Lecture Notes in Computer Science 16381

Founding Editors

Gerhard Goos

Juris Hartmanis

Editorial Board Members

Elisa Bertino, *Purdue University, West Lafayette, IN, USA*

Wen Gao, *Peking University, Beijing, China*

Bernhard Steffen, *TU Dortmund University, Dortmund, Germany*

Moti Yung, *Columbia University, New York, NY, USA*

The series Lecture Notes in Computer Science (LNCS), including its subseries Lecture Notes in Artificial Intelligence (LNAI) and Lecture Notes in Bioinformatics (LNBI), has established itself as a medium for the publication of new developments in computer science and information technology research, teaching, and education.

LNCS enjoys close cooperation with the computer science R & D community, the series counts many renowned academics among its volume editors and paper authors, and collaborates with prestigious societies. Its mission is to serve this international community by providing an invaluable service, mainly focused on the publication of conference and workshop proceedings and postproceedings. LNCS commenced publication in 1973.

Huazhong Liu · Shadi Ibrahim · Thomas Rauber
Editors

Algorithms and Architectures for Parallel Processing

25th International Conference, ICA3PP 2025
Zhengzhou, China, October 30 – November 2, 2025
Proceedings, Part I

 Springer

Editors
Huazhong Liu
Hainan University
Haikou, China

Shadi Ibrahim
Inria
Rennes, France

Thomas Rauber
University of Bayreuth
Bayreuth, Germany

ISSN 0302-9743　　　　　　ISSN 1611-3349 (electronic)
Lecture Notes in Computer Science
ISBN 978-981-95-8398-0　　ISBN 978-981-95-8399-7 (eBook)
https://doi.org/10.1007/978-981-95-8399-7

This Springer imprint is published by the registered company Springer Nature Singapore Pte Ltd.
The registered company address is: 152 Beach Road, #21-01/04 Gateway East, Singapore 189721, Singapore

If disposing of this product, please recycle the paper.

Preface

On behalf of the Conference Committee, we welcome you to the proceedings of the 2025 International Conference on Algorithms and Architectures for Parallel Processing (ICA3PP 2025), which was held in Zhengzhou, China from October 30 – November 2, 2025. ICA3PP 2025 was the 25th in this series of conferences (started in 1995) that are devoted to algorithms and architectures for parallel processing. ICA3PP is now recognized as the main regular international event that covers the many dimensions of parallel algorithms and architectures, encompassing fundamental theoretical approaches, practical experimental projects, and commercial components and systems. This conference provides a forum for academics and practitioners from countries around the world to exchange ideas for improving the efficiency, performance, reliability, security, and interoperability of computing systems and applications.

A successful conference would not be possible without the high-quality contributions made by the authors. This year, ICA3PP received a total of 543 submissions from authors in various countries and regions. Each paper underwent a single-blind review process by at least three reviewers. Based on rigorous peer reviews by the Program Committee members and reviewers, 158 regular papers were selected to be presented at the conference, giving an acceptance rate of 29%. We also accepted 104 research papers and 48 special session papers. In addition to the contributed papers, distinguished scholars were invited to give keynote lectures, providing us with the recent developments in diverse areas in algorithms and architectures for parallel processing and applications.

We would like to take this opportunity to express our sincere gratitude to the Program Committee members and 305 reviewers for their dedicated and professional service. We highly appreciate the General Chairs, Didier El Baz, Zizhong Chen, and Gudula Runger, for their help and support in organizing the conference. Then we would like to express our appreciation to the four program vice-chairs, Rong Gu, Deze Zeng, Jiawei Huang, and Zhe Peng, for their leadership in managing the various tracks of the conference. We would also like to give our sincere thanks to the Local Chairs, Jihong Ding, Yi Jiang, Song Zhang, Zhicheng Liu, Chunfeng Du, Zecan Yang, and Xin Nie, for their hard work and commitment to the success of the conference. Our heartfelt thanks also go to the Workshop Chair Shaojun Zou for his efforts in organizing the wonderful special sessions and enriching the technical discourse. We are also deeply appreciative of the Publicity Chairs, Jia Hu, Jing Guo, and Wenxuan Zhang, for their tireless efforts in promoting the conference. We are so grateful to the publication chairs, Binbin Zhou and Lan Yan, and the publication assistants for their dedicated support in overseeing the publication process. Thanks also go to Registration Chairs Yuan Tian and Tingting Han, and Web Chairs Jiawei Wang and Ren Li. Our deepest thanks also go to the Steering Chairs, Yang Xiang, Wanlei Zhou, Yi Pan, Laurence T. Yang, and Weijia Jia, for their continuous leadership and support throughout the planning and execution phases of the conference. We must also say "thank you" to all the volunteers who helped us at various stages of this conference. Moreover, we were so honored to have many renowned scholars be part of this conference. Finally, we would like to thank all speakers, authors, and participants for their great contribution to and support for the success of ICA3PP 2025!

October 2025

Huazhong Liu
Shadi Ibrahim
Thomas Rauber

Organization

General Chairs

Didier El Baz	LAAS-CNRS, France
Zizhong Chen	University of California, Riverside, USA
Gudula Runger	Chemnitz University of Technology, Germany

Program Chairs

Huazhong Liu	Hainan University, China
Shadi Ibrahim	Inria, France
Thomas Rauber	University of Bayreuth, Germany

Program Vice-chairs

Rong Gu	Nanjing University, China
Deze Zeng	China University of Geosciences, Wuhan, China
Jiawei Huang	Central South University, China
Zhe Peng	Hong Kong Polytechnic University, China

Local Chairs

Jihong Ding	Hainan University, China
Yi Jiang	Hainan University, China
Song Zhang	Zhengzhou University, China
Zhicheng Liu	Zhengzhou University, China
Chunfeng Du	Zhengzhou University, China
Zecan Yang	Zhengzhou University, China
Xin Nie	Zhengzhou University, China

Workshop Chair

Shaojun Zou	Hainan University, China

Publicity Chairs

Jia Hu	University of Exeter, UK
Jing Guo	Hainan University, China
Wenxuan Zhang	Hainan University, China

Publication Chairs

Binbin Zhou	Hangzhou City University, China
Lan Yan	Hainan University, China

Registration Chairs

Yuan Tian	Hainan University, China
Tingting Han	Hainan University, China

Web Chairs

Jiawei Wang	University of Warwick, UK
Ren Li	Hainan University, China

Steering Committee

Yang Xiang	Deakin University, Australia
Wanlei Zhou	City University of Macau, China
Yi Pan	Shenzhen University of Advanced Technology, China
Laurence T. Yang	Zhengzhou University, China
Weijia Jia	Beijing Normal University, China

Program Committee

Adnan Anwar	Deakin University, Australia
Sen Bai	Tsinghua University, China
Wen Bai	Hubei University, China
Selcuk Baktir	American University of the Middle East, Kuwait

Nicola Bena	University of Milan, Italy
Ran Bi	Northeastern University, China
Yanling Bu	Nanjing University of Aeronautics and Astronautics, China
Massimo Cafaro	University of Salento, Italy
Miao Cai	Nanjing University of Aeronautics and Astronautics, China
Zhicheng Cai	Nanjing University of Science and Technology, China
Chenhong Cao	University of Science and Technology of China, China
Zhilei Chai	Jiangnan University, China
Sudip Chakraborty	Valdosta State University, USA
Xiaolin Chang	Beijing Jiaotong University, China
Zhaoxin Chang	Institut Polytechnique de Paris, France
Bo Chen	Shenzhen University, China
Chao Chen	Chongqing University, China
Huijie Chen	Beijing University of Technology, China
Jiaoyan Chen	Wuhan University of Science and Technology, China
Jing Chen	Hainan University, China
Liying Chen	Hunan Normal University, China
Long Chen	Guangdong University of Technology, China
Longbiao Chen	Xiamen University, China
Ning Chen	Soochow University, China
Quan Chen	Guangdong University of Technology, China
Simin Chen	University of Texas at Dallas, USA
Ting Chen	University of Electronic Science and Technology of China, China
Xiang Chen	Zhejiang University, China
Ying Chen	Hunan Normal University, China
Yuzhu Chen	Changsha Social Work College, China
Zi Chen	Wuhan University of Technology, China
Zixuan Chen	China Telecom, China
Francesco Colace	University of Salerno, Italy
Jin Cui	Northwest University, China
Jinhua Cui	Huazhong University of Science and Technology, China
Lin Cui	Jinan University, China
Weihao Cui	Shanghai Jiao Tong University, China
Yuan Tian	Hainan University, China
Gianni D'Angelo	University of Salerno, Italy
Haipeng Dai	Nanjing University, China

Fan Dang	Beijing Jiaotong University, China
Xia Deng	Guangzhou University, China
Xiaojun Deng	Hunan University of Technology, China
Dian Ding	Shanghai Jiao Tong University, China
Kai Dong	Southeast University, China
Pingping Dong	Hunan Normal University, China
Xiaogang Dong	Jiujiang University, China
Bingqian Du	Huazhong University of Science and Technology, China
Haizhou Du	Shanghai University of Electric Power, China
Xin Du	Zhejiang University, China
Yang Du	Soochow University, China
Cheng Dai	Sichuan University, China
Christian Esposito	University of Salerno, Italy
Weibei Fan	Nanjing University of Posts and Telecommunications, China
Xin Fan	Beijing Forestry University, China
Xinggang Fan	Zhejiang University of Technology, China
Chunrong Fang	Nanjing University, China
Xing Fang	Xiamen University, China
Libo Feng	Yunnan University, China
Yunyun Feng	University of Science and Technology of China, China
Zhiwei Feng	Northeastern University, China
Shaojing Fu	National University of Defense Technology, China
Xiong Fu	Nanjing University of Posts and Telecommunications, China
Chengxi Gao	Shenzhen Institutes of Advanced Technology, Chinese Academy of Sciences, China
Shang Gao	Hong Kong Polytechnic University, China
Tianchong Gao	Southeast University, China
Yifan Gao	Victoria University of Wellington, New Zealand
Harald Gjermundrød	University of Nicosia, Cyprus
Xiaoli Gong	Nankai University, China
Chaojie Gu	Zhejiang University, China
Chen Gu	Hefei University of Technology, China
Yunfei Gu	Shanghai Jiao Tong University, China
Zonghua Gu	Umeå University, Sweden
Zhitao Guan	North China Electric Power University, China
Shuai Guo	Qingdao University of Technology, China
Tao Guo	Shenzhen University, China

Yongming Han	Beijing University of Chemical Technology, China
Zengyi Han	Dalian Maritime University, China
Lin He	Tsinghua University, China
Tengjiao He	Jinan University, China
Yi He	University of CSU, China
Zhenli He	Yunnan University, China
Peng Hou	Soochow University, China
Jinbin Hu	Changsha University of Science and Technology, China
Pengfei Hu	Shandong University, China
Shihong Hu	Hohai University, China
Yusheng Hua	Nanjing University of Posts and Telecommunications, China
Kuochan Huang	National Taichung University of Education, Taiwan
Qinlong Huang	Beijing University of Posts and Telecommunications, China
Ran Huang	Victoria University of Wellington, New Zealand
Mingtao Ji	Hohai University, China
Juncheng Jia	Soochow University, China
Guiyuan Jiang	Ocean University of China, China
Shan Jiang	Sun Yat-sen University, China
Wenhui Jiang	Jiangxi University of Finance and Economics, China
Xianliang Jiang	Ningbo University, China
Lanju Kong	Shandong University, China
Peter Kropf	University of Neuchâtel, Switzerland
Laurent Lefevre	Inria, France
Bei Li	China Unicom, China
Bo Li	Victoria University, Australia
Borui Li	Southeast University, China
Chao Li	Shanghai Jiao Tong University, China
Dagang Li	Macau University of Science and Technology, China
Guangli Li	Institute of Computing Technology, Chinese Academy of Sciences, China
Hu Liang	Qilu University of Technology, China
Jiahui Li	Jilin University, China
Jingzong Li	City University of Hong Kong, China
Liying Li	Nanjing University of Science and Technology, China
Meng Li	Hefei University of Technology, China

Nannan Li	Tsinghua University, China
Ning Li	Universidad Politécnica de Madrid, Spain
Wenjuan Li	Education University of Hong Kong, China
Wenjun Li	Hunan Normal University, China
Xiaolu Li	Huazhong University of Science and Technology, China
Yan Li	Hunan University of Science and Technology, China
Zhiqiang Li	A10 Networks, USA
Teng Liang	Peng Cheng Laboratory, China
Xueqin Liang	Xidian University, China
Wenmin Lin	Hangzhou Normal University, China
Yijing Lin	Beijing University of Posts and Telecommunications, China
Zhen Ling	Southeast University, China
Bo Liu	Zhengzhou University, China
Dajiang Liu	Chongqing University, China
Hongyan Liu	Fuzhou University, China
Jia Liu	Nanjing University, China
Jianchun Liu	University of Science and Technology of China, China
Jianwei Liu	Changsha University, China
Shaojie Liu	Hainan University, China
Tao Liu	National Supercomputer Center in Jinan, China
Xuxun Liu	South China University of Technology, China
Ying Liu	National University of Defense Technology, China
Yuchen Liu	Beihang University, China
Xiaojuan Lu	Central South University, China
Lailong Luo	National University of Defense Technology, China
Luyao Luo	Soochow University, China
WangQing Luo	Hunan University of Science and Technology, China
Hongwu Lv	Harbin Engineering University, China
Sheng Ma	National University of Defense Technology, China
Shunmei Meng	Nanjing University of Science and Technology, China
Weizhi Meng	Lancaster University, UK
Yinbin Miao	Xidian University, China
Priyadarsi Nanda	University of Technology Sydney, Australia
Xuefeng Ni	Hunan University, China

Xintao Niu	Nanjing University, China
Shujie Pang	Guangdong University of Technology, China
Kai Peng	Huaqiao University, China
Yaqiong Peng	Hunan University, China
Chiara Pero	University of Salerno, Italy
Nikolaos Pitropakis	American College of Greece, Greece
Vincenzo Piuri	Università degli Studi di Milano, Italy
Lingjun Pu	Nankai University, China
Hui Qi	Changchun University of Science and Technology, China
Ji Qi	Chinese Academy of Sciences, China
Jiacheng Qu	Hainan University, China
Zhengwei Qi	Shanghai Jiao Tong University, China
Yan Qiao	Hefei University of Technology, China
Yuben Qu	Nanjing University of Aeronautics and Astronautics, China
Zhihao Qu	Hohai University, China
Vinayakumar Ravi	Prince Mohammad Bin Fahd University, Saudi Arabia
Shenyuan Ren	Beijing Jiaotong University, China
Fariza Sabrina	Central Queensland University, Australia
Jun Shao	Zhejiang Gongshang University, China
Chao Shen	Xi'an Jiaotong University, China
Fanfan Shen	Nanjing Audit University, China
Han Shen	Xi'an Technological University, China
Hang Shen	Nanjing Tech University, China
Houqiang Shen	Changsha University of Science and Technology, China
Zhishu Shen	Wuhan University of Technology, China
Fang Shi	South China Agricultural University, China
Qingyu Shi	Hunan University of Technology and Business, China
Wanxin Shi	Fudan University, China
Na Song	Putian University, China
Qingyu Song	Xiamen University, China
Milos Stojmenovic	Singidunum University, Serbia
Qiwei Sun	Zhengzhou Police University, China
Lizhuang Tan	Qilu University of Technology, China
Weiqiang Tan	Guangzhou University, China
Chaogang Tang	China University of Mining and Technology, China
Hailun Tang	Tianjin University, China

Huijun Tang	Durham University, UK
Jianzhi Tang	Soochow University, China
Xing Tang	Wuhan University of Technology, China
Xiaoqiang Teng	Beijing Technology and Business University, China
Yun Teng	Hunan University of Science and Technology, China
Xiang Tian	Qilu University of Technology, China
Ying Wan	Southeast University, China
Bin Wang	Nanchang Hangkong University, China
Chenyang Wang	Shenzhen University, China
En Wang	Jilin University, China
Guijuan Wang	Qilu University of Technology, China
Jingzhou Wang	Soochow University, China
Kai Wang	Harbin Institute of Technology, China
Tian Wang	Jiangsu College of Finance and Accounting, China
Yingjie Wang	Yantai University, China
Zhijie Wang	Chongqing University, China
Zhu Wang	Northwestern Polytechnical University, China
Kaimin Wei	Jinan University, China
Jiaxin Wu	Guangdong University of Technology, China
Jinling Wu	Zhejiang Normal University, China
Wei Wu	Hainan Institute of Zhejiang University, China
Yalan Wu	Guangdong University of Technology, China
Zhongying Wu	Hainan University, China
Junxu Xia	National University of Defense Technology, China
Nian Xia	Nanjing Normal University, China
Qiufen Xia	Dalian University of Technology, China
Haolong Xiang	Nanjing University of Information Science and Technology, China
Zhengzhe Xiang	Hangzhou City University, China
XiongJiang Xiao	Hainan University, China
Mande Xie	Zhejiang Gongshang University, China
Jia Xu	Nanjing University of Posts and Telecommunications, China
Mengwei Xu	Beijing University of Posts and Telecommunications, China
Pin Xu	Hainan University, China
Shujiang Xu	Shandong Computer Science Center, China
Xianghao Xu	Nanjing University of Science and Technology, China

Yueshen Xu	Xidian University, China
Xingfu Yan	South China Normal University, China
Bo Yang	Northwestern Polytechnical University, China
Bo Yang	Northwest A&F University, China
Chen Yang	Beijing Institute of Technology, China
Wen Yang	China Three Gorges University, China
Wenjun Yang	University of Victoria, Canada
Yubo Yang	Hunan University of Science and Technology, China
Biyuan Yao	Hainan University, China
XiaoXue Yin	Hainan University, China
Long Yuan	Wuhan University of Technology, China
Min Yuan	Nanjing Normal University, China
Hengshan Yue	Jilin University, China
Rongfei Zeng	Northeastern University, China
Yue Zeng	Nanjing University of Science and Technology, China
Xiangping Bryce Zhai	Nanjing University of Aeronautics and Astronautics, China
Hao Zhang	Harbin Institute of Technology, China
Jiapeng Zhang	Hunan University, China
Jinquan Zhang	Guangdong University of Technology, China
Lianming Zhang	Hunan Normal University, China
Qingyang Zhang	Anhui University, China
Shuhui Zhang	Shandong Computer Science Center, China
Xianwei Zhang	Sun Yat-sen University, China
Yin Zhang	University of Electronic Science and Technology, China
Yong Zhang	Shenzhen Institutes of Advanced Technology, Chinese Academy of Sciences, China
Yongqin Zhang	Hainan University, China
Yuhao Zhang	Tianjin University, China
Yun Zhang	Hunan Normal University, China
Yuhui Zhang	Chinese Academy of Sciences, China
Pinlong Zhao	Hangzhou Dianzi University, China
Shengrong Zhao	Qilu University of Technology, China
Tianfang Zhao	Jinan University, China
Yu Zhao	Chongqing University of Posts and Telecommunications, China
Yunfeng Zhao	Tianjin University, China
Zhihui Zhao	Taiyuan University of Technology, China
Tianyue Zheng	Southern University of Science and Technology, China

Xiao Zheng	Anhui University of Technology, China
Ruiting Zhou	Southeast University, China
Zhou Zhou	Changsha University, China
Rongbo Zhu	Huazhong Agricultural University, China
Weitao Zou	Northeast Forestry University, China
Xiaolin Zou	Zhaoqing University, China
Sen Liu	Fudan University, China
Weijie Liu	National University of Defense Technology, China
Chang Ruan	Changsha Unviersity of Science and Technology, China
Xinchen Wan	ByteDance, USA
Wei Wang	National University of Defense Technology, China
Linyi Han	Tianjin University, China
Ye Liu	Donghua UniversityChina
Jinquan Nie	University of Chinese Academy of Sciences, China
Le Qin	Jiangxi Normal University, China
Xinping Rao	Jiangxi Normal University, China
Wei Yang	Jiangxi Normal University, China
Jiamin Deng	Hong Kong Polytechnic University, China
Hongfei Huang	Hangzhou Dianzi University, China
Xin He	Nanjing University of Posts and Telecommunications, China
Youhuizi Li	Hangzhou Dianzi University, China
Jinhai Liu	Henan Normal University, China
Yafei Shi	Henan Normal University, China
Yukun Shi	Hangzhou Dianzi University, China
Muyu Deng	China Unicom, China
Xixi Sun	Hangzhou Dianzi University, China
Baofu Wu	Hangzhou Fanjia Technology Co. Ltd, China
Wenjian Xu	Zhejiang University of Science and Technology, China
Weiqi Yue	Hangzhou Dianzi University, China
Tingjun Zhao	Zhejiang University of Technology, China
Leilei Zheng	Hangzhou Dianzi University, China
Yuan Liu	Xiangtan University, China
Xia Ou	Zhejiang University of Technology, China
Yuntao Su	Hunan University of Technology and Business, China
Taotao Yu	Zhejiang University of Technology, China
Wenchao Dong	Central China Normal University, China

Hanrui Gao	Zhengzhou University, China
Xi Kong	Henan University, China
Wenhui Xu	Henan Normal University, China
Hongwei Yu	Beihang University, China

Contents

Regular Papers

Parallel and Distributed Architectures

Software Systems and Programming Models

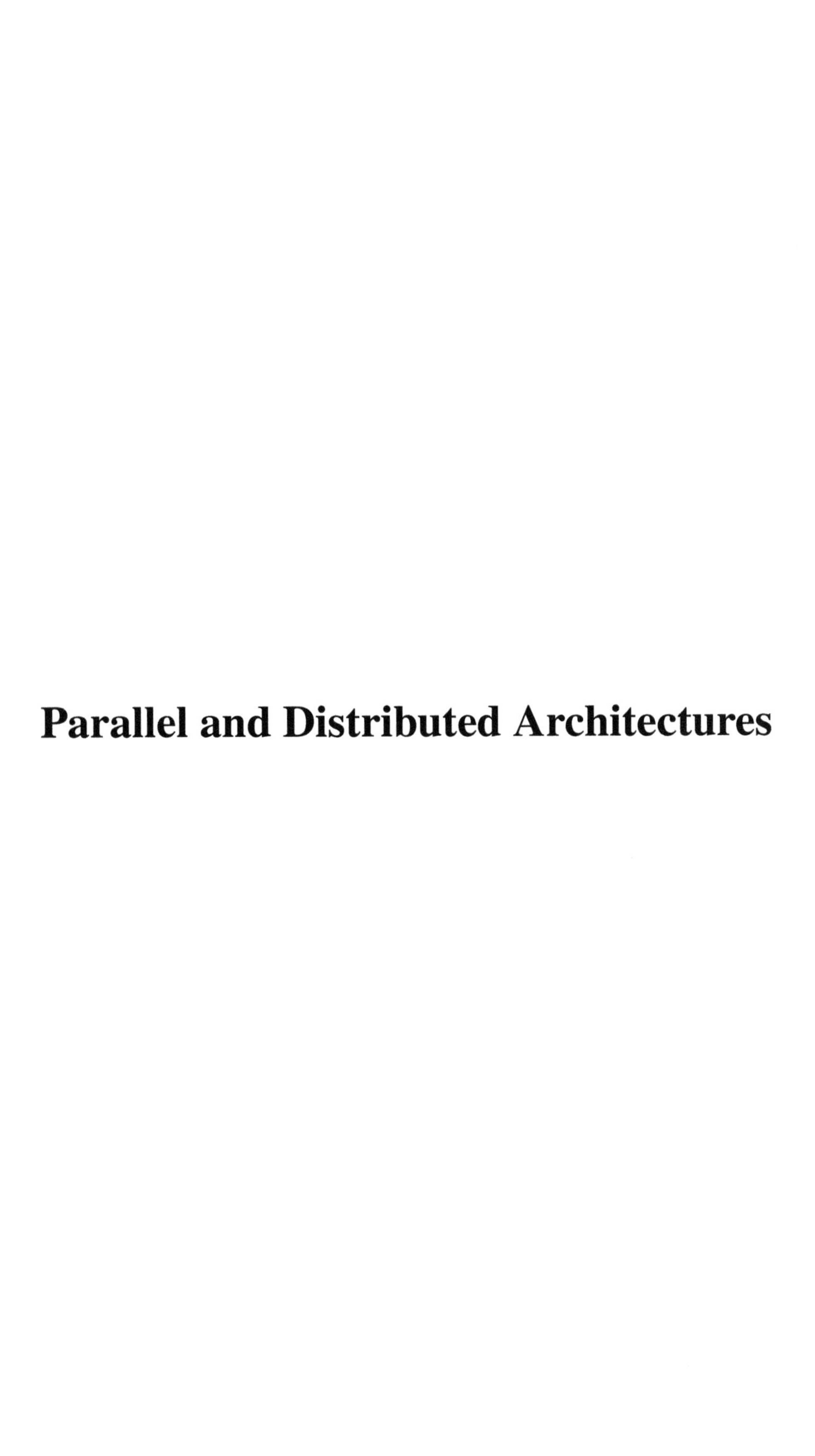

Parallel and Distributed Architectures

Streamline: A High-Parallelized BFT Consensus Protocol for Blockchain

Guoze Li, Yuchao Zhang[✉], Xiaofeng He, and Xiaotian Wang

Beijing University of Posts and Telecommunications, Beijing, China
{liguozebupt,yczhang,xiaofenghe,wangxiaotian}@bupt.edu.cn

Abstract. The rapid adoption of blockchain technology, exemplified by Bitcoin and Ethereum, has intensified the need for high-performance systems. Existing optimization approaches primarily focus on two directions: simplifying consensus protocols and parallelizing consensus processes. Simplified protocols reduce consensus overhead but remain constrained by single-leader inefficiencies, particularly in large-scale networks where throughput degrades rapidly. Parallelization enhances throughput but frequently introduces complexity and tightly coupled architectures that lack dynamic adaptability, limiting adaptability.

To address these limitations, we propose Streamline, a functionally decoupled multi-leader consensus protocol for permissioned blockchains that decouples transaction distribution from consensus processes. By integrating horizontal parallelism for concurrent consensus instances and vertical decoupling of data distribution, Streamline eliminates single-leader bottlenecks and supports dynamic modularity. Its layered design enables flexible workload adjustments, such as real-time load balancing under varying node resources, ensuring robust adaptability without compromising security.

Compared to the latest consensus protocols, our approach demonstrates stable scalability and adaptability, incurring lower performance degradation as the network scales while maintaining more stable throughput and latency. Experimental results show that Streamline's performance is comparable to state-of-the-art parallel consensus protocols and effectively delays network saturation; under skewed load, its throughput degrades by only 10%.

Keywords: Byzantine fault tolerance · permissioned blockchain · distributed ledgers consensus · multi-leader consensus · load balancing

1 Introduction

Blockchain technology, powering systems like Bitcoin and Ethereum, relies on distributed consensus to ensure all nodes agree on a shared ledger. In trusted

This work was supported by the National Natural Science Foundation of China under Grant 62172054, the Beijing Nova Program under Grant 2023140, and the National Key R&D Program of China under Grant 2024YFB2906900.

settings, Crash Fault Tolerant (CFT) systems handle node failures, while State Machine Replication (SMR) ensures consistency across nodes. In untrusted environments, Byzantine Fault Tolerant (BFT) systems extend SMR to maintain security and progress despite malicious nodes, making them ideal for applications like financial transactions [11], healthcare [2], and gaming [6].

At the heart of BFT systems lies the consensus mechanism, enabling nodes to agree on the content and total order of a sequence of operations. In the most common partial-synchronous network model [7], BFT consensus typically relies on a single leader node responsible for broadcasting blocks and coordinating different nodes to reach agreement [3,20]. However, the limited computational and bandwidth resources of a single node inherently create a performance bottleneck in BFT consensus systems. This issue is exacerbated when malicious or faulty nodes are elected as leaders, leading to prolonged stalls in consensus and further scalability issues. For commercial applications requiring high throughput (e.g., financial transactions processing thousands of transactions per second), these limitations become particularly problematic.

To address the performance bottleneck caused by the single leader, existing work has explored simplified BFT protocols [4,10,13,17]. These approaches simplify coordination or consensus pipelining to enhance efficiency. However, their reliance on a single leader restricts performance in demanding scenarios. For example, in high-frequency trading requiring thousands of transactions per second, these protocols struggle to maintain speed and reliability. To overcome these limitations, researchers have developed parallelized protocols that distribute tasks across multiple nodes [8,15,19,21]. However, these parallel designs often rely on static protocol structures that are difficult to adjust after deployment. Their lack of modularity makes them rigid, complicating extensions or modifications to meet new requirements without affecting the entire system. In real-world scenarios, systems often need dynamic adjustments to meet changing requirements, which static designs struggle to support. For instance, when node resources vary, static protocols require redesign to adjust workloads securely while adapting to new requirements. Similarly, when operators deploy different consensus protocols for nodes to meet varying security needs across scenarios, these static multi-leader designs often necessitate a complete system reconstruction.

The primary challenge, therefore, lies in overcoming the inefficiencies and bottlenecks of consensus protocols while ensuring a loosely coupled framework that facilitates seamless adjustments, optimizations, and adaptability to diverse environments.

In this paper, we introduce Streamline, a multi-leader BFT consensus protocol that decouples transaction distribution from consensus, mitigating the single-leader bottleneck and enhancing modularity. Unlike prior solutions, Streamline offers robust dynamic adjustment capabilities through its layered design, enabling flexible adaptation to varying deployment needs. Taking load balancing as an example, when node resources change, it supports real-time workload redistribution without compromising security or requiring extensive system changes.

Experimental results show that Streamline not only achieves comparable or even superior throughput compared to recent consensus protocols in regular environments, but more importantly, it demonstrates effective scalability, maintaining stable latency and consistent performance without significant degradation even under unbalanced loads or resource constraints.

Overall, this paper makes the following key contributions:

1. We present Streamline, an efficient BFT consensus that integrates horizontal multi-leader parallelism with vertical decoupling of broadcasting and consensus, simplifying the protocol structure and enhancing modularity, scalability and adaptability for diverse environments. We conduct rigorous theoretical analysis to ensure the security properties of the protocol are preserved while achieving these improvements.
2. We introduce two broadcast primitives for transaction distribution and inter-consensus coordination, enabling multi-dimensional parallelism and enhancing system efficiency across different stages of the consensus process.
3. We evaluate our prototype, demonstrating comparable or even superior throughput compared to recent consensus protocols and showcasing scalability with reasonable latency even under unbalanced load conditions.

The remainder of this paper is organized as follows: Sect. 2 reviews existing approaches for improving consensus protocol efficiency and discusses their limitations. Section 3 defines the system model and security assumptions. Section 4 details our system design, including core protocols, mechanisms and system workflow. Section 5 presents a theoretical analysis of system security and performance. Section 6 compares Streamline with recent BFT consensus solutions. Finally, Sect. 7 concludes the paper with key contributions and future directions.

2 Related Work

Blockchain technology is still in its early stages and faces significant challenges, particularly in improving consensus efficiency. Early public blockchain systems, primarily employing the Proof-of-Work (PoW) consensus mechanism, suffer from severe inefficiencies. For instance, Bitcoin processes only 7 transactions per second (TX/s), and Ethereum achieves 14 TX/s [18], far below the thousands of TX/s required for commercial applications like high-frequency trading, payment processing, and supply chain management. With the introduction of hybrid consensus [9] and permissioned blockchains [1], BFT consensus algorithms such as PBFT [3] have gained research attention. However, scalability remains a significant challenge for applying BFT consensus in large-scale blockchain systems. Traditional BFT consensus typically exhibits high communication complexity and computational overhead, limiting its ability to handle increasing workloads and node counts. Consequently, existing approaches to improving BFT consensus performance can be categorized into protocol simplification and parallelization.

Protocol simplification focuses on reducing consensus overhead. EBFT [17] introduces an egalitarian block generation mechanism, eliminating the leader role to enhance resilience against targeted attacks and reduce failover complexity. Streamlet [4] distills recent consensus advancements into a streamlined framework, providing a simpler yet effective alternative to PBFT and Paxos.

Despite the progress made by protocol simplification in reducing consensus overhead, it still faces bottlenecks inherent in single-leader architectures [8,12]. Therefore, more efforts have shifted towards parallelization strategies to further enhance performance.

Parallelization, on the other hand, is further divided into vertical and horizontal strategies. Vertical parallelism enhances transaction processing by reducing serial dependencies. For example, EPaxos [16] eliminates the single-leader bottleneck by allowing any node to propose, enabling parallel execution. PigPaxos [5] optimizes communication via aggregation and relays, improving throughput. Stratus [8] introduce a transaction broadcast primitive that decouples payload distribution from consensus, balancing network load and alleviating leader bottlenecks. Horizontal parallelism improves inter-node parallelism. ISS [19] adopts sequential broadcast to allow nodes to engage in multiple consensus instances while maintaining consistency, breaking single-leader constraints.

However, despite these improvements in existing works, both protocol simplification and parallelization approaches exhibit inherent limitations. Simplified protocols reduce consensus overhead but remain constrained by single-leader architectures, centralizing decision-making and causing bottlenecks under high-latency or contentious workloads. Malicious or delayed leaders further disrupt consensus, limiting scalability and robustness.

Similarly, vertical parallelism optimizes transaction processing workflows but fails to eliminate the fundamental limitations of single-leader designs. When the leader encounters stalls or behaves maliciously, the entire protocol can suffer disruptions, reducing system responsiveness and scalability. Horizontal multi-leader parallelism improves throughput but often lacks the ability to dynamically adjust its structure. Many designs employ static protocol structures that cannot be easily modified after deployment, limiting flexibility and adaptability in varying deployment contexts. In cases requiring load balancing or traffic control, these protocols often necessitate substantial architectural changes, posing challenges for scalability and long-term maintainability.

Recognizing the shortcomings in flexibility and adaptability of existing methods, this paper introduces a highly parallelized function-decoupled multi-leader consensus protocol that separates the distribution process from the consensus process, called Streamline.

3   System Model

We consider a system comprising a fixed set of $n = 3f + 1$ replicas, identified by indices $i \in [n]$, where $[n] = \{1, 2, \ldots, n\}$. Up to f of these replicas may exhibit Byzantine faults, while the remaining replicas are assumed to function correctly. For brevity, we assume that the Byzantine replicas are coordinated by a single adversary who has full access to their internal state, including any cryptographic keys they hold.

3.1 Network Model

Communication between replicas is modeled as point-to-point, authenticated, and reliable. Specifically, a message sent by a correct replica is guaranteed to be received. In our protocol, the term "broadcast" refers to the process whereby a correct sender transmits the identical message individually to every replica (including itself). We adopt the partial synchrony model, in which there exists an unknown Global Stabilization Time (GST) after which all messages between correct replicas are guaranteed to be delivered within a known bounded delay.

3.2 Security Assumptions

Our protocol relies on threshold signature schemes for authentication. In a (k, n)-threshold signature scheme, all replicas share a public key, while each replica holds its distinct private key. The i-th replica can generate a partial signature, denoted by $\sigma_i(m)$, for any given message m. A valid digital signature $\sigma(m)$ can then be obtained by combining partial signatures from any set I of k replicas, i.e., where $|I| = k$. Any replica may verify $\sigma(m)$ using the common public key and a verification function $\texttt{Verify}(m, \sigma(m))$, which must return true when the combined signature is valid. We assume that if an adversary has access to the signature oracles of fewer than k faulty replicas, the probability of producing a valid signature for a new message m is negligible. Throughout this work, we set the threshold parameter to $k = 2f + 1$, as this is the minimum number required to guarantee agreement in the presence of up to f Byzantine faults.

In addition, we assume the availability of a collision-resistant hash function h, which maps inputs of arbitrary length to fixed-length outputs. The collision resistance property ensures that it is computationally infeasible for an adversary to find distinct messages m and m' such that $h(m) = h(m')$, thereby allowing $h(m)$ to serve as a unique identifier for m.

We further assume that clients refrain from submitting duplicate transaction requests, except in cases where timeouts occur due to blockages caused by Byzantine nodes. Additionally, each client is assigned a unique identity and maintains a strictly increasing transaction sequence number; these identifiers are typically enforced through a combination of public key signatures and timestamps, thereby ensuring both the uniqueness of the client and the correct ordering of its transactions.

4 Design of Streamline

Our design is based on a parallel multi-consensus instance architecture combined with a decoupled mechanism that separates data distribution from the consensus process, facilitating flexible load balancing. This section is organized into four subsections: the system framework, the design of parallel mechanism on consensus process, the design of parallel mechanism on distribution process, and the load balancing strategy.

4.1   System Framework

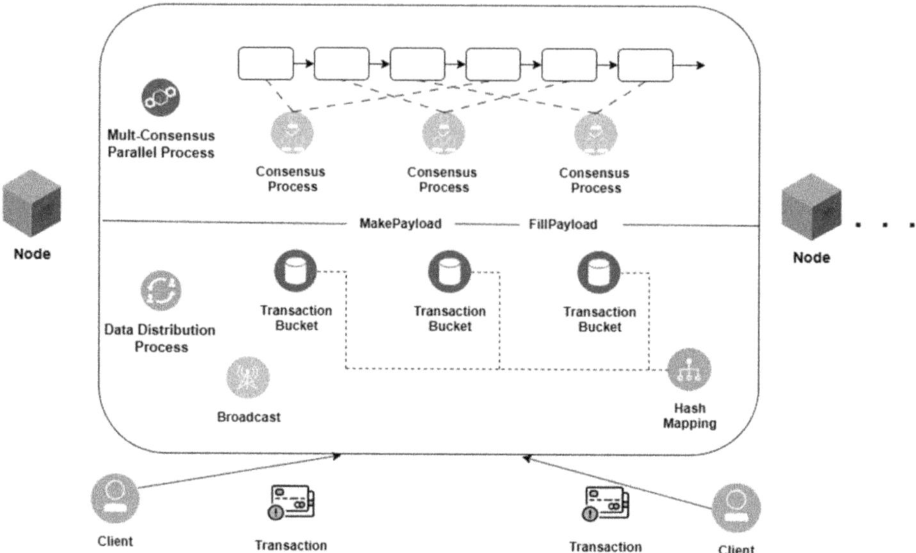

Fig. 1. The workflow of Streamline comprises two concurrent processes: multi-consensus parallel execution and decoupled transaction distribution, carried out by consensus nodes in response to client requests.

The Streamline framework consists of two main processes: the multi-consensus parallel process and the data distribution process, as shown in Fig. 1. In our system, client nodes are responsible solely for generating and sending transaction requests to the network, while nodes (also referred to as consensus nodes) participate in the network to process, propagate, and reach consensus on transactions.

In the multi-consensus parallel process, nodes employ a divide-and-conquer strategy, partitioning the log to instantiate multiple consensus instances. Logs are synchronized and backed up at epoch ends to ensure consistency and security.

In the data distribution process, transaction distribution is decoupled from consensus. Transactions are aggregated into buckets using a hash mapping function and subsequently broadcast for dissemination. Should a node be missing any payload, it proactively retrieves the necessary data from its peers, while an integrated load-balancing strategy optimizes resource utilization.

These two processes run concurrently, ensuring seamless operation with minimal interference. In cases where no transactions are available, the consensus protocol proceeds with empty commits. Moreover, if payloads are missing, a dedicated retrieval thread is initiated to guarantee uninterrupted transaction dissemination.

Algorithm 1: Limited Broadcast Protocol

---

1 **Variables:**
2   init-threshold, σ, sn, m, $Seqs$, M

3 **Procedure** LB-Broadcast(sn, m)
4   Send $\langle$LB-CAST $\mid (sn, m)\rangle$ to all nodes

5 **Procedure** LB-Deliver(sn, m)
6   Commit m at $\log[sn]$
7   LB-INIT($sn + 1$)

8 **Procedure** LB-INIT(sn)
9   Send $\langle$LB-INIT $\mid sn\rangle$ to all nodes

10 **Upon receiving** $\langle$LB-INIT $\mid sn\rangle$ **at node** i:
11   **if** $i = \sigma$ **then**
12   init-threshold $\leftarrow$ init-threshold $+ 1$
13   **if** $init\text{-}threshold = 2f + 1$ **then**
14   m $\leftarrow$ select message from M
15   LB-Broadcast(sn, m)
16   init-threshold $\leftarrow 0$

17 **Upon receiving** $\langle$LB-CAST $\mid (sn, m)\rangle$:
18   LB-Deliver(sn, m)

19 **Upon timeout for** sn:
20   LB-Deliver(sn, nil)
21   trigger view-change

4.2   Parallel Mechanism Design on Consensus Process

Limited Broadcast. Our multi-consensus instance parallel process relies on a new Byzantine broadcast primitive called limited broadcast. Limited broadcast is a variant of Byzantine ordered broadcast, with the core idea being to achieve consensus in limited rounds for transactions.

In a limited broadcast, only a specific sender node σ can assign a sequence number sn from a sequence set $Seqs$ to a message m from a message set M and initiate a consensus round on that message through the broadcast operation $\langle LB - CAST \mid (sn, m)\rangle$ (Algorithm 1, Line 10). Receiving nodes complete the consensus round by delivering the message m with sequence number sn through the event $\langle LB - DELIVER \mid (sn, m)\rangle$ (Algorithm 1, Line 17).

If a sender node is silent (i.e., no LB-CAST message is received within a timeout period), nodes deliver a null value using $\langle LB - DELIVER \mid (sn, nil)\rangle$ to continue the consensus instance (Algorithm 1, Line 19).

A limited broadcast instance $LB(\sigma, S, M)$ is initialized with the $\langle LB - INIT\rangle$ operation (Algorithm 1, Line 8) and has the following properties:

– Integrity: If a correct node delivers $\langle LB - DELIVER \mid (sn, m)\rangle$ (where $m \neq nil$), σ must have broadcast $\langle LB - CAST \mid (sn, m)\rangle$.

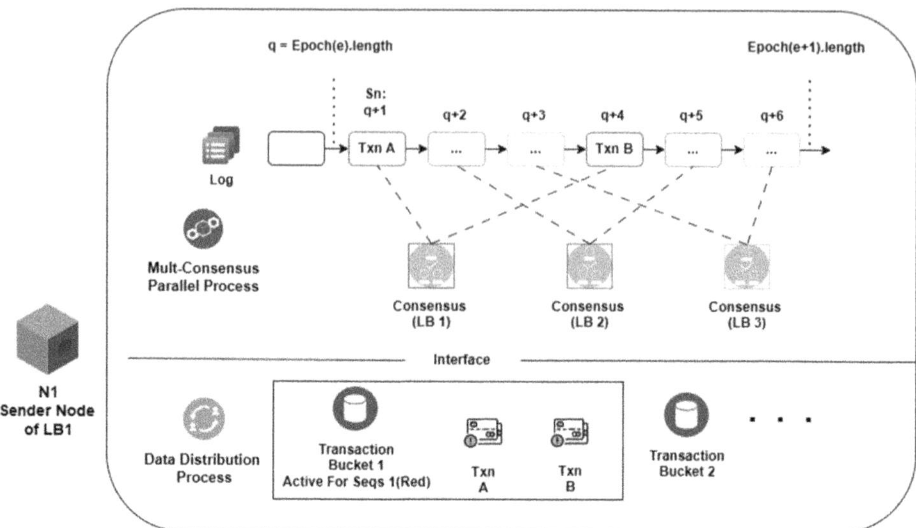

Fig. 2. StreamLine Multi-Consensus Instance Diagram. In epoch e, the log is partitioned into three subsequences via round-robin. For the LB1 instance, the sender is Node 1, the sequence set corresponds to the first subsequence, and the information set for consensus input is derived from Transaction Bucket 1.

– Agreement: If exist two correct nodes, which both deliver
  $\langle LB - DELIVER|(sn, m)\rangle$ and $\langle LB - DELIVER|(sn, m')\rangle$, then $m = m'$,
  meaning these messages are consistent.
– Termination: Every correct node eventually delivers
  $\langle LB - DELIVER|(sn, m)\rangle$ for each sn, where m is from M or is nil.
– Eventual Progress: If nodes deliver nil, some nodes will suspect σ after
  $\langle LB - INIT\rangle$.

Limited Broadcast (LB) ensures that every correct node will eventually complete consensus for each sequence number sn, a fault detector described by [14] and the mechanism allowing nodes to consume sequence numbers by delivering null values nil. We will continue with limited broadcast and achieve the dividing part.

Divide-and-Conquer Strategy. Using the aforementioned limited broadcast primitive, we can achieve consensus on a finite set of transactions within a bounded number of rounds. This provides a foundational idea for parallelizing the consensus process.

Divide Phase

At the start of each epoch, we decompose the consensus process into multiple limited broadcast instances, thereby enabling nodes to participate in several parallel consensus instances. To instantiate each consensus instance, three essential components must be allocated: the sender node, the log subsequence, and the transaction buckets.

First, each consensus node maintains a local log in which entries are identified by sequence numbers sn. To enable parallelism, these sequence numbers are partitioned into non-overlapping subsets (referred to as Seqs) using a round-robin method. By selecting the range from $Epoch(e).Length$ to $Epoch(e+1).Length$, we create a finite set of sequence subsets available for the consensus instances in that epoch.

Next, transactions are mapped into buckets using a hash function based on their payload digest. This structure, termed transaction buckets, distinguishes the transactions that each node should propose and prevents redundant consensus. Each log subsequence (Seq) is then assigned a set of transaction buckets (Buckets) also via a round-robin method, ensuring that each sender node proposes only the transactions contained in its allocated buckets.

Finally, the sender node for each limited broadcast instance is determined by an existing leader selection algorithm, *Leader-Policy()*. The set Leaders(e), computed for epoch e, contains every sender node l that will act as a leader in a consensus instance.

Taking Fig. 2 as an example, it illustrates the scenario in epoch e where Node 1 participates in multiple consensus instances – specifically, three instances – while serving as the leader in the LB1 consensus instance. Initially, the log is divided into six sequence numbers based on the range from $Epoch(e).Length$ to $Epoch(e+1).Length$. A round-robin method is then used to further partition these log entries into three subsequences: $Seq\{q+1, q+4\}$ (displayed in red), $Seq\{q+2, q+5\}$ (blue), and $Seq\{q+3, q+6\}$ (orange). In the current epoch e, Node 1 is selected as one of the leader nodes and takes on the sender role in the LB1 consensus instance. For LB1, its designated sender employs the subsequence $Seq\{q+1, q+4\}$, and the corresponding information set M is derived from Transaction Bucket 1, which contains transactions such as Txn A and Txn B; these transactions are ultimately proposed and committed by Node 1 at log positions $q+1$ and $q+4$, respectively. The other nodes follow a similar process.

Through this dividing method, we ensure liveness: over an infinite execution period, at least one node will (1) not be suspected by all correct nodes (as ensured by the properties of the sequencer broadcast) and (2) eventually be allocated every transaction bucket.

Merge Phase

At the end of each epoch e, nodes merge consensus results, synchronize logs, and create checkpoints by broadcasting signed messages containing the highest sequence number in subsequence $S_n(e)$ and the Merkle root of committed entries. Once $2f+1$ signatures are collected, a stable checkpoint forms, enabling nodes with missing entries to catch up by fetching logs from peers. After all subsequences commit, nodes advance to epoch $e+1$, ensuring no duplicate transactions and maintaining ledger consistency in an asynchronous environment.

Algorithm 2: Transaction Distribution Broadcast Protocol

---

1 **Variables:**
2   m (transaction), acks[] (ACK counter), stable_messages[] (stable messages), $q = f + 1$

3 **Procedure** *TDB-Broadcast(m)*
4   Send $\langle$TDB-CAST $\mid m, \text{node.id}\rangle$ to all nodes

5 **Procedure** *TDB-ACK(m, σ)*
6   Send $\langle$TDB-ACK $\mid m.id\rangle$ to σ

7 **Procedure** *MakePayload()*
8   Select m from *stable_messages*
9   Return $\langle m.digest, m.proof \rangle$

10 **Procedure** *FillPayload*(proposal)
11   digest $\leftarrow$ proposal.m.digest
12     $m \leftarrow$ Retrieve corresponding m from storage by digest
13   Return m

14 **Upon receiving** m :
15   Validate m
16   *TDB-Broadcast(m)*

17 **Upon receiving** $\langle$TDB-CAST $\mid m, \sigma\rangle$ **at node** i:
18   Validate m
19   *TDB-ACK(m, σ)*

20 **Upon receiving** $\langle$TDB-ACK $\mid m.id\rangle$ **at sender** σ:
21   acks[$m.id$] $\leftarrow$ acks[$m.id$] $+ 1$
22   if $acks[m.id] \geq q$ then
23     $\mid$ $m.proof \leftarrow$ threshold-signature(acks[$m.id$], m)
24     $\mid$ *stable_messages*.add(m)

25 **Upon consensus proposal phase:**
26   Propose *MakePayload()*

27 **Upon consensus commit phase:**
28   Commit *FillPayload()*

4.3   Parallel Mechanism Design on Distribution Process

Transaction Distribution Broadcast. The distribution process relies on a Byzantine broadcast primitive, Transaction Distribution Broadcast (TDB), to ensure efficient message delivery. This broadcast ensures nodes eventually receive relevant transactions without waiting for missing loads during consensus.

When node σ receives a transaction m from a client, it broadcasts $\langle$TDB-CAST $\mid m\rangle$ (Algorithm 2, Line 3). Upon receiving this broadcast, other nodes validate m and invoke the *TDB-ACK* procedure by sending back $\langle$TDB-ACK $\mid m.id\rangle$ to σ (Algorithm 2, Line 5). After the sending node σ receives at least $q = f + 1$ ACKs for m (Algorithm 2, Line 24), it uses these to generate proof for m via a threshold signature tool. This proof ensures that m has been

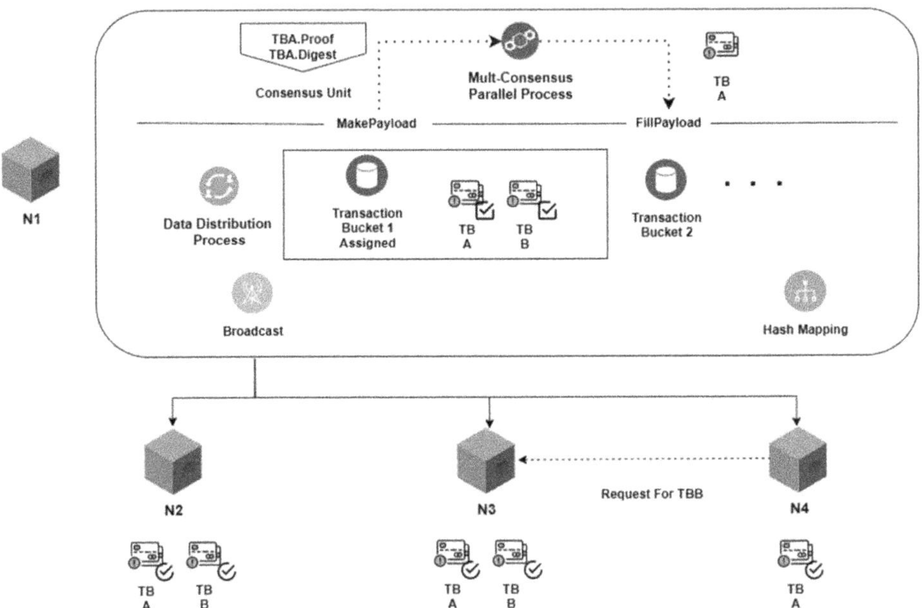

Fig. 3. Node manages transaction buckets locally, each with a transaction pool running TDB protocol. As sending node, it can use the MakePayload to produce consensus payload.

delivered to at least one correct node in the network and guarantees that m can be reliably retrieved under the fault model.

Once proof is generated, the message transitions from a pending state to a stable state and is stored in *stable_messages*. During the consensus proposal phase, the *MakePayload* procedure converts stable messages into consensus load units $\langle m.digest, m.proof \rangle$ (Algorithm 2, Line 7), which are then proposed as the consensus content. During the consensus commit phase, nodes reconstruct the full transaction m using the *FillPayload* procedure, which retrieves m from local storage by its digest (Algorithm 2, Line 10).

We propose two interfaces for consensus:

- *MakePayload* **interface:** This interface converts a certain number of transactions m that have completed the TDB process into consensus load units $\langle m.digest, m.proof \rangle$ for consensus use.
- *FillPayload* **interface:** This interface restores the consensus load units $\langle m.digest, m.proof \rangle$ back to the full transaction m using locally stored load information.

Transaction Logic. Based on the broadcast described above, we can summarize the transaction processing logic into two parts: distribution logic and retrieval logic.

Transaction Distribution Logic

Each node has several transaction buckets for storing transaction requests. When a transaction arrives and is mapped into a certain bucket, is aggregated in the bucket. Once the bucket meets the predefined threshold (by count or load), the transactions are batched into a transaction distribution block $TB\langle digest, payload\rangle$, where $digest$ is the hash digest of the block and $payload$ is the original transaction request information. The sending node then broadcasts $\langle$TDB-CAST$\|TB\rangle$ to other nodes.

Upon collecting enough ACKs, the sending node generates proof for the TB using a threshold signature tool and a stable transaction distribution block $STB\langle digest, payload, proof\rangle$ is formed. This stable transaction distribution block, now containing proof, is ready for consensus. The sending node can extract stable transaction distribution blocks via the MakePayload interface when the corresponding transaction bucket is assigned and use them as consensus content.

During the consensus process, only the digest $digest$ and the proof $proof$ generated during distribution are used as consensus payload units $\langle STB.digest, STB.proof\rangle$ to represent the original transaction information.

Transaction Recall Logic

Ideally, at the end of a consensus round, nodes use digests in the consensus payload units $\langle STB.digest, STB.proof\rangle$ to retrieve the corresponding transactions from local storage and restore them by using the FillPayload interface, then commit them to the ledger. However, in special cases, such as network fluctuations resulting in some nodes failing to receive a transaction block, the sending node can still generate $proof$ if it gets enough ACKs from other nodes. This block is then included in the consensus process, and nodes missing it can request the block upon receiving the consensus proposal from other nodes that signed the ACK. If the payload arrives before the end of consensus, the node uses the FillPayload interface to restore it via the FillPayload interface and commit it to the ledger. In the worst-case scenario, network failures or Byzantine nodes may prevent the request from reaching the intended node, and the node fails to restore the payload by the end of the consensus process. Still without blocking the process, the node will keep retrying to retrieve the missing block with a timeout retry mechanism using a new thread. Since the parallel consensus protocol uses sequences for submission, log commitment does not require strict ordering, so as long as the node retrieves the payload before the view timeout in that epoch, the impact on consensus is minimal. In Byzantine systems where malicious nodes are fewer than one-third, there's only about 1% chance of failure after four attempts, with ten retries nearly guaranteeing payload retrieves before a view change.

Figure 3 illustrates how Node 1 distributes transactions TB1 and TB2. Node 1 first collects transaction requests in Transaction Bucket 1 and aggregates them into TB1 and TB2, which are then broadcast to Node 2, Node 3, and Node 4 via the TDB protocol. However, Node 4 fails to receive TB2. Upon receiving TB1 or TB2, nodes return ACKs to Node 1 (represented by circular check marks). Once enough ACKs are collected, Node 1 generates proof, allowing them to transition into a stable state and preparing them for consensus (indicated by square check marks). During the consensus proposal phase, Node 1 uses *MakePayload* to include only the proof and digest of TB1 instead of its full payload. After consensus, *FillPayload* retrieves the full payload from storage for log commitment. For TB2, since Node 4 lacks the full payload, it requests it from other nodes (e.g., Node 3) using the proof and digest in the consensus proposal. Once retrieved, the payload is committed to the log.

4.4 Load Balancing

Load Estimation Strategy. In our design, the consensus content consists only of a hash and related proofs. Therefore, load balancing is not based on the consensus content itself but rather on the data structure used in the distribution process–Transaction distribution Blocks (TB). Based on this, we use the number of transaction distribution blocks sent by a node in the recent past as a load measurement indicator. A sliding window mechanism is employed to effectively measure the recent distribution rate. We maintain a fixed-size window queue while recording the timestamps of each broadcast of transaction distribution blocks. By using the earliest and current timestamps, along with the queue length, the recent distribution rate of transaction distribution blocks can be computed. Using parameters such as the distribution threshold and the node's bandwidth configuration, the node's load status can be inferred.

Load Forwarding Strategy. In our design, load forwarding by a node does not disrupt the consensus process. When a node forwards the load, it first determines which nodes to forward to. To prevent the selection of Byzantine nodes, which may cause forwarding failures and further congest transaction requests, we employ a Random-K strategy for selecting forwarding nodes, which involves randomly selecting K nodes from a set of candidates to perform a task, helping to distribute load and reduce the risk of selecting malicious nodes. Upon receiving a load forwarding request, the selected node broadcasts on behalf of the original sender. Other nodes then reply to the original sender with an acknowledgment (ACK) upon receiving the related transaction distribution block.

5 Theoretical Analysis

5.1 Security Analysis

Our model extends the partially synchronous Byzantine fault-tolerant (BFT) framework by introducing multi-consensus instance parallelism and decoupled transaction distribution to enhance system performance. To ensure that these optimizations do not compromise consistency or liveness, we analyze their impact separately.

First, we examine the effect of decoupling transaction distribution from consensus. In our approach, transactions are pre-distributed before the consensus phase, using only transaction digests during consensus instead of full payloads. Since this mechanism does not interfere with the consensus process itself, it preserves system consistency. Moreover, transactions are assigned to buckets via a hash mapping function and distributed in a round-robin manner. Given a bounded confirmation time, every valid transaction is guaranteed to reach at least one correct node and be proposed for consensus, preventing Byzantine nodes from indefinitely delaying transactions. This ensures system liveness.

Next, we analyze the impact of multi-consensus instance parallelism. This design allows multiple independent consensus instances to execute concurrently, each adhering to the standard BFT consensus protocol, thereby maintaining consistency. To synchronize the global state, a checkpoint mechanism is introduced at the end of each epoch, ensuring that all nodes reach agreement on committed transactions. Additionally, as long as the leader selection algorithm guarantees eventual participation–where every correct node is chosen as a leader within a bounded number of rounds–transactions will eventually be proposed and committed by a non-malicious leader. This guarantees system liveness.

In summary, our system design preserves both consistency and liveness while introducing optimizations to improve efficiency.

5.2 Performance Analysis

We evaluate the impact of our parallelization strategies on system performance from two key aspects: end-to-end transaction processing latency and overall system throughput.

First, we assess their effect on transaction processing latency. Decoupling transaction distribution from consensus shifts transaction handling to an independent broadcast phase, introducing additional message propagation steps, such as *TDB-CAST* and *TDB-ACK*, as well as aggregation delays for forming transaction distribution blocks (TBs). These steps slightly increase latency compared to non-parallel systems. However, multi-consensus instance parallelism introduces no extra steps beyond *epoch checkpoint synchronization*, allowing nodes to engage in multiple instances, maximizing resource utilization and minimizing idle time. As a result, while transaction distribution introduces some additional delay, the parallel consensus mechanism mitigates waiting times and optimizes resource efficiency. In scenarios with low resource utilization, this parallelism reduces average transaction processing latency.

Next, we evaluate the impact on system throughput. Transaction distribution broadcasting allows transactions to be disseminated before consensus begins, enabling the consensus process to focus solely on protocol execution while reducing communication overhead. Additionally, shifting the transaction distribution burden from a *single leader node* to the entire network prevents bandwidth bottlenecks at the leader. This transforms transaction processing from a *burst mode* centered on a single node to a *balanced mode* distributed across multiple nodes, significantly enhancing throughput. Likewise, multi-consensus instance parallelism enables nodes to participate in multiple consensus instances concurrently. When one instance enters a waiting phase, nodes can continue advancing others, thereby accelerating overall transaction commitment. Similar to the transaction distribution mechanism, this approach alleviates computational and communication bottlenecks by evenly distributing workloads across the network. Consequently, our system achieves a notable improvement in throughput through better resource utilization. Moreover, in real-world environments where network resources are unevenly distributed or client requests exhibit locality, our approach dynamically balances the workload across the network, making it more adaptable to complex and unbalanced conditions.

6 Experimental Evaluation

6.1 Experimental Setup

We implemented our protocol prototype, Streamline, in approximately 3000 lines of Go code and conducted experiments on 33 Alibaba Cloud ECS `e-c1m2.large` servers. Each server was equipped with a dual-core CPU and 4 GB of memory, running Ubuntu Server 20.04. All blockchain nodes and clients were deployed on separate instances. The system was tested in a wide-area network (WAN) environment, where each node had a 100 Mbps bandwidth and communicated via public IPs to evaluate system performance.

In the workload design, a client node was deployed on a separate cloud server instance, sending transaction requests to randomly selected blockchain nodes at a fixed rate of 6144 transactions per second. At this rate, we confirmed that the computational and network resources of all nodes were nearly saturated. In the scalability performance evaluation, workloads were evenly distributed across the network, whereas in the adaptive performance evaluation, we introduced an unbalanced load environment to evaluate the system's adaptability to complex network conditions.

For protocol configuration, we selected four blockchain consensus protocols as baselines for comparison, each representing a distinct direction in consensus protocol research. Streamlet [4] and EBFT [17] exemplify the approach of simplifying the consensus process, focusing on reducing protocol complexity and communication rounds. In our experiments, the reproduced implementation of EBFT [17] is denoted as LBFT. Stratus [8] represents the strategy of optimizing consensus through procedural parallelization, enhancing processing efficiency via structural improvements. ISS [19] adopts a multi-leader parallel consensus design, leveraging concurrent processing to further improve system performance. To ensure a fair comparison, we maintained consistency in common parameters across all protocols while optimizing protocol-specific parameters accordingly.

6.2 Scalability Performance Evaluation

In this experiment, we evaluate the scalability differences among various consensus protocols. The experiment is conducted in a WAN environment, where we test our protocol, Streamline, alongside four comparative protocols: ISS [19], Stratus [8], Streamlet [4], and LBFT [17]. The transaction request size is set

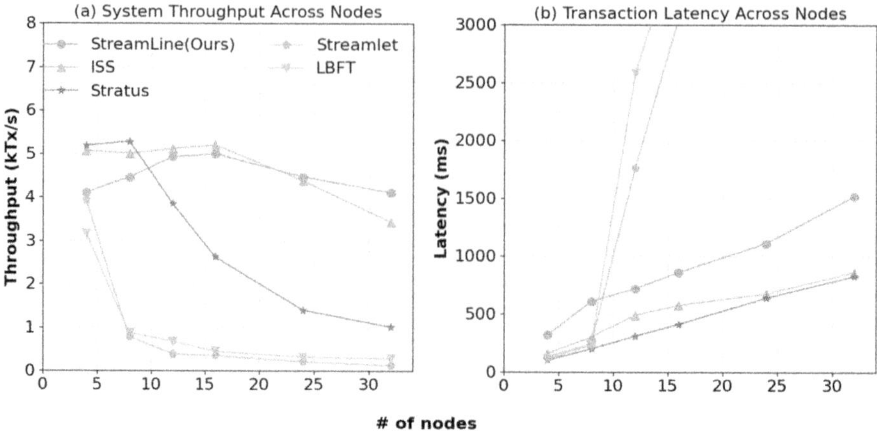

Fig. 4. Throughput and latency evaluation of scalability differences among consensus protocols.

to 512 bytes, with each node running on an independent cloud server instance and communicating via a 100 Mbps public network. To analyze scalability under different network sizes, we test with 4, 8, 12, 16, 24, and 32 nodes, and measure the throughput and latency. The results are presented in Fig. 4.

Throughput Analysis. Figure 4(a) illustrates the change in transaction throughput as the number of nodes increases. Overall, as the network scales, transaction throughput declines for all protocols. In single-leader protocols (LBFT and Streamlet), throughput stagnates when the number of nodes exceeds 16, indicating a significant limitation in handling increased workload. Among the three parallelized protocols (Streamline, ISS, and Stratus), throughput is initially comparable. However, as the network expands, different parallelization strategies exhibit distinct behaviors. ISS and Streamline, which employ multi-leader consensus, sustain or even increase throughput as the system scales, only experiencing a notable decline beyond 16 nodes. This is because multi-leader protocols distribute the consensus workload, effectively balancing network pressure as the system scales. In contrast, Stratus employs a shared memory pool for process parallelization. However, since it still relies on a single-leader architecture, the coordination overhead increases with network size, leading to a sharper decline in throughput in larger-scale deployments.

Latency Analysis. Figure 4(b) presents transaction processing latency as the number of nodes increases. In general, latency rises with network size. Protocols designed with parallel processing–Streamline, ISS, and Stratus–exhibit significantly lower latencies compared to simplified consensus protocols–LBFT and Streamlet. As the number of nodes increases, LBFT and Streamlet experience a sharp rise in transaction latency, indicating that their processing capacities have reached their limits, making further scaling impractical. In contrast, parallelized protocols ISS and Stratus benefit from reduced transaction waiting times, resulting in lower latencies. However, Streamline exhibits slightly higher latency due to additional processing steps, including transaction bucket aggregation, broadcast distribution, and allocation to consensus nodes, introducing moderate queuing overhead. This is acceptable, as the latency curve for Streamline grows more gradually with scale. This indicates that the overhead primarily stems from its processing pipeline rather than a scalability bottleneck in protocol design.

These results demonstrate the significant impact of parallelization on protocol scalability. Streamline incorporates a multi-dimensional parallelization design, allowing it to maintain performance comparable to state-of-the-art consensus protocols while exhibiting greater stability in large-scale network environments. By dynamically adjusting allocation based on node waiting times and real-time resource utilization, the system effectively balances workload, preventing single-node bottlenecks.

6.3   Adaptive Performance Evaluation

In this experiment, we evaluate the adaptability of Streamline in real-world environments, particularly under varying load conditions and increasing transaction sizes. While most existing protocols lack dynamic load balancing capabilities, Streamline incorporates a built-in adaptive mechanism. To contextualize its effectiveness, we compare Streamline with ISS [19]–a structurally similar multi-leader consensus protocol that does not implement load balancing. Notably, enabling such functionality in ISS would require significant architectural redesign, as verified through code inspection. The experiment is conducted in a WAN environment, with all nodes running on independent cloud instances. It consists of two parts: load balancing evaluation and load handling evaluation, as illustrated in Figs. 5 and 6.

Load Balancing Evaluation. Figure 5 presents the impact of uniform and skewed load distributions on the performance of Streamline and ISS. We maintain a fixed transaction size of 512 bytes and evaluate both protocols under two traffic scenarios: uniform load, where requests are evenly distributed, and skewed load, where requests are concentrated. In the skewed load scenario, 25% of the nodes are bandwidth-limited to 50 Mbps, and 70% of requests are directed to a single node, simulating a severe imbalance. Experimental results show that Streamline exhibits only a modest performance degradation–throughput drops by around 10% when switching from uniform to skewed load. In contrast, ISS suffers a severe decline, with throughput reduced to approximately half of its baseline. In terms of latency, both protocols experience increases under skewed load. Streamline's average latency doubles compared to the uniform case, while ISS exhibits a roughly threefold latency increase. Moreover, ISS reaches a saturation point at 32 nodes, where its latency surges due to accumulated network congestion–indicating the lack of adaptability in static load handling. This indicates that Streamline's adaptive load balancing mechanism effectively mitigates performance bottlenecks. ISS, lacking such a mechanism, exhibits early signs of saturation, especially under skewed load. When a node becomes overloaded or bandwidth-limited, Streamline dynamically redistributes its workload

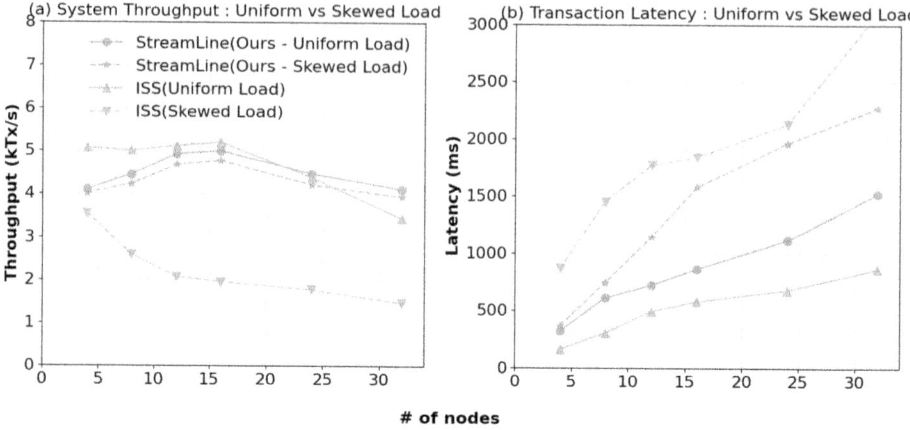

Fig. 5. Evaluation of Streamline's load balancing under bandwidth constraints and uneven request distributions.

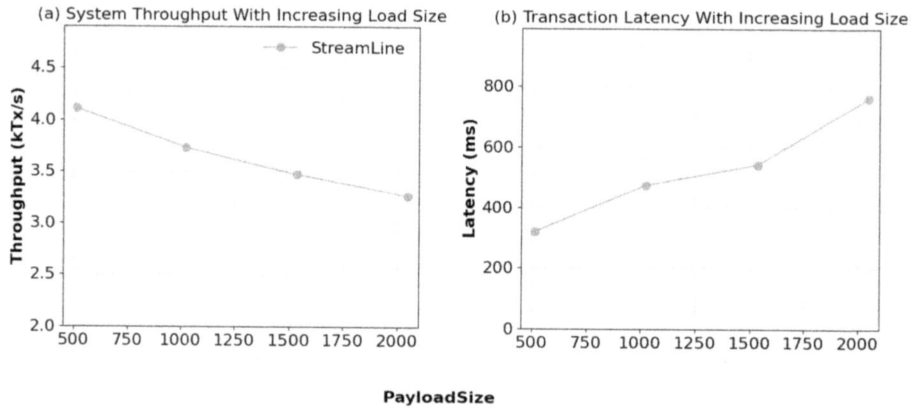

Fig. 6. Evaluation of Streamline's performance with increasing transaction sizes from 512B to 2000B.

to other nodes, maintaining stable consensus execution and preventing system-wide congestion. Furthermore, due to Streamline's modular and parallelized process design, its load balancing mechanism can be easily fine-tuned using lightweight scheduling algorithms, providing high flexibility and scalability.

Load Handling Evaluation. Figure 6 illustrates the impact of increasing transaction sizes on system performance. We analyze system performance under increasing transaction sizes using a 4-node WAN setup, where transaction size is gradually increased from 512 bytes to 2000 bytes. The results show that larger transaction sizes negatively impact system performance, with throughput decreasing to 87% of the original level and latency increasing to 2.37× of the baseline when the transaction size reaches four times the initial value. Notably, the impact on throughput is relatively minor, whereas latency is significantly affected. This phenomenon can be attributed to the TDB broadcast mechanism, which processes transaction digests instead of full transaction payloads during consensus. As a result, increasing transaction size has minimal impact on the consensus process, allowing the system to maintain relatively stable throughput even under high-load conditions.

7   Conclusion

We propose Streamline, an extensible and efficient BFT consensus protocol that integrates horizontal multi-leader parallelism with vertical decoupling of data distribution and consensus. By adopting multi-dimensional parallelism, Streamline reduces bottlenecks and enhances system scalability. Experimental results demonstrate that Streamline achieves competitive throughput compared to existing multi-leader protocols while maintaining stable performance under unbalanced load conditions. Its dynamic adjustment capability ensures efficient resource utilization across diverse network environments. Future work will explore hierarchical protocol designs to further enhance scalability and adaptability in large-scale deployments.

References

1. Androulaki, E., et al.: Hyperledger fabric: a distributed operating system for permissioned blockchains. In: Proceedings of the Thirteenth EuroSys Conference. EuroSys '18, ACM, New York, NY, USA (2018). https://doi.org/10.1145/3190508.3190538
2. Azaria, A., Ekblaw, A., Vieira, T., Lippman, A.: MedRec: using blockchain for medical data access and permission management. 2016 2nd International Conference on Open and Big Data (OBD), pp. 25–30 (2016). https://doi.org/10.1109/OBD.2016.11
3. Castro, M., Liskov, B.: Practical byzantine fault tolerance. In: Proceedings of the Third Symposium on Operating Systems Design and Implementation, pp. 173–186. USENIX Association, USA (1999)
4. Chan, B.Y., Shi, E.: Streamlet: textbook streamlined blockchains. In: Proceedings of the 2nd ACM Conference on Advances in Financial Technologies, AFT '20, pp. 1–11. Association for Computing Machinery, New York, NY, USA (2020). https://doi.org/10.1145/3419614.3423256
5. Charapko, A., Ailijiang, A., Demirbas, M.: PigPaxos: devouring the communication bottlenecks in distributed consensus. In: Proceedings of the 2021 International Conference on Management of Data, pp. 235–247. Association for Computing Machinery (2021)
6. Di, S.U., Michalas, A.: Blockchain in online gaming: a survey of solutions and opportunities. Int. J. Comput. Games Technol. **2017**, 1–8 (2017). https://doi.org/10.1155/2017/5323168
7. Dwork, C., Lynch, N., Stockmeyer, L.: Consensus in the presence of partial synchrony. J. ACM, 288–323 (1988)

8. Gai, F., Niu, J., Beschastnikh, I., Feng, C., Wang, S.: Scaling blockchain consensus via a robust shared mempool (2022)
9. Gilad, Y., Hemo, R., Micali, S., Vlachos, G., Zeldovich, N.: Algorand: scaling Byzantine Agreements for cryptocurrencies. In: Proceedings of the 26th Symposium on Operating Systems Principles, pp. 51–68 (2017)
10. Gueta, G.G., et al.: SBFT: a scalable and decentralized trust infrastructure. In: 2019 49th Annual IEEE/IFIP International Conference on Dependable Systems and Networks (DSN), pp. 568–580. IEEE (2019)
11. Guo, Y., Wang, C., Shao, Y.: Blockchain application and outlook in the banking industry. Financ. Innov. **2**(1), 24 (2016). https://doi.org/10.1186/s40854-016-0034-9
12. Gupta, S., Hellings, J., Sadoghi, M.: Revisiting consensus protocols through wait-free parallelization. arXiv preprint arXiv:1908.01458, August 2019. https://arxiv.org/abs/1908.01458
13. Kotla, R., Alvisi, L., Dahlin, M., Clement, A., Wong, E.: Zyzzyva: speculative Byzantine fault tolerance. In: Proceedings of Twenty-First ACM SIGOPS Symposium on Operating Systems Principles, pp. 45–58 (2007)
14. Malkhi, D., Reiter, M.: Unreliable intrusion detection in distributed computations. In: Proceedings of the 10th IEEE Workshop on Computer Security Foundations, p. 116. IEEE Computer Society (1997)
15. Miller, A., Xia, Y., Croman, K., Shi, E., Song, D.: The honey badger of BFT protocols. In: Proceedings of the 2016 ACM SIGSAC Conference on Computer and Communications Security, pp. 31–42 (2016)
16. Moraru, I., Andersen, D.G., Kaminsky, M.: Egalitarian Paxos. In: ACM Symposium on Operating Systems Principles (2012)
17. Niu, J., et al.: EBFT: simplifying BFT consensus through egalitarianism (2023). https://arxiv.org/abs/2012.01636
18. Poon, J., Dryja, T.: Scaling bitcoin to billions of transactions per day. In: 2016 IEEE Symposium on Security and Privacy (SP), pp. 3–37 (2016). https://doi.org/10.1109/SP.2016.10
19. Stathakopoulou, C., Pavlovic, M., Vukolić, M.: State machine replication scalability made simple. In: Proceedings of the Seventeenth European Conference on Computer Systems, pp. 17–33. Association for Computing Machinery (2022). https://doi.org/10.1145/3492321.3519579
20. Yin, M., Malkhi, D., Reiter, M.K., Gueta, G.G., Abraham, I.: HotStuff: BFT consensus with linearity and responsiveness. In: Proceedings of the 2019 ACM Symposium on Principles of Distributed Computing, pp. 347–356. Association for Computing Machinery, New York, NY, USA (2019)
21. Yu, D., Wang, J., Li, L., Jiang, W., Liu, C.: RecAGT: shard testable codes with adaptive group testing for malicious nodes identification in sharding permissioned blockchain. In: Tari, Z., Li, K., Wu, H. (eds.) International Conference on Algorithms and Architectures for Parallel Processing, pp. 398–418. Springer, Singapore (2023). https://doi.org/10.1007/978-981-97-0859-8_24

PliKOS: Pre-warming Serverless Functions Under Pulsed Loads

Tengtao Xiao[1], Zhengong Cai[1], Chi Zhang[1], Pu Zhang[1], and Bowei Yang[2(✉)]

[1] School of Software Technology, Zhejiang University, Ningbo, Zhejiang, China
{tengtaox,cstcaizg,puzhang,zhangchi_se}@zju.edu.cn
[2] School of Aeronautics and Astronautics, Zhejiang University,
Hangzhou, Zhejiang, China
boweiy@zju.edu.cn

Abstract. Cold start latency is a fundamental challenge in serverless computing. The alternating burst and non-burst phases of function invocations complicate pre-warming, and existing strategies often lack phase-aware modeling, leading to limited effectiveness and high idle resource cost, with poor trade-offs between responsiveness and efficiency. PliKOS is a phase-aware pre-warming strategy designed for pulsed load patterns. It uses I-DBSCAN to identify burst phases from invocation intervals, avoiding full-sequence prediction and reducing overhead. It also adjusts the number of pre-warmed instances based on execution time, request rate, and latency objectives. PliKOS reduces cold start rates by over 95%, lowers idle resource cost by 19%, shortens service time by 13%, and cuts prediction error and overhead by up to 86.4% and 91.4%. By modeling pulsed workloads with phase awareness, PliKOS offers an efficient and practical solution for cold start mitigation in serverless environments.

Keywords: Serverless Computing · Cold Start · Pre-warming Optimization · Pulsed Load Modeling

1 Introduction and Background

Serverless computing has recently become a key cloud trend due to flexible resource management and high scalability [6,10]. Function-as-a-Service (FaaS), a core serverless model, offers event-driven execution for efficient, convenient development [7,13]. Keep-alive strategies reduce latency by maintaining function availability. However, such strategies introduce additional resource and cost overheads, and with limited server resources, setting an appropriate keep-alive duration is critical for minimizing cold starts and cost.

The growing diversity of workloads makes tuning keep-alive and pre-warming strategies more complex [20,23,25]. Existing keep-alive strategies are generally classified into three categories: (1) Fixed duration strategies, such as the default settings of cloud providers like AWS and OpenWhisk [1,2,5,8,22]; (2) Dynamic strategies based on data statistics [20,26]; and (3) Predictive models that adjust the keep-alive duration dynamically based on historical invocation patterns, considering the frequency and trends of function calls [11,12,19].

© The Author(s), under exclusive license to Springer Nature Singapore Pte Ltd. 2026
H. Liu et al. (Eds.): ICA3PP 2025, LNCS 16381, pp. 23–36, 2026.
https://doi.org/10.1007/978-981-95-8399-7_2

Existing strategies struggle with heterogeneous invocation patterns that alternate between bursts and idle periods. Fixed and predictive methods often fail to balance cold start latency and resource overhead. Short keep-alive durations cause frequent cold starts during bursts, while long durations waste resources during idle phases. To address these issues, we propose the following key contributions:

1. Pulsed Load Pattern Modeling. We observe that many serverless functions exhibit invocation patterns composed of alternating burst and non-burst phases, which we define as the Pulsed Load Pattern. To capture this structure, we introduce a lightweight clustering method that identifies burst phases based on the density of invocation intervals. This phase-aware modeling enables efficient pre-warming without relying on full-range sequence prediction, thereby reducing overhead and improving accuracy.

2. Dynamic Instance Pre-Warming for Burst Phases. We design a dynamic pre-warming strategy tailored to burst phases. It adjusts the number of warm instances based on each function's execution time, request rate, and service-level latency objectives. During burst phases, PliKOS provisions sufficient instances to meet performance targets under concurrent workloads. During non-burst periods, idle instances are released early to reduce resource consumption. This phase-aware strategy enables effective cold start mitigation while maintaining cost efficiency across diverse functions.

2   Observation and Motivation

In this chapter, we present several key observations that serve as a motivation and entry point for our proposed improvements. This analysis lays the foundation for our proposed improvements, motivating a shift towards a more adaptive strategy.

2.1   Differences in Keep-Alive Function

In serverless architectures, cold starts refer to the latency incurred when a function is reactivated after remaining idle for an extended period, which becomes more severe under high concurrency. To mitigate this, keep-alive strategies are commonly used to temporarily retain the function environment, enabling faster startups [3,9].

Different functions respond differently to varying keep-alive durations. While increasing keep-alive time generally reduces cold start rates, the benefit diminishes with longer durations, while idle cost rises sharply. Figure 1 illustrates these trends. A fixed-duration strategy fails to handle this diversity, indicating the need for dynamic, per-function keep-alive tuning.

2.2   Pulsed Load Pattern

Through our observation of serverless function invocation traces, we have identified a recurring invocation pattern, which we refer to as the Pulsed Load Pattern, where the invocation frequency exhibits clear cyclical fluctuations. Figure 2 reveals that function invocations exhibit a pulsed pattern with alternating burst and idle periods, confirming the prevalence of cyclic invocation rhythms.

Specifically, during the non-burst phase, functions are triggered rapidly and continuously, with invocation intervals ranging from seconds to minutes. This often occurs when the system needs to handle a large number of concurrent requests, such as during peak traffic periods or batch task execution. Subsequently, the system transitions into a non-burst phase during which the frequency of function invocation drops significantly. The intervals between requests become much longer compared to the concentrated invocation phase, ranging from several minutes to hours, or even longer. Idle periods occur during low usage, e.g., off-hours or post-task.

This behavior, driven by user patterns and workload characteristics, challenges conventional keep-alive strategies and highlights the need for phase-aware scheduling.

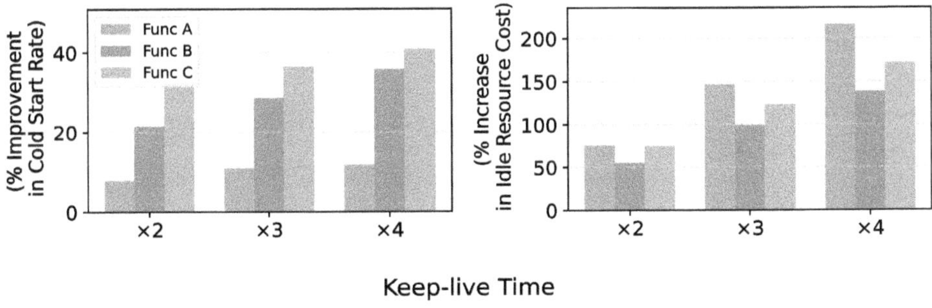

Fig. 1. Cold start improvement and idle resource cost increase under varying keep-alive durations.

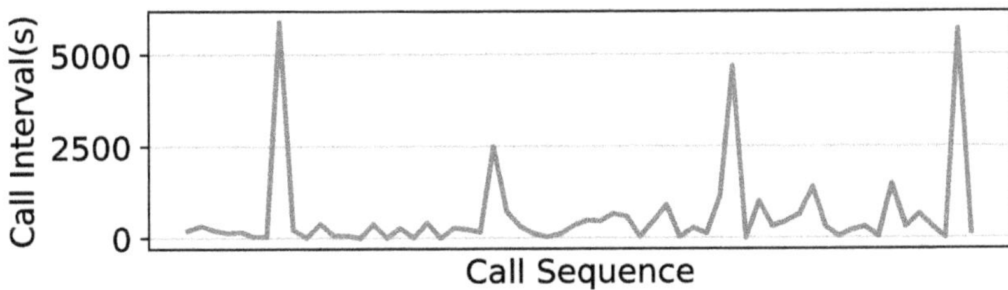

Fig. 2. Pulsed invocation pattern with alternating burst and non-burst phases.

2.3   Function Invocation Predictor

Existing research has primarily focused on improving prediction accuracy through various modeling approaches [19,20], or on mitigating the cost of keep-alive failures by leveraging heterogeneous server resources. However, unpredictable invocation behavior in serverless systems makes full-sequence forecasting difficult and costly.

In the case of pulsed load patterns, there exists a substantial gap between the invocation intervals in non-burst and non-burst phases, and this extreme disparity makes accurate prediction more difficult. Moreover, fine-grained prediction of invocation intervals during the burst phase is often not cost-effective. Given the burst frequency of function invocations in this phase, it's more practical to cover burst periods with keep-alive than predict every call.

Based on these observations, we model the pulsed load pattern in serverless function invocations and propose a novel keep-alive optimization strategy tailored to this behavior.

3   PliKOS: Key Ideas and Design

PliKOS consists of two key components: Invocation Predictor and Keep-Alive Duration Setter (Fig. 3).

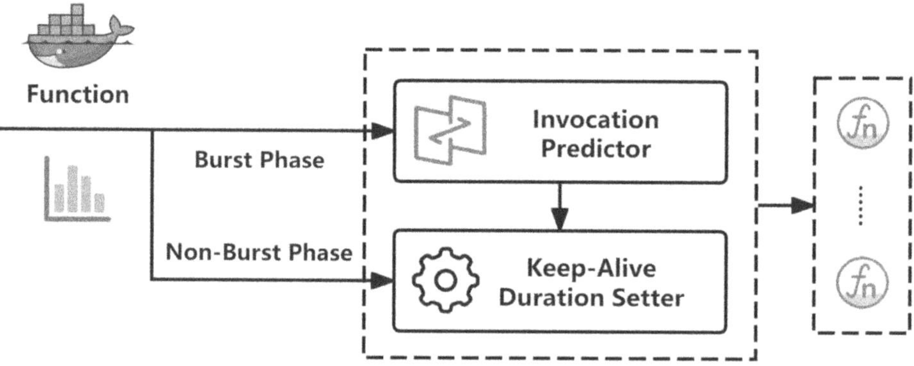

Fig. 3. Overview of the PliKOS system design.

3.1   Invocation Predictor

According to tracing data from Microsoft Azure, nearly 98% of serverless function invocations exhibit some form of periodicity [19]. However, the intervals between consecutive invocations often show high volatility and uncertainty, which makes interval prediction methods based on probability histograms or traditional time series models less effective in such scenarios. For example, Wild adopts an ARIMA model for predicting function invocation intervals, while icebreaker enhances this by introducing harmonic modeling to better capture periodic patterns [19,20].

Modeling the entire invocation sequence ignores the sharp contrast between burst and idle phases. During bursts, frequent calls make interval prediction costly and redundant; during idle periods, sparse data increases prediction errors. Hence, full-range predictors incur high overhead without much gain. In contrast, during burst phases, we prefer to maintain a number of warm function instances using keep-alive strategies to improve hit rates, rather than relying on continuous interval prediction.

Therefore, our research shifts focus to modeling the pulsed load behavior of serverless function invocations, with the goal of identifying and forecasting the onset of burst phases. PliKOS employs an Interval-based DBSCAN (I-DBSCAN), an unsupervised clustering algorithm based on function call interval density, to identify burst call stages. Unlike traditional timestamp-based clustering approaches, I-DBSCAN operates directly on the interval features between adjacent invocations, avoiding the impact of temporal drift and sparse noise on clustering quality. To improve discrimination, I-DBSCAN adopts a two-dimensional density-based strategy that jointly considers the temporal position and the local invocation density of each function call. Specifically, for each invocation in the time series, we define the feature vector as:

$$\mathbf{x}_i = [t_i, \Delta t_i] \tag{1}$$

where t_i denotes the timestamp of the i-th invocation, and $\Delta t_i = t_{i+1} - t_i$ represents the interval between the current and next invocation. Both dimensions are normalized, and the DBSCAN neighborhood radius is set to the 95th percentile of the invocation interval distribution. With this configuration, I-DBSCAN can effectively identify compact, high-density clusters of invocations over time, where each cluster naturally corresponds to a distinct burst invocation phase.

After identifying burst phases, PliKOS extracts their start times and uses Facebook Prophet to forecast the next burst onset. The predicted window then guides instance pre-warming and keep-alive duration decisions.

3.2   Keep-Alive Duration Setter

In the pulsed load invocation pattern of serverless functions, the burst phase typically accounts for the vast majority of requests within each cycle. Therefore, reducing the cold start rate hinges on optimizing the function hit rate during this burst phase. Given that serverless functions generally have short execution durations, with studies showing that approximately 50% of functions complete in less than one second [20], subsequent requests during the burst phase can often reuse recently executed function instances that are still in a warm state. Based on this observation, we first need to determine how many function instances should be pre-warmed. The following formula illustrates how we compute this value:

$$N_{\text{warm}} = \left\lceil R \cdot T_{\text{exec}} \cdot \left(1 - \frac{SLO}{T_{\text{cold}} + T_{\text{exec}}}\right) \right\rceil \tag{2}$$

R represents the average request rate (QPS) during the target time window; T_{exec} denotes the average execution time of the function; T_{cold} is the cold start latency, which can take a long time relative to the function execution [25]; SLO is the maximum allowable response time threshold.

The first term, $R \cdot T_{\text{exec}}$, indicates the baseline number of instances required under steady-state conditions to handle a continuous stream of incoming requests. The adjustment factor $\left(1 - \frac{SLO}{T_{\text{cold}} + T_{\text{exec}}}\right)$ reflects the proportion of requests at risk of violating the SLO due to cold starts. It quantifies the extent to which pre-warming is needed to compensate for the additional latency introduced by cold starts.

As the sum $T_{\text{cold}} + T_{\text{exec}}$ becomes larger relative to the SLO, the likelihood of violating the latency target increases, and thus the necessity for pre-warming grows accordingly. Finally, N_{warm} is rounded up to ensure the result is an integer number of instances.

Similar to most FaaS platforms and existing keep-alive strategies, an appropriate keep-alive duration must also be determined for function instances to ensure that subsequent requests can be served by warm containers. The keep-alive duration is set to the 95th percentile of burst-phase intervals, balancing coverage and efficiency. During non-burst periods, early termination of idle instances helps reduce unnecessary overhead with minimal impact on user experience. Because the number of cold starts caused is relatively small compared to the overall volume of service requests, making this trade-off acceptable.

Algorithm 1: I-DBSCAN

Input: Dataset $D = \{x_i = (t_i, \Delta t_i)\}$; neighborhood radius ε; minimum points $MinPts$

Output: Cluster labels $C = \{c_1, c_2, ..., c_n\}$

1 Sort D by t_i in ascending order;
2 Normalize each $x_i = (t_i, \Delta t_i)$ to obtain X;
3 Initialize all points as unvisited and cluster label $k \leftarrow 0$;
4 **foreach** *point $p \in X$* **do**
5   **if** *p is visited* **then**
6   continue;
7   Mark p as visited;
8   $N_\varepsilon(p) \leftarrow \{q \in X : d(p,q) \leq \varepsilon\}$;
9   **if** $|N_\varepsilon(p)| < MinPts$ **then**
10   Label p as noise;
11   **else**
12     $k \leftarrow k + 1$;
13   Assign cluster label k to p;
14   **foreach** *point $q \in N_\varepsilon(p)$* **do**
15   **if** *q is not visited* **then**
16   Mark q as visited;
17         $N_\varepsilon(q) \leftarrow \{r \in X : d(q,r) \leq \varepsilon\}$;
18   **if** $|N_\varepsilon(q)| \geq MinPts$ **then**
19           $N_\varepsilon(p) \leftarrow N_\varepsilon(p) \cup N_\varepsilon(q)$;
20   **if** *q has no cluster label* **then**
21   Assign cluster label k to q;

---

4 Experiment and Analysis

4.1 Experimental Setup

Workload and Metric Explanation
To simulate realistic workloads, we use Azure traces containing two weeks of invocation data and select representative applications from ServerlessBench [27], covering diverse workloads (e.g., image processing, analytics), help ensure evaluation under real-world conditions.

We evaluate PliKOS using three key metrics:

Service Time: Average time from request initiation to completion, including execution and delays from cold starts or scheduling.

Cold Start Rate: Percentage of requests requiring cold starts due to unavailable warm instances. High rates imply suboptimal instance management.

Idle Resource Cost: Time wasted keeping inactive instances alive, reflecting resource overhead without useful work.

PliKOS is deployed on an x86-64 Linux machine equipped with sixteen 36-core Intel(R) Xeon(R) Platinum 8374C CPUs and 128 GB of memory.

4.2 Identification of Pulsed Load Invocation Patterns

Modeling invocation patterns is key to designing efficient keep-alive strategies. In this section, we present the advantages of using the I-DBSCAN algorithm to identify the burst and non-burst invocation phases of functions. As illustrated in Fig. 4, I-DBSCAN detects burst phases via time and density features. Specifically, I-DBSCAN relies on two key parameters:

eps, which defines the neighborhood radius for clustering. We determine its value by normalizing the 95th percentile of call intervals within a sliding time window.

min_samples, which specifies the minimum number of requests required to form a valid cluster. This parameter prevents excessive resources from being allocated to very small clusters and is typically set based on system load or empirical thresholds. In our experiments, we set ***min_samples*** to 5.

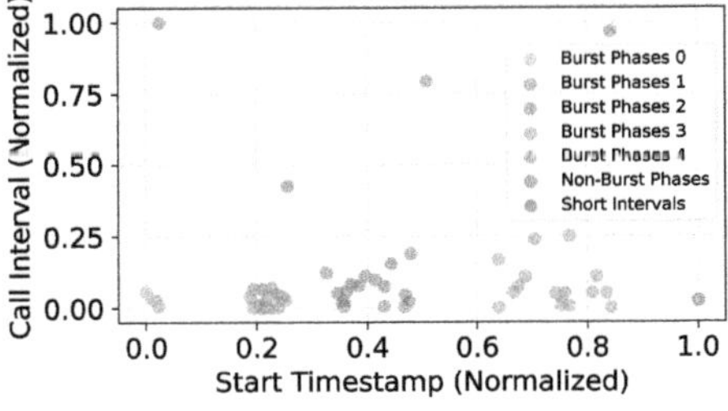
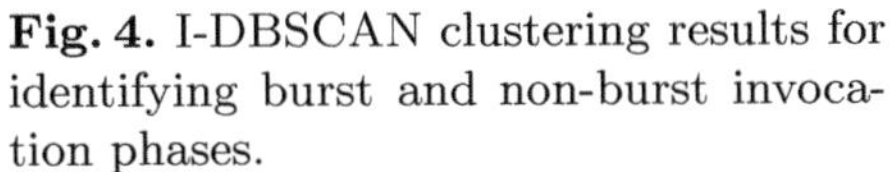

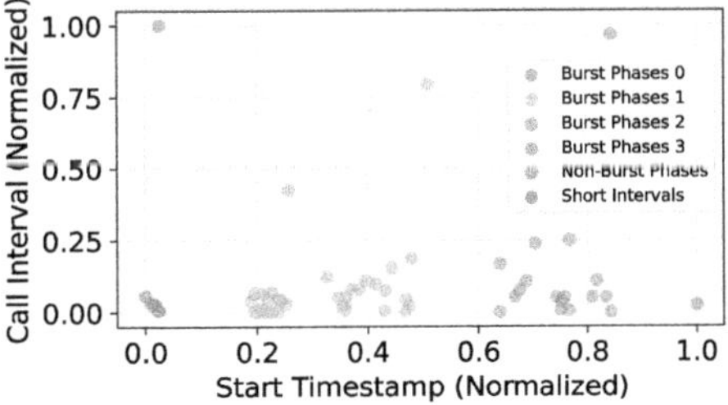

Fig. 4. I-DBSCAN clustering results for identifying burst and non-burst invocation phases.

Fig. 5. Clustering accuracy improved with combined time and density features.

Through I-DBSCAN clustering, the High Phases in the figure clearly indicate several distinct burst invocation stages. These phases not only exhibit tight temporal proximity in invocation time but also have significantly shorter call intervals. Non-Burst phases, marked as noise, reflect sparse, irregular invocations outside burst intervals. Short Intervals highlight some transient dense invocation bursts, which, due to their small scale, are often not considered sustained burst phases. For each burst phase, I-DBSCAN also determines the earliest and latest invocation times, which are essential for optimizing subsequent prewarming and scheduling strategies.

The advantages of I-DBSCAN are multifaceted. Firstly, it can autonomously identify burst and non-burst phases without supervision, eliminating the need for manual intervention or complex model training. Secondly, it automatically identifies and excludes non-burst phases by treating them as outliers, focusing only on the burst clusters. Thirdly, the parameterization of min_samples allows the algorithm to adapt to different system load scenarios, flexibly controlling the minimum cluster size and enhancing the adaptability of keep-alive strategies. Compared to deep learning-based clustering methods, I-DBSCAN achieves significantly faster convergence with a typical complexity of $O(N\log N)$, making it well-suited for large-scale, real-time load analysis in serverless environments.

We also analyzed the advantages of combining invocation time and call density as two complementary dimensions. Experimental results show that burst phases are characterized not only by temporal proximity in request sequences but also by significantly shorter call intervals. Figure 5 illustrates that relying solely on the call sequence for clustering forms some dense regions in the time sequence, but it fails to capture differences in call density. As a result, some regions are misidentified as burst phases despite large call intervals. Additionally, the lack of density information introduces more short gray segments, reducing the overall accuracy of burst phase identification.

4.3   Burst Invocation Prediction

Our strategy is based on identifying burst invocation phases and leveraging the start time points of these phases to proactively pre-warm function instances. In contrast to most existing pre-warming strategies that rely on continuously predicting the entire invocation sequence, these traditional methods not only face significant challenges in accuracy but also result in unnecessary resource overhead.

We believe that for the concentrated requests within burst phases, the key is to provide sufficient available instances and an appropriate keep-alive duration, rather than performing full-range prediction for every request. As long as the start points of burst phases can be accurately detected, targeted pre-warming can effectively minimize the latency caused by cold starts while significantly reducing the additional resource overhead introduced by unnecessary predictions.

Figure 6 compares the prediction accuracy of burst phase start times identified by Pulsed Load Modeling against the full-range invocation sequence prediction method. Experimental results demonstrate that, across different data scales,

PliKOS can significantly reduce prediction time, with an average reduction of 83.6% and a maximum of 91.4%. Meanwhile, it also substantially improves prediction accuracy, with the average MAPE reduced by 75.6% and up to 86.4% in the best case (Fig. 7).

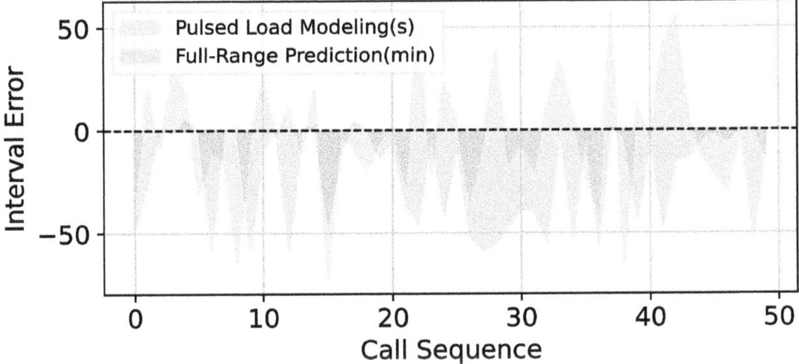

Fig. 6. Prediction accuracy improvement through burst phase identification.

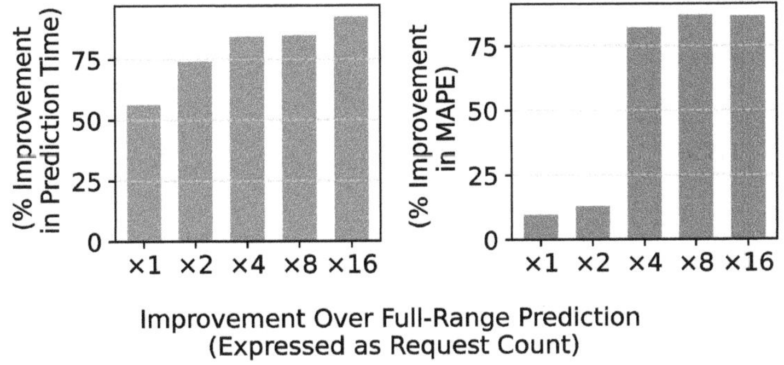

Fig. 7. Pulsed load modeling improves prediction accuracy and reduces overhead.

4.4   Instance Pre-warming for Burst Phases

During bursts, dynamic pre-warming ensures low latency and meets SLOs. Beyond conventional pre-warming mechanisms, PliKOS further considers concurrent demand by dynamically adjusting the number of warm instances based on phase-specific load. This allows the system to minimize instance retention time while ensuring compliance with SLOs, thereby reducing overall system overhead.

For functions with longer execution times, active instances are occupied for extended periods under high concurrency. Without timely scaling, this leads to request queuing and cold starts, significantly increasing response latency. PliKOS dynamically estimates the minimum number of warm instances required by incorporating request rate, execution time, and cold start latency, enabling elastic scheduling. As shown in Fig. 8, PliKOS consistently maintains lower average service time as execution time increases, achieving up to 15–25% improvement over OpenWhisk and outperforming baselines such as Wild and ACMP. This benefit stems from its proactive awareness of resource contention, allowing it to pre-allocate instances before phase transitions.

Moreover, cold start delays vary significantly across platforms and environments. Without explicitly modeling cold start costs, retention strategies can cause service time inflation and elevated cold start rates. In another set of experiments (Fig. 9), PliKOS demonstrates strong resilience under high cold start cost scenarios—reducing the cold start rate by over 90% compared to OpenWhisk, while maintaining superior response time relative to Wild and ACMP.

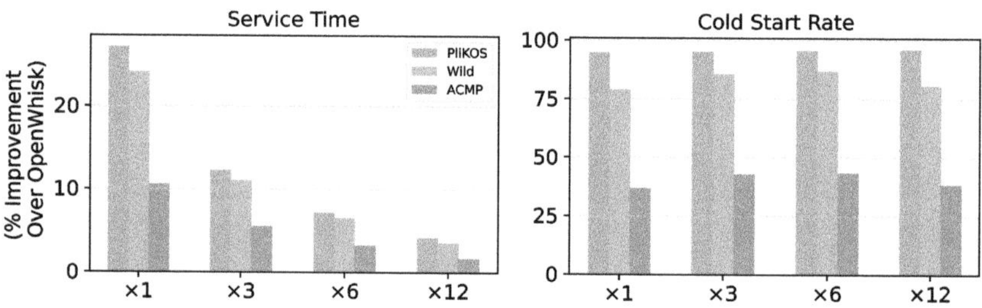

Fig. 8. PliKOS is effective across different execution times of functions.

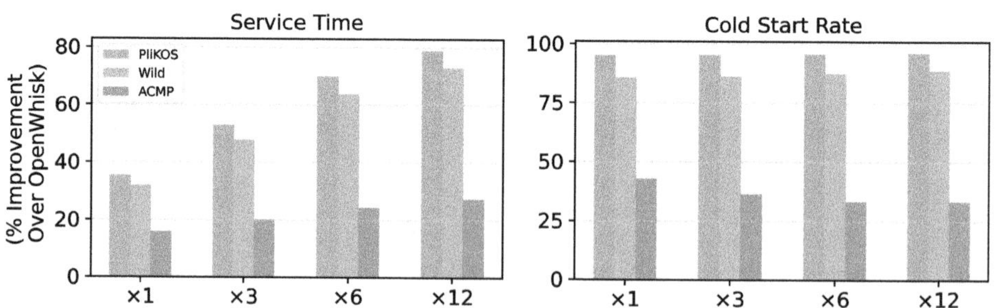

Fig. 9. PliKOS is effective across different cold start latencies.

4.5   Impact of PliKOS

We evaluate PliKOS using Azure traces against three baselines using two weeks of function invocation traces from the Azure Functions platform. We examined several key performance metrics, including average service time, cold start rate, and idle resource cost, and compared PliKOS with representative baseline strategies. As shown in Fig. 10, PliKOS achieved significant improvements across all three metrics.

PliKOS reduced the average service time by approximately 13% compared to the default OpenWhisk strategy, while lowering the cold start rate by over 95%. Additionally, by precisely controlling the number of warm instances and their keep-alive durations, PliKOS reduced idle resource costs by around 19%.

These gains result from PliKOS's pulsed-aware, resource-adaptive design, which identifies burst phases through invocation pattern modeling and proactively prewarms instances ahead of time. Unlike traditional approaches that rely on continuous prediction across the full invocation sequence, PliKOS focuses on burst periods, enabling more targeted and efficient prewarming that significantly reduces cold starts. In high-concurrency scenarios, PliKOS further incorporates cold start latency, function execution time, and request rate into its model to dynamically adjust the number of prewarmed instances. This ensures service level objectives (SLOs) are met while minimizing unnecessary resource usage.

Overall, PliKOS achieves superior cold start mitigation, lower latency, and better resource efficiency than existing strategies.

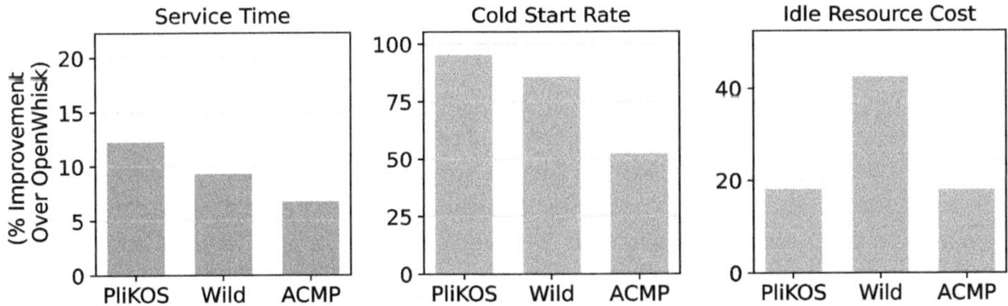

Fig. 10. PliKOS improves service time, cold start rate, and idle resource cost.

5   Related Work

5.1   Invocation Characteristics

Numerous studies have analyzed serverless workload patterns from user and provider perspectives, analyzing trends in serverless computing through representative benchmarks on the user side and workload traces on the provider side.

On the user side, these studies typically run benchmarks to capture usage patterns, while on the provider side, they examine workload traces [20,23,25,27].Key characteristics include scheduling effects, invocation patterns, and I/O trends [14,15,17]. Some works analyze function invocation chains and contextual dependencies, particularly invocation chains and contextual dependencies. For example, SAND pre-warms subsequent calls using sandboxing, and Shen et al. mine dependencies to reduce cold starts [4,21]. These efforts overlook phase-level invocation dynamics. In contrast, our work models burst and non-burst phases for finer-grained pre-warming.

5.2   Cold Start Mitigation

Cold start mitigation remains a key challenge for performance–cost trade-offs. To address the cold start problem of functions, prior solutions fall into two categories: fast startup and keep-alive optimization. For fast startup, researchers explored pre-initialized containers and memory sharing [18,24]. For keep-alive tuning, prediction-based methods dominate [11,12,16,19].

While these predictions have improved keep-alive strategies, they often struggle in serverless environments with highly variable invocation intervals and diverse workloads. Moreover, full-range prediction approaches incur high computational costs and are prone to error under pulsed workloads. PliKOS addresses these issues by avoiding full-sequence modeling and instead focusing on identifying and predicting burst phase start times, which reduces overhead and improves robustness.

6   Conclusion

PliKOS is a pre-warming optimization strategy tailored to pulsed load patterns. It leverages the I-DBSCAN algorithm to identify burst phases, avoiding full-sequence prediction and reducing computational overhead, and dynamically adjusts the number of pre-warmed instances based on execution time, request rate, and latency objectives. Compared to existing methods, PliKOS reduces cold start rates by over 95%, shortens service time by 13%, and lowers idle resource costs by 19%. It also reduces prediction error and computational overhead by up to 86.4% and 91.4%, respectively, offering an efficient and practical solution for cold start mitigation in serverless computing through phase-aware pulsed load modeling.

References

1. AWS lambda limits. https://docs.aws.amazon.com/lambda/latest/dg/gettingstarted-limits.html. Accessed 25 June 2025
2. Apache openwhisk: Open source serverless cloud platform (2020). https://openwhisk.apache.org/. Accessed 25 June 2025

3. Agarwal, S., Rodriguez, M.A., Buyya, R.: A reinforcement learning approach to reduce serverless function cold start frequency. In: 2021 IEEE/ACM 21st International Symposium on Cluster, Cloud and Internet Computing (CCGrid), pp. 797–803. IEEE (2021)
4. Akkus, I.E., et al.: Sand: towards high-performance serverless computing. In: 2018 USENIX Annual Technical Conference (USENIX ATC 2018), pp. 923–935 (2018)
5. Alliaume, E., Le Roux, B.: Cold start/warm start with AWS lambda. OCTO Talks (2018). https://blog.octo.com/en/cold-startwarm-start-with-aws-lambda/. Accessed 22 July 2021
6. Back, T., Andrikopoulos, V.: Using a microbenchmark to compare function as a service solutions. In: Service-Oriented and Cloud Computing: 7th IFIP WG 2.14 European Conference, ESOCC 2018, Como, Italy, 12–14 September 2018, Proceedings 7, pp. 146–160. Springer (2018)
7. Castro, P., Ishakian, V., Muthusamy, V., Slominski, A.: The rise of serverless computing. Commun. ACM **62**(12), 44–54 (2019)
8. Cui, Y.: How long does AWS lambda keep your idle functions around before a cold start (2018)
9. Daw, N., Bellur, U., Kulkarni, P.: Xanadu: mitigating cascading cold starts in serverless function chain deployments. In: Proceedings of the 21st International Middleware Conference, pp. 356–370 (2020)
10. Hellerstein, J.M., et al.: Serverless computing: one step forward, two steps back. arXiv preprint arXiv:1812.03651 (2018)
11. HoseinyFarahabady, M.R., Zomaya, A.Y., Tari, Z.: A model predictive controller for managing QOS enforcements and microarchitecture-level interferences in a lambda platform. IEEE Trans. Parallel Distrib. Syst. **29**(7), 1442–1455 (2017)
12. Hoseiny Farahabady, M.R., Taheri, J., Tari, Z., Zomaya, A.Y.: A dynamic resource controller for a lambda architecture. In: 2017 46th International Conference on Parallel Processing (ICPP), pp. 332–341. IEEE (2017)
13. Jangda, A., Pinckney, D., Brun, Y., Guha, A.: Formal foundations of serverless computing. Proc. ACM Program. Lang. **3**(OOPSLA), 1–26 (2019)
14. Klimovic, A., Litz, H., Kozyrakis, C.: Selecta: heterogeneous cloud storage configuration for data analytics. In: 2018 USENIX Annual Technical Conference (USENIX ATC 2018), pp. 759–773 (2018)
15. Klimovic, A., Wang, Y., Kozyrakis, C., Stuedi, P., Pfefferle, J., Trivedi, A.: Understanding ephemeral storage for serverless analytics. In: 2018 USENIX Annual Technical Conference (USENIX ATC 2018), pp. 789–794 (2018)
16. Kumari, A., Sahoo, B.: ACPM: adaptive container provisioning model to mitigate serverless cold-start. Clust. Comput. **27**(2), 1333–1360 (2024)
17. Lin, X.C., Gonzalez, J.E., Hellerstein, J.M.: Serverless boom or bust? An analysis of economic incentives. In: 12th USENIX Workshop on Hot Topics in Cloud Computing (HotCloud 2020) (2020)
18. Mohan, A., Sane, H., Doshi, K., Edupuganti, S., Nayak, N., Sukhomlinov, V.: Agile cold starts for scalable serverless. In: 11th USENIX Workshop on Hot Topics in Cloud Computing (HotCloud 2019) (2019)
19. Roy, R.B., Patel, T., Tiwari, D.: Icebreaker: warming serverless functions better with heterogeneity. In: Proceedings of the 27th ACM International Conference on Architectural Support for Programming Languages and Operating Systems, pp. 753–767 (2022)
20. Shahrad, M., et al.: Serverless in the wild: characterizing and optimizing the serverless workload at a large cloud provider. In: 2020 USENIX annual technical conference (USENIX ATC 2020), pp. 205–218 (2020)

21. Shen, J., Yang, T., Su, Y., Zhou, Y., Lyu, M.R.: Defuse: a dependency-guided function scheduler to mitigate cold starts on FAAS platforms. In: 2021 IEEE 41st International Conference on Distributed Computing Systems (ICDCS), pp. 194–204. IEEE (2021)
22. Shillaker, S., Pietzuch, P.: FAASM: lightweight isolation for efficient stateful serverless computing. In: USENIX ATC, vol. 20, pp. 419–433 (2020)
23. Wang, A., Chang, S., Tian, H., et al.: FaaSNet: scalable and fast provisioning of custom serverless container runtimes at Alibaba cloud function compute. In: USENIX ATC, vol. 21, pp. 443–457 (2021)
24. Wang, K.-T.A., Ho, R., Wu, P.: Replayable execution optimized for page sharing for a managed runtime environment. In: Proceedings of the Fourteenth EuroSys Conference 2019, pp. 1–16 (2019)
25. Wang, L., Li, M., Zhang, Y., Ristenpart, T., Swift, M.: Peeking behind the curtains of serverless platforms. In: 2018 USENIX Annual Technical Conference (USENIX ATC 2018), pp. 133–146 (2018)
26. Yang, Y., Zhao, L., Li, Y., Zhang, H., et al.: INFless: a native serverless system for low-latency, high-throughput inference. In: ASPLOS, pp. 768–781 (2022)
27. Yu, T., et al.: Characterizing serverless platforms with serverlessbench. In: Proceedings of the 11th ACM Symposium on Cloud Computing, pp. 30–44 (2020)

VPFU: A Bit-Serial Architecture for Energy-Efficient Acceleration of Ultra-low Precision DNNs

Cheng Xu, Yibai He$^{(\boxtimes)}$, and Lu Wang

National Innovation Institute of Defense Technology,
Academy of Military Sciences, Beijing, China
heyibai100@163.com

Abstract. Quantizing deep neural networks (DNNs) to ultra-low precision (<4 bits) with mixed-precision strategies—employing variable bit-widths across layers—effectively alleviates storage bottlenecks and accuracy degradation for on-device edge deployment. However, the low-power constraints of edge devices persistently limit the full exploitation of inference performance. To address this challenge, this paper proposes a Variable-Precision Fusion Unit (VPFU) leveraging bit-serial computation to enable dynamic ultra-low-precision multiply-accumulate operations through logical fusion of Basic Blocks, achieving high parallelism while maintaining low power consumption. Capitalizing on the intrinsic channel-level parallelism of convolutional layers, we further design a VPFU-based 2D array architecture that significantly enhances convolutional computation efficiency. We implement the VPFU in RTL and synthesize it at a 45 nm technology node. Using cycle-accurate simulations on selected convolutional layers from four representative models, we evaluate the performance of the VPFU array. Experimental results show that compared to Bit Fusion's Fusion Unit, the proposed VPFU achieves a 33.6% area reduction and a 35.16% power reduction, while delivering an average 1.33× speedup. When the precision is reduced to 1-bit, the average speedup further increases to 2.14×.

Keywords: Quantized neural networks · Ultra-low precision · Bit-serial computation · Variable-precision computing

1 Introduction

With the continuous development of neural network models, both their depth and the number of neurons have increased dramatically, with some model parameters exceeding the hundred-billion threshold [1]. Trained models are typically deployed on edge devices (e.g., smartphones, embedded sensors, and intelligent cameras) to execute tasks such as object recognition [2] and speech recognition [3]. However, these devices face significant challenges in accommodating large-scale neural network models due to their low-power designs and limited

H. Liu et al. (Eds.): ICA3PP 2025, LNCS 16381, pp. 37–57, 2026.
https://doi.org/10.1007/978-981-95-8399-7_3

storage resources [4]. Specifically, the high computational demands of model inference and frequent memory access requirements severely constrain practical application performance. In this context, the co-optimization of model compression techniques and hardware acceleration has emerged as a prominent research focus.

Quantization, as a core technique for model compression, effectively reduces computational demands and memory footprint by decreasing the bit precision of weights and activations, while maintaining model inference accuracy [5]. Recent years have witnessed rapid advancements in quantization techniques. Studies have shown that models quantized to extremely low precisions below 4 bits can achieve results comparable to their full-precision counterparts [6–10,16]. Moreover, mixed-precision quantization strategies, which dynamically assign different quantization bit-widths to convolutional layers, further mitigate the accuracy degradation caused by uniform ultra-low-precision quantization [5].

Meanwhile, dedicated hardware architectures for accelerating deep neural networks have also made remarkable progress. The Tensor Processing Unit (TPU) [11], based on a systolic array structure, supports 8/16-bit computations and significantly enhances the efficiency of large-scale matrix operations. DaDianNao [12] employs a fixed bit-width design capable of performing 256 multiplications per cycle, optimizing convolutional computation efficiency, though it lacks support for variable-precision operations and consequently suffers from limited flexibility. Among quantization-specific hardware architectures, Stripes [13] and BitMoD [14] employ bit-serial architectures, while Bit Fusion [15] adopts a dynamic fusion bit-width adjustment mechanism, both achieving reduced hardware power consumption while maintaining inference accuracy. Nevertheless, existing architectures still exhibit limitations in supporting extremely low variable-precision (sub-4-bit) quantized models, particularly regarding the computational efficiency of multiply-accumulate (MAC) operations at ultra-low bit widths, which remains to be optimized.

This paper observes that quantized neural networks with extremely low precision exhibit the following key characteristics:

1) Computational features: MAC operations in convolutional layers account for over 90% of the total computational workload during model inference [15].
2) Quantization properties: Both weights and activations can be quantized to ultra-low-bitwidth (<4 bits) without significant accuracy degradation [6–10].
3) Necessity of mixed precision: To mitigate the accuracy loss caused by uniformly quantizing weights and activations to extremely low precision, mixed-precision quantization can be adopted [5].
4) Parallelism and data reuse: The independence among convolutional kernels enables channel-wise parallel computation, while the reuse of input or weight data reduces memory access overhead.

Based on the above observations, this paper proposes a variable-precision fused unit (VPFU) for ultra-low-bitwidth operations to accelerate convolutional layers with extremely low precision. Specifically, the contributions of this paper are summarized as follows:

- **Variable-Precision Multiply-Accumulate Operations at Ultra-low Bit-Widths.** This work introduces a Bit-serial algorithm to design a Variable-Precision Functional Unit (VPFU) supporting ultra-low bit-widths. The VPFU enables any combination of input and weight precisions at 1, 2, or 4 bits. It is designed to precisely match the bit-width requirements of convolution layers in extremely low mixed-precision quantized DNNs, maximizing computational unit utilization and minimizing resource waste.
- **Channel-Level Parallel VPFU Array Architecture.** To effectively exploit the abundant parallelism inherent in convolution layers, this work designs a channel-level parallel array architecture based on the VPFU. This architecture significantly enhances the parallelism of convolution operations, thereby improving execution efficiency.
- **Adaptive Dynamic Dataflow Selection Algorithm.** As a single operation of the VPFU array cannot produce a complete output feature value, multiple computation dataflows exist depending on the prioritized output feature being computed, with significant variations in their data reuse efficiency. This work quantifies the data reuse efficiency of different dataflows and proposes an adaptive dynamic dataflow selection algorithm to maximize data reuse efficiency.

We implemented the VPFU in Verilog and synthesized it using a 45 nm process. Performance evaluation was conducted on partial convolution layers from MobileNet [17], GoogLeNet [18], YoloTiny [19], and ResNet [20] models. Experimental results demonstrate that the VPFU array achieves an average 5.78× speedup compared to a traditional systolic array (Scale-Sim [21]) under identical configurations. Furthermore, the VPFU reduces area by 33.60% and power consumption by 35.16% versus the Fusion Unit within the state-of-the-art DNN accelerator, Bit Fusion [15]. Under equivalent area and clock frequency constraints, the VPFU array delivers an average 1.33× speedup over Bit Fusion. This speedup increases to an average 2.14× when data precision is further reduced to 1-bit.

2   Variable-Precision Fusion Unit and Array Architecture

2.1   Theory of Bit-Serial Computation

In low-precision dot-product computations, bit-serial algorithms exhibit higher computational efficiency [22]. The theoretical foundation of these algorithms is derived from the following formula:

$$\sum_{i=0}^{I-1} w_i \times a_i = \boldsymbol{w} \cdot \boldsymbol{a} = \sum_{n=0}^{N-1} \sum_{m=0}^{M-1} 2^{n+m} popcount(\boldsymbol{w_n} \wedge \boldsymbol{a_m}) \tag{1}$$

Let I denote the dimension of $\boldsymbol{w}$ and $\boldsymbol{a}$; N and M denote their element bit-widths, respectively. The *popcount* function is used to count the number of 1s in a binary representation. We observed that the dot-product operation between two vectors can be decomposed into a sequence of bitwise AND operations, *popcount* operations, shift operations, and accumulation operations.

2.2   Basic Block Architecture of Bit-Serial Computation

The three fundamental operations in bit-serial algorithm—vector bitwise AND operations, *popcount* operations, and shift operations—are abstracted as a bit-serial computation Basic Block (BB), as illustrated in Fig. 1(a).

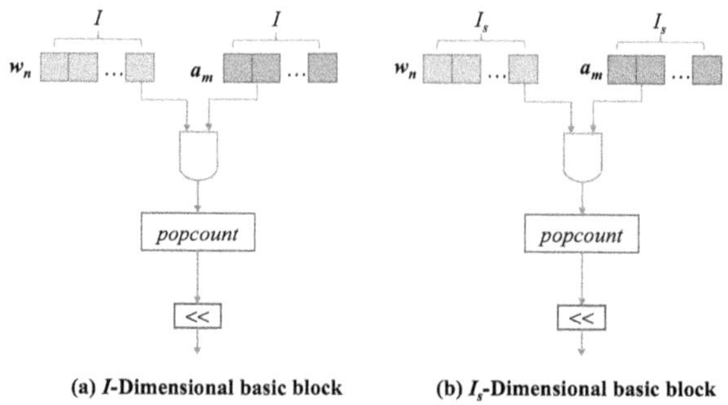

(a) *I*-Dimensional basic block   (b) I_s-Dimensional basic block

Fig. 1. Bit-serial computation Basic Block (BB).

It can be observed that both $\boldsymbol{w_n}$ and $\boldsymbol{a_m}$ are I-dimensional vectors with 1-bit precision elements. Therefore, according to the bit-serial algorithm, computing the dot product between $\boldsymbol{w}$ (with N-bit precision) and $\boldsymbol{a}$ (with M-bit precision) using bit-serial computation BBs requires $M \times N$ such BBs.

In practical convolution operations, the dimension I of the two vectors– representing the number of MAC operations required to compute a single output feature–is typically large. Directly using operand vectors of length I as inputs to the BBs fails to meet the design requirements. To address this, the MAC operations must be decomposed, as formulated in (2):

$$
\boldsymbol{w} \cdot \boldsymbol{a} = \sum_{i=0}^{I-1} w_i \times a_i = \sum_{i=0}^{I_s-1} w_i \times a_i + \sum_{i=I_s}^{2I_s-1} w_i \times a_i +
$$

$$
\cdots + \sum_{i=(k-1)I_s}^{kI_s-1} w_i \times a_i + \sum_{i=kI_s}^{I-1} w_i \times a_i \tag{2}
$$

$$
= \boldsymbol{w}0 \cdot \boldsymbol{a}0 + \cdots + \boldsymbol{w}(k-1) \cdot \boldsymbol{a}(k-1) + \boldsymbol{w}k \cdot \boldsymbol{a}k
$$

As derived from (2), the dot product of an I-dimensional vector can be decomposed into $k+1$ groups of vector dot products. Except for the last group, each group involves vectors of dimension I_s, where $I_s \ll I$. Consequently, the dot-product computation for each group can be independently executed I_s-dimensional BBs, as illustrated in Fig. 1(b).

The decomposition of an I-dimensional vector dot product into a summation of I_s-dimensional vector dot products—for computation on BBs with an operand bit-width of I_s—can be categorized into three cases. Firstly, when $I = I_s$, no decomposition is required, and the vector can be directly partitioned for computation on the BBs. Secondly, when $I < I_s$, the vector must be padded to I_s. For instance, consider $I = 3$ and $I_s = 4$; the dot-product computation using bit-serial BBs requires padding, as formulated in (3), where the padded elements are set to $w_3 = a_3 = 0$.

$$MAC = w_0a_0 + w_1a_1 + w_2a_2 = w_0a_0 + w_1a_1 + w_2a_2 + w_3a_3 \qquad (3)$$

Thirdly, when $I > I_s$, decomposition is required. For example, given $I = 8$ and $I_s = 4$, As shown in (4), the 8-dimensional vector dot product can be transformed into the sum of two sets of 4-dimensional vector dot products, enabling the use of basic blocks for dot product computation. All other cases can be comprehensively addressed through a combination of the three scenarios mentioned above.

$$MAC = \sum_{i=0}^{7} w_i \times a_i = \sum_{i=0}^{3} w_i \times a_i + \sum_{i=4}^{7} w_i \times a_i \qquad (4)$$

2.3 Variable-Precision Fusion Unit

As indicated above, using $M \times N$ BBs, we can implement the dot product operation between two I_s-dimensional vectors with bit-widths of M and N respectively, which essentially corresponds to I_s sets of MAC operations. When given a fixed number of bit-serial computational BBs, these blocks can be internally reconfigured through logic fusion to achieve variable-precision MAC computations.

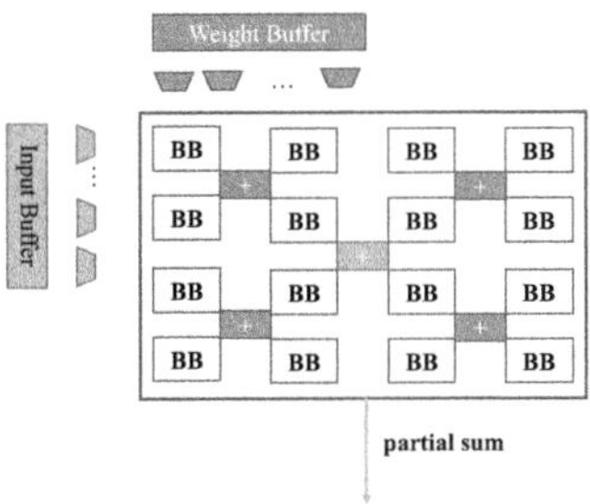

Fig. 2. Variable-precision fused computing unit with 16 BBs.

Variable-Precision Fusion Unit Architecture. As illustrated in Fig. 2, 16 BBs are arranged in a 4×4 physical array structure and interconnected via adder trees. The array is externally equipped with two buffers—a weight buffer and an input buffer—to supply operands to all BBs within the array. Together, the basic block array and buffers constitute a VPFU. Within the fused unit, a configurable number of BBs are logically grouped to form processing elements

(PEs). A smaller number of BBs per PE corresponds to lower arithmetic precision support but enables higher parallelism by accommodating more PEs within the VPFU. Conversely, increasing the number of BBs per PE enhances computational precision at the expense of reduced PE count and parallelism in the VPFU.

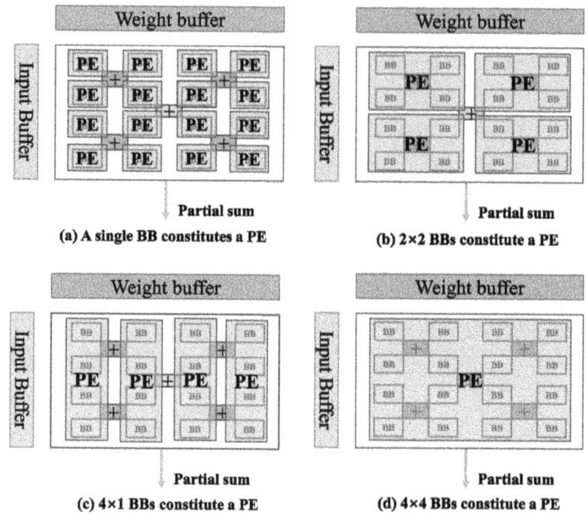

Fig. 3. PEs are logically formed using varying numbers of BBs.

The Fusion Unit Implements Variable-Precision Operations. Variable-precision MAC operations can be achieved through logical fusion of varying numbers of BBs. Figure 3 illustrates four distinct approaches for logically fusing bit-serial computational basic blocks. In configuration (a), a single BB constitutes a PE, where each PE executes MAC operations with 1-bit input data precision and 1-bit weight precision. The VPFU integrates 16 such PEs, achieving maximum inter-PE parallelism and delivering $16 \times I_s$ MAC operation groups. However, this configuration operates with the minimum bit-width (1-bit) for both input data and weights, ultimately generating a partial sum output.

Configuration (b) employs logical fusion of 2×2 BBs to form a PE capable of processing I_s operation groups with 2-bit input data and 2-bit weight precision. The VPFU contains 4 such PEs, enabling $4 \times I_s$ MAC operation groups while producing a single partial sum.

In configuration (c), 1×4 BBs are fused into a PE that processes I_s operation groups with 4-bit input data precision and 1-bit weight precision. The VPFU similarly incorporates 4 PEs, achieving $4 \times I_s$ MAC operation groups and generating a partial sum output.

Finally, configuration (d) integrates 16 BBs into a single PE supporting I_s operation groups with 4-bit input data and 4-bit weight precision. This VPFU contains only one PE, resulting in the lowest inter-PE parallelism but operating at maximum bit-width (4-bit) for both input data and weights. This hierarchical

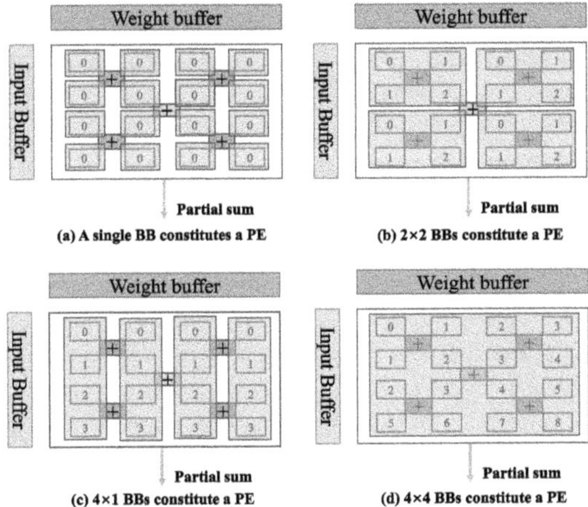

Fig. 4. Shift conditions of basic blocks under different logic fusion approaches (numbers in the BBs indicate shift counts).

design demonstrates the inherent trade-off between computational parallelism and operand precision across different architectural configurations. Beyond these four configurations, the VPFU can support MAC operations with heterogeneous bit-width combinations (e.g., 1-bit/4-bit, 2-bit/4-bit for input/weight) through logical fusion of varying numbers of BBs.

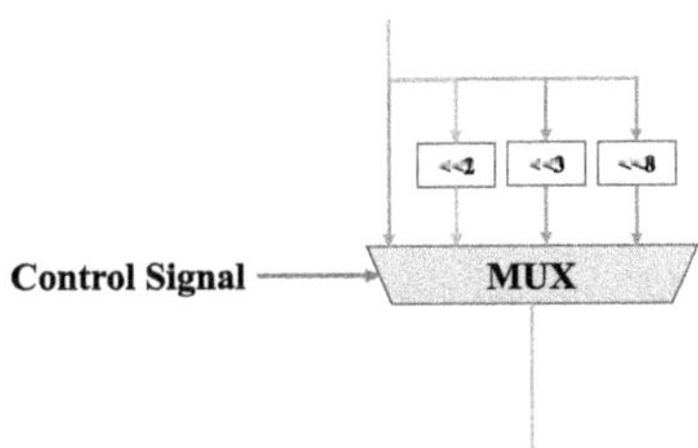

Fig. 5. The BB incorporates multiple shifters and utilizes multiplexers to enable diverse shifting functionalities.

Logical Fusion in the VPFU Is Achieved Through Shift Operations.
The concept of logical fusion is realized through the shifting of BBs. Figure 4 illustrates the shifting configurations of individual basic blocks within the PE when operating on data of varying precisions. Therefore, each BB must incorporate a configurable shifter supporting multiple shift amounts, with the correct value selected via multiplexers. As exemplified by the four cases in Fig. 4, the bottom-right BB in the VPFU requires shift amounts of 0, 2, 3, and 8 bits (as detailed in Fig. 5), though practical implementations may demand additional shift configurations.

The computational capacity of the VPFU is intricately linked to the input data bit-width N and the weight data bit-width M. As discussed above, $M \times N$ basic blocks constitute a PE, where each PE can perform I_s sets of MAC operations. Consequently, in a VPFU comprising 16 basic blocks, the maximum computational throughput I_{max} is expressed as:

$$I_{max} = \frac{16}{MN} \times I_s \tag{5}$$

2.4   Array Design Based on Variable-Precision Fusion Units

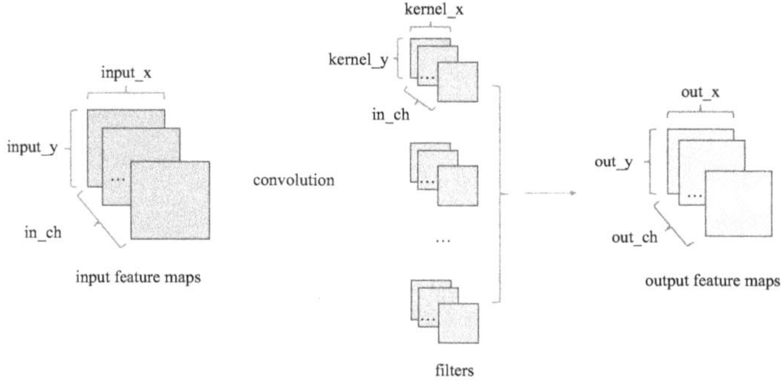

Fig. 6. Convolutional layer.

Convolution Operations and Array Structures. The generic convolutional layer is illustrated in Fig. 6 The input feature map adopts a three-dimensional structure comprising height, width, and channels. The convolution kernel groups are organized as a four-dimensional structure (*quantity* × *channels* × *height* × *width*). Each kernel performs a convolution computation with the overlapping region of the input feature map through a sliding window operation, generating a corresponding feature value in the output feature map. This process collectively produces an output feature map, while multiple independent kernels convolve with the input feature map to generate multiple output feature maps (the number of output feature maps corresponds to the number of convolution kernels). Since there are no data dependencies between the computations of different kernels, the computation process can be fully parallelized. Leveraging this characteristic and building upon VPFUs, a two-dimensional array has been designed, as shown in Fig. 7.

The array is organized into H rows and W columns of VPFUs. As shown in the structure, input buffers are positioned at the edge of each row, where each row of VPFUs shares a common input buffer, thereby enabling the input buffer to simultaneously broadcast data to all VPFUs within the same row. Each VPFU is equipped with an independent weight buffer for localized parameter

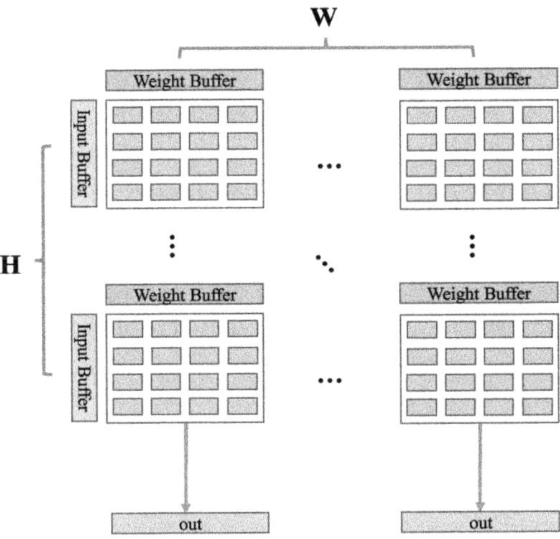

Fig. 7. Variable-precision fusion unit array architecture.

storage. An output buffer is located at the base of the array, which aggregates and accumulates computational results vertically along each column of VPFUs. The final accumulated outputs are stored in the output buffer for subsequent processing stages.

3 Implementation of Convolutional Layer Operations on the VPFU Array

3.1 Convolution Operation Data Flows and Data Reuse

The convolution operation can be equivalently implemented through vector reorganization and matrix multiplication. By leveraging the im2col (image-to-column) algorithm, each sliding window of the convolution kernel over the input feature map is expanded into a row vector to form the input matrix, while each convolution kernel is flattened into a column vector to construct the weight matrix. The matrix multiplication between the input matrix and the weight matrix generates an output matrix, which is subsequently reshaped into the output feature map.

In the VPFU array architecture, the data flows of the convolution operation can be analyzed through a matrix multiplication model. We first define the following key parameters: The number of columns in the input matrix equals the number of rows in the weight matrix, representing the number of MAC operations required for computing a single output feature value, as formulated in (6):

$$Num_{mac} = kernel\_x \times kernel\_y \times in\_ch \qquad (6)$$

The number of rows in the input matrix equals that of the output matrix, representing the total number of activation values (or elements) within an output feature map, as defined in (7):

$$Num_{one\_feature} = out\_x \times out\_y \tag{7}$$

The number of columns in the weight matrix equals that of the output matrix, indicating both the number of convolution kernel groups and the number of channels in the output feature map. In the VPFU array architecture designed in Fig. 7, the capability to compute one output feature value per cycle is determined by the number of array rows H. Given that the maximum computational capacity of a single fused unit is I_{max}, as defined in (5), the total number of multiply-accumulate (MAC) operations required for the array to compute one output feature value per cycle is expressed as:

$$I_{once} = I_{max} \times H \tag{8}$$

Under typical conditions, the number of MAC operations required to compute an output feature value, denoted as Num_{mac}, exceeds I_{once}. Depending on the prioritization of computational sequences, three distinct data flows can be categorized.

The first data flow prioritizes the computation of the current output feature value (Present First, PF), meaning it completes the calculation of one output feature value before moving to the next. This workflow is described in Table 1.

Table 1. The PF Data Flow

Data Flow 1: Priority is given to computing the current output feature value.

1: **for** $input\_row \leftarrow 0$ to $num\_one\_feature$ **do**
2:     **for** $weight\_col\_index \leftarrow 0$ to $\frac{out\_ch}{W}$ **do**
3:         **for** $input\_col\_index \leftarrow 0$ to $\frac{num\_mac}{I_{once}}$ **do**
4:             **for** $input\_col\_offset \leftarrow 0$ to I_{once} **do** // **Loop unrolling**
5:             **for** $weight\_col\_offset \leftarrow 0$ to W **do**   // **Loop unrolling**
6:                 $input\_col = input\_col\_index \times I_{once} + input\_col\_offset$
7:                 $weight\_col = weight\_col\_index \times W + weight\_col\_offset$
8:                 $out\_feature\_mat[input\_row][weight\_col] +=$
         $input\_mat[input\_row][input\_col] \times weight\_mat[input\_col][weight\_col]$

In the PF data flow, the innermost loop variable, $input\_col\_index$, drives the dynamic update of the $input\_col$ variable within each clock cycle. This design results in updates to both input feature data and weight data during every computational cycle, thereby preventing temporal data reuse—i.e., the repeated utilization of the same data across consecutive time intervals. However, due to the multi-channel parallel processing mechanism employed by the computing array, input data is shared across multiple fused computing units via a centralized buffer. This inherently enables spatial data reuse, where a single data element is distributed to multiple computing units simultaneously within the same clock cycle, significantly improving data throughput.

The second data flow prioritizes the computation of output feature values within the same channel but across different spatial positions (Next First, NF). During this process, the partial sums of the current output feature values are preserved to enable prioritized computation of subsequent values within the same channel. This workflow is formally described in Table 2.

Table 2. The NF Data Flow

Data Flow 2: Prioritize computing output activations at different spatial positions within the same channel.

1: **for** $weight\_col\_index \leftarrow 0$ to $\frac{out\_ch}{W}$ **do**

2:     **for** $input\_col\_index \leftarrow 0$ to $\frac{num\_mac}{I_{once}}$ **do**

3:         **for** $input\_row \leftarrow 0$ to $num\_one\_feature$ **do**

...

In the NF data flow, the innermost loop variable $input\_row$ is updated iteratively during each cycle. This design ensures that each loop iteration computes output feature values at distinct spatial positions within the same channel. Crucially, only the row index of the input feature matrix is updated per iteration, while the weight data remains unchanged throughout the loop cycles, thereby enabling temporal data reuse of weight data.

Table 3. The CF Data Flow

Data Flow 3: Prioritize computing output features across different channels at the same spatial position.

1: **for** $input\_row \leftarrow 0$ to $num\_one\_feature$ **do**

2:     **for** $input\_col\_index \leftarrow 0$ to $\frac{num\_mac}{I_{once}}$ **do**

3:         **for** $weight\_col\_index \leftarrow 0$ to $\frac{out\_ch}{W}$ **do**

...

The third data flow prioritizes the computation of output feature values across different channels but at the same spatial position (Channel First, CF). During this process, the partial sums of the currently computed output feature values are preserved to enable prioritized computation of subsequent channels' values at the identical spatial position. This workflow is formally described in Table 3.

In the CF data flow, the parameter $weight\_col$, controlled by the innermost loop variable $weight\_col\_index$, is updated during each iteration of the innermost loop. This design ensures that only the $weight\_col$ parameter changes per iteration, while the input data participating in the computation remains unchanged.

Consequently, weight data is dynamically updated, whereas input data achieves temporal reuse.

In summary, distinct data reuse patterns for input data and weight data emerge across different data flows, as systematically summarized in Table 4.

Table 4. Data Flow Reuse Pattern

Data Flow	Weight Data Reuse Pattern	Activation Reuse Pattern
The PF Data Flow	No Reuse	Spatial Data Reuse
The NF Data Flow	Temporal Data Reuse	Spatial Data Reuse
The CF Data Flow	No Reuse	Spatial Data Reuse and Temporal Data Reuse

3.2 Dynamic Dataflow Selection Algorithm

Due to differences in data reuse patterns across data flow strategies, their data reuse efficiency exhibits significant variations under the hardware constraints of specific convolutional layer dimensions and fused computing unit arrays. For a given problem scale, identifying a more efficient data flow strategy enhances the computational efficiency of the array. The PF data flow strategy, where neither weights nor input data exhibit temporal reuse, demonstrates markedly lower data reuse efficiency compared to the NF and CF strategies. Consequently, it is not discussed further in this analysis.

First, we define the evaluation criteria for reuse efficiency. Consider a complete convolutional layer (as illustrated in Fig. 6) executed on a fused computing unit array of a specific scale (as shown in Fig. 7). The total amount of data loaded into the weight buffer and the input buffer is used as the metric for evaluating data reuse efficiency, formalized in (9):

$$
\begin{aligned}
Count_{data} &= Count_{Input} \times (I_{max} \times N) + Count_{Weight} \times (I_{max} \times M) \\
&= I_{max} \times (Count_{Input} \times N + Count_{Weight} \times M)
\end{aligned}
\tag{9}
$$

In this context, $Count_{Input}$ and $Count_{Weight}$ represent the total number of data loads for W input buffers and $W \times H$ weight buffers, respectively.

To compare the reuse efficiency of the two data flows, common factors can be neglected, yielding an equivalent comparison formula as shown in (10).

$$
Count_{data} = Count_{Input} \times N + Count_{Weight} \times M
\tag{10}
$$

The total number of clock cycles required to compute the convolutional layer in Fig. 6 using the array architecture in Fig. 7 is:

$$
Count_{total} = \left\lceil \frac{kernel\_x \cdot kernel\_y \cdot in\_ch}{I_{once}} \right\rceil \cdot out\_x \cdot out\_y \cdot \left\lceil \frac{out\_ch}{W} \right\rceil
\tag{11}
$$

the term $\left\lceil \frac{kernel\_x \cdot kernel\_y \cdot in\_ch}{I_{once}} \right\rceil$ denotes the number of clock cycles required to compute a single output feature value within one column of fused computation units. Consequently, $\left\lceil \frac{kernel\_x \cdot kernel\_y \cdot in\_ch}{I_{once}} \right\rceil \cdot out\_x \cdot out\_y$ represents the total clock cycles needed to generate one output feature map. Meanwhile, $\left\lceil \frac{out\_ch}{W} \right\rceil$ corresponds to the parallel computation of W output feature maps.

For the NF data flow, where weights are temporally reused (as shown in Table 2), all weight buffers are reloaded once every $out\_x \cdot out\_y$ clock cycles. Consequently, the total number of load operations for all weight buffers is:

$$Count_{Weight\_nf} = \frac{Count_{total}}{out\_x \cdot out\_y} \cdot W \cdot H \tag{12}$$

since the input data lacks temporal reuse and must be loaded every clock cycle, the total number of load operations for all input buffers is:

$$Count_{Input\_nf} = H \times Count_{total} \tag{13}$$

substituting (12) and (13) into (10) yields the total data load volume for all buffers:

$$\begin{aligned}
Count_{data_nf} = H \cdot {} & \left\lceil \frac{kernel_x \cdot kernel_y \cdot in_ch}{I_{once}} \right\rceil \\
& \cdot \left\lceil \frac{out_ch}{W} \right\rceil \cdot (out_x \cdot out_y \cdot N + W \cdot M)
\end{aligned} \tag{14}$$

For the CF data flow, where input data is temporally reused (as shown in Table 3), the input data buffers are reloaded once every $\left\lceil \frac{out\_ch}{W} \right\rceil$ clock cycles. Consequently, the total number of load operations for all input data buffers is:

$$Count_{Input\_cf} = \frac{Count_{total}}{\left\lceil \frac{out\_ch}{W} \right\rceil} \cdot H \tag{15}$$

similarly, since the weights lack temporal reuse, the total number of load operations for all weight buffers is:

$$Count_{Weight\_cf} = W \cdot H \times Count_{total} \tag{16}$$

substituting (15) and (16) into (10) yields the total data load volume for all buffers:

$$\begin{aligned}
Count_{data_cf} = H \cdot {} & \left\lceil \frac{kernel_x \cdot kernel_y \cdot in_ch}{I_{once}} \right\rceil \\
& \cdot out_x \cdot out_y \cdot \left(N + \left\lceil \frac{out_ch}{W} \right\rceil W \cdot M \right)
\end{aligned} \tag{17}$$

Subtracting (17) from (14) yields the following.

$$d = Count_{data\_nf} - Count_{data\_cf}$$
$$= H \cdot \left\lceil \frac{kernel\_x \cdot kernel\_y \cdot in\_ch}{I_{once}} \right\rceil$$
$$\cdot \left[\left\lceil \frac{out\_ch}{W} \right\rceil \cdot (out\_x \cdot out\_y \cdot N + W \cdot M) \right.$$
$$\left. - out\_x \cdot out\_y \cdot (N + \left\lceil \frac{out\_ch}{W} \right\rceil \cdot W \cdot M) \right] \tag{18}$$

The sign of the difference d is determined by the difference factor:

$$d' = \left\lceil \frac{out\_ch}{W} \right\rceil \cdot (out\_x \cdot out\_y \cdot N + W \cdot M)$$
$$- out\_x \cdot out\_y (N + \left\lceil \frac{out\_ch}{W} \right\rceil \cdot W \cdot M)$$
$$\approx (\frac{N}{W} - M) \cdot out\_x \cdot out\_y \cdot out\_ch$$
$$+ M \cdot out\_ch - N \cdot out\_x \cdot out\_y \tag{19}$$

To simplify the selection of data flows, we perform a qualitative analysis. When $\frac{N}{W} - M \leq 0$, we consider $d' < 0$, indicating that the total data load volume of the NF data flow is smaller than that of the CF data flow. Consequently, the NF data flow is selected, prioritizing the computation of output feature values at different spatial positions within the same channel. Conversely, when $\frac{N}{W} - M > 0$, we assume $d' > 0$, and the CF data flow is chosen, prioritizing the computation of output feature values across different channels at the same spatial position.

4   Evaluation

This section evaluates the area and power consumption of the VPFU, as well as the performance of the VPFU array in executing convolutional layers, the trade-off between data precision and performance, and the dynamic data flow selection algorithm. We configure the parameter $I_s = 4$ to guide the RTL implementation and synthesis of the VPFU, and further leverage this configuration for systematic evaluation of the VPFU array.

Table 5. Benchmarks

Model	Convolutional Layer Name	Input Feature Map Dimensions	Kernel Dimensions	Strides
GoogLeNet	$Inc4e\_5 \times 5$	$14 \times 14 \times 32$	$5 \times 5 \times 32 \times 128$	1
	$Inc5b\_5 \times 5$	$7 \times 7 \times 32$	$5 \times 5 \times 32 \times 128$	1
	$Inc5b\_5 \times 5red$	$7 \times 7 \times 832$	$1 \times 1 \times 832 \times 48$	1
MobileNet	FC	$1 \times 1 \times 1024$	$1 \times 1 \times 1024 \times 1000$	1
	$Conv4$	$112 \times 112 \times 64$	$3 \times 3 \times 64 \times 1$	2
	$Conv6$	$56 \times 56 \times 128$	$3 \times 3 \times 128 \times 1$	1
	$Conv9$	$28 \times 28 \times 128$	$1 \times 1 \times 128 \times 256$	2
	$Conv14$	$14 \times 14 \times 512$	$3 \times 3 \times 512 \times 1$	1
	$Conv17$	$14 \times 14 \times 512$	$1 \times 1 \times 512 \times 1$	1
	$Conv24$	$14 \times 14 \times 512$	$3 \times 3 \times 512 \times 1$	2
Yolotiny	$Conv3$	$104 \times 104 \times 32$	$3 \times 3 \times 32 \times 64$	1
	$Conv4$	$52 \times 52 \times 64$	$3 \times 3 \times 64 \times 128$	1
Resnet	$Conv2\_2a$	$56 \times 56 \times 64$	$3 \times 3 \times 64 \times 64$	1
	$Conv5\_s$	$14 \times 14 \times 256$	$1 \times 1 \times 256 \times 512$	2
	$Conv5\_1b$	$7 \times 7 \times 512$	$3 \times 3 \times 512 \times 512$	1

4.1 Methodology

Benchmarks. As summarized in Table 5, this study comprehensively evaluates the computational capabilities of the VPFU array using selected convolutional layers from four representative models: GoogLeNet [18], MobileNet [19], Yolotiny [17], and ResNet [16]. These models are widely adopted in critical application domains such as object detection and image classification. The chosen layers encompass diverse computational patterns, including data-intensive convolutional layers and fully connected layers, ensuring a systematic evaluation of the VPFU array's performance across varying operational scenarios.

Hardware Cost Evaluation Methodology. The VPFU was implemented using RTL-Verilog and validated for functional correctness through RTL simulation. It was synthesized with the Synopsys Design Compiler (DC) tool using an open-source 45 nm process library, generating area and power consumption reports to evaluate the hardware cost of the VPFU array.

Performance Evaluation Methodology. The performance of the VPFU array under two distinct dataflow patterns was evaluated using the Scale-Sim [20] cycle-accurate simulator. Scale-Sim provides precise measurements of computational unit latency, off-chip memory access latency, and cache access cycles. The simulation cycle counts were validated against RTL implementations to enhance the credibility of the simulation results. To ensure fair comparison with systolic arrays and Bit Fusion–neither of which incorporates dynamic dataflow selection–this study exclusively employs the NF dataflow for VPFU array performance evaluation. For algorithm validation, the execution efficiency of NF

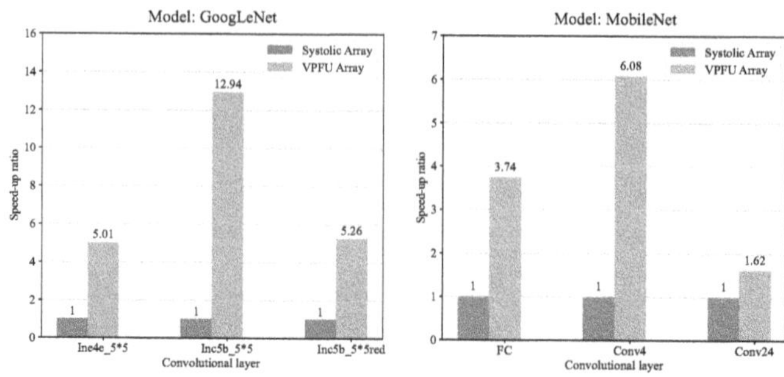

Fig. 8. Performance comparison between systolic array dataflow and VPFU array dataflow. The systolic array has a 4×4 configuration, while the VPFU array is 2×2.

and CF dataflows is compared independently to verify the correctness of the proposed selection mechanism.

Table 6. Area and Power Consumption Comparison Between Fusion Unit and VPFU

Computational unit	Fusion Unit	VPFU
Technology library	FreePDK45	
Clock Frequency	100 MHz	
Area(um^2)	2493.483994	1655.584002
Power(uW)	214.8148	139.2819

4.2 Experimental Results

Comparison with Systolic Arrays. To ensure a fair comparison, we scale the array sizes such that the total computational capacity remains equivalent. Since each PE in a systolic array performs only one MAC operation per cycle, whereas a VPFU can execute 4 MAC operations at 4-bit precision, we configure the systolic array with 4× the number of PEs compared to VPFUs. Under this iso-computational configuration, we evaluate both architectures across six convolutional layers from two representative models using the Scale-Sim simulator, comparing their execution times to assess relative performance. Results are shown in Fig. 8. The analysis demonstrates that the VPFU array achieves an average speedup of 7.74× over conventional systolic architectures in selected convolutional layers of the GoogLeNet model, and 3.81× acceleration in corresponding layers of the MobileNet model.

Comparison with Bit Fusion. Bit Fusion also supports variable-precision MAC operations with a minimum granularity of 2 bits, accommodating 2-, 4-,

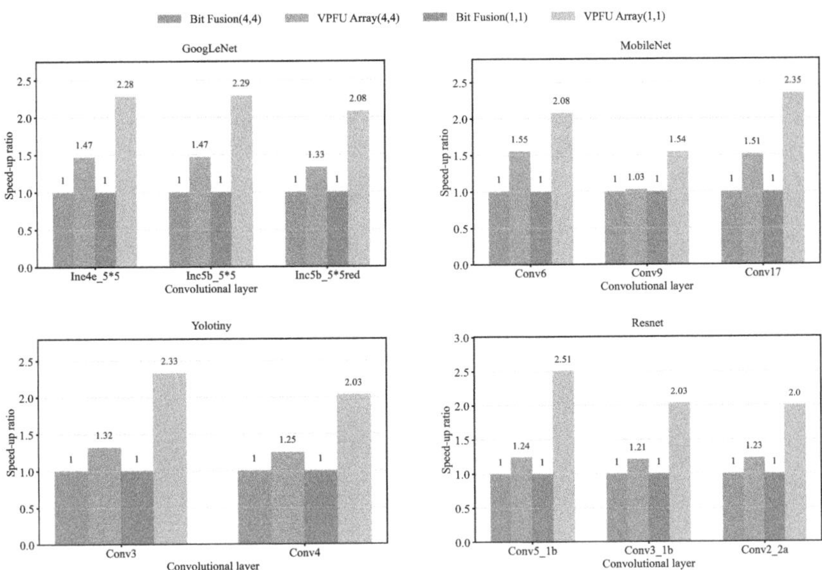

Fig. 9. Performance comparison between BitFusion dataflow and VPFU array dataflow. For the GoogLeNet and MobileNet comparative analysis, the Bit Fusion and VPFU arrays are configured at 6×6 and 6×9 dimensions, respectively. In the Yolotiny and ResNet evaluations, the array scales are set to 10×10 and 12×12.

and 8-bit precisions. For a comparative analysis of area and power consumption, RTL implementations of Bit Fusion's Fusion Unit and the proposed VPFU were synthesized using the DC tool with an open-source 45 nm process library. The results are summarized in Table 6. As indicated in the table, the VPFU achieves a 33.60% reduction in area and a 35.16% decrease in power consumption compared to the Fusion Unit. Under iso-area conditions between the Bit Fusion and VPFU arrays, we performed performance evaluation using the Scale-Sim simulator, with the results presented in Fig. 9.

The results indicate that when operating with both input data and weight data at 4-bit precision, the VPFU array demonstrates a significant performance improvement over selected convolutional layers within the GoogLeNet, MobileNet, Yolotiny, and ResNet models. Experimental data reveal that compared to the Bit Fusion architecture, the VPFU array achieves average speedups of $1.42\times$, $1.36\times$, $1.29\times$, and $1.23\times$ on these four models, respectively. The performance disparity becomes more pronounced when the computational precision is further reduced to 1 bit: the VPFU array attains increased average speedups of $2.22\times$ (GoogLeNet), $1.99\times$ (MobileNet), $2.18\times$ (Yolotiny), and $2.18\times$ (ResNet) on the same convolutional layers. This performance advantage originates from the fundamental difference in computational granularity between the two architectures: Bit Fusion is constrained by its minimum computational granularity of 2 bits, necessitating the use of 2-bit multipliers even for 1-bit precision operations, thereby resulting in suboptimal hardware resource utilization. Conversely, the VPFU array, leveraging a bit-serial computational architecture, achieves a minimum computational granularity of 1 bit. Consequently, this fine-grained computational capability enables maximal exploitation of hardware potential in ultra-low-precision computing scenarios.

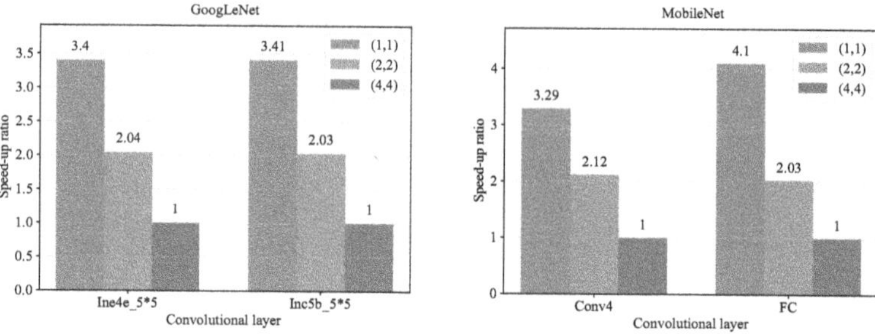

Fig. 10. Comparative analysis of NF dataflow execution efficiency across data precision levels. The VPFU array adopts a 4×4 configuration.

The Trade-Off Between Data Precision and Performance. This study quantitatively analyzes the computational characteristics of the VPFU: as derived in (5), its operational parallelism exhibits an inverse relationship with data precision, whereby increased precision reduces operations per unit time. To evaluate the impact of this trade-off on convolutional computing performance, comparative experiments were conducted on the VPFU array's execution efficiency across varying precision configurations, with detailed results presented in Fig. 10.

Experimental results demonstrate that reducing data precision from 4-bit to 2-bit and 1-bit configurations achieves average speedup ratios of $2.06\times$ and $3.55\times$, respectively, in the VPFU array. This trend aligns with the theoretically predicted inverse relationship between computational precision and acceleration performance.

Evaluation of Dynamic Dataflow Selection Algorithm. To validate the effectiveness of the dynamic dataflow selection algorithm, we compare execution times of NF and CF dataflow patterns implemented via Scale-Sim under varying array dimensions and data precision configurations. This analysis verifies whether the algorithm accurately selects the dataflow pattern with superior execution efficiency. The results are summarized in Fig. 11. Experimental results demonstrate that the dynamic dataflow selection algorithm consistently identifies performance-optimal dataflow patterns, thereby validating its correctness and enhancing the computational efficiency of the VPFU array in executing convolutional layers.

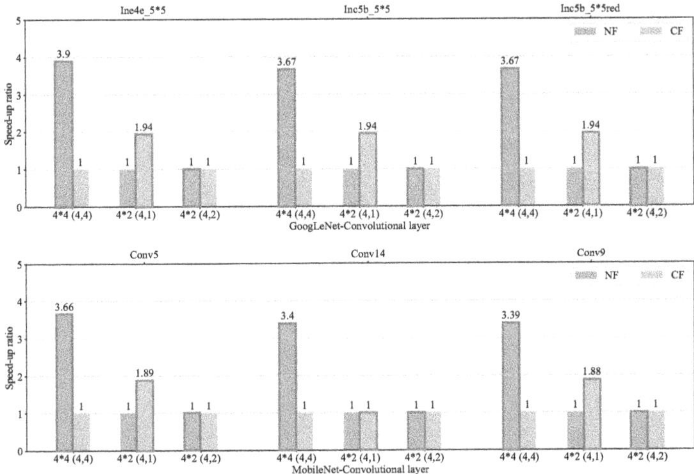

Fig. 11. Under varying array scales (W*H) and data precision (input precision, weight precision) configurations, the execution of identical convolutional layers using NF dataflow versus CF dataflow. Performance-superior dataflows selected by the algorithm are highlighted in red boxes. (Color figure online)

5   Conclusion

Ultra-low-bitwidth quantized neural networks effectively alleviate storage bottlenecks on resource-constrained edge devices. To address their stringent low-power requirements, this work proposes the Variable-Precision Fusion Unit (VPFU), an ultra-low-bitwidth arithmetic unit enabling efficient MAC operations for dynamically variable-precision data. Capitalizing on the inherent parallelism in DNNs, we design a channel-parallel array architecture based on VPFUs for efficient DNN inference. An adaptive dynamic dataflow selection algorithm is also introduced to maximize data reuse efficiency.

The VPFU was implemented in Verilog and synthesized using Synopsys Design Compiler with an open-source 45 nm PDK. Synthesis results demonstrate significant reductions in area and power consumption for the VPFU compared to computational units in state-of-the-art accelerators. Cycle-accurate simulations further reveal that the VPFU array delivers substantial acceleration for DNN inference workloads versus advanced accelerator architectures.

References

1. Guo, D., Zhang, H., et al.: DeepSeek-R1 incentivizes reasoning in LLMs through reinforcement learning. Nature **645**(8081), 633–638 (2025)
2. Nurul Achmadiah, M., Ahamad, A., et al.: Energy-efficient fast object detection on edge devices for IoT systems. IEEE Internet Things J. **12**(5), 16681–16694 (2025)

3. Zhang, Y., Sun, S., Ma, L.: Tiny transducer: a highly-efficient speech recognition model on edge devices. In: 2021 IEEE International Conference on Acoustics, Speech and Signal Processing (ICASSP), pp. 6024–6028. IEEE (2021)

4. Dupuis, T., et al.: Sparq: a custom RISC-v vector processor for efficient sub-byte quantized inference. In: 2023 21st IEEE Interregional NEWCAS Conference (NEWCAS), pp. 1–5 (2023)

5. Gholami, A., Kim, S., Dong, Z., et al.: A survey of quantization methods for efficient neural network inference. Low-Power Comput. Vis., 291–326 (2022)

6. Yao, K., Li, W.: Re-quantization based binary graph neural networks. Sci. China (Inf. Sci.) **67**(07), 160–171 (2024)

7. Xu, Y., Han, X., Yang, Z., et al.: Onebit: towards extremely low-bit large language models. In: Proceedings of the 38th International Conference on Neural Information Processing Systems, pp. 66357–66382 (2024)

8. Liu, Y., Wen, J., Wang, Y., et al.: VPTQ: extreme low-bit vector post-training quantization for large language models. In: 2024 Conference on Empirical Methods in Natural Language Processing, pp. 8181–8196 (2024)

9. Kim, H., Kim, Y., Ryu, S., et al.: BitSplit-Net: multi-bit deep neural network with bitwise activation function. arXiv preprint arXiv:1903.09807 (2019)

10. Zhang, H., et al.: BitNet: scaling 1-bit transformers for large language models. arXiv preprint arXiv:2310.11453 (2023)

11. Jouppi, N.P., Young, C., Patil, N., et al.: In-datacenter performance analysis of a tensor processing unit. In: 2017 ACM/IEEE 44th Annual International Symposium on Computer Architecture (ISCA), pp. 1–12 (2017)

12. Chen, Y., et al.: DaDianNao: a machine-learning supercomputer. In: 2014 47th Annual IEEE/ACM International Symposium on Microarchitecture (MICRO), pp. 609–622 (2014)

13. Judd, P., Albericio, J., Hetherington, T., Aamodt, T.M., Moshovos, A.: Stripes: bit-serial deep neural network computing. In: 2016 49th Annual IEEE/ACM International Symposium on Microarchitecture (MICRO), pp. 1–12 (2016)

14. Chen, Y., et al.: BitMoD: bit-serial mixture-of-datatype LLM acceleration. In: 2025 IEEE International Symposium on High Performance Computer Architecture (HPCA), pp. 1082–1097 (2025)

15. Sharma, H., et al.: Bit fusion: bit-level dynamically composable architecture for accelerating deep neural network. In: 2018 ACM/IEEE 45th Annual International Symposium on Computer Architecture (ISCA), pp. 764–775 (2018)

16. Hubara, I., Courbariaux, M., Soudry, D., El-Yaniv, R., Bengio, Y.: Quantized neural networks: training neural networks with low precision weights and activations. J. Mach. Learn. Res. **8**, 154–234 (2018)

17. Howard, A.G.: MobileNets: efficient convolutional neural networks for mobile vision applications. arXiv preprint arXiv:1704.04861 (2017)

18. Szegedy, C., et al.: Going deeper with convolutions. In: Proceedings of the IEEE Conference on Computer Vision and Pattern Recognition, pp. 1–9 (2015)

19. Redmon, J., Divvala, S., Girshick, R., Farhadi, A.: You only look once: unified, real-time object detection. In: 2016 IEEE Conference on Computer Vision and Pattern Recognition (CVPR), pp. 779–788 (2016)

20. He, K., Zhang, X., Ren, S., Sun, J.: Deep residual learning for image recognition. In: Proceedings of the 2016 IEEE Conference on Computer Vision and Pattern Recognition (CVPR), pp. 770–780 (2016)
21. Samajdar, A., et al.: SCALE-sim: systolic CNN accelerator simulator. arXiv preprint arXiv:1811.02883 (2019)
22. Won, J., et al.: ULPPACK: fast sub-8-bit matrix multiply on commodity SIMD hardware. Proc. Mach. Learn. Syst. **4**, 52–63 (2022)

3D-DP: A Practical DRL-Based Data Prefetcher with a Dynamic Prefetch Degree and Direction

Yafei Zhao[1,2], Jizeng Wei[1,2(✉)], Shuangsheng Li[1,2], and Yaogong Yang[3]

[1] School of Cyber Security, Tianjin University, Tianjin, China
`weijizeng@tju.edu.cn`
[2] College of Intelligence and Computing, Tianjin University, Tianjin, China
[3] Phytium Technology Co., Ltd., Tianjin, China

Abstract. Data prefetching is an essential technique that bridges the memory wall by speculatively loading data to hide memory access latency. Although Reinforcement Learning (RL)-based prefetchers such as Pythia show significant potential, they remain limited by state representations, high hardware overhead, and the inability to dynamically coordinate the prefetch degree and direction. To address these limitations, we propose 3D-DP, a novel prefetcher based on Deep Reinforcement Learning (DRL). 3D-DP replaces large Q-tables with a lightweight fully connected neural network. This approach allows for the direct processing of high-dimensional state features, alleviating the state aliasing issues introduced by hash-based indexing. It also addresses slow-start issues thanks to the network's ability to generalize across parameters. 3D-DP also introduces an adaptive control mechanism that dynamically determines the optimal prefetch degree for each memory page based on its access history and current system load. This is complemented by a low-overhead algorithm for predicting prefetch direction, enabling fine-grained coordination of both the amount and direction of prefetched data. For practical hardware implementation, we also propose 3D-DP-TQ, a quantized version of 3D-DP that leverages ternary quantization to compress the storage overhead to just 4.92 KB. Our evaluations using the SPEC CPU 2006 and 2017 benchmarks show that 3D-DP outperforms state-of-the-art prefetchers, achieving performance improvements of 10.2%, 7.6%, and 5.2% over Pythia, MLOP, and Berti, respectively. Notably, 3D-DP-TQ retains 95.5% of 3D-DP's performance and incurs a substantially lower hardware cost, demonstrating an effective trade-off between performance and efficiency.

Keywords: Hardware Data Prefetching · Deep Reinforcement Learning · Microarchitecture

1 Introduction

The performance gap between processors and the main memory, known as the memory wall, represents a critical bottleneck in modern computer systems.

H. Liu et al. (Eds.): ICA3PP 2025, LNCS 16381, pp. 58–82, 2026.
https://doi.org/10.1007/978-981-95-8399-7_4

While caches mitigate this gap, the long latency of cache misses continues to hinder performance. Data prefetching addresses this challenge by predicting a program's memory access patterns and loading data into the cache before it is explicitly requested, effectively hiding memory latency and improving processor performance.

Recently, thanks to the powerful decision-making capabilities demonstrated by Machine Learning (ML), a new wave of high-performance, ML-based prefetchers [4,12,17,21,24] has been proposed. Recent work on Pythia [4], in particular, has demonstrated the significant potential of Reinforcement Learning (RL) in the data prefetching domain. Pythia's design offers two key advantages. First, its use of online training provides excellent dynamic adaptability. By continuously learning from real-time interactions with the memory system, Pythia can accurately capture dynamically changing access patterns during program execution–a capability beyond the reach of models trained offline. Second, from a practical standpoint, many contemporary ML-based prefetchers incur storage overheads of tens of megabytes or more. In contrast, Pythia's modest hardware footprint makes it a much more feasible solution for integration into real systems.

However, Pythia and prior prefetchers still exhibit several key shortcomings:

Limited State Representation. The state space a prefetcher encounters at runtime is vast; a 64-bit program counter (PC) alone constitutes 2^{64} possible states. In sharp contrast, Pythia's Q-table contains only 128 entries, limiting it to representing just 128 distinct states. To overcome this disparity, Pythia hashes the full-state vector to an index within its small table. This approach, however, inevitably leads to hash collisions, where multiple distinct states map to the same entry. The resulting state aliasing degrades the quality of the prefetcher's decisions. Furthermore, Pythia suffers from a slow-start problem. When a new, previously unseen state is encountered, its corresponding Q-table entry is empty. Consequently, the prefetcher is initially unable to make an informed decision and must wait for the entry to be populated through subsequent learning, which delays its effectiveness for new access patterns.

Excessive Hardware Overhead. As a value-based RL method, Pythia's Q-learning algorithm relies on a Q-table to store the expected value of every state–action pair. A key drawback is that the size of this table scales directly with the product of the state and action spaces. Consequently, even though Pythia aggressively prunes these spaces, which optimizes its implementation, the resulting 24 KB Q-table still imposes a significant hardware overhead.

Inadequate Control of Prefetch Degree. The prefetch degree the number of cachelines fetched at once—is a key parameter for controlling a prefetcher's aggressiveness. An aggressive strategy can improve prefetch coverage but risks cache pollution and wasted memory bandwidth if the prefetched data is unused. A conservative strategy may achieve high accuracy but often fails to hide latency effectively, limiting performance gains. As illustrated in Fig. 1, on the 429.mcf-184B, performance (IPC) speedup over a no-prefetch baseline initially rises with the prefetch degree but declines sharply past an optimal point. Therefore, dynamically managing this trade-off in prefetcher aggressiveness is crucial for maximizing performance.

However, existing designs fail to effectively control this trade-off. Most designs set the prefetch degree to a fixed value or offer a few static options. For example, BOP [18] disables itself by setting the degree to zero when the accuracy drops below a certain threshold. Footprint-based prefetchers like Bingo [3] try to fetch all addresses within a predicted memory region, a strategy that ignores real-time system conditions. Issuing a high-degree prefetch when the memory bandwidth is already heavily contended is clearly suboptimal.

Ignoring Prefetch Direction. The prefetch direction (e.g., forward or backward) determines the address sequence that should be fetched when the prefetch degree is greater than one. For a base address x and a degree of 3, a forward prefetch would fetch $\{x, x + 1, x + 2\}$, while a backward prefetch would fetch $\{x, x - 1, x - 2\}$. The prefetch direction reflects the program's memory traversal trend.

Coordinating the prefetch direction and degree is essential for optimizing efficiency. An incorrect direction prediction, especially with a high degree, not only leads to failure in obtaining useful data, but also severely pollutes the cache with useless data. However, most prior prefetchers do not predict the prefetch direction dynamically or accurately. Pythia, for example, fixes the direction to forward, limiting its effectiveness for programs with backward or bidirectional access patterns.

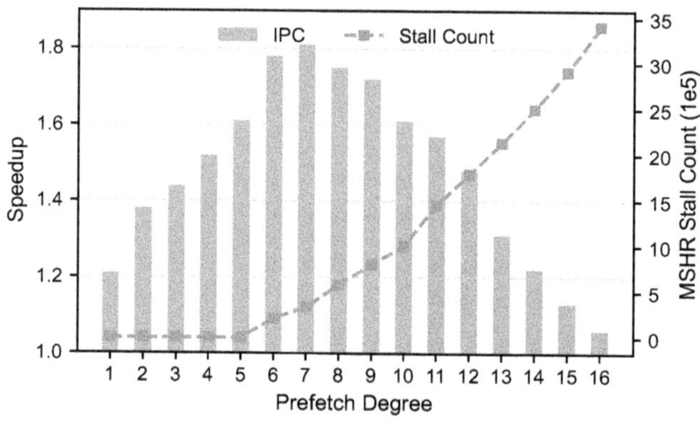

Fig. 1. The impact of prefetch degree on IPC and MSHR stall count.

To address these shortcomings, we propose 3D-DP, a novel prefetcher based on Deep Reinforcement Learning (DRL). 3D-DP replaces the large Q-table of traditional methods with a lightweight, fully connected neural network, enabling more effective prefetching decisions to be made. This approach offers several advantages. First, we match the network's input layer dimension to the state vector, allowing it to directly process complex state features and capture deep correlations within data access patterns. Second, the neural network approximates the entire Q-value function with a compact set of parameters, which significantly reduces storage requirements and overcomes the "curse of dimensionality" inherent to Q-table in vast state spaces. Finally, through parameter sharing, the network generalizes from learned experiences to make intelligent prefetching decisions for new, unseen states.

Recognizing that complex neural networks introduce significant overhead in hardware, we also propose 3D-DP-TQ, a more practical and lightweight variant of 3D-DP with a minor performance trade-off. 3D-DP-TQ applies 2-bit integer quantization (ternarization) to the neural network, significantly reducing both parameter storage overhead and computational complexity.

We dynamically adjust the prefetch degree on a per-page basis, considering both the page's access patterns and the current memory system load. Under high MSHR and memory bandwidth utilization, the prefetch degree is reduced to mitigate the penalties of over-prefetching. By tuning the degree for each page individually, our approach prevents mutual interference between access patterns, leading to more precise prefetching.

Our prefetch direction prediction algorithm is based on the observation that most pages exhibit a stable, unidirectional access pattern. Leveraging this, we determine the prefetch direction by simply comparing the current access address to the initial access address within that page. This lightweight method is effective and does not require significant hardware resources or complex decision logic.

The main contributions of this paper are as follows:

- We propose 3D-DP, a DRL-based prefetcher using a lightweight network to approximate the Q-table for enhanced decision-making. We also introduce 3D-DP-TQ, a ternarized version of 3D-DP that achieves a minimal footprint of 4.92 KB by significantly reducing parameter size and complexity.
- We introduce a novel prefetch degree prediction mechanism that dynamically controls prefetching aggressiveness by integrating historical access patterns with real-time system feedback (e.g., MSHR occupancy, bandwidth pressure).
- We propose a low-overhead prefetch direction prediction mechanism and demonstrate that its synergy with prefetch degree prediction significantly boosts efficiency.
- An extensive evaluation shows that 3D-DP outperforms state-of-the-art prefetchers, including Pythia (by 10.2%), MLOP (by 7.6%), and Berti (by 5.2%). The highly compact 3D-DP-TQ retains 95.5% of 3D-DP's performance.

2   Related Work

2.1   Traditional Spatial and Temporal Prefetchers

Traditional hardware prefetchers exploit the principle of memory access locality. Spatial prefetchers [3, 5, 15, 16, 18, 19, 23, 26] leverage spatial locality by predicting that memory regions physically near the current access will be subsequently requested. These prefetchers, which include offset-, delta-, and footprint-based designs, are characterized by simple hardware implementation, low storage overhead, and effectiveness in reducing compulsory misses. For example, BOP [18] prefetches the address $X + D$ upon access to address X, where D is a dynamically learned offset. However, the fundamental limitation of spatial prefetchers is their reliance on physically contiguous memory accesses, which renders them ineffective for data structures with irregular access patterns, such as linked lists, trees, and graphs. Their predictive range is further restricted by the common constraint that they cannot cross 4 KB physical page boundaries. In contrast, temporal prefetchers [1,2,9,27,28] exploit temporal locality by recording and replaying historical memory access sequences to predict future requests. Common approaches are based on a Global History Buffer (GHB) [20] or large pattern-matching tables. Temporal prefetchers excel at capturing complex, spatially non-contiguous access patterns. For instance, Domino [2] replays chains of past miss addresses, while MISB [28] maps physically disjoint addresses into a contiguous "structure address" space to linearize irregular access streams. The primary drawbacks of temporal prefetchers are their significant metadata storage overhead and high computational complexity, as they require large history structures to capture long-range dependencies.

2.2   Neural Network Prefetchers

The advent of machine learning has spurred interest in neural network prefetchers [8,12,14,17,22,24,31], which can capture complex, non-linear correlations in memory access patterns that elude traditional prefetchers. For instance, the LSTM-based prefetcher, Delta-LSTM [12], frames prefetching as a classification problem and achieves high prediction accuracy, but its reliance on learning address deltas renders it ineffective for irregular access patterns. To address such challenges, Voyager [24] introduced a hierarchical neural model that decomposes address prediction into two more manageable sub-problems: page prediction and offset prediction. This decomposition significantly shrinks the model's output space. Voyager further resolves the "offset aliasing" problem–where identical offsets on different pages create ambiguity–by incorporating a page-aware, attention-based offset embedding layer.

Despite this progress, the practical deployment of neural prefetchers faces two primary challenges: a vast output space and prohibitive computational cost. In a 64-bit system, the sheer number of unique addresses makes any classification model that predicts absolute addresses intractable. The computational overhead is equally daunting; a single prediction from Voyager requires approximately 81 million floating-point operations (FLOPs), inducing an inference latency of up to 18,000 ns. This latency far exceeds the practical time budget for a hardware prefetcher, negating the benefit of hiding memory latency.

2.3 RL-Based Prefetchers

RL-Based prefetchers [4,10,21,30] possess the key advantage of adapting to complex workloads with potentially less overhead than neural network-based methods. An example of an RL-based prefetcher is the Context Prefetcher [21], which employs a Contextual Bandits model. By using a high-level program context, such as the program counter and branch history, it identifies data correlations from program semantics to address irregular access patterns. Similarly, Pythia [4] models prefetching as a Markov Decision Process (MDP), where states are defined by program features and actions represent prefetch offsets and degrees. It learns online via the SARSA algorithm, using a reward mechanism based on prefetch timeliness and accuracy. A common challenge for these RL approaches, however, is the high complexity and storage overhead required to maintain action values across a vast number of states.

Our work differs from prior work in several ways. First, we propose 3D-DP, a prefetcher based on Deep Reinforcement Learning that replaces the conventional Q-table with a compact neural network. This design overcomes key limitations of Q-table-based methods like Pythia, such as state aliasing and slow-start issues, while enabling generalization to unseen access patterns. We also introduce 3D-DP-TQ, a highly compressed variant that uses ternarization to achieve a minimal hardware footprint of only 4.92 KB. Second, we introduce a novel mechanism to dynamically adjust prefetching aggressiveness. Unlike prior approaches that use a fixed prefetch degree, our method predicts and adapts the degree for each memory page individually. This decision is informed by both historical access patterns and real-time system feedback, including MSHR occupancy and memory bandwidth utilization. Third, we propose a low-overhead algorithm that dynamically predicts the memory access direction (forward or backward). This enables our prefetcher to effectively coordinate the prefetch degree with the access direction—a capability largely absent in previous prefetchers, which typically assume a fixed forward direction.

3 Method

3.1 Data Prefetching Using Deep Reinforcement Learning

As illustrated in Fig. 2, we formulate data prefetching as a Deep Reinforcement Learning problem. We model the interaction between the prefetching agent and the memory system environment as an MDP. The agent learns an optimal prefetching policy by observing the system state, taking a prefetch action, and receiving a corresponding reward from the environment at each step. The agent's objective is to learn a policy that maximizes the cumulative discounted reward, thereby enabling it to issue accurate and timely prefetches.

State. We represent the state as a k-dimensional vector composed of the system-level features listed in Table 1. A fundamental requirement for RL-based prefetching is to establish a meaningful closed-loop interaction with the memory system, where the agent's actions influence the subsequent state it observes.

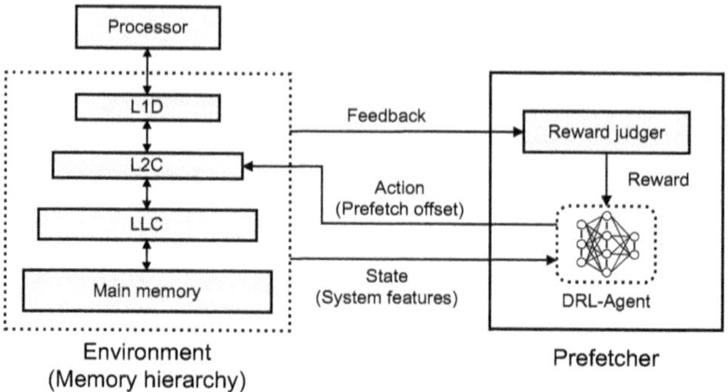

Fig. 2. The Deep Reinforcement Learning framework for 3D-DP.

However, prior methods (e.g., Pythia) primarily construct the state from static program features, such as the instruction address (PC) and load addresses. These features are largely invariant to the prefetcher's actions; whether the prefetch is accurate or inaccurate has no bearing on the state representation for the next decision. This insensitivity breaks the crucial "state–action–reward" feedback loop essential to the RL framework.

To address this limitation, we augment the state representation with dynamic, system-level feedback features, such as memory bandwidth utilization, historical prefetch accuracy, and the program's instructions per cycle (IPC). By incorporating these runtime-sensitive metrics, the state becomes responsive to the outcomes of previous prefetching actions. For example, an effective prefetch action that improves program performance will lead to a measurable increase in IPC, which is then captured as a change in the state vector at the subsequent timestep. This design enables the RL agent to learn a more accurate and adaptive policy by tightly coupling its decisions to system-level feedback.

Action. The agent's action is defined as selecting a prefetch offset—the delta between the predicted and demanded cacheline addresses. For a system with 4 KB pages and 64B cachelines, the valid offset range within a page is $[-63, 63]$, totaling 127 possible actions. However, such a large action space is impractical and inefficient.

First, leveraging our dynamic prefetch direction mechanism, the agent can implicitly determine the prefetch direction (forward or backward). As a result, we can safely restrict the action space to non-negative offsets $[0, 63]$, effectively halving its size. Second, we observe that large prefetch offsets are inefficient for two primary reasons. On one hand, data prefetched with a large offset has a longer latency until it is potentially used, increasing the risk of eviction before use and thus reducing prefetch timeliness. On the other hand, since prefetching is confined within a physical page, a large offset is more likely to generate an address that crosses a page boundary, causing the request to be discarded and resulting in a loss of coverage.

Table 1. System-level features used in the state vector.

Features	Description
PC	The program counter of the load instruction that triggered the prefetch
Load Address	The physical memory address accessed by the load instruction
Page Footprint	The footprint of all addresses within the same memory page as the load address
Last 4 offsets	The page offsets of the last four memory accesses within the same page
Prefetch accuracy	The accuracy of previous prefetches
L2 Cache hit rate	The hit rate of the L2 Cache
Bandwidth occupancy	Memory bandwidth utilization
IPC	The instructions per cycle executed by the processor

Based on these insights, we refine the action space by discarding all offsets in the [32, 63] range. From the remaining [0, 31] range, we empirically select the 16 most frequently used offsets as the final action space for 3D-DP.

Reward. The reward scheme defines the prefetcher's objective. To guide the agent's behavior with high precision, we define a fine-grained reward scheme with the following components:

- **Accurate and Timely** (R_{AT}): A prefetch is successful, arriving in the cache before the processor requests it.
- **Accurate but Late** (R_{AL}): Demand occurs before the prefetched line is filled.
- **Accurate but Premature** (R_{AP}): A prefetched line is evicted before it is ever used.
- **Inaccurate Prefetch** (R_{IN}): A prefetched line is evicted without ever being used.
- **No Prefetch** (R_{NP}): The agent chooses not to issue a prefetch.
- **Page Boundary Crossing** (R_{BC}): The prefetch address crosses a physical page boundary, causing the request to be discarded.
- **Prefetch Hit** (R_{PH}): The agent issues a prefetch for data that is already in the cache.

3.2   Detailed Design of 3D-DP

Figure 3 shows a high-level overview of 3D-DP, which is composed of a Prediction module and a Training module.

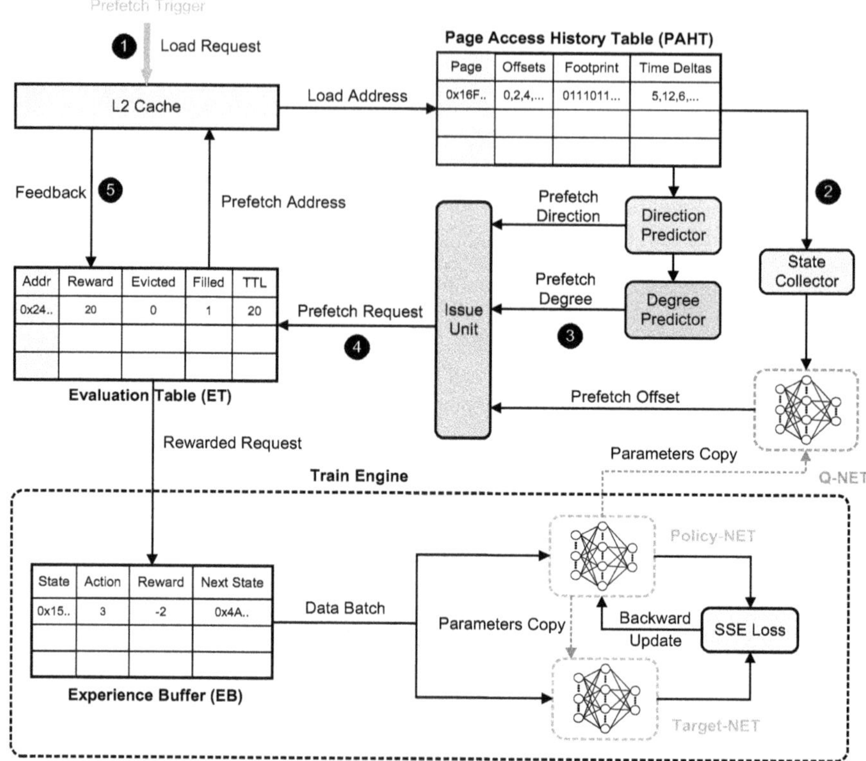

Fig. 3. Microarchitecture and data flow of the 3D-DP prefetcher.

Prediction. A complete prefetch cycle involves predicting the prefetch direction, degree, and offset. Based on these predictions, the Issue Unit generates and sends prefetch requests, which are then evaluated by an Evaluation Table (ET) to assign rewards. The Page Access History Table (PAHT) is the primary data structure in 3D-DP. It tracks the page access history to provide the state information required for prediction.

1. **Prefetch Trigger.** A demand request to the L2 Cache triggers the 3D-DP prefetching process. 3D-DP queries the PAHT using the page number of the demand address. If the page is found (a hit), the corresponding entry is updated with new offset and footprint data. If not (a miss), a new entry is created. When the PAHT reaches capacity, entries are evicted using the Least Recently Used (LRU) policy.
2. **State Collection.** Following the PAHT lookup, the State Collector determines the current system state. To prevent numerical instability resulting from using 64-bit features (e.g., PC, memory address) as direct neural network inputs, 3D-DP hashes each 64-bit feature to 32 bits, then splits them into four 8-bit sub-features. These are combined to form a 16-dimensional state vector, which serves as the input for the prediction modules.

3. **Prefetch Decision.** The state vector is fed into the Q-Network (Q-NET) to predict the prefetch offset. Concurrently, independent predictors determine the prefetch degree and direction (details of the prediction procedures are given in Sect. 3.3). These three predictions–offset, degree, and direction— collectively form the final prefetch decision.

4. **Prefetch Issuance.** The Issue Unit combines the prefetch offset, degree, and direction to generate the final prefetch address. It then sends a prefetch request to both the L2 Cache and the ET. Before dispatching the request, 3D-DP performs a page boundary check. Any request that crosses a 4 KB page boundary is considered invalid and is immediately discarded. It is then assigned a reward of R_{BC} without occupying ET resources.

5. **Prefetch Evaluation.** Prefetch requests are inserted into the ET to await a delayed reward assignment. Each ET entry contains the prefetch address (Addr), a reward value, an evicted bit, a filled bit, and a Time-To-Live (TTL) counter. When a prefetched line is filled into the L2 Cache, the ET is queried and the filled bit of the corresponding entry is set. If the data is already resident in the cache at fill time (e.g., a prefetch hit), a reward of R_{PH} is assigned. Similarly, if the prefetched data is later evicted from the L2 Cache, the entry's evicted bit is set. When the L2 Cache receives a demand load, it checks the ET for a matching prefetch. If found, a reward is assigned based on the state of the evicted and filled bits:
   - If the evicted bit is set, it indicates that the prefetch was correct but evicted prematurely. In this case, a reward of R_{AP} is assigned.
   - If the filled bit is set, the prefetch is considered both accurate and timely, and a reward of R_{AT} is assigned.
   - If neither bit is set, the prefetch is accurate but the data has not yet arrived, indicating that the prefetch was untimely. A reward of R_{AL} is assigned.

The ET uses a TTL mechanism to identify useless prefetches. The TTL of each entry is periodically decremented. If an entry's TTL reaches zero, the prefetch is deemed useless as it has not been accessed in time, and a reward of R_{IN} is assigned. A reward of R_{IN} is also assigned when an entry is evicted from a full ET.

Training and Network Update. The training process of 3D-DP is illustrated by the Train Engine in Fig. 3. Prefetch requests that retire from ET are stored in the Experience Buffer (EB) as training samples. Once the EB accumulates a full batch of N samples (where N is the train batch size), it triggers a training iteration of the neural network.

To enhance training stability, 3D-DP employs Double DQN, an improved version of the Deep Q-Network (DQN) algorithm. The architecture decouples training and inference into separate paths to prevent mutual interference. Training is performed exclusively on a replica of the inference network (Q-NET), termed the Policy-NET. This ensures that the parameters of the primary Q-NET remain stable during inference. After every K training iterations, the weights from the

Policy-NET are synchronized to the Q-NET to update the parameters used for prediction.

This decoupled design provides two key benefits. First, it ensures continuous prefetch operation without incurring prediction stalls or performance degradation due to frequent weight updates. Second, it reduces hardware complexity by avoiding long-latency memory accesses on the critical inference path.

3.3 Prefetch Degree and Direction Prediction

Algorithm 1 presents our proposed prefetch degree prediction algorithm. This design considers several key factors to dynamically adjust the aggressiveness of the prefetching strategy, adapting to varying program behaviors and system conditions.

Spatial Locality of Page Accesses. For highly compact and sequential access patterns, employing a large prefetch degree can pre-load the required data with high accuracy, yielding significant performance gains. Conversely, for irregular patterns such as random access, it is imperative to use a minimal, or even zero, prefetch degree to avoid wasting memory bandwidth and polluting the cache by prefetching useless data.

Our algorithm is based on the core idea that a request issued by the prefetcher essentially covers a contiguous data region, defined by a starting prefetch offset and a length equal to the prefetch degree. Therefore, we construct a historical reference region to evaluate the compactness of recent accesses. Specifically, if the current prefetch direction is forward, the reference region is defined as shown in line 5; otherwise, it is as shown in line 7, where D_{max} is the maximum prefetch degree. The algorithm quantifies the compactness of the current memory access pattern by calculating the number of distinct offsets within the last D_{max} accesses that fall into this historical reference region (line 11). This count is then used as the new prefetch degree (line 14).

Temporal Locality of Page Accesses. An analysis of program memory access history reveals that in some applications, consecutive accesses to the same page are separated by significant time intervals. For instance, in the 607.cactuBSSN_s, over 85% of page revisit intervals exceed several hundred timesteps, where one timestep is defined as a single L2 Cache access.

Such patterns are common in scientific computing workloads. Consider a General Matrix–Vector Multiplication (GEMV) operation performed column-wise. In this approach, each step of the computation uses one column of matrix A and one element of vector B. The entire column from A is scaled by the single element from B and accumulated into the output vector. Consequently, the access to an element in vector B is separated from the access to the next element by a time interval proportional to the matrix dimension M. For a large M, even if a prefetcher fetches subsequent data for vector B in a timely manner, this data is likely to be evicted from the cache long before it is referenced, rendering the prefetch ineffective.

Therefore, for an access pattern with high spatial locality but poor temporal locality, aggressive prefetching can degrade system performance. Motivated by this observation, we introduce a throttling mechanism based on temporal locality. When our algorithm detects that a page's revisit interval exceeds a predefined threshold T, it sets the prefetch degree for that page to zero (line 17), thus suppressing this class of ineffective prefetches.

MSHR Stalls. An overly aggressive prefetching strategy can overwhelm critical memory system resources, leading to severe MSHR (Miss Status Holding Register) stalls. Prefetch requests and demand misses from upper-level caches must contend with the limited MSHR entries in the lower-level cache. When these entries become saturated, new memory requests are blocked, which directly causes CPU stalls and degrades performance.

We observe that even with throttling mechanisms based on temporal and spatial locality, certain memory-intensive workloads (e.g., 429.mcf) still suffer from significant MSHR stalls due to prefetching. As shown in Fig. 1, the speedup over a no-prefetch baseline increases with the prefetch degree, reaching its peak at a degree of 7 before declining. Beyond this point, IPC drops sharply, while MSHR stalls rise rapidly once the degree exceeds 5. This highlights that overly aggressive prefetching can overwhelm MSHRs, causing the performance cost to outweigh the benefits.

To address this, we introduce a feedback mechanism that dynamically adjusts the maximum prefetch degree, D_{max}, based on recent MSHR stall behavior. If stalls are detected, D_{max} is reduced to relieve the pressure on MSHRs. If no stalls occur over time, D_{max} is gradually increased to explore higher performance potential.

Prefetch Direction Prediction. The memory access pattern within a page often reveals the program's overall access direction. A sequential pattern produces a consistent offset sequence (e.g., $\{1, 1, 1, 1, 1, 1\}$), clearly indicating forward access. In contrast, a quasi-sequential pattern (e.g., $\{-1, 1, -2, 1, -2, 1\}$) exhibits frequent sign changes, making the instantaneous direction ambiguous. However, prefetchers aim to capture the overall access trend rather than precisely predict each individual direction. Despite its fluctuations, the quasi-sequential pattern shares a forward-moving trend similarly to the sequential pattern.

Therefore, our proposed direction prediction mechanism primarily targets two access patterns with distinct macroscopic trends: (1) sequential, which has a stable access direction, and (2) quasi-sequential, which has a fluctuating instantaneous direction but a stable overall trend. We employ a simple, low-overhead method to predict the prefetch direction. The direction is determined by comparing the cumulative offset of the current access relative to the initial access within the page. A positive cumulative offset results in the prediction of a forward direction, while a non-positive offset results in the prediction of a backward direction.

Algorithm 1. 3D-DP's Prefetch Degree Prediction Algorithm

Input: The page number P of the load address, the time delta threshold T, and the maximum prefetch degree D_{max}.

Output: The prefetch degree D.

1: $O \leftarrow \text{GetLastKOffsets}(P, D_{max})$
2: $X \leftarrow \text{GetLastOffset}(P)$
3: $Gos \leftarrow \emptyset$ ▷ Initialize an empty set for matching offsets
4: **if** $\text{GetPrefetchDir}(P) == \text{FORWARD}$ **then**
5:      $R \leftarrow \{X - D_{max} + 1, ..., X - 1, X\}$
6: **else**
7:      $R \leftarrow \{X, X + 1, ..., X + D_{max} - 1\}$
8: **end if**
9: **for** $i = 0$ **to** D_{max} **do**
10:      **if** $O[i] \in R$ **then**
11:          $\text{Insert}(Gos, O[i])$ ▷ Add historical offsets that overlap with the region
12:      **end if**
13: **end for**
14: $D \leftarrow \text{SizeOf}(Gos)$ ▷ Initial degree is the number of overlapping offsets
15: $Td \leftarrow \text{GetLastKTimeDeltas}(P, 5)$
16: **if** $\text{AVG}(Td) \ ¿ \ T$ **then**
17:      $D \leftarrow 0$ ▷ If average access time is too high, cancel prefetch
18: **end if**
19: **return** D

3.4 The Design of 3D-DP-TQ

To improve the hardware feasibility of the 3D-DP prefetcher, we propose 3D-DP-TQ, a variant that significantly reduces hardware overhead with **ternary quantization**. While the original 3D-DP already employs a lightweight Q-NET, its direct hardware implementation remains costly.

Quantizing the Q-NET. To efficiently implement the fully connected neural network as a hardware prefetcher, we leverage quantization. Inspired by WAGE [29], we quantize the Q-NET's weights (W), activations (A), gradients (G), and errors (E) to 2-bit, 8-bit, 8-bit, and 8-bit fixed-point integers, respectively. For the Q-NET's weights, we adopt a ternary quantization strategy, mapping each weight parameter to a value in the set $\{-1, 0, +1\}$.

This quantization scheme simplifies computation. The product of an input x and a quantized weight w_q is determined as follows: if $w_q = +1$, the result is x; if $w_q = -1$, the result is $-x$; and if $w_q = 0$, the result is 0. These operations are highly hardware-friendly, implemented using only multiplexers (Muxs) and integer adders, thereby eliminating the need for area-intensive and high-latency Multiply–Accumulate (MAC) units.

Moreover, we further reduce the action space of 3D-DP-TQ by selecting only the eight most frequently used offsets, which effectively reduces the size of the

Q-NET model. The Q-NET configurations for 3D-DP and 3D-DP-TQ are shown in Table 5.

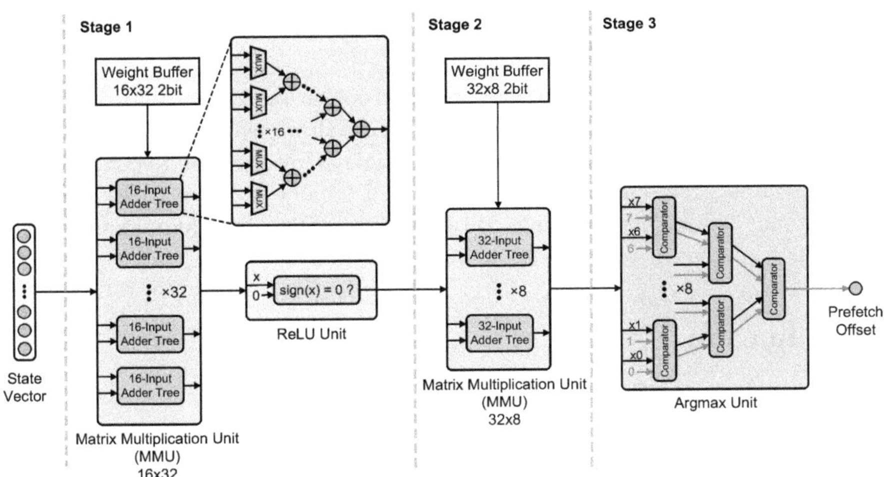

Fig. 4. A three-stage pipeline for the Q-NET hardware implementation.

Hardware Implementation of Q-NET. We propose a pipelined hardware implementation for Q-NET that balances prediction latency with timing performance to minimize hardware overhead.

The timeliness of data prefetching is critical: a prefetch request must be issued several cycles before the actual demand to effectively hide memory latency. Consequently, it is crucial to minimize prediction latency. At the same time, the prefetcher must not become the processor's critical path, as this would limit the overall clock frequency.

Performing Q-NET inference in a single clock cycle would introduce a long combinational logic path, creating a new timing bottleneck. To address this issue, we design a three-stage pipelined hardware implementation for Q-NET, as illustrated in Fig. 4.

Stage 1 is centered on a 16×32 Matrix Multiplication Unit (MMU). The MMU contains 32 16-input adder trees, each calculating the inner product of the state vector and a corresponding column of 16 ternarized $(-1, 0, 1)$ weights from the Weight Buffer. To minimize access latency, the 192 B Weight Buffer is implemented as a register array. Within each adder tree, multiplication is performed by a parallel array of MUXs that select and output either $-x$, 0, or x based on the weight's value. A four-level ($\lceil \log_2 16 \rceil$) adder tree then reduces the 16 operands. The MMU output passes through a ReLU unit, implemented efficiently by checking the sign bit and zeroing negative values.

Stage 2 is organized similarly to the Stage 1. It contains a 32×8 Weight Buffer and a corresponding MMU to produce eight prediction values.

Stage 3 implements an Argmax unit to select the action with the highest predicted value. This unit uses a three-level comparator tree ($\lceil \log_2 8 \rceil$) to identify the maximum value and its index. Isolating Argmax into its own stage balances pipeline latency and prevents bottlenecks.

As shown in Fig. 3, the full prefetch flow comprises three phases: (1) State Construction (one cycle), (2) Offset Prediction (three cycles), and (3) Prefetch Issue (one cycle), resulting in a total prediction latency of five cycles. This latency ensures prefetches can be issued in a timely manner to hide memory access latency while maintaining compatibility with high-frequency processor operation.

4   Evaluation

4.1   Experimental Setup

We evaluate all prefetchers using the trace-driven ChampSim [11] simulator, a widely adopted tool for microarchitectural research in areas such as branch prediction, cache replacement policies, and data prefetching. Table 2 summarizes the key simulation parameters used in our evaluation. For each simulation, we warm up the cache and associated structures using the first 50 million instructions, followed by a detailed simulation of the next 200 million instructions.

Workloads. To demonstrate the broad applicability of our proposed DRL-based prefetcher, we evaluate its performance using a diverse set of 83 traces from multiple workload suites. The workloads used for our evaluation include SPEC CPU 2006 [13], SPEC CPU 2017 [7], PARSEC [6], and Ligra [25]. We exclude traces with fewer than 10 million L2 Cache accesses, since on such traces, all prefetchers exhibit performance similar to a baseline with no prefetching, making it difficult to demonstrate their acceleration effects.

Prefetchers. We compare the performance of 3D-DP and 3D-DP-TQ with that of three state-of-the-art prefetchers: Pythia [4], MLOP [23], and Berti [19]. Pythia is a state-of-the-art RL-based prefetcher that dynamically adjusts its prefetching policy based on historical memory access behavior. MLOP is a high-performance offset prefetcher that achieved excellent results in the 3rd Data Prefetching Championship (DPC-3). Berti is a recently proposed lightweight, pattern-based prefetcher that is adept at capturing complex memory access patterns. The parameter configurations for 3D-DP, 3D-DP-TQ, and the three competing prefetchers are shown in Table 3.

Table 2. Simulation parameters.

Core	6-wide issue, 4-wide retire, out-of-order execution, 2.5 GHz
ROB/IQ/LQ/SB	512/204/192/114 entries
Branch Predictor	TAGE-SC-L
L1-I Cache	32 KB, 8-way, 4-cycle hit latency, 16 MSHRs
L1-D Cache	48 KB, 8-way, 5-cycle hit latency, 16 MSHRs
L2 Cache	512 KB, 8-way, 12-cycle hit latency, 32 MSHRs
LLC	4 MB, 16-way, 36-cycle hit latency, 64 MSHRs
DRAM	8 GB, 100-cycle access latency

Table 3. Configuration of evaluated prefetchers.

Prefetchers	Configurations	Overhead
Pythia	2 features, 2 vaults, 3 planes, 16 actions	25.5 KB
MLOP	128-entry AMT, 500-update, 16-degree	8 KB
Berti	8-set, 16-way History Table, 16-entry Delta Table, 16-entry PQ/MSHR, virtual address training	2.55 KB
3D-DP	See Table 5 for hardware configuration and Table 4 for parameters	8.73 KB
3D-DP-TQ	See Table 5 for hardware configuration and Table 4 for parameters	4.92 KB

4.2 Performance of 3D-DP

Overall Performance. Figure 5 shows the performance of the four prefetchers across various workloads. On average, 3D-DP improves performance by 83.1% over a no-prefetching baseline, outperforming Pythia, MLOP, and Berti by 10.2% (1.66×), 7.6% (1.70×), and 5.2% (1.74×), respectively.

3D-DP's advantage is particularly pronounced in the computation-intensive SPEC CPU 2006 and 2017 workloads, where it improves performance by 76.8% and 108.9%, respectively, significantly outperforming all other prefetchers. Figure 7 further illustrates the performance comparison between 3D-DP and Pythia, the state-of-the-art RL-based prefetcher, on SPEC CPU workloads. On several memory-intensive traces, 3D-DP achieves a performance improvement of over 10%, and notably, it outperforms Pythia by up to 94.3% on 429.mcf-217B.

In 429.mcf-217B, 87.3% of page accesses exhibit quasi-sequential patterns, where the access deltas alternate between large positive and negative values without clear regularity. This irregularity hampers delta-based prefetchers like Pythia. Moreover, 93.1% of accesses are backward, while Pythia only supports forward prefetching, further degrading its accuracy. In contrast, 3D-DP dynam-

Table 4. Parameters of 3D-DP and 3D-DP-TQ.

Parameters	Value
Action list	3D-DP: $\{-8, -6, -4, -3, -1, 0, 1, 2, 3, 4, 5, 6, 8, 10, 12, 16\}$ 3D-DP-TQ: $\{-4, -2, 0, 1, 2, 3, 4, 8\}$
Reward values	$R_{NP} = -1$, $R_{AP} = 3$, $R_{AT} = 20$, $R_{AL} = 2$, $R_{IN} = -23$, $R_{BC} = -3$, $R_{PH} = -1$
$N/K/T/D_{\max}$	32/10/560/16

ically adjusts both the prefetch degree and direction, allowing it to better adapt to such complex patterns and achieve higher accuracy and performance.

3D-DP also performs well on the Ligra graph processing workload, achieving a 1.90× speedup. In the PARSEC workload, memory access patterns are more regular and sequential, resulting in high cache hit rates and limited potential for prefetching improvement. In this scenario, Berti's aggressive approach slightly outperforms 3D-DP, achieving a 1.49× speedup compared to 3D-DP's 1.44×.

Accuracy. As shown in Fig. 9a, 3D-DP achieves the highest prefetch accuracy. On average, 3D-DP achieves 86.6% accuracy, significantly outperforming Pythia (85.3%) and slightly surpassing MLOP (86.5%) and Berti (86.3%).

3D-DP's superior accuracy is due to its design, using independent, per-page access histories in its PAHT. This approach prevents interference between different pages' memory access patterns, allowing for more precise, fine-grained prefetching. The benefit of this approach is most evident on the SPEC CPU 2006 workload, where Pythia's accuracy drops to 76.8%—markedly lower than 3D-DP's accuracy of 82.3%. Pythia uses global access history, which struggles to identify the correct pattern when multiple complex patterns interfere with

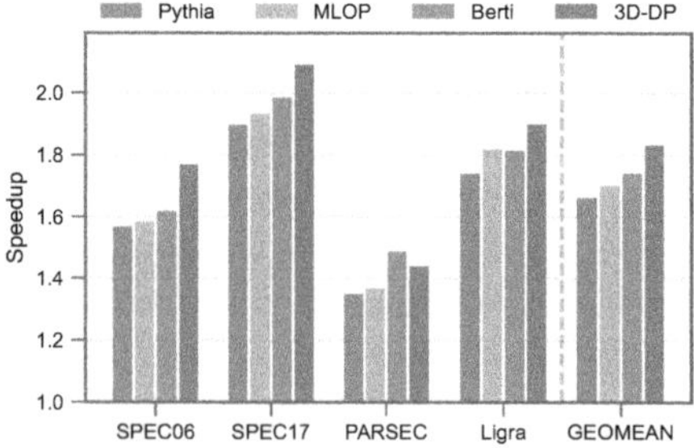

Fig. 5. Performance comparison of 3D-DP against state-of-the-art prefetchers across multiple workloads.

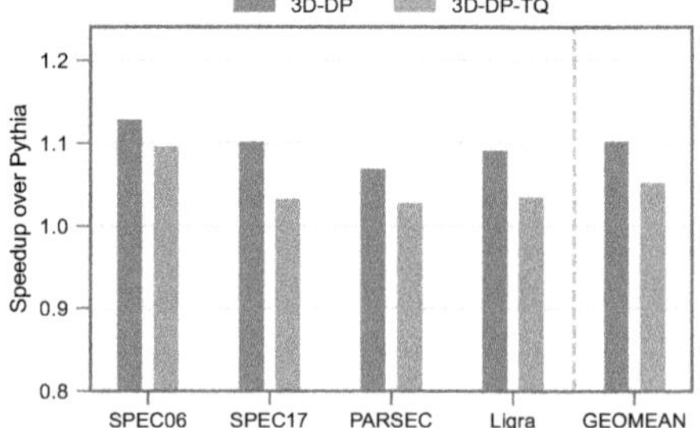

Fig. 6. Relative performance of 3D-DP and its quantized version, 3D-DP-TQ, normalized to the Pythia prefetcher.

each other. Conversely, 3D-DP's per-page mechanism isolates these patterns, providing stronger interference immunity and thus higher accuracy.

Timeliness. Prefetch timeliness indicates whether data is fetched and placed into the cache before the processor requests it. Accordingly, correct prefetches can be classified as either timely or late. Only timely prefetches effectively mitigate the performance penalty of cache misses. In contrast, late prefetches, similarly to incorrect ones, lead to cache pollution and wasted memory bandwidth. Therefore, an efficient prefetcher must not only achieve high accuracy but also ensure the timeliness of its prefetch requests.

Figure 9a illustrates the timeliness of different prefetchers, where the green portion of each bar represents timely prefetches and the orange portion represents late prefetches. The figure shows that 3D-DP achieves high timeliness. On average, 3D-DP's late prefetches constitute only 1.5% of its total prefetch requests, significantly lower than Pythia (3.9%), MLOP (2.4%), and Berti (3.0%). With a comparable overall prefetch accuracy, this high timeliness enables 3D-DP to achieve a higher effective prefetch accuracy. 3D-DP's timely prefetch accuracy is 84.8%, outperforming Pythia (81.6%) and Berti (82.6%), and is comparable to that of MLOP (84.2%).

MLOP achieves high prefetch timeliness through its Multi-Lookahead mechanism, which determines the optimal prefetch offset from various lookahead distances. 3D-DP also prioritizes prefetch timeliness. In 3D-DP's Reinforcement Learning reward function, a late but correct prefetch receives a negative reward similar to that of an incorrect prefetch. This policy incentivizes the prefetcher to issue timely and effective prefetch requests, thereby improving not only accuracy but also the practical utility of its prefetches and overall system performance.

Coverage and Overprediction. Prefetch coverage is defined as the fraction of L2 Cache misses eliminated by the prefetcher. Overprediction is defined as the fraction of incorrect and late prefetches relative to all L2 Cache load requests. Figure 9b presents the coverage and overprediction results across workloads.

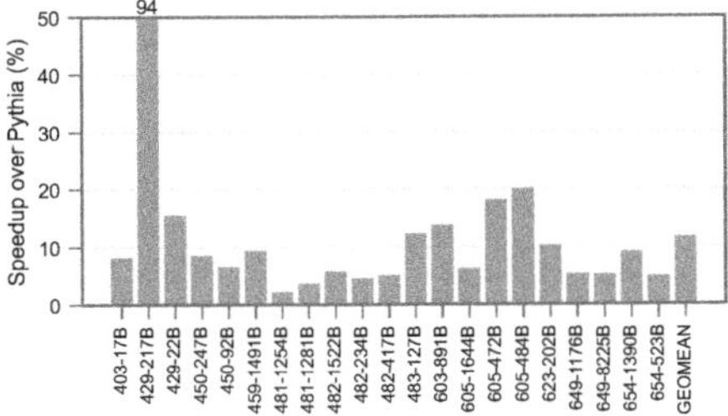

Fig. 7. Speedup of 3D-DP over Pythia on SPEC CPU workloads.

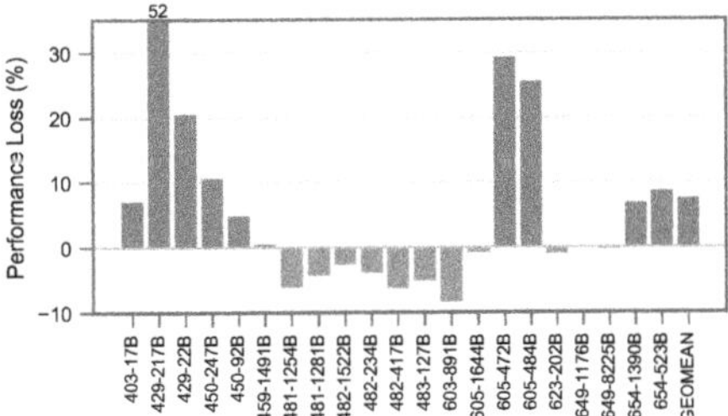

Fig. 8. Performance impact of disabling dynamic direction prediction.

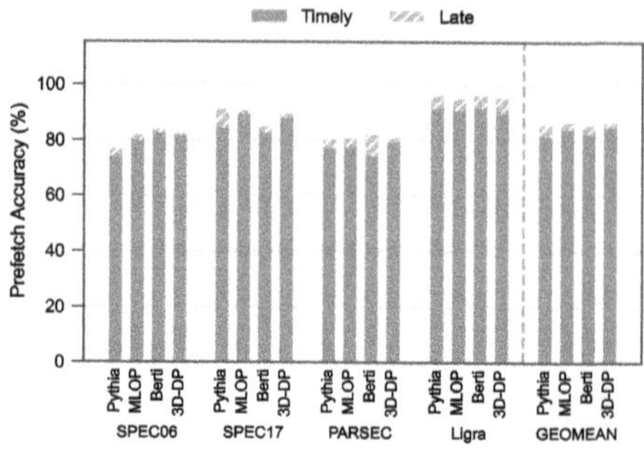

(a) Prefetch accuracy and timeliness.

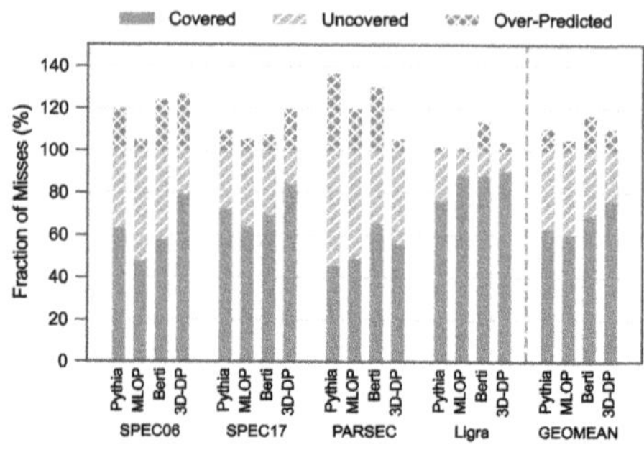

(b) Prefetch coverage and overprediction.

Fig. 9. Performance comparison of four prefetchers. (a) Prefetch accuracy, broken down into timely and late prefetches. (b) Prefetch coverage and degree of overprediction.

On average, 3D-DP achieves 76.3% coverage—20.5%, 26.3%, and 9.8% higher than Pythia, MLOP, and Berti, respectively. Despite its significantly higher coverage than Pythia and MLOP, 3D-DP's overprediction rate of 10.6% is comparable to Pythia's (10.4%) and higher than MLOP's (5.4%), but substantially lower than Berti's (16.7%). 3D-DP's superior performance is due to its dynamic prefetch degree control mechanism, which adaptively balances coverage and accuracy based on program behavior. When 3D-DP detects regular memory access patterns and the system load is low, it adopts a more aggressive policy, increasing the prefetch degree to improve coverage. Conversely, when memory access patterns are difficult to predict, it switches to a conservative mode, reducing the number of prefetches to prevent performance degradation due to overprediction.

4.3 Understanding the Benefit of Predicting Prefetch Degree and Direction

To better understand 3D-DP's prefetching behavior, Fig. 10 shows its prefetch direction distribution and average prefetch degree on the SPEC CPU 2006 and 2017 workloads. A key observation is that 3D-DP effectively adapts its prefetch degree based on workload characteristics. For instance, 603.bwaves_s-891B exhibits a highly regular memory access pattern, with 99.2% of its memory pages accessed sequentially 64 times and excellent temporal locality (an average access interval of four timesteps). Consequently, 3D-DP adaptively selects a large average prefetch degree of 13.3.

In contrast, 403.gcc-17B has a more complex, quasi-sequential access pattern and poor temporal locality; 57.2% of its pages have an average access interval exceeding 500 timesteps. In response, 3D-DP appropriately chooses a smaller average prefetch degree of 4.8 to prevent performance degradation from overly aggressive prefetching.

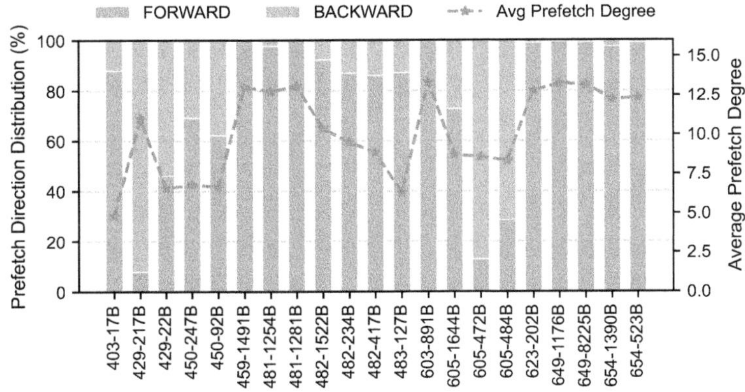

Fig. 10. 3D-DP's adaptive prefetching behavior on SPEC CPU workloads.

Figure 8 highlights the importance of dynamic direction prediction by showing the performance impact of fixing the prefetch direction to forward. Disabling dynamic direction prediction degrades 3D-DP's performance by 7.4% on average. The impact is particularly severe for 429.mcf-217B, which suffers a 52% performance loss because many of its pages are accessed backward. This incorrect prefetch direction, combined with a large prefetch degree, leads to a significant drop in prefetch accuracy and coverage. Conversely, for workloads such as 459.GemsFDTD-1491B, where memory access is almost exclusively forward, fixing the prefetch direction causes only a negligible 0.4% performance loss. For 482.sphinx3-417B, whose complex access patterns make direction prediction challenging, a fixed forward prefetch direction improves performance by 6.3%.

4.4 3D-DP-TQ Evaluation

In this section, we evaluate 3D-DP-TQ's performance and analyze the reasons for its performance degradation compared to 3D-DP.

Performance of 3D-DP-TQ. Figure 6 compares the speedup of 3D-DP and 3D-DP-TQ over Pythia across various workloads. On average, 3D-DP-TQ achieves 95.5% of 3D-DP's performance. Although slightly less effective than 3D-DP, 3D-DP-TQ still improves upon Pythia by 6.2%.

Performance Loss Analysis. 3D-DP-TQ is a practical prefetcher whose design centers on two lightweight strategies: low-precision integer quantization for the Q-NET and a simplified action space. While these strategies effectively reduce hardware overhead, they are the primary sources of 3D-DP-TQ's performance degradation compared to 3D-DP.

Using 403.gcc-17B as an example, Fig. 11 illustrates the predicted offset distribution for both 3D-DP and 3D-DP-TQ. Low-precision quantization significantly decreases the Q-NET's accuracy in fitting Q-values, which leads to an increase in incorrect prefetch offsets. 3D-DP-TQ learns only the general pattern of the offset distribution, losing detailed information and thereby reducing prefetch accuracy.

The smaller action space also negatively impacts performance. 3D-DP uses the 16 most frequent offsets, while 3D-DP-TQ reduces this to 8 to further minimize Q-NET overhead. Figure 12 shows that reducing the action space from an ideal configuration to 16 causes only a 0.9% drop in performance, but further reduction to 8 leads to a 2% decrease. If the action space is reduced below 4, performance degrades sharply. In the extreme case of using only one offset, performance drops to 85.4% of the optimal value.

4.5   Hardware Overhead

Table 5 shows the storage overhead of 3D-DP and 3D-DP-TQ. 3D-DP requires 8.73 KB of storage overhead. The hardware structures, including PAHT, ET, EB, and Q-NET, consume most of this storage, with PAHT and Q-NET weight storage being the dominant components. 3D-DP-TQ reduces Q-NET weight storage to 0.19 KB by applying low-precision quantization and using fewer neurons. Consequently, the total storage overhead of 3D-DP-TQ is only 4.92 KB.

To evaluate the hardware overhead of 3D-DP-TQ, we implemented its Register-Transfer Level (RTL) design in SystemVerilog and synthesized it using Synopsys Design Compiler with the FreePDK 15nm technology library. The synthesis reveals that 3D-DP-TQ's core logic consumes only 0.17 mm of the area and 25.19 mW of the dynamic power.

Table 6 compares this overhead with three mainstream commercial processors. The area and power consumption of 3D-DP-TQ are minimal compared to these processors. Specifically, they account for only 0.32% and 0.67% of the area and power, respectively, of a 4-core Ryzen Embedded V1605B; 0.57% and 0.23%

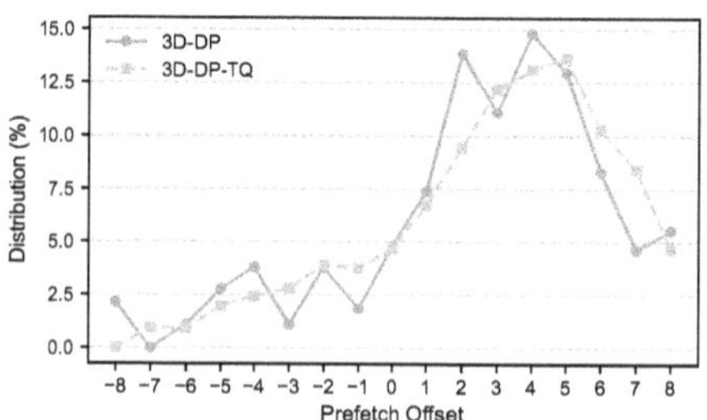

Fig. 11. Comparison of predicted prefetch offset distributions for 3D-DP and 3D-DP-TQ.

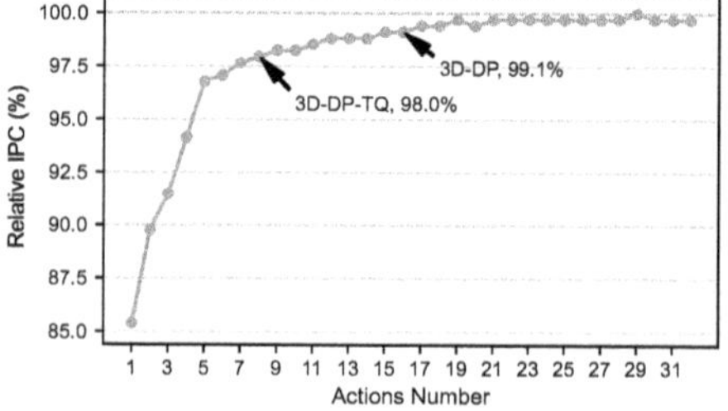

Fig. 12. The effect of action space size (number of prefetch offsets) on relative IPC.

of the area and power of a 6-core Ryzen 5 5500; and 0.94% and 0.52% of the area and power of a 32-core EPYC 7452. We also performed a Static Timing Analysis (STA) on the synthesized netlist. Due to the fine-grained pipelining of the Q-NET inference logic, 3D-DP-TQ's critical path delay is only 0.21 ns, enabling an operating frequency up to 4.6 GHz. This frequency is higher than the maximum frequency of all three processors in Table 6, allowing 3D-DP-TQ to be clocked at the speed of the L2 Cache or memory controller without becoming a system bottleneck.

In summary, 3D-DP-TQ represents an efficient and practical hardware design that can be integrated into modern processors with minimal area and power overhead, and without impacting the overall processor frequency.

Table 5. Storage overhead of 3D-DP (and 3D-DP-TQ in parentheses).

Structure	Description	Storage
PAHT	128-entry. Each entry: 16-bit hashed page, 16×8-bit offsets, 64-bit footprint, 4×5-bit time deltas	3.56 KB
ET	32-entry. Each entry: 16-bit hashed address, 6-bit reward, 1-bit evicted, 1-bit filled, 6-bit TTL	0.12 KB
EB	32-entry. Each entry: $2 \times 16 \times 8$-bit state, 2×4-bit action, 6-bit reward	1.05 KB
Q-NET	Hidden layer weight: $16 \times 32 \times 32$-bit($16 \times 32 \times 2$-bit), no bias. Output layer weight: $32 \times 16 \times 32$-bit($32 \times 8 \times 2$-bit), no bias	4.00 (0.19) KB
Total		8.73 (4.92) KB

Table 6. Overhead of 3D-DP-TQ compared to commercial processors.

Overhead compared to real systems	Area	Power	Frequency
4-core, Ryzen Embedded V1605B, 15W TDP	0.32%	0.67%	<4.6 GHz
6-core, Ryzen 5 5500, 65W TDP	0.57%	0.23%	<4.6 GHz
32-core, EPYC 7452, 155W TDP	0.94%	0.52%	<4.6 GHz

5   Conclusions

In this study, we introduced 3D-DP, a DRL-based data prefetcher. By replacing traditional Q-table with a lightweight neural network, 3D-DP effectively mitigates the state aliasing and slow-start issues that are common in prior RL-based prefetchers. Furthermore, 3D-DP incorporates novel, low-overhead mechanisms to dynamically predict both the prefetch degree and direction, allowing it to adapt its aggressiveness in response to shifting memory access patterns and system load. For practical hardware implementation, we proposed 3D-DP-TQ, a quantized version of 3D-DP whose storage overhead is reduced to only 4.92 KB via ternary weight quantization. Our detailed hardware analysis confirms that 3D-DP-TQ's area, power, and timing allow for its seamless integration into mainstream commercial processors. Our evaluations demonstrate that 3D-DP outperforms the state-of-the-art prefetchers Pythia, MLOP, and Berti by 10.2%, 7.6%, and 5.2%, respectively. Remarkably, the highly efficient 3D-DP-TQ retains 95.5% of 3D-DP's full performance, showcasing an excellent trade-off between performance and hardware cost.

Acknowledgments. This study was funded by the Natural Science Foundation of Tianjin Municipality (grant 23JCYBJC01770).

Disclosure of Interests. The authors have no competing interests to declare that are relevant to the content of this article.

References

1. Ainsworth, S., Mukhanov, L.: Triangel: a high-performance, accurate, timely on-chip temporal prefetcher. In: 2024 ACM/IEEE 51st Annual International Symposium on Computer Architecture (ISCA), pp. 1202–1216. IEEE (2024)
2. Bakhshalipour, M., Lotfi-Kamran, P., Sarbazi-Azad, H.: Domino temporal data prefetcher. In: 2018 IEEE International Symposium on High Performance Computer Architecture (HPCA), pp. 131–142. IEEE (2018)
3. Bakhshalipour, M., Shakerinava, M., Lotfi-Kamran, P., Sarbazi-Azad, H.: Bingo spatial data prefetcher. In: 2019 IEEE International Symposium on High Performance Computer Architecture (HPCA), pp. 399–411. IEEE (2019)
4. Bera, R., Kanellopoulos, K., Nori, A., Shahroodi, T., Subramoney, S., Mutlu, O.: Pythia: a customizable hardware prefetching framework using online reinforcement learning. In: 2021 54th IEEE/ACM International Symposium on Microarchitecture (MICRO), pp. 1121–1137 (2021)
5. Bera, R., Nori, A.V., Mutlu, O., Subramoney, S.: DSPatch: dual spatial pattern prefetcher. In: Proceedings of the 52nd Annual IEEE/ACM International Symposium on Microarchitecture (MICRO), pp. 531–544 (2019)
6. Bienia, C., Kumar, S., Singh, J.P., Li, K.: The parsec benchmark suite: characterization and architectural implications. In: Proceedings of the 17th International Conference on Parallel Architectures and Compilation Techniques (PACT), pp. 72–81 (2008)

7. Bucek, J., Lange, K.D., v. Kistowski, J.: SPEC CPU2017: next-generation compute benchmark. In: Companion of the 2018 ACM/SPEC International Conference on Performance Engineering, pp. 41–42 (2018)

8. Duong, Q., Jain, A., Lin, C.: A new formulation of neural data prefetching. In: 2024 ACM/IEEE 51st Annual International Symposium on Computer Architecture (ISCA), pp. 1173–1187. IEEE (2024)

9. Fu, G., Xia, T., Luo, Z., Chen, R., Zhao, W., Ren, P.: Differential-matching prefetcher for indirect memory access. In: 2024 IEEE International Symposium on High-Performance Computer Architecture (HPCA), pp. 439–453. IEEE (2024)

10. Gerogiannis, G., Torrellas, J.: Micro-armed bandit: lightweight & reusable reinforcement learning for microarchitecture decision-making. In: Proceedings of the 56th Annual IEEE/ACM International Symposium on Microarchitecture (MICRO), pp. 698–713 (2023)

11. Gober, N., et al.: The championship simulator: architectural simulation for education and competition. arXiv preprint arXiv:2210.14324 (2022)

12. Hashemi, M., et al.: Learning memory access patterns. In: International Conference on Machine Learning (ICML), pp. 1919–1928. PMLR (2018)

13. Henning, J.L.: SPEC CPU2006 benchmark descriptions. SIGARCH Comput. Archit. News **34**, 1–17 (2006)

14. Jia, L., Mcmahon, J.P., Gudaparthi, S., Singh, S., Balasubramonian, R.: Pathfinder: practical real-time learning for data prefetching. In: Proceedings of the 29th ACM International Conference on Architectural Support for Programming Languages and Operating Systems (ASPLOS), pp. 785–800 (2024)

15. Jiang, S., Yang, Q., Ci, Y.: Merging similar patterns for hardware prefetching. In: 2022 55th IEEE/ACM International Symposium on Microarchitecture (MICRO), pp. 1012–1026. IEEE (2022)

16. Kim, J., Pugsley, S.H., Gratz, P.V., Reddy, A.N., Wilkerson, C., Chishti, Z.: Path confidence based lookahead prefetching. In: 2016 49th Annual IEEE/ACM International Symposium on Microarchitecture (MICRO), pp. 1–12. IEEE (2016)

17. Liu, Y., Tziantzioulis, G., Wentzlaff, D.: Building efficient neural prefetcher. In: Proceedings of the International Symposium on Memory Systems, pp. 1–12 (2023)

18. Michaud, P.: Best-offset hardware prefetching. In: 2016 IEEE International Symposium on High Performance Computer Architecture (HPCA), pp. 469–480. IEEE (2016)

19. Navarro-Torres, A., Panda, B., Alastruey-Benedé, J., Ibáñez, P., Viñals-Yúfera, V., Ros, A.: Berti: an accurate local-delta data prefetcher. In: 2022 55th IEEE/ACM International Symposium on Microarchitecture (MICRO), pp. 975–991. IEEE (2022)

20. Nesbit, K.J., Smith, J.E.: Data cache prefetching using a global history buffer. In: 10th International Symposium on High Performance Computer Architecture (HPCA), pp. 96–96. IEEE (2004)

21. Peled, L., Mannor, S., Weiser, U., Etsion, Y.: Semantic locality and context-based prefetching using reinforcement learning. In: Proceedings of the 42nd Annual International Symposium on Computer Architecture (ISCA), pp. 285–297 (2015)

22. Peled, L., Weiser, U., Etsion, Y.: A neural network prefetcher for arbitrary memory access patterns. ACM Trans. Archit. Code Optim. (TACO) **16**(4), 1–27 (2019)

23. Shakerinava, M., Bakhshalipour, M., Lotfi-Kamran, P., Sarbazi-Azad, H.: Multi-lookahead offset prefetching. In: The Third Data Prefetching Championship (DPC-3), pp. 1–4 (2019)

24. Shi, Z., Jain, A., Swersky, K., Hashemi, M., Ranganathan, P., Lin, C.: A hierarchical neural model of data prefetching. In: Proceedings of the 26th ACM International Conference on Architectural Support for Programming Languages and Operating Systems (ASPLOS), pp. 861–873 (2021)
25. Shun, J., Blelloch, G.E.: Ligra: a lightweight graph processing framework for shared memory. In: Proceedings of the 18th ACM SIGPLAN Symposium on Principles and Practice of Parallel Programming, pp. 135–146 (2013)
26. Vavouliotis, G., Chacon, G., Alvarez, L., Gratz, P.V., Jiménez, D.A., Casas, M.: Page size aware cache prefetching. In: 2022 55th IEEE/ACM International Symposium on Microarchitecture (MICRO), pp. 956–974. IEEE (2022)
27. Wu, H., Nathella, K., Pusdesris, J., Sunwoo, D., Jain, A., Lin, C.: Temporal prefetching without the off-chip metadata. In: Proceedings of the 52nd Annual IEEE/ACM International Symposium on Microarchitecture (MICRO), pp. 996–1008 (2019)
28. Wu, H., Nathella, K., Sunwoo, D., Jain, A., Lin, C.: Efficient metadata management for irregular data prefetching. In: Proceedings of the 46th International Symposium on Computer Architecture (ISCA), pp. 449–461 (2019)
29. Wu, S., Li, G., Chen, F., Shi, L.: Training and inference with integers in deep neural networks. ArXiv **abs/1802.04680** (2018)
30. Zhang, P., Kannan, R., Srivastava, A., Nori, A.V., Prasanna, V.K.: Resemble: reinforced ensemble framework for data prefetching. In: SC22: International Conference for High Performance Computing, Networking, Storage and Analysis, pp. 1–14. IEEE (2022)
31. Zhang, P., Srivastava, A., Brooks, B., Kannan, R., Prasanna, V.K.: RAOP: recurrent neural network augmented offset prefetcher. In: Proceedings of the International Symposium on Memory Systems, pp. 352–362 (2020)

HRE-Store: A Hierarchical and Scalable Storage Architecture for Permissioned Blockchains

Yang Liu(✉), XiangYu Cui, FangChao Tian, Feng Wang, Han Li, and Min Zhang

School of Information Science and Engineering, Henan University of Technology, Zhengzhou 450001, China
{liu_yang,fangchao.tian}@haut.edu.cn,
{cuixy,lihan,2023930881}@stu.huat.edu.cn, wfmail@sina.com

Abstract. With the widespread adoption of blockchain technology, its inherent data storage bottleneck has become increasingly prominent. Traditional blockchain systems typically rely on full replication storage strategies, resulting in high storage overhead and poor scalability, which in turn limits their applicability in key domains such as the Internet of Things (IoT), finance, and supply chains. Although current approaches—such as off-chain storage and data sharding—have alleviated some of the storage pressure on blockchain systems, they still fall short of fundamentally overcoming the scalability constraints imposed by full replication. In contrast, erasure coding technologies, by leveraging data partitioning and redundant encoding mechanisms, provide a fundamental breakthrough to these limitations and have thus attract significant attention. However, existing erasure-code-based storage schemes continue to face challenges such as high encoding computational complexity, significant communication overhead, and difficulties in dynamic node adjustment. To address these challenges, this paper proposes HRE-Store, a blockchain storage scheme based on a hierarchical storage architecture. This design employs a hierarchical storage framework and partitioned encoding mechanism to decouple global encoding tasks, thereby reducing encoding complexity. A lightweight indexing mechanism is constructed to minimize communication overhead, and a minimum threshold mechanism for nodes is introduced to simplify the process of dynamic node adjustment. Theoretical analysis and experimental results demonstrate that, in large-scale node environments, HRE-Store achieves approximately $64\times$ and $3\times$ improvements in computational efficiency compared to the classical BFT Store and the latest PartitionChain coding scheme, respectively. In terms of read performance, latency is reduced by an average of approximately 90% and 80%, respectively, while in terms of system stability, data recovery time is reduced by approximately 87% and 55%, respectively.

Keywords: Blockchain · Storage Scalability · Erasure Coding · Partitioning Design

1   Introduction

Due to its decentralized nature, tamper resistance, and transparency [1], blockchain technology demonstrates unique application value and technical advantages in scenarios such as proof-of-existence, data provenance, and trustworthy transactions among multiple competing entities [2–7]. However, mainstream blockchain systems currently adopt a full-replication data storage strategy [8], where all nodes are required to store the complete blockchain ledger. For example, in the Bitcoin system, the data volume per node has reached approximately 650 GB, with around 12,000 daily active full nodes, resulting in a total storage footprint of about 7.8 PB, growing at a rate of 2.4 TB per day. In the Ethereum system, individual nodes store up to 2 TB of data, with approximately 9,000 daily active full nodes, leading to a total storage volume exceeding 17.6 PB and a daily increase of around 6.2 TB [9]. While this redundant full-replication strategy enhances system fault tolerance, it significantly restricts the practical deployment of blockchain systems. Consequently, the storage scalability bottleneck has become a critical issue that urgently needs to be addressed.

To tackle this problem, researchers have proposed various storage optimization methods. For instance, off-chain storage solutions based on the InterPlanetary File System (IPFS) and Distributed Hash Tables (DHT) aim to alleviate on-chain storage burdens by offloading blockchain data to external systems [10,11]. The Simplified Payment Verification (SPV) mechanism [12], proposed by Bitcoin's creator Satoshi Nakamoto, allows light nodes to store only block headers to reduce storage overhead. Sharding techniques [13,14] divide on-chain data into multiple shards stored across different sub-networks, enabling partial data storage. However, within each sub-network, all nodes still store the same shard data.

While these methods alleviate storage pressure to varying extents, they do not fundamentally overcome the limitations of full-replication strategies. To further improve blockchain storage scalability, researchers have explored integrating Erasure Coding (EC) with the Practical Byzantine Fault Tolerance (PBFT) consensus protocol, leading to storage-optimized schemes such as BFT-Store [15] and PartitionChain [16].

BFT-Store was the first to introduce Reed-Solomon (RS) coding into permissioned blockchain systems. By adopting an $RS(n - 2f, n)$ scheme, it distributes $n-2f$ original shards and $2f$ redundant shards across multiple nodes. Any $n-2f$ shards suffice to reconstruct the original data, thus significantly reducing per-node storage overhead while ensuring recoverability. The encoding and decoding complexities of this scheme are $O(T^2 \cdot N^2)$ and $O(T^2 \cdot N^3)$, respectively. Furthermore, BFT-Store uses Threshold Signatures (TS) to verify shard correctness, but its reliance on a Trusted Third Party (TTP) contradicts the decentralized ethos of blockchain. Additionally, the system requires global re-encoding upon node changes, incurring high computational costs.

To further optimize performance, PartitionChain builds upon BFT-Store's strengths by proposing a shard-level optimized coding strategy. It divides block data into multiple first-level shards and applies RS encoding independently to each. This reduces the encoding and decoding complexities to $O(T \cdot C \cdot N)$ and

$O(T \cdot C \cdot N^2)$, respectively. To eliminate TTP dependency, PartitionChain introduces the Certificateless Aggregate Signature (CLAS) scheme, allowing signature updates without global re-encoding when nodes change, thereby lowering update overhead.

Although current EC-based storage schemes effectively break the storage bottleneck of full-replication strategies, they still face several practical challenges. Specifically, encoding and decoding efficiency requires further improvement, communication overhead remains substantial, and under frequent node dynamics, there is a lack of flexible and efficient mechanisms to maintain data availability and structural consistency.

To this end, this paper proposes HRE-Store (Hierarchical Redundancy Encoding Store), a storage optimization scheme tailored for permissioned blockchain systems. Building upon the architectural strengths of BFT-Store and PartitionChain, HRE-Store introduces a hierarchical storage framework, partitioned encoding mechanism, lightweight indexing, and a minimum node threshold strategy. These enhancements collectively address key limitations of existing erasure coding schemes, including high encoding complexity, substantial communication overhead, and the difficulty of maintaining system performance in dynamic network environments.

The main contributions of this paper are as follows:

- HRE-Store innovatively introduces a partitioning mechanism to parallelize encoding tasks and reduce computational complexity. Unlike BFT-Store, where the global RS encoding imposes heavy computational loads on a small subset of nodes, and PartitionChain, which still relies on network-wide coordination despite its first-level sharding, HRE-Store evenly distributes original data shards across multiple partitions. Each partition performs RS encoding independently. This significantly enhances encoding parallelism, reduces the computational load on individual nodes, and improves overall system efficiency and scalability.
- HRE-Store is the first to construct a two-tier lightweight indexing structure combining partition-level Bloom filters [17] and inter-partition digest negotiation to optimize communication efficiency. Existing schemes often require global broadcasting or multiple negotiation rounds for shard location. In contrast, HRE-Store maintains a Bloom filter within each partition to quickly check for local shard presence. If the query misses, it utilizes synchronized inter-partition Bloom digests to forward requests toward the most likely target partition. This strategy avoids unnecessary broadcasts and redundant transfers, significantly reducing communication costs while ensuring high query accuracy and enhancing data retrieval efficiency.
- HRE-Store proposes a novel minimum node threshold mechanism based on redundancy tolerance to maintain data availability and structural stability under dynamic node conditions. Unlike BFT-Store, which triggers global re-encoding upon every node change, and PartitionChain, which still maintains a global encoding structure, HRE-Store implements a monitoring mechanism for the minimum number of effective nodes. Only when the number of active nodes falls below the RS fault tolerance threshold does the system initi-

ate data recovery and encoding reconstruction. This mechanism significantly reduces unnecessary reconfiguration operations and enhances the system's adaptability and robustness in the face of frequent node joins, departures, or failures.

2   Preliminaries

This section introduces the fundamental assumptions and symbol definitions on which this study is based.

2.1   Assumptions

- Among a total of n nodes, at most f nodes may exhibit Byzantine behavior—meaning they may behave arbitrarily, including sending incorrect data, acting maliciously, or becoming unresponsive, where $f = \left\lfloor \frac{(n-1)}{3} \right\rfloor$. Within each partition, the number of Byzantine nodes does not exceed $f_p = \left\lfloor \frac{(n_p-1)}{3} \right\rfloor$, where n_p denotes the number of nodes in the partition.
- This study adopts a partially synchronous network model, in which message delivery may be temporarily delayed due to network latency but will eventually reach the intended destination. Communication channels, both intra-partition and inter-partition, are assumed to be secure, ensuring that messages are neither tampered with nor lost.

2.2   Notation

To clearly describe the system logic and algorithms, the basic symbols used in this paper are defined in Table 1.

Table 1. The Notations Used in This Paper

Symbol	Description
T	Size of the original data block
R	Node reputation value
N	Total number of nodes in the system
n	Total number of encoded data shards
k	Number of original shards; minimum required for data reconstruction
$n - k$	Number of redundant shards; maximum tolerable shard loss
p	Number of partitions
N_p	Number of nodes in each partition
n_p	Total number of shards stored in each partition
k_p	Number of original shards in each partition
$N_{\min}$	Minimum node threshold

3   System Design

This section details the design of HRE-Store.

3.1   Hierarchical Structuring

HRE-Store classifies all nodes into three tiers based on their Node Capability Score R, which is periodically updated. The score R is computed by integrating a node's storage capacity S, computational capability C, and historical task success rate H, as follows:

$$R = \alpha S + \beta C + \gamma H \tag{1}$$

where α, β, and γ are weighting factors that balance the contribution of each component. Based on their R values, nodes are categorized into:

- **Core Tier**: $R \geq 0.7$, responsible for computation-intensive tasks and storing original data shards.
- **Extension Tier**: $0.4 \leq R < 0.7$, responsible for storing redundant shards and assisting with data recovery.
- **Edge Tier**: $R < 0.4$, assigned to lightweight tasks and storing redundant shards.

This mechanism ensures that high-performance nodes are prioritized for critical tasks, while lower-capacity nodes handle auxiliary functions, improving resource allocation and adaptability.

3.2   Partitioning

After hierarchical structuring, the system divides the N total nodes evenly into p partitions, each with $N_p = \frac{N}{p}$ nodes (Fig. 1). Partitioning follows load-balancing principles to ensure that node tiers are evenly distributed, supporting efficient task scheduling and parallel processing.

Effective partitioning in large-scale blockchain storage reduces both computational complexity and communication overhead while maintaining fault tolerance. However, the number of partitions must balance:

- **Computation Efficiency**: More partitions reduce workload per group and lower encoding/decoding complexity.
- **Communication Overhead**: Excessive partitions increase inter-partition communication cost.

To strike this balance, we propose a dynamic partition adjustment formula:

$$p = \min\left(\frac{N}{E}, p_{\max}\right) \tag{2}$$

where E is the expected node count per partition, and p_{max} is the maximum allowed number of partitions. This allows p to adapt to changes in N, ensuring performance scalability while limiting overhead.

Each partition operates independently, performing local storage, encoding, decoding, and recovery tasks in parallel. Despite the intra-partition focus, inter-partition cooperation is possible when recovering lost data or handling node failures.

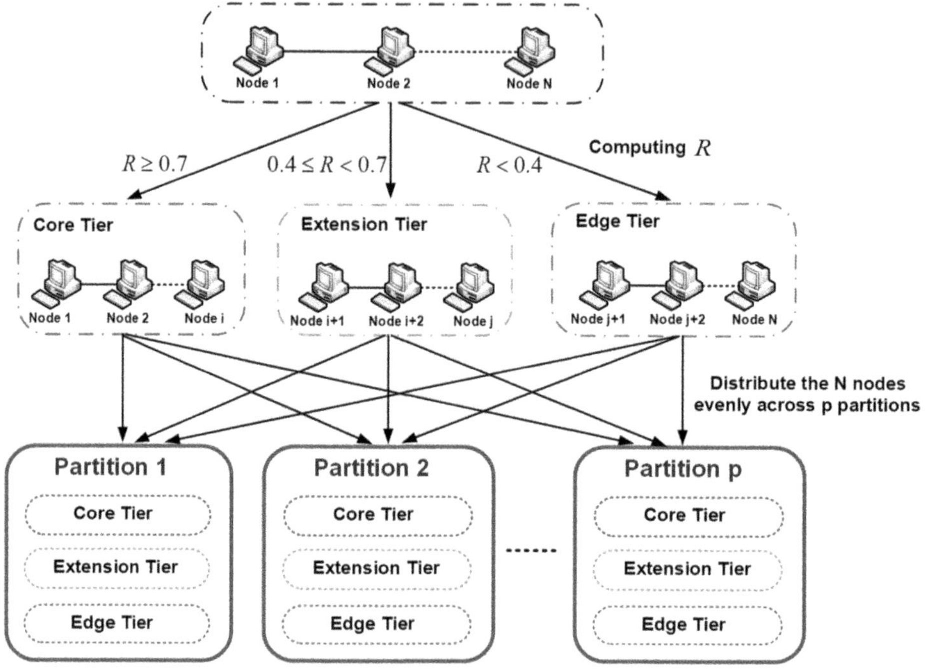

Fig. 1. System Node Architecture.

3.3 Data Storage

HRE-Store adopts Reed-Solomon (RS) coding to ensure fault tolerance and leverages hierarchical and partitioned structures for shard distribution (Fig. 2). The storage process proceeds as follows:

1. An original block $B(h)$ with unique hash h is divided into k equal-sized original shards.
2. These k shards are distributed across p partitions.
3. Each partition independently encodes its received shards to generate $n - k$ redundant shards.

4. Original shards are stored in Core Tier nodes; redundant shards are assigned extension-tier and edge-tier nodes.

Original shards contain the raw data and can be directly accessed, while redundant shards are used for recovery. If data loss occurs, any k valid shards (original or redundant) suffice for full recovery via RS decoding.

Algorithm 1 details the shard encoding and distribution process based on node hierarchy, partitioning, and RS coding.

Algorithm 1. Hierarchical partitioning and data storage

1: **Input:** Data block $B(h)$, NodeSet N, RS(k, n)
2: **Output:** Encoded shards distributed across all nodes
3: **for** each node $n \in N$ **do**
4:     $R \leftarrow \alpha \cdot \text{StorageAbility} + \beta \cdot \text{ComputeAbility} + \gamma \cdot \text{TacSuccessRate}$
5:     **if** $R \geq 0.7$ **then**
6:         Assign n to **CoreTier**
7:     **else if** $R \geq 0.4$ **then**
8:         Assign n to **ExtensionTier**
9:     **else**
10:         Assign n to **EdgeTier**
11:     **end if**
12: **end for**
13: $P \leftarrow \lfloor |N|/\text{PartitionSize} \rfloor$
14: Initialize empty list **Partitions** of size p
15: **for** $i = 0$ to $|N| - 1$ **do**
16:     $pid \leftarrow i \bmod p$
17:     Append $N[i]$ to **Partitions**$[pid]$
18: **end for**
19: Split $B(h)$ into k shards: $S[0], \ldots, S[k-1]$
20: **for** $i = 0$ to $k - 1$ **do**
21:     $pid \leftarrow i \bmod P$
22:     Append $S[i]$ to **Partitions**$[pid]$.OriginalShards
23: **end for**
24: **for** each partition P in **Partitions do**
25:     $RSout \leftarrow \text{RSEncode}(k, n)$
26:     Assign $RSout[0 : k - 1]$ to CoreTier nodes in p
27:     Assign $RSout[k : n - 1]$ to ExtensionTier $\cup$ EdgeTier nodes in p
28:     Store all assigned shards to respective nodes
29: **end for**

3.4   Lightweight Indexing Mechanism

To improve shard lookup efficiency and reduce inter-partition communication, HRE-Store introduces a two-level Bloom filter-based indexing mechanism:

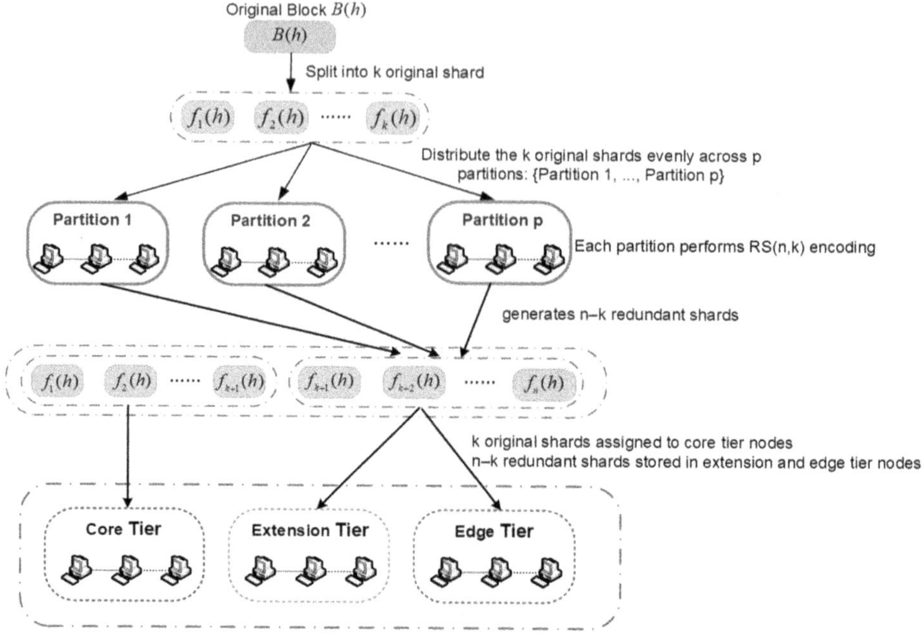

Fig. 2. Data Storage Process.

Bloom Filter Initialization. Each partition configures its Bloom filter based on the number of stored shards and a target false positive rate. The bit array size m and number of hash functions h are determined by:

- m depends on shard count n and false positive rate P:

$$m = -\frac{n \cdot \ln P}{(\ln 2)^2} \tag{3}$$

- h is derived from m and n:

$$h = \frac{m}{n} \cdot \ln 2 \tag{4}$$

Intra-Partition Filter Construction. Each partition builds its Bloom filter (BF_i) by inserting the identifiers of all locally stored shards.

Inter-Partition Digest Synchronization. To avoid global broadcasts during queries, each partition periodically broadcasts a compressed digest of its (BF_i) to neighboring partitions or core nodes. A Bloom filter update threshold θ is defined as 10% of a partition's shard count. Once updates exceed θ, a new digest is broadcast to maintain consistency.

This design balances query accuracy and system overhead, enabling fast and reliable shard discovery at scale. Algorithm 2 outlines the initialization and dynamic update process of the lightweight index.

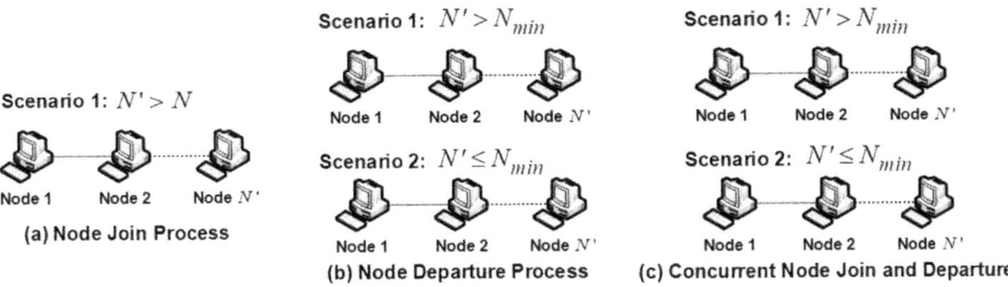

Fig. 3. Node Count Variations.

Table 2. Performance Comparison

		BFT-Store	PartitionChain	HRE-Store
Storage Overhead		$O(1)$	$O(1)$	$O(1)$
Computational Efficiency	Encoding	$O(T^2 \cdot N^2)$	$O(T \cdot C \cdot N)$	$O\left(\frac{N \cdot T}{p^2}\right)$
	Decoding	$O(T^2 \cdot N^3)$	$O(T \cdot C \cdot N^2)$	$O\left(\frac{N^2 \cdot T}{p^3}\right)$
Additional Communication Cost	Shard Distribution	$O(T \cdot N)$	$O(T \cdot C \cdot N)$	$O\left(\frac{N \cdot T}{k}\right)$
	Data Recovery	$O(k \cdot T)$	$O(T \cdot C)$	$O\left(\frac{T}{p}\right)$
	Dynamic Adjustment	$O(T \cdot N)$	$O(T \cdot C \cdot N)$	$O\left(\frac{N \cdot T}{k \cdot p}\right)$
Read Performance		$O(N)$	$O(C \cdot N)$	$O(1)$

4.1   Storage Overhead

In HRE-Store, storage overhead is optimized through $RS(k, n)$ erasure coding, where k denotes the number of original shards and n the total number of shards. Each data block is divided into k original shards and $n - k$ redundant shards, with each shard of size $\frac{T}{k}$. The total storage overhead is:

$$k \cdot \frac{T}{k} + (n - k) \cdot \frac{T}{k} \approx \frac{n \cdot T}{k} \tag{5}$$

Since both k and n dynamically adjust with the number of nodes, the storage overhead remains independent of the system size N. Thus, storage cost is constant in scale, i.e., $O(1)$.

4.2   Computational Efficiency

Computational efficiency reflects the cost and performance of the data encoding and decoding process. HRE-Store leverages partition-based parallelism to significantly reduce complexity and improve encoding speed.

Encoding. Assume the system is divided into p partitions, each with $N_p = \frac{N}{p}$ nodes. Original shards are evenly distributed across partitions, and RS encoding is independently performed in each. Let each partition handle k_p original shards and generate n_p total shards. The encoding involves applying a generator matrix G of size $n_p \times k_p$ to a data matrix of size $\frac{T}{k}$. The complexity per partition is:

$$O(G \cdot \frac{T}{k}) = O(n_p \cdot k_p \cdot \frac{T}{k}) = O\left(\frac{n}{p} \cdot \frac{k}{p} \cdot \frac{T}{k}\right) \approx O\left(\frac{N \cdot T}{p^2}\right) \tag{6}$$

Decoding. Data recovery is possible as long as $n \geq k$. Decoding involves inverting a $n_p \times k_p$ matrix G, with complexity $O(k_p^2)$, followed by matrix multiplication to reconstruct the original data, with complexity $O\left(k_p^3 \cdot \frac{T}{k}\right)$. Therefore, total decoding complexity is:

$$O(k_p^2 + k_p^3 \cdot \frac{T}{k}) = O\left(\left(\frac{k}{p}\right)^2 + \frac{k^2 \cdot T}{p^3}\right) \approx O\left(\frac{k^2 \cdot T}{p^3}\right) \approx O\left(\frac{N^2 \cdot T}{p^3}\right) \tag{7}$$

The dominant term is typically the matrix multiplication step, particularly for large data blocks.

4.3   Additional Communication Overhead

Additional communication overhead refers to extra data transferred between nodes during storage, recovery, or reinitialization. It includes:

- **Shard Distribution:** Cost of transmitting encoded shards during initial storage or reinitialization.
- **Data Recovery:** Communication required to collect shards for recovering lost data.
- **Dynamic Adjustment:** Overhead due to changes in node count requiring RS parameter updates and shard redistribution.

Shard Distribution. In the initial storage phase, n shards of size $\frac{T}{k}$ are distributed to N nodes, yielding total communication cost:

$$O\left(\frac{n \cdot T}{k}\right) \approx O\left(\frac{N \cdot T}{k}\right) \tag{8}$$

HRE-Store's hierarchical and partitioned design reduces global broadcasts, thereby minimizing this cost.

Data Recovery. Thanks to partitioning, data recovery is localized. Given p partitions, recovery within a partition incurs:

$$O\left(\frac{k}{p} \cdot \frac{T}{k}\right) = O\left(\frac{T}{p}\right) \tag{9}$$

This ensures that recovery-related communication does not scale with total node count.

Dynamic Adjustment. When node count changes, RS parameters are reconfigured and shards redistributed. If only a single partition is affected, communication is localized. Global reinitialization is triggered only when $N \leq N_{\min}$ (the minimum number of nodes to maintain RS tolerance). Communication per partition:

$$O\left(\frac{n}{p} \cdot \frac{T}{k}\right) = O\left(\frac{n \cdot T}{k \cdot p}\right) \tag{10}$$

Thus, partitioning helps contain communication within partitions under normal conditions, significantly reducing global traffic.

Read Performance. HRE-Store's hierarchical architecture enhances read performance. Original shards are stored on core-tier nodes, allowing direct access without decoding under normal conditions, greatly reducing latency. If a core node fails, extension and edge nodes use redundant shards for recovery.

Partitioning also contributes to read efficiency by limiting data lookup to specific partitions. Independent decoding within a partition minimizes recovery latency and related overhead.

In addition, HRE-Store employs Bloom filters for rapid shard localization. Each query checks hash-based Bloom filters to identify the corresponding storage node with $O(1)$ time complexity. This dramatically reduces lookup latency and improves scalability, especially in large-node environments.

5   Experimental Evaluation

To comprehensively evaluate the performance of HRE-Store, we conducted a series of simulation experiments and compared it against full replication, BF-Store, and PartitionChain. The evaluation focuses on four aspects: storage overhead, computational efficiency, read performance, and system stability. Node counts were set to 4, 8, 16, 32, and 64. Each node stores one shard generated via RS coding with a redundancy ratio of $n = 2k$. Nodes are categorized into three tiers: core (50%), extension (30%), and edge (20%). The minimum node threshold $N_{\min}$ is set to the RS fault-tolerance limit k. The number of partitions p is dynamically adjusted following our proposed strategy to ensure balanced distribution. Experimental settings are summarized in Table 3.

Table 3. Experimental Parameter Settings

N	p	(k, n)	Core Tier	Extension Tier	Edge Tier
4	1	$k = 2, n = 4$	2	1	1
8	1	$k = 4, n = 8$	4	2	2
16	2	$k = 8, n = 16$	8	5	3
32	4	$k = 16, n = 32$	16	10	6
64	8	$k = 32, n = 64$	32	19	13

5.1 Storage Overhead

To evaluate storage overhead, we fixed the block size at 1MB and tested each scheme under increasing numbers of nodes. Only a single block was used for clarity.

As shown in Fig. 4, the full replication scheme requires each node to store the entire block, resulting in linear growth in storage with node count. In contrast, BFT-Store, PartitionChain, and HRE-Store significantly reduce storage burden per node through erasure coding. Their total storage costs remain constant, unaffected by the number of nodes, thus overcoming the scalability bottlenecks inherent in full replication.

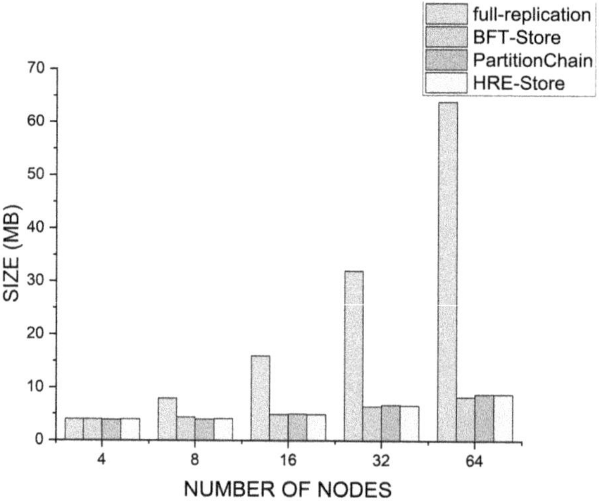

Fig. 4. Comparison of Storage Overhead.

5.2 Computational Efficiency

We evaluate encoding and decoding efficiency, as well as system throughput. Since full replication does not involve coding, it is excluded from this section.

As depicted in Fig. 5, HRE-Store significantly outperforms the others due to its partition-based parallelism. At $N = 64$, HRE-Store achieves approximately $2\times$ higher encoding efficiency than PartitionChain and $64\times$ higher than BFT-Store. For decoding, it outperforms PartitionChain and BFT-Store by $3\times$ and $64\times$, respectively. In terms of throughput, HRE-Store achieves $2\times$ the throughput of PartitionChain and $218\times$ that of BFT-Store. These results demonstrate HRE-Store's superior computational efficiency, especially in medium to large-scale networks ($N \geq 16$).

5.3   Read Performance

Read performance is critical for responsive query handling, especially under high concurrency. We compared the read latency of HRE-Store, PartitionChain, and BFT-Store.

As illustrated in Fig. 6, HRE-Store consistently achieves the lowest read latency. Its parallel partition execution and Bloom filter-based indexing significantly reduce inter-partition communication and lookup delays. At $N = 64$, the average read latency of HRE-Store is 424 ms—only one-fifth of PartitionChain's and one-tenth of BFT-Store's—making it well-suited for high-throughput, low-latency blockchain applications.

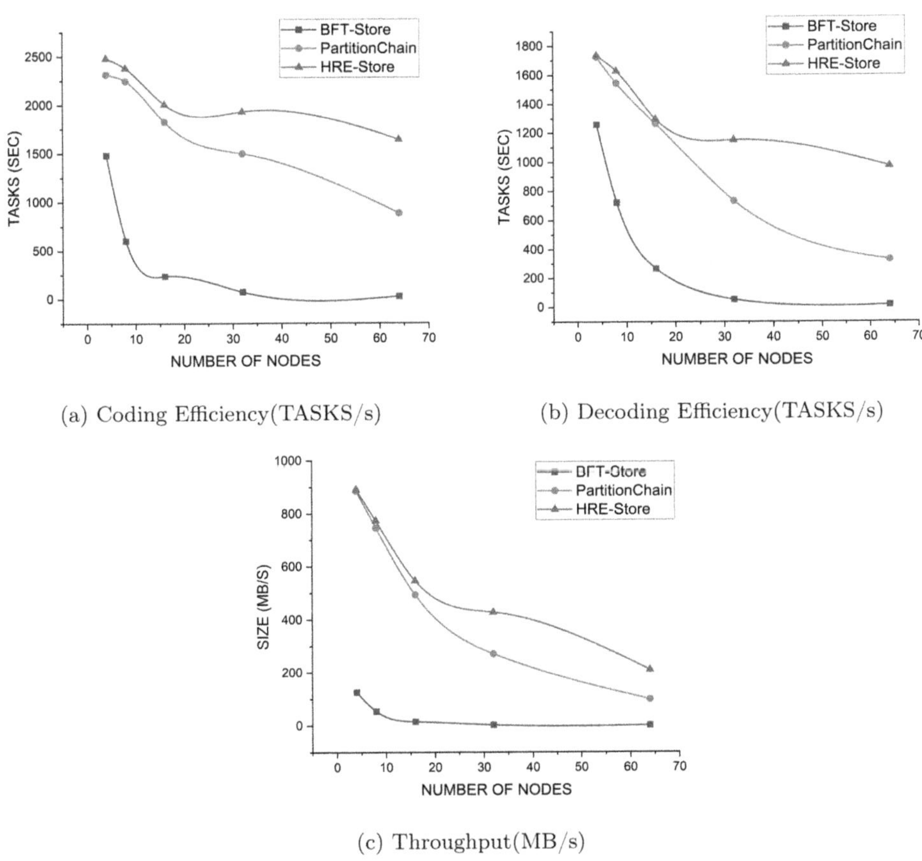

Fig. 5. Computational Efficiency.

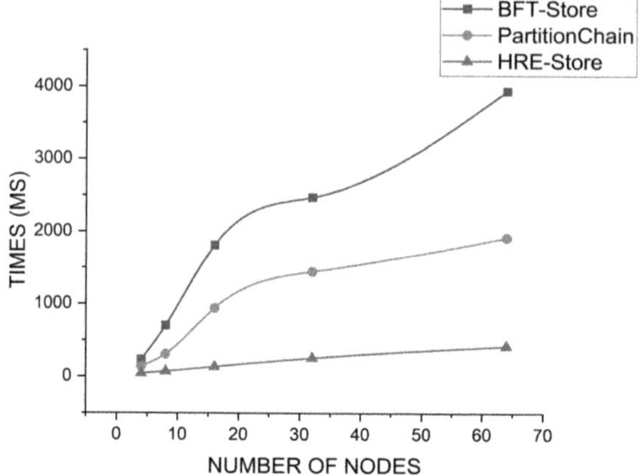

Fig. 6. Read Response Time (ms).

5.4   System Stability

System stability measures the ability to maintain availability in dynamic environments such as node failures, departures, or reinitialization.

Figure 7 presents recovery and reconfiguration times across three scenarios: node failure, node exit, and system reinitialization. In the failure scenario, HRE-Store leverages intra-partition recovery and inter-partition redundancy coordination to significantly reduce recovery latency. At $N = 64$, recovery time is 1284 ms—55% and 87% lower than PartitionChain and BFT-Store, respectively. When $N \leq N_{min}$, HRE-Store triggers reinitialization with a latency of 1672 ms, reducing overhead by 62% and 85% compared to PartitionChain and BFT-Store. These results highlight HRE-Store's robustness and fault tolerance under dynamic conditions.

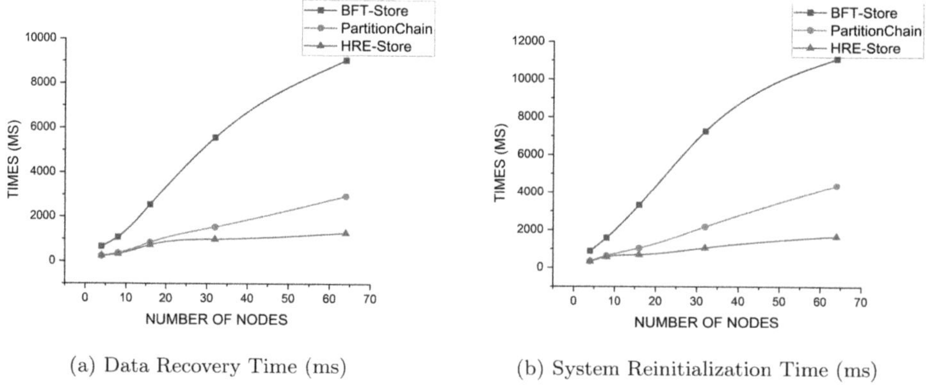

(a) Data Recovery Time (ms)   (b) System Reinitialization Time (ms)

Fig. 7. System Stability.

6   Conclusion

This paper presents HRE-Store, an efficient storage optimization scheme for permissioned blockchains, designed to address the scalability and overhead challenges of full replication. HRE-Store significantly reduces per-block storage overhead while preserving data availability. By incorporating a tiered storage architecture based on node performance and reputation, HRE-Store assigns nodes to core, extension, and edge tiers for differentiated task allocation. This enhances resource efficiency and reduces communication costs. A partitioned RS coding mechanism divides the network into partitions, each independently performing encoding/decoding tasks, thereby reducing computational complexity. Combined with a Bloom filter-based indexing mechanism, HRE-Store supports rapid localization and recovery of both original and redundant shards, minimizing inter-partition communication. To ensure availability under dynamic conditions (e.g., node joins, departures, or failures), HRE-Store introduces a minimum node threshold mechanism. When node count falls below this threshold, the system automatically triggers partition reconstruction and data recovery. Experimental results show that HRE-Store outperforms BFT-Store and PartitionChain in terms of computational overhead, read performance, and system stability—especially in large-scale, heterogeneous blockchain environments—demonstrating superior scalability and fault tolerance.

References

1. Belotti, M., Božić, N., Pujolle, G., Secci, S.: A vademecum on blockchain technologies: when, which, and how. IEEE Commun. Surv. Tutor. **21**(4), 3796–3838 (2019)
2. Gan, Q., Lau, R.Y.K.: Trust in a 'trust-free' system: blockchain acceptance in the banking and finance sector. Technol. Forecast. Soc. Change **199**, 123050 (2024)
3. Liang, W., Liu, Y., Yang, C., Xie, S., Li, K., Susilo, W.: On identity, transaction, and smart contract privacy on permissioned and permissionless blockchain: a comprehensive survey. ACM Comput. Surv. **56**(12), 1–35 (2024)
4. Wang, M., Yang, F., Shan, F., Guo, Y.: Blockchain adoption for combating remanufacturing perceived risks in a reverse supply chain. Transp. Res. Part E Logist. Transp. Rev. **183**, 103448 (2024)
5. Chen, X., Cheng, Q., Yang, W., Luo, X.: An anonymous authentication and secure data transmission scheme for the Internet of Things based on blockchain. Front. Comput. Sci. **18**(3), 183807 (2024)
6. Si, H., Li, W., Wang, Q., Cao, H., Bacao, F., Sun, C.: A secure cross-domain interaction scheme for blockchain-based intelligent transportation systems. PeerJ Comput. Sci. **9**, e1678 (2023)
7. Zhang, Q., et al.: Blockchain-based asymmetric group key agreement protocol for internet of vehicles. Comput. Electr. Eng. **86**, 106713 (2020)
8. Heo, J.W., Ramachandran, G.S., Dorri, A., Jurdak, R.: Blockchain data storage optimisations: a comprehensive survey. ACM Comput. Surv. **56**(7), 1–27 (2024)
9. Blockchair Blockchain Explorer. https://blockchair.com/bitcoin/charts/blockchain-size. Accessed 24 June 2025

10. Zheng, Q., Li, Y., Chen, P., Dong, X.: An innovative IPFS-based storage model for blockchain. In: 2018 IEEE/WIC/ACM Int. Conf. on Web Intelligence (WI), pp. 704–708. IEEE, New York (2018)
11. Hassanzadeh-Nazarabadi, Y., Küpçü, A., Özkasap, Ö.: LightChain: scalable DHT-based blockchain. IEEE Trans. Parallel Distrib. Syst. **32**(10), 2582–2593 (2021)
12. Nakamoto, S.: Bitcoin: A Peer-to-Peer Electronic Cash System, 1st edn. (2008). https://bitcoin.org/bitcoin.pdf
13. Luu, L., Narayanan, V., Zheng, C., Baweja, K., Gilbert, S., Saxena, P.: A secure sharding protocol for open blockchains. In: Proc. of the 2016 ACM SIGSAC Conf. on Computer and Communications Security, pp. 17–30. ACM, New York (2016)
14. Dang, H., Dinh, T.T.A., Loghin, D., Chang, E.C., Lin, Q., Ooi, B.C.: Towards scaling blockchain systems via sharding. In: Proc. of the 2019 Int. Conf. on Management of Data, pp. 123–140. ACM, New York (2019)
15. Qi, X., Zhang, Z., Jin, C., Zhou, A.: A reliable storage partition for permissioned blockchain. IEEE Trans. Knowl. Data Eng. **33**(1), 14–27 (2020)
16. Du, Z., Pang, X., Qian, H.: PartitionChain: a scalable and reliable data storage strategy for permissioned blockchain. IEEE Trans. Knowl. Data Eng. **35**(4), 4124–4136 (2021)
17. Luo, L., Guo, D., Ma, R.T., Rottenstreich, O., Luo, X.: Optimizing Bloom filter: challenges, solutions, and comparisons. IEEE Commun. Surv. Tutor. **21**(2), 1912–1949 (2019)

Reputation-Based Secure Node Allocation and Dynamic Monitoring for Blockchain Sharding

Linlin Zhang[1], Shengrong Deng[1], Wenshou Wu[1], Xuehua Bi[2], and Kai Zhao[3]([✉])

[1] School of Software, Xinjiang University, Ürümqi, Xinjiang, China
{zllnadasha,wuws}@xju.edu.cn
[2] College of Medical Engineering and Technology, Xinjiang Medical University, Ürümqi, Xinjiang, China
[3] School of Computer Science and Technology, Xinjiang University, Ürümqi, Xinjiang, China
zhawkk@xju.edu.cn

Abstract. Sharding technology is considered the most promising solution to address the scalability issues of blockchain. Most existing sharding techniques focus on performance improvement but fail to consider the categories of nodes when assigning them to shards. This may lead to malicious clustering in the same shard, reducing the security of the sharded blockchain. This paper introduces the random forest model for classifying nodes and proposes the ANC (Assign Nodes by Category) algorithm, which isolates malicious nodes and distributes the remaining nodes to different shards. During the process of transaction execution, we monitor nodes to detect whether malicious behavior has occurred, enhancing system security. Experimental results demonstrate that the node classification result achieves 97% accuracy and indirectly increases the safety of Blockchain. Compared to the baseline models, the proposed solution is superior in terms of throughput, latency, and security.

Keywords: Blockchain · Sharding · Node classification · Node allocation

1 Introduction

Sharding technology is a prioritized solution for enhancing blockchain scalability. Some existing studies [1] primarily focus on node allocation to reduce overhead, shard reshuffling to ensure security, and accelerating block confirmation to improve the efficiency of transaction processing. Among these, shard security research mainly concentrates on shard reorganization. For instance, solutions like tMPT [2] and S-Store [3] modify the state tree to obtain efficient and low overhead in shard reorganization progress, thereby safeguarding system security. Approaches such as CoChain [4] and Benzene [5] design multi-shard supervision schemes that adopt Trusted Execution Environments (TEE) and multi-shard

H. Liu et al. (Eds.): ICA3PP 2025, LNCS 16381, pp. 101–116, 2026.
https://doi.org/10.1007/978-981-95-8399-7_6

voting mechanisms for node monitoring, ensuring system security. Additionally, other schemes [6–8] randomly compose each shard from nodes, relying on the probability of limiting the allocation of malicious nodes to shards.

The above research work enhances the security of sharded blockchains during operation and promotes sharding technology. However, these schemes have limitations. For example, randomly assembling nodes into shards or not considering node attribute categories may increase the risk of malicious nodes clustering in the same shard. As shown in Fig. 1a, when over 1/3 of malicious nodes cluster in a single shard, the correctness of transaction processing is compromised. Similarly, an accumulation of lazy nodes (low-activity nodes) slows down transaction processing, degrading overall blockchain performance.

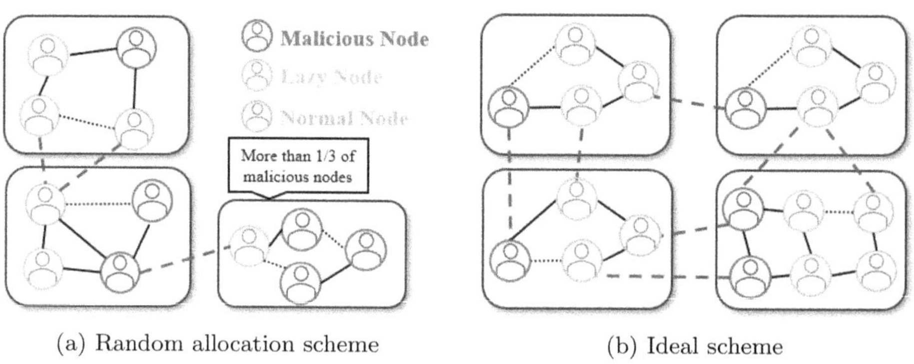

(a) Random allocation scheme   (b) Ideal scheme

Fig. 1. Comparison of allocation schemes: (a) random allocation and (b) ideal scheme

This work addresses these issues through node identification and disposal of nodes, using classification results as the basis for node allocation to ensure controlled quantities of malicious and lazy nodes in a shard. In addition, the paper's objective is to reduce shard system damage risk via early malicious node handling, to build a secure, efficient sharded blockchain system. The ideal node distribution, as shown in Fig. 1b, ensures that no shard contains more than 1/3 malicious nodes. This paper focuses on node identification, malicious node handling, and node behavior monitoring in sharded blockchains to enhance system security. The key contributions are as follows:

- **Firstly**, a reputation-based node classification method is designed to classify nodes in the Ethereum dataset.
- **Secondly**, we propose a node processing algorithm that isolates identified malicious nodes and distributes lazy and normal nodes to shards based on predefined node ratios and allocation strategies to ensure system security.
- **Finally**, we develop a node monitoring algorithm to track node behavior in real-time, preventing sudden surges of malicious nodes in any single shard.

The remainder of this paper is organized as follows: Sect. 2 reviews related work on sharding technologies; Sect. 3 details the design of the proposed algorithms and models; Sect. 4 analyzes the model's effectiveness and security; Sect. 5

presents experimental results and performance evaluation; Sect. 6 concludes the paper and discusses future directions.

2   Related Work

2.1   The Development of Sharding Technology

Sharding technology originated from distributed databases and is used to solve the problem of insufficient scalability in traditional databases. However, as a new generation of Internet technology, blockchain is facing the problem of insufficient scalability in enterprise applications. For this reason, many scholars utilize the concept of sharding technology in distributed databases as a reference point for conducting relevant research. Elastico [6], as the first-generation sharding technology solution, has achieved efficient parallel processing of transactions and promoted the development of blockchain technology. Omniledger [7] is based on the first-generation sharding scheme and utilizes PBFT consensus and a VRF-based leader selection algorithm to address the issues of high consensus latency and low resource utilization. In addition, the Monoxide [9] scheme first proposed the Relay consensus mechanism, which addresses the challenge of cross-shard transactions in the sharding domain.

Based on previous work, current research on blockchain sharding technology primarily focuses on performance improvement and resource utilization, including the efficient processing of cross-shard transactions and maintaining shard load balancing. Some state-of-the-art solutions, such as Pyramid [10] proposing hierarchical consensus, BrokerChain [11] proposing account splitting and graph splitting algorithms, and GridDB [12] proposing off-chain operations, have improved the efficiency and security of cross-shard transactions. In addition, TxAllo [13], LB-Chain [14], and Sliver [15] further improve the utilization of system resources. These works have made outstanding contributions to promoting the widespread application of blockchain technology.

2.2   The Current Research Status of Shard Security

Increasing the number of shards to achieve higher processing efficiency can easily lead to security issues. For example, under the condition of limited nodes, partitioning into smaller shards makes these shards more vulnerable to 51% attacks. Regarding the difficulties, some scholars are currently conducting relevant research work. The tMPT [2] and S-Store [3] schemes scramble nodes in shards through a shard reshuffle to ensure security. To reduce the latency of large-scale sharding and reshuffling, the above solutions propose a new MPT tree structure to lower the cost of data synchronization. CoChain [4] proposed a multi-shard supervision scheme, allowing shards to monitor other shards, ensuring system security even as the number of shards continues to increase. Benzene [5] adopts a consensus protocol based on a TEE and multi-shard voting to enhance the system's fault tolerance and reduce transaction latency. The above solution provides security guarantees for high-performance sharded blockchains, further promoting the development of blockchains.

2.3   Research the Status of Node Processing Schemes

Node allocation refers to the strategic distribution of nodes across different shards to enable concurrent transaction processing, thereby enhancing system performance. Traditional sharding protocols [6–8] typically adopt random node allocation, where each shard is formed through random node combinations, relying on probabilistic distribution to disperse malicious nodes across shards.

Huang et al. proposed the CLPA [19] scheme, which took into account the situation of cross-shard transactions when allocating nodes and assigning nodes to shards with more frequent transactions, thereby reducing the Cross-Shard Transactions Ratio (CSTR). Zhang et al.'s trust-aware model [17] allocates nodes based on trust levels to achieve balanced trust distribution across shards. Xu et al.'s category-based approach [18] that proportionally distributes nodes of different types to maintain shard security.

The above-mentioned sharding schemes have enhanced the performance and security of the blockchain, yet these schemes still have certain limitations. For instance, during the node allocation period, if the attribute characteristics of the nodes are not taken into account, it may lead to malicious nodes being included in the same shard, thereby affecting the security of the system. To optimize the node allocation process, we defined a reputation value range to evaluate node categories and used a random forest model to determine node classification. Based on the classification results, isolate malicious nodes and then allocate them to shards according to the strategy to avoid lazy node aggregation. This method effectively prevents clusters of malicious or lazy nodes.

3   Design of Algorithm and Model

We extract the "From" and "To" addresses from the Ethereum dataset as nodes to construct a node relationship network $G = \{p_1, p_2, \ldots, p_i\}$, where p_i represents the shard for indexed address i and p_i consists of node set V_i and transaction set E_i ($p_i = \{V_i, E_i\}$). Based on the node network G, this paper designs different reputation value ranges for nodes. Then, the Random Forest model is used to determine node categories, and account allocation is performed based on the obtained node classification results. Finally, a node behavior monitoring scheme is designed to monitor node behavior, which serves as an evaluation basis for their classification. The specific process is shown in Fig. 2. The detailed procedures for each stage are as follows.

3.1   Design of Node Categories Based on Reputation Values

We propose a reputation-based node classification method to distinguish node types. Firstly, we collected real Ethereum transaction data from Xblock (a dataset sharing platform aimed at promoting the healthy development of blockchain and data research) as our experimental dataset, extracting *source* and *destination address fields* as node addresses. Then, we extracted three categories of node features from the dataset: basic features (whether double-spending occurred), transaction statistical features (Number of transactions, transaction failure rate, cross-shard ratio), and transaction-derived features (Total transaction amount, total Gas consumption, contract execution ratio). Each node was assigned a feature vector based on these characteristics.

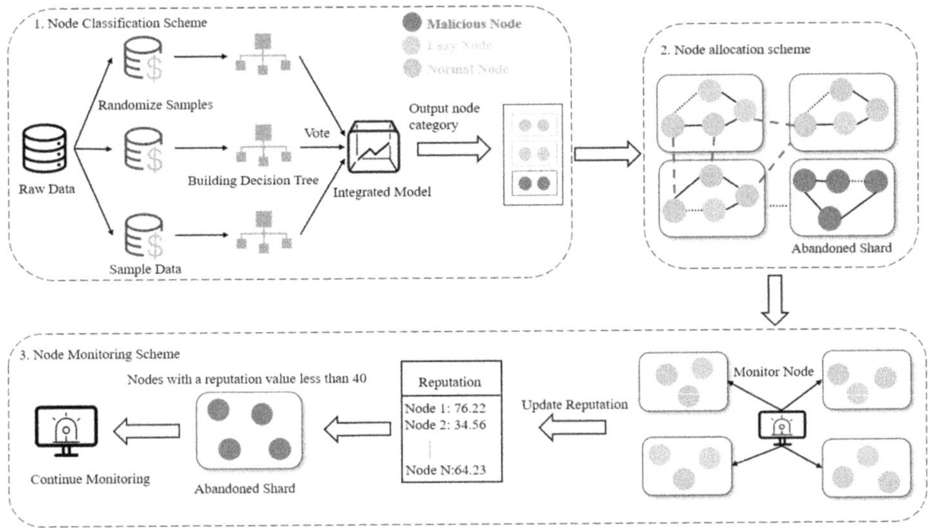

Fig. 2. Model Architecture

After calculating node reputation values, we observed that some nodes exhibiting frequent inactivity, frequent transaction failures, or double-spend behavior consistently displayed low reputation values. This paper confirms the reputation values of node categories through mathematical statistical analysis methods. During this period, it is found that the reputation value range of most nodes is $(60, 100]$, while a small portion fell into $[0, 40]$ or $(40, 60]$. Based on these findings, we classified nodes into three categories:

- **Malicious nodes** V_m: Reputation $\in [0, 40]$ or existing double-spending (class label 0);
- **Lazy nodes** V_l: Reputation $\in (40, 60]$ without double-spending (class label 1);
- **Normal nodes** V_n: Reputation $\in (60, 100]$ (class label 2).

In the initialization phases, all nodes were assigned a default reputation value of 100 (class label 2) to facilitate subsequent node detection processes.

3.2 Design of Node Classification Method

Hardware resources limit the training of node classification models. For instance, training large-scale Graph Neural Networks (GNNs) and Transformers requires high-performance GPUs, making such tasks difficult to handle for average blockchain users. To address this challenge, our study employs a lightweight Random Forest model for node classification. This model effectively captures nonlinear relationships between features while maintaining good classification performance for our complex task.

The workflow consists of three stages: data preparation, processing, model training, and evaluation. To ensure reproducibility, we fixed the random seed (`random_state=42`) throughout our experiments. The complete algorithm is presented in Algorithm 1.

Algorithm 1. Random Forest Model for Determining Node Categories

---

Require: Node set N; Random seed $S = 42$;
Ensure: Prediction label set y_{test}
1: // **Data Preparation**
2: **for** each $node_i \in N$ **do**
3:     $f_i \leftarrow \phi(node_i) \in \mathbb{R}^d$ $\triangleright$ f is the eigenvector, ϕ is the feature function
4:     $y_i \leftarrow \begin{cases} 0, & \text{if } 0 < R_i \leq 40 \wedge D_i \geq 1 \\ 1, & \text{if } 40 < R_i \leq 60 \wedge D_i = 0 \\ 2, & \text{if } 60 < R_i \leq 100 \wedge D_i = 0 \end{cases}$
5: **end for**
6: // **Data Processing**
7: Feature segmentation and via (1) eliminate differences in feature dimensions.
8: Divide the training set and testing set
9: // **Model Training**
10: Model $\leftarrow$ RandomForest($n\_trees = 100, criterion = gini, random\_state = S$)
11: $y_{\text{test}} \leftarrow$ Model.predict($X_{\text{test\_std}}$)
12: **return** y_{test}

- **Data Preparation:** We iterate the node set to extract feature vectors for each node, assign category labels based on predefined rules, and store both features and labels in separate lists while preserving the original indices for data partitioning tracking.
- **Data Processing:** This stage involves handling missing feature values, injecting Gaussian noise, and standardizing features to enhance model robustness. Specifically, Z-score standardization (Eq. 1) is applied to the training set features to eliminate scale differences across features.

$$X_{\text{scaled}} = \frac{X - \mu}{\sigma} \tag{1}$$

- **Model Training and Evaluation:** The Random Forest classifier aggregates predictions from multiple decision trees (using majority voting to determine node classes), mitigating overfitting risks inherent in single trees while efficiently processing high-dimensional features.

3.3 Node Processing Scheme Design

We found through investigation that some classic sharding algorithms [9] obtain shard indices by modulo node addresses and assign nodes to the shard. However, this approach does not consider the transaction relationships between nodes and the attribute categories of nodes, which can easily lead to a high CSTR, malicious nodes, or lazy node aggregation. Therefore, based on the attribute categories of nodes and the transaction behavior characteristics of nodes, the refined ANC (Allocation based on Node Category) algorithm is designed. This algorithm assigns nodes to different shards based on their category results, balancing the types of nodes in each shard and avoiding aggregation of lazy and malicious nodes.

Specifically, we obtain the determined node category results (Pre-Nodes) as the initial data source for node allocation and define the criteria for node allocation to shards. To ensure security, we assign all identified malicious nodes to a discarded shard, without any processing on the nodes in that shard; To achieve efficiency, the proposed method evenly distributes the identified normal nodes and lazy nodes to different shards, ensuring that the normal nodes in each shard always have an advantage and avoiding lazy nodes clustering in one shard, which affects the overall progress of transaction processing. However, in the process of sharding, some factors such as the frequency of transactions between nodes and the current load size of sharding also need to be considered. Therefore, calculation formulas for CSTR and shard load are defined to evaluate the performance of node allocation algorithms. Among them, the calculation of CSTR is shown in (2), which is based on the ratio of the total number of cross-shard transactions to the total number of shard transactions, where $T_{i,j}$ is the number of transactions between node i and j; $v_{j,k} = 1$ indicates that node j belongs to shard k, while $v_{j,k} = 0$ indicates that it does not belong. The calculation formula for shard load is shown in (3), where T_i is the transaction volume of node i.

$$\text{CSTR} = \frac{\text{CST}_{\text{total}}}{\text{CST}_{\text{total}} + \text{IST}_{\text{total}}} = \frac{\sum\limits_{i=1}^{N}\sum\limits_{j=1}^{N} T_{i,j} \times (1 - v_{i,k}) \times v_{j,k}}{\sum\limits_{i=1}^{N}\sum\limits_{j=1}^{N} T_{i,j} \times (v_{j,k} + v_{i,k})}, \quad \forall k \in P, i, j \in V \quad (2)$$

$$\text{Load}_k - \sum_{i=1}^{N} V_{i,k} \times T_i, \quad \forall k \in S \tag{3}$$

The algorithm traverses Pre-Nodes to obtain the current node $v \in V$ that needs to be allocated. Then, it calculates the Score obtained by assigning v to different shards and selects the one with the highest Score as the `target_shard`. The Score calculation method is shown in (4), where w_1: Cross-shard transaction weight, w_2: Load weight, w_3: Category weight, category (p_i) represents the proportion of normal nodes in shard p_i. ($w_1 + w_2 + w_3 = 1$) can be dynamically adjusted according to the real-time status of the system to maintain system balance. For example, increasing the weight of w_2 when the system load is high.

$$\text{Score}(v, p_i) = w_1 \times (1 - \text{CSTR}(v, p_i)) + w_2 \times (1 - \text{Load}(p_i)) + w_3 \times \text{category}(p_i) \quad (4)$$

Subsequently, the highest scoring shard is selected as the target shard p_i, and the nodes are assigned to the shards using the formula shown in (5). The algorithm execution process is shown in 2.

$$\text{target}_{p_i} = \arg\max_{p_i \in P} \text{Score}(v, p_i) \quad (5)$$

Algorithm 2. Assign Nodes Based on Classification Results

Require: Classification set Pre-Nodes; Shard set P; Weight coefficient W_i;
Ensure: Node allocation result Map
1: //**1. Isolation and processing of malicious nodes**
2: $p_{\text{isolate}} \in P$ ▷ Create isolation shard
3: **for** each v in Pre_Nodes **do**
4:     **if** v.category == Malicious **then**
5:         $\text{Map}[v] = p_{\text{isolate}}$
6:         $p_{\text{isolate}} = 0$ ▷ Mark isolation shard as inactive
7:     **end if**
8: **end for**
9: //**2. Dynamic allocation of other nodes**
10: $V = \text{Pre\_Nodes} \setminus \text{Malicious}$ ▷ Filter out malicious nodes
11: **while** $V \neq \emptyset$ **do**
12:     Via (4) calculate the scores of nodes in different shards
13:     Via (5) select the highest score as target_{p_i}
14:     $\text{Map}[v] = \text{target}_{p_i}$ ▷ Allocate node to shard
15: **end while**
16: **return** Map

3.4 Design of Node Monitoring Scheme

We propose a node monitoring scheme for real-time detection of node behavior in shards, further enhancing the security of the system. Specifically, during the transaction execution process, monitor whether the node initiates a double-spending transaction, whether the node's reputation value is less than or equal to 40, and so on. The calculation of the reputation value of a node includes four parameters: Transaction Failure Rate (FR): Transactions with low success rates may be malicious node behavior; CSTR: A high proportion of cross-shard transactions may disrupt shard balance; transaction activity (TA): Frequent transaction behavior may be brushing orders, and very few transaction behaviors may be cold nodes; Double-spending attack frequency (DSA): If double-spending behavior is detected, it is directly considered a malicious node.

Based on the above four parameters, formula (6) calculates the node's reputation value, where $\alpha + \beta + \gamma + \delta = 100$. In this experiment, we referred to the setting of reputation value weight parameters by some scholars and made corresponding adjustments based on the dataset, where $\alpha = 10$, $\beta = 20$, $\gamma = 10$, and $\delta = 60$ (Because the double flower attack poses the greatest threat to the system, the highest penalty weight is assigned).

$$\text{Reputation} = 100 - (\alpha \cdot \text{FR} + \beta \cdot \text{CSTR} + \gamma \cdot \text{IS} + \delta \cdot \text{DSA}) \tag{6}$$

When malicious behavior is detected on a node, the corresponding reputation is deducted. If the reputation value of the node is less than or equal to 40, it is removed from the source shard, reassigned to the discard shard, and isolated. If the node value is greater than 40 but less than or equal to 60, it is considered a lazy node and requires focused monitoring.

This paper proposes a combination scheme of "time decay + positive excitation" to solve the problem of node misjudgment, ensuring that the reputation value of nodes remains within a reasonable range during long-term operation. According to formula (7), for each epoch period, the historical penalty records of the node decay exponentially, and the reputation value can be partially restored, where $\lambda \in (0, 1)$ is the decay coefficient that controls the recovery speed.

$$\text{Reputation}_{t+1} = \min(100, \text{Reputation}_t + \lambda \cdot (100 - \text{Reputation}_t)) \tag{7}$$

According to formula (8), when a node successfully processes a transaction or participates in cross-shard collaboration, its reputation value is increased. η is the incentive weight, which is related to the importance of behavior (such as ordinary transaction $\eta = 0.1$, critical verification $\eta = 0.5$); Valid_Actions (weighted sum of compliance behaviors completed by nodes during the statistical period).

$$\text{Reputation} = \min(100, \text{Reputation} + \eta \cdot \text{Valid\_Actions}) \tag{8}$$

Our design combines reward-punishment mechanisms with dynamic adjustments. This approach curbs malicious behavior while providing compliant nodes with reputation repair channels. Thus, it enhances both system fairness and security. The specific implementation process of the plan is shown in Algorithm 3.

4   Model Analysis

This section analyzes the effectiveness of the proposed node classification scheme and the security of the sharded blockchain system. By integrating cryptographic protocols, consensus layer reinforcement, and dynamic reputation mechanisms, a multi-layer defense architecture is constructed to deal with diverse attack vectors, ensuring a secure and efficient operating environment for the system.

Algorithm 3. Node Behavior Monitoring

Require: node set $V = \{v_1, v_2, \ldots, v_n\}$, transaction set $T = \{t_1, t_2, \ldots, t_m\}$, weight
    parameters $\alpha, \beta, \gamma, \delta$ ($\alpha + \beta + \gamma + \delta = 100$), FR, CSR, TA, DSA, λ, η, k
Ensure: Malicious node list V_M, lazy node list V_L
1: // **Calculate node reputation and monitor behavior**
2: **for** each node $v \in V$ **do**
3:     Via (6) calculate current reputation value R_v
4:     **if** k mod $10 == 0$ **then**   ▷ Executed every 10 epochs
5:         Via (7) apply time decay to R_v
6:         Via (8) apply positive incentives to R_v
7:     **end if**
8:     **if** $R_v \leq 40$ **then**
9:         V_M.append(v)   ▷ Add to malicious list
10:       migrate_and_isolate(v)   ▷ Isolate malicious node
11:    **else if** $40 < R_v \leq 60$ **then**
12:       V_L.append(v)   ▷ Add to lazy list
13:       set_monitoring_level(v, HIGH)   ▷ Increase monitoring
14:    **else**
15:       maintain_normal_status(v)   ▷ Normal operation
16:    **end if**
17: **end for**
18: **return** V_L, V_M

4.1 Node Classification Effectiveness

According to the results of node category recognition, the random forest model
is efficient in determining node categories. It utilizes multiple decision trees for
classification and regression, effectively processing high-dimensional data and
making it an ideal choice for node classification. When determining the category
of nodes, it was found that the number of nodes did not change significantly when
using datasets of 5K and 10K, but the category of nodes changed significantly.
This is because, as the dataset increases, the number of nodes does not show a
significant increase; rather, the existing nodes handle an increasing number of
transactions. Therefore, the more transactions a node has, the more prominent
the features used for training, and the more accurate the recognition results will
be.

4.2 Security

The solution we have constructed can resist security problems such as double-
spending attacks, Sybil attacks, Byzantine attacks, and privacy breaches. Firstly,
we design reputation penalty and node eviction mechanisms to enforce the
uniqueness of blockchain transactions by maintaining global transaction ledger
synchronization across shards. The verification is as follows: Let $H(tx)$ be the
hash value of transaction tx, T be the set of all submitted transactions, and for
any new transaction tx', the verification function satisfies:

$$\forall tx \in T, H(tx') \neq H(tx) \implies \text{is\_double\_spend}(tx') = \text{False}. \tag{9}$$

A Sybil-resistant consensus participation mechanism is implemented by binding RSA-secured identities to reputation thresholds. Specifically: Each node registers a unique RSA key pair (pk, sk) during network initialization, where pk serves as its immutable identity; Before joining consensus, a smart contract verifies whether the node's reputation satisfies $reputation(n) > 50$ (Formula 10). Nodes failing this check are excluded from the consensus pool C; For any node $n \in N_{\mathrm{sybil}}$ detected initiating fake identities, its reputation is decayed exponentially:

$$reputation(n)_{t+1} = \lambda \cdot reputation(n)_t \quad (\lambda = 0.7 \text{ per epoch}) \tag{10}$$

Nodes triggering $reputation(n) \leq 50$ within $T_{\max} = 2$ blocks are deactivated for $\Delta T = 10$ epochs.

Finally, in response to Byzantine attacks, the PBFT consensus protocol and view-changing mechanisms are adopted to improve the system's fault tolerance. Specifically, using a view-changing protocol can address the issue of master node failure. The PBFT consensus protocol can tolerate malicious nodes and must satisfy $3f + 1 \leq |N|$. The liveness verification is as follows: According to Castro et al.'s study [20], if the number of Byzantine nodes does not exceed f, the PBFT protocol can guarantee: $Latency \leq 2\Delta + O(|N|)$, where Δ is the network delay.

5   Performance Evaluation

Based on Python, this experiment constructed a prototype of a blockchain sharding system. On this basis, protocols such as node classification, node type determination, and node behavior monitoring were designed. This paper selects the Ethereum transaction dataset to evaluate throughput (TPS), transaction confirmation delay (TCL), and CSTR. In addition, the experimental method is also adapted to other blockchain datasets. The source and parameter Settings of the dataset for this experiment are as follows:

The 10K transaction dataset was selected to simulate real transaction situations. The experimental shard size is set to $S = \{2, 4, 8, 16\}$, PBFT is used as the shard consensus protocol, and batch processing transactions are specified as 200.

The hardware resource situation includes: RTX4060, a 24-core CPU, and 32 GB of video memory. In addition, we use Random and Monoxide [9] as baseline models, where Random uses a scheme of randomly assigning accounts to shards; Monoxide converts node addresses into decimal numbers and performs modulo operations on the number of shards to obtain the positions that nodes will be assigned to shards.

5.1   Node Categories Identify Results

We use datasets of 5K and 10k for model training and prediction, with a ratio of 7:3 between the training and testing sets. As shown in Fig. 3a, the 5K dataset exhibits a clear distribution of malicious (red), lazy (blue), and normal (green) nodes. The node characteristics in the dataset can effectively distinguish the type of nodes. In addition, Fig. 3b shows the use of the random forest model to determine node categories, which rarely makes errors in determining node categories, indicating the superiority of using this model. To comprehensively evaluate the performance of the model, this paper used a 10K dataset to repeat the above experiments, and the results are shown in Fig. 4a and 4b. The more transaction data, the more comprehensive the features of the nodes, and the more accurate the recognition of node categories. The experimental results show that this method can effectively distinguish malicious nodes, lazy nodes, and normal nodes, providing a reliable tool for analyzing the behavior of network nodes.

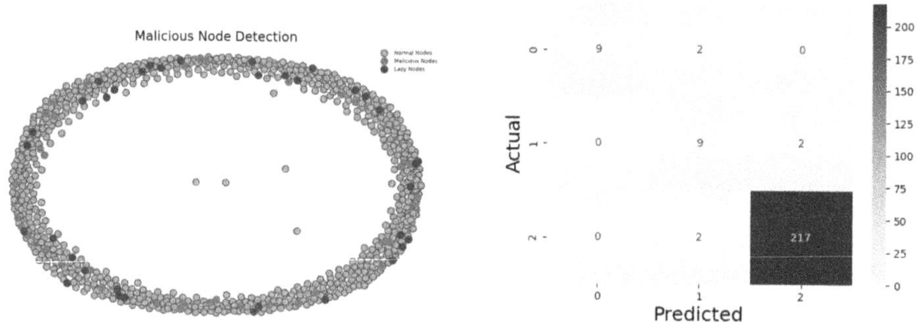

(a) 5K Dataset Node Category Diagram

(b) 5K Dataset Node category recognition heatmap

Fig. 3. 5K Dataset Node Category Analysis: (a) diagram of node categories and (b) recognition heatmap (Color figure online)

5.2   Throughput

We use 10K Ethereum datasets to simulate the process of real transactions, with 200 transactions processed per batch. Throughput changes are compared with baseline models under different shard quantities, with shard quantities ranging from $S = \{2, 4, 8, 16\}$. As shown in Fig. 5a, with the continuous increase of the number of shards, the performance of the shard scheme also improves almost linearly. However, the proposed ANC scheme has the best performance. Compared with solutions such as Random and Monoxide, ANC's performance has improved by about 10% under the same hardware configuration and dataset. This is because the Random and Monoxide schemes have a certain degree of randomness in node allocation and have not fully evaluated the characteristic relationships between nodes. The ANC scheme selects the optimal shard of nodes from a global perspective by introducing transaction relationships between

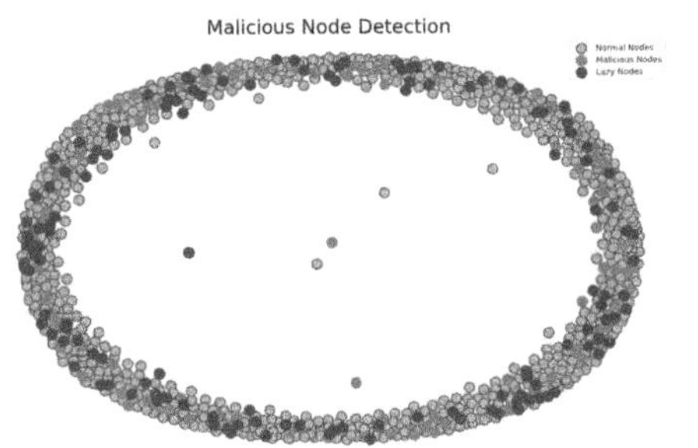

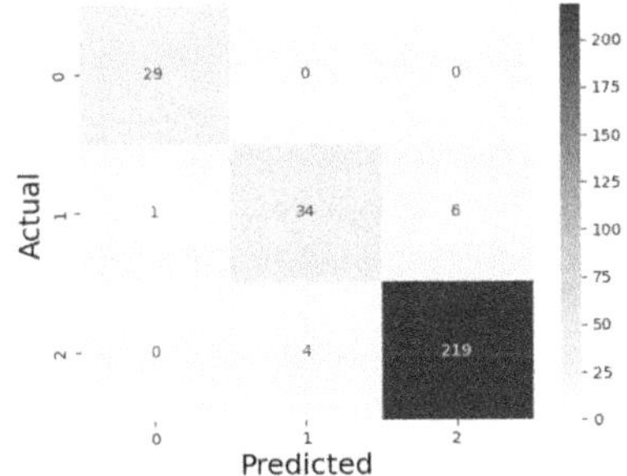

(a) 10K Dataset Node Category Diagram   (b) 10K Dataset Node category recognition heatmap

Fig. 4. 10K Dataset Node Category Analysis: (a) diagram of node categories and (b) recognition heatmap (Color figure online)

nodes, node class, and node load to avoid getting stuck in locally optimal solutions. In addition, the reasonable allocation of nodes to shards can improve the resource utilization of the system, thereby enhancing its performance to a certain extent. Finally, we also investigated the TPS situation under different transaction injection speeds, as shown in Fig. 6b. Increasing the injection speed of transactions can significantly improve the overall performance of the scheme, among which the ANC scheme still has the best performance.

5.3 Cross-Shard Transaction Ratio

Cross-shard transaction refers to transactions conducted between nodes in two different shards. Compared to intra-shard transactions, cross-shard transaction consumes more resources overhead. We explore the changes in the CSTR under different shard quantities. As shown in Fig. 5b, the overall trend of the CSTR increases with the increase of shard quantity. However, when $S>=8$, the CSTR of ANC, Random, and Monoxide schemes approach 0.86. This is because in a limited data set, when the number of shards increases to a certain extent, transactions within each shard can be considered as cross-shard transactions. Therefore, when the number of shards reaches a certain threshold, the fluctuation of the CSTR is very small. To comprehensively evaluate the system, we also investigated CSTR under different injection speeds. As shown in Fig. 6a, the effect of injection speed on CSTR is not significant. Despite the increase in the number of shards leading to an increase in CSTR, ANC still maintains its overall performance advantage through load balancing and malicious node isolation.

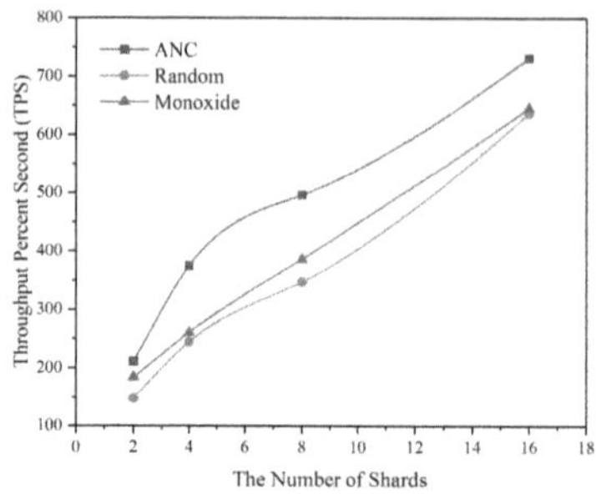

(a) TPS changes of different schemes

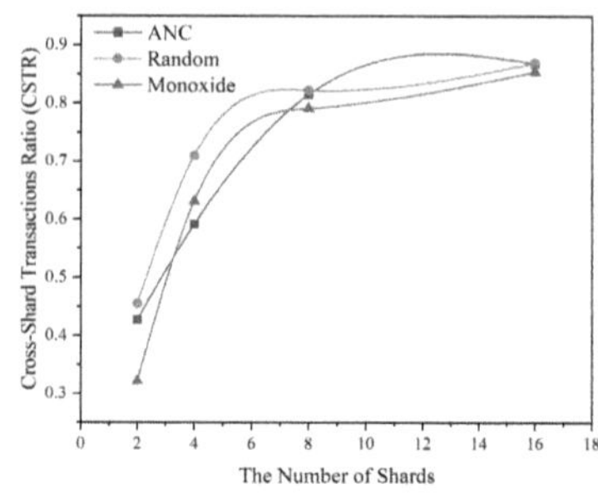

(b) CSTR changes of different schemes

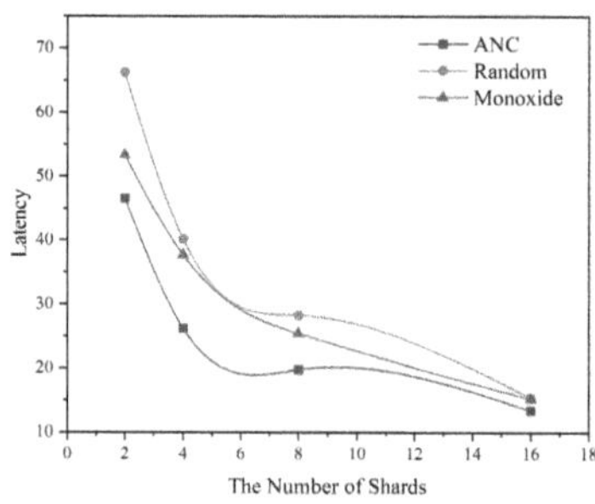

(c) TCL changes of different schemes

Fig. 5. Overall performance comparison of schemes under different numbers of shards

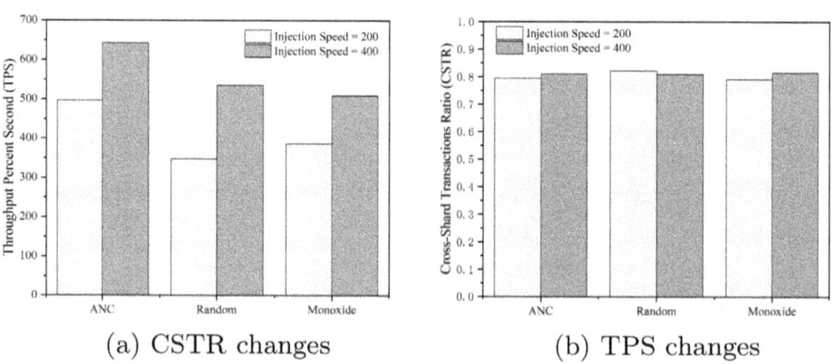

(a) CSTR changes

(b) TPS changes

Fig. 6. Changes in CSTR and TPS under different transaction injection speeds

5.4 Transaction Latency

We define transaction latency as the total time from transaction initiation to completion. In our system, the overall trend of transaction latency decreases as the number of shards increases, due to improved system parallelism and higher TPS processing speed. As shown in Fig. 5c, the latency of our proposed ANC scheme is lower than both the Random and Monoxide baseline models (ANC_Latency = 13.41 vs Random_Latency = 15.40 and Monoxide_Latency = 15.20), demonstrating its superior TPS performance. However, with a fixed dataset size, when the shard count exceeds a certain threshold, the latency difference between ANC and baseline models becomes negligible. This occurs because while sharding improves efficiency, it simultaneously introduces additional resource overhead for cross-shard transactions. Excessive sharding can lead to substantial overhead costs, necessitating the dynamic adjustment of shard numbers based on actual data size.

6   Conclusion

This paper proposes a node allocation scheme based on node attribute categories, which uses a Random Forest model to classify nodes and designs a refined account allocation algorithm (ANC) based on node classification results. The ANC algorithm considers both the transaction frequency between nodes and the load of shards, proportionally distributing lazy nodes and normal nodes across shards to indirectly improve the sharded blockchain performance and security. Subsequently, this paper proposes a node behavior monitoring scheme to detect potential category changes during transaction processing, thereby improving node classification accuracy. Compared with traditional solutions, our system model demonstrates higher security and performance.

Future work will investigate dynamic weight adjustment mechanisms to address the limitations of fixed weights (e.g., $\alpha = 10$, $\beta = 20$) in adapting to dynamic attack scenarios. While the current approach to isolating malicious nodes relies on centralized decision-making, subsequent research will explore staking penalties or DAO voting mechanisms for determining malicious node punishments. In addition, further analysis of Ethereum transaction data characteristics will also be conducted to enhance the sharding system's security.

Acknowledgments. This material is based upon work supported by the National Natural Science Foundation of China under Grant No. 62366052, Natural Science Foundation of Xinjiang Uygur Autonomous Region (No. 2022D01C429, No. 2022D01C427), Key R&D Program of Xinjiang Uygur Autonomous Region (No. 2022B03023, No.2022B01046), Sichuan Province Joint Fund Project of China No. 25QYCX0103.

References

1. Yang, Q., et al.: The state-of-the-art and promising future of blockchain sharding. IEEE Commun. Mag. 4(63), 192–198 (2024)
2. Huang, H., Zhao, Y., Zheng, Z.: tMPT: reconfiguration across blockchain shards via trimmed Merkle_Patricia Trie. In: 31st International Proceedings of the IEEE/ACM Symposium on Quality of Service, pp. 1–10. IEEE (2023)

3. Qi, X.: S-store: a scalable data store towards permissioned blockchain sharding. In: IEEE Conference on Computer Communications, pp. 1978–1987. IEEE (2022)
4. Li, M., et al.: CoChain: High concurrency blockchain Sharding via consensus on consensus. In: Proceedings of the IEEE Conference on Computer Communications, pp. 1–10. IEEE (2023)
5. Cai, Z., et al.: Benzene: Scaling blockchain with cooperation-based Sharding. IEEE Trans. Parallel Distrib. Syst. **34**(2), 639–654 (2022)
6. Luu, L., et al.: A secure Sharding protocol for open blockchains. In: Proceedings of the 2016 ACM SIGSAC Conference on Computer and Communications Security, pp. 17–30. ACM (2016)
7. Kokoris-Kogias, E., et al.: Omniledger: a secure, scale-out, decentralized ledger via Sharding. In: 2018 IEEE Symposium on Security and Privacy, pp. 583–598. IEEE (2018)
8. Zamani, M., et al.: RapidChain: scaling blockchain via full Sharding. In: Proceedings of the 2018 ACM SIGSAC Conference on Computer and Communications Security, pp. 931–948. ACM (2018)
9. Wang, J., Wang, H.: Monoxide: scale out blockchains with asynchronous consensus zones. In: 16th USENIX Symposium on Networked Systems Design and Implementation, pp. 95–112. NSENIX (2019)
10. Hong, Z., Guo, S., Li, P., Chen, W.: Pyramid: a layered Sharding blockchain system. In: Proceedings of the IEEE Conference on Computer Communications, pp. 1–10. IEEE (2021)
11. Huang, H., Peng, X., Zhan, J., et al.: BrokerChain: a cross-shard blockchain protocol for account/balance-based state Sharding. In: Proceedings of the IEEE Conference on Computer Communications, pp. 1968–1977. IEEE (2022)
12. Hong, Z., et al.: GriDB: scaling blockchain database via Sharding and off-chain cross-shard mechanism. arxiv preprint arxiv:2407.03750 (2024)
13. Zhang, Y., Pan, S., Yu, J.: Txallo: dynamic transaction allocation in sharded blockchain systems. In: 39th International Conference on Data Engineering, pp. 721–733. IEEE (2023)
14. Li, M., et al.: LB-Chain: load-balanced and low-latency blockchain Sharding via account migration. IEEE Trans. Parallel Distrib. Syst. **34**(10), 2797–2810 (2023)
15. Tao, Y., Li, B., Li, B.: On Sharding across heterogeneous blockchains. In: 39th International Conference on Data Engineering, pp. 477–489. IEEE (2023)
16. Li, X., Luo, H., Duan, J.: Security analysis of Sharding in blockchain with PBFT consensus. In: 4th International Conference on Blockchain Technology, pp. 9–14. ACM (2022)
17. Zhang, P., et al.: Optimized blockchain Sharding model based on node trust and allocation. IEEE Trans. Netw. Service Manag. **20**(3), 2804–2816 (2023)
18. Xu, Y., Huang, Y.: An N/2 Byzantine node tolerate blockchain Sharding approach. In: 35th Annual ACM Symposium on Applied Computing, pp. 349–352. IEEE (2020)
19. Li, C., Huang, H., Zhao, Y., et al.: Achieving scalability and load balance across blockchain shards for state Sharding. In: proceedings of the International Symposium on Reliable Distributed Systems, pp. 284–294. IEEE (2022)
20. Castro, M., et al.: Practical byzantine fault tolerance. In: OsDI, vol. 99, pp. 173–186 (1999)

A Parallel Implementation of ChaCha20 on MT-3000 Heterogeneous Multi-zone Processor

Yongtao Luo[1,2], Jie Liu[1,2], Tun Li[3], and Chunye Gong[3,4](✉)

[1] Laboratory of Digitizing Software for Frontier Equipment, National University of Defense Technology, Changsha 410073, China
`{luoyongtao,liujie}@nudt.edu.cn`
[2] National Key Laboratory of Parallel and Distributed Computing, National University of Defense Technology, Changsha 410073, China
[3] College of Computer Science and Technology, National University of Defense Technology, Changsha 410073, China
`tunli@nudt.edu.cn, gongchunye@126.com`
[4] National Supercomputer Center in Tianjin, Tianjin 300457, China

Abstract. In the latest Transport Layer Security (TLS) v1.3 protocol, ChaCha20 and the Advanced Encryption Standard (AES) are the only supported encryption algorithms. ChaCha20, a variant of Salsa20, is widely recognized for its simplicity, robust security, and compatibility. Given the sensitive nature of much of the data processed on supercomputers remotely, efficient encryption is paramount. However, encrypting large-scale data can be time-consuming. To address this challenge, we propose a parallel implementation of the ChaCha20 algorithm on the MT-3000 heterogeneous multi zone processor. By leveraging its unique architectural features, we employ various optimization techniques, including multilevel parallel strategies for heterogeneous multi-zone architecture, Direct Memory Access (DMA) for efficient data transmission, and Single-Instruction Multiple Data (SIMD) for parallel data processing within the acceleration core. Experimental results demonstrate the effectiveness of our parallel scheme, achieving a throughput of up to 168.11 GB/s on a single MT-3000 processor, and finally scaling to 512 acceleration clusters with good scalability.

Keywords: MT-3000 · ChaCha20 Algorithm · Heterogeneous Parallelism · SIMD

1 Introduction

With the rapid development of Internet technology and the advent of the big data era, data has become an important factor in the development of various industries [1,2]. Data sharing, integration, and value mining have become its main development trends. At the same time, the security issues surrounding big

H. Liu et al. (Eds.): ICA3PP 2025, LNCS 16381, pp. 117–134, 2026.
https://doi.org/10.1007/978-981-95-8399-7_7

data have become increasingly prominent, and encryption of network transmission data is an essential method for protecting data security [3]. Therefore, there is an urgent need to accelerate encryption speed.

ChaCha is a family of stream cipher algorithms, a variant of Salsa20 [4,5], developed by Daniel J. Bernstein in 2008. The ChaCha family includes the 20-round cipher ChaCha20, the 12-round cipher ChaCha12, and the 8-round cipher ChaCha8. The number of operation rounds determines the security level, with more rounds providing higher security but slower performance. ChaCha20 is used to replace RC4 [6] in OpenSSL for secure Internet communications. Google initially implemented HTTPS (TLS/SSL) traffic between the Chrome browser (Android mobile version) and Google's website. It is now widely used in areas such as TLS [7], OpenSSH [8], and Virtual Private Networks (VPNs) [9]. In the latest TLS v1.3, the protocol has been simplified while enhancing security, retaining only the AES and ChaCha20 algorithms [10]. Compared to other encryption algorithms, numerous security studies have demonstrated its effectiveness in protecting against a wide range of attacks, such as post-quantum attacks, differential attacks, fault attacks, timing attacks, and side-channel attacks [11–13]. Although AES is more widely used, ChaCha20 is significantly faster than AES in software-only implementations. On systems without dedicated AES hardware, it is approximately three times faster [4,14]. In recent years, ChaCha20 has been adopted as an alternative to the AES algorithm to increase encryption/decryption speed. From the perspective of encryption/decryption efficiency, reference [15] provides a detailed analysis of whether to use ChaCha20 or AES in different scenarios.

In supercomputers, much of the data is sensitive, such as nuclear simulation data, drug screening data, and weather forecasting data, and must be protected by encryption. Recently, many encryption algorithms, such as AES, ChaCha20, and SHA256, have been deployed in the Sunway TaihuLight Supercomputer for encrypting sensitive data [16–18]. However, there are few security studies on the MT-3000 multi-zone processor [19,20], which is autonomously designed and implemented by the National University of Defense Technology and used in next-generation exascale prototype supercomputers [21]. To protect the data on these next-generation exascale prototype supercomputers, we deploy the ChaCha20 stream cipher for data encryption for the first time. Furthermore, to achieve higher encryption efficiency, we have designed a parallel ChaCha20 algorithm that is closely integrated with the MT-3000 architecture. Specifically, this paper makes the following contributions.

- We propose a multi-level data parallelism approach, including process-level parallelism through the Message Passing Interface (MPI), thread-level parallelism through a low-overhead and high-availability heterogeneous programming interface (hThreads) [22], and intra-core data parallelism through SIMD, to fully utilize the performance of the MT-3000.
- To fully utilize SIMD instructions within the acceleration core, we propose a SIMD-friendly data layout for the ChaCha20 algorithm. This performance outperforms other schemes implemented on CPUs using NEON instructions.

- We implement a parallel version of the ChaCha20 algorithm for data encryption on the MT-3000 and evaluate it thoroughly. To the best of our knowledge, we are the first to accelerate ChaCha20 on the MT-3000, providing a new option for encrypting sensitive data on supercomputers. The results show that MT-ChaCha20 can achieve a throughput of 168.11 GB/s on a single MT-3000 processor and can be scaled to 512 acceleration clusters with good scalability.

The remainder of this paper is organized as follows. Section 2 introduces the standard ChaCha20 stream cipher algorithm, the architecture of the MT-3000 multi-zone processor, and reviews some related work on ChaCha20. In Sect. 3, we specifically describe the implementation of ChaCha20 on the MT-3000. In Sect. 4, we evaluate the actual performance of the parallel ChaCha20 implementation on the next-generation exascale prototype supercomputer platform and compare it with other related work on different platforms. Additionally, we have also implemented the ChaCha20 algorithm on a CPU using NEON instructions. Finally, we conclude this paper in Sect. 5.

2   Background and Related Work

In this section, we begin with a brief overview of ChaCha20 algorithm. Then, we introduce the MT-3000 heterogeneous multi-zone platform and software stack. Finally we provide a brief review of related work.

2.1   ChaCha20 Algorithm

ChaCha20 is a high-speed stream cipher, which has 20 rounds. The initial state of ChaCha20 encryption consists of four parts, including a 128-bit constant ($Constant$), a 256-bit key (Key), a 32-bit count ($Count$) and a 96-bit random number ($Nonce$) [4]. There are arranged into a 4 * 4 32-bit initial matrix (I_m) as follows:

$$I_m = \begin{bmatrix} Constant[0] & Constant[1] & Constant[2] & Constant[3] \\ Key[0] & Key[1] & Key[2] & Key[3] \\ Key[4] & Key[5] & Key[6] & Key[7] \\ Counter & Nonce[0] & Nonce[1] & Nonce[2] \end{bmatrix} \tag{1}$$

The basic operation of the ChaCha20 algorithm is called the quarter-round (QR), which mainly operates on four 32-bit unsigned integers. The specific arithmetic operations of the QR include addition($+$), XOR ($\oplus$) and left rotate ($ROTL^n$). The ChaCha20 block function applies ten double-rounds to update the matrix state, wherein each double-round consists eight quarter-rounds. The state matrix (O_m) is initialized with the values of I_m. The result is stored in the same position as the state taken by the input when the rounds finalize. Finally, by adding the I_m and O_m, the key-stream is obtained, and at the same time the counter adds one to generate the next I_m. Algorithm **1** presents the process of keystream generation.

As shown in Fig. 1, ChaCha20 encrypts in 512-bit blocks. The keystream is generated independently of the plaintext. When the keystream is updated, the counter value is incremented by one and the other parameters remain unchanged. Any data that is shorter than 512 bits will be processed according to its actual length without padding. The decryption process is similar to the encryption process, as shown below:

$$ciphertext = plaintext \oplus keystream \tag{2}$$

$$plaintext = ciphertext \oplus keystream \tag{3}$$

Algorithm 1. Keystream generation

Input: $Constant \in (0,1)^{128}, Key \in (0,1)^{256}, Counter \in (0,1)^{32}, Nonce \in (0,1)^{96}$
Output: $keystream \in (0,1)^{512}$
Begin
\#define $QR(a,b,c,d)$ {
    $a = a + b; d = a \oplus d; d = ROTL^{16}(d)$
    $c = c + d; b = b \oplus c; b = ROTL^{12}(b)$
    $a = a + b; d = a \oplus d; d = ROTL^{8}(d)$
    $c = c + d; b = b \oplus c; b = ROTL^{7}(b)$
}
$I_m = init(Constant, Key, Counter, Nonce)$
$O_m = I_m$
for $i = 20; i > 0; i = i - 2$ **do**
    $QR(O_m[0,4,8,12])$
    $QR(O_m[1,5,9,13])$
    $QR(O_m[2,6,10,14])$
    $QR(O_m[3,7,11,15])$
    $QR(O_m[0,5,10,15])$
    $QR(O_m[1,6,11,12])$
    $QR(O_m[2,7,8,13])$
    $QR(O_m[3,4,9,14])$
end for
$keystream = serialize(I_m + O_m)$
return $keystream$
End

2.2   The MT-3000 Processor

MT-3000 is a heterogeneous multi-zone processor for high performance computing (HPC) autonomously designed and implemented by the National University of Defense Technology for the next-generation exascale prototype supercomputer [21]. MT-3000 consists of a general purpose (GP) zone and four acceleration (ACC) zone clusters. The GP zone contains 16 general purpose CPU cores with L1 and L2 caches. Each cluster contains 24 acceleration arrays, an on-chip global

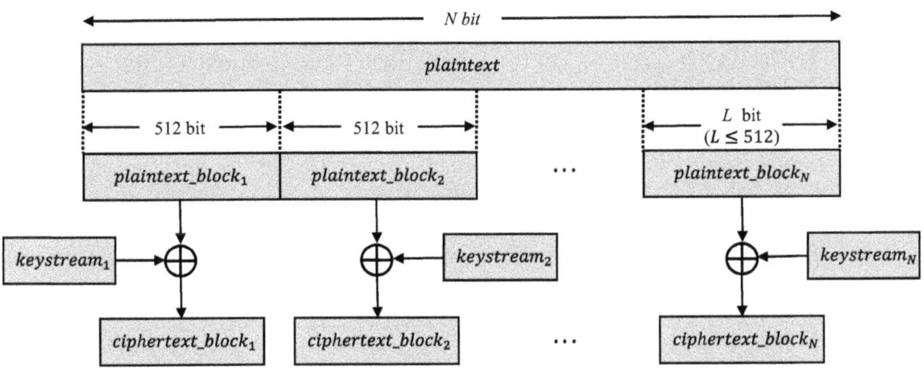

Fig. 1. ChaCha20 stream cipher process.

shared memory (GSM) with 6 MB capacity and 307 GBps bandwidth, and DDR4 with 32 GB capacity and a total bandwidth of 51 GBps. Each acceleration array contains a control memory (CM) with 64 KB, and an array memory (AM) with 768 KB. The AM is a vector memory, and can support two load/stores to each accelerator core at most. The acceleration array establishes a robust foundation for the efficient exploitation of data-level parallelism (DLP), a prevalent form of parallelism in many HPC workloads. Since the MT-3000 adopts multiple levels of memory storage, data movement between these different memories is through dual channel direct memory access (DMA). In addition the GP zone can access four DDR4 memory at the same time, while the ACC cluster can only access the corresponding DDR4 memory. Figure 2 shows the overall architecture of the MT-3000.

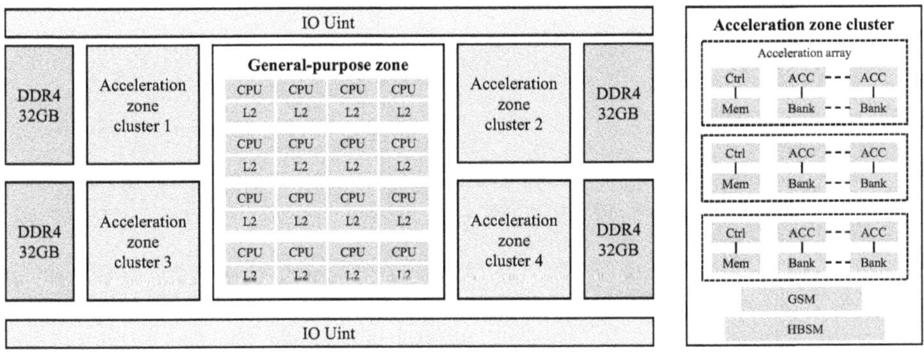

Fig. 2. The overall architecture of MT-3000.

To facilitate user programming on MT-3000 platform, a feature-rich and high-performance software stack is provided. A standard Linux operating system (OS) handles the interactions between CPUs and ACC clusters via device driver. The compiler system (M3CC and libMT) translates C-like code into binaries running on MT-3000. In addition, two programming interfaces are provided to avoid user handling the underlying hardware, including the heterogeneous threads (hThreads) and MOCL3. The programming implementation in this paper is based on hThreads. The hThreads is a low-overhead and high-availability multi-thread model for MT-3000 programming [22]. It not only provides a POSIX-like thread interface to manage the acceleration arrays, but also offers user-friendly wrappers of sophisticated hardware functional interfaces such as DMA interfaces and hardware synchronization interfaces.

Figure 3 illustrates the hThreads runtime conceptual framework, where the module above the dotted line represents the user interface, while the internal implementing functional modules are below the dotted line. Communication between the GP zone and the ACC zones occurs through the use of interrupts and shared memory [21,22].

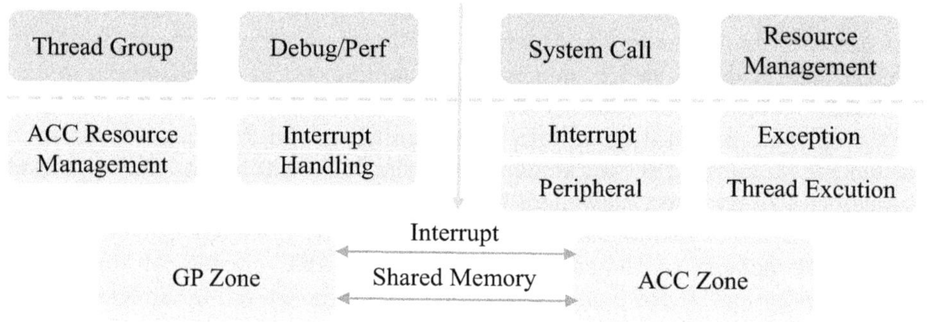

Fig. 3. The implementation modules of hThreads.

2.3   Related Work

Since the introduction of ChaCha20 in 2008 [4], a variety of optimization techniques have been used to improve the performance of the cipher. Here, we will introduce some related work briefly.

In terms of vectorization, Goll et al. [23] presented a software optimization for ChaCha20 with AVX2 instructions. The implementation performed at 1.43 cycles per byte on the Intel Haswell microarchitecture (on a 4 KB message). In addition, by comparing the numbers of SSE (128-bit), AVX2 (256-bit), and AVX512 (512-bit) instructions count, the authors predicted the performance could be doubled with AVX512 vectorization.

In terms of software refactoring, Zinzindohoué et al. [24] presented HACL*, a verified cryptographic library written in the F* programming language that implements the ChaCha20 encryption algorithms, and other message authentication, hash functions, and signature algorithms. The results show that the scheme performs as well as the C implementation in OpenSSL and libsodium on the 64-bit platforms. Almeida et al. [25] presented a new cryptographic implementations framework, which is based on the Jasmin framework. The scheme use vectorized instructions, compiler optimization, and other low-level optimizations to maximize performance benefits. The results show that the non-vectorized code is 2 times faster than HACL*, and the vectorized code is about 10x faster.

In terms of High Performance Computing (HPC) hardware platforms, Ma et al. [26] presented a lightweight cryptographic customized SoC for the security of IoTs (Internet of Things) devices, based on a 32-bit VexRiscv processor. They implemented PRINCE, PRESENT-80, ChaCha20 and SHA3-512 in the Artix-7 XC7A100T FPGA board. The throughput of ChaCha20 can achieve 662 Mbps that runs at 75Mhz, and the power is just 0.008 W. Wang et al. [17] presented a hybrid CPU/GPU optimization scheme of ChaCha20 stream cipher. On Intel Xeon 5218R, the throughput can achieve up to 18.32 GB/s when there is on data reuse. On Nvidia GTX-3090, the maximum throughput is 211.41 GB/s. Considering the interconnected collaboration between CPU and GPU, the actual throughput is 9.86 GB/s, reaching 87.76% of the theoretical bandwidth. Cai et al. [17] presented a parallel implementation of ChaCha20 stream cipher on Sunway TaihuLight Supercomputer. They used DMA for memory access, SIMD technique for key stream generation and dynamic optimization scheme for different input data sizes. The throughput is 32.43 GB/s on a single SW26010 processor, and the parallel efficiency is above 99.9% when extended to 1024 core groups.

Finally, MT-3000, a high-performance chip on the supercomputer, there is no relevant research on ChaCha20 using the processor yet.ÂăIn our work, we implement a parallel ChaCha20 algorithm with high throughput and scalability.

3   ChaCha20 Implementation on MT-3000

3.1   Programming Model

Based on the heterogeneous multi-zone architecture of MT-3000, MPI and hThreads are the programming models that used in the heterogeneous parallel implementation ChaCha20 algorithm on the next-generation exascale prototype supercomputer. Since the four ACC clusters of the MT-3000 can not form a shared memory system, each cluster processes all tasks of a process independently. This means four processes are assigned to four CPU cores within each computing node. The hThreads abstracts each acceleration array as a thread and treats multiple acceleration arrays as a thread group for unified management on the GP zone. Thus, 24 acceleration arrays are controlled by 24 thread-like within each ACC cluster, and each thread can perform the same task or a completely different task. In the scheme of this paper, each acceleration array performs the same task. Similar to CUDA, OpenCL and other heterogeneous programming models, the hThreads consists of host APIs and device APIs, and provides managements of device, program images, shared resources, thread parallelism, synchronization, and data transfers.

To summarize, four processes are created on each computing node, corresponding to four CPU cores. Each process manages an ACC cluster, and then 24 thread-like processes are created corresponding to 24 ACC arrays. SIMD instructions are used for data parallelism within each ACC array. Thereby ultimately accelerating the parallel computation of ChaCha20 and enabling collaborative computation of GP zone and ACC zone clusters. Figure 4 shows the overview of the multi-level parallelism of ChaCha20.

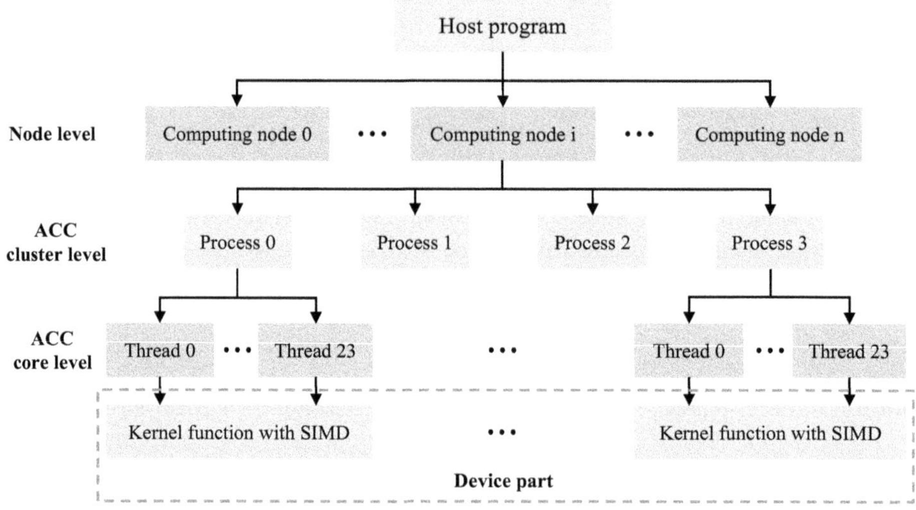

Fig. 4. Overview of the multi-level parallelism of ChaCha20 implementation.

3.2   Process-Level and Thread-Level Parallelism

The heterogeneous ChaCha20 algorithm adopts an equal distribution strategy for task computation, and each ACC zone cluster only encrypts the assigned plaintext. Firstly, we assume that the number of processes is N and the total size of the plaintext is D, that means plaintext is equally divided into the size of D/N. Finally, the data transfers between different processes are through the MPI interface. However, the overhead of transfer between the GP zone and ACC zone cluster is expensive, and the data transfers should be minimized.

In the hThreads programming model, the ACC cluster is responsible for performing computationally intensive tasks. Each ACC array on an ACC cluster is abstracted as a thread. That means each thread is uniquely mapped to an ACC array, and has unique thread ID. To further improve computation efficiency, threads will perform computing tasks in SIMD, and different threads can perform different computing tasks. The data transfers between different threads are through shared memory.

Due to the unique hybrid memory hierarchy structure of MT-3000, it is important to investigate the data transfer strategy carefully. That involves reducing memory access operations during computation to minimize access overhead and achieve high performance. According to the computational flow of the ChaCha20 algorithm in Sect. 2.1, only the counter parameter differs in each initial matrix during the key-stream generation process, while the other parameters remain unchanged. Consequently, in the parallel ChaCha20 algorithm, constant parameters are broadcast by the master process, while counter values are calculated independently in each ACC array.

In fact, there are two methods for data transfer between the GP zone (DDR4 memory) and ACC cluster (AM/CM/GSM). The ACC array can access data either from the three-level (DDR4-GSM-AM/SM) memory hierarchy, or directly from the DDR4 memory by DMA. When performing vectorizationÂăcalculations, the data in AM is further loaded into the vector registers and saved to AM when the calculation is complete. In the first case, the ACC array access the data items through the DDR4-GSM-AM/SM hierarchy. Due to the limited 6 MB capacity of the GSM, the amount of data that can be stored at one time is small. Moreover, our experience with High Performance Computing Gradient (HPCG) benchmark optimization is that the storage of data in the GSM does not significantly improve performance. In the second case, the ACC array can directly access data items from DDR4 memory by DMA. Such a direct memory access pattern is also efficient and provides high bandwidth. In this paper, we adopt the second way for plaintext transfer between DDR4 memory and AM/SM and each ACC array uses DMA to fetch data blocks as large as possible. Figure 5 shows the memory acccss on thc CPU and acceleration side.

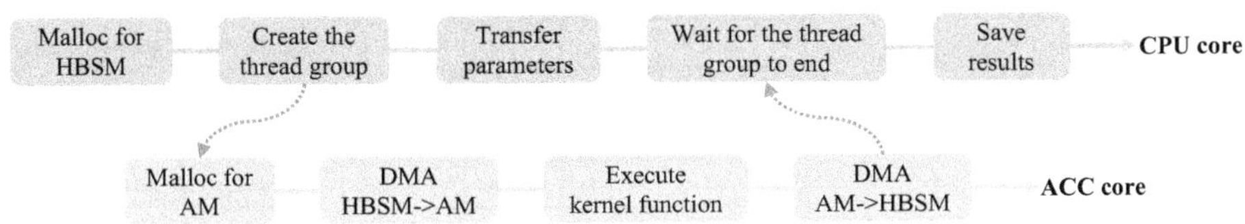

Fig. 5. Memory access on the CPU and ACC side.

3.3   Data Parallelism

In Sect. 3.1, the multi-level parallelism strategy utilizes SIMD instructions to implement multiple-data parallelism within each ACC array. However, implementing parallel ChaCha20 within a core is challenging for the following reasons. Firstly, ChaCha20 encrypts 512 bits of data at a time, while the SIMD instructions have a length of 1024 bits. Secondly, the MT-3000 compiler does not support automatic vectorization. As a result, we need to explicitly load data from AM to registers for computation, then explicitly store the data back to AM when the computation is complete, and finally transfer the data back to DDR4 through DMA. To address these challenges, we rearrange the data to achieve data parallelism within the core. The implementation details are described below.

ChaCha20 encrypts a 512-bit message in two stages. Firstly, the key stream is generated and secondly, the plaintext block performs XOR operation with the 512-bit key stream. During the 512-bit key stream generation process, the parallel acceleration cannot fully exploit the SIMD capability of the 1024-bit vector register because the 20 rounds (10 iterations) depend on the sequence. To fully utilize the parallelization capabilities of SIMD, we have presented a SIMD-friendly data layout, which simultaneously loads 32 initial matrices (32 × 512-bit) for processing.

As shown in Fig. 6, the initial matrices are independent of each other. This allows us to process the 32 initial matrices as a batch and allocate 16 1024-bit registers to ensure that the data from each batch will be stored in the same register. By rearranging the data structure in this way, we can take advantage of SIMD instructions to accelerate the generation of the key stream.

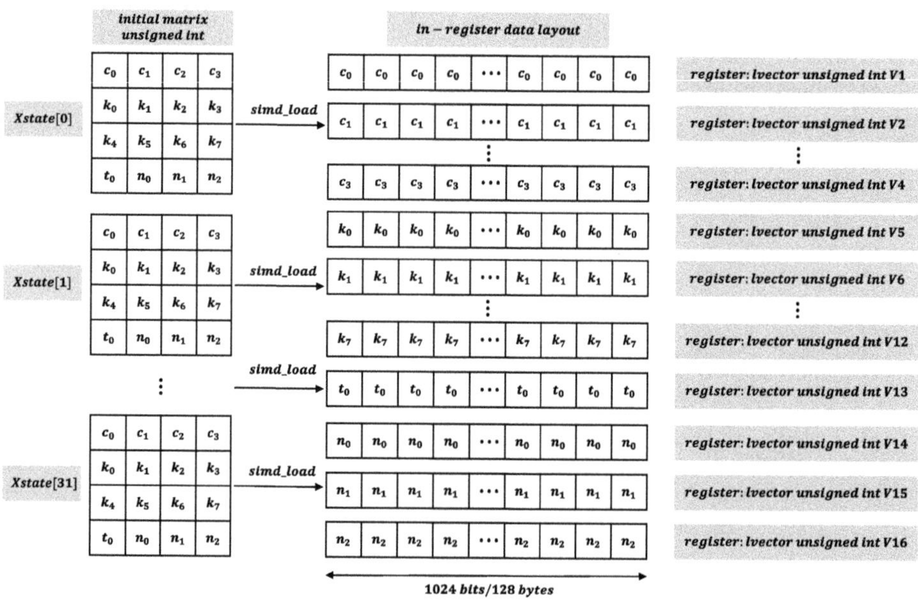

Fig. 6. Demonstration of SIMD-friendly data layout. 32 initial matrices are loaded into 16 1024-bit registers.

As shown in Algorithm **2**, we have implemented vectorization of the QR operation. The QR_SIMD follows the same logic as the previous operation in Sect. 2.1, with the exception that the operation on four 32-bit unsigned integers is now carried out on four 1024-bit vector variables.

Algorithm 2. The parallel ChaCha20 cipher algorithm with SIMD

Input: 32 initial matrices $I_m[32][16]$, 32 512-bit plaintext
Output: 32 512-bit ciphertext
Begin
lvector unsigned int a,b,c,d, VEC_X,VEC_Y
char keystream, plaintext, ciphertext
$ROTL\_SIMD^n(x) = vec\_or(vec\_shfll((n),(x)), vec\_shflr((32-n),(x)))$
#define $QR\_SIMD(a,b,c,d)$ {
   $a = vec\_add(a,b); d = vec\_xor(a,d); d = ROTL\_SIMD^{16}(d)$
   $c = vec\_add(c,d); b = vec\_xor(b,c); b = ROTL\_SIMD^{12}(b)$
   $a = vec\_add(a,b); d = vec\_xor(a,d); d = ROTL\_SIMD^{8}(d)$
   $c = vec\_add(c,d); b = vec\_xor(b,c); b = ROTL\_SIMD^{7}(b)$
}
$simd\_load\ I_m[32][16]\ into\ VEC\_X[16]$
$VEC\_Y = VEC\_X$
for $i = 20; i > 0; i = i - 2$ **do**
   $QR\_SIMD(VEC\_Y[0,4,8,12])$
   $QR\_SIMD(VEC\_Y[1,5,9,13])$
   $QR\_SIMD(VEC\_Y[2,6,10,14])$
   $QR\_SIMD(VEC\_Y[3,7,11,15])$
   $QR\_SIMD(VEC\_Y[0,5,10,15])$
   $QR\_SIMD(VEC\_Y[1,6,11,12])$
   $QR\_SIMD(VEC\_Y[2,7,8,13])$
   $QR\_SIMD(VEC\_Y[3,4,9,14])$
end for
$VEC\_X = vec\_add(VEC\_X, VEC\_Y)$
$store\ VEC\_X\ into\ I_m[32][16]$
for $i = 0; i < 32; i = i + 1$ **do**
   $u32\ to\ u8(I_m[i], keystream)$
   $ciphertext = keystream \oplus plaintext$
end for
return $ciphertext$
End

---

4 Performance Evaluation

We implement ChaCha20 with a multi-level programming model, with MPI to launch 4 processes on 4 ACC clusters of a single chip, and hThreads APIs to launch 24 thread-like to the 24 ACC arrays within one ACC cluster. Data transmission between DDR4 and AM is through DMA access. All the experimental code is implemented with the C programming language and compiled by mpi-x compiler mixed with FT-M3000 compiler. We evaluate the performance on a single MT-3000 processor, and also evaluate the scalability of the parallel version ChaCha20. The system configurations is listed in Table 1.

Table 1. System configurations

Item	Value
Processor	MT-3000 multi-zone processor
Operation system	Ubuntu 20.04.2 LTS
Instruction set	FT-M3000 Instruction set
Compilers	gcc-9.3.0, FT-M3000 compiler
MPI compiler	mpi-x (based on MPICH 3.4.2)
CPUs per node	16
Clusters per node	4
Arrays per cluster	24

4.1   Performance on a Single ACC Array

Firstly, we implement ChaCha20 with different optimization methods on a single CPU core and a single ACC array, respectively. Figure 7 shows the experimental performance of four different implementations of ChaCha20 running on MT-3000 with different data sizes (from 48 KB to 1.5 GB), including the original serial implementation running on a CPU of the GP zone, the optimized implementation running on a CPU with SIMD (NEON 128-bit instruction), the serial implementation running on an ACC array, the optimized implementation running on an ACC array with SIMD (1024-bit instruction).

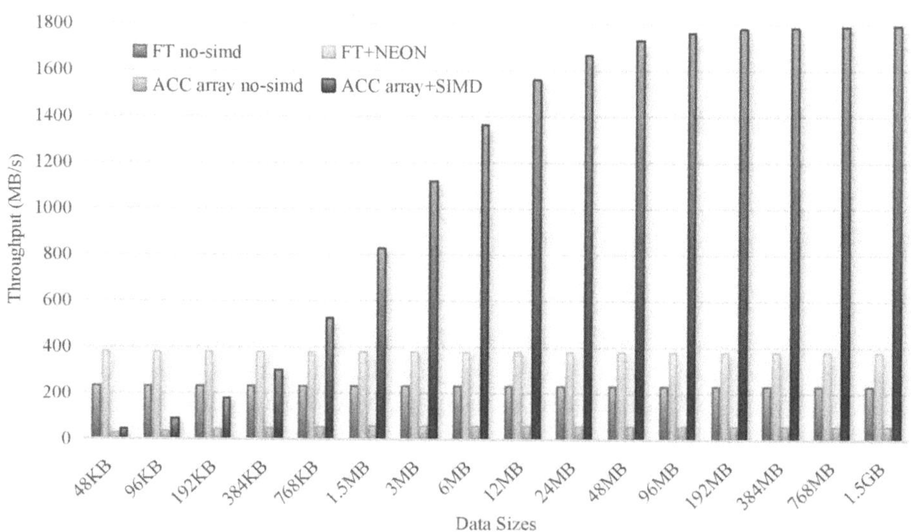

Fig. 7. Performance of ChaCha20 on a single core under different input data sizes.

Taking the serial implementation without SIMD as a baseline, it can be seen that the effect of SIMD instruction optimization is significant. Due to the architecture and frequency, the performance of a single CPU is significantly higher than a single ACC core. For 128-bit vectorization of NEON, it is theoretically possible to accelerate 32-bit calculation by 4 times, but the actual acceleration is just about 1.6 times. However, for 1024-bit vectorization of MT-3000, the actual acceleration is up to 30.6 times, which is close to the theoretical 32 times acceleration. This is due to a combination of the special architecture of the ACC array and kernel instruction optimization. For input data, more than 48 KB, the throughput of the CPU does not change with the increase of data. This is because one-time encrypted data is only 256 bytes within the NEON instruction, and the CPU computation does not require data transfer. The throughput of the ACC array becomes larger with the increase of data. The maximum computational performance of ACC array is achieved when the input data is more than 192 MB, and the throughput levels off at 1778.28 MB/s.

4.2 Performance on Multiple Clusters

In the multiple processes test, input data are generated independently in each process. This is because the network bandwidth is about 11.7 GB/s through MPI communication [28]. If we adopt the method of distributing input data equally to each process by the master process, the network bandwidth will be the main bottleneck. Consequently, the throughput of multiple parallel processes will be probably lower than that of a single process.

Before the scalability test, we firstly show the overall performance of the parallel ChaCha20 on MT-3000. Figure 8 shows the throughput of a single MT-3000 processor (4 ACC clusters) with various data sizes ranging from 192 KB to 6 GB. When the data is greater than 384 MB, the throughput increases at a slower rate. The maximum throughput is approximately 168.11 GB/s.

For weak scaling test, each ACC cluster encrypts 1.5 GB plaintext independently. Each MT-3000 processor contains 4 ACC clusters as a computing node, and a single ACC cluster can achieve 42.03 GB/s throughput. When expanded to 512 ACC clusters (128 computing nodes), the maximum throughput can reach up to 21512.52 GB/s. Taking the throughput of a single ACC cluster as a baseline, the parallel efficiency can still be 99% when scaling to 512 ACC clusters. It implies almost linear scalability. Table 2 lists the maximum throughput under different ACC clusters, and Fig. 9 shows the weak scalability of MT-ChaCha20 on multiple ACC clusters.

4.3 Comparison with Other Works

In this section, we will compare the performance of the parallel ChaCha20 algorithm on MT-3000 with some other related works on different hardware platforms. The results are shown in Table 3. Firstly, we compare our scheme with the implementation on the high performance chip SW26010, which is used on the Sunway TaihuLight Supercomputer. The ChaCha20 implemented on SW26010 achieves a throughput of 32.43 GB/s. Since SW26010 (2016 introduced) is a relatively old chip, it is not as good as MT-3000 in terms of architecture, so the performance is lower than the scheme in this paper. The newly introduced SW26010-pro (2023 introduced) has a 4x performance increase, but there is no relevant implementation yet.

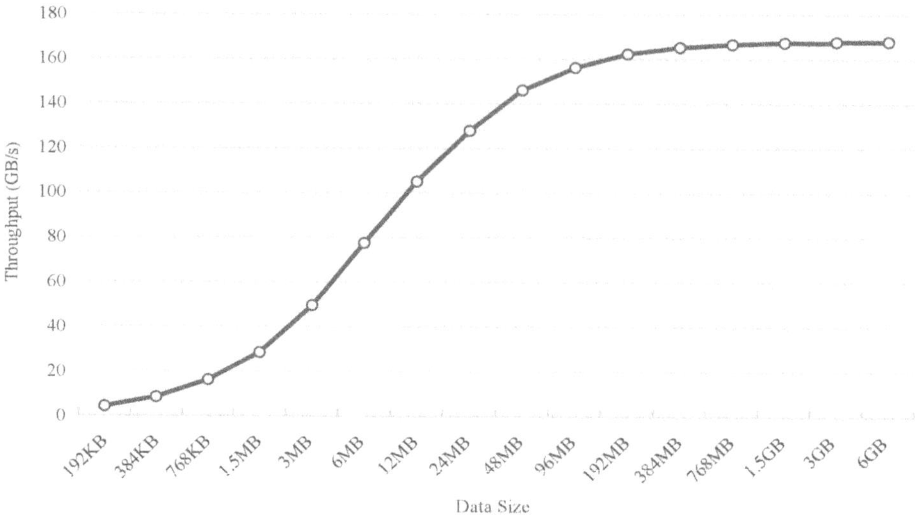

Fig. 8. Performance of ChaCha20 under different data sizes on a single MT-3000 processor.

Table 2. Throughput of parallel ChaCha20 on a single ACC cluster and multiple ACC clusters

Number of ACC clusters	Throughput (GB/s)
1	42.03
4	168.11
8	336.23
16	672.4
32	1344.89
64	2689.76
128	5379.59
256	10759.26
512	21512.52

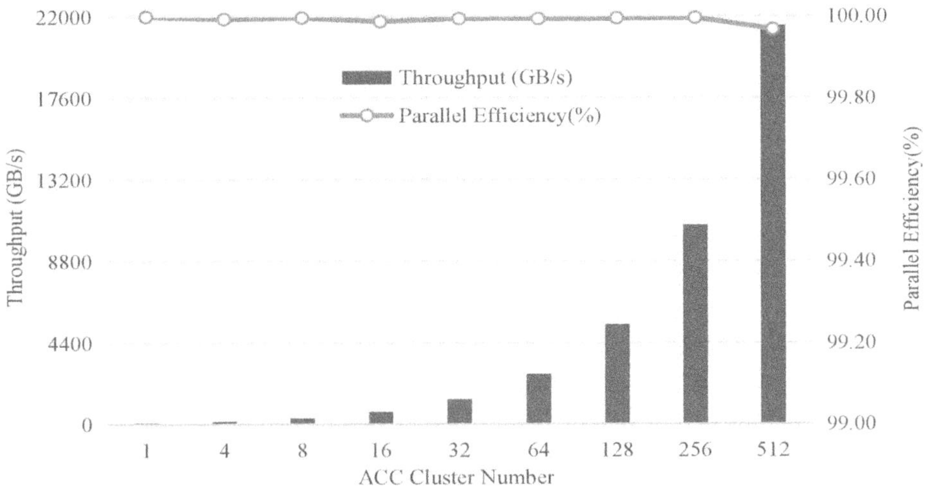

Fig. 9. Parallel weak scaling efficiency and throughput for 1.5GB of data per ACC cluster.

At present, there are many works about the implementation of ChaCha on CPU, GPU, and FPGA. The throughput of ChaCha20 on Intel Xeon Gold 5128R is 18.32 GB/s. The highest throughput of ChaCha8 on FPGA is 21.87 GB/s, but ChaCha20 performance is several times lower. In terms of GPU parallel implementation, the throughput of this paper is higher than that of the GTX-3070 and lower than that of the GTX-3090. This is because the GTX-3090 has 10496 CUDA cores with 384-bit bit-width, while the GTX-3070 only has 5888 CUDA cores with 256-bit bit-width. More cores mean more data can be computed in parallel, and larger bit widths mean higher bandwidth. Both of these affect the parallel performance of the GPU. However, the hybrid CPU/GPU throughput is only 9.85 GB/s, much lower than the GPU-only implementation. The bottleneck of the hybrid CPU/GPU parallel implementation is mainly in the bandwidth of PCIE.

Furthermore, in some scenarios where superior encryption efficiency is required, ChaCha20 is often used instead of the AES algorithm.ÂăWe also list some related works of AES. Since the algorithm of AES is more complex, the performance is lower than ChaCha20 with the same hardware, e.g. SW26010. Even on a more advanced GPU Tesla V100, the throughput of AES is 173.06 GB/s, which is lower than ChaCha20's throughput on a GTX 3090 of 211.41 GB/s. The throughput of AES on MT-3000 is not listed in Table 3 because the work has not yet been completed.

In summary, the parallel ChaCha20 scheme for data encryption on MT-3000 is highly efficient. It demonstrates the feasibility and promise of optimizing encryption algorithms on MT-3000. Since the hardware characteristics of each test platform are different and some are older models. Moreover, the ChaCha20 algorithm involves only addition and logic operations, without complex multiplication and division operations. It does not fully reflect the performance of the hardware devices. So, the maximum throughput is only a reference.

Table 3. Comparison of MT-ChaCha20 with other works on a single device

Algorithm	Work	Computing components	Throughput
ChaCha20	[29], 2016	Radeon HD 6850	16.0 GB/s
ChaCha8	[30], 2019	Virtex-7 XC7VX485T	21.87 GB/s
ChaCha20	[27], 2021	Intel Xeon 5218R	18.32 GB/s
ChaCha20	[27], 2021	Nvidia GTX-3070	119.32 GB/s
ChaCha20	[27], 2021	Nvidia GTX-3090	211.41 GB/s
ChaCha20	[27], 2021	Hybrid 5218R/3090	9.86 GB/s
ChaCha20	[17], 2022	SW26010	32.43 GB/s
AES-ECB	[31], 2019	Nvidia GTX-1080	100.65 GB/s
AES-ECB	[31], 2019	Nvidia Tesla P100	104.07 GB/s
AES-ECB	[31], 2019	Nvidia Tesla V100	173.06 GB/s
AES-ECB	[16], 2020	SW26010	13.50 GB/s
ChaCha20	This paper	MT-3000	168.11 GB/s

5   Conclusion

In this paper, we propose a parallel ChaCha20 implementation on the MT-3000 heterogeneous multi-zone processor, which is designed for next-generation exascale prototype supercomputers. Experimental results demonstrate that a single MT-3000 processor achieves a peak throughput of 168.11 GB/s. When scaled to 512 ACC clusters, MT-ChaCha20 attains a throughput of 21,512.52 GB/s while maintaining parallel efficiency above 99%.ÂăCompared to parallel implementations of ChaCha20 on other high-performance devices, including FPGA, GPU, and SW26010, our solution delivers 1.4-17× higher performance, except for the parallel implementation on the GTX-3090. Furthermore, our scheme scales efficiently across more compute nodes with near-linear scalability.

Finally, we plan to port additional encryption algorithms to the MT-3000 platform, enabling a high-efficiency, secure data transmission framework.

Acknowledgments. This work is partially supported by grants from the National Key Research and Development Program of China (2021YFB0300101), and National Natural Science Foundation of China (62032023, 42104078, 6190241).

References

1. Manyika, J., et al.: Big Data: The Next Frontier for Innovation, Competition, and Productivity. McKinsey Global Institute, San Francisco, CA, USA (2011)
2. Chen, C.L., Zhang, C.: Data-intensive applications, challenges, techniques and technologies: a survey on Big Data. Inf. Sci. **275**, 314–347 (2014)
3. Rawat, D.B., Doku, R., Garuba, M.: Cybersecurity in big data era: from securing big data to data-driven security. IEEE Trans. Serv. Comput. **14**(6), 2055–2072 (2019)
4. Bernstein, D.J.: ChaCha, a variant of Salsa20. In: Workshop Record of SASC, vol. 8, no. 1, pp. 3–5 (2008)
5. Bernstein, D.J.: The Salsa20 family of stream ciphers. In: New Stream Cipher Designs: The eSTREAM Finalists, pp. 84–97 (2008)
6. Paul, G., Maitra, S.: RC4 Stream Cipher and Its Variants, 1st edn. CRC Press, Boca Raton (2011)
7. Langley, A., Chang, W., Mavrogiannopoulos, N., Strombergson, J., Josefsson, S.: ChaCha20-Poly1305 cipher suites for transport layer security (TLS). RFC Editor, United States (2016)
8. Mclaren, P., Buchanan, W.J., Russell, G., Tan, Z.: Deriving ChaCha20 key streams from targeted memory analysis. J. Inf. Secur. Appl. **48**, 102372 (2019)
9. Lackorzynski, T., Köpsell, S., Strufe, T.: A comparative study on virtual private networks for future industrial communication systems. In: Proceeding of 15th IEEE International Workshop on Factory Communication Systems (WFCS), pp. 1–8 (2019)
10. Rescorla, E.: The transport layer security (TLS) protocol version 1.3. RFC Editor, United States (2018)
11. Baksi, A., Bhasin, S., Breier, J., Jap, D., Saha, D.: A survey on fault attacks on symmetric key cryptosystems. ACM Comput. Surv. **55**(4), 1–34 (2022)
12. Dilip Kumar, S.V., Patranabis, S., Breier, J., Mukhopadhyay, D., Bhasin, S., Chattopadhyay, A.: A practical fault attack on ARX-like ciphers with a case study on ChaCha20. In: Proceeding of Workshop on Fault Diagnosis and Tolerance in Cryptography (FDTC), pp. 33–40 (2017)
13. Dey, S., Sarkar, S.: Proving the biases of salsa and ChaCha in differential attack. Des. Codes Crypt. **88**, 1827–1856 (2020)
14. Nir, Y.: ChaCha20 and Poly1305 for IETF Protocols. RFC Editor, United States (2018)
15. Wang, Z., Chen, H., Wu, W.: Client-aware negotiation for secure and efficient data transmission. Energies **13**(21), 5777 (2020)
16. Li, L., et al.: Efficient AES implementation on Sunway TaihuLight supercomputer: a systematic approach. J. Parallel Distrib. Comput. **138**, 178–189 (2020)
17. Cai, W., Chen, H., Wang, Z., Zhang, X.: Implementation and optimization of ChaCha20 stream cipher on Sunway TaihuLight supercomputer. J. Supercomput. **78**(3), 4199–4216 (2022)
18. Wang, Z., Dong, X., Kang, Y., Chen, H.: Parallel SHA-256 on SW26010 many-core processor for hashing of multiple messages. J. Supercomput. **79**(2), 2332–2355 (2023)
19. Luo, Y., et al.: MT-office: parallel password recovery program for office on domestic heterogeneous multi-core processor. CCF Trans. High Perform. Comput. **5**(3), 231–244 (2023)

20. Luo, Y., Liu, J., Xiao, T., Gong, C.: Parallel Implementation of SHA256 on multi-zone heterogeneous systems. In: Proceeding of IEEE International Conference on Parallel & Distributed Processing with Applications (ISPA), pp. 416–422. IEEE (2023)
21. Lu, K., et al.: MT-3000: a heterogeneous multi-zone processor for HPC. CCF Trans. High Perform. Comput. 4(2), 150–164 (2022)
22. Fang, J., et al.: Programming bare-metal accelerators with heterogeneous threading models: a case study of matrix-3000. Front. Inf. Technol. Electron. Eng. 24(4), 509–520 (2023)
23. Goll, M., Gueron, S.: Vectorization on ChaCha stream cipher. In: Proceeding of 11th International Conference on Information Technology: New Generations, pp. 612–615 (2014)
24. Zinzindohoué, J.K., Bhargavan, K., Protzenko, J., Beurdouche, B.: HACL*: a verified modern cryptographic library. In: Proceeding of the ACM SIGSAC Conference on Computer and Communications Security (CCS), pp. 1789–1806 (2017)
25. Almeida, J.B., Barbosa, M., Barthe, G., Grégoire, B., Koutsos, A., Laporte, V.: The last mile: high-assurance and high-speed cryptographic implementations. In: Proceeding of the IEEE Symposium on Security and Privacy (SP), pp. 965–982 (2020)
26. Ma, K.M., Le, D.H., Pham, C.K., Hoang, T.T.: Design of an SoC based on 32-Bit RISC-V processor with low-latency lightweight cryptographic cores in FPGA. Future Internet 15(5), 186 (2023)
27. Wang, Z., Chen, H., Cai, W.: A hybrid CPU/GPU scheme for optimizing ChaCha20 stream cipher. In: Proceeding of the IEEE Intl Conf on Parallel & Distributed Processing with Applications (ISPA), pp. 1171–1178. IEEE (2021)
28. Li, R., Liu, J., Zhang, G., Gong, C., Yang, B., Liang, Y.: An efficient heterogeneous parallel algorithm of the 3D MOC for multizone heterogeneous systems. Comput. Phys. Commun. 292, 108806 (2023)
29. Velea, R., Gurzău, F., Mărgărit, L., Bica, I., Patriciu, V.V.: Performance of parallel ChaCha20 stream cipher. In: Proceeding of the IEEE 11th International Symposium on Applied Computational Intelligence and Informatics (SACI), pp. 391–396. IEEE (2016)
30. Pfau, J., Reuter, M., Harbaum, T., Hofmann, K., Becker, J.: A hardware perspective on the ChaCha ciphers: scalable Chacha8/12/20 implementations ranging from 476 slices to bitrates of 175 Gbit/s. In: Proceeding of the 32nd IEEE International System-on-Chip Conference (SOCC), pp. 294–299 (2019)
31. Hajihassani, O., Monfared, S.K., Khasteh, S.H., Gorgin, S.: Fast AES implementation: a high-throughput bitsliced approach. IEEE Trans. Parallel Distrib. Syst. 30(10), 2211–2222 (2019)

INF-DRAM: An In-Memory Prefetching DRAM Architecture

Hairui Zhu, Haitao Du, Zhongguang Xu, and Yi Kang[✉]

University of Science and Technology of China,
Hefei 230026, Anhui, People's Republic of China
ykang@ustc.edu.cn

Abstract. DRAM remains the primary main memory in CPU and GPU systems for its cost-effectiveness, density, and speed. However, prior work has shown that DRAM bandwidth often cannot reach the theoretical maximum in real-world applications, primarily due to row-buffer conflicts. This paper introduces INF-DRAM, a new DRAM architecture that converts unused internal bandwidth into accessible external bandwidth via an in-memory prefetching mechanism. By exploiting the wide internal bandwidth to preload anticipated data into dedicated prefetch buffers, INF-DRAM mitigates row-switching overhead caused by memory interference. Accessing data in the prefetch buffers is not constrained by most standard DRAM operations (e.g., activation, precharge, refresh), as these accesses bypass the physical banks. Simulation results show that INF-DRAM achieves an average 6% performance improvement (up to 21%) and a 1.2% energy efficiency gain on CPU workloads from SPEC 2006/2017, with only 0.159% area overhead. On GPUs, INF-DRAM reduces latency by 32.6% for end-to-end single-token inference on Llama 3.2-1B compared to standard HBM.

Keywords: DRAM bandwidth · Memory interference · Memory architecture

1 Introduction

In conventional CPU and GPU architectures, as the performance of compute units continues to accelerate, critical workloads such as graph processing and Large Language Model (LLM) training and inference increasingly face memory-bound bottlenecks, which is often referred to as the "memory wall" issue. As the predominant main memory technology, DRAM suffers from significantly higher latency and lower bandwidth for random accesses compared with sequential ones. This limitation comes from the fundamental DRAM access mechanism: transferring a row of data from capacitor cells to the row-buffer incurs significant latency and blocks subsequent accesses. When the requested data is not present in the row-buffer (termed row-buffer conflict), the access suffers extended service delay. In multicore systems, memory interference from different applications disrupts access locality, increases the frequency of row-buffer conflicts, and degrades effective bandwidth [1]. To mitigate this, modern DRAM architectures exploit bank-level parallelism by employing multiple independent banks to overlap activation and precharge latencies. The number of banks has increased across DRAM generations. For example, DDRx DRAM roughly doubles the bank count each generation, from 4 in DDR to 32 in DDR5 [2–6]. Although increased bank-level parallelism improves access concurrency and reduces row-buffer conflicts, adding more banks duplicates peripheral circuitry and causes substantial area overhead [7]. As a result, prior work has explored alternative approaches to reducing row-buffer conflicts without increasing bank count, which can be categorized into three classes: scheduling, mapping, and reorganizing.

Approaches based on scheduling [8,9] enhance row-buffer utilization to maximize bandwidth efficiency through request reordering. Such techniques are area-intensive, as comparisons across all requests have to be completed within one cycle in the Memory Controller (MC). With continued threads spawning from many cores per chip, memory scheduling algorithms are ineffective at solving this problem at low complexity [10]. Approaches based on mapping [11,12] allocate dedicated memory resources for threads to mitigate memory interference-induced contention. But this often limits memory resources for each thread and becomes inefficient when the number of threads is large. Reorganizing-based approaches [7,13–18] expand row-buffer capacity or enhance row-buffer granularity control by integrating extra storage/control components. However, most of them require: 1) modifications to the memory data array, introducing design complexity and potential reliability issues; 2) additional commands for controls; 3) coarse-grained static partitioning (e.g., 1/2 row, 1/8 row).

In this paper, we present an architecture of **IN**-memory pre**F**etching DRAM – INF-DRAM, which operates at atomic DRAM operation granularity. INF-DRAM enables prefetching without modifying the memory data array and without introducing additional management commands in the DRAM access sequence. Our key objective is to leverage idle bank-level bandwidth for in-memory prefetching to bypass latency-intensive DRAM core operations. The impact of memory interference is then mitigated as the prefetch in INF-DRAM preserves partial page locality.

Our key contributions are as follows:

1) We analyze LLC-filtered memory access sequences across multiple applications and characterize their specific access patterns.
2) We propose INF-DRAM with an in-memory prefetching mechanism that achieves reduced row-buffer conflict rates and improved bandwidth utilization, and we implement it with low area overhead;
3) Simulations of INF-DRAM show a significant latency reduction compared to standard HBM in GPU systems for inference on Llama 3.2-1B and a good average performance gain compared to standard DDR4 in CPU systems for SPEC 2006/2017 workloads.

2   Background

2.1   DRAM Organization

The DRAM memory system features a hierarchical architecture comprising channels, ranks, bankgroups, and banks. As shown in Fig. 1, each chip contains four

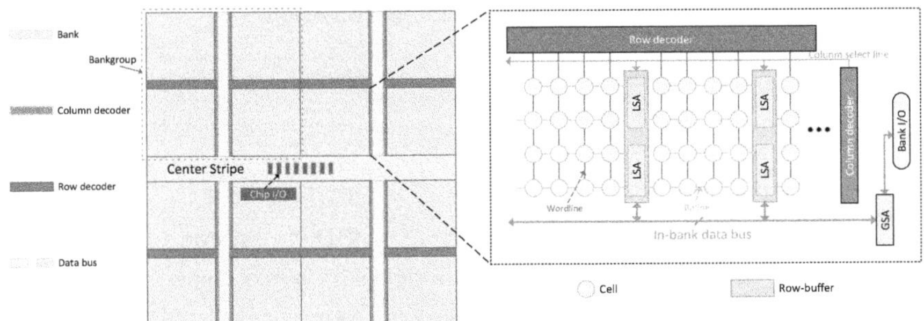

Fig. 1. Organization of a DRAM chip with 4×4 banks

bankgroups, with each bankgroup consisting of four banks. These four banks within the group share certain critical resources, including the internal bus connected to the chip I/O interface. Each bank operates as an independent memory array, representing the smallest parallelizable unit in DRAM architecture. Inside a bank, data cells are organized in a two-dimensional matrix structure with multiple rows and columns. Modern DRAM designs typically contain thousands of rows and columns within a single bank. To address the challenges of excessively large decoding structures, each bank is further partitioned into smaller sub-arrays and MATs (Memory Array Tiles), which then serve as fine-grained decoding units.

2.2 DRAM Operation

We use a read request to illustrate a typical DRAM access. It should be noted that DRAM cannot process access requests autonomously; it operates by receiving explicit commands from the MC. First, the MC issues an ACT (activate) command that includes a bank ID and row address. This activates the corresponding row decoder, which opens the wordline of a specified row. The data of the row is then transferred from the capacitor cells into the local SA (sense amplifier), i.e., the row-buffer. Next, the MC issues a RD (read) command with the bank ID and column address. The corresponding column decoder selects the data from the row-buffer and routes it through the internal bus to the chip I/O, and finally to the MC.

For a write request, the data follows the same path as in a read request, but in the opposite direction.

2.3 DRAM Timing Constrains

DRAM devices provide a set of timing parameters that restrict the intervals between commands to ensure each command is executed correctly. The timing parameters used in this article are summarized in Table 1.

Table 1. DRAM timing constrains

Symbol	Parameter	Note
tRCD	row to column delay	ACT -> RD/WR
tRAS	short for row access strobe	ACT -> PRE
tRP	row precharge	PRE -> ACT
tCL	column latency	RD -> data on I/O
tCCDS	short column to column delay	diff bankgroups
tCCDL	long column to column delay	same bankgroup

3   Motivation

In standard DRAM, each bank allows only one row to be activated at a time. The data from this row is latched in the row-buffer. Access to a bank generates different command sequences based on whether or not the target data are already in the row-buffer. These different command sequences result in different access delays for row-buffer hits, misses, or conflicts [19].

Row-buffer conflicts incur the highest access latency and disrupt DRAM's pipelined operation. During a row-buffer conflict, the row replacement process forces the impacted bank into an extended period of inaccessible status, introducing substantial pipeline bubbles. Although modern DRAM architectures employ multi-bank parallelism to hide such bubbles, row-buffer conflicts remain a major bottleneck that limits the bandwidth of many real-world DRAM devices well below their theoretical peaks. This problem becomes more severe in multithreaded systems, where interference among memory requests from multiple threads leads to a sharp increase in row-buffer conflict rates [18]. In worst-case scenarios, when multiple threads access different rows within the same bank in turn, each memory request triggers a row conflict even when individual threads exhibit good spatial locality in their memory access patterns. Consequently, conventional DRAM architectures demonstrate poor efficiency when handling such workloads.

To mitigate issues caused by memory access interference, prior research has proposed various solutions, which can be categorized into scheduling, mapping, and reorganizing. The works based on scheduling and mapping are designed through off-chip approaches without modifying DRAM architectures. Scheduling approaches optimize the service order of memory access requests to improve system throughput, by clustering [9] or ranking [20]. Works based on mapping try to separate threads in accordance by allocating memory resources in a parallel-accessible way. Different mapping methods use different memory resources such as cache banks [21], DRAM channels [12], DRAM banks [11]. However, as the processor core count scales, these approaches face increasing challenges of cost and implementation complexity.

During the last decade, reorganization-based techniques have become mainstream. The main idea is to create more row-buffers or divide the row-buffer into finer-grained units. A bitline-direction sub-bank scheme was proposed in [16], in which a physical bank is divided into multiple horizontal sub-array groups and each sub-array group is utilized as a sub-bank for interleaving. By inserting a dedicated set of row address latches to sub-banks, each sub-bank can hold an active wordline, i.e., the row-buffer. A more intuitive approach of multi row-buffers is proposed in [15], in which a certain number of independent buffers are added, and one buffer per thread or application is allocated to achieve isolation. Sub-bank in wordline direction is proposed that a wordline is divided into multiple mini wordline sections [7,14,17]. The row-buffer is divided into finer-grained sub-row-buffers so that one whole row-buffer can be allocated to different threads. Ke et al. [18] combine the more row-buffers approach with the finer-grained row-buffer approach. They add several VRBs (victim row-buffers) per bank, and also copy part of row-buffer (e.g. 1/8 row-buffer) into VRBs when the bank is precharged. However, existing reorganizing research approaches face one or more limitations, as follows: First, modifications to the DRAM data array increase design complexity, as DRAM data arrays maintain structural regularity to preserve scalability and uniform access patterns. Second, row-buffer management and switching require additional commands and introduce latency. Finally, the retained data blocks remain coarse-grained and unavoidably contain a large amount of unused data, leading to additional energy consumption.

To address the limitations of prior approaches, we propose a novel in-DRAM prefetching architecture called INF-DRAM. It mitigates memory interference through a prefetching mechanism at the granularity of DRAM atomic operation, without extra control commands or modifications to the DRAM data array. Unlike conventional off-DRAM prefetching schemes [22–24], INF-DRAM does not introduce extraneous DRAM requests that would waste DRAM I/O bandwidth and exacerbate interference in multithreaded workloads.

4   Architecture

This section illustrates the architecture of our proposed in-DRAM prefetching mechanism. While we demonstrate the design using DDRx as an example, the approach can be directly applied to LPDDRx, HBMx, and other DRAM variants because our solution does not require any modification to the core DRAM array.

4.1   Prefetch Mechanism

First, we introduce two new DRAM commands for prefetching: **RDP** (read with preparation) triggers a normal read and migrates data from an adjacent column to the prefetch buffer inside DRAM; **RDF** (read fetch) reads data directly from the prefetch buffer. Each RDP command triggers two data transfers. As shown in Fig. 2, RDP1 first moves DATA1 to the internal bus, followed by transferring FETCH1 to the prefetch buffer. Normal and prefetched data from different banks can be transferred in parallel by leveraging bank-level bandwidth. For example, while Bank1 sends DATA2 for RDP2, Bank0 simultaneously transfers FETCH1

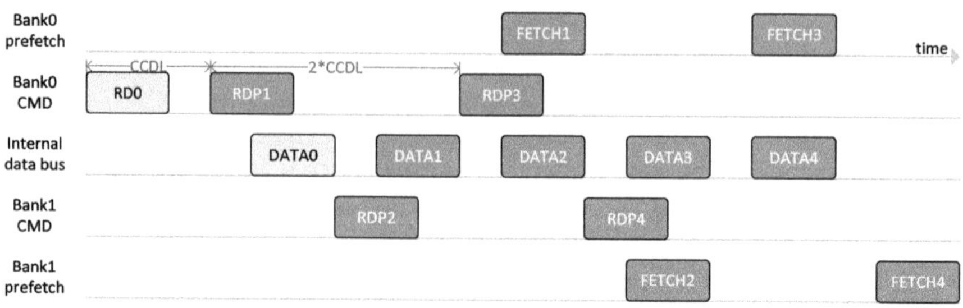

Fig. 2. Schematic diagram of performing RDP operations

for RDP1. A CAM is added to the MC to record addresses of prefetched data. In subsequent accesses, if a read request hits the CAM, the MC will send a RDF command with the corresponding index over the C/A bus to access the data directly from the prefetch buffer.

INF-DRAM handles write requests identically to standard DRAM without involving prefetched data. To maintain coherence, a dirty bit is added to each prefetch CAM entry. During write command execution, the CAM is probed and the matching entry's dirty bit is set to 1. Dirty entries are invalidated to prevent the MC from generating RDF for stale data in the prefetch buffer.

The prefetch buffer can be implemented on-chip in DRAM using one of the following approaches:

1) Approach 1 - Part of DRAM space: This approach reserves one or multiple rows in DRAM to store the prefetched data directly in adjacent rows, which is similar to some in-DRAM cache designs, such as TL-DRAM [25]. This implementation requires DRAM to support simultaneous activation of multiple rows to avoid row-buffer conflicts when accessing the prefetch data array and the regular data array alternately.

2) Approach 2 - SRAM within DRAM banks: Embedded SRAM is added to each DRAM bank to store prefetched data, which is similar to the approaches like TRB (Threads row-buffer) [15] and VRB (Victim row-buffer) [18]. However, this approach requires an independent Prefetch Manager per bank, which increases overhead. It is also ineffective to balance uneven prefetch demands across banks since low-demand banks underutilize their prefetch buffers, while high-demand banks have capacity shortage issues.

3) Approach 3 - SRAM at chip I/O: A single SRAM and management module per chip better adapts to imbalanced prefetch demands across banks by using a unified buffer. However, modern DRAM uses shared internal buses within bankgroups, which allow only one bank per group to transmit data simultaneously. Consequently, prefetch operations on any bank in a group block all other banks in that group from responding to normal read/write or prefetch commands.

4) Approach 4 - SRAM alongside bank-to-I/O data paths: This approach places SRAM along the internal bus between banks and chip I/O, with dedicated

data links between banks and their corresponding prefetch buffers. This design closely resembles [26]. This method partially mitigates uneven prefetch demands across banks by leveraging bank-level internal bandwidth. Prefetch data can be transferred to the prefetch buffer concurrently with transferring normal data over internal buses.

The physical placement of the Prefetch Manager does not fundamentally impact the effectiveness of the proposed prefetch mechanism. Since Approach 4 can be implemented with low area overhead and minor impact on normal memory accesses, while allowing placement outside the DRAM bank's data array, we adopt this approach to present our prefetch architecture in this work.

To ensure the effectiveness of prefetching, the selection of prefetch data is critical. To select the right prefetch data, we first analyze memory access sequences from selected SPEC CPU2006/2017 benchmarks. The minimum offset between a memory request and its preceding 16 requests (MinOffset) is extracted. The scope of MinOffset is intentionally set to values of 1–7, as data with larger strides are more likely to be mapped into different DRAM channels, ranks, or banks during address translation [27]. As shown in Fig. 3, the most frequent MinOffset observed across most applications is 1. For applications such as 519.lbm and 520.omnetpp, poor spatial locality or large memory access intervals often result in most MinOffset being categorized as "other". In such cases, fixed-stride prefetching is largely ineffective. As for alternative approaches, such as variable-stride prefetching, the information needed for stride prediction is filtered out by the LLC (last-level cache), which can lead to inaccurate predictions. Furthermore, such methods require additional commands to dynamically adjust prefetch strides; therefore, more overhead is needed. Consequently, we adopt a fixed stride of 1 as the address offset in INF-DRAM.

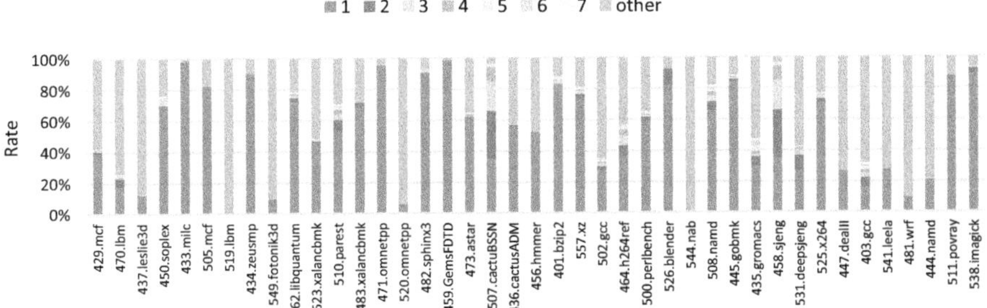

Fig. 3. Minimum offset rate between each memory request and its preceding 16 requests of SPEC 2006/2017

4.2 Hardware

The overall architecture of INF-DRAM is illustrated in Fig. 4. Compared to a standard DRAM, the key modifications are highlighted in orange-red, including:

I/O gating at bank interfaces, a Prefetch Manager containing prefetch buffers positioned adjacent to internal buses, and a Prefetch Controller integrated within the MC. Details of these modifications are described in the following.

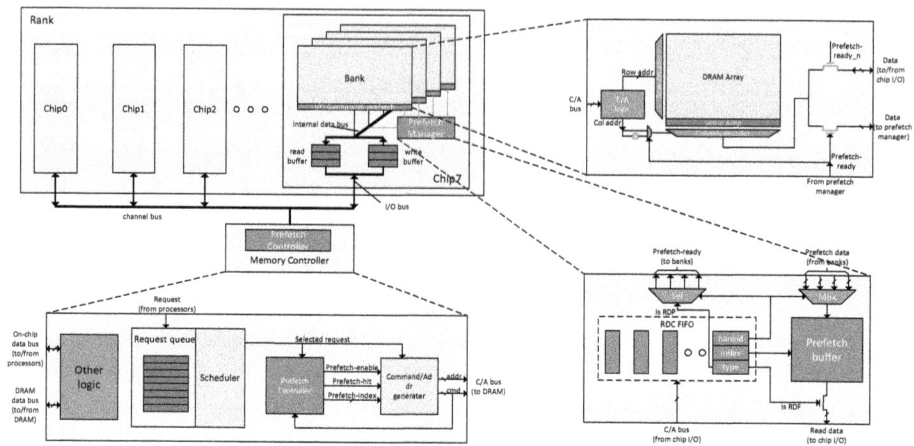

Fig. 4. Overall architecture of INF-DRAM

Bank I/O Gating. To enable prefetching, extra data gating circuits within DRAM banks are added to allow the bank to direct the data either to the internal bus or to the Prefetch Manager. The INF-DRAM bank introduces two modifications: 1) a dedicated path to the Prefetch Manager, and 2) an adder that generates column addresses for prefetching based on those used in standard DRAM read operations.

The data path is selected by the prefetch-ready signal from the Prefetch Manager: When the prefetch-ready is inactive, the column decoder uses the original column address, and data is routed normally to the internal bus; When the prefetch-ready is active, the column decoder switches to use the offset column address while redirecting data to the Prefetch Manager.

Prefetch Manager. This part primarily consists of a prefetch buffer, an RDC (read command) FIFO, and control logic for managing input/output paths. The RDC FIFO receives commands and addresses from the C/A bus, where each entry contains three fields: type, index, and bankid. When an RDP command is received, a new entry with type = RDP and bankid is enqueued to the RDC FIFO; when an RDF command is received, a new entry with type = RDF and its index is enqueued, where the index serves as an access pointer to the prefetch buffer. The Prefetch Manager does not take any other commands.

Following each RDP command, and after a tCCDL delay, the Prefetch Manager issues a prefetch-ready signal to the corresponding bank identified by the bankid, forcing the bank into prefetch data transfer mode. This tCCDL delay ensures that the prefetch operation is handled as a standard bank read operation so that it complies with the timing constraints of internal bank reads. The prefetched data from the target bank is then transferred and stored in the prefetch buffer. In the INF-DRAM implementation in this work, the prefetch buffer employs a round-robin replacement policy that prioritizes recent prefetched data for higher prefetch hit rates. This implementation is motivated by the following two factors: 1) Prefetched blocks are unlikely to be reaccessed soon, as memory requests to DRAM are filtered by the LLC, which reduces repeated access to the same block in short time windows. 2) Recently prefetched data is more likely to be accessed due to temporal locality.

For RDF operations, the prefetch buffer transfers the data block at the specified index to the chip I/O. To align the latency of RDF operations with that of standard RD operations over the internal bus, an additional delay is introduced at the buffer's output interface. This adjustment compensates for the inherently lower access latency of SRAM compared to DRAM. As a result, the ordering of data responses from RDF, RD, and RDP operations is preserved.

Prefetch Controller. Inside the MC, a Prefetch Controller is inserted between the scheduler and the C/A (command/address) generation logic. When the scheduler selects a request, the Prefetch Controller generates three signals: 1) Prefetch-enable determines whether prefetching should be applied to the request. This signal dynamically enables/disables prefetch to reduce energy overhead in cases where prefetch is ineffective. For example, the controller maintains a prefetch hit rate table. Prefetching is disabled when the hit rate falls below a threshold (e.g., 5%), and re-enabled after a fixed interval. 2) Prefetch-hit indicates whether or not the selected request hits an existing prefetched entry. The Prefetch Controller uses a CAM to track the addresses of previously issued prefetches from RDP, and performs a lookup to determine hit status. 3) Prefetch-index specifies the location in the prefetch buffer where the matched prefetched data resides, in the event of a prefetch hit.

The C/A generator then issues the command and address based on the signals provided by the Prefetch Controller and the request selected by the scheduler.

4.3   Overall Flowchart

Figure 5 shows the workflow of INF-DRAM. During each MC cycle, after the scheduler selects a memory request, the Prefetch Controller first determines whether prefetching is enabled for the request. If prefetching is not enabled, command generation and DRAM operations remain identical to those in standard DRAM. If prefetching is enabled, the Prefetch Controller then checks for prefetch-hit. If a prefetch hit occurs, the C/A generator issues an RDF command carrying the prefetch-index on the C/A bus. Upon receiving the RDF command, the Prefetch Manager transmits the corresponding data to the chip I/O. In the case of a miss, the C/A generator produces standard C/A signals, but replaces the RD command with an RDP command when applicable. For non-RDP commands, INF-DRAM responses match standard DRAM behaviors. For RDP commands, the target DRAM bank first sends the read data to the chip I/O. Afterward, the Prefetch Manager issues a prefetch-ready signal to the bank. The bank then transmits the prefetched data to the Prefetch Manager for storage in the prefetch buffer. The Prefetch Controller also updates the prefetch CAM when RDP commands are issued and sets the dirty bits on WR commands.

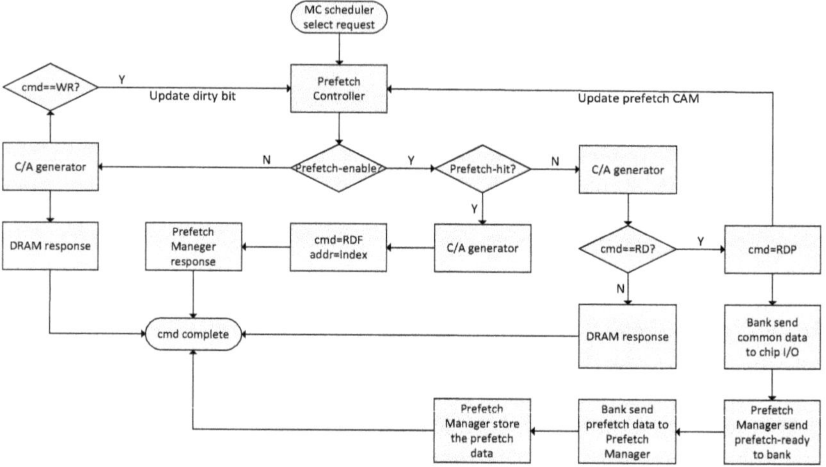

Fig. 5. Overall flowchart of INF-DRAM

5 Evaluation Methodology

We use Ramulator [28], an open-source cycle-accurate DRAM simulator, as the baseline memory simulator and add support for INF-DRAM. Since INF-DRAM introduces additional selection logic in the datapath, we model the corresponding delay using the 0.07 ns delay of an AND2X2 gate from FREEPDK45 [29]. This process technology exhibits a similar feature pitch to the 22 nm DRAM peripheral region [30,31]. This delay accounts for 11% of a DDR4-3200 clock cycle (0.625ns). To conservatively model this delay, we add one cycle to tCL. Additionally, to prevent contention between standard read and prefetch operations to the same bank, we increase the timing constraint from tCCDL to 2tCCDL for subsequent reads to the same bank after an RDP, as RDP triggers the transmission of prefetched data immediately after the standard data.

We conduct comparative experiments with VRB [18], a state-of-the-art in-DRAM technique for mitigating memory interference without modifying the memory array. VRB copies a portion of the row-buffer into an additional buffer at 1/8 row-buffer granularity before row closure. In our evaluation, VRB is idealized by assuming instantaneous row-buffer copying upon receiving a bank precharge command, while omitting additional commands for block selection and replacement.

Simulations for CPU Systems: The experimental configuration of the CPU is detailed in Table 2. We select applications from SPEC 2006 [32] and SPEC 2017 [33] to form 20 4-threaded workloads. The workloads are grouped into 5 classes based on memory usage intensity. Performance metrics are gathered during the execution interval of 100M to 200M instructions for each application in our workloads, with the first 100M instructions used for warming up. Weighted speed-up [34] is adopted as the final evaluation criterion.

Table 2. CPU Systems evaluation configuration

Processor	core num	4 out-of-order
	frequency	3.2 GHz
	inst window	128
	n-wide issue	4
Cache	cache block size	64B
	L1/L2/LLC size	32 KB/256 KB/8 MB
	latency	4/16/47
	n-way	8-ways
MC	scheduler	FRFCFS
	read/write queue size	32
	row policy	open
DRAM	standard	DDR4
	frequency	1600 MHz
	channel/rank/bankgroup/bank	1/1/4/4
	CL-RCD-RP	22-22-22
INF-DRAM	CL-RCD-RP	**23**-22-22

Simulations for GPU Systems: To evaluate the impact of INF-DRAM on memory interference under massively parallel threads, we add GPU-based inference for LLM (large-language-model) load into our experiments. Our setup employs AccelSim [35] + Ramulator with configurations based on the NVIDIA V100 GPU. The configuration is shown in Table 3. The workloads consist of multidimensional matrix operations derived from the parameter dimensions of Llama 3.2 1B/3B [36] and DeepSeek-R1 1.5B/7B [37]. To measure latency and throughput, we simulate the complete single-token inference of Llama 3.2-1B, as each token follows the same process in an LLM.

Table 3. GPU Systems evaluation configuration

GPU	model	V-100
	SM Count	80
	frequency	1132 MHz
	cache block size	32 sectored
	L2 cache size	6 MB
base DRAM	standard	HBM
	channel/rank/bankgroup/bank	32/1/4/4
	frequency	800 MHz
	scheduler	FRFCFS
	CL-RCD-RP	12-12-12
INF-DRAM	CL-RCD-RP	**13**-12-12

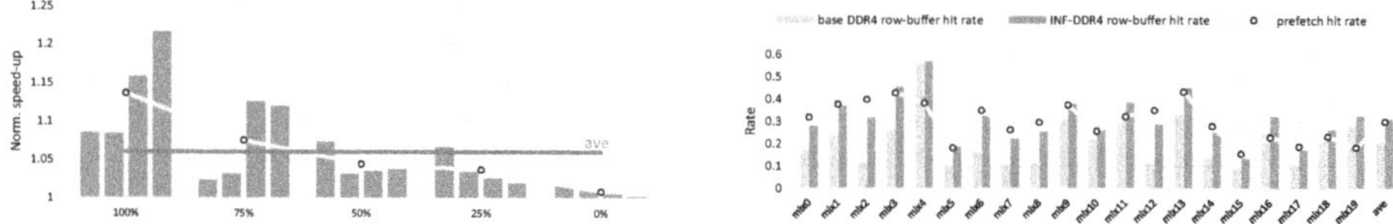

Fig. 6. Results of baseline DDR4 and INF-DDR4. Left is the performance of INF-DDR4 normalized to baseline DDR4, right is the row-buffer hit rate and prefetch hit rate

6   Evaluation Results

6.1   Performance of CPU Systems

In Fig. 6, INF-DDR4 refers to a DDR4 DRAM built with INF-DRAM architecture. As shown on the left of Fig. 6, for the 4-thread workloads selected from SPEC 2006 and SPEC 2017, INF-DDR4 achieves an average performance improvement of 6% (up to 21%) over baseline DDR4, and 13% on average for memory-intensive applications. The right of Fig. 6 presents the row-buffer hit rates of baseline DDR4 and INF-DDR4. It also shows the prefetch hit rate in INF-DDR4, where the prefetch hit rate is defined as the percentage of RDF commands among all reads. Since each RDF must pair with a preceding RDP, the upper bound of prefetch hit rate is 50%. INF-DDR4 achieves an average prefetch hit rate of approximately 30%, along with a roughly 10% improvement in row-buffer hit rate compared to baseline DDR4. For most mixes, the row-buffer hit rate enhancement exhibits a positive correlation with the prefetch hit rate. However, for workloads like mix4 and mix9, despite their prefetch hit rates approaching 40%, the row-buffer hit rate improvements are limited to only 1.2% and 6.8%, respectively, resulting in relatively low performance gains. We attribute this to lower spatial interference in these mixes. Actually, even under the baseline DDR4 configuration, row-buffer locality is already well preserved for these mixes. In such cases, a higher proportion of ACT commands serve multiple accesses, and prefetching must hit all accesses associated with an ACT command to skip the ACT.

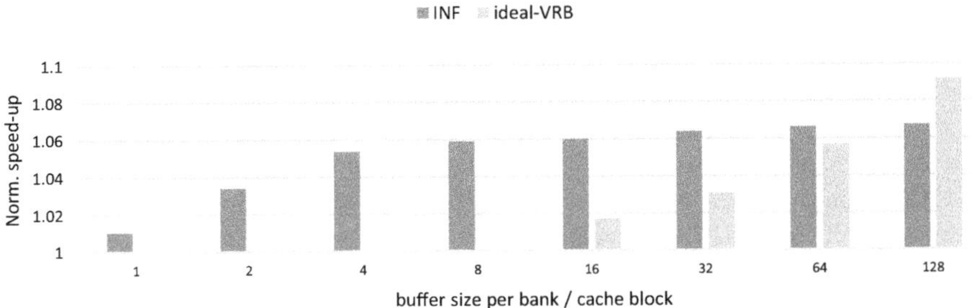

Fig. 7. Average normalized performance of INF-DDR4 and VRB-DDR4 with different buffer size

As shown in Fig. 7, INF-DDR4 outperforms 64-block VRB-DDR4 at the configuration of 8 blocks per bank, while demonstrating performance saturation with increasing buffer sizes. This phenomenon stems from closely spaced accesses to adjacent blocks in most applications, where modest buffer sizes already maximize INF benefits. Although VRB-DDR4 has a higher performance potential, its 1/8-row granularity demands greater intra-bank bandwidth for low-latency row-buffer copying, as well as more complex buffer-management mechanisms. In contrast, INF-DRAM offers lower implementation complexity and better practicality.

6.2   Performance of GPU Systems

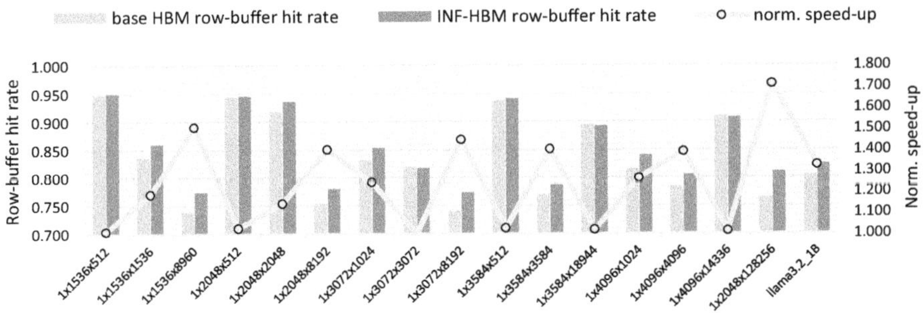

Fig. 8. Row-buffer hit rate of baseline HBM and INF-HBM, and normalized performance of INF-HBM

Figure 8 shows row-buffer hit rates of both baseline HBM and INF-HBM, and normalized performance of INF-HBM, where INF-HBM refers to an HBM device built with INF-DRAM architecture. For vector-matrix multiplications of different dimensions, INF-HBM exhibits variable performance gains. The performance gains are inversely correlated with the baseline HBM's row-buffer hit rate and positively correlated with the row-buffer hit rate enhancement of INF-HBM over the baseline HBM. Lower row-buffer hit rates in the baseline HBM, which indicate heavier memory interference, give INF-HBM greater room for optimization to reduce row-buffer conflicts and thus deliver higher speed-up. For the $1 \times 2048 \times 128256$ vector-matrix multiplication task derived from Llama 3.2-1B, INF-HBM achieves a 4.8% row-buffer hit rate improvement and a 71% performance gain. Across the complete single-token inference of the Llama 3.2-1B model, INF-HBM achieves a 32.6% overall speed-up.

Beyond the data shown in Fig. 8, from more experiments we have the following observations: 1) For matrix-matrix multiplications, INF-HBM shows almost no performance improvement over baseline HBM, as over 90% row-buffer hit rates are obtained. We attribute this to high data reuse in matrix-matrix multiplications that results in lower memory-request frequency and smaller address offset, both making the row-buffer conflicts less. In contrast, vector-matrix multiplications exhibit low data reuse and are more susceptible to row-buffer conflicts caused by memory interference. 2) All experimental results, including both vector-matrix/matrix-matrix multiplications and full Llama 3.2-1B single-token inference, demonstrate prefetch hit rates consistently approaching 50%. This means that almost all prefetched data trigger prefetch hits.

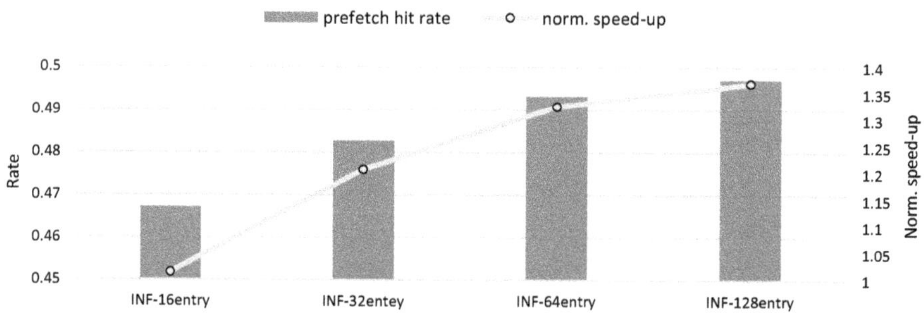

Fig. 9. The affect of prefetch buffer size per bankgroup to Llama 3.2-1B single-token inference

As shown in Fig. 9, the prefetch buffer size significantly impacts Llama 3.2-1B inference performance, but only marginally affects prefetch hit rates. The INF-16entry configuration achieves a 46.7% prefetch hit rate yet only a 1.4% performance gain, indicating that most prefetch hits overlap with row-buffer hits; therefore, memory accesses in this case have the same latency as in the no-prefetch scenarios. On the other hand, larger prefetch buffers introduce additional prefetch hits that cover data not in row-buffers, so more row-buffer conflicts are mitigated and higher performance gains are achieved. With a small buffer, prefetched data with the potential to mitigate row-buffer conflicts is frequently evicted before being requested, due to the high number of prefetched blocks per row. We therefore adopt 64 entries per bank group as the default since the benefit of increasing buffer size becomes little after the data buffer size exceeds 64 entries.

7 Discussions of Overhead

INF-DRAM incorporates additional circuitry into standard DRAM, resulting in increased area and power consumption. In this section, we quantify the overheads using an 8 Gb DDR4-x8-3200 device as the baseline.

7.1 Area

The three modified components are analyzed individually as follows.

The additional components introduced by INF-DRAM within each bank are shared across the entire bank. These additions are limited to the column logic and are much smaller than the original DRAM column decoders. According to prior work [14], column decoders occupy only $0.002/37.129$ of the total chip area—less than 0.01%. Therefore, the added column logic contributes negligible area overhead.

To evaluate the Prefetch Manager component, we synthesized it using Nangate45 [38] as the 45 nm logic process exhibits feature size similar to 22 nm DRAM peripheral region [30]. For a 64-entry configuration, the synthesized area is 24,562 μm^2. With four bank groups per chip, the total area becomes 98,248 μm^2, which corresponds to 0.159% of a 22 nm 8 Gb DDR4 chip (61.73 mm^2) [31].

On the MC side, each prefetch CAM entry includes a 7-bit column address, 16-bit row address, 4-bit bank address, 1-bit dirty bit, and a 6-bit prefetch-index. This results in 34 bits per entry. With 64×4 entries, the total CAM size is 8,704 bits (1.06 KB). Prior work shows that a 56 KB SRAM incurs only 0.02% area overhead [39]. Hence, the area overhead of the 1 KB CAM is negligible.

7.2 Power

INF-DRAM behaves identically to standard DRAM for most commands. The only exceptions are the RDP and RDF commands, which introduce additional data transfer and selection. This subsection first analyzes the energy consumption of individual RDP and RDF operations, followed by an evaluation of overall DRAM energy efficiency.

On the bank side, RDP commands fetch an additional burst of data from the memory array compared to RD operations. The primary energy overhead occurs between the local SAs and the global SAs. Prior work [17] reports that this segment accounts for approximately 43% of the energy in a single burst read. In contrast, RDF commands bypass this path, avoiding intra-bank data movement. The additional selection circuits introduced in INF-DRAM banks contribute negligible power consumption, as their scale and switching activity are lower than those of the column decoder. According to prior DRAM power breakdown studies [14], the column decoder itself contributes less than 0.1% of total energy consumption.

Table 4. Energy comparison of baseline DDR4 and INF-DDR4

State&Cmd	baseline DDR4	INF-DDR4
background-idle	44 mW	47.5 mW
background-act	68.4 mW	71.5 mW
ACT-PRE	369 pJ	369 pJ
RD	336 pJ	336 pJ
RDP	-	483 pJ
RDF	-	187 pJ
WR	297 pJ	297 pJ

Energy consumption of the Prefetch Manager is evaluated at the netlist level. In standby mode with no input activity, power consumption measures 793 µW. To quantify the per-command energy overhead of RDP and RDF operations, average power was measured under randomized data patterns over 100 consecutive command bursts. It consumed 5394 µW during RDP and 2269 µW during RDF.

Following Micron's *Calculating Memory Power for DDR4 SDRAM* methodology [40] and using parameters from Micron's *DDR4 datasheet* [41], the resulting energy breakdown is presented in Table 4. Only VDD consumption is considered, as it dominates DRAM power and INF-DRAM introduces no additional VPP overhead.

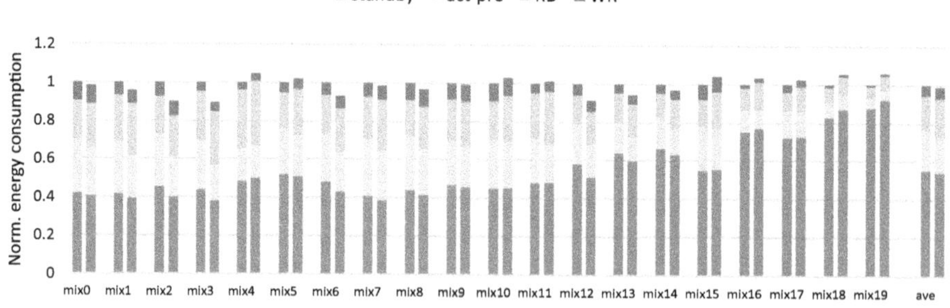

Fig. 10. Normalized energy consumption of baseline DDR4 (left) and INF-DDR4 (right)

For all 4-thread CPU experiments, command traces were collected for both baseline DDR4 and INF-DDR4 operations. These traces were used to generate DRAM energy consumption profiles, as shown in Fig. 10. The vertical axis represents normalized energy consumption per access request, relative to baseline DDR4. Results show that INF-DDR4 improves energy efficiency by 1.2% across 20 mixed workloads. For most workloads, INF-DDR4 consumes more power on read operations because of the existence of useless prefetching. However, overall energy efficiency of INF-DDR4 is improved due to higher performance and increased row-buffer hit rate. For extremely non-memory-intensive workloads (e.g., mix15–mix19), INF-DDR4 exhibits reduced energy efficiency with negligible performance improvement. This degradation stems mainly from the Prefetch Manager's standby power. The limitation can be mitigated using memory-intensive-aware power gating, which dynamically disables prefetch circuitry and its power domain under low memory-intensity conditions.

In GPU-based vector-matrix multiplication and end-to-end LLM inference, INF-HBM is expected to achieve greater energy efficiency improvements over standard HBM than INF-DDR4 over standard DDR4. This claim is supported by three observations: 1) vector-matrix multiplication and LLM inference are highly memory-intensive; 2) a prefetch hit rate near 50% ensures that most prefetched data are used efficiently, minimizing energy waste from over-prefetch; and 3) performance gains are more substantial, so that the standby power overhead is reduced.

8   Conclusion

This paper presents INF-DRAM, an in-memory prefetching architecture that enhances effective I/O bandwidth by leveraging underutilized bank-level bandwidth. INF-DRAM achieves an average 6% performance improvement on SPEC 2006/2017 benchmarks and a 32.6% speedup in GPU-based LLM inference, with only 0.159% area overhead. The core DRAM architecture and timing behavior are largely preserved, as modifications of INF-DRAM are confined to selective data-path multiplexing. Consequently, it can be easily integrated into other DRAM architectures.

References

1. Udipl, A.N., Murallmanohar, N., Chatterjee, N., Balasubramonian, R., Davis, A., Jouppi, N.P.: Rethinking dram design and organization for energy-constrained multi-cores. In: Proceedings of the 37th Annual International Symposium on Computer Architecture, pp. 175–186 (2010)
2. JEDEC: Double Data Rate (DDR) SDRAM Specification (JESD79) (2000)
3. JEDEC: DDR2 SDRAM Specification (JESD79-2) (2003)
4. JEDEC: DDR3 SDRAM Specification (JESD79-3) (2007)
5. JEDEC: DDR4 SDRAM Specification (JESD79-4) (2012)
6. JEDEC: DDR5 SDRAM Specification (JESD79-5) (2020)

7. Lym, S., Ha, H., Kwon, Y., Chang, C.K., Kim, J., Erez, M.: ERUCA: efficient dram resource utilization and resource conflict avoidance for memory system parallelism. In: 2018 IEEE International Symposium on High Performance Computer Architecture (HPCA), pp. 670–682. IEEE (2018)

8. Kim, Y., Han, D., Mutlu, O., Harchol-Balter, M.: ATLAS: a scalable and high-performance scheduling algorithm for multiple memory controllers. In: HPCA-16 2010 The Sixteenth International Symposium on High-Performance Computer Architecture, pp. 1–12. IEEE (2010)

9. Kim, Y., Papamichael, M., Mutlu, O., Harchol-Balter, M.: Thread cluster memory scheduling: exploiting differences in memory access behavior. In: 2010 43rd Annual IEEE/ACM International Symposium on Microarchitecture, pp. 65–76. IEEE (2010)

10. Ausavarungnirun, R., Chang, K.K.W., Subramanian, L., Loh, G.H., Mutlu, O.: Staged memory scheduling: achieving high performance and scalability in heterogeneous systems. ACM SIGARCH Comput. Archit. News **40**(3), 416–427 (2012)

11. Jeong, M.K., Yoon, D.H., Sunwoo, D., Sullivan, M., Lee, I., Erez, M.: Balancing dram locality and parallelism in shared memory CMP systems. In: IEEE International Symposium on High-Performance Computer Architecture, pp. 1–12. IEEE (2012)

12. Muralidhara, S.P., Subramanian, L., Mutlu, O., Kandemir, M., Moscibroda, T.: Reducing memory interference in multicore systems via application-aware memory channel partitioning. In: Proceedings of the 44th Annual IEEE/ACM International Symposium on Microarchitecture, pp. 374–385 (2011)

13. Chatterjee, N., et al.: Architecting an energy-efficient dram system for GPUs. In: 2017 IEEE International Symposium on High Performance Computer Architecture (HPCA), pp. 73–84. IEEE (2017)

14. Zhang, T., Chen, K., Xu, C., Sun, G., Wang, T., Xie, Y.: Half-DRAM: a high-bandwidth and low-power dram architecture from the rethinking of fine-grained activation. ACM SIGARCH Comput. Archit. News **42**(3), 349–360 (2014)

15. Herrero, E., Gonzalez, J., Canal, R., Tullsen, D.: Thread row buffers: improving memory performance isolation and throughput in multiprogrammed environments. IEEE Trans. Comput. **62**(9), 1879–1892 (2012)

16. Kim, Y., Seshadri, V., Lee, D., Liu, J., Mutlu, O.: A case for exploiting subarray-level parallelism (SALP) in dram. ACM SIGARCH Comput. Archit. News **40**(3), 368–379 (2012)

17. O'Connor, M., et al.: Fine-grained dram: energy-efficient dram for extreme bandwidth systems. In: Proceedings of the 50th Annual IEEE/ACM International Symposium on Microarchitecture, pp. 41–54 (2017)

18. Gao, K., Fan, D., Wu, J., Liu, Z.: Decoupling contention with victim row-buffer on multicore memory systems. In: 2015 IEEE International Parallel and Distributed Processing Symposium Workshop, pp. 454–463. IEEE (2015)

19. Jacob, B., Wang, D., Ng, S.: Memory Systems: Cache, DRAM, Disk. Morgan Kaufmann (2010)

20. Ghose, S., Lee, H., Martínez, J.F.: Improving memory scheduling via processor-side load criticality information. In: Proceedings of the 40th Annual International Symposium on Computer Architecture, pp. 84–95 (2013)

21. Lin, J., Lu, Q., Ding, X., Zhang, Z., Zhang, X., Sadayappan, P.: Gaining insights into multicore cache partitioning: Bridging the gap between simulation and real systems. In: 2008 IEEE 14th International Symposium on High Performance Computer Architecture, pp. 367–378. IEEE (2008)

22. Michaud, P.: Best-offset hardware prefetching. In: 2016 IEEE International Symposium on High Performance Computer Architecture (HPCA), pp. 469–480. IEEE (2016)
23. Lee, Y., Kim, S.: CLAP: clustered look-ahead prefetching for energy-efficient dram system. IEEE Trans. Very Large Scale Integr. (VLSI) Syst. 24(5), 1770–1782 (2015)
24. Li, M., Zhang, Q., Ren, Y., Xie, Z.: Integrating prefetcher selection with dynamic request allocation improves prefetching efficiency. In: 2025 IEEE International Symposium on High Performance Computer Architecture (HPCA), pp. 204–216 (2025)
25. Lee, D., Kim, Y., Seshadri, V., Liu, J., Subramanian, L., Mutlu, O.: Tiered-latency dram: a low latency and low cost dram architecture. In: 2013 IEEE 19th International Symposium on High Performance Computer Architecture (HPCA), pp. 615–626. IEEE (2013)
26. Hidaka, H., Matsuda, Y., Asakura, M., Fujishima, K.: The cache DRAM architecture: a dram with an on-chip cache memory. IEEE Micro 10(2), 14–25 (2002)
27. Wang, M., Zhang, Z., Cheng, Y., Nepal, S.: DRAMDig: a knowledge-assisted tool to uncover dram address mapping. In: 2020 57th ACM/IEEE Design Automation Conference (DAC), pp. 1–6. IEEE (2020)
28. Kim, Y., Yang, W., Mutlu, O.: Ramulator: a fast and extensible DRAM simulator. IEEE Comput. Archit. Lett. 15(1), 45–49 (2015)
29. Stine, J.E., et al.: FreePDK: an open-source variation-aware design kit. In: 2007 IEEE International Conference on Microelectronic Systems Education (MSE'07), pp. 173–174. IEEE (2007)
30. Sung, M., et al.: Gate-first high-K/metal gate dram technology for low power and high performance products. In: 2015 IEEE International Electron Devices Meeting (IEDM), pp. 26–6. IEEE (2015)
31. Li, S., et al.: Scope: a stochastic computing engine for dram-based in-situ accelerator. In: 2018 51st Annual IEEE/ACM International Symposium on Microarchitecture (MICRO), pp. 696–709. IEEE (2018)
32. Henning, J.L.: SPEC CPU2006 benchmark descriptions. ACM SIGARCH Comput. Archit. News 34(4), 1–17 (2006)
33. Standard Performance Evaluation Corporation: SPEC CPU 2017 (2017)
34. Snavely, A., Tullsen, D.M.: Symbiotic jobscheduling for a simultaneous multi-threaded processor. In: Proceedings of the Ninth International Conference on Architectural Support for Programming Languages and Operating Systems, pp. 234–244 (2000)
35. Khairy, M., Shen, Z., Aamodt, T.M., Rogers, T.G.: Accel-sim: an extensible simulation framework for validated GPU modeling. In: 2020 ACM/IEEE 47th Annual International Symposium on Computer Architecture (ISCA), pp. 473–486. IEEE (2020)
36. Grattafiori, A., Dubey, A., Jauhri, A., et al.: The Llama 3 herd of models (2024). https://arxiv.org/abs/2407.21783
37. DeepSeek-AI: DeepSeek-R1: incentivizing reasoning capability in LLMs via reinforcement learning (2025). https://arxiv.org/abs/2501.12948
38. Knudsen, J.: Nangate 45nm open cell library. CDNLive, EMEA, pp. 1–21 (2008)
39. Zhou, R., Ahmed, S., Roohi, A., Rakin, A.S., Angizi, S.: DRAM-Locker: a general-purpose dram protection mechanism against adversarial DNN weight attacks. In: 2024 Design, Automation & Test in Europe Conference & Exhibition (DATE), pp. 1–6. IEEE (2024)
40. Micron Technology, Inc.: Calculating Memory Power for DDR4 SDRAM (2017)
41. Micron Technology, Inc.: DDR4 SDRAM Datasheet (2015)

PQIns: Pipeline-Driven Application-Specific Instruction-Set Architecture for Hybrid Post-quantum Cryptography Acceleration

Danni Wang[1], Sibo Gong[1], Sizhao Li[1,2]([✉]), Guisheng Yin[1],
Hechang Chen[3], and Yue Cao[1,2]

[1] College of Computer Science and Technology, Harbin Engineering University,
Harbin, China
{wangdanni,gongsibo,sizhao.li,yinguisheng,cscaoyue}@hrbeu.edu.cn
[2] Key Laboratory of Information Secrecy and Protection Technology,
Ministry of Industry and Information Technology, Harbin, China
[3] School of Artificial Intelligence, Jilin University, Changchun, China
chenhc@jlu.edu.cn

Abstract. Modern Post-Quantum Cryptography (PQC) is represented by the latest standardized Kyber (ML-KEM) and Dilithium (ML-DSA) algorithms from NIST. Nevertheless, significant on-chip data movement overhead inherent in PQC hardware critically constrains computational efficiency. To address these issues, we propose an application-specific instruction set architecture (ISA) extension optimized for PQC systems termed *PQIns*. This architecture achieves enhanced performance-resource efficiency through a configurable four-stage pipeline (fetch-decode-execute-feedback) employing distinct instruction classes: configuration directives, execution operations, and flow-control commands. Moreover, this framework integrates specialized accelerators comprising an expanded 8-way parallel processing for rejection sampling and an η-adaptive parallel Centered Binomial Distribution (CBD) sampler, a high-throughput 320-bit Keccak engine, and a unified modular arithmetic units and a customized data flow finite state machine (FSM) for number theoretic transform (NTT). Finally, system-level optimizations further incorporate execution-phase reordering with pipeline staggering, alongside memory coalescing and register, file enhancements to decouple computation from bandwidth constraints. Evaluated on Xilinx KCU105 platform, *PQIns* used $15K/30K$ LUTs, $11K/18K$ FFs, $10/14$ DSPs and $14/20$ BRAMs respectively in ML-KEM/DSA, at $200\,\mathrm{MHz}$ clock frequency. Compared with the state-of-the-art other works, the performance improves $10\%(\pm4\%)$ and $5\%(\pm4\%)$. As a novel ISA extension natively supporting both ML-KEM/DSA parameter sets, this work establishes an extensible hardware-software co-design framework for lattice-based cryptosystems, bridging quantum-resistant algorithmic complexity and hardware efficiency.

Keywords: Post-quantum cryptography · Instruction extension ·
Data flow · Hybrid architecture · Multi-precision computing

1   Introduction

Quantum computing poses an existential threat to conventional public-key cryptosystems (*e.g.*, RSA, ECC) by enabling polynomial-time solutions to mathematically hard problems, specifically integer factorization and discrete logarithm problems, that currently underpin their security [20]. This vulnerability necessitates an urgent migration toward quantum-resistant cryptographic standards. In 2025, NIST standardized lattice-based primitives including Kyber (ML-KEM) for key encapsulation and Dilithium (ML-DSA) for digital signatures as quantum-resistant alternatives [15,16].

However, recent research reveals that merely simplifying computational fails to overcome hardware efficiency bottlenecks. A validated methodology employing a loosely-coupled acceleration architecture with unified instruction control to streamline workflows and improve efficiency [5]. In ASICs, it employs custom instruction sets and FSM controllers [23], while FPGA implementations utilize LUTs and BRAM for reconfigurable, security-adaptive hardware [7].

Owing to significant computational requirements, critical PQC modules such as Keccak, sampling, and NTT are commonly accelerated via instruction set extensions for enhanced parallelism [13,26] or multi-instruction cooperation [11,14]. However, the absence of heterogeneous coordination among these modules in a unified pipeline results in weak hardware-algorithm coupling, causing low resource utilization and diminished computational efficiency [6,21]. Meanwhile, although an efficient instruction architectures enhance chip performance, inadequate hardware-algorithm dataflow integration still limits PQC efficiency. In current hardware implementations targeting ML-KEM/DSA algorithms, most implementations construct parallel task scheduling frameworks to improve memory utilization [1,2,25]. Additionally, multi-sized FIFO buffers improve memory utilization and pipeline latency [3,18,19,24]. While reducing memory access time, these do not fundamentally resolve inherent data redundancy, necessitating further research into dataflow scheduling and path optimization.

Building upon foundational research, this paper introduces a pipelined instruction set architecture specifically engineered to bridge the persistent algorithmic-hardware efficiency gap in ML-KEM and ML-DSA implementations (*i.e.* **PQIns**). The architecture systematically addresses PQC computational complexity through four contributions:

- We propose a 12-bit instruction set architecture optimized for ML-KEM/DSA. Its configuration, execution, and flow-control instructions enable a four-stage pipeline for parameterized security adaptation. Hardware abstraction and controllers ensure efficiency while minimizing resource redundancy.
- To accelerate ML-KEM/DSA under instruction control, we optimized the key modules. Rejection sampling was enhanced with extended range and 8-way

parallelism. CBD flexibility was improved via extensible combinatorial design. Keccak latency was reduced through data path widening. Unified reusable arithmetic units and custom FSMs were designed for efficient NTT operation and data flow control.
- Our algorithmic analysis of ML-KEM/DSA dataflow patterns informs a customized optimization strategy. By partitioning computational workflows and strategically allocating clock cycles, achieves latency reduction. Experimental validation shows that, compared with the state-of-the-art works, our instruction set architecture reduces the clock cycles in each stage of the ML-KEM by $1.1\times$ and the ML-DSA reduces by $1.5\times$, respectively.

The paper is logically organized as follows: Sect. 2 provides a high-level overview of ML-KEM and ML-DSA, along with core modules. Section 3 presents the motivation. Section 4 outlines the central contributions of this work. Section 5 describes the dataflow optimization strategy. Section 6 evaluate performance and compares with existing works. Finally, Sect. 7 draws the main conclusions.

2   Background

2.1   ML-KEM

ML-KEM is an IND-CCA2-secure key exchange protocol, which offers parameter sets: ML-KEM-512/768/1024, corresponding to NIST security 1, 3, and 5. ML-KEM is defined over the polynomial quotient ring $R_q = \mathbb{Z}_q[x]/x^n + 1$, where $n{=}256$ and the modulus $q = 3329$ which comprises three core algorithms: *KeyGen*, *Enc*, and *Dec*. Among them, *KeyGen* samples a polynomial matrix A, an error vector e, and a secret vector s to generate a public-private key pair (pk, sk). The public key $pk = A||t$ and private key $sk = A||s||t$ are derived from the relationship $t = As + e$, where arithmetic operations are performed q. *Enc* utilizes pk to produce a random session key K and its corresponding ciphertext encapsulation c. This process involves generating a random vector r, small error vectors e_1 and polynomial e_2, followed by computing the ciphertext components $c_1 = A^T r + e_1$ and $c_2 = t^T r + e_2 + K$, which collectively form $c = c_1||c_2$. *Dec* employs sk and c to compute K through the operation $c_2 - s^T c_1$, ensuring correct session key reconstruction under bounded error conditions.

2.2   ML-DSA

ML-DSA is a lattice-based digital signature scheme. It defines three security levels: ML-DSA-44/65/87, corresponding to NIST Levels 2, 3, and 5. The scheme operates on the polynomial quotient ring R_q, where the modulus $q = 8380417$. ML-DSA comprises three core cryptographic primitives: *KeyGen*, *Sign*, and *Verify*. Among them, *KeyGen* samples a polynomial matrix A, error vectors s_1, and a secret vector s_2 to generate a public-private key pair (pk, sk), where the public key $pk = A||t$, the secrect key $sk = A||s_1||s_2||t$, and $t = As_1 + s_2$. *Sign* utilizes secret key to produce a signature σ for message m. Specifically, first generate a random polynomial y and a specific hash c of the message m, then compute $w = Ay$ and $z = y + cs_1$, and hash w and m into $\hat{c}$. Signature $\sigma = z || \hat{c}$. Finally, $Verify$ employs the public key to validate σ as an authentic signature for m. Firstly, $w' = Az - ct$ is calculated and corrected to remove small errors, and then $\hat{c}'$ is derived by hashing using m. If the signature is valid there is $\hat{c}' = \hat{c}$.

2.3   Related Modules

Rejection Sampling. Rejection sampling probabilistically screens values in lattice-based cryptography. ML-KEM uses it in $KeyGen$ to build polynomial matrices A. ML-DSA applies it during signature generation to prevent key leakage: after computing candidate signature $z = y + cs_1$, if any coefficient in z exceeds $\gamma_1 - \beta$ or low-order bits of $Az - ct$ exceed $\gamma_2 - \beta$, signing restarts. This dual-check ensures signature correctness and long-term key security.

CBD. Center Binomial Distribution (CBD) is the core sampling mechanism for ML-KEM, generating polynomial coefficients for vectors s, e, and r. Sampling B_η (where $\eta \in \{2, 3\}$) involves: collecting 2η random bits $(a_1, ..., b_\eta)$, then outputting $\sum_{i=1}^{\eta} (a_i - b_i)$ within $[-\eta, \eta]$.

Keccak. SHA-3 hash functions (SHA3-256/512, SHAKE-128/256) are core symmetric primitives in ML-KEM and ML-DSA, leveraging the Keccak-f[1600] permutation with θ, ρ, π, χ, and ι transformations via sponge construction. Hardware optimizations minimizes computational cycles and resource overhead. This achieves high-throughput execution while maintaining cross-platform adaptability, essential for efficient PQC.

NTT. Number Theoretic Transform (NTT) accelerates polynomial multiplication by operating in finite fields, reducing complexity to $O(n \log n)$. Defined over ring $R_q = \mathbb{Z}_q[x]/x^n + 1$ with $n = 2^k$ $(k \in \mathbb{Z}^+)$ and modulus $q \equiv 1 \bmod n$. NTT transforms a polynomial into point-value representation via forward NTT $\hat{a}_j = \sum_{i=0}^{n-1} a_i \omega^{ij} \bmod q$, inverse NTT (INTT) $a_i = n^{-1} \sum_{j=0}^{n-1} \hat{a}_j \omega^{-ij} \bmod q$, and computes as $\mathbf{c} = \text{INTT}(\text{NTT}(\mathbf{a}) \circ \text{NTT}(\mathbf{b}))$. In the hardware design, coefficient-wise multiplication and shared Cooley-Tukey/Gentleman-Sande butterfly operations is employed to enhance efficiency.

3   Motivation

3.1   Ineffective Heterogeneous Module Coordination

The core cryptographic primitives involved in ML-KEM and ML-DSA require different computational resources and data access modes. While instruction set extensions (*e.g.*, dedicated NTT instructions) exist for individual modules, they operate in isolation. The lack of a unified pipeline architecture to coordinate these heterogeneous modules leads to poor coupling, severe resource underutilization, and module coordination failure. Attempts at multicore parallelization merely avoid, rather than resolve, this architectural dissonance.

3.2 Persistent Data Redundancy and Rigid Scheduling

Current hardware optimizations focus on reducing memory access latency or parallel task scheduling. However, they ignore the fundamental problem of data redundancy inherent in PQC workflows. Redundant memory accesses and rigid execution sequences consume significant computation time, wasting bandwidth and energy. In addition, the significant parameter differences between ML-KEM and ML-DSA exacerbate this challenge, as most architectures lack sufficient flexibility to dynamically reconfigure data paths and control logic between these different security levels. The ensuing problem is that the computational efficiency of ML-KEM/ML-DSA is severely degraded, while mainstream solutions lack holistic data flow optimization.

4 System Architecture

This work presents a custom instruction set architecture specifically designed for ML-KEM and ML-DSA (Fig. 1), integrating a hardware-software co-design that enables instruction-level configuration to dynamically adapt to different security levels and operational modes. The hybrid architecture employs reusable functional modules like Keccak and NTT to significantly reduce area and power overhead. By unifying the execution of both algorithms, the design effectively balances performance and flexibility within a single hardware framework.

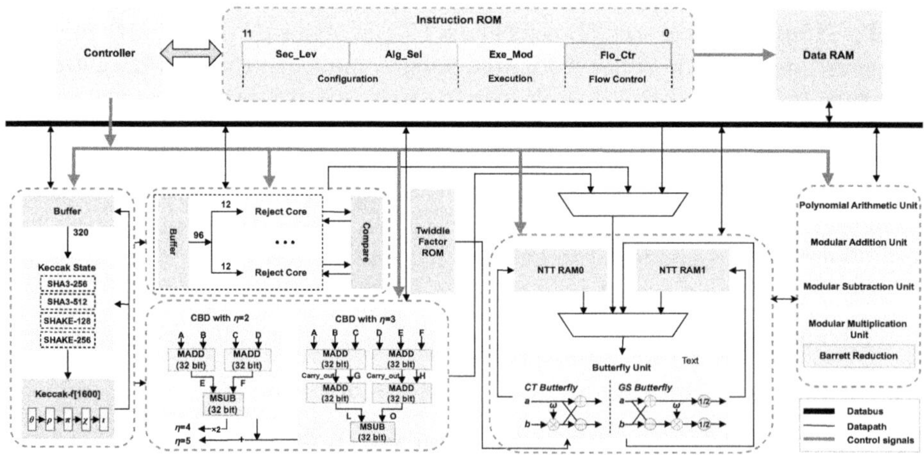

Fig. 1. Overall architecture diagram of post-quantum cryptography chip.

The **PQIns** architecture implements a hierarchical instruction set stored in a 12-bit instruction ROM, addressing macro-operation sequencing at the algorithmic level and incorporating a micro-instruction pipeline at the operational level. These instructions are structured into four components: security level designation (**Sec_Lev**), algorithm selection (**Alg_Sel**), execution mode configuration (**Exe_Mod**), and flow control management (**Flo_Ctr**).

During module reuse operations, the controller dynamically dispatches algorit-hm-specific control signals, which carry distinct configuration parameters, to dedicated processing units including Keccak, rejection sampling, CBD sampler, NTT, and polynomial arithmetic modules. Processed data outputs are then routed through a unified Databus interface to the Data RAM, enabling seamless execution of diverse algorithm operations.

4.1   Instruction-Set Definition and Coding

As illustrated in Fig. 2, this design employs adaptive encoding schemes that dynamically reconcile the competing demands of instruction density and execution parallelism, while maintaining strict timing determinism through the ROM's fixed-latency access characteristics, enabling precise coordination between instruction decoding and execution units.

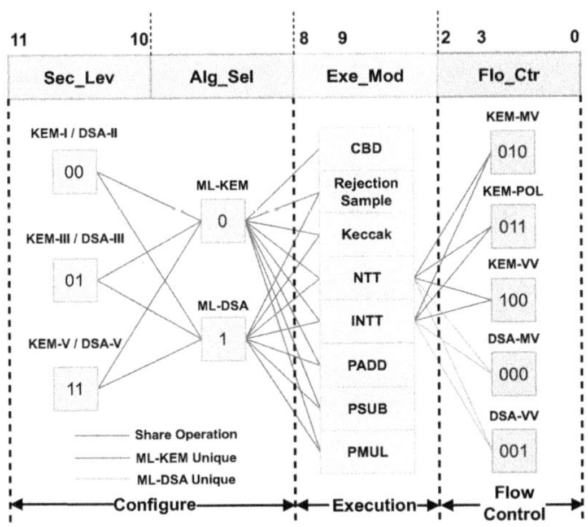

Fig. 2. Three-phase structure: configuration (parameter initialization), execution (arithmetic/logic operations), and flow control (NTT/INTT) components.

Configuration Instructions. The configuration instructions enable dynamic security level switching and algorithm selection, supporting parameter adjustments across polynomial dimensions and modulus spaces. Architecturally, these instructions occupy bits 9–11, the upper 2 bits select the security levels, while the least significant 1 bit specifies the target algorithm.

Execution Instructions. Our executing instructions directly map algorithmic workflows to hardware execution, abstracting cryptographic primitives into dedicated hardware-accelerated operations. Focused on critical computational modules shared by both standards. These instructions occupy 6 bits with upper 4 bits encoding operational functions and lower 2 bits refining operand specifications. As detailed in Table 1, while most instructions utilize two operands, Keccak operations require three operands to accommodate its wider state management needs.

Flow Control Instructions. Specialized flow control instructions manage dataf-low scheduling during polynomial multiplication operations, particularly optimizing NTT-based computations, where M, V, and POL denote matrix, vector, and polynomial data types respectively. This domain-specific approach dynamically adapts execution flow to computational requirements while significantly reducing controller scheduling overhead through operation-aware hardware sequencing.

Table 1. Execution instructions description.

Instruction	Code_1	Code_2	Description
Rej_Sample	0000	00	Rejection sampling and configure the number of reject cores
Rej_Comp	0000	01	Compare the rejected values and decide whether to loop again
CBD2	0001	00	CBD sampling of $\eta = 2$
CBD3	0001	01	CBD sampling of $\eta = 3$
Keccak_XOR3	0010	00	Rotate 1 bit to the left and perform the XOR operation
Keccak_XOR2	0010	01	Perform the XOR operation and rotate the result by x bits to the left
Keccak_REV	0010	10	Performe the left rotation and split components operate bitwise XOR
Keccak_CHI	0010	11	Execute the χ steps in Keccak
NTT	0011	00	Forwar NTT operation
INTT	0100	00	Inverse NTT operation
PADD	0101	00	Modular add operation
PSUB	0110	00	Modular subtract operation
PMUL	0111	00	Modular multiplication operation

4.2 *PQIns* Pipeline

Our ***PQIns*** architecture implements a specialized 4-stage instruction pipeline comprising: Instruction Fetch (***IF***), Decode (***ID***), Execute (***EX***), and Wait Feedback (***WF***) stages as illustrated in Fig. 3.

The ***IF*** stage employs an innovative dual-phase prefetch mechanism utilizing a true dual-port ROM (12-bit width, 512-depth), which concurrently retrieves the current instruction while prefetching subsequent instructions. This design effectively eliminating fetch-related structural hazards and reduces pipeline latency by decoupling current execution from next-instruction anticipation.

In the ***ID*** instruction fields undergo lexical decomposition into operation codes and operand specifiers, with proactive operand pre-staging into data memory buffers to mitigate access latency.

The ***EX*** stage serves as the computational core, generating time-critical control signals to activate specialized arithmetic units and implementing dedicated data/address bus arbitration to optimize operand routing. This phase requires strict signal sequencing within single-clock boundaries and efficient resource allocation to maintain throughput.

The ***WF*** monitors completion signals from execution units and processes exception, implementing a tag-based dependency tracking to resolve data hazards dynamically. For cryptographic algorithms requiring iterative computation with intermediate result dependencies, this synchronization mechanism prevents pipeline stalls by confirming operand readiness before releasing instructions.

4.3 Instruction Accelerating

Rejection Sampling. To reduce the rejection probability during parallel ML-KEM operations, we expand the rejection sampling range. Specifically, we follow the algorithmic pattern to extract the sampled data from the hash value generated by Keccak and extend the sampling range from the original interval $[0, q-1]$ to $[0, k^*q - 1]$, where k is a small positive integer, selected by a pre-experimental

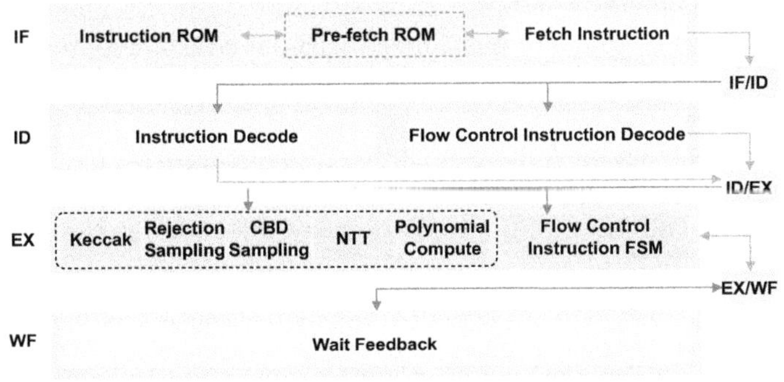

Fig. 3. 4-stage instruction pipeline for *PQIns* architecture.

traversal with the highest probability of receiving. Simultaneously, we use modular arithmetic to reduce the expanded values back to the predetermined range. This approach can be applied for any prime modulus q. It effectively lowers the rejection probability.

Based on this optimized design, our rejection sampling instruction supports eight parallel operations, which applies to both ML-KEM and ML-DSA algorithms. Furthermore, each sampling operation requires checking if values fall within the right range. To handle this, we designed the Rej_Comp instruction that triggers a jump if any sampled value exceeds this range. Specifically, it causes execution to loop back to the start of the sampling instruction sequence, as illustrated in Fig. 4. This jump-and-loop mechanism ensures that in both lattice-based cryptographic algorithms, every generated random number satisfies the required range constraints during parallel sampling.

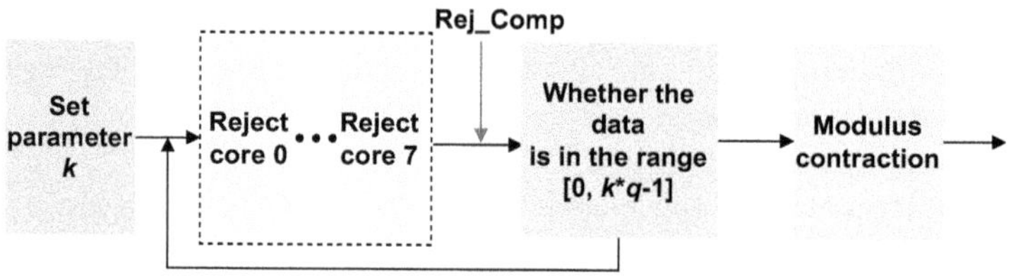

Fig. 4. Rejection sampling encoding process.

CBD. The CBD2 instruction execution begins with rejection sampling acquiring the lower 32-bit data segment, which undergoes bit-level partitioning into 32 discrete components as shown in Figure 5. Each component is zero-extended to 4 bits through the concatenation of three trailing zero bits, an optimization that eliminates the need for supplementary adder circuits in the sampling unit. The extended data is subsequently divided into four 32-bit vectors (A, B, C, D), where A is arithmetically combined with B, and C with D, generating intermediate values E and F that are guaranteed without overflow potential. The final computational stage performs $E - F$, with the resultant value stored within a 256-bit memory space.

The CBD3 instruction follows a similar processing flow to CBD2. The key difference is that CBD3 processes the lower 48 bits. This data is divided into six 32-bit blocks and then performed similarly addition and subtraction. This method enables CBD3 to support sampling across a larger range.

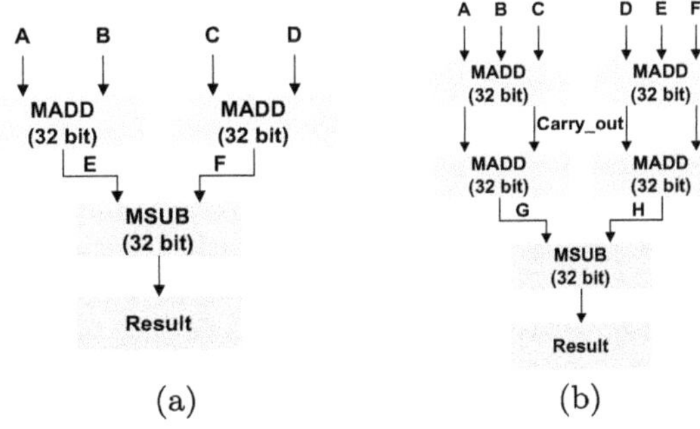

(a)   (b)

Fig. 5. CBD sampling dataflow, (a) $\eta = 2$ and (b) $\eta = 3$.

Furthermore, sampling for other η values can be achieved through a combinatorial approach using both CBD2 and CBD3 instructions. For instance, $\eta = 4$ can be implemented by combining two CBD2 instructions. This method significantly enhances the adaptability and functionality of the instruction set.

Keccak. The conventional Keccak hardware implementation requires approximately 13,000 clock cycles to complete a full cryptographic operation. To address computational latency, we extended the data path width from 256 bits to 320 bits, thereby enabling parallel processing of five 64-bit data per cycle. This datapath optimization is complemented by a novel set of application-specific instructions with the detailed in Fig. 6.

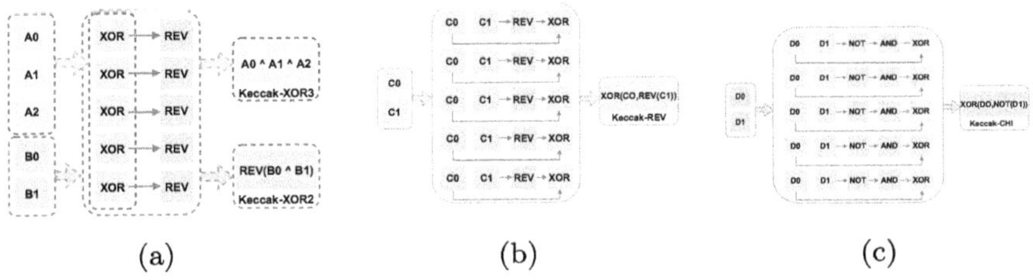

(a)   (b)   (c)

Fig. 6. Keccak instructions, (a) Keccak-XOR2 and Keccak-XOR3 operation, (b) left rotation and XOR (Keccak_REV), and (c) executing with Keccak_CHI.

The implemented Keccak instruction set employs a three-operand format. Functionally categorized into four types: $Keccak\_XOR3$ rotates operands $A0$ and $A1$ left by 1 bit before XOR with $A2$, accelerating the θ-step computation. $Keccak\_XOR2$ performs vectorized XOR operations between $B0$ and $B1$ followed by configurable rotations, targeting the ρ-step. Simultaneously, $Keccak\_REV$ processes dual operands $(C0, C1)$ through lane segmentation, rotation, and bitwise XOR, effectively consolidating the θ and ρ steps, where

$C1$ encodes θ and rotation parameters embed ρ offsets. Finally, $Keccak\_CHI$ completes the computation pipeline by accepting partitioned data from $D0$ and $D1$ and applying the non-linear χ-step transformation.

Modular Arithmetic for NTT. The modular arithmetic instructions include modular addition (PAdd), modular subtraction (PSub), and modular multiplication (PMul) (Fig. 7).

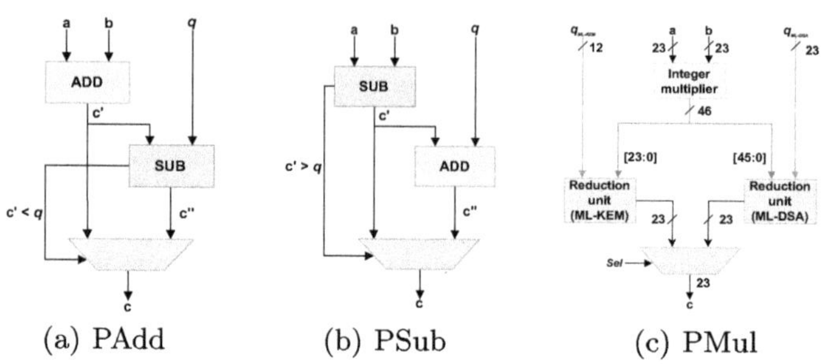

Fig. 7. Modular arithmetic computing units.

PAdd and PSub can share hardware via functional reuse, but performing a butterfly operation in single cycle requires executing both operations simultaneously. This necessitates dedicated PAdd and PSub instructions for parallel computation. Specifically, the operation corresponding to the PAdd instruction is shown in Fig. 7(a). Operands are input into an unsigned integer adder, producing sum $c' = a + b$. Next, an unsigned subtraction operation is performed to obtain $c'' = c' - q$. If $c' < q$, then $c = c'$. Otherwise, $c = c''$. Since unsigned arithmetic is used, this conditional logic preserves the relevant data bits excluding the highest bit. For PSub, as depicted in Fig. 7(b), similarly computes $c' = a - b$. Then, an unsigned addition operation yields $c'' = c' + q$. If $c' < q$, then $c = c''$. Otherwise, $c = c'$. Moreover, both PAdd and PSub instructions discards the highest bit to guarantee the result c remains within the range $[0, q - 1]$.

The PMul instruction executes modular multiplication in two stages. First, a shared 23×23-bit multiplier computes a×b, producing a 46-bit result that supports both ML-DSA (full precision) and ML-KEM (23-bit output), avoiding redundant multiplier hardware. Second, Barrett reduction replaces division-instensive operations with shift-add sequences. As Fig. 7(c) illustrated, a unified multiplexer selects the algorithm-specific result, minimizing area overhead while meeting their distinct computational demands.

Flow Control for NTT. The flow control instructions are used in polynomial-vector computation (POL), matrix (M) - vector (V) multiplication and vector-vector multiplication. ML-KEM controls state transitions using three instructions: KEM_POL, KEM-MV, and KEM-VV. ML-DSA utilizes only two types: DSA-MV and DSA-VV.

Figure 8 and Fig. 9 show the FSMs for ML-KEM and ML-DSA, respectively. Taking a typical operation $t = As_1 + s_2$ in ML-DSA. Firstly, s_2 is written into NTT-RAM0. Then, to compute As_1, the elements $A_{[0][i]}$ and $s_{1[i]}$ are transferred to NTT-RAM0 and NTT-RAM1. Their corresponding state transitions are IDLE → WRITE0 → IDLE and IDLE → WRITE1 → NTT1 → PWM → INTT0 → MACC → IDLE. MACC is used to perform polynomial addition and write back to RAM0. Once loading completed, the product of A and s_1 can be computed. The product process follows the state sequence: IDLE → WRITE0 → PWM → INTT0 → MACC → IDLE. Finally, the computed result is written back into NTT-RAM0.

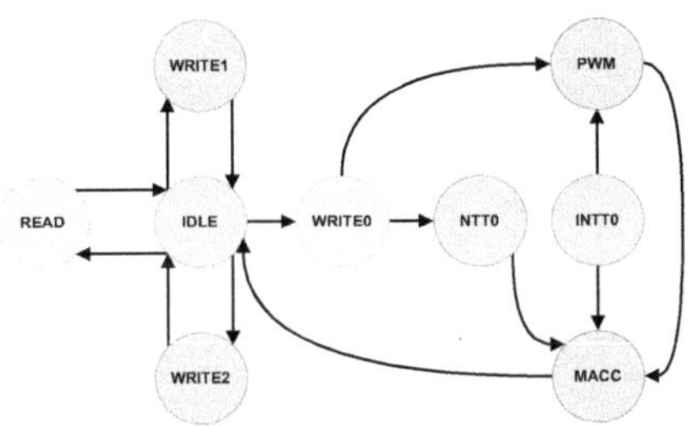

Fig. 8. ML-KEM states machine.

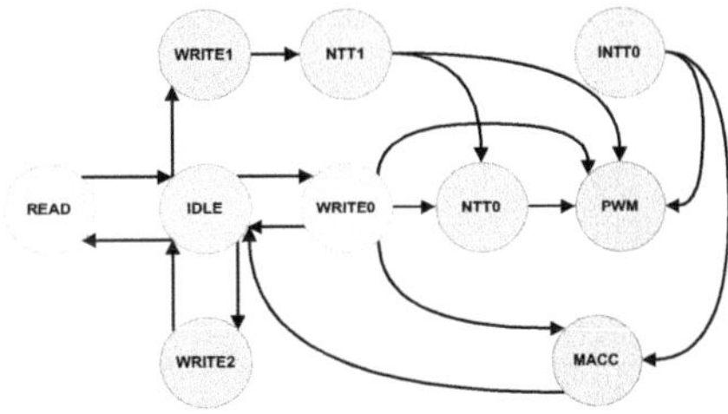

Fig. 9. ML-DSA states machine.

5 Data Flow Optimisation

In hardware implementations, optimizing data flow is critical for enhancing system performance. This work conducts algorithmic-level analysis and optimization of data flow for essential stages of ML-KEM and ML-DSA.

5.1 ML-KEM

We first analyze *KenGen* stage. The modulus q governs the construction of matrix A. For noise polynomials(s and e), the noise vector converts to the NTT domain before matrix initiates. During PWM, polynomial vector from A and corresponding vector feed into the multiplier(Fig. 10(a)). Functional modules align vertically, while the horizontal axis represents clock cycles. Box lengths indicate execution cycle ranges rather than precise temporal proportions.

For *Enc* stage, two polynomial multiplications and additions are required to obtain u and v which are encoded, compressed, and concatenated to form the ciphertext c. As Fig. 10(c) shows, the noise polynomials r, e_1, and e_2 are generated while the public key pk is decoded to obtain the polynomial $\hat{t}$. Once r sampling is complete, it undergoes NTT to yield $\hat{r}$ which undergoes PWM with $\hat{t}$. The result is written back into the storage space occupied by $\hat{t}_{[0]}$. Similarly, the result between A and r is written back into r's space. e_1 and e_2 are temporarily stored in buffer. After PWM, INTT transforms the result. Then output data are added vector-wise to e_2. The synthesized data stream flows into buffer where the polynomial addition is performed with the message M. After the corresponding bits are added in sequence, the data stream is fed into the compression module and encoded into c. The ciphertext output order is c_2 first, then c_1.

The critical data flow for Dec comes from decrypting and decompressing sk and c (Fig. 10(b)). First, c is decoded to extract the polynomial, which is used for NTT. Then sk is decoded to obtain $\hat{s}$ and added to each element of $\hat{u}$. $\hat{s} \circ \hat{u}$ is continued to perform INTT with results buffered temporarily. After v is decompressed, its polynomial coefficients are padded and stored similarly. Once buffers are full, reading data into the module addition unit, and the results are sequentially input $Compress$ to get the decrypted message m. s and e are written to the dedicated storage units, and A is stored temporarily.

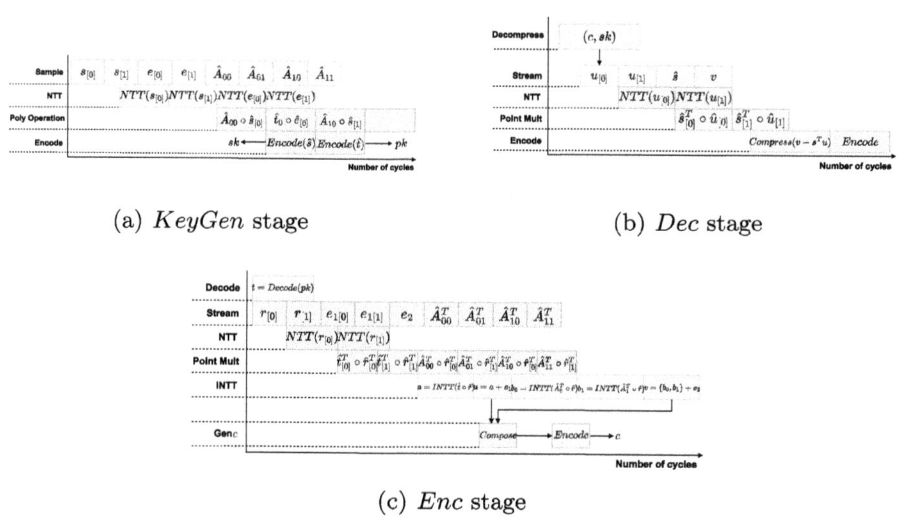

(a) KeyGen stage   (b) Dec stage

(c) Enc stage

Fig. 10. Dataflow optimization for ML-KEM.

5.2 ML-DSA

Figure 11(a) illustrates the data flow design for $KenGen$ stage. The Keccak operation unit is active throughout the whole cycle. The vectors s_1, s_2, and matrix A employ sequential generation mechanism. This timing strategy significantly reduce the pipeline pause probability. This phase primarily involves the storage and retrieval operations for s_1 and s_2, as well as the storage processes for t_0 and t_1. The intermediate result f transmitted to the pre-stage buffer of the Power2Round module where undergoes modular addition and decomposition operations with s_2.

In the ML-DSA algorithm, *Sign* stage demonstrates the highest computational complexity and time consumption, with its primary bottleneck lying in the rejection sampling loop. To address this issue, we analysis the *MakeHint* algorithm in the baseline and propose our approach that only requires two conditional checks on h to determine the corresponding bit. The function $MakeHint(r, w_1, a)$ outputs 1 or 0, indicating the result.

The rejection sampling loop divides two phases. After optimizing the *MakeHint* procedure, the first phase (Fig. 11(b)) primarily generates the challenge polynomial c and vector polynomials w_1, w_2, which form the basis for rejection condition evaluation. The second phase (Fig. 11(c)) involves the generation of z and h, along with signature validity verification. If any parameter fails to meet the predefined criteria, the system immediately terminates and reverts to the first phase for regenerating masking vectors.

Finally, *Verify* stage consists of a triple-AND logic decision. As shown in Fig. 11(d), it first checks the infinity norm of z while counting the set bits in h. If conditions are satisfied, Keccak and NTT initiated synchronously. The value μ is generated using the private key tr, reducing computation cycles. Then, using pre-computed z, c and t_0 by NTT, the calculation of $\hat{A}\hat{z} - ct_1 \cdot 2^d$ proceeds as follows: intermediate result $\hat{f}$ (representing $\hat{A}\hat{z}$) is stored; $\hat{c}$ and t_1 are loaded in parallel into the modular multiplier. The output from this unit is added to $\hat{f}$ to form polynomial $\hat{g}$ in one clock cycle without extra storage. After cyclic modular operations, INTT is applied to obtain $\hat{g}$, which enters $UseHint$ to reconstruct $w_1{}'$. Finally, μ participates in a hash loop to produce $\hat{c}'$, which is compared with the input signature to complete verification. Notably, only accesses for $\hat{A}\hat{z}$ and ct_1 require memory. All other operations exist purely within the data flow.

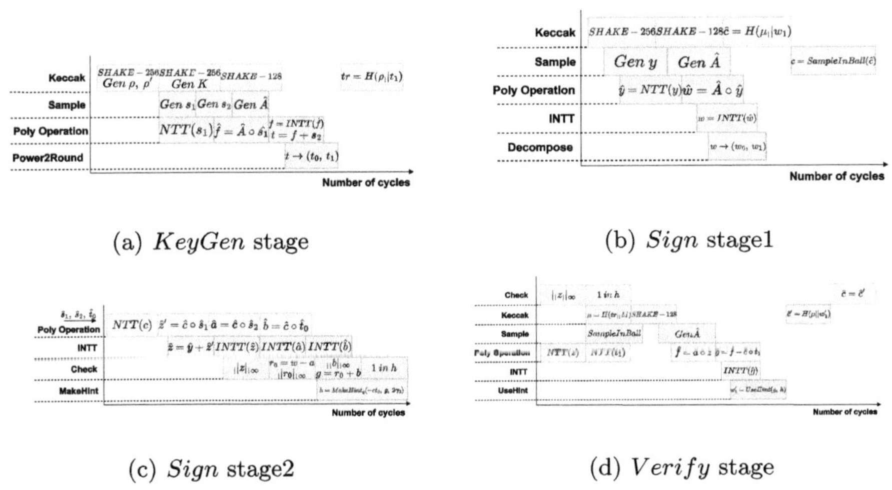

(a) *KeyGen* stage   (b) *Sign* stage1

(c) *Sign* stage2   (d) *Verify* stage

Fig. 11. Dataflow optimization for ML-DSA.

6   Evaluations

6.1   Experimental Setup

The proposed design of ML-KEM and ML-DSA have been simulated, synthesized and implemented with Vivado 2019.2 design suite on Xilinx KCU105 kit, where the FPGA chip is xcku040-ffva1156-2-e with Verilog HDL. All of building blocks implemented in hardware. We compare the resource consumption with some state -of-the-art designs to demonstrate the advantages of our design.

6.2   Resource Proportion

The overall resource consumption of this design is shown in Fig. 12. It utilizes 53K/242K LUTs, 32K/484K FFs, 24/1920 DSPs, and 29/600 BRAMs, which accounted for 21.9%, 6.6%, 1.2% and 4.8% of individual resources. Additionally, for the instruction set control unit, ML-KEM/DSA require only 1 BRAM as instruction storage. For ML-KEM, LUT usage 15K, FF usage 11K, DSP usage 10 and BRAM usage 14. However, ML-DSA consumes more resources due to the larger volume of algorithmic parameters. Specifically, 30K LUTs, 18K FFs, 14 DSPs, and 20 BRAMs are used, with shares of 56.6%, 56.2%, 58.3%, and 68.0% of the various resources used, respectively. Analysis reveals the combined BRAM usage of ML-DSA and ML-KEM (34) exceeds the total number of BRAM used overall acount (29). This is due to ML-KEM and ML-DSA share certain BRAMs.

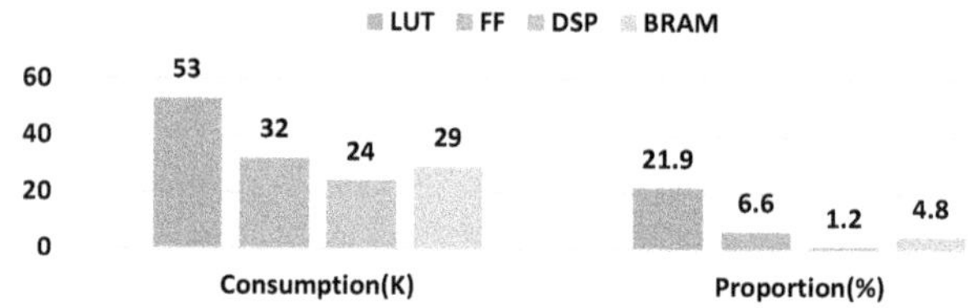

Fig. 12. Overall resource consumption.

6.3   Performance Results

Figure 13 present the execution performance metrics (clock cycle counts and latency (in μs) for the functional modules of the ML-KEM and ML-DSA, respectively, for our design on Xilinx KCU105 platform running with 200 MHz system clock frequency. The quantitative analysis results show that the execution time of the core operations of the ML-KEM-512 algorithm is $14.82\mu s$ for key generation, $21.34\mu s$ for encryption, and $29.82\mu s$ for decryption. Similarly, the execution time of the core operations of the ML-DSA-44 algorithm is $48.74\mu s$ for key generation, $112.9\mu s$ for signature generation, and $48.62\mu s$ for signature verification, respectively.

Notably, the signature generation stage of ML-DSA-44 exhibits higher latency compared to its key generation and verification stages, attributable to its inherent computational complexity. Conversely, the decryption stage of ML-KEM-512 requires slightly more time than its encryption stage. In order to systematically evaluate the performance advantages of our design, a comparative performance analysis against related work will be conducted.

6.4   Comparison with Related Works

Comparison with Instruction-Set Designs in Literature. Since instructions play a holistic regulatory role in the application of algorithms, the quantitative results alone are usually ambiguous. Consequently, this section analyses and compares the instruction design approaches available in the related literature.

In [4], a flexible ML-KEM accelerator uses a custom instruction set with 20 high-level codes, each comprising a 5-bit opcode and two 10-bit operand addresses. However, this paper specifically addresses hash operations. Meanwhile, [1] introduces a unified ML-KEM/ML-DSA architecture with a dual-issue instruction set that enables parallel execution, decoding instructions via decimal opcodes to activate crypto modules. The work in [13], focuses on efficient polynomial arithmetic, employing 7-bit instructions that encode security level, algorithm, opcode, and mode control to maximize hardware reuse across both algorithms. In [7], instructions are designed 38-bit control words and seven 9-bit pointers; synchronized internal/external sets enhance parallelism, though latency-driven scheduling may lead to pipeline imbalance and underutilization.

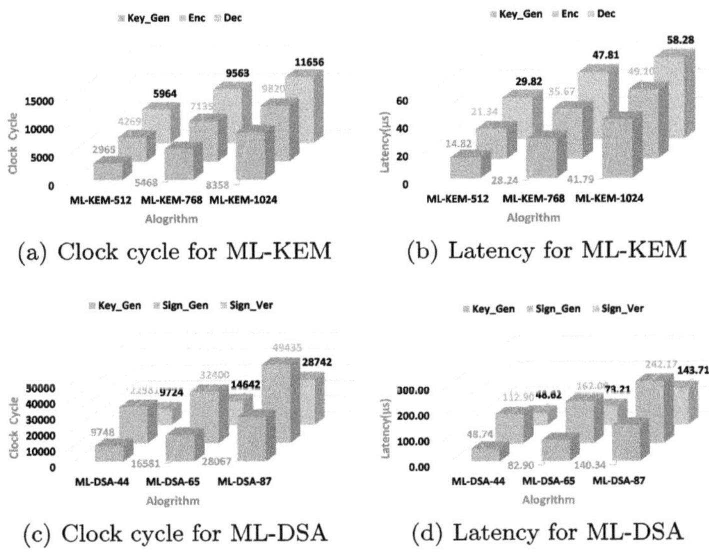

(a) Clock cycle for ML-KEM   (b) Latency for ML-KEM

(c) Clock cycle for ML-DSA   (d) Latency for ML-DSA

Fig. 13. ML-KEM and ML-DSA execution performance.

In contrast, as shown in Table 2, our work demonstrates advantages in algorithm compatibility, instruction efficiency, and performance innovation compared to related work. First, our design overcomes the limitations of single-algorithm implementations such as [12] (ML-DSA-only) and [7] (ML-KEM-only), achieving efficient dual-algorithm support. Second, our 12-bit instruction achieves greater conciseness relative to [7, 12] and extensibility than [13]. Finally, we innovatively propose a 4-stage instruction pipeline that eliminates clock-cycle waste caused by sequential execution or module-switching delays.

Table 2. Comparison table for instruction design.

Work	ML-KEM	ML-DSA	Platform	Instruction bit	Pipeline Design for Instructions
[4]	✔	✘	Artix-7	$5^B + 2 \times 10^B$	✘
[1]	✔	✔	Ultrascale+	16°	✘
[13]	✔	✔	UltraScale+ and Artix-7	7^B	✘
[12]	✘	✔	UltraScale+	32^B	✘
[7]	✔	✘	Artix-7	$38^B + 7 \times 9^B$	✘
PQIns	✔	✔	UltraScale+	12^B	✔

^{B} Number of instruction bits in binary.

^{O} Number of instruction codes in decimal.

Comparison with ML-KEM and ML-DSA Designs in Literature. When implementing the proposed architecture in hardware, we selected Ultrascale+ as the target platform and operated frequency of approximately 200 MHz. Some other works in the literature were optimized for Artix-7 or others. While the resource consumption typically remains relatively stable across different FPGA technologies. Our design is evaluated against a selection of representative recent works, particularly those employing instruction control and pipeline design, to facilitate a direct and meaningful comparison. Furthermore, to assess the enhancement brought by our instruction set and pipeline design, we measure the overall performance of the algorithms.

In Table 3, we present FPGA implementation results for ML-KEM. [10] proposed an FPGA-based, fully dedicated hardware design for ML-KEM, employing a modular design methodology to implement the encryption process on a single FPGA. However, it is evident that this design consumes significantly more resources; our design demonstrates superior performance across various aspects. [4] introduced an instruction-based FPGA acceleration scheme for ML-KEM. They merged preprocessing into the NTT algorithm to enhance transformation throughput and mitigated absorption latency by parallelizing the Keccak core operations. The pure hardware implementation of ML-KEM presented in [24] maximized hardware resource utilization through optimizations in sampling and

NTT-related computations. They also devised a strategy to minimize memory usage by employing shared memory blocks, thereby reducing memory requirements. [17] proposed a compact hash module for padding, hashing, and data storage, alongside an NMI-NTT architecture. Furthermore, they optimized the dataflow between modules to improve parallelism and reduce execution time. The lower operating frequencies reported in [4,24], and [17] can be attributed to their use of Artix-7 FPGAs, which represent a lower-tier technology compared to our UltraScale+ platform. Consequently, their designs exhibit notably inferior performance relative to ours.

[7] operates at 1.6× the clock frequency but delivers approximately 1.3× lower performance compared to our implementation. [8] incorporates k double butterfly units. Its pipelined design enables the control circuitry to skip pipeline flush cycles whenever possible, achieving low resource consumption alongside high performance at elevated frequencies. In [18], FIFO buffers are employed for NTT computation and pipeline staging to balance different pipeline phases, resulting in 0 BRAM usage reported in their resource utilization. [9] implements multiple polynomial parallelism via double butterflies algorithm and a high-speed pipelined architecture. It can be seen that in resource consumption is about half of outs, it requires slightly higher clock cycles.

In Table 4, we present FPGA implementation results for ML-DSA from the literature. Compared to [25], the clock frequency we used is 2.1×. Although our design is lower in frequency than [3], it reduces in resource usage by 1.8×, 1.6×, 1.1× and 1.5×, respectively. [22]'s hardware acceleration scheme for ML-DSA designs a conflict-free memory mapping and fast modulo multiplication scheme for polynomial multiplication, and synergistic multifunctional units for hashing and sampling. The increased complexity of the design results in a better performance of the scheme at frequencies lower than 1.3× of this design. But still it is higher than the present design by more than 2× in latency. A unified cryptoprocessor integrating ML-DSA and Saber is proposed in [2], which although it has the same clock frequency as the one we use, we outperform it by about 2.3× with a lower number of clock cycles and higher performance than that work. Aikata et al. [1] designed a unified ML-DSA and ML-KEM cryptochip, not only proposing a new polynomial computation unit, but also designing an efficient and reusable Keccak. However, our design not only supports both ML-KEM and ML-DSA algorithms as well, but also proposes a dedicated instruction-set architecture and analyses the data flow at the algorithmic level for overall optimization. Although this work operates at a higher frequency than this design, the performance is reduced by a factor of about 1.6× compared to it.

Table 3. Comparison table for ML-KEM implementation.

Security level	Work	Platform	Resources LUT/FF/ DSP/BRAM	Frequency (MHz)	Cycles(kCCs) KeyGen/ Enc/Dec	Performance(μs)
1	[10]	Aritix-7	85K/-/204/202	155	-/49.1/68.9	–
	[4]	Artix-7	18K/5K/6/15	115	4/7/10	148.0
	[24]	Artix-7	8K/5K/2/3	161	3.8/5.1/6.7	95.2
	[8]	US$^+$	10K/9K/4/6	450	2.1/3.3/4.5	21.9
	[9]	Artix-7	7K/4K/2/3	200	3.0/4.0/5.3	66.5
	[7]	US$^+$	9K/7K/4/6	311	2.5/3.7/4.6	34.7
	[17]	Artix-7	6K/4K/2/4	185	3.3/4.5/6.1	75.1
	PQIns	**US**	**15K/11K/10/1**	**200**	**3.0/4.3/6.0**	**66.0**
3	[10]	Aritix-7	104K/-/164/202	155	-/77.5/102.2	–
	[4]	Artix-7	16K/6K/9/16	115	7/10/14	209.0
	[24]	Artix-7	8K/5K/2/3	161	6.3/7.9/10.0	149.1
	[8]	US$^+$	11K/11K/6/9	450	2.7/3.9/5.0	25.8
	[9]	Artix-7	7K/4K/2/3	200	5.1/6.6/7.9	94.7
	[7]	US$^+$	9K/7K/4/6	311	4.1/5.6/6.9	53.4
	[17]	Artix-7	6K/4K/2/4	185	5.6/7.1/9.2	118.4
	PQIns	**US**	**15K/11K/10/1**	**200**	**5.5/7.2/9.6**	**111.7**
5	[10]	Virtex-7	127K/-/292/202	155	-/107.1/135.6	–
	[4]	Artix-7	16K/6K/12/17	115	10/14/18	286.0
	[24]	Artix-7	8K/5K/2/3	161	9.4/11.3/13.9	212.3
	[8]	US$^+$	12K/12K/8/11	450	3.6/4.8/6.0	31.8
	[9]	Artix-7	7K/4K/2/3	200	8.0/9.0/11.2	142.5
	[7]	US$^+$	9K/7K/4/6	311	6.3/8.2/9.9	78.5
	[17]	Artix-7	6K/4K/2/4	185	8.5/10.1/12.9	170.3
	PQIns	**US**	**15K/11K/10/1**	**200**	**8.4/9.9/11.7**	**149.2**

From the comparison of the above work, the dedicated instruction set architecture design supporting ML-DSA and ML-KEM designed in this paper outperforms most of the work in terms of performance, and according to the data flow optimisation design, the computational data of each stage of the algorithm is used right in the middle of the path, which reduces the access to the storage space, and further enhances the efficiency of the system. Overall, the dedicated instruction set architecture and data flow optimisation proposed in this paper can improve the performance of the system and can achieve a good balance between resources and performance.

Table 4. Comparison table for ML-DSA implementation.

Security level	Work	Platform	Resources LUT/FF/ DSP/BRAM	Frequency (MHz)	Cycles(kCCs) KeyGen/ Sign/Verify	Performance(μs)
2	[25]	Artix-7	30K/22K/10/11	96.9	4.2/28.1/4.5	–
	[3]	US	54K/29K/16/29	256	4.9/29.9/6.6	162.0
	[22]	Zynq-7000	19K/8K/10/17	159	7.8/52.1/7.7	424.0
	[2]	US	-/-/-/-	200	14.2/30.4/15.1	297.7
	[1]	US	-/-/-/-	270	14.6/31.7/15.5	228.4
	PQIns	**US**	**30K/18K/14/20**	**200**	**9.8/22.6/9.8**	**210.3**
3	[25]	Artix-7	30K/22K/10/11	96.9	5.9/44.7/6.2	–
	[3]	US	54K/29K/16/29	256	8.3/49.5/9.8	264.0
	[22]	Zynq-7000	20K/9K/10/17	159	13.0/89.3/11.3	714.0
	[2]	US	19K/10K/4/24	200	23.0/47.5/25.6	479.3
	[1]	US	23K/10K/4/24	270	23.7/48.8/26.2	340.8
	PQIns	**US**	**30K/18K/14/20**	**200**	**16.6/32.4/14.7**	**381.1**
5	[25]	Artix-7	30K/22K/10/11	96.9	8.8/49.0/9.1	–
	[3]	US	54K/29K/16/29	256	14.1/55.1/14.7	327.0
	[22]	Zynq-7000	19K/8K/10/17	159	20.2/93.8/15.9	816.0
	[2]	US	-/-/-/-	200	38.9/68.5/45.8	765.2
	[1]	US	-/-/-/-	270	39.8/70.2/46.7	578.3
	PQIns	**US**	**30K/18K/14/20**	**200**	**28.1/49.5/28.8**	**526.2**

7   Conclusion

This paper introduces **_PQIns_**, a novel instruction set architecture (ISA) extension designed to accelerate Post-Quantum Cryptography (PQC) operations. **_PQIns_** integrates a configurable four-stage pipeline with specialized hardware accelerators, including an expanded 8-way parallel rejection sampler, and an η-adaptive parallel Center Binomial Distribution (CBD) sampler, a high-throughput Keccak engine, a customized data flow FSMs for NTT. This co-designed architecture significantly enhances hardware efficiency by minimizing computational overhead and maximizing resource utilization. As an innovative ISA extension natively supporting both ML-KEM and ML-DSA algorithms, **_PQIns_** establishes an extensible hardware-software co-design paradigm for lattice-based cryptosystems. This work bridges the critical gap between quantum-resistant algorithmic complexity and practical hardware efficiency, providing a scalable foundation for next-generation PQC implementations.

Acknowledgments. This work is supported by the National Key R&D Program of China (No. 2022YFB4400703), the National Key R&D Program of Heilongjiang Province (2024ZXDXA12) and the Fundamental Research Funds for the Central Universities (3072025YC603).

References

1. Aikata, A., Mert, A.C., Imran, M., Pagliarini, S., Roy, S.S.: KaLi: a crystal for post-quantum security using Kyber and Dilithium. IEEE Trans. Circuits Syst. I Regul. Pap. **70**(2), 747–758 (2023). https://doi.org/10.1109/TCSI.2022.3219555
2. Aikata, A., Mert, A.C., Jacquemin, D., Das, A., Matthews, D., Ghosh, S., Roy, S.S.: A unified cryptoprocessor for lattice-based signature and key-exchange. IEEE Trans. Comput. /textbf72(6), 1568–1580 (2023). https://doi.org/10.1109/TC.2022.3215064
3. Beckwith, L., Nguyen, D.T., Gaj, K.: High-performance hardware implementation of crystals-Dilithium. In: 2021 International Conference on Field-Programmable Technology (ICFPT), pp. 1–10 (2021). https://doi.org/10.1109/ICFPT52863.2021.9609917
4. Bisheh-Niasar, M., Azarderakhsh, R., Mozaffari-Kermani, M.: Instruction-set accelerated implementation of crystals-kyber. IEEE Trans. Circuits Syst. I Regul. Pap. **68**(11), 4648–4659 (2021). https://doi.org/10.1109/TCSI.2021.3106639
5. Brohet, M., Valencia, F., Regazzoni, F.: Invited paper: instruction set extensions for post-quantum cryptography. In: 2023 IEEE/ACM International Conference on Computer Aided Design (ICCAD), pp. 1–6 (2023). https://doi.org/10.1109/ICCAD57390.2023.10323931
6. Carril, X., et al.: Hardware acceleration for high-volume operations of crystals-kyber and crystals-dilithium. ACM Trans. Reconfigurable Technol. Syst. **17**(3) (2024). https://doi.org/10.1145/3675172
7. Cui, Y., Chen, J., Ni, Z., Zhang, Z., Wang, C., Liu, W.: Instruction-based high-performance hardware controller of crystals-kyber with balanced resource utilization. IEEE Trans. Circuits Syst. I Regul. Pap. **72**(5), 2394–2407 (2025). https://doi.org/10.1109/TCSI.2025.3547799
8. Dang, V.B., Mohajerani, K., Gaj, K.: High-speed hardware architectures and FPGA benchmarking of crystals-kyber, NTRU, and saber. IEEE Trans. Comput. **72**(2), 306–320 (2023). https://doi.org/10.1109/TC.2022.3222954
9. Guo, W., Li, S.: Split-radix based compact hardware architecture for crystals-kyber. IEEE Trans. Comput. **73**(1), 97–108 (2024). https://doi.org/10.1109/TC.2023.3320040
10. Huang, Y., Huang, M., Lei, Z., Wu, J.: A pure hardware implementation of crystals-kyber PQC algorithm through resource reuse. IEICE Electronics Express **17**(17), 20200234–20200234 (2020). https://doi.org/10.1587/elex.17.20200234
11. Li, D., Pakala, A., Yang, K.: MeNTT: a compact and efficient processing-in-memory number theoretic transform (NTT) accelerator. IEEE Trans. Very Large Scale Integration (VLSI) Syst. **30**(5), 579–588 (2022). https://doi.org/10.1109/TVLSI.2022.3151321
12. Li, X., Lu, J., Liu, D., Li, A., Yang, S., Huang, T.: A high speed post-quantum crypto-processor for crystals-dilithium. IEEE Trans. Circuits Syst. II Express Briefs **71**(1), 435–439 (2024). https://doi.org/10.1109/TCSII.2023.3304416
13. Lu, J., et al.: An efficient and configurable hardware architecture of polynomial modular operation for crystals-kyber and dilithium. In: 2024 IEEE 67th International Midwest Symposium on Circuits and Systems (MWSCAS), pp. 29–32 (2024). https://doi.org/10.1109/MWSCAS60917.2024.10658892
14. Mu, J., Ren, Y., Wang, W., Hu, Y., Chen, S., Chang, C.H.: Scalable and conflict-free NTT hardware accelerator design: methodology, proof, and implementation. IEEE Trans. Comput. Aided Des. Integr. Circuits Syst. **42**(5), 1504–1517 (2022). https://doi.org/10.1109/TCAD.2022.3205552

15. National Institute of Standards and Technology: Module-lattice-based digital signature standard. In: Federal Information Processing Standards Publication (FIPS) NIST FIPS 204 (2024). https://doi.org/10.6028/NIST.FIPS.204
16. National Institute of Standards and Technology: Module-lattice-based key encapsulation mechanism standard. In: Federal Information Processing Standards Publication (FIPS) NIST FIPS 203 (2024). https://doi.org/10.6028/NIST.FIPS.203
17. Nguyen, T.H., Dam, D.T., Duong, P.P., Kieu-Do-Nguyen, B., Pham, C.K., Hoang, T.T.: Efficient hardware implementation of the lightweight crystals-kyber. IEEE Trans. Circuits Syst. I Regul. Pap. **72**(2), 610–622 (2025). https://doi.org/10.1109/TCSI.2024.3443238
18. Ni, Z., Khalid, A., Kundi, D.e.S., O'Neill, M., Liu, W.: Hpka: A high-performance crystals-kyber accelerator exploring efficient pipelining. IEEE Trans. Comput. **72**(12), 3340–3353 (2023). https://doi.org/10.1109/TC.2023.3296899
19. Ricci, S., et al.: Implementing CRYSTALS-dilithium signature scheme on FPGAs In: The 16th International Conference on Availability, Reliability and Security (ARES), pp. 1–11 (2021). https://doi.org/10.1145/3465481.3465756
20. Shor, P.W.: Polynomial-time algorithms for prime factorization and discrete logarithms on a quantum computer. SIAM J. Comput. **26**(5), 1484–1509 (1997). https://doi.org/10.1137/s0097539795293172
21. Wang, C., Gao, M.: SAM: a scalable accelerator for number theoretic transform using multi-dimensional decomposition. In: 2023 IEEE/ACM International Conference on Computer Aided Design (ICCAD), pp. 1–9 (2023). https://doi.org/10.1109/ICCAD57390.2023.10323744
22. Wang, T., Zhang, C., Cao, P., Gu, D.: Efficient implementation of dilithium signature scheme on FPGA SoC platform. IEEE Trans. Very Large Scale Integr. (VLSI) Syst. **30**(9), 1158–1171 (2022). https://doi.org/10.1109/TVLSI.2022.3179459
23. Wang, T., Zhang, C., Zhang, X., Gu, D., Cao, P.: Optimized hardware-software co-design for kyber and dilithium on RISC-V SoC FPGA. IACR Trans. Cryptographic Hardw. Embedded Syst. **3**, 99–135 (2024). https://doi.org/10.46586/tches.v2024.i3.99-135
24. Xing, Y., Li, S.: A compact hardware implementation of CCA-secure key exchange mechanism crystals-kyber on FPGA. IACR Trans. Cryptographic Hardw. Embedded Syst. **2**, 328–356 (2021). https://doi.org/10.46586/tches.v2021.i2.328-356
25. Zhao, C., et al.: A compact and high-performance hardware architecture for crystals-dilithium. IACR Trans. Cryptographic Hardw. Embedded Syst. **2022, 1**, 270–295 (2022). https://doi.org/10.46586/tches.v2022.i1.270-295
26. Zhao, Y., Xie, R., Xin, G., Han, J.: A high-performance domain-specific processor with matrix extension of RISC-V for module-LWE applications. IEEE Trans. Circuits Syst. I Regul. Pap. **69**(7), 2871–2884 (2022). https://doi.org/10.1109/TCSI.2022.3162593

Federated Incremental Learning Method Based on Memory Replay and Self-distillation Technology

Guizhi Miao[1], Haimin Hong[2], Yiying Zhang[1(✉)], Xiaoliang Mo[2], and Yeshen He[2]

[1] College of Artificial Intelligence, Tianjin University of Science and Technology, Tianjin, China
97313284@tust.edu.cn
[2] China Gridcom Co., Ltd,
1st Floor, No. 13, Hualian Industrial Zone, Xinshi Community, Dalang Street, Longhua District
Shenzhen, China

Abstract. Federated learning (FL), as a distributed machine learning paradigm, can realize multi-party collaborative training models while protecting user privacy and security. However, traditional federated learning methods usually assume that the data distribution of participants is static, ignoring the characteristics of dynamic expansion of data scale and increasing categories in actual scenarios. This limitation causes the model to face the problem of catastrophic forgetting in the process of incremental learning, that is, the introduction of new knowledge will overwrite the memory of old knowledge. To address this challenge, this paper proposes a novel "memory-enhanced adaptive distillation" (MEAD) method. MEAD uses the scores of new categories predicted by the current model to enrich the scores of old categories of the historical model, and adjusts the scale difference so that the sum of the scores of new and old categories is 1, constructing composite knowledge for self-distillation. In addition, this method also adds a memory replay module and uses a generative adversarial network (GAN) to generate samples of old categories.By using memory replay technology to retain information of old categories, it also uses generated samples to indirectly review historical tasks to reduce catastrophic forgetting. Experimental results show that MEAD reduces the average forgetting rate by 6.49%–24.27% compared to mainstream methods on multiple benchmark datasets, and improves the global accuracy by 0.37%–20.37%.

Keywords: Federated Learning · Incremental Learning · Catastrophic Forgetting · Self-distillation · Memory Playback

1 Introduction

With the growing demand for data privacy protection, FL has become a key technical paradigm in privacy-sensitive scenarios due to its distributed collaborative training characteristics [1]. This innovative framework adopts a distributed computing architecture, allowing participating terminals to complete model training by exchanging encrypted intermediate parameters without sharing original data, thereby ensuring that the data of the participants is always retained locally. This not only fundamentally avoids the privacy leakage risks caused by the cross-domain circulation of sensitive data, but also makes cross-institutional collaboration possible.

Although existing federated learning frameworks have made significant progress in privacy protection and data collaboration, their designs are mostly based on the strong assumption of static data distribution, that is, the client's data distribution and category space remain unchanged during the training process. This situation is difficult to cope with the incremental learning needs in actual scenarios where the data scale is constantly expanding and the categories are dynamically evolving. It causes the model to frequently encounter catastrophic forgetting problems during the incremental learning process [2]. Catastrophic forgetting (CF) is a core challenge in machine learning, especially continuous learning and incremental learning. It refers to the phenomenon that when a model learns new tasks or new data, its performance on old tasks or old data significantly decreases due to excessive bias in parameter adjustment towards new knowledge [3]. Its essence is that the neural network loses its key representation ability for historical tasks during the dynamic optimization process. For example, adding a new classification model for category B causes the accuracy of the original category A to drop sharply from 90% to 50%.

To address this challenge, Federated Class Incremental Learning (FCIL) has become a new research direction [4]. Its core goal is to maintain high-precision recognition of old categories while learning new categories in a distributed environment. Existing FCIL methods mostly use self-distillation technology, which uses the soft labels of historical models to constrain the optimization of the current model. Although some progress has been made in alleviating forgetting, there are still many shortcomings. For example, directly superimposing the scores of new and old categories in knowledge distillation may lead to scale inconsistency and affect training stability. The GLFC method is a typical example. It directly connects the old category logic of the historical model with the new logic of the current model, resulting in the non-normalization of the new and old probability distributions and a shift in the optimization direction [5].

The "Memory Enhancement and Adaptive Distillation" (MEAD) method proposed in this paper improves the performance of FCIL by introducing memory replay and a novel self-distillation technique. Compared with existing methods, the novelty of MEAD lies in combining the new category scores of the current model with the old category scores of the historical model, adjusting the proportion of the old category scores and normalizing the total score. Specifically, the old category scores of the historical model are multiplied by (1-the sum of the new category scores of the current model), while the new category scores directly use the output of the current model. This ensures that the sum of all category scores is 1, avoids scale differences, and significantly improves the ability to retain knowledge. In addition, MEAD analyzes the communication cost and computational complexity to ensure the feasibility of practical deployment.

The main contributions of this paper are as follows:

1. The MEAD method achieves scale normalization and optimizes the distillation method of knowledge transfer by adjusting the coefficients and combining the old

category scores of the historical model with the new category scores of the current model.

2. A memory replay module is introduced to generate pseudo samples of old categories based on GAN to enhance the retention of historical knowledge
3. Regularize the model parameters to prevent overfitting or excessive forgetting of old knowledge. Our method not only designs a dynamically expanded fully connected layer to adapt to the increase in new task categories, but also retains the parameters of the old fully connected layer to avoid training from scratch.

2 Related Work

2.1 Federated Class-Incremental Learning

In real-world scenarios, the amount and categories of data will gradually increase over time. Traditional federated learning methods have difficulty in maintaining the memory of old knowledge while protecting privacy and coping with data heterogeneity [6]. FCIL integrates the core technology of quasi-incremental learning and further adapts to the needs of distributed environments through a client-server collaboration mechanism. FCIL has formed three major technical schools: regularization method, architecture expansion method, and replay method.

Regularization methods preserve old task knowledge by constraining model parameter updates. For example, Elastic Weight Consolidation (EWC) uses an information matrix to quantify parameter importance and imposes stronger regularization penalties on key parameters to reduce forgetting [7]; Architecture extension methods dynamically adjust the model structure to allocate independent capacity for new tasks and avoid overwriting old knowledge. For example, FedCL balances model capacity and communication overhead through client-side adaptive parameter extension and server-side dynamic aggregation [8]; Replay methods are another important strategy to combat forgetting, which implements knowledge review by storing old data or its representation. For example, Gradient Context Memory (GEM) constrains the direction of the new task gradient to avoid conflict with the old task [9]. However, these methods still suffer from the problem of scale inconsistency when fusing new and old category scores, and the reliance on old data may be limited by storage capacity.

2.2 Memory Replay

The basic idea of the memory replay technique is to utilize stored old samples or generated pseudo samples to enable the model to review historical knowledge when learning new tasks, thereby maintaining the ability to remember old categories.

Traditional memory replay methods usually rely on storing a small amount of real samples. However, in order to overcome the limitation of limited storage space, generative methods have gradually become a new direction for memory replay [10]. Some studies have attempted to combine memory replay with federation mechanisms. FedRep proposed a virtual replay method based on feature distillation, which circumvents privacy issues by transmitting feature representations instead of original data [11]. However, this method has a strong dependence on feature extractors and fails to fully utilize the potential of generative models. The MEAD method aims to further improve the retention rate of old knowledge in FCIL by introducing an improved generative memory replay mechanism, providing a more practical solution for dynamic incremental learning.

2.3 Knowledge Distillation and Knowledge Transfer

Knowledge distillation (KD) is a technique for optimizing model performance through knowledge transfer. Its core is to transfer the knowledge of complex models into lightweight models to improve generalization ability or achieve efficient deployment [12].

Traditional knowledge distillation faces significant challenges in dynamic incremental scenarios. First, most methods are based on the assumption of static category distribution and are difficult to adapt to dynamic environments with increasing categories and knowledge interference between tasks [13]. Second, existing methods often lack sophisticated designs when integrating new and old knowledge. For example, the FedWa method uses a fixed weight parameter to linearly combine the classification loss and the distillation loss [14]. This simple weighting method is difficult to effectively coordinate the conflict between the new and old tasks and may aggravate catastrophic forgetting. Similarly, although GLFC integrates new and old category features through weighted binary cross entropy loss, its limited ability to capture complex semantic associations between categories is likely to limit the robustness of knowledge transfer. These limitations indicate that traditional knowledge distillation faces the dual challenges of efficiently integrating new and old knowledge and suppressing forgetting in dynamic incremental scenarios.

3 Problem Definition

Consider a federated learning system consisting of K clients, each client $k \in \{1, 2, ..., K\}$ collects and stores a local private dataset D^k. The data arrives gradually in T tasks, where the tth task (denoted as T^t, $t = 1, 2, ..., T$) introduces new data D_t^k, so the total data set for client k can be denoted as $D^k = \bigcup_{t=1}^{T} D_t^k$. There may be category overlaps between task datasets, but new tasks often contain unseen categories, causing the data category set to expand over time. Therefore, task division simulates dynamic scenarios, and the number of categories varies significantly between tasks. For example, Task 1 introduces 4 categories and Task 2 introduces 10 categories.

Due to the limited storage and computing power of the client, each client cannot retain all historical data. Assume that when processing a task T^t, client k can only retain a small number of samples from the previous t-1 tasks, denoted as D_{t-old}^k, where the number of samples is m, and $m \ll \sum_{i=1}^{t-1} | D_i^k |$. These reserved samples are usually extracted from historical data through selection strategies (such as random sampling or representative sampling) to assist in the training of the current task. At the same time, the current task data D_t^k and the retained old data D_{t-old}^k together constitute the training set of client k at the time of the task T^t.

The goal of federated incremental learning is to train a global model W^S and a local model W^k for each client, achieving good performance on all tasks through a client-server collaboration mechanism. Specifically, the global model is updated collaboratively by the server with the goal of minimizing the global loss function $J^S(W^S)$, which is defined as the weighted expectation of the client's local loss:

$$\min_{W^S} J^S\left(W^S\right) = E_k\left[\frac{|D_t^k|}{\sum_{i=1}^K |D_t^i|} J^k\left(W^k\right)\right]|D_t^k| \tag{1}$$

Among them, $\sum_{i=1}^K |D_t^i|$ is the total current data of all clients, and the weight $\frac{|D_t^k|}{\sum_{i=1}^K |D_t^i|}$ reflects the contribution ratio of each client to the global model. $J^k(W^k)$ is the local loss function of client k, which needs to consider both the learning of new data and the knowledge retention of old data in the current stage.

In addition, FCIL also needs to satisfy the dynamic constraints of the task category set expanding over time, which may lead to conflicts between new and old knowledge. Therefore, the optimization goal of FCIL is not only to improve the accuracy of the global model on the current task, but also to minimize the forgetting of old tasks.

4 Method

As shown in Fig. 1., MEAD combines two modules, memory playback and adaptive distillation, to achieve efficient fusion and retention of new and old knowledge while protecting privacy.

4.1 Memory Playback Module

In the sample generation phase, MEAD introduces the generative adversarial network GAN. GAN consists of a generator G and a discriminator D. The training process of the game between the two is shown in Fig. 2.

The generator G learns old categories from the data feature distribution of the historical model, takes random noise z as input, and generates realistic pseudo data samples in an attempt to deceive the discriminator;The discriminator D is responsible for distinguishing real old samples from generated samples. The optimization goal of GAN follows the classic adversarial training paradigm:

$$\min_{G}\max_{D} E_{X_{old}}\left[\log D(X)\right]E_z\left[\log(1 - D(G(z)))\right] \tag{2}$$

Where E represents the expected operation. In order to avoid generating generalized or confusing samples and improve the unique representation ability of the generated samples for the target data category, the output of the historical model is used as a conditional input during training, and the generated pseudo samples are combined with the current task data D_t^k to form an extended training set for subsequent model updates.

To ensure the reliability of generated samples, we use intra-class variance and pseudo-label accuracy as evaluation metrics. Intra-class variance reflects the degree of dispersion of generated samples in the data feature space. Lower variance means more concentrated samples and higher quality. Pseudo-label accuracy measures category consistency by

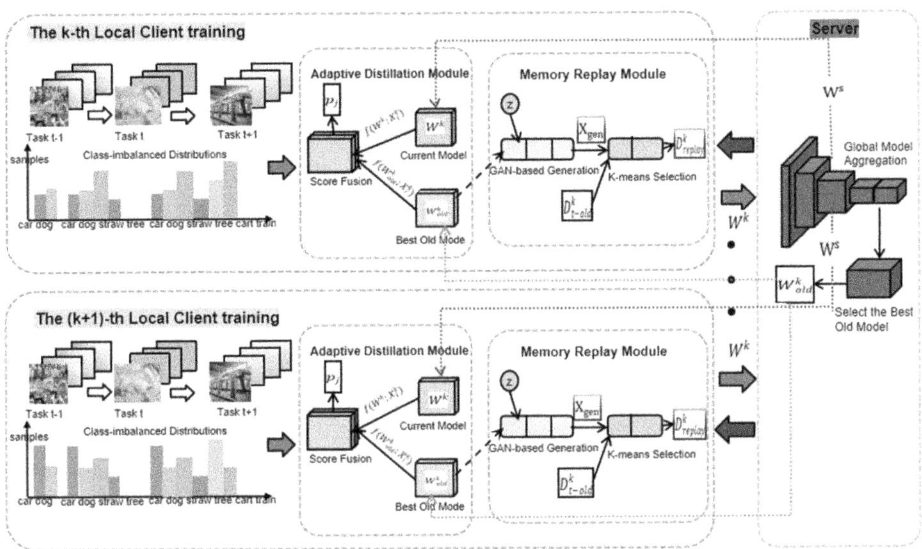

Fig. 1. The overall framework of the MEAD method. MEAD fuses new and old knowledge through an adaptive distillation module to generate a fusion score p_j. In the memory replay module, random noise z and historical model W^k_{old} are used to generate sample X_{gen} through GAN. $D^k_{t-old} \cup X_{gen}$ Obtain replay set through K-means selection D^k_{replay}. Perform local training on the client and update the local model W^k. The server aggregates and updates the global model W^s and selects the optimal historical model W^k_{old} to feed back to the client to achieve balanced learning of new and old tasks.

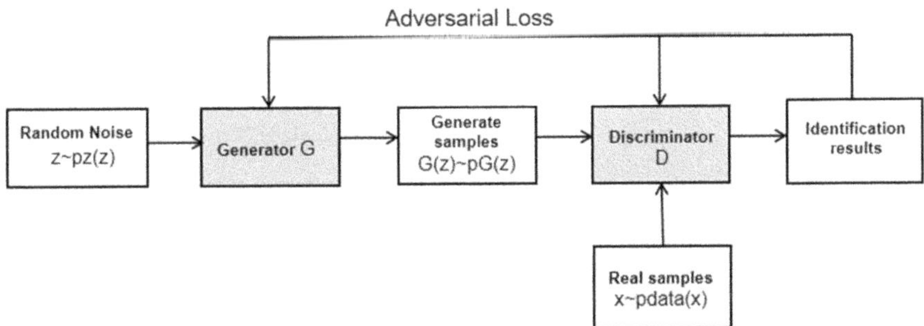

Fig. 2. Schematic diagram of GAN

classifying generated samples using a historical model and calculating the matching rate between pseudo-labels and true categories. Furthermore, a verification mechanism is introduced, using the discriminator D to screen generated samples. Because a too low threshold may introduce noisy samples and affect model stability, a too high threshold may result in insufficient sample diversity and weaken the playback effect. Therefore, based on empirical trade-offs, the confidence threshold is set to 0.8. Only samples with a confidence level above 0.8 are retained for playback to reduce the risk of noise and bias.

In the sample selection stage, due to the limited storage capacity of edge devices, MEAD needs to select the most representative samples from old data D^k_{t-old} and generated samples X_{gen} to construct replay set D^k_{replay}. Specifically, the K-means clustering algorithm is used to cluster the sample feature space of $D^k_{t-old} \cup X_{gen}$ and extract samples corresponding to m cluster centers. These samples cover the main distribution patterns of the old categories, providing rich and representative historical information for subsequent training, helping the model to effectively use historical knowledge when performing new tasks.

4.2 Adaptive Distillation Module

The adaptive distillation module guides the current model training through the knowledge of historical models to achieve the coordinated fusion of new and old categories. Its core is to dynamically adjust the category scores to ensure the consistency of probability distribution. When dynamically expanding the fully connected layer, the parameters of the new categories are initialized using Xavier uniform initialization, while the parameters of the old categories remain unchanged. To reduce the interference of initialization on the prediction of the old categories, we introduce a regularization term in the loss function (see Eq. 8) to constrain the update direction of the new parameters and avoid excessive deviation from the feature space of the old categories.

Self-Distillation Framework: MEAD uses the output of historical models W^k_{old} as soft labels to guide the training of the current model W^k. Assume that in Task T^t, the total number of data categories is $g + h$, where g is the number of old categories and h is the number of new categories. The historical model only has prediction capabilities for the old categories $\{1,...,g\}$, while the current model needs to cover all categories $\{1,...,g + h\}$. Self-distillation compares the outputs of the two models, constraining the predictions for the old categories to be as consistent as possible with the historical model, while learning the representation of the new categories.

Fusion of New and Old Scores: MEAD constructs a unified fusion score $p_j (j = 1, ..., g + h)$ by processing the probability distribution of new and old categories separately to achieve seamless integration of new and old knowledge. Assume that the total number of categories for the current task is $g + h$. For the input sample X^k_i, the current model outputs $logits f(W^k; X^k_i)$, covering all $g + h$ categories; the historical model outputs $logits f(W^k_{old}; X^k_i)$, only covering the old category g. The goal is to generate a fusion score p_j that satisfies $\sum_{j=1}^{g+h} p_j = 1$.

For the new category score, since the new category has not been seen by the historical model, the predicted probability of the current model is directly used:

$$p_j = \frac{\exp\left(f(W^k; X^k_i)_j/t\right)}{\sum_{m=1}^{g+h} \exp(f(W^k; X^k_i)_m/t)}, j \in \{g + 1, \ldots, g + h\} \tag{3}$$

Among them, $f(W^k; X_i^k)_j$ is the logit value of the j-th category, τ is the temperature parameter, which controls the smoothness of the probability distribution. The denominator is the normalization factor of the current model on all categories.

For the old category scores, the historical model's predictions for the old categories need to be adjusted to fit the expanded category space. MEAD redistributes the old class probabilities by introducing an adjustment factor:

$$p_j = \frac{\exp\left(f\left(W_{old}^k; X_i^k\right)_j / \tau\right)}{\sum_{m=1}^{g} \exp\left(f\left(W_{old}^k; X_i^k\right)_m / \tau\right)} \cdot \left(1 - \sum_{c=g+1}^{g+h} \frac{\exp\left(f\left(W^k; X_i^k\right)_c / \tau\right)}{\sum_{m=1}^{g+h} \exp\left(f\left(W^k; X_i^k\right)_m / \tau\right)}\right),$$
$$j \in \{1, \ldots, g\}$$

(4)

Among them, the first term is the softmax probability of the historical model on the old category. The second term is the complement of the sum of the new category probabilities, which serves as an adjustment coefficient to ensure that the old category scores are consistent with the new category scores.

4.3 Optimization Goals

The MEAD method achieves balanced learning of new and old knowledge by jointly optimizing multiple loss functions of client k, effectively suppressing catastrophic forgetting.

Classification Loss: The classification loss is for the classification task of the current task data D_t^k and the playback set D_{replay}^k, ensuring the model's predictive ability on new and old categories. The formula is:

$$L_{cls}^k = \frac{1}{\left|D_t^k \cup D_{replay}^k\right|} \sum_{(X,y) \in D_t^k \cup D_{replay}^k} - \log\left(\frac{\exp\left(f(W^k, X)_y\right)}{\sum_{m=1}^{g+h} \exp(f(W^k; X)_m)}\right)$$

(5)

Among them, $f(W^k; X)_y$ represents the value of the current model for the sample in the true category y, and the denominator is the normalization factor for all categories. The classification loss ensures the learning effect of the current task by optimizing the classification accuracy of the model on new and old data, while maintaining the memory of old categories with the help of the replay set.

The Memory Replay Loss: It targets the classification consistency of the generated samples X_{gen}, ensuring the quality of the generated data and the model's ability to remember old categories. The formula is:

$$L_{rep}^k = \frac{1}{\left|X_{gen}\right|} \sum_{X_{gen}} - \log\left(\frac{\exp\left(f(W^k; X_{gen})_{y_{gen}}\right)}{\sum_{m=1}^{g+h} \exp(f(W^k; X_{gen})_m)}\right)$$

(6)

Among them, X_{gen} is the pseudo labels for the generated samples are generated by the historical model W_{old}^k prediction (for example $y_{gen} = \text{argmax} f\left(W_{old}^k; X_{gen}\right)$), ensuring that the category information X_{gen} is consistent with the historical knowledge.

Loss from Distillation: The self-distillation loss is based on the fusion score p_j, and through knowledge transfer, it constrains the output of the current model to be consistent with historical knowledge, thereby alleviating the interference of new task learning on old tasks. The formula is:

$$L_{dist}^k = \frac{1}{\left|D_t^k \cup D_{replay}^k\right|} \sum_{(X,y) \in D_t^k \cup D_{replay}^k} \sum_{j=1}^{g+h} p_j \cdot \log\left(\frac{p_j}{\frac{\exp\left(f(w^k;X)_j/t\right)}{\sum_{m-1}^{++h} \exp\left(f(w^k;X)_m/t\right)}}\right) \tag{7}$$

Among them, p_j is the fusion score, τ is the temperature parameter, and the probability distribution of the current model is calculated in exponential form.

Total Loss Function: The weighted combination of the losses is the total loss function for client k:

$$L^k(W^k) = L_{cls}^k + \beta L_{dist}^k + \lambda L_{rep}^k \tag{8}$$

Where β is a hyperparameter that controls the weight of the distillation loss, balances the knowledge transfer from the historical model, and is used to adjust the contribution of self-distillation. The optimization process is performed locally on the client using stochastic gradient descent. The updated local model is uploaded to the server, which then updates the global model using weighted averaging.

Algorithm 1 MEAD Algorithm Flow

1: **Initialization phase:**
2: Server initializes global model W^S distribute to K clients $W^k, k \in \{1, 2, \ldots, K\}$
3: Each client k let W^S copy as local model W^k and initialize the history model W_{old}^k
4: Client k initialize data storage D_{t-old}^k. Used to store historical data (due to storage limitations, only samples are retained) m samples are retained)
5: The server uses the Xavier initialization method to initialize parameters
6: **for** each task $T^u \in \{T^1, T^2, \ldots, T^U\}$ **do**
7:     **Task processing phase:**
8:     Client k receive current task data D_t^k perform local training in combination with historical data
9:     **if** $T^u = T^1$ **then**
10:         Client k Optimize $J_{CE}(W^k)$
11:     **else**
12:         $D_{t-old}^k \leftarrow Subset(D_{(t-1)-old}^k \cup D_t^k)$. Storage of restricted data
13:         $W_{old}^k \leftarrow W^k$, Expand W^k accommodate the new category.
14:         **while** T^u not finished **do**
15:             Client k Optimize $J^k(W^k)$, with memory playback (generate X_{gen}, search D_{replay}^k) and knowledge distillation
16:             Clinet k upload W^k Sever.
17:             The server calculates the weighted average:
18:             $W^S \leftarrow \sum_{k=1}^{K} \frac{|D^k|}{\sum_{l=1}^{K} |D^l|} W^k$
19:             Server broadcasts W^S to client, the client updates W^k
20:         **end while**
21:     **end if**
22: **end for**

---

5 Experiment

5.1 Experimental Setup

To comprehensively evaluate the performance of the proposed MEAD method for federated incremental learning, we conduct experiments on multiple datasets, baseline methods, and evaluation metrics.

Datasets: We selected three widely used benchmark datasets: SVHN, CIFAR-10, and EMNIST. These datasets are chosen to simulate the real-world scenarios of different class distributions and task complexities in FCIL.

Experimental Environment: The experiment was conducted under the federated learning framework, and 20 clients were set up to simulate a distributed environment. Set dirichlet_alpha = 0.5 for non-independent identically distributed (Non-IID) partitioning to simulate data heterogeneity in the real world. The model was trained using the stochastic gradient descent (SGD) optimizer and mixed precision training. The learning rate is 0.01, the momentum is 0.9, temperature_old and temperature_new are set to 5 and 3 respectively, each client performs 10 local iterations in each round of communication, the number of training rounds is 70, the total number of training rounds of GAN is 10, and the frequency of experience replay is 2. The regularization loss loss_re is calculated by the regularization_loss method, which may be used to constrain model parameters. Use dynamically expanded fully connected layers to adapt to the increase in categories of new tasks, retain the parameters of the old fully connected layers, and avoid training from scratch.

Task Setting: We divide the CIFAR-10, SVHN, and EMNIST datasets into three-task scenarios. CIFAR-10 and SVHN learn [3, 4] new categories on Task 1, Task 2, and Task 3, respectively; EMNIST learns [15, 15, 17] new categories respectively.

Comparison with Baseline Methods: We compared MEAD with the current mainstream incremental federated learning methods FedWa, GLFC, and ICARL, and used four key indicators to evaluate the learning and memory capabilities of the model: global accuracy Acc1, global accuracy Acc3, response time, and average forgetting rate. The global accuracy is used to measure the accuracy of the model's predictions; the response time reflects how quickly the model responds to new data or changes in operating status; and the average forgetting rate evaluates the degree to which the model forgets old task knowledge when learning new tasks.

5.2 Performance Comparison

The experimental comparison results of MEAD with three mainstream methods, GLFC, ICARL and FedWa, are shown in Table 1. Under limited experimental conditions, although the accuracy of the MEAD method on the new task Task3 is slightly lower than that of other methods, its global accuracy Acc1 and Acc3, two key indicators, are higher than GLFC, ICARL, and FedWa. More importantly, the average forgetting rate and task-specific forgetting rate of MEAD are much lower than those of the other three methods. On the simulated distributed resource SVHN dataset, the average forgetting rate is only 16.54%, which shows that MEAD can effectively alleviate catastrophic forgetting when learning new tasks, while its poor performance in few tasks is due to the need to retain old category information and learn new categories. MEAD's outstanding performance is due to its dual mechanisms of memory replay and adaptive distillation, which work together to improve the model's ability to learn and retain new and old knowledge.

Table 1. Performance comparison of the MEAD method with three baseline algorithms on three incremental tasks on three datasets.

Dataset	Method	T^1Acc	T^2Acc	T^3Acc	T^2ForgRate	T^3ForgRate	Global Acc1	Global Acc3	Avg Forg Rate
CIFAR-10	GLFC	62.55	64.86	67.97	29.48	30.83	40.72	76.96	31.41
	ICARL	61.08	74.60	74.87	33.03	47.50	41.78	75.31	39.28
	FedWa	65.85	73.56	74.30	27.36	46.53	41.02	76.38	40.11
	MEAD	66.54	60.10	58.23	6.88	41.38	45.25	80.63	24.92
SVHN	GLFC	66.23	71.18	46.93	44.27	24.24	42.79	74.06	36.38
	ICARL	74.98	74.54	71.43	53.79	30.55	45.52	69.50	40.81
	FedWa	70.69	75.63	55.63	44.22	27.94	43.55	75.46	39.57
	MEAD	70.72	50.68	49.49	20.05	6.94	56.01	81.96	16.54
EMNIST	GLFC	29.98	40.85	72.24	30.31	43.26	49.31	68.74	38.20
	ICARL	74.69	75.56	75.89	43.92	28.45	69.31	80.92	35.59
	FedWa	50.37	59.83	63.17	39.69	40.97	62.47	77.92	42.53
	MEAD	73.29	71.57	66.34	19.81	28.19	69.68	85.40	20.51

5.3 Communication Efficiency

Training Efficiency: Experiments,To verify the training efficiency of the MEAD method, we compared it with the baseline methods GLFC,ICARL and FedWa on the EMNIST dataset.,The experimental setting includes 20 clients.Table 2 shows the results of each round of training time. It can be seen that the average training time of MEAD is 175 seconds, which is not slower than other methods under the premise of increasing training accuracy and reducing forgetting rate.This is due to the mixed precision training and dynamically expanded fully connected layer design adopted by MEAD, which reduces unnecessary computational overhead while maintaining data privacy protection.

Table 2. Time per training round using 20 clients on the EMNIST dataset.

Method	GLFC	ICARL	FedWa	MEAD
Training time	175	194	165	179

Convergence Evaluation: To verify the convergence of the proposed MEAD framework, we compare the training loss rate of MEAD with ICARL, GLFC and FedWa on the dataset SVHN in Figure 3.It can be seen from Fig. 3. that the convergence effect of MEAD after 100 rounds of training is higher than that of the other three methods, and the convergence speed is faster. This indicates that MEAD can converge faster and more stably in dynamic incremental learning scenarios while maintaining a balanced learning of new and old knowledge.

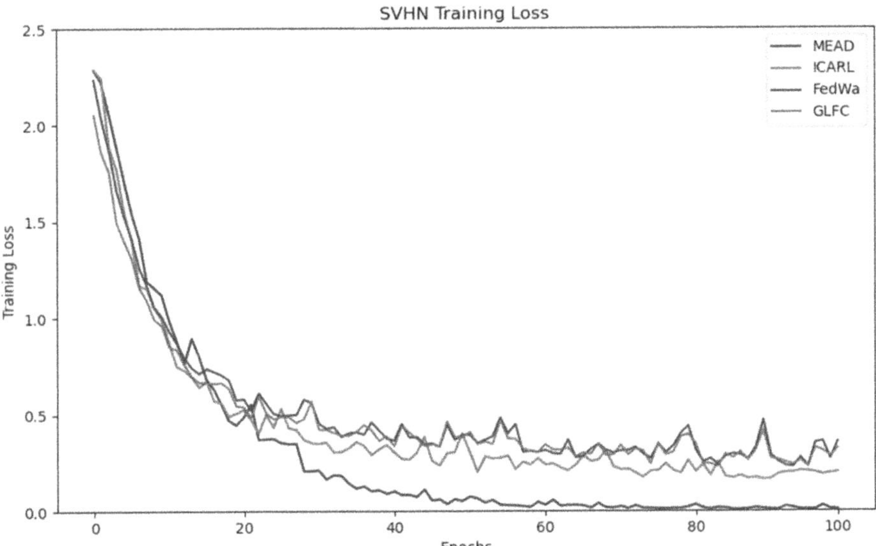

Fig. 3. Comparison of training loss convergence of MEAD, ICARL, FedWa, and GLFC on the SVHN dataset

5.4   Ablation Experiment

To explore the contributions of each key component of MEAD, we conduct ablation experiments to evaluate the impact on performance by removing modules one by one. MEAD w/o GAN, MEAD w/o ADM, and MEAD w/o GAN + ADM represent the performance without using the GAN module, without using the self-distillation technology, and without using the GAN module and the self-distillation technology, respectively. The experimental results are shown in Table 3. Compared with our method MEAD, the accuracy of MEAD w/o GAN, MEAD w/o ADM, and MEAD w/o GAN + ADM is significantly reduced by 2.67% ~ 28.59%. The forgetting rate is significantly increased by 11.48% ~ 28.14%.Ablation experiments verify the effectiveness of the collaborative work of various modules and show that these modules are crucial for training global class incremental models.

The distillation coefficient in the adaptive distillation module controls the weight of the distillation loss in the total loss, affecting the balance between new and old task knowledge. To explore the impact on MEAD performance, we set it to {0, 3, 8, 10} and evaluate its impact on the global accuracy Acc1, global accuracy Acc3, and average forgetting rate. The experimental results are shown in Table 4. We can see that the global accuracy Acc1 and Acc3 first decrease and then increase, while the forgetting rate shows a continuous downward trend. Experiments have verified that adaptive distillation can effectively coordinate the learning of new and old knowledge in federated incremental learning, and reasonable adjustment of the distillation coefficient can optimize model performance.

Table 3. The impact of the memory playback module on key indicators of MEAD method.

Module	Global Acc1	Global Acc3	Avg Forg Rate
MEAD w/o GAN	54.37	79.29	33.60
MEAD w/o ADM	49.71	63.89	28.02
MEAD w/o GAN + ADM	29.35	53.37	44.68
MEAD	56.01	81.96	16.54

Table 4. Effects of different distillation β coefficients on key indicators of MEAD method.

β	Global Acc1	Global Acc3	Avg Forg Rate
0	53.84	79.35	36.24
3	46.39	77.21	29.03
8	55.28	80.91	19.97
10	56.01	81.96	16.54

6 Conclusion and Future Work

This paper proposes an innovative federated incremental learning framework MEAD, which effectively alleviates the catastrophic forgetting problem of FCIL when facing the dynamic growth of data categories through memory replay and self-distillation techniques. The memory replay module generates synthetic samples of old categories through GAN and selects representative samples for replay, ensuring that the model can continuously review historical knowledge. The self-distillation module fuses the predictions of the current model with the best historical model to generate a balanced score, guiding the model to achieve smooth knowledge transfer between new and old categories. Under Non-IID, MEAD performs well in global accuracy, response time, average forgetting rate, and convergence, outperforming mainstream baseline methods.

Although MEAD has achieved significant performance improvements, there is still room for improvement. Future research can introduce more advanced generative models (such as diffusion models) to replace GAN to further improve the quality of generated samples and enhance the effect of memory playback. At the same time, more efficient adaptive distillation strategies can be explored to better adapt to complex and changing operating environments.

References

1. Wu Z, He T, Sun S, et al.: Federated Class-Incremental Learning with New-Class Augmented Self-Distillation. arXiv preprint arXiv:2401.00622, (2024).
2. Lesort, T., Lomonaco, V., Stoian, A., et al.: Continual learning for robotics: definition, framework, learning strategies, opportunities and challenges. Inf. Fusion. **58**, 52–68 (2020)
3. van de Ven G M, Soures N, Kudithipudi D.: Continual learning and catastrophic forgetting. arXiv preprint arXiv:2403.05175, (2024).
4. Yang, X., Yu, H., Gao, X., et al.: Federated continual learning via knowledge fusion: a survey. IEEE Trans. Knowl. Data Eng. **36**(8), 3832–3850 (2024)
5. Dong, J., Wang, L., Fang, Z., et al.: Federated class-incremental learning. In: Proceedings of the IEEE/CVF Conference on Computer Vision and Pattern Recognition, pp. 10164–10173 (2022)
6. Wang, N., Li, Y., Zhang, H., et al.: Data-Free Federated Class Incremental Learning with Diffusion-Based Generative Memory. In: arXiv preprint arXiv:2405.17457, pp. 1–15. arXiv, Online (2024).
7. Kirkpatrick, J., et al.: Overcoming catastrophic forgetting in neural networks. Proc. Natl. Acad. Sci. (PNAS). **114**(13), 3521–3526 (2017)
8. Liu, Z., Wu, F., Wang, Y., et al.: FedCL: federated contrastive learning for multi-center medical image classification. Pattern Recogn. **143**, 109739 (2023)
9. Lopez-Paz, D., Ranzato, M.'.A.: Gradient episodic memory for continual learning. Adv. Neural Inf. Proc. Syst. (NeurIPS), 6470–6479 (2017)
10. Shin, H., Lee, J.K., Kim, J., et al.: Continual learning with deep generative replay. Adv. Neural Inf. Proces. Syst. **30** (2017)
11. Li, Y., Li, Q., Wang, H., et al.: Towards efficient replay in federated incremental learning. In: Proceedings of the IEEE/CVF Conference on Computer Vision and Pattern Recognition, pp. 12820–12829 (2024)
12. Gou, J., Yu, B., Maybank, S.J., et al.: Knowledge distillation: a survey. In: Int. J. Comput. Vis. (IJCV), 129(6), pp. 1789–1819. Springer, Berlin (2021).
13. Chen, Z., Wang, T., Li, D., et al.: Mitigating catastrophic forgetting in online continual learning by modeling previous task interrelations via Pareto optimization. In: International Conference on Machine Learning (ICML), pp. 12345–12356. PMLR, Online (2024)
14. Kim, J., Lee, S., Park, T., et al.: Weighted averaging federated learning based on example forgetting events in label imbalanced non-IID. In: Appl. Sci., 13(9), pp. 5487. MDPI, Basel (2023).

SpikeEAR: Low-Power Neuromorphic Auditory System for Real-Time Scene Analysis on FPGA

Yiwei Si[1], Hao Yu[1], Yuhang Zhu[1], Sizhao Li[1,2]([envelope]), Zechao Liu[1,2], Yongrui Zhang[3], and Di Gao[4]

[1] College of Computer Science and Technology, Harbin Engineering University, Harbin, China
`sizhao.li@hrbeu.edu.cn`
[2] Key Laboratory of information Secrecy and Protection Technology, Ministry of Industry and Information Technology, Harbin, China
[3] College of Mechanical and Electrical Engineering, Harbin Engineering University, Harbin, China
[4] Hangzhou Institute for Advanced Study, University of Chinese Academy of Sciences, Zhejiang, China

Abstract. Traditional audio processing systems are constrained by the challenges of high power consumption and high latency. To address these challenges, this paper proposes SpikeEAR, an end-to-end spiking auditory system that implements a Sensing-Computing Integration architecture on a single FPGA. SpikeEAR achieves ultra-low end-to-end latency by seamlessly integrating a biomimetic cochlea, utilizing a novel noniterative adaptive encoding scheme, with a deeply pipelined Spiking Neural Network (SNN) accelerator. We implement the system on a Xilinx Zynq platform to evaluate the latency and power consumption. SpikeEAR achieves a 95.69 % recognition accuracy on a subset of the Google Speech Commands dataset, with an end-to-end latency of only 256.95 µs. This latency represents a $4 \sim 7\times$ improvement over recent FPGA-based solutions. This high-performance operation is achieved within a total power budget of 1.25 W. The overall performance validates the SCI architecture as a superior approach for constructing high-performance, low-power intelligent auditory systems for edge devices.

Keywords: Spiking Neural Network · Sensing-Computing Integration · Neuromorphic Auditory System · FPGA · Low Latency

1 Introduction

Neuromorphic computing, a paradigm inspired by the structure and function of biological neural systems, opens new avenues for efficient information processing. By emulating biological sensory mechanisms, these fields offer solutions to the efficiency bottlenecks of traditional computing architectures [22]. The biological auditory system exemplifies this potential. It achieves millisecond-level

© The Author(s), under exclusive license to Springer Nature Singapore Pte Ltd. 2026
H. Liu et al. (Eds.): ICA3PP 2025, LNCS 16381, pp. 190–209, 2026.
https://doi.org/10.1007/978-981-95-8399-7_11

spike coding and spatiotemporal integration in complex acoustic environments, providing valuable insights for tackling modern challenges in power consumption and real-time processing. This system inherently integrates sensing and computing, with information acquisition and processing tightly coupled both physically and temporally.

In contrast, mainstream audio processing techniques, particularly in automatic sound classification (ASC), depend on a frame-based approach. This method involves converting audio into Mel spectrograms [24], followed by processing with convolutional neural network (CNN) [20] or recurrent neural networks (RNN) [1]. Despite its success, this paradigm has notable drawbacks. Continuous spectral analysis demands significant computational resources and power, limiting deployment on resource-constrained edge devices. Additionally, its processing latency hinders instantaneous responses to sudden acoustic events.

To address these limitations, we propose a neuromorphic approach that mirrors biological mechanisms, achieving seamless Sensing-Computing Integration for audio signals. The biological cochlea transforms sound waves into sparse neural spike sequences, which a spiking neural network (SNN) processes in an event-driven, "no-event, no-computation" manner [17]. As a third-generation neural network model [25], SNN offer low power consumption and high real-time performance through their asynchronous, event-driven nature [11,22]. Combining a "silicon cochlea" with an SNN classifier thus represents an optimal strategy for next-generation auditory systems.

Significant progress has been made in this domain. At the sensing front-end, Spiketrum [2] efficiently converts audio into biomimetic spike sequences, while a MEMS-based cochlea [15] adjusts sensing in real time. At the computational back-end, a multi-FPGA auditory system [5] demonstrates robustness in noisy environments, and a lightweight SNN processor [16] enables real-time analysis on FPGA. However, these efforts often remain fragmented, focusing either on sensing or computation without deep integration. This paper addresses this gap by designing an end-to-end neuromorphic auditory system, SpikeEAR, optimized for FPGA. It processes the full pipeline from sensory coding to decision-making in parallel, achieving low latency, high energy efficiency, and robust recognition accuracy for real-time sound event recognition in smart environments [11].

In response to this research gap, the core research question of this paper is: can we design and implement a complete end-to-end digital neuromorphic Sensing-Computing Integrated system, which processes the full pipeline from sensory coding to cognitive decision-making in parallel, while achieving unprecedented low latency and high energy efficiency on resource-constrained platforms like FPGA, while ensuring high recognition accuracy? To answer this question, we propose a solution designed specifically for efficient operation on FPGA, aimed at solving the real-time sound event recognition problem in intelligent environments. The motivation of this research lies in advancing the concept of Sensing-Computing Integration from a theoretical model to a hardware-validated practical application.

To this end, our work presents three key contributions aimed at systematically addressing the challenges mentioned above:

1. **A Neuromorphic Cochlea with Adaptive Event Generation and Dynamic Thresholding.** A lightweight, non-iterative adaptive thresholding mechanism is applied to Gammatone filter banks. This optimizes the hardware-fidelity trade-off to produce high-quality, sparse event streams.
2. **A Parallel SNN Accelerator with Co-located Compute and Storage.** A parallel SNN architecture with inter-layer pipelining and intra-layer parallelism is proposed. By storing synaptic weights in on-chip Block RAMs (BRAM), the design achieves tight compute-storage coupling, thus mitigating memory bottlenecks and maximizing event throughput.
3. **An End-to-End System Embodying "Sensing-Computing Integration".** The primary contribution is a complete, FPGA-validated System-on-Chip (SoC) blueprint. It realizes a true "Sensing-Computing Integration" by seamlessly co-designing the spiking cochlea front-end and the SNN back-end within a unified framework.

The remainder of this paper is organized as follows: Sect. 2 reviews the background and research motivation. Section 3 details the core methods. Section 4 presents the FPGA hardware architecture design of the system. Section 5 showcases comprehensive evaluation results. Section 6 concludes the paper.

2   Background and Motivations

2.1   Biologically Inspired Cochlear Model

Achieving low-power, real-time audio processing on edge devices is a persistent challenge, for which the biological cochlea provides an energy-efficient model. The cochlea avoids energy-intensive Fourier Transforms by using its physical structure to decompose sound, which Inner Hair Cells (IHCs) then convert into sparse, information-rich neural spikes [17]. This event-driven approach is key to next-generation smart sensors.

Inspired by this, early "Silicon Cochleas" employed Cascaded Filter Architectures [15]. However, their sequential processing inherently introduces signal latency and amplifies quantization noise, hindering applications that demand high temporal precision [29,30].

Parallel Filter-Bank Architectures solve the latency and noise accumulation problems by processing all channels concurrently. These architectures commonly use the Gammatone filter, which effectively mimics the human auditory response [31]. Nonetheless, state-of-the-art implementations like Spiketrum [2] introduce a new bottleneck: to maintain accuracy, they employ complex, iterative algorithms. This computational burden compromises the real-time performance and hardware efficiency that the parallel architecture was meant to achieve.

2.2   FPGA Implementation of SNN

FPGAs, with their inherent parallelism and reconfigurability, have become an ideal hardware platform for implementing SNN. To simulate large-scale networks on resource-constrained hardware, TDM is a widely adopted strategy.

This method utilizes a small number of Processing Elements (PEs) to serially update the states of numerous neurons over time, thereby enabling effective hardware resource sharing. Early research validated the feasibility of this approach as far back as 1998 [27]. Following a similar line of thought, Guo et al. ran a complete network using only 25 physical cores [9], and Luo et al. also reused computational hardware through pipelining to reduce resource consumption [18]. However, the serial nature of TDM limits a system's throughput and real-time capabilities when processing high-concurrency spike streams.

In contrast to TDM, massively parallel architectures prioritize extreme real-time performance. Such schemes assign a dedicated PE to each neuron (or a small group of neurons) in the network, thereby maximizing computational concurrency. For instance, the modern SiBrain architecture achieves multi-dimensional parallelism across space, time-steps, and channels through its Sparse Spatio-Temporal Parallel Processing Element Array (S^2TP-PE Array) [4]. To effectively support high-concurrency processing, researchers often employ techniques like parallel adder trees. In ReplaceNet, designed by Hwang et al. (2023) for neural rehabilitation applications, each Integrate-and-Fire (LIF) unit in the SNN hardware has a one-to-one correspondence with a replaced biological neuron, ensuring high throughput and real-time replacement capabilities via parallel operations [12]. For more general image processing tasks, the DeepFire architecture proposed by Aung et al. (2021) achieves state-of-the-art throughput of tens of thousands of frames per second (kFPS) on convolutional SNN by deploying hundreds of parallel compute cores and optimizing the data pipeline across the entire network [3]. Beyond general parallel strategies, some studies have explored specialized parallel schemes for specific network topologies. For example, the work by Ogaki et al. ingeniously maps the spatial locality of biological neuronal connections to a stencil computation pattern on hardware, creating a true "dataflow" pipeline that significantly enhances the simulation efficiency of networks with nearest-neighbor connectivity in the Hodgkin-Huxley model [19].

2.3   Motivation

To overcome the power and latency limitations of traditional audio processing, we propose an event-driven neuromorphic auditory system integrating a silicon cochlea (sensing) with a SNN. The cochlea's sparse spike output is the native input for an SNN, creating a naturally efficient, event-driven pipeline.

However, a key challenge exists. As discussed in Sect. 2.1, current high-fidelity cochlea models, while using hardware-friendly parallel filter banks, rely on slow and resource-intensive iterative algorithms. This conflicts with the goal of an efficient integrated system. Our first objective is therefore to design a lightweight, non-iterative event generation mechanism for the filter-bank architecture, seeking a better trade-off between classification accuracy and system efficiency.

Furthermore, overall system performance depends on the back-end. The high throughput spike data from our proposed fast cochlea would be bottlenecked by conventional, serial TDM SNN implementations, negating any latency gains. Our second objective, guided by system-level co-design, is to replace the

TDM paradigm. We will develop a massively parallel FPGA accelerator for the SNN, featuring a tightly-coupled, deeply pipelined architecture to preserve high-throug-hput and low-latency processing across the entire system.

3   Methods

This section details the bio-inspired, event-driven preprocessing module of the SpikeEAR system. The module's core function is to convert a raw audio signal into an information-dense, spatio-temporally sparse stream of spike events, mimicking the response mechanisms of the biological auditory pathway. The methodology is presented in two stages: first, bio-inspired frequency decomposition and energy envelope extraction; and second, the dynamic spike encoding and adaptive thresholding mechanism, which is the core innovation of this study.

3.1   Bio-inspired Frequency Decomposition and Envelope Extraction

To emulate cochlear frequency selectivity, we implement a parallel bank of M fixed-coefficient, second-order Infinite Impulse Response (IIR) filters, analogous to Gammatone models, to decompose the input signal x[n]. This efficient, hardware friendly design avoids the computational overhead and non-determinism of adaptive filters.

The filter bank's center frequencies are distributed on the Equivalent Rectangular Bandwidth (ERB) [21]scale to mimic human hearing. This yields higher resolution at lower frequencies, which are critical for speech perception, and coarser resolution at higher frequencies. The filtering in each channel m follows the standard second-order difference equation [13]:

$$y_m[n] = b_{m,0}x[n] + b_{m,1}x[n-1] + b_{m,2}x[n-2] - a_{m,1}y_m[n-1] - a_{m,2}y_m[n-2] \quad (1)$$

where $\{b_{m,k}\}$ and $\{a_{m,k}\}$ are the constant feedforward and feedback coefficients that define the desired frequency response. The output, $y_m[n]$, represents the basilar membrane's vibrational response within a specific frequency band.

Following frequency decomposition, the next stage emulates the function of IHCs by extracting the temporal energy envelope from each channel's output. This mimics the IHC transduction process, including its rectification and temporal integration properties. Operationally, this involves half-wave rectification of the filtered signal (approximated by its absolute value), followed by a first-order IIR low-pass filter to extract the energy envelope, $e_m[n]$. This is described by [23]:

$$e_m[n] = (1 - \alpha) \cdot |y_m[n]| + \alpha \cdot e_m[n-1] \quad (2)$$

where the smoothing factor, $\alpha \in [0, 1]$, controls the trade-off between the envelope's temporal fidelity and its smoothness. This filtering effectively demodulates the signal, removing the high-frequency carrier while preserving the slowly varying envelope that carries salient temporal information.

The parallel architecture with fixed-coefficient filters avoids the latency accumulation of cascaded designs and is well-suited for FPGA implementation. This approach provides a deterministic and resource-efficient foundation for the subsequent spike generation module, to which the resulting set of energy envelopes, $\{e_m[n]\}$, is passed as direct input.

3.2 Adaptive Encoding and Dynamic Thresholding

After obtaining the energy envelope $e_m[n]$ for each frequency channel, we introduce a novel, stateful processing module that emulates the behavior of auditory neurons. The objective of this module is to convert the continuous, analog-like energy envelopes into discrete, digital spike events. This process is executed independently for every channel and at each time step in an online, streaming fashion.

Composite Event Detection Criteria. A core innovation of our method is the replacement of simple spike generation schemes based on fixed thresholds with a more biologically plausible, adaptive encoding mechanism. Instead of a single criterion, an event is triggered by a composite condition that integrates sensitivity to three distinct acoustic features. Local Peak Detection: This criterion captures instantaneous high points in the signal envelope, designed specifically to encode high-frequency transients with steep attack and decay slopes, such as plosives or percussive clicks. Significant Onset Detection: This criterion identifies the beginning of new acoustic events by monitoring the rapid rate of increase in energy. It is crucial for segmenting the audio stream and marking the boundaries of speech or music. Sustained Activity Detection: This criterion ensures that following an onset, the sustained content of a sound, such as its pitch or texture, is not disregarded, thereby preserving the steady-state portions of the signal. Events are triggered by a logical OR of three detectors ('is_peak', 'is_onset', 'is_sustained'), ensuring sensitivity to diverse acoustic dynamics, from sharp transients to sustained tones (Algorithm 1, lines 4–8). This method produces a spike train that encodes both the sound's presence and its temporal structure, yielding a richer representation than single-threshold approaches.

Dynamic Thresholding and Refractory Period. For robust encoding in variable environments, such as fluctuating Signal-to-Noise Ratios and volume levels, we implement an independent dynamic threshold for each channel (Algorithm 1, lines 9–15), governed by a synergistic adaptation and recovery mechanism. This mechanism features an activity-driven fast adaptation, where upon event detection, the threshold is immediately adjusted upward in relation to the current signal energy $e_m[n]$. This rapid gain control prevents a strong signal from causing spike saturation and allows the system to remain responsive to subsequent energy changes. Complementing this is a slow recovery process. At every time step, regardless of event activity, the threshold decays exponentially at a fixed rate γ_{decay} toward a predefined minimum baseline value θ_{min}.

This "leaky memory" allows the system to gradually regain its sensitivity during quiet intervals. The update rule is governed by the equation:

$$\theta_m[n] = \max(\theta_m[n-1] \cdot \gamma_{decay}, \theta_{min}) \tag{3}$$

Here, $\theta_m[n]$ and $\theta_m[n-1]$ are the thresholds for channel m at the current and previous time steps, respectively, γ_{decay} is a constant close to 1, and θ_{min} is a floor value that ensures the system maintains basic sensitivity after long periods of silence. Finally, to ensure the temporal sparsity of the output event stream, we incorporate a refractory period from neuroscience, which suppresses a channel's firing for a brief interval after it generates a spike. This prevents redundant encoding of a single, continuous acoustic feature. The module's final output is a sparse, robust, and information-rich spatiotemporal spike pattern, providing an ideal input for efficient processing by subsequent SNN.

Algorithm 1. Dynamic Event Generation and Adaptive Encoding

Require: Envelope signals $\{e_m[n]\}_{m=1}^{M}$; parameters $\gamma_{decay}, \theta_{min}$
Ensure: Event stream
1: Initialize threshold $\theta_m[0] \leftarrow \theta_{min}$ for all m
2: **for** $n = 1$ to $N - 1$ **do**
3:    **for** $m = 1$ to M **do**
4:       // **Step 1: Check Composite Event Conditions**
5:       $is\_peak \leftarrow (e_m[n-1] > e_m[n]) \wedge (e_m[n-1] > e_m[n-2])$
6:       $is\_onset \leftarrow (e_m[n] - e_m[n-2]) > (0.5 \cdot \theta_m[n-1])$
7:       $is\_sustained \leftarrow e_m[n] > (1.5 \cdot \theta_m[n-1])$
8:       $has\_event \leftarrow (is\_peak \vee is\_onset \vee is\_sustained) \wedge (e_m[n] > \theta_{min})$
9:       // **Step 2: Generate Event and Update Threshold**
10:      **if** has_event and no recent event on channel m **then**
11:         Emit event $\langle m, n, 1.0 \rangle$
12:         $\theta_m[n] \leftarrow \max(\theta_m[n-1], 0.5 \cdot e_m[n])$ {Activity-driven adaptation}
13:      **else**
14:         $\theta_m[n] \leftarrow \theta_m[n-1]$
15:      **end if**
16:      // **Step 3: Apply Exponential Decay**
17:      $\theta_m[n] \leftarrow \max(\theta_m[n] \cdot \gamma_{decay}, \theta_{min})$
18:    **end for**
19: **end for**
20: **return** Event stream

---

4   SpikeEAR: Design and Implementation

SpikeEAR, the end-to-end neuromorphic auditory system proposed in this paper, is founded on the core idea of emulating the complete biological auditory pathway, from perception to cognition. The system is conceptualized as a deep processing pipeline, illustrated in Fig. 1, which clearly shows the progressive transformation of information from continuous, redundant raw audio signals into discrete, sparse spike events. The pipeline proceeds as follows: (a) A raw audio signal is received as input. (b) This signal is decomposed into multiple frequency channels by a parallel array of cochlear models. (c) Spike signals are then independently generated within each channel. (d) An encoder module integrates the sparse spikes from all channels, (e) formatting them into a unified event data stream. (f) This event stream, rich in spatio-temporal information, is fed into a SNN backend for real-time pattern recognition. (g) The final classification decision is represented by the firing of neurons in the network's output layer.

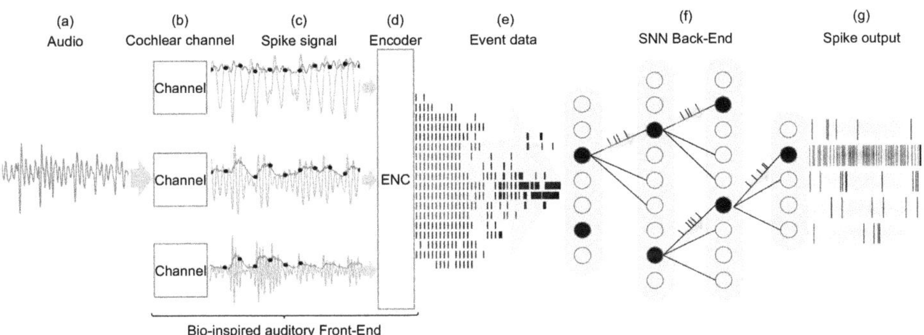

Fig. 1. The conceptual diagram of the end-to-end neuromorphic processing pipeline of SpikeEAR.

To efficiently implement this conceptual model in hardware, we designed a heterogeneous SoC architecture, shown in Fig. 2. This architecture integrates a Processing System (PS) with Programmable Logic (PL) to separate the control plane from the computation plane. In this design, the processor-based PS is responsible for high-level task management and external memory control, while all computationally intensive neuromorphic processing tasks are offloaded to the PL and executed efficiently by our custom-designed IP cores.

The PL side serves as the core computational unit of the system and is primarily composed of two main modules: a Parallel Bionic Cochlea Model and an SNN Engine. Specifically, the cochlea model consists of a parallel array of EGNs. Each EGN, acting as a basic unit, converts the continuous audio signal into a discrete stream of spike events through filtering, rectification, and comparison with an adaptive threshold. The SNN Engine, in turn, is constructed from an array of PEs, where each PE implements a Leaky LIF neuron model. These PEs are responsible for the spatio-temporal integration of weighted input spikes and will fire a spike and reset their state when their membrane potential exceeds a threshold. For inter-module communication, a high-speed AXI-Stream bus transfers the event stream between the two core modules. Concurrently, the PS configures and monitors the status of these modules via a standard AXI bus and utilizes a DMA controller to coordinate data exchange between the PL and external memory.

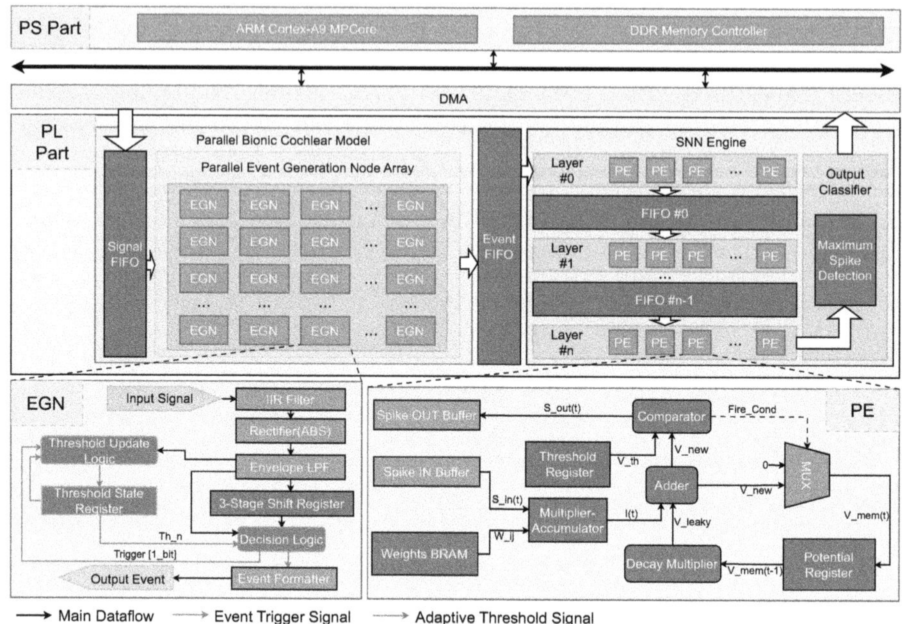

Fig. 2. The overall hardware architecture of the SpikeEAR system.

4.1   Parallel Bionic Cochlea Model

The Parallel Bionic Cochlea Model constitutes the core "sensory" front-end of the entire neuromorphic system. Its primary function is to emulate the biological cochlea, converting continuous raw audio signals into a sparse stream of spike events that encode rich spatio-temporal information in real-time and with high efficiency. As depicted in the left half of Fig. 2, the model's top-level hardware architecture is a large-scale parallel array composed of multiple identical **Event Generation Nodes (EGNs)**. To achieve maximum data throughput and minimum processing latency, this array is fully unrolled in hardware. This design choice means that each EGN has its own dedicated, physically distinct computational and storage resources. Consequently, a single audio input sample can be broadcast to all nodes and processed in parallel within the same clock cycle, which lays the foundation for the real-time performance of the entire system.

Each EGN is a sophisticated, deeply pipelined processing unit, whose microarchitecture is detailed in Fig. 2. This pipeline precisely executes the complete workflow from frequency decomposition to event encoding. Data enters the first stage of the pipeline, the IIR filter, from a Signal Buffer. This is a fourth-order Gammatone IIR filter, whose passband characteristics are carefully designed to mimic the frequency selectivity of different locations along the biological cochlear basilar membrane. For the hardware implementation, we construct this fourth-order filter by cascading two second-order Biquad sections, a structure that strikes a good balance between resource utilization and numerical stability. The coefficients for all filters are pre-computed based on the ERB scale and stored as constants in on-chip memory, thereby avoiding complex online computations.

Following frequency decomposition, the band-passed signal stream is fed into the next stage: the Envelope Detector. The core function of this module is to extract the energy features of the signal, a process analogous to how IHCs in the biological system convert mechanical vibrations into neural signals. Specifically, the signal first undergoes half-wave rectification to preserve its energy information and is then passed through a first-order IIR low-pass filter. This low-pass filter smooths out high-frequency fluctuations, thereby extracting the slowly varying envelope of the signal's energy, which we denote as the instantaneous energy E_t. This continuously varying energy envelope serves as the direct input for the subsequent event generation stage.

The final stage of the pipeline is centered around a core dynamic feedback loop, which consists of a unified Hierarchical Decision Logic and an Adaptive Threshold Logic. Departing from conventional methods that use a fixed threshold, this design introduces a sophisticated, two-stage event decision process, as illustrated in Fig. 3.

The first stage is **Stage 1: Instantaneous Candidate Detection**, utilizes a combinational logic datapath to identify potential sound events. This stage concurrently executes three parallel detection paths—Sustained Energy, Energy Onset, and Peak Detection—to capture diverse acoustic features. As illustrated in Fig. 3, the outputs of these paths are first aggregated via an OR gate and then gated by a noise signal (when Envelope(n) > MIN_THRESH), producing a single-bit Candidate Event Flag.

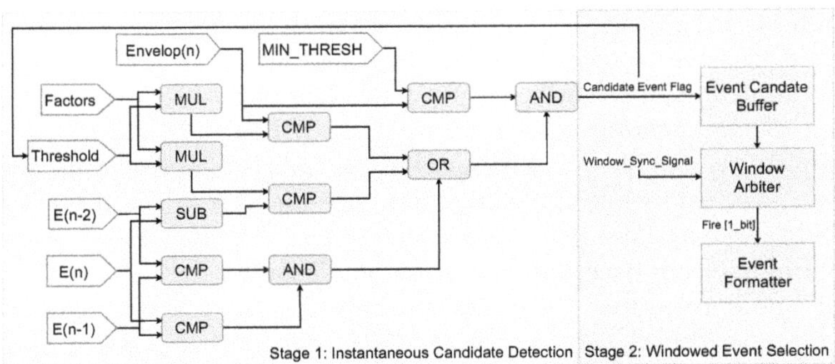

Fig. 3. The Adaptive Thresholding Unit.

This flag initiates **Stage 2, Windowed Event Selection**. Instead of immediate processing, candidate metadata is buffered. A periodic Window_Sync_Signal then triggers a Winner-Take-All (WTA) arbiter to select the most significant candidates within the buffer, issuing a Fire signal for each winner to generate a final output event. Critically, the system employs a dual-path control mechanism.

The Candidate Event Flag provides immediate feedback to the adaptive threshold logic for rapid micro-dynamic adaptation. Simultaneously, the WTA-based selection regulates the final event rate, ensuring output sparsity. This decoupling of rapid threshold adaptation from final event generation significantly enhances the encoding's signal-to-noise ratio (SNR) and robustness.

4.2 SNN Engine

The SNN Engine serves as the "cognitive" core of the system. It is a hardware accelerator specifically designed to efficiently process streams of spike data and perform the final classification task. The engine's computation is based on the discrete-time LIF neuron model, whose dynamics are described by Equations (4) through (6).

$$V[t + 1] = \beta V[t] + I[t + 1] \tag{4}$$

$$S[t + 1] = \Theta(V[t + 1] - V_{\text{th}}) \tag{5}$$

$$V[t + 1] \leftarrow V[t + 1] \cdot (1 - S[t + 1]) + V_{\text{reset}} \cdot S[t + 1] \tag{6}$$

The primary design challenge for the SNN Engine is to process in real-time the sparse yet highly concurrent event stream generated by the front-end bionic cochlea, which uses the adaptive threshold encoding described in Sect. 3.2. To meet these high-throughput and low-latency demands, we adopted two key optimization strategies: a deeply pipelined architecture and a co-design of memory and computation. These strategies ensure that the processing capability of the SNN Engine seamlessly matches the output rate of the front-end cochlea.

Our deeply pipelined architecture (right half of Fig. 2) is designed to overcome the serial processing bottleneck of traditional SNN accelerators, which often rely on TDM. In our architecture, the n layers of the SNN are directly mapped onto n physically sequential and functionally independent hardware processing modules. Inter-layer communication is buffered by $n - 1$ FIFOs, which enables task-level parallelism. This pipelined mechanism allows subsequent layers to process an existing event stream while preceding layers are simultaneously processing a new one, thereby dramatically increasing event processing efficiency and preventing the SNN Engine from becoming a system bottleneck.

Furthermore, to address the limitations imposed by memory access latency on the real-time performance of large-scale SNNs, we adopted a design that tightly couples memory and computation. The core of the SNN Engine consists of a large-scale array of LIF Processing Elements, with all synaptic weights pre-stored in BRAMs located in close proximity to their respective PEs. This design offers two major advantages. First, the BRAMs support single-cycle parallel access to a large number of weights, providing the PE array with a data bandwidth that far exceeds that of off-chip memory. Second, the physical coupling of computation and storage eliminates time-consuming off-chip bus transactions, significantly reducing both data access latency and power consumption.

Through deep pipelining and memory-computation co-design, the SNN Engine can process the highly concurrent event streams from the front-end cochlea in real-time and with high energy efficiency. This design not only breaks through the performance bottlenecks of traditional SNNs but also provides a solid hardware foundation for the seamless integration of our end-to-end neuromorphic system.

5   Evaluations

5.1   Evaluation of the Cochlea

To evaluate the performance of our Cochlea model and to validate its capability as a robust and efficient auditory front-end for speech recognition in noisy environments, we conducted a series of comparative experiments.

We employed a consistent SNN with a single hidden layer as the backend classifier to assess the Cochlea model configured with different parameters. The experiments were performed on the public Speech Commands dataset, from which we selected five word categories for training and prediction: 'go', 'happy', 'on', 'off', and 'no'. The objective of this section is to determine an optimal network configuration that balances high recognition accuracy with computational efficiency through a systematic evaluation of key preprocessing parameters, such as time resolution and the number of filter bank channels.

First, we investigated the combined impact of time resolution and the environmental SNR on word classification accuracy. The results of this five-class classification experiment are presented in Fig. 4. We generated multiple test sets with SNRs ranging from $-5\,\mathrm{dB}$ to $20\,\mathrm{dB}$ by mixing varying levels of Additive White Gaussian Noise (AWGN) into the original audio.

As can be observed from the figure, the optimal choice of time resolution is closely correlated with the SNR level. Under high-SNR conditions (e.g., $20\,\mathrm{dB}$), a finer time resolution of $1\,\mathrm{ms}$ achieves the highest classification accuracy by capturing richer phonetic details. In low-SNR environments (e.g., $0\,\mathrm{dB}$), however, the optimal time resolution shifts to $5\,\mathrm{ms}$. This phenomenon reveals a core trade-off in selecting the time resolution: an excessively high resolution (e.g., $0.5\,\mathrm{ms}$) increases the model's sensitivity to noise, whereas an overly coarse resolution (e.g., $30\,\mathrm{ms}$) leads to a loss of effective information due to insufficient sampling. Both scenarios result in a degradation of classification performance. Considering the model's generalization capability across different noisy environments, we ultimately selected $5\,\mathrm{ms}$ as the fixed time resolution. This choice aims to strike a balance between robustness and precision, ensuring that the model maintains stable performance under varying SNR conditions.

Next, we evaluated the effect of the number of filter bank channels on classification accuracy, as shown in Fig. 5. The experiment compared our proposed adaptive threshold method from Sect. 3.2 against two baseline fixed-threshold methods. The first fixed-threshold method (Fixed, Mean of Max/Min) is a common baseline strategy in the field of event detection [26]. It determines a global threshold by calculating the mean of the peak-to-trough amplitude of the signal

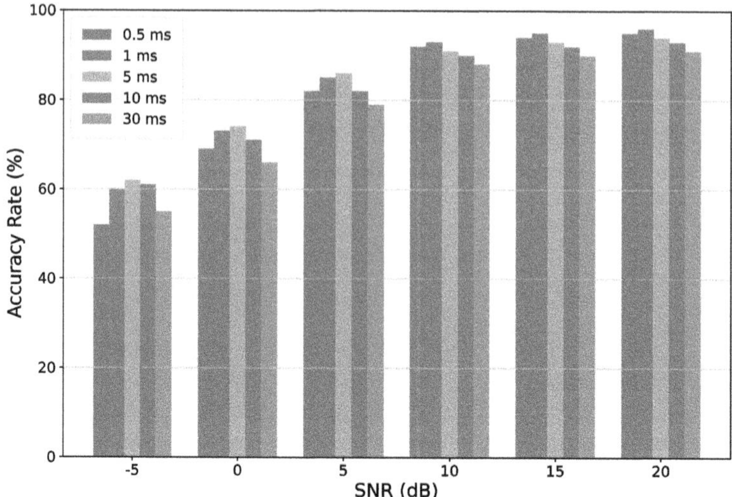

Fig. 4. Comparison of classification accuracy.

within windows across the entire time series, thereby making simple judgments based on the signal's dynamic range. For a more comprehensive comparison, we also introduced a second fixed-threshold method (Fixed, Mean+Std Dev), which is widely used as a standard baseline in bio-signal processing [28]. Its threshold is set to the mean of the signal energy plus a specific multiple of its standard deviation.

The results clearly demonstrate that for all tested channel counts, our proposed adaptive threshold method significantly outperforms both fixed-threshold schemes. Under the adaptive thresholding strategy, the classification accuracy increases monotonically with the number of channels, reaching its peak performance at 128 channels. Although the two fixed-threshold methods also exhibit a similar growth trend, a considerable performance gap remains between them and the adaptive method, validating the superiority of our adaptive approach. Notably, the accuracy gains for all methods begin to diminish significantly after the channel count exceeds 48. Specifically, for the top-performing adaptive method, increasing the number of channels from 48 to 128 yields an accuracy improvement of less than one percentage point. Considering that more channels imply a greater computational load and higher hardware resource consumption, we conclude that 48 channels represent an optimal trade-off between classification accuracy and computational efficiency, as it achieves near-saturating recognition performance without excessively increasing computational complexity.

Based on the preceding experimental analysis, the final parameter configuration for the signal preprocessing pipeline in this study was determined as follows: All incoming audio signals are uniformly resampled to $f_s = 16\,\mathrm{kHz}$. Subsequently, the signal is fed into a filter bank composed of M=48 parallel

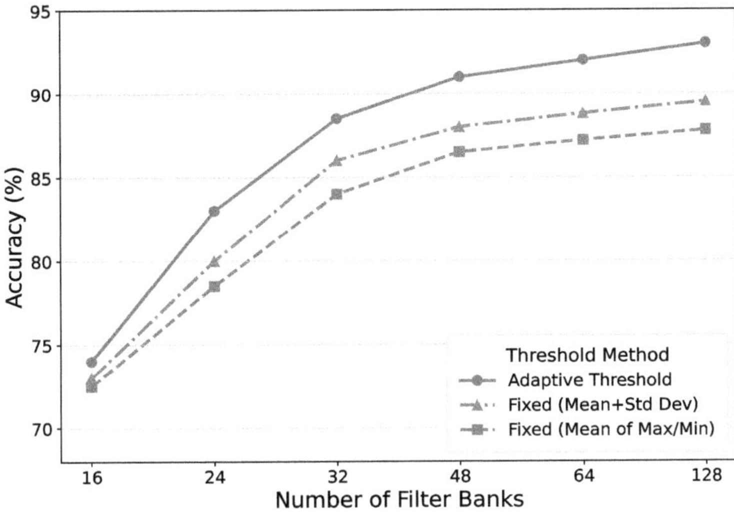

Fig. 5. Comparison of Classification Accuracy for Different Threshold Methods.

fourth-order (K=2) Gammatone filters to emulate the frequency decomposition function of the cochlea.

5.2 System-Level Evaluation

In this subsection, we conduct a system-level evaluation of SpikeEAR. All experiments were conducted on the benchmark dataset previously described in Sect. 5.1. For the network structure, we chose to use a structure with two hidden layers, each with 128 neurons. The network was trained using the snnTorch framework [6] by jointly learning the membrane potential decay constant (β), the firing threshold, and the synaptic weights. For each 1-second audio sample, the *SNN* processes a 48-dimensional feature vector using a 20 ms sliding window. This window length is chosen based on the shortest perceivable auditory event in humans (approximately 10–40 ms), adhering to bio-inspired design principles.

For network training, 80% of this subset was utilized, augmented with additive noise to achieve a SNR of 20 dB. For the FPGA implementation, we converted SpikeEAR to a fixed-point representation. Since the cochlea module outputs a "1" only upon event detection, its integer bit requirement is minimal. This allows for allocating more bits to the fractional part, thereby preserving computational precision while leveraging end-to-end fixed-point arithmetic to enhance processing speed. Ultimately, a 16-bit fixed-point format was adopted, achieving a recognition accuracy of 95.69% on this subset.

For hardware implementation, SpikeEAR was described in Verilog and subsequently synthesized and routed using the *Xilinx Vivado Design Suite* [7]. The precise power consumption was obtained post-route from the suite's built-in *Power Report* tool. The synthesis results confirm that the design meets tim-

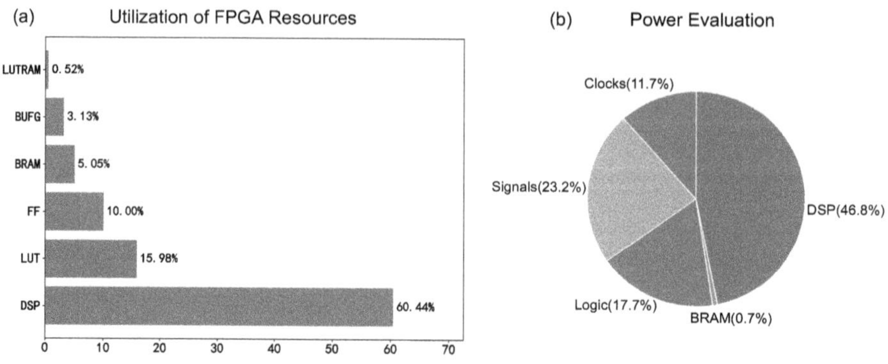

Fig. 6. Resources Report.

ing constraints at 137.63 MHz. The system's power distribution is illustrated in Fig. 6.

Resource Utilization (Fig. 6-a). Among the on-chip FPGA resources, DSP48s account for the highest proportion (60.44 %). This is directly related to the extensive use of parallel Multiply-Accumulate (MAC) operations within SpikeEAR's *SNN* inference core. Following this are LUTs and FFs, with utilization rates of 15.98 % and 10.00 %, respectively. These are primarily consumed by control logic, finite-state machines (FSMs), and small-scale register files. The utilization of BRAMs is relatively low for two main reasons: 1) The total storage required for weights and FIFOs is only about 15 kB, which can be easily implemented using a small number of 18k/36k BRAMs. 2) Small arrays, such as those for neuron membrane potentials and thresholds, are preferentially mapped by the synthesizer to distributed RAM, further reducing the demand for dedicated BRAM blocks.

Power Breakdown (Fig. 6-b). The power consumption of DSP logic follows at 46.8 %, corresponding to the intensive MAC computations mentioned above. This is followed by signal routing (23.2 %) and general-purpose logic (17.7 %). Overall, both the power hotspots and the resource distribution profile are highly consistent with the workload characteristics of SpikeEAR's "front-end filtering + parallel SNN" architecture, validating the feasibility and efficiency of the hardware accelerator in terms of resource utilization and power management.

Figure 7 visualizes the end-to-end processing pipeline of SpikeEAR for five distinct audio commands. The figure illustrates the layer-by-layer transformation of each command from a raw audio waveform into sparse spike trains by the cochlea, followed by progressive feature abstraction in the SNN's hidden layers. The classification result is clearly demonstrated in the output layer: for each input, the neuron corresponding to the correct class fires intensely while all other neurons are suppressed. For instance, processing the word "Happy" triggers a dense spike sequence exclusively in the designated 'Happy' neuron. This visu-

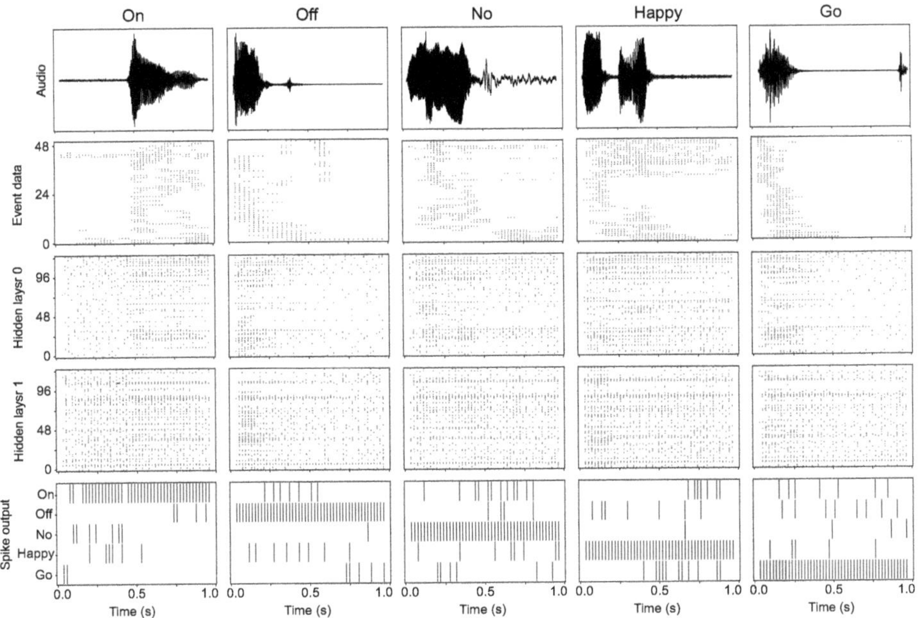

Fig. 7. Classification results for samples from each audio category.

alization provides compelling evidence of the system's end-to-end effectiveness, from robust feature encoding to accurate spike-based classification.

To better assess the performance of the SpikeEAR front-end and back-end, we independently perform a comprehensive evaluation of the Cochlear Model and SNN Engine. The comparison with existing electronic cochlea solutions is presented in Table 1, while the performance of the SNN Engine is benchmarked against other state-of-the-art FPGA accelerators in Table 2.

First, we evaluated the front-end of the system. In Table 1, we compare the key metrics of our proposed electronic cochlea against existing VLSI and FPGA solutions. The primary advantage of our design is its power efficiency. Our implementation consumes only 535 mW, a figure that represents a greater than twofold reduction compared to other solutions on similar FPGA platforms (1260 mW and 1382 mW). While dedicated VLSI implementations inherently offer lower power consumption, such as the 0.055 mW achieved in an advanced 0.18 μm process, our work demonstrates exceptional optimization for a reconfigurable 28nm FPGA platform. Regarding the number of channels, although our 48 channels are fewer than in some other works, this configuration is a deliberate design trade-off to balance performance and resource consumption while preserving classification accuracy, as justified in Sect. 5.1. This work sets a new benchmark for power

Table 1. Comparison with Existing Electronic Cochleas

Specs.	VLSI-based		FPGA-based		
	[8] (2005)	[31] (2016)	[30] (2018)	[2] (2025)	This work
Process Technology	0.5 μm	0.18 μm	28 nm	28 nm	28 nm
Architecture based	CMOS	CMOS	Cyclone-5	Artix-7	Zynq-7000
Filter Type	Parallel	Parallel	Cascade	Parallel	Parallel
Channel Number	100	64	70	120	48
Frequency Range (Hz)	200–20,000	8,000–20,000	Up to 22,050	20–8,000	20–8,000
Power Consumption (mW)	1.7	0.055	1,260	1,382	535

Table 2. Benchmarking the SNN Engine against Similar Works

Ref.	FIX Models	Date Precision	FPGA Freq. (MHz)	LUT	DSP	FPGA Family	Latency (μ s)	Throughput (GOPS)
[14] (2016)	S-LIF	FLO	132	15137	400	Spartan-6	2260	48.67
[10] (2020)	S-LIF	FLO	100	56230	64	Virtex-6	1100	22.81
[32] (2022)	LIF	FIX	100	80172	0	Kintex-7	1210	12.96
[4] (2024)	LIF	FIX	200	87870	1000	Virtex-7	1960	8.02
This work	LIF	FIX	137	39180	287	Zynq-7000	255	89.11

efficiency on reconfigurable platforms, taking an important step toward narrowing the efficiency gap with dedicated ASICs.

$$\text{Throughput (GOPS)} = \frac{N_{\text{mult}} + N_{\text{add}}}{T_{\text{latency}} \times 10^9} \tag{7}$$

We prioritized throughput in our evaluation, as it is a more comprehensive and reliable performance metric than latency for hardware accelerators handling diverse and continuous data streams. Our system's back-end SNN engine sets a new performance benchmark with a throughput of 89.11 GOPS (Table 2), substantially outperforming other FPGA-based accelerators. For the throughput calculation, we define a single operation (OP) as a multiplication or an addition (7). Notably, compared to [4], our design delivers over 11 times the throughput while utilizing fewer than half the LUTs and one-third of the DSPs. Such high computational efficiency is a direct result of our novel fixed-point neuron optimization and highly parallel architecture. Furthermore, the system exhibits a minimal inference latency of just 255 μs, which is 4 to 7 times faster than competing millisecond-scale accelerators. In terms of power, the SNN engine itself consumes 765 mW.

The integrated SpikeEAR system achieves a recognition accuracy of 95.69 % with an average end-to-end latency of 256.95 μs. This performance is achieved with a total on-chip power consumption evaluated at 1.25 W (which consists of a 535 mW front-end electronic cochlea and a 765 mW back-end SNN engine), not including the additional 1.56 W consumed by the ZynqPS side. These results validate the end-to-end functionality of our proposed system, demonstrating a successful balance between sensing and computation for real-time auditory processing applications.

6   Conclusion

In this work, we presented SpikeEAR, a neuromorphic auditory system that embodies the SCI principle by unifying a bionic cochlea and a SNN accelerator on a single digital fabric. Hardware validation on a Xilinx ZC706 platform demonstrates a compelling performance profile: SpikeEAR achieves 95.69 % accuracy on the Google Speech Commands subset and an end-to-end latency of only $256.95\,\mu s$, a $4 \sim 7\times$ reduction compared to recent FPGA accelerators. Notably, these results are achieved with a total on-chip accelerator power of 1.25 W. This unique combination of microsecond-level latency and a competitive power budget on a reconfigurable platform validates the SCI paradigm as an effective strategy for developing high-performance intelligent systems at the edge.

Acknowledgments. This work is supported by the National Key R&D Program of China (No. 2022YFB4400703), the National Key R&D Program of Heilongjiang Province (2024ZXDXA12) and the Fundamental Research Funds for the Central Universities (3072025YC602).

References

1. Al-Selwi, S.M., et al.: RNN-LSTM: from applications to modeling techniques and beyond—systematic review. J. King Saud Univ. Comput. Inf. Sci. 102068 (2024)
2. Alsakkal, M.A., Wijekoon, J.: Spiketrum: an FPGA-based implementation of a neuromorphic cochlea. IEEE Transactions on Circuits and Systems I: Regular Papers (2025). preprint, accessed 2024
3. Aung, M.T.L., Qu, C., Yang, L., Luo, T., Goh, R.S.M., Wong, W.F.: DeepFire: acceleration of convolutional spiking neural network on modern field programmable gate arrays. In: 2021 31st International Conference on Field-Programmable Logic and Applications (FPL), pp. 28–32. IEEE (2021)
4. Chen, Y., Ye, W., Liu, Y., Zhou, H.: Sibrain: a sparse spatio-temporal parallel neuromorphic architecture for accelerating spiking convolution neural networks with low latency. IEEE Transactions on Circuits and Systems I, Regular Papers (2024)
5. Deng, B., Fan, Y., Wang, J., Yang, S.: Auditory perception architecture with spiking neural network and implementation on FPGA. Neural Netw. **165**, 31–42 (2023)
6. Eshraghian, J.K., et al.: Training spiking neural networks using lessons from deep learning. Proc. IEEE **111**(9), 1045–1082 (2023)
7. Feist, T.: Vivado design suite. White Paper **5**(30), 24 (2012)
8. Fragnière, E.: A 100-channel analog CMOS auditory filter bank for speech recognition. In: ISSCC. 2005 IEEE International Digest of Technical Papers. Solid-State Circuits Conference, 2005. pp. 140–589. IEEE (2005)
9. Guo, W., Yantır, H.E., Fouda, M.E., Eltawil, A.M., Salama, K.N.: Towards efficient neuromorphic hardware: unsupervised adaptive neuron pruning. Electronics **9**(7) (2020)

10. Gupta, S., Vyas, A., Trivedi, G.: FPGA implementation of simplified spiking neural network. In: 2020 27th IEEE International Conference on Electronics, Circuits and Systems (ICECS), pp. 1–4. IEEE (2020)

11. Han, J., Li, Z., Zheng, W., Zhang, Y.: Hardware implementation of spiking neural networks on FPGA. Tsinghua Sci. Technol. **25**(4), 479–486 (2020)

12. Hwang, S., et al.: ReplaceNet: real-time replacement of a biological neural circuit with a hardware-assisted spiking neural network. Front. Neurosci. **17**, 1161592 (2023)

13. Islam, R., Tarique, M.: Investigating the performance of gammatone filters and their applicability to design cochlear implant processing system. Designs **8**(1), 16 (2024)

14. Kiselev, I., Neil, D., Liu, S.C.: Event-driven deep neural network hardware system for sensor fusion. In: 2016 IEEE International Symposium on Circuits and Systems (ISCAS), pp. 2495–2498. IEEE (2016)

15. Lenk, C., et al.: Neuromorphic acoustic sensing using an adaptive microelectromechanical cochlea with integrated feedback. Nat. Electron. **6**(5), 370–380 (2023)

16. Leone, G., et al.: A Tiny RISC-V-Controlled SNN Processor for Real-Time Sensor Data Analysis on Low-Power FPGAs. IEEE Transactions on Circuits and Systems I, Regular Papers (2024)

17. Liu, S.C., van Schaik, A., Minch, B.A., Delbruck, T.: Asynchronous binaural spatial audition sensor with $2 \times 64 \times 4$ channel output. IEEE Trans. Biomed. Circuits Syst. **8**(4), 453–464 (2014)

18. Luo, J., Coapes, G., Degenaar, P., Yamazaki, T., Mak, T., Tin, C.: A real-time silicon cerebellum spiking neural model based on FPGA. In: 2014 International Symposium on Integrated Circuits (ISIC), pp. 276–279. IEEE (2014)

19. Ogaki, M., Sato, Y.: Hodgkin-huxley-based neural simulation with networks connecting to near-neighbor neurons. In: 2021 IEEE 32nd International Conference on Application-specific Systems, Architectures and Processors (ASAP), pp. 109–116. IEEE (2021)

20. Purwono, P., Ma'arif, A., Rahmaniar, W., Fathurrahman, H.I.K., Frisky, A.Z.K., ul Haq, Q.M.: Understanding of convolutional neural network (CNN): a review. Int. J. Robot. Control Syst. **2**(4), 739–748 (2022)

21. Radha, K., Bansal, M., Pachori, R.B.: Automatic speaker and age identification of children from raw speech using SincNet over ERB scale. Speech Commun. **159**, 103069 (2024)

22. Rathi, N., et al.: Exploring neuromorphic computing based on spiking neural networks: algorithms to hardware. ACM Comput. Surv. (CSUR) **55**(12), 1–49 (2023)

23. Roonizi, A.K.: Digital IIR filters: effective in edge preservation? Signal Process. **221**, 109492 (2024)

24. Shan, S., Liu, J., Wu, S., Shao, Y., Li, H.: A motor bearing fault voiceprint recognition method based on Mel-CNN model. Measurement **207**, 112408 (2023)

25. Szczerek, W.J., Podobas, A.: A quarter of a century of neuromorphic architectures on FPGAs – an overview. arXiv preprint arXiv:2502.20415v2 (2025)

26. Terranova, F., et al.: Windy events detection in big bioacoustics datasets using a pre-trained convolutional neural network. Sci. Total Environ. **949**, 174868 (2024)

27. Waldemark, J.T.A., Lindblad, T., Lindsey, C.S., Waldemark, K.E., Öberg, J., Millberg, M.: Pulse coupled neural network implementation in FPGA. In: Rogers, S.K., Fogel, D.B., Bezdek, J.C., Bosacchi, B. (eds.) Applications and Science of Neural Networks, Fuzzy Systems, and Evolutionary Computation. SPIE Proceedings, vol. 3390, pp. 392–402 (1998)

28. Wang, L.H., et al.: A novel real-time threshold algorithm for closed-loop epilepsy detection and stimulation system. Sensors **25**(1), 33 (2024)
29. Xu, Y., Perera, S., Bethi, Y., Afshar, S., van Schaik, A.: Event-driven spectrotemporal feature extraction and classification using a silicon cochlea model. Front. Neurosci. **17**, 1125210 (2023)
30. Xu, Y., et al.: A FPGA implementation of the CAR-FAC cochlear model. Front. Neurosci. **12**, 198 (2018)
31. Yang, M., Chien, C.H., Delbruck, T., Liu, S.C.: A 0.5 V 55 μW 64×2 channel binaural silicon cochlea for event-driven stereo-audio sensing. IEEE J. Solid-State Circuits **51**(11), 2554–2569 (2016)
32. Ye, W., Chen, Y., Liu, Y.: The implementation and optimization of neuromorphic hardware for supporting spiking neural networks with MLP and CNN topologies. IEEE Trans. Comput. Aided Des. Integr. Circuits Syst. **42**(2), 448–461 (2022)

KiRa: A Unified Memory Architecture for Efficient Post-quantum Cryptographic Algorithms

Zhiming Hu[1], Tianlin Liu[1], Xinhua Wang[1], Sizhao Li[1,2]($\boxtimes$), Xiaojing Fu[1,2], and Qiuliang Li[3]

[1] College of Computer Science and Technology, Harbin Engineering University, Harbin, China
sizhao.li@hrbeu.edu.cn
[2] Key Laboratory of information Secrecy and Protection Technology, Ministry of Industry and Information Technology, Harbin, China
[3] Shanghai Jian Qiao University, Shanghai, China

Abstract. As quantum computing poses a significant threat to traditional cryptographic systems, the need for post-quantum cryptographic (PQC) solutions has become urgent. Implementing these lattice-based algorithms on resource-constrained platforms, such as embedded systems and IoT devices, presents challenges. These challenges stem from the high computational and memory demands of the algorithms. Existing hardware designs often focus on accelerating specific computational tasks, like the Number Theoretic Transform (NTT), but are usually tailored to individual algorithms, making them inefficient for multi-algorithm systems. In this work, we address these challenges by designing **KiRa**: A Unified Memory Architecture for PQC that supports both CRYSTALS-Dilithium and CRYSTALS-Kyber at all NIST security levels. The architecture allows core components, such as Keccak, NTT, and sampling modules, to be shared across both Dilithium and Kyber algorithms. A uniform control logic and dataflow enable seamless switching between the two algorithms without duplicating resources. To support high-throughput polynomial operations, KiRa includes a dual-port memory design that efficiently handles polynomial data, and features a high-throughput NTT engine compatible with both algorithms. We implement and test the design on a Zynq UltraScale+ MPSoC FPGA platform. The results demonstrate a significant reduction in logic usage, memory access frequency, and dynamic power consumption, while improving latency for signing and decryption operations. Our results show that KiRa achieves a 23.8% reduction in area, a 17.3% improvement in latency, and a 72.1% reduction in dynamic power, proving its potential as an efficient and scalable solution for post-quantum cryptographic hardware.

Keywords: Lattice-based lgorithms · Unified Memory Architecture · Memory Access Optimization · Near-Memory Computing · FPGA Implementation

1   Introduction

As quantum computing technology advances rapidly, traditional public-key cryptographic systems like RSA [1] and ECC are becoming vulnerable. Shor's algorithm [2] and advancements in quantum hardware make it possible to solve problems such as integer factorization and elliptic curve discrete logarithms in polynomial time. This poses a significant threat to the core security of current cryptographic algorithms.

To respond to this threat, the global cryptographic community has focused on developing post-quantum cryptography (PQC) [3]. In 2016, the National Institute of Standards and Technology (NIST) launched the PQC standardization process [4].

After several evaluation rounds, NIST recommended lattice-based algorithms like CRYSTALS-Dilithium [5] and CRYSTALS-Kyber [6] as the main candidates for digital signatures and key encapsulation [7]. These algorithms offer high security and efficient performance, making them leading choices for future secure systems. Several software implementations of CRYSTALS-Dilithium and CRYSTALS-Kyber have already been analyzed, and vulnerabilities discovered have been patched in the subsequently released versions. However, the evaluation of hardware implementations has only just begun. Applying these complex lattice-based algorithms on resource-limited platforms, such as embedded systems or IoT devices, remains a significant challenge.

Most hardware implementations [8,9] today focus on speeding up computation-heavy modules like the Number Theoretic Transform (NTT) or polynomial multiplication [11]. But these designs are often customized for a single algorithm, and their memory architecture cannot be reused across different schemes.

Fixed-function memory designs lack flexibility and cannot support both Dilithium and Kyber in the same hardware system. Additionally, duplicated storage and repeated memory access lead to high power consumption and large area costs. This is especially problematic for low-power battery-based devices. Thus, a general-purpose, energy-efficient memory architecture is needed for PQC hardware.

Our Contributions: In this paper, we presents KiRa: A Unified Memory Architecture for Efficient Post-Quantum Cryptographic Algorithms, which efficiently supports both Dilithium and Kyber algorithms at all NIST security levels on the same hardware:

1. We designed a modular and reusable architecture where core components like Keccak, NTT, and sampling modules are shared across both algorithms. There is a uniform control logic and dataflow that allow Dilithium and Kyber to switch operations without duplicating resources;
2. To support high-throughput polynomial operations, we adopt a dual-port, conflict-free storage design at the same time. The architecture combines lightweight butterfly units and near-memory mapping logic. the polynomial storage architecture consists of a control unit, address generation and mapping units, polynomial RAM with four banks, a twiddle factor ROM, and

a radix-4 NTT engine. The system is designed for in-place, high-throughput polynomial transforms, compatible with both Dilithium and Kyber.
3. We test our architecture on a Zynq UltraScale+ MPSoC FPGA platform. The design supports full algorithm flows for Dilithium and Kyber. Compared with traditional shared designs of Dilithium and Kyber, KiRa uses 37890 LUTs (57.7%), 15753 DFFs (60.9%), 20 DSPs (62.5%), and 15 BRAMs (50.0%) in FPGA. our design achieves a highest 69% improvement in memory access of NTT stage, consumes 23.8% less area, achieves a 17.3% improvement in latency, reduces dynamic power by 72.1%.

The rest of this paper is organized as follows. Section 2 gives a background on Dilithium and Kyber algorithm, then reviews previous work related to Dilithium and Kyber hardware implementations. Section 3 presents the challenges of designing a unified memory Architecture for both two algorithms. Section 4 presents the overall architecture design and dual-port, conflict-free storage design of KiRa. Section 5 summarizes experimental results. Section 6 concludes the paper.

2   Background

Lattice-based cryptographic schemes have gained increasing prominence in the post-quantum cryptography (PQC) standardization process. As a result, Dilithium and Kyber have emerged as representative algorithms and primary targets for practical hardware implementations. To enable hardware support for both algorithms, it is essential to comprehend their algorithmic workflows and structures. Additionally, an in-depth analysis of the data characteristics generated during their computational routines and their storage requirements is necessary. This foundational analysis provides the basis for Integrating those two algorithm and designing efficient memory architectures.

2.1   Dilithium Algorithm

Dilithium is a post-quantum secure digital signature scheme based on the hardness of the Short Integer Solution (SIS) problem in lattice-based cryptography [24]. It is one of the final candidates in the NIST post-quantum cryptography standardization project [17]. Dilithium provides security levels suitable for different applications, and its parameter sets offer strong security guarantees against quantum attacks. It has three main operations:

KeyGen(ρ). The private key consists of two vectors s_1 and s_2, both independently sampled from a binomial distribution B_η. The private key sk is composed as $sk = (s_1, s_2)$. The public key contains a matrix A and a vector t. The matrix A is sampled from a uniform distribution and has a size of $k \times k$, used to construct the signature. The generation of matrix A is done using an extendable output function (XOF), which generates each column based on a random seed. The vector t in the public key is calculated as $t = A \cdot s_1 + s_2$. The final public key is $pk = (A, t)$, consisting of the matrix A and the vector t.

Sign(sk, m). A signature is generated by the private key for a given message m. First, a random vector y is selected and then a challenge value c is computed, which is generated by the message m and a hash of the public key information. The challenge value c and the private key s_1 are used to compute the intermediate value $\hat{s} = A \cdot s_1 + s_2$ using a number-theoretic transform (NTT). A candidate signature z is then generated, which is derived from pk and is computed using rejection sampling to ensure its validity. The final signature σ consists of multiple values ensuring the validity and security of the signature.

Verify(pk, m, σ). The verifier uses the public key $pk = (A, t)$ and the signature $\sigma = (z_1, z_2)$ to verify the message m. First, the verifier computes the challenge value c, which is derived from the message m and the public key pk. Then, the verification formula $c \cdot t = \hat{s} + z$ is checked to verify the signature. If the condition is satisfied, the signature is valid; otherwise, the signature is invalid. Through efficient operations like number-theoretic transform (NTT), Dilithium ensures high efficiency in the signature and verification processes.

There are currently two typical full hardware designs for Dilithium. In [10], a high-performance design for Dilithium version 2 [25] is described. The authors report area for individual modules instantiated for level 3 and performance for security levels 1–4. They utilized a 2×2 NTT butterfly arrangement to calculate two layers of the NTT at a time. This NTT module is duplicated several times within the key generation, sign, and verify modules to improve performance through parallelization. Another implementation [26], describes a midrange implementation for version 3.1 of Dilithium, which focuses on achieving the best performance possible with reasonable resource utilization. This design consists of three top-level modules, each capable of performing all operations at a single security level, as well as individual modules that only perform a single operation at a single security level.

2.2 Kyber Algorithm

Kyber is an IND-CCA2-secure key encapsulation mechanism (KEM) based on the hardness of the Module Learning With Errors (Module-LWE) problem, which is a generalization of the standard LWE problem over polynomial rings [13,16]. It is one of the finalists selected in the third round of the NIST post-quantum cryptography standardization project [17]. Kyber provides three different security levels, each corresponding to a different parameter set (Kyber512, Kyber768, and Kyber1024). These align with NIST security levels 1, 3, and 5, respectively. These parameter sets are carefully designed to balance efficiency, bandwidth, and post-quantum security. The Kyber cryptosystem is notable for its simplicity, strong theoretical foundations, and suitability for both hardware and software implementations. The parameter list for Kyber is shown in the table.

Kyber operates over a module $\mathbb{Z}_q[x]/(x^n + 1)$, where $q = 3329$ and $n = 256$. It utilizes the number-theoretic transform (NTT) to accelerate polynomial arithmetic [21]. The central idea of Kyber is to use noisy linear algebra operations over small secret polynomials and publicly shared matrices to generate and recover a shared key, it also has three main opeartions:

KeyGen(ρ). The key generation function samples secret polynomials s and e from a centered binomial distribution χ_η, and generates a public matrix $A \in R_q^{k \times k}$ using a public seed ρ. The public key is computed as $t = A \cdot s + e$, where multiplication is done using efficient NTT-based operations. The secret key is the vector s, and the public key is $pk = (\rho, t)$. Additionally, Kyber stores a hashed version of the public key and a random value z in the secret key to provide CCA security via the Fujisaki-Okamoto transform.

Enc(pk, m, μ). To encrypt, the sender uses the public key pk, a message m, and a uniformly random seed μ. Using μ, the sender samples ephemeral secret polynomials r, e_1, e_2 from the same distribution χ_η, and derives a matrix A as in KeyGen. The ciphertext is computed as $u = A^T \cdot r + e_1$, and $v = t^T \cdot r + e_2 + \text{Encode}(m)$, where the addition of noise ensures semantic security. The ciphertext $ct = (u, v)$ is then compressed for transmission.

Dec(sk, ct). The receiver uses the secret key s to compute the shared secret from the ciphertext. Specifically, it computes $v' = v - u \cdot s$ to recover the encoded message. This result is decoded and verified against expected hashes to ensure ciphertext integrity and correctness. Due to the Fujisaki-Okamoto transform, invalid ciphertexts do not leak information, ensuring full IND-CCA2 security.

There are a few researchs that have implemented kyber separately on FPGAs. In [21] a hardware implementation of CRYSTALS-Kyber in [19] FPGA is presented which takes 10 times less clock cycles than the ARM Cortex-M4 implementation of CRYSTALS-Kyber from. In [23] an even smaller and faster hardware implementation of CRYSTALS-Kyber is demonstrated. With suitably designed pipelines and well-optimized architecture, this implementation is capable of executing the decapsulation procedure of any version of CRYSTALS-Kyber in 14,000 clock cycles while consuming only 7,500 LUTs.

2.3 NTT and Polynomial Multiplication

Number Theoretic Transform (NTT) is a modular analog of the Fast Fourier Transform (FFT), designed for efficient polynomial multiplication over finite fields [11,15]. It is a fundamental building block in lattice-based post-quantum cryptographic schemes, offering reduced computational complexity from $\mathcal{O}(n^2)$ to $\mathcal{O}(n \log n)$, where n is the degree of the polynomial.

In both Dilithium and Kyber, polynomials are defined over the ring $\mathbb{Z}_q[x]/(x^n + 1)$, where q is a prime modulus that supports an n-th primitive root of unity ω [12,13]. Given a polynomial $a(x) = (a_0, a_1, \ldots, a_{n-1})$, the forward NTT transforms it into:

$$A_k = \sum_{j=0}^{n-1} a_j \cdot \omega^{jk} \quad \text{mod } q \tag{1}$$

The inverse NTT (INTT) restores the original polynomial using:

$$a_j = n^{-1} \sum_{k=0}^{n-1} A_k \cdot \omega^{-jk} \quad \text{mod } q \tag{2}$$

where n^{-1} denotes the modular inverse of n modulo q. Twiddle factors ω^i are typically precomputed and stored in memory to accelerate the transformation.

NTT is commonly implemented using an iterative butterfly structure. Radix-2 architectures require $\log_2 n$ stages with $n/2$ butterfly operations per stage, whereas radix-4 architectures reduce the number of stages to $\log_4 n$, improving throughput and reducing control overhead. Arithmetic operations during NTT, such as modular multiplications, are usually implemented using Montgomery or Barrett reduction to avoid costly divisions.

3   Motivation

3.1   Challenge 1: Differences in Algorithmic Structures and Data Forms

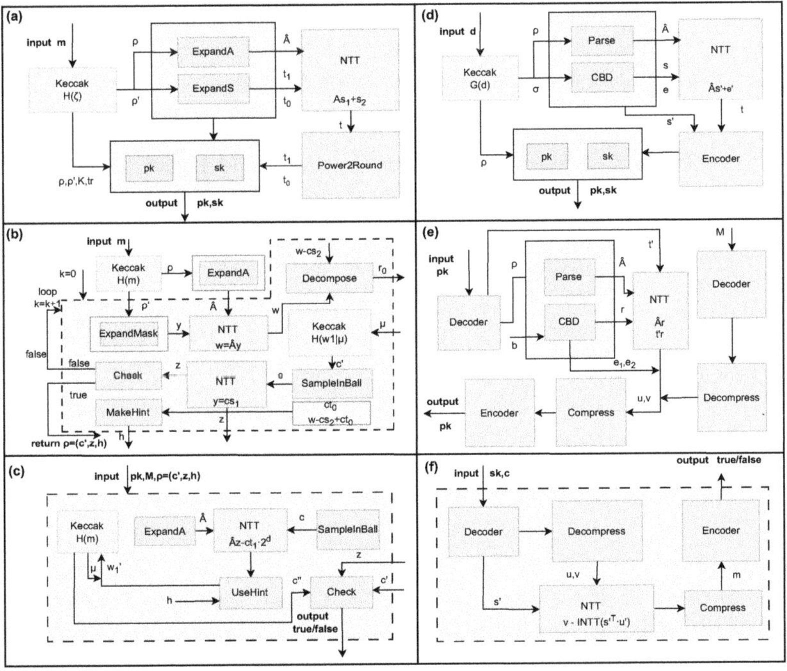

Fig. 1. The Algorithm Flow and Data Flow for Dilithium and Kyber. Figures (a), (b), and (c) show the Key Generation, Signature, and Verification in Dilithium, respectively. Figures (d), (e), and (f) show the Key Generation, Encryption, and Decryption in Kyber, respectively.

Figure 1 shows the algorithm and data flow for Dilithium and Kyber. The two schemes differ significantly in terms of algorithmic structure. Dilithium, as a signature scheme, involves more complex logic with multiple steps. These steps include rejection sampling [20], which introduces control flow divergence. In contrast, Kyber, a key encapsulation scheme, uses simpler operations. It avoids rejection sampling and relies on centered binomial distributions for polynomial generation.

The data structures also vary: Dilithium requires larger polynomial coefficients and multiple intermediate buffers (such as z, w, and c) to store and process data, leading to higher memory usage and more complex memory management. Kyber, on the other hand, uses smaller polynomial coefficients and simpler data structures, making its memory management more efficient and predictable.

Table 1 compares the data lengths for both schemes. Dilithium requires more memory due to larger coefficients (e.g., 1313 and 2636 bits for public and secret keys in Dilithium 2), while Kyber's coefficients are smaller (e.g., 800 and 1592 bits for Kyber 512). This makes Kyber's memory footprint scale more linearly with the security level, while Dilithium's complexity increases with higher security levels.

In summary, Dilithium's complex data structures and higher memory usage require more sophisticated memory management strategies, while Kyber is more efficient and straightforward in terms of memory usage.

Table 1. Data both exists in Dilithium and Kyber [5–7]

Unit	DIff.core	A	S	pk	sk	m	e	H
Dilithium	Dilithium 2	1313	2636	1313	2636	33	3490	48
	Dilithium 3	1992	4096	1992	4096	33	3598	48
	Dilithium 5	2892	4594	2892	4594	33	4592	48
Kyber	Kyber 512	800	1592	800	1592	32	769	52
	Kyber 768	1164	2409	1164	2409	32	1098	32
	Kyber 1024	1509	5169	1509	5169	32	1508	32

3.2 Challenge 2: Resource Contention and Access Overhead in Multi-Mode Architectures

Although only one scheme executes at a time, the memory subsystem must support the full data paths of both algorithms due to integration requirements, leading to an "integrated but non-shareable" problem. Resources are provisioned for peak load but cannot be efficiently shared across modes. In compute-intensive operations like NTT and high-dimensional matrix multiplication, large volumes of polynomial data must move between memory and arithmetic units, causing bandwidth congestion and increased latency. Memory contention further arises when intermediate data—such as Kyber's (r, e_1, e_2) and Dilithium's (z, h, c)—must coexist within limited address space.

The situation is complicated by differences in data formats. Dilithium uses larger polynomial coefficients and buffers than Kyber, increasing memory management complexity. Its rejection sampling and norm checking also require higher precision and more frequent memory access. In contrast, Kyber features simpler data flow and smaller coefficients. The memory system must reconcile these differences and dynamically adjust layout and access patterns during algorithm switching to avoid conflicts and resource shortages. When multiple intermediate values coexist in memory, contention and latency worsen. Thus, the memory subsystem must manage diverse access patterns while ensuring parallel task execution without bottlenecks.

3.3 Opportunities: Shared Components and Structural Similarities

Dilithium and Kyber share key features that enable unified memory design, as shown in Fig. 1. Both operate on fixed-degree polynomials (typically 256) and use the Number Theoretic Transform (NTT) for fast modular multiplication. They also employ Keccak-based hashing (SHAKE128/256) for key derivation, challenge computation, and pseudorandom number generation [27]. Their shared ring structure $\mathbb{Z}_q[x]/(x^n + 1)$ supports standardized encoding and data widths, improving implementation efficiency.

Common operations—such as NTT/INTT, polynomial accumulation, modular reduction, and binomial sampling—exhibit similar access and computation patterns [18], enabling the design of parameterized, reusable memory interfaces and scheduling logic. With proper abstraction and dynamic address mapping, a unified memory subsystem can support both algorithms, fostering hardware co-design that balances flexibility, efficiency, and post-quantum readiness.

Leveraging these shared elements reduces hardware complexity and cost—crucial for systems demanding high throughput and low power. A unified design eliminates the need for separate hardware paths, enabling faster computation and more efficient resource use, and paving the way for future post-quantum cryptographic standards.

4 KiRa: DESIGN and Implementation

4.1 Overall Architecture Design Overview

Section 2 presents the algorithmic flow of CRYSTALS-Dilithium and CRYSTALS-Kyber. Section 3 discusses the implementation challenges and opportunities of supporting both schemes in a shared hardware platform.

To support both algorithms efficiently, we design **KiRa**: a unified memory architecture for efficient post-quantum cryptographic algorithms. Figure 2 illustrates the high-level architecture proposed in this design.

KiRa adopts a modular hardware structure where critical units, such as Keccak, samplers, and the polynomial arithmetic unit, are reused across both schemes, and these critical units are equipped with dedicated storage units like Sampler Ram, Keccak Register, Polk RAM and Twiddle Factor ROM. Among

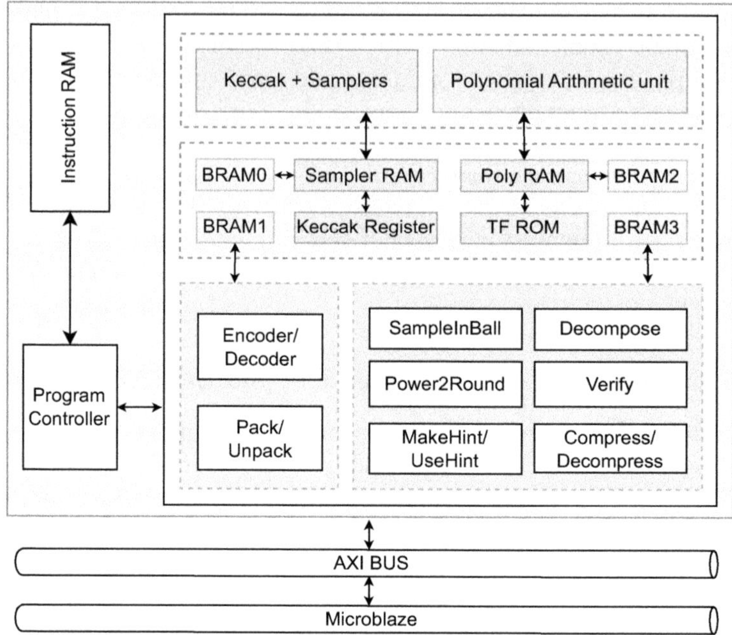

Fig. 2. High-level Architecture of KiRa.

those modules, the modules in the blue box are unique to Kyber, and the modules in the green box are unique to Dilithium. The controller uses instructions to flexibly choose which operation to schedule, and the BRAM blocks are logically partitioned and mapped to the required data structures in both the Dilithium and Kyber algorithms, according to their execution phases. The Fig. 3 shows how the data in the Dilithium algorithm is mapped to the storage address in the BRAM at different stages of the algorithm. This memory organization—combined with dynamic address mapping and dataflow scheduling optimization which will be discussed in the following sections enables adaptive switching across different algorithms within a unified memory architecture.

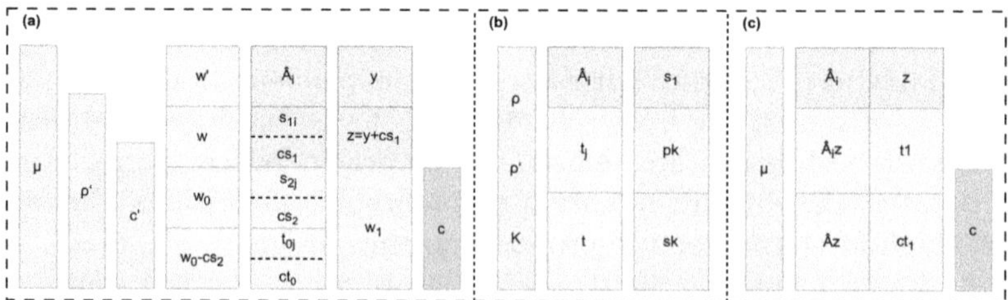

Fig. 3. Memory Address Mapping in BRAM with Dilithium Algorithm. (a) Key Generation; (b) Signature; (c) Verification.

The more detailed top-level architecture partially shown in Fig. 4.In the following subsections, we present each key component in detail.

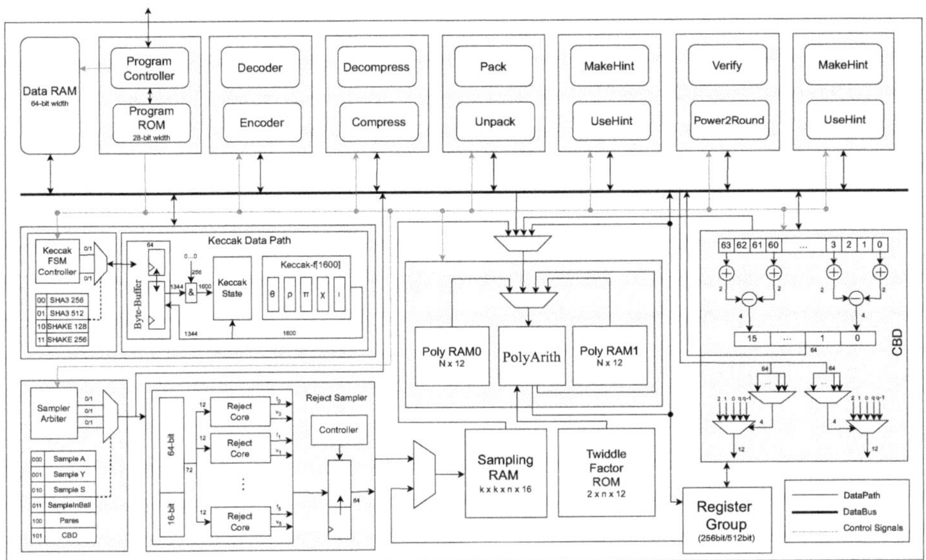

Fig. 4. Detailed KiRa Top-Level Architecture that Supports both CRYSTALS-Dilithium and CRYSTALS-Kyber at All NIST Security Levels.

Program Controller and Instruction ROM. At the core of the system is a program controller that decodes 28-bit wide instructions from the ROM, specifying an operation type, source/destination operands, and control signals. This approach allows for flexible execution of all Dilithium and Kyber protocol steps without modifying the underlying datapath. The controller manages synchronization, parallel scheduling, and data movement among modules. Instructions are defined for operations including sampling, polynomial multiplication, compression, encoding, Keccak hashing, and hint-related steps required in Dilithium.

Keccak Hash and PRNG Core. A unified Keccak module, shared by both schemes, supports SHA3-256, SHA3-512, SHAKE-128, and SHAKE-256. The hash function mode is selected through a control signal from the Keccak controller. The module operates in a pipelined manner with an internal 1600-bit state and 64-bit I/O buffers. Keccak outputs are consumed directly by the samplers, eliminating intermediate buffering. The same core is reused for generating random coins, matrix expansion, and hashing operations in both schemes.

Configurable Sampling Module. A dedicated sampling unit is designed to support all sampler types required by Dilithium and Kyber. This includes uniform sampling, CBD-based binomial sampling (with $\eta = 2$ or 3), rejection sampling, and SampleInBall used by Dilithium. The sampler is tightly coupled with the Keccak output and contains parallel logic units for each mode. An arbiter inside the sampler selects the sampling variant based on the current instruction and activates the corresponding submodule. The sampled polynomials are stored in Poly RAM blocks for further NTT processing.

Dual-Port NTT Memory and Polynomial Architecture. The polynomial arithmetic engine performs NTT/INTT and pointwise multiplication over $\mathbb{Z}_q[x]/(x^n+1)$. A radix-4 butterfly unit is reused for all transform and multiplication steps. To support high-throughput access, the architecture uses two BRAM banks (RAM0 and RAM1) in ping-pong mode. Data are read from one bank and written to the other during each stage of the transform. The Poly RAM is logically partitioned to store different polynomial types and is organized to enable coefficient-aligned access patterns. Further details on the polynomial storage and mapping strategy are described in Sect. 4.2.

Pointwise Multiplication and Data Packing. A unified pointwise multiplication unit is integrated with the NTT engine. It supports coefficient-wise multiplication for Kyber and linear polynomial multiplication for Dilithium. Control logic configures the arithmetic datapath to switch between operations. In post-processing, encoding and compression units convert polynomials into bit-packed or compressed ciphertexts. These units are implemented using shift and mask logic for low-latency performance. They also support decompression and decoding paths during key recovery and signature verification.

Shared Data Bus and Global Memory. All functional modules are connected through a shared 64-bit data bus managed by the controller. A unified memory interface ensures consistent data flow between Keccak, sampler, polynomial units, and the main memory. Registers are used as staging buffers for instructions and sampled coefficients. The memory system supports both fixed-length and variable-length polynomial formats, ensuring compatibility with Dilithium and Kyber parameter sets.

4.2 Polynomial Storage Architecture

To support high-throughput polynomial operations, we adopt a dual-port, conflict-free storage design at the same time. The architecture combines lightweight butterfly units and near-memory mapping logic. As shown in Fig. 5, the polynomial storage architecture consists of a control unit, address generation and mapping units, polynomial RAM with four banks, a twiddle factor ROM, and a radix-4 NTT engine. The system is designed for in-place, high-throughput polynomial transforms, compatible with both Dilithium and Kyber.

The control unit monitors the current computation stage—initialization, NTT, pointwise multiplication, or INTT—and produces control signals for memory access and address mode. It drives the address generator and orchestrates memory interactions.

At each stage of the NTT, the address generator produces logical addresses based on the transform structure. These addresses are converted to physical memory locations by the address mapping unit. To ensure that multiple coefficients accessed in the same butterfly cycle do not conflict, a static mapping scheme is used.

The polynomial RAM is divided into four banks: RAM0, RAM1, RAM2, and RAM3. Each polynomial contains 256 coefficients, evenly distributed across the banks. As shown in Fig. 5(a), four coefficients are accessed per butterfly cycle, and the mapping logic ensures they fall into distinct banks to avoid collisions.

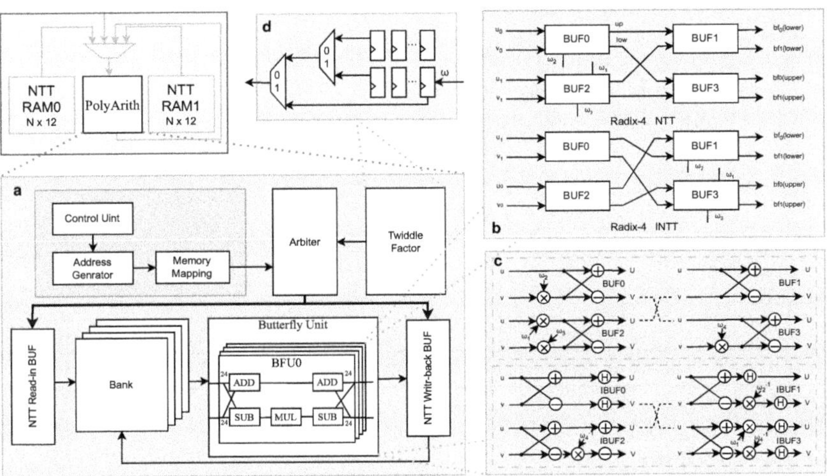

Fig. 5. Overall Polynomial Storage Architecture. (a) represents memory mapping unit and polynomail arithmetic logic unit; (b) show the scheduling of each butterfly unit in the modes of radix-4 NTT/INTT; (c) illustrates the computational process of butterfly unit; (d) demonstrates how twiddle factors coordinate the scheduling with the Arbiter.

The address generation algorithm is described in Algorithm 1. A full radix-4 NTT has four stages. At stage s, there are 4^s rounds and each round involves $256/4^{s+1}$ butterfly operations. The addresses produced are logical indexes for computation.

The physical address is determined by a conflict-free mapping strategy. After generation, each address is split into a bank index and offset using Algorithm 2. The lower two bits select the bank. The upper bits give the offset. Figure 6 directly shows how coefficients are distributed across banks. After each butterfly operation is completed, the result is directly written back to the same memory region .This method ensures that all four addresses in one butterfly fall into distinct banks and eliminates read-after-write (RAW) and write-after-read (WAR) hazards across NTT stages.

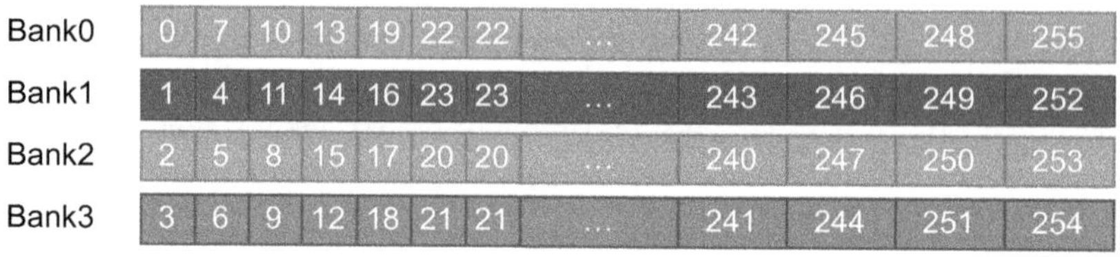

Bank0	0	7	10	13	19	22	22	...	242	245	248	255
Bank1	1	4	11	14	16	23	23	...	243	246	249	252
Bank2	2	5	8	15	17	20	20	...	240	247	250	253
Bank3	3	6	9	12	18	21	21	...	241	244	251	254

Fig. 6. Storage sequence of polynomial coefficients in Bank.

Algorithm 1. Address Generation Algorithm

1: **Input:** R, N $\triangleright$ R is radix, N is number of coefficients
2: **for** $s \leftarrow \log_R N - 1$ **downto** 0 **do**
3:      $r \leftarrow R^s$
4:      **for** $k \leftarrow 0$ **to** $\frac{N}{R^s} - 1$ **do**
5:          **for** $j \leftarrow 0$ **to** $r - 1$ **do**
6:              $\text{addr}_0 \leftarrow R \cdot k \cdot r + j$
7:              $\text{addr}_1 \leftarrow R \cdot k \cdot r + j + r$
8:              $\text{addr}_2 \leftarrow R \cdot k \cdot r + j + 2r$
9:              $\text{addr}_3 \leftarrow R \cdot k \cdot r + j + 3r$
10:          **end for**
11:      **end for**
12: **end for**

Algorithm 2. Address Mapping Algorithm

1: **Input:** a_0, a_1, a_2, a_3
2: **Output:** $(BI_0, BA_0), \ldots, (BI_3, BA_3)$
3: **for** $i \leftarrow 0$ **to** 3 **do**
4:      $BI_i \leftarrow a_i \bmod 4$
5:      $BA_i \leftarrow a_i / 4$
6: **end for**

During butterfly operations, the required twiddle factors are retrieved from a precomputed ROM, with each indexed by stage and round, as shown in Fig. 5(d). The control unit fetches the correct value from the index.

To support both schemes, the memory format adapts accordingly: each word holds a single 24-bit coefficient for Dilithium, and two 12-bit coefficients of the same parity (even or odd) for Kyber. This supports Kyber's incomplete NTT structure with even–odd separation and merge stages. It also improves efficiency in data packing and parallel access. This architecture provides a unified and efficient memory organization for both schemes, ensuring safe, parallel access to coefficients with minimal conflict and enabling in-place NTT execution under a shared logic design.

5   Experiments

5.1   Experimental Setup

All experiments were carried out on a Xilinx ZCU102 board, which integrates a Zynq UltraScale+ MPSoC. We mapped the proposed KiRa cryptoprocessor to the programmable logic and verified the full Dilithium and Kyber functionality under Vivado 2020.2. The synthesis, place and route, and timing closure used worst-case PVT at 200 MHz. Static and toggle-rate-aware dynamic power were estimated with the Xilinx Power Estimator. Post-route simulation streamed real

Table 2. FPGA Implimentation Results For KiRa

Unit	Comp. Core	S-1	S-2	LUT	FF	DSP	BRAM
Dilithium	Memory	–	–	0	0	0	3×6
	Keccak	✓	✓	5483×3	2451×3	0	0
	SampleA	✓	✓	1593	617	0	0
	SampleS	✓	✓	1655	375	0	0
	SampleY	✓	✓	2020	529	0	0
	SampleC	✓	✓	2056	863	0	0
	NTT	✓	✓	4691×2	1157×2	16	0
	Decompose	✓	–	774	538	0	0
	Pow2Round	–	✓	1553	523	0	0
	Pack	✓	–	1802	1285	0	0
	Unpack	✓	–	715	428	0	0
	MakeHint	–	✓	2264	730	0	0
	UseHint	–	✓	4273	1799	0	0
	Refresh	✓	–	729	522	0	0
	VerifyEq.	✓	–	455	316	0	0
Kyber	Memory	–	–	0	0	0	3×4
	Keccak	✓	✓	5483×3	2451×3	0	0
	Encode	-	✓	2237	870	0	0
	Decode	✓	–	1688	621	0	0
	SampleA	✓	✓	1593	617	0	0
	SampleS	✓	✓	1655	375	0	0
	NTT	✓	✓	4691×2	1157×2	16	0
	Com./Decom.	–	✓	722	367	0	0
	Verify	–	✓	387	116	0	0
	CMOV	–	✓	207	90	0	0
	COPY	–	✓	144	59	0	0
KiRa	Memory	–	–	0	0	0	3×4
	Keccak	✓	✓	5483×3	2451×3	0	0
	NTT	✓	✓	4691×2	1157×2	16	0
	Multiplier	–	–	3531	1157×2	4	3
	Porg.Contr.	–	–	3569	372	0	0
	Total			37,890	15,753	20	15

test vectors through the complete encrypt/decrypt and sign/verify chains. Byte-level buffers are realized with distributed LUT-RAM FIFOs, while polynomial memories use a four-bank BRAM organization. This design allows the same netlist to switch between Dilithium and Kyber micro-programmes at runtime without reconfiguration.

5.2   Area and Performance Results of KiRa

Table 2 presents the detailed utilization of each building blocks in KaLi for Ultra-Scale+ ZCU102 platform. ComPared with traditional Dilithium and Kyber, KiRa uses 37890 LUTs (57.7%), 15753 DFFs (60.9%), 20 DSPs (62.5%), and 15 BRAMs (50.0%).

Table 3. Performance Results for KiRa in FPGA

Operation	Dilithium-2 Kyber-512		Dilithium-3 Kyber-768		Dilithium-5 Kyber-1024	
	Cycle	μs	Cycle	μs	Cycle	μs
Dil.Gen	5684	28.42	8064	40.32	15846	79.23
Dil.Sign	7346	36.73	10652	53.26	18325	91.63
Dil.Verify	6478	32.39	8793	43.97	17213	86.07
Kyb.KeyGen	3758	18.79	5215	26.07	6832	34.16
Kyb.Encaps	5842	29.21	8825	44.13	10812	54.06
Kyb.Decaps	5133	25.66	8266	41.3	8932	44.66

Table 3 presents the average cycle count and latency for Dilithium and Kyber operations at a 200 MHz clock frequency. In the best-case scenario, where a valid signature is successfully generated after the first loop iteration, the CCA-secure operations for Dilithium-2 take the following time: key generation (21.63 μs), signature generation (32.96 μs), and signature verification (19.36 μs). For Kyber-512, the corresponding times are: key generation (15.62 μs), encapsulation (21.70 μs), and decapsulation (17.30 μs).

5.3   Comparison with Dilithium-only and Kyber-only Work

We compare the performance of the KiRa architecture with two other designs: the traditional Kyber implementation and the shared DK design, which integrates Dilithium and Kyber without the optimizations of near-memory computation and conflict-free address mapping. Table 4 presents a comparison of KiRa with these two designs in terms of resource consumption, power efficiency, and energy situation. The data is taken from the average values of the Kyber algorithm at each security level.

Table 4. Comparison with Kyber-only and shared DK design in Kyber mode

Architecture	Slices	BRAM	Static Pow. (mW)	Dyn. Pow. (mW)	Area (mm^2)	Mem. Acc. (M)	EFF. (M/W)	Energy (nW/mm^2)
Kyber-only [21]	21400	30	220.1	539.5	2.45	2.31	3.04	310.1
Shared DK design [28]	18670	21	210.4	326.0	2.12	1.85	3.95	253.0
KiRa	**16320**	**15**	**207.8**	**150.5**	**1.83**	**1.43**	**5.11**	**195.8**

KiRa outperforms the traditional design by reducing slices, BRAM usage, and dynamic power consumption, achieving 23.8% less area, 72.1% lower dynamic power, and a 17.3% improvement in latency. The total energy per Kyber encapsulation is reduced from 310.1 nJ to 195.8 nJ, showing a significant improvement in energy efficiency.

Table 5. Average Power, Memory Accesses, and Time Share of Each Substage in Dilithium Signature

Substage	Power (mW)	Mem.Acc. (M)	Time (%)	Power (mW)	Mem.Acc. (M)	Time (%)
	KiRa			**Dilithium-Only** [22]		
Sampling	91.1	0.34	22.2	178.5	0.58	20.0
NTT	113.9	0.42	27.4	225.0	0.71	32.0
MakeHint	70.2	0.23	15.0	137.4	0.39	14.0
UseHint	45.3	0.13	8.5	88.0	0.22	10.0
Decompose	49.4	0.18	11.8	97.0	0.31	11.0
Pack	24.5	0.12	7.8	49.0	0.20	7.0
Unpack	18.2	0.11	7.2	36.5	0.19	6.0

We also evaluated the performance of KiRa in the signature path under the Dilithium mode. Table 5 presents the average power consumption, memory accesses, and time proportion for each operation in the signature path, including Sampling, NTT, MakeHint, UseHint, Decompose, Pack, and Unpack. Also, the data is taken from the average values of the signature path at each Dilithium security level.

As shown in Fig. 7, KiRa reduces power consumption by approximately 50%. This improvement is likely due to KiRa using fewer LUTs, FFs, and BRAMs. The highest power consumption occurs in the NTT stage, Compared with the traditional Dilithium, KiRa's power consumption is reduced from 225 nW to 113.9 nW. In terms of memory access, KiRa demonstrates significant improvement. The NTT stage has the largest improvement, with the number of memory accesses reduced by 69%, from 0.71M to 0.42M. And the time share has been reduced from 32% to 27.4%. This shows that Polynomial Storage Architecture plays an important role.

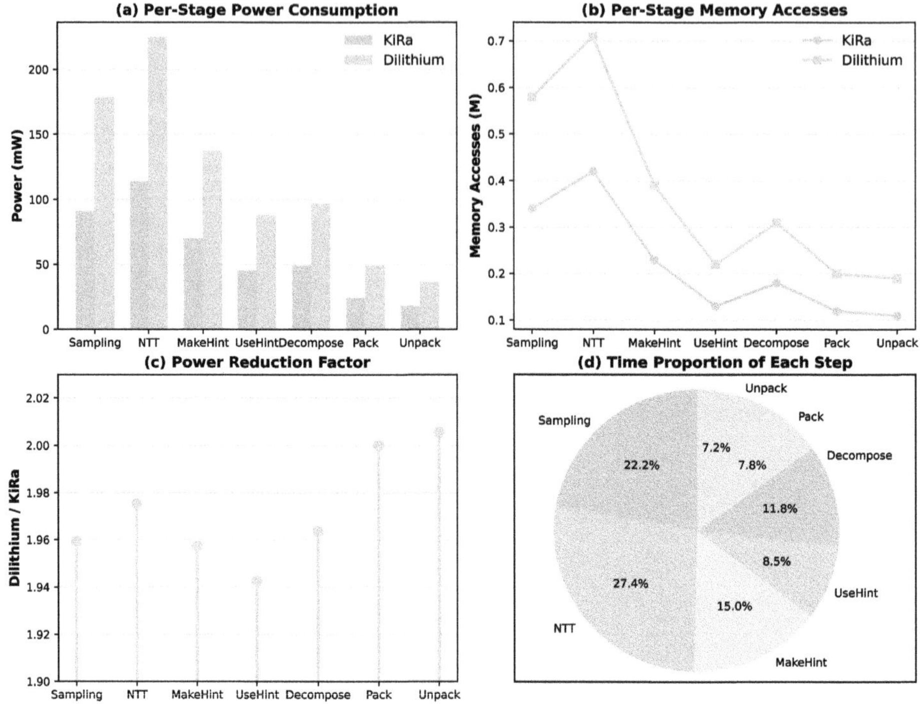

Fig. 7. Comparison with Dilithium-only Design in Dilithium Mode.

Overall, the KiRa architecture achieves superior performance in both energy efficiency and resource utilization, fully supporting all security levels of Dilithium and Kyber in a single bitstream. It reduces the total energy consumption by half compared to traditional implementations while maintaining lower memory access and latency.

6   Conclusion

We presents **KiRa**: A Unified Memory Architecture for Efficient Post-Quantum Cryptographic Algorithms, which efficiently supports both Dilithium and Kyber algorithms on the same hardware. To achieve this, we designed a modular and reusable architecture where core components like Keccak, NTT, and sampling modules are shared across both algorithms. We introduced a uniform control logic and dataflow that allow Dilithium and Kyber to switch operations without duplicating resources.

And We designed a dual-port, conflict-free memory architecture for high-throughput polynomial operations in Dilithium and Kyber. The architecture ensures efficient parallel access by mapping coefficients to four memory banks and avoiding memory conflicts during NTT computations. This design improves memory access efficiency, enabling in-place NTT execution with minimal overhead.

The complete system was implemented and tested on a Zynq UltraScale+ MPSoC FPGA platform. Experimental results confirm that KiRa architecture can fully support the complete algorithm flow of both Dilithium and Kyber with functional correctness and timing reliability. ComPared with traditional or shared designs of Dilithium and Kyber, KiRa uses 37890 LUTs (57.7%), 15753 DFFs (60.9%), 20 DSPs (62.5%), and 15 BRAMs (50.0%) in FPGA. Our design achieves a highest 69% improvement in memory access of NTT stage, consumes 23.8% less area, achieves a 17.3% improvement in latency, reduces dynamic power by 72.1%.

In conclusion, KiRa demonstrates an effective and feasible method to integrate multiple lattice-based cryptographic algorithms into a single memory architecture. It achieves high resource reuse, low power consumption, and reduced latency, providing a strong reference design for post-quantum cryptographic hardware platforms. This work lays a foundation for future secure chips that support flexible algorithm switching and efficient hardware consolidation.

Acknowledgments. This work is supported by the National Key R&D Program of China (No. 2022YFB4400703), the National Key R&D Program of Heilongjiang Province (2024ZXDXA12) and the Fundamental Research Funds for the Central Universities (3072025YC602).

References

1. Blake, I.F., Gao, X., Mullin, R.C., Vanstone, S.A., Yaghoobian, T.: Elliptic Curve Cryptosystems. In: Menezes, A.J. (ed.) Applications of Finite Fields, pp. 151–171. Springer, Boston, MA (1993). https://doi.org/10.1007/978-1-4757-2226-0_8
2. Shor, P.W.: Algorithms for quantum computation: discrete logarithms and factoring. In: Proceedings 35th Annual Symposium on Foundations of Computer Science, pp. 124–134. IEEE (1994). https://doi.org/10.1109/SFCS.1994.365700
3. Arute, F., Arya, K., Babbush, R., et al.: Quantum supremacy using a programmable superconducting processor. Nature **574**, 505–510 (2019). https://doi.org/10.1038/s41586-019-1666-5
4. NIST Homepage, https://csrc.nist.gov/projects/, last accessed 2025/6/28
5. Ducas, L., Kiltz, E., Lepoint, T., Lyubashevsky, V., Schwabe, P., Seiler, G., Stehlé, D.: CRYSTALS-Dilithium: A Lattice-Based Digital Signature Scheme. IACR Trans. Cryptogr. Hardw. Embed. Syst. **2018**(1), 238–268 (2018). https://doi.org/10.13154/tches.v2018.i1.238-268
6. Bos, J., Ducas, L., Kiltz, E., Lepoint, T., Lyubashevsky, V., Schanck, J.M., Schwabe, P., Seiler, G., Stehlé, D.: CRYSTALS - Kyber: A CCA-Secure Module-Lattice-Based KEM. In: 2018 IEEE European Symposium on Security and Privacy (EuroS&P), pp. 353–367. IEEE (2018). https://doi.org/10.1109/EuroSP.2018.00032
7. pq-crystals Homepage, https://pq-crystals.org/index.shtml, last accessed 2025/6/28

8. Kamucheka, T., Fahr, M., Teague, T., Nelson, A., Andrews, D., Huang, M.: Power-based Side Channel Attack Analysis on PQC Algorithms. Cryptology ePrint Archive, Paper 2021/1021 (2021). https://eprint.iacr.org/2021/1021
9. Rodriguez, R.C., Bruguier, F., Valea, E., Benoit, P.: Correlation Electromagnetic Analysis on an FPGA Implementation of CRYSTALS-Kyber. Cryptology ePrint Archive, Paper 2022/1361 (2022). https://doi.org/10.1109/PRIME58259.2023.10161764, https://eprint.iacr.org/2022/1361
10. Ricci, S., Malina, L., Jedlicka, P., Smekal, D., Hajny, J., Cibik, P., Dobias, P.: Implementing CRYSTALS-Dilithium Signature Scheme on FPGAs. Cryptology ePrint Archive, Paper 2021/108 (2021). https://eprint.iacr.org/2021/108
11. Nicholson, P.J.: Algebraic theory of finite Fourier transforms. J. Comput. Syst. Sci. $5(5)$, 524–547 (1971). https://doi.org/10.1016/S0022-0000(71)80014-4
12. Zhuang, S., Zhang, L., Lai, Q.: Heuristic Ideal Obfuscation Based on Evasive LWR. Cryptology ePrint Archive, Paper 2024/392 (2024). https://eprint.iacr.org/2024/392
13. Shirase, M.: Reduction of Search-LWE Problem to Integer Programming Problem. Cryptology ePrint Archive, Paper 2023/1162 (2023). https://eprint.iacr.org/2023/1162
14. Dagdelen, Ö., Fischlin, M., Gagliardoni, T.: The Fiat–Shamir Transformation in a Quantum World. In: Sako, K., Sarkar, P. (eds) Advances in Cryptology - ASIACRYPT 2013. ASIACRYPT 2013. Lecture Notes in Computer Science, vol 8270. Springer, Berlin, Heidelberg (2013). https://doi.org/10.1007/978-3-642-42045-0_4, https://doi.org/10.1007/978-3-642-42045-0_4
15. Mert, A.C., Karabulut, E., Öztürk, E., Savaş, E., Aysu, A.: An Extensive Study of Flexible Design Methods for the Number Theoretic Transform. IEEE Trans. Comput. $71(11)$, 2829–2843 (2022). https://doi.org/10.1109/TC.2020.3017930
16. Bos, J., Ducas, L., Kiltz, E., Lepoint, T., Lyubashevsky, V., Schanck, J.M., Schwabe, P., Seiler, G., Stehlé, D.: CRYSTALS – Kyber: a CCA-secure module-lattice-based KEM. Cryptology ePrint Archive, Paper 2017/634 (2017). https://doi.org/10.1109/EuroSP.2018.00032, https://eprint.iacr.org/2017/634
17. Moody, D., Alagic, G., Apon, D., Cooper, D., Dang, Q., Kelsey, J., Liu, Y., Miller, C., Peralta, R., Perlner, R., Robinson, A., Smith-Tone, D., Alperin-Sheriff, J.: Status Report on the Second Round of the NIST Post-Quantum Cryptography Standardization Process. NIST Interagency/Internal Report (NISTIR), National Institute of Standards and Technology, Gaithersburg, MD (2020). https://doi.org/10.6028/NIST.IR.8309, https://doi.org/10.6028/NIST.IR.8309 last accessed 2025/6/28
18. Scott, M.: A Note on the Implementation of the Number Theoretic Transform. In: O'Neill, M. (ed.) Cryptography and Coding, pp. 247–258. Springer International Publishing, Cham (2017). https://doi.org/10.1007/978-3-319-71045-7_13
19. Botros, L., Kannwischer, M.J., Schwabe, P.: Memory-Efficient High-Speed Implementation of Kyber on Cortex-M4. Cryptology ePrint Archive, Paper 2019/489 (2019). https://eprint.iacr.org/2019/489
20. Bisheh-Niasar, M., Azarderakhsh, R., Mozaffari-Kermani, M.: High-Speed NTT-based Polynomial Multiplication Accelerator for CRYSTALS-Kyber Post-Quantum Cryptography. Cryptology ePrint Archive, Paper 2021/563 (2021). https://eprint.iacr.org/2021/563
21. Huang, M.: A pure hardware implementation of CRYSTALS-KYBER PQC algorithm through resource reuse. IEICE Electronics Express. Institute of Electronics, Information and Communications Engineers (IEICE), $17(2020)$.https://doi.org/10.1587/ELEX.17.20200234

22. Beckwith, L., Nguyen, D.T., Gaj, K.: High-Performance Hardware Implementation of CRYSTALS-Dilithium. In: 2021 International Conference on Field-Programmable Technology (ICFPT), pp. 1–10. IEEE (2021). https://doi.org/10.1109/ICFPT52863.2021.9609917
23. Xing, Y., Li, S.: A compact hardware implementation of CCA-secure key exchange mechanism CRYSTALS-KYBER on FPGA. IACR Trans. Cryptogr. Hardw. Embed. Syst., pp. 328–356 (2021). https://doi.org/10.13154/tches.v2021.i1.328-356
24. Lyubashevsky, V.: Fiat-Shamir with Aborts: Applications to Lattice and Factoring-Based Signatures. In: Sako, K., Sarkar, P. (eds.) Advances in Cryptology – ASIACRYPT 2009. ASIACRYPT 2009. Lecture Notes in Computer Science, vol 8270, pp. 598–616. Springer, Berlin, Heidelberg (2009). https://doi.org/10.1007/978-3-642-42045-0_4, https://doi.org/10.1007/978-3-642-42045-0_4
25. Ducas, L., Lepoint, T., Lyubashevsky, V., Schwabe, P., Seiler, G., Stehlé, D.: CRYSTALS – Dilithium: Digital Signatures from Module Lattices. Cryptology ePrint Archive, Paper 2017/633 (2017). https://eprint.iacr.org/2017/633
26. Land, G., Sasdrich, P., Güneysu, T.: A Hard Crystal - Implementing Dilithium on Reconfigurable Hardware. Cryptology ePrint Archive, Paper 2021/355 (2021). https://eprint.iacr.org/2021/355
27. National Institute of Standards and Technology: FIPS PUB 202: SHA-3 Standard: Permutation-Based Hash and Extendable-Output Functions. Aug. 2015. https://doi.org/10.6028/NIST.FIPS.202
28. Bisheh-Niasar, M., Azarderakhsh, R., Mozaffari-Kermani, M.: Instruction-Set Accelerated Implementation of CRYSTALS-Kyber. IEEE Trans. Circuits Syst. I Regul. Pap. **68**(11), 4648–4659 (2021). https://doi.org/10.1109/TCSI.2021.3106639

A GCN Accelerator with Unified Architecture

Meng Wu[1,2], Mingyu Yan[1,2(✉)], Lei Deng[3], Wenming Li[1,2],
Zhimin Zhang[1,2], Xiaochun Ye[1,2], and Dongrui Fan[1,2]

[1] SKLP, Institute of Computing Technology, Chinese Academy of Sciences,
Beijing, China
{wumeng,yanmingyu,liwenming,zzm,yexiaochun,fandr}@ict.ac.cn
[2] University of Chinese Academy of Sciences, Beijing, China
[3] Tsinghua University, Beijing, China
leideng@mail.tsinghua.edu.cn

Abstract. Graph convolutional networks (GCNs) are prevalent in graph learning. Hybrid architecture, comprised of interconnected engines, is widely used in accelerator designs for GCNs to handle their hybrid execution patterns. However, its inherent inefficiency arises from local bottlenecks in each engine, driven by fluctuating compute resource requirements, thereby causing low overall compute utilization. In this work, we propose a novel GCN accelerator with unified architecture called uFlowGCN, which is in-cooperated with reduction-tree-based computing unit design and tag-driven scheduling. uFlowGCN enables the efficient parallel execution of hybrid execution patterns on a unified hardware substrate, allowing for dynamic allocation of compute resources on the fly to alleviate local bottlenecks. Compared to the optimal design of hybrid architecture, uFlowGCN achieves up to 4.3× speedup and 95.3%~99.5% compute utilization.

Keywords: Graph Convolutional Neural Network · Hardware Accelerator · Unified Architecture

1 Introduction

Graph convolutional networks (GCNs) have become the premier paradigm for analyzing graph data, finding widespread applications in real-world scenarios. GCNs have infiltrated into the data centers at Google and Alibaba, and have been widely used in many real-world applications including real-time estimated time of arrivals in Google Maps, content recommendations in Pinterest, spam review detection in Alibaba, as well as many others [1,2].

GCNs comprise the aggregation and combination phases, which exhibit distinct, even opposed compute and memory access patterns, resulting in hybrid execution patterns [3–10]. To tackle these hybrid execution patterns, numerous previous and state-of-the-art work designs GCN accelerators using hybrid

H. Liu et al. (Eds.): ICA3PP 2025, LNCS 16381, pp. 230–246, 2026.
https://doi.org/10.1007/978-981-95-8399-7_13

architecture [7,9,11–18] due to its pipelined execution and low design complexity. For example, previous work proposes HyGCN [7], a GCN accelerator based on the hybrid architecture that builds individual engines for the two distinct phases, respectively, and constructs inter-engine dataflow to enable inter-phase fusion. HyGCN achieves higher efficiency than the software framework running on NVIDIA V100 GPU.

However, hybrid architecture inherently suffers from inefficiency: the local bottleneck in each engine caused by varying compute resource requirements results in low overall compute utilization. Redesign alone cannot address its inherent weakness.

In this work, we propose a novel GCN accelerator using unified architecture.

- We abstract the computational graphs of GCNs as a graph operator and design a novel processing element (PE). This PE is flexible enough to support various operations of GCNs and easy enough to exploit various parallelisms.
- We propose a tag-driven dataflow scheduling, which leverages GCN-semantics-aware tags to drive and wire dataflow between graph operators, offering two execution modes to efficiently perform the two distinct phases on PEs. Additionally, these two modes can run simultaneously to establish the dataflow between the two phases.
- Building upon the aforementioned PE and scheduling, we introduce a unified architecture named uFlowGCN to accelerate GCNs. uFlowGCN allows the two distinct phases to efficiently perform in the unified hardware substrate in parallel and enables compute resources to be dynamically allocated to each phase in balance.
- We evaluate uFlowGCN using a microarchitectural simulation. Compared to the optimal design of hybrid architecture, uFlowGCN achieves up to $4.3\times$ speedup and $95.3\%\sim99.5\%$ compute utilization.

2   Background and Related Work

This section introduces the background and related work.

2.1   Background

GCNs. GCNs mainly consist of the aggregation and combination phases, as shown in Fig. 1. In the aggregation phase, the feature vector of each vertex is updated by aggregating the feature vectors of its neighboring vertices. In the consequent combination phase, the feature vector of each vertex is then transformed into a new one using the same multi-layer perceptron (MLP). The two phases are usually executed iteratively according to the number of layers [19–21]. After several iterations (also called layers), each vertex is represented

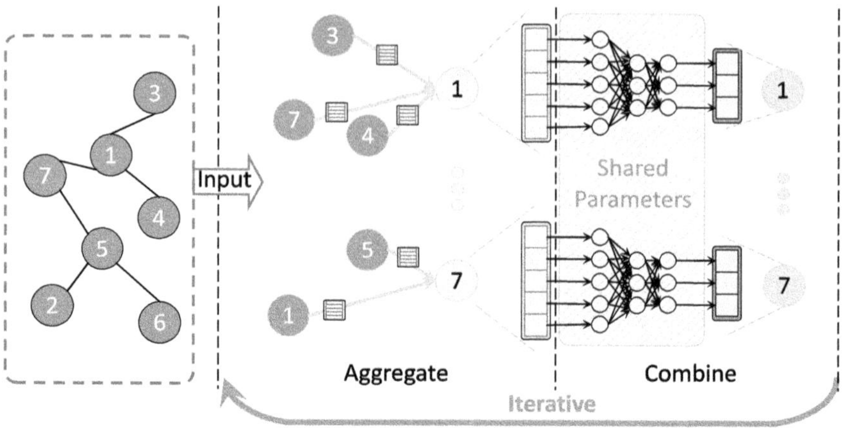

Fig. 1. Illustration of GCNs.

by its final feature vector, which can be used for downstream tasks. Typically, the k-th iteration of GCNs can be generally formulated as

$$a_v^k = \mathbf{Aggregate}\big(h_u^{(k-1)} : u \in N_v \cup \{v\}\big),$$

$$h_v^k = \mathbf{Combine}\big(a_v^k\big),$$

where N_v is the number of neighboring vertices of v, a_v^k and h_v^k are the aggregated feature vector and feature vector of v at the k-th iteration, respectively. Specifically, the aggregate function aggregates the feature vectors from source neighbors at the $(k-1)$-th iteration into one single a_v^k, and the combine function further transforms a_v^k to generate h_v^k using an MLP. The length of feature vector is determined by the input dataset or the number of the MLP's output neurons. The parameters such as weights of the MLP are shared among vertices. The activation function in the MLP can be the widely-used ReLU function in deep learning. Moreover, a sampling technique can be used to reduce the number of neighbors when performing the aggregation phase [22].

Hybrid Execution Patterns. These two phases demonstrate distinct, even contrasting compute and memory access patterns, leading to hybrid execution patterns [3–10]. The computational graph and memory access pattern in the aggregation phase are irregular due to the divergent amount of neighbors and scattered memory locations of neighboring feature vectors across different vertices. In contrast, the computational graph and memory access pattern are regular and identical across vertices in the combination phase because all vertices share the same MLP, but the combination phase suffers from intensive matrix-vector multiplications (MVMs) with shared parameters.

2.2   Related Work

To handle hybrid execution patterns, recent efforts commonly design GCN accelerators based on hybrid architecture, delivering significant performance improvements compared to general-purpose processors and prior art [7,9,11–18,23–25]. HyGCN [7], as prior art, introduces a hybrid architecture featuring separate acceleration engines for the aggregation phase (G-Engine) and the combination phase (NN-Engine), positioned side by side to enable inter-engine pipeline. The G-Engine employs vector compute units, whereas the NN-Engine is constructed using a systolic array [26]. Based on the hybrid architecture of HyGCN [7], SGCN [17], a state-of-the-art (SOTA) GCN accelerator, employs a GCN-friendly compression format to minimize off-chip memory traffic, introduces microarchitectures for seamless handling of compressed features, and utilizes sparsity-aware cooperation for improved locality management. FlowGNN [16], another state-of-the-art accelerator, designs a versatile dataflow architecture for accelerating graph neural networks (GNNs), applicable to a broad spectrum of message-passing GNNs. Like HyGCN [7], FlowGNN comprises two acceleration engines. Moreover, a vertex queue is interposed between these engines, facilitating inter-engine pipeline. Similarly, many other designs share the same design principles as the above accelerators [9,11–15,18,23,27,28].

Some GCN accelerators utilize multiplier-accumulator (MAC) arrays or MAC vector units to construct a unified architecture. However, such designs encounter issues such as limited flexibility, shallow pipeline depth, or insufficient parallelism exploitation. For instance, AWB-GCN [29], as prior art, represents the computation of the first GCN model [21] as two consecutive sparse matrix multiplications, executed through MAC arrays. Despite being a unified architecture, it lacks flexibility and cannot accommodate many GCN models [16]. Additionally, it fails to exploit vertex-level parallelism. Similarly, GROW [30] adopts the same computation representation as AWB-GCN, utilizing MAC vector units to build a unified architecture. However, it cannot realize dataflow between computing units like systolic arrays and cannot stream compute operations [26].

Table 1. Benchmark datasets.

Dataset	#Vertex	#Feature	#Edge	Average Degree
IMDB-BIN (IB)	2,647	136	28,624	11
COLLAB (CL)	12,087	492	1,446,010	120
Pubmed (PB)	19,717	500	88,648	4
Reddit (RD)	232,965	602	114,615,892	492

3   Motivation

Limitation of State-of-the-Art. Accelerator designs using hybrid architecture, exemplified by SOTA designs like SGCN [17] and FlowGNN [16], inherently suffer from low overall compute utilization due to local bottlenecks in each engine, where insufficient compute resources hinder performance improvement.

To demonstrate this, we evaluate an on-chip implementation of hybrid architecture, where all data is accommodated in on-chip buffers[1] This excludes the DRAM to focus solely on addressing low compute utilization. This implementation closely resembles SOTA designs such as SGCN [17], where the G-Engine and NN-Engine comprise SIMD-8 units and 8×8 systolic arrays, respectively. For the GCN models, we employ GCN [21], GraphSage (GSC) [20], and GINConv (GIN) [19]. The aggregate function for these models is accumulation, and the MLP structure in the combine function is $|a_v^{k-1}|-|h_v^k|$, $|a_v^{k-1}|-|h_v^k|$, and $|a_v^{k-1}|-|h_v^k|-|h_v^k|$, respectively. Table 1 lists datasets. Figure 2 shows results. The y-axis and x-axis represent the peak performance of G-Engine and NN-Engine in tera-operations per second (TOPS), corresponding to various configurations of hybrid architecture.

In Fig. 2(a), it is evident that local bottlenecks occur in different engines across various configurations, even on the same dataset and model. Moreover, these bottlenecks vary among different models, datasets, and $|h_v^k|$, even in the same configuration. These variations arise from the fluctuating compute resource requirements across different scenarios. Figure 2(b) gives how local bottlenecks can easily lead to low overall compute utilization. The root cause lies in the fixed allocation of computing resources in the two engines of the hybrid architecture, which fails to accommodate the diverse compute demands of different models, datasets, and $|h_v^k|$.

Opportunity for Designing Unified Architecture. We discover that the primary operations in the aggregation and combination phases can be depicted using a reduction-tree-style dataflow graph. This finding presents an opportunity to develop an efficient and versatile unified architecture for GCN acceleration. The key operation of the aggregation is to aggregate feature vectors from the neighboring vertices, where the computation is dominated by element-wise reduction (i.e., accumulation, or max, or min), as shown in Fig. 3(a). The key operation of the combination is matrix-vector multiplication with a regular number of aggregated feature elements, which can be abstracted as the dot product, as shown in Fig. 3(b). Thus, the output result of each element in the feature vectors can be generated through a reduction-tree-style dataflow graph, which is unified no matter in the aggregation or combination phases. In this study, we term this reduction-tree-style dataflow graph in Fig. 3(c) as graph operators, specifically, an 8-input reduction binary tree. The compute graphs of both phases can be formed using the graph operator, choosing between sum (or max, or min)

---

[1] Graph partition can be used to split the graph into multiple parts, each of which is stored in on-chip buffers, similar to graph analytics [31,32]..

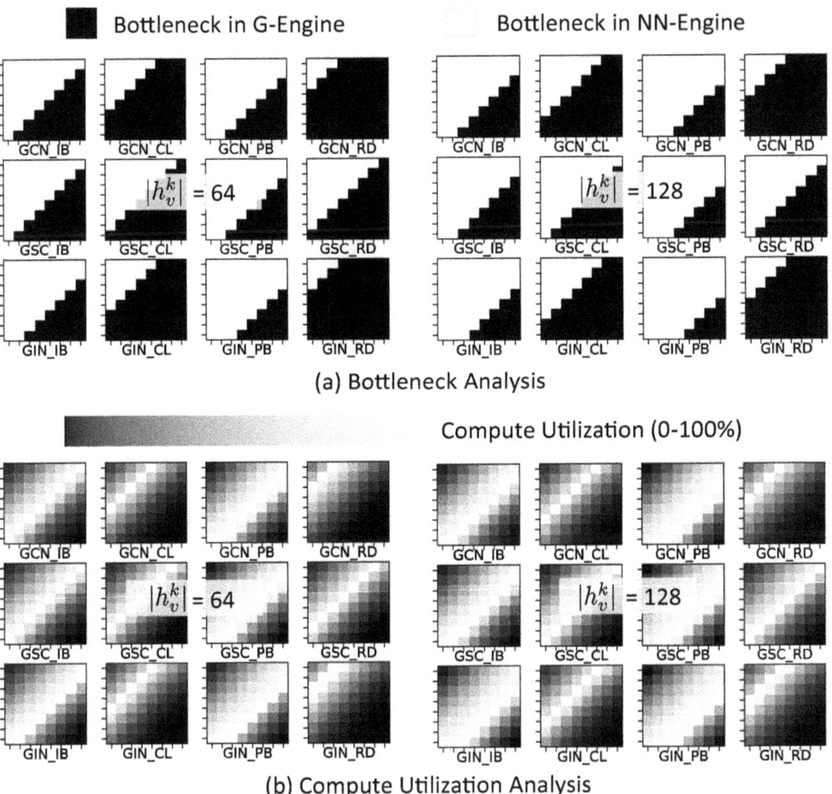

Fig. 2. Inefficiency Analysis: (a) local bottleneck in the engine, i.e., a shortage of compute resources and (b) computation utilization across different TOPS of G-Engine (y-axis = $\{\frac{1}{32}, \frac{1}{16}, \frac{1}{8}, \frac{1}{4}, \frac{1}{2}, 1, 2, 4\}$) and NN-Engine (x-axis = $\{\frac{1}{4}, \frac{1}{2}, 1, 2, 4, 8, 16, 32\}$) on different models, datasets, and $|h_v^k|$.

or multiplication operations, along with determining the number of graph operators and their inputs.

4 Architecture Design of uFlowGCN

This section introduces uFlowGCN with two key components: 1) a graph-operator-based processing element and 2) tag-driven dataflow scheduling.

4.1 Architecture Overview

To address the issue of hybrid architecture, a novel unified architecture is proposed, namely uFlowGCN, helping consolidate total compute resources and prevent resource isolation. The overall architecture of uFlowGCN consists of the

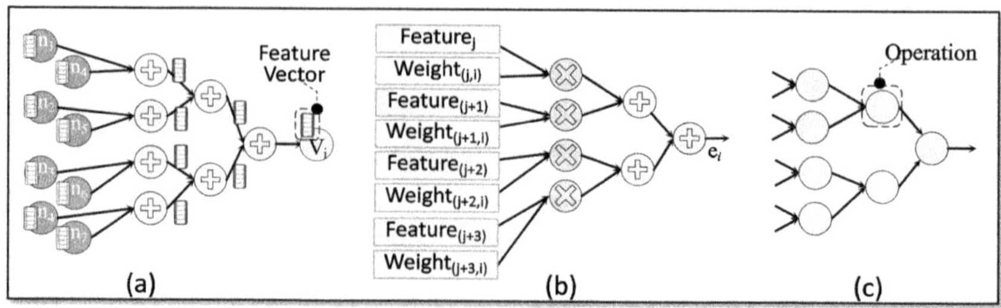

Fig. 3. Extraction of reduction-tree-based graph operator: (a) computation for V_i in the aggregation phase; (b) computation of element e_i in the combination phase; (c) unified graph operator (8-input reduction binary tree).

following two parts. **1)** *Graph-operator-based Processing Element (gPE)* (**Section** 4.2) is proposed to execute the graph operator of both the irregular computational graph of the aggregation phase and the regular computational graph of the combination phase. **2)** *Tag-driven Dataflow Scheduling* (**Section** 4.3) is proposed to leverage GCN-semantics-aware tags to drive dataflows between graph operators in the gPE, providing two execution modes to respectively execute the aggregation and combination phases. This allows the two distinct execution phases to simultaneously perform on the same physical design, enables compute resources to be dynamically allocated to each phase in balance, and realizes the inter-phase pipeline.

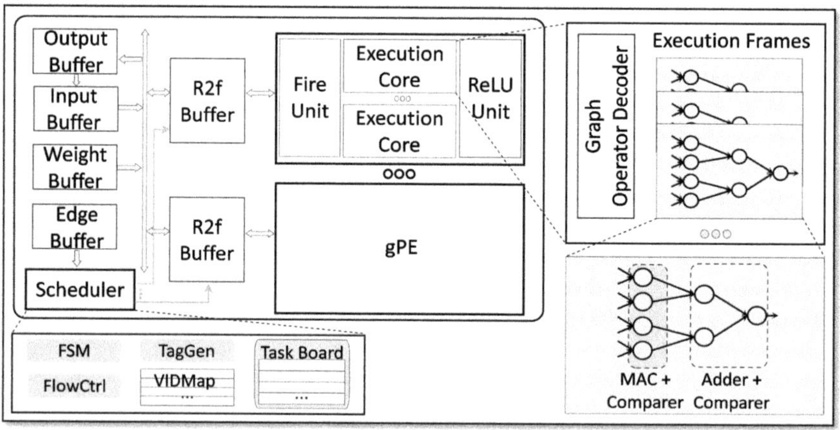

Fig. 4. Hardware Design of uFlowGCN.

The hardware design is shown in Fig. 4, comprising a tag-driven dataflow scheduler, eDRAM scratchpad memory (SPM)-based buffers, and graph-operator-based PEs (gPEs). The scheduler has a finite state machine (FSM), a tag generator (TagGen), a flow controller (FlowCtrl), a task board, and a vertex ID remap table (VIDMap). The FSM partitions the computational graph into graph operators. The TagGen generates the tag for each graph operator. The FlowCtrl builds the inputs for graph operators and drives dataflow between graph operators by tag. Task board registers each computational graph into a task entry. The VIDMap remaps the original vertex ID of the processing vertex to a new one with fewer bits. The R2f (ready-to-fire) buffer stores the input of the graph operator. The input and output buffers respectively hold input and output feature vectors. The weight buffer holds the MLP parameters. The edge buffer stores the edge lists for processing vertices. The gPE links to an R2f buffer, featuring a fire unit, four execution cores, and a ReLU unit. The fire unit fires the graph operator in the R2f buffer to each execution core. The ReLU executes bias addition and activation.

4.2   Graph-Operator-Based Processing Element

We design the gPE to directly execute the graph operator, providing flexibility to perform various operations in GCNs and facilitating efficient exploitation of parallelisms, thereby enhancing execution efficiency over individual operations.

Design and Flexibility. The gPE mainly includes execution core and fire unit, as shown in the right of Fig. 4. Execution core in gPE is built by a graph operator decoder and eight execution frames. The decoder is used for decoding graph operators and setting the input for each frame. Each frame is a binary tree with eight inputs and seven nodes, which directly performs a graph operator. Each input node has a MAC, and a comparer, while the other nodes each only have an adder and a comparer. We define four opcodes for frame, i.e. 00, 01, 10, and 11, for the 8-input sum, max, min, and MAC trees, respectively. Thus, frame can be reconfigured to execute sum/max/min reduction or dot product, which is flexible enough to support the computation of GCNs. Besides, although the number of input operands of each frame is eight, it also supports a range from two to seven, where fire unit complements a binary tree with eight inputs by padding constant values (e.g. 0, 1, or $\pm\infty$). After padding, fire unit fires the graph operator to execution cores. Note that, using the 8-input execution frame is to avoid too much padding and reduce the design complexity of gPE. All frames in the same core run in a lock-step fashion.

Parallelism Exploitation. The gPE is sufficient to exploit the following parallelisms.

1) Intra-frame Parallelism: The computation of the same layer in the binary tree of the frame is data independent, thus all of the intra-layer computational nodes can be processed in parallel.

2) Inter-frame Parallelism: All frames inside the same core can be processed in parallel due to the same execution behavior with the unified binary-tree execution dataflow.

3) Streaming: The dataflow inside the frame naturally allows streaming, where a batch of graph operators can be processed in a pipeline manner.

Exploiting these parallelisms improves computation utilization and overall performance.

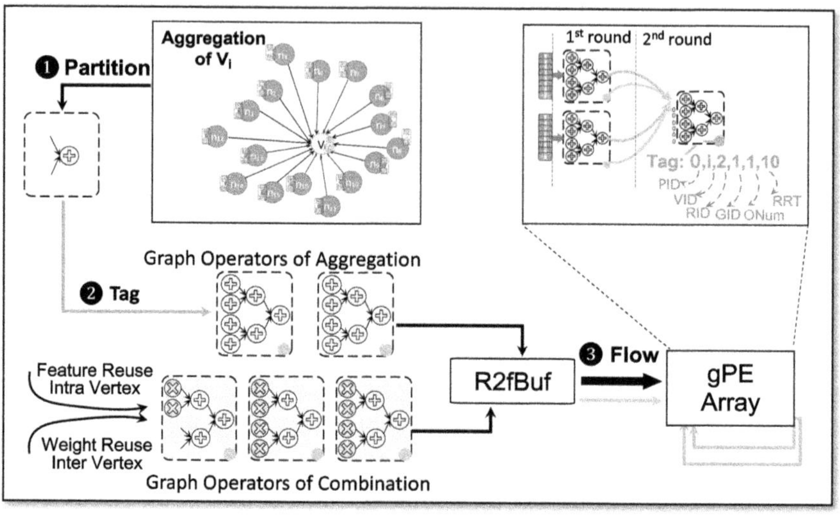

Fig. 5. Tag-driven dataflow scheduling.

4.3   Tag-Driven Dataflow Scheduling

The key idea to perform irregular and regular computational graphs and meet the variable compute resource requirements is to schedule a practical dataflow atop graph operators. Thus, we propose a tag-driven dataflow scheduling tailored to GCNs, which leverages GCN-semantics-aware tags to drive dataflows between graph operators, providing two execution modes to efficiently execute the aggregation and combination phases, respectively. Moreover, these two modes can run simultaneously to pipeline these two phases. Besides, support for large computational graphs is provided.

Scheduling Procedure. The scheduling involves *Partition*, *Tag*, and *Flow*, all managed by the scheduler. To illustrate the scheduling procedure, we employ an example of aggregation for V_i with 16 neighbors, as depicted in Fig. 5. The aggregation operation is accumulation.

❶ *Partition.* The compute graph is partitioned into multiple graph operators by the FSM. The partitioning principle aims to maximize the effectual inputs in each sub-graph. For example, as given in Fig. 5, the 16-input accumulation compute graph can be partitioned into three graph operators with eight inputs, eight inputs, and two inputs, respectively. Before partition, the scheduler registers the current compute graph into a task entry of the task board and produces a task to process its operation. The task entry manages the task status, which has five segments: phase ID (PID), vertex ID (VID), #Total Neighbors (i.e., the total number of neighbors), #Fired Neighbors (i.e., the number of fired neighbors), #Completed Neighbors (i.e., the number of completed neighbors). When PID = 0, the task handles the aggregation; otherwise, it does the combination. To reuse the task board in the combination phase, each element of the aggregated feature vector and weight row is treated as a "neighbor" (see Fig. 3(a) and (b)).

❷ *Tag.* The TagGen assigns a tag to each graph operator, which is crucial for linking multiple graph operators to construct the original compute graph. Each tag comprises six segments: PID, VID, round ID (RID), graph operator ID (GID) in the current round, the number of outputs (ONum) of the current round, and residual repetition time (RRT). The number of rounds corresponds to the number of graph operator levels required to reduce the original compute graph. ONum of the current round equals the number of inputs (#Inputs) of the next round. Given the number of inputs of a round, its number of outputs can be calculated by ONum = $\lceil$#Inputs/8$\rceil$, and GID ranges from 1 to ONum. When ONum equals one, RID stops increasing. RRT denotes the number of repetitive executions of the same compute graph. In the aggregation, the initial value of RRT corresponds to the number of elements in the input feature vector, while in the combination, the initial value of RRT is the number of elements in the output feature vector.

❸ *Flow.* Driven by the tags, the dataflow between the graph operators flows through three steps controlled by the FlowCtrl, with the graph operators executed by gPEs. 1) Input Flow: The input data flowing into the graph operators of the first round is addressed using the VID and GID. *First*, the R2f buffers read the three tags of the first round to check if their graph operators lack input data (i.e., 16 10-element feature vectors) and send data requests to other buffers. The data request is generated using VID and GID. *Second*, the input data returns to the R2f buffers slice by slice. *Finally*, the fire unit complements a binary tree with eight inputs for each graph operator by padding zero and fires them to execution cores. 2) Linking Flow: The dataflow between the graph operators is linked by tags. The intermediate results of the current round flow to the graph operators of the next round according to VID, RID, and GID. These graph operators are executed by gPEs when all input operands are ready. 3) Repetitive Flow: The overall dataflow is allowed to flow repetitively until RRT equals zero. For example, the dataflow between all graph operators flows 10 times for the calculation of 10 elements.

Execution Modes of Unified Architecture. By scheduling the graph operators at run-time, we offer the aggregation and combination modes to efficiently handle the irregular computational graph of the aggregation phase and the regular computational graph of the combination phase, respectively.

- *Aggregation Mode.* The computational graphs depicting the aggregation phase with various types of operations and different numbers of neighboring vertices can be processed as outlined in Sect. 4.3. Different computational graphs only determine how many trees (i.e., graph operators) and how many effectual inputs of each tree are needed, and what the operations in each tree node are. Thus, we omit it here for simplicity.
- *Combination Mode.* To exploit the high-degree data reusability in the combination phase, the combination mode is built by enhancing the scheduling procedure of the aggregation mode. The distinction between these two modes lies in the *Input Flow* step. Firstly, the same aggregated feature vector is repeatedly reused intra-vertex, flowing into the same graph operators in the first round repetitively until RRT equals zero, while replacing weight rows. This facilitates the calculation of different output elements by reusing the aggregated features. Secondly, the weight matrix is reused inter-vertex. For example, Fig. 6 illustrates the computational graph of the combination, which is based on MAC graph operators. This example sets VID=1, VID=2, #Elements_in=20 (i.e., the sum of the lengths of the aggregated feature vector and weight row), and #Elements_out=4 (i.e., the length of the output feature vector). The input operands include both the aggregated feature data and weight rows. To calculate different output elements by reusing the aggregated features, the aggregated feature elements continuously flow four times into the same graph operators. While weight rows cannot be reused when processing a single vertex, they are reused across different vertices. Additionally, ReLU unit conducts additional processing on the final results.

Inter-phase Fusion. These two modes can also run simultaneously to fuse these two phases. Within the processing of a vertex, the combination mode closely follows the aggregation mode, thus the intermediate aggregated feature vector from the aggregation mode can be processed in-place by the combination mode without extra memory overhead. To further boost the inter-phase fusion, the generations of resulting feature elements (i.e., aggregated feature elements or output feature elements) for a single vertex are distributed to a portion of the gPE array and processed in parallel. Moreover, to overlap scheduling and execution, multiple vertices are fired to the gPE array in each cycle. Compared to the design using hybrid architecture, our design enables inter-phase fusion with more parallelism.

Support for Large Computational Graph. Limited by the capacity of R2f buffers, uFlowGCN is unable to process extremely large computational graph at once. Specially, in the aggregation mode, if the number of neighbors is too large to store all tags and input operands in R2f buffers, the task for the large graph is split into multiple sub-tasks. Each sub-task is executed as above, while the intermediate results from previous sub-tasks should be accumulated into the current sub-task. For example, Fig. 7(a) and (b) respectively present the execution flow of a task for the aggregation of 125 neighbors and the evolution of the

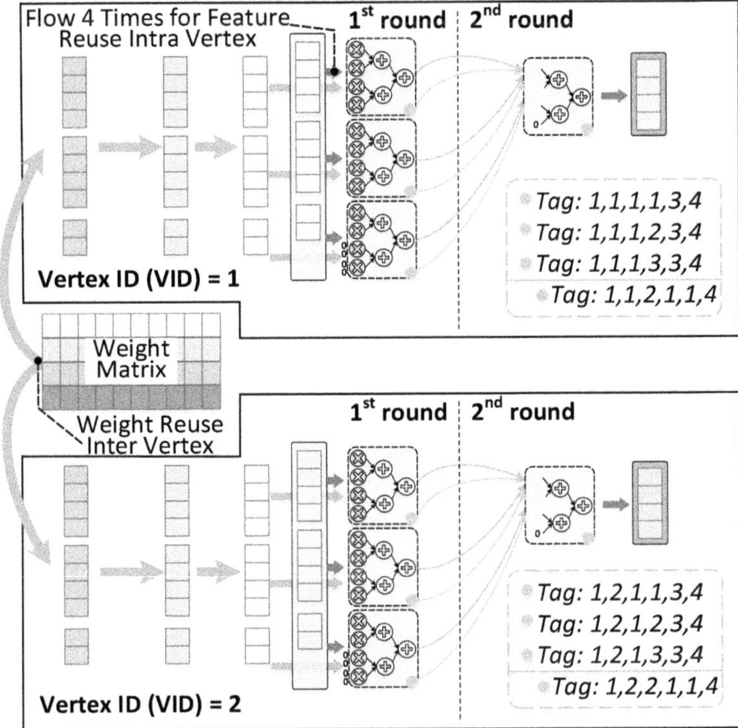

Fig. 6. Illustration of combination mode.

task entry, where each sub-task handles a 32-input accumulation computational graph.

5   Evaluation Results

Evaluation Setup. The models, datasets, and hybrid architecture used in the evaluation are the same as described in Sect. 3. To assess their performance, we develop cycle-accurate simulators for both the hybrid architecture and uFlowGCN. To focus the improvement of compute utilization, both are set up as on-chip implementations, ensuring sufficient on-chip buffer capacity for accommodating the total data. The Synopsys Design Compiler is used in conjunction with the TSMC 12nm standard VT library for the synthesis of the RTL of the scheduler and gPE array. The critical path delay is less than 0.9 ns, allowing uFlowGCN to comfortably operate at a clock frequency of 1 GHz.

Results. To ensure a fair comparison with the hybrid architecture, we conducted experiments to find the optimal configuration for the hybrid architecture. The experimental results are shown in Fig. 8, where each cell represents

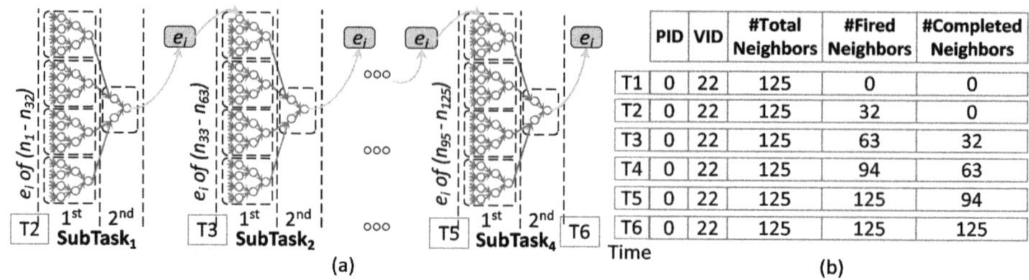

	PID	VID	#Total Neighbors	#Fired Neighbors	#Completed Neighbors
T1	0	22	125	0	0
T2	0	22	125	32	0
T3	0	22	125	63	32
T4	0	22	125	94	63
T5	0	22	125	125	94
T6	0	22	125	125	125

Fig. 7. Execution flow of the aggregation phase in the case with a large number of neighbors: (a) task split; (b) evolution of the task entry.

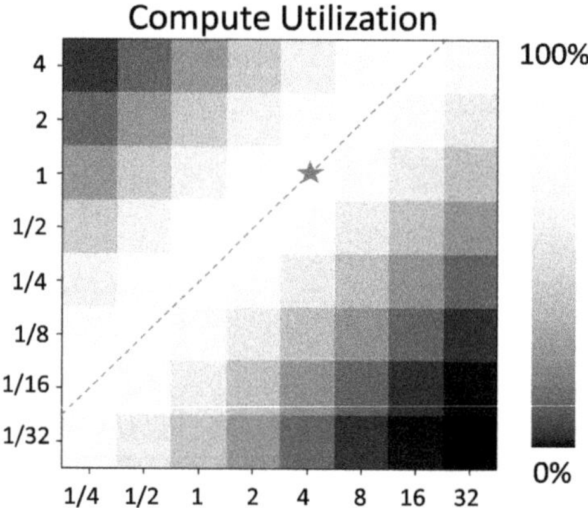

Fig. 8. Optimal configuration of hybrid architecture.

the average compute utilization across all datasets, models, and $|h_v^k|$. The optimal configuration is selected from the configurations marked with a star along the diagonal representing the highest compute utilization rate, i.e., 1 TOPS for G-Engine and 4 TOPS for NN-Engine. Accordingly, we set uFlowGCN to 5 TOPS.

Figure 9 illustrates that uFlowGCN achieves a notable speedup, reaching up to 4.3× compared to the optimal configuration of hybrid architecture. This improvement is primarily attributed to the dynamic allocation of total compute resources to each phase as required, effectively mitigating local bottlenecks. The modest improvement noted in the IB and PB datasets is due to the fact that the compute resources needed for the aggregation phase are minimal, whereas those for the combination phase are substantial, given the low average vertex degrees. Similarly, the improvement on GSC is modest as neighboring vertices are sampled and consequently reduced in number.

Figure 10 shows that the compute utilization of uFlowGCN is high, ranging from 95.3% to 99.5%, in stark contrast to the optimal configure of hybrid architecture, which exhibits a much lower utilization rate of only 24%. This stark difference underscores the substantial performance enhancement offered

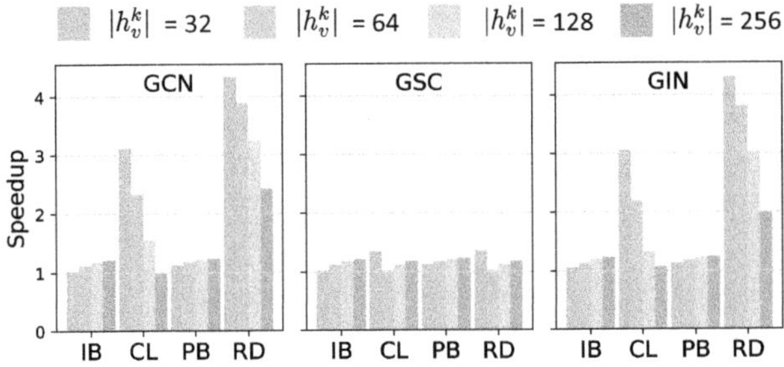

Fig. 9. Speedup to optimal configuration of hybrid architecture across various models, datasets, and $|h_v^k|$.

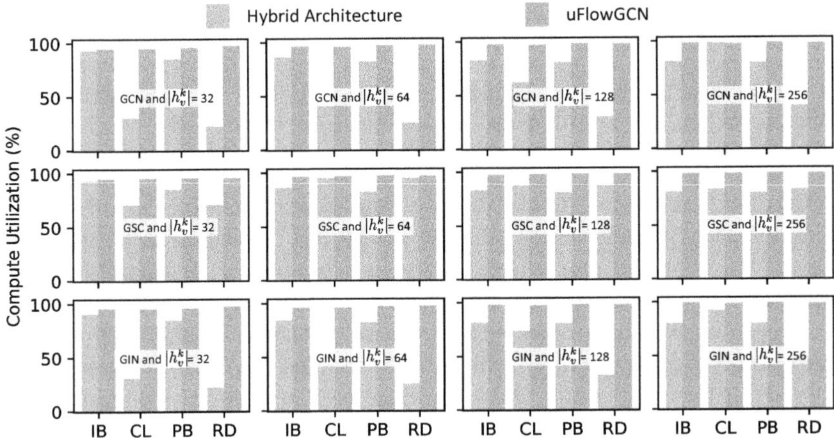

Fig. 10. Compute utilization under optimal configuration of hybrid architecture and uFlowGCN across various models, datasets, and $|h_v^k|$.

by uFlowGCN. The heightened compute utilization in uFlowGCN is mainly attributed to several factors: 1) dynamic allocation of compute resources as required, 2) inter-phase fusion, 3) leveraging various parallelism, and 4) overlapping of scheduling and execution. Note that the compute utilization of the optimal configuration of the hybrid architecture on the CL and RD datasets under the GCN and GIN models is lower compared to the others. This difference is due to these two datasets having a higher average degree, making them more prone to exhausting compute resources during aggregate operations. However, under the GSC, the compute utilization is relatively higher due to the sampling of neighboring vertices, which reduces the compute resource requirements of the aggregate operations.

Figure 11 shows that uFlowGCN introduces certain invalid compute operations, primarily due to operand padding, with the majority arising during the aggregate operations owing to the irregular number of neighboring vertices. The occurrence of these operations is significantly higher in IB and PB datasets com-

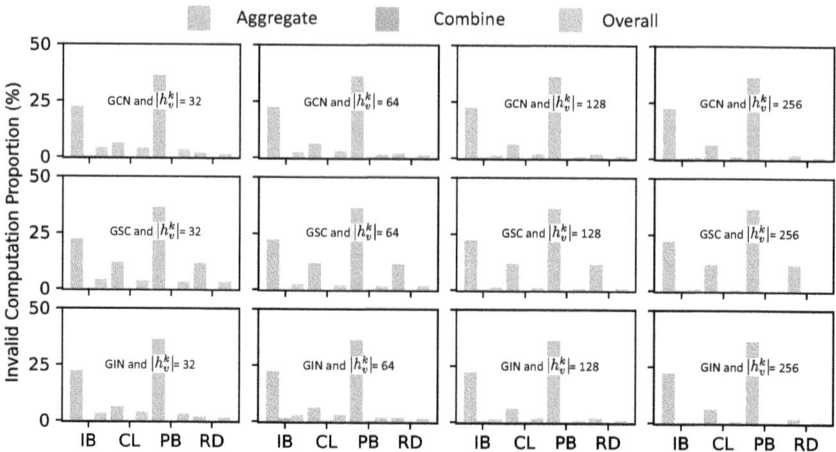

Fig. 11. Invalid computation proportion of uFlowGCN for the aggregate operation, combine operation, and overall across various models, datasets, and $|h_v^k|$.

pared to others, attributed to the smaller average vertex degree in these datasets. For example, the average vertex degree in the PB dataset is four, which is less than eight. However, as the vertex degree increases to 11, the number of invalid computations in aggregate operations decreases to below 25%, and this reduction becomes even more pronounced with higher degrees. Furthermore, this additional computational overhead remains minimal overall, as the combine operations constitute the majority of computations and necessitate minimal padding.

Table 2. Layout characteristics under 5 TOPS, 1 GHz, and 12 nm.

	Power	Area
Compute	5.28 W	2.57 mm^2
Control	0.01 W	0.01 mm^2

Table 2 presents the power and area metrics of uFlowGCN, with a flip rate set to 0.85. In terms of the compute component, its power and area are primarily influenced by numerous 32-bit fixed-point arithmetic logic units. Conversely, for the control component, factors such as tag implementation, task board, and flow control predominantly determine its power and area characteristics. Notably, the power and area of the control component are smaller compared to those of the compute component, accounting for less than 1% of the total.

6   Conclusion

This work proposes uFlowGCN to tackle the hybrid execution patterns in GCN acceleration. It allows the two distinct phases to perform in the unified hardware substrate in parallel and enables compute resources to be dynamically allocated to each phase in balance, alleviating the local bottlenecks in hybrid architecture.

Acknowledgments. This work was supported by the National Natural Science Foundation of China (Grant No. 62202451), CAS Project for Young Scientists in Basic Research (Grant No. YSBR-029), CAS Project for Youth Innovation Promotion Association, and Beijing Nova Program (Grant No. 20250484774).

References

1. Lin, H., et al.: A comprehensive survey on distributed training of graph neural networks. Proceedings of the IEEE (2023)
2. Wu, Z., Pan, S., Chen, F., Long, G., Zhang, C., Philip, S.Y.: A comprehensive survey on graph neural networks. IEEE Transactions on Neural Networks and Learning Systems (2020)
3. Yan, M., et al.: Characterizing and understanding GCNs on GPU. IEEE Computer Architecture Letters (2020)
4. Zhang, Z., Leng, J., Ma, L., Miao, Y., Li, C., Guo, M.: Architectural implications of graph neural networks. IEEE Computer Architecture Letters (2020)
5. Yan, M., et al.: Characterizing and understanding HGNNs on GPUs. IEEE Comput. Archit. Lett. **21**(2), 69–72 (2022)
6. Han, D., Yan, M., Ye, X., Fan, D.: Characterizing and understanding HGNN training on GPUs. ACM Trans. Archit. Code Optim. **22**(1), 1–25 (2025)
7. Yan, M., et al.: HyGCN: a GCN accelerator with hybrid architecture. In: 2020 IEEE International Symposium on High Performance Computer Architecture (HPCA) (2020)
8. Chen, D., et al.: MetaNMP: leveraging cartesian-like product to accelerate HGNNs with near-memory processing. In: Proceedings of the 50th Annual International Symposium on Computer Architecture (2023)
9. Xue, R., et al.: HiHGNN: accelerating HGNNs through parallelism and data reusability exploitation. IEEE Transactions on Parallel and Distributed Systems (2024)
10. Wu, M., Yan, M., Li, W., Ye, X., Fan, D., Xie, Y.: Survey on characterizing and understanding GNNs from a computer architecture perspective. IEEE Transactions on Parallel and Distributed Systems (2025)
11. Zeng, H., Prasanna, V.: GraphACT: accelerating GCN training on CPU-FPGA heterogeneous platforms. In: Proceedings of the 2020 ACM/SIGDA International Symposium on Field-Programmable Gate Arrays, pp. 255–265 (2020)
12. Stevens, J.R., Das, D., Avancha, S., Kaul, B., Raghunathan, A.: GNNerator: a hardware/software framework for accelerating graph neural networks. In: 2021 58th ACM/IEEE Design Automation Conference (DAC). IEEE (2021)
13. Zhang, B., Zeng, H., Prasanna, V.: Hardware acceleration of large scale GCN inference. In: 2020 IEEE 31st International Conference on Application-specific Systems, Architectures and Processors (ASAP), pp. 61–68. IEEE (2020)
14. Zhong, K., et al.: CoGNN: an algorithm-hardware co-design approach to accelerate GNN inference with minibatch sampling. IEEE Trans. Comput. Aided Des. Integr. Circuits Syst. **42**(12), 4883–4896 (2023)

15. Kiningham, K., Levis, P., Ré, C.: GRIP: a graph neural network accelerator architecture. IEEE Trans. Comput. **72**(4), 914–925 (2023)
16. Sarkar, R., Abi-Karam, S., He, Y., Sathidevi, L., Hao, C.: FlowGNN: a dataflow architecture for real-time workload-agnostic graph neural network inference. In: 2023 IEEE International Symposium on High-Performance Computer Architecture (HPCA), pp. 1099–1112 (2023)
17. Yoo, M., Song, J., Lee, J., Kim, N., Kim, Y., Lee, J.: SGCN: exploiting compressed-sparse features in deep graph convolutional network accelerators. In: 2023 IEEE International Symposium on High-Performance Computer Architecture (HPCA), pp. 1–14 (2023)
18. Chen, C., Li, K., Li, Y., Zou, X.: ReGNN: a redundancy-eliminated graph neural networks accelerator. In: 2022 IEEE International Symposium on High-Performance Computer Architecture (HPCA). IEEE (2022)
19. Xu, K., Hu, W., Leskovec, J., Jegelka, S.: How powerful are graph neural networks? In: International Conference on Learning Representations (2018)
20. Hamilton, W., Ying, Z., Leskovec, J.: Inductive representation learning on large graphs. Adv. Neural Inf. Process. (NeurIPS) **30** (2017)
21. Kipf, T.N., Welling, M.: Semi-supervised classification with graph convolutional networks. CoRR (2016)
22. Liu, X., et al.: Sampling methods for efficient training of graph convolutional networks: a survey. IEEE/CAA J. Automatica Sinica **9**(2), 205–234 (2021)
23. Auten, A., Tomei, M., Kumar, R.: Hardware acceleration of graph neural networks. In: 2020 57th ACM/IEEE Design Automation Conference (DAC), pp. 1–6. IEEE (2020)
24. Li, H., et al.: Hardware acceleration for GCNs via bidirectional fusion. IEEE Comput. Archit. Lett. **20**(1), 66–69 (2021)
25. Xue, R., et al.: SiHGNN: leveraging properties of semantic graphs for efficient HGNN acceleration. IEEE Trans. Comput.-Aided Design Integr. Circuits Syst. 1–1 (2025)
26. Jouppi, N.P., et al.: In-datacenter performance analysis of a tensor processing unit. In: Proceedings of the 44th Annual International Symposium on Computer Architecture, pp. 1–12 (2017)
27. Xue, R., et al.: GDR-HGNN: a heterogeneous graph neural networks accelerator frontend with graph decoupling and recoupling. In: DAC '24: 61st ACM/IEEE Design Automation Conference (2024)
28. Sun, G., et al.: Multi-node acceleration for large-scale GCNs. IEEE Trans. Comput. **71**(12), 3140–3152 (2022)
29. Geng, T., et al.: AWB-GCN: a graph convolutional network accelerator with run-time workload rebalancing. In: 2020 53rd Annual IEEE/ACM International Symposium on Microarchitecture (MICRO) (2020)
30. Hwang, R., et al.: GROW: a row-stationary sparse-dense GEMM accelerator for memory-efficient graph convolutional neural networks. In: 2023 IEEE International Symposium on High-Performance Computer Architecture (HPCA), pp. 42–55 (2023)
31. Ham, T.J., Wu, L., Sundaram, N., Satish, N., Martonosi, M.: Graphicionado: a high-performance and energy-efficient accelerator for graph analytics. In: 2016 49th Annual IEEE/ACM International Symposium on Microarchitecture (MICRO), pp. 1–13. IEEE (2016)
32. Yan, M., et al.: Alleviating irregularity in graph analytics acceleration: a hardware/software co-design approach. In: Proceedings of the 52nd Annual IEEE/ACM International Symposium on Microarchitecture, pp. 615–628 (2019)

ACIM: Approximate Computing Based Ising Machine with Dynamic-Ratio Parallel Annealing

Yingzhe Luo[1,2], Zhi Wang[1,2(✉)], Wenya Deng[1,2], Xuhui Li[1,2], Yi Liu[1,2], and Yang Guo[1,2]

[1] College of Computer Science and Technology, National University of Defense Technology, Changsha, China
{luoyingzhe23,zhiwang,dengwenya,guoyang}@nudt.edu.cn
[2] Key Laboratory of Advanced Microprocessor Chips and Systems, National University of Defense Technology, Changsha, China

Abstract. Combinatorial optimization problems (COPs) hold significant application value in logistics, chip design, resource allocation and so on. Ising machines demonstrate unique advantages in solving COPs through physical annealing processes. However, existing Ising architectures face two key challenges: first, current annealing strategies struggle to balance parallelism and solution quality, with existing parallel annealing methods often trapped in local optima while conventional serial methods exhibit prohibitively slow convergence; second, the hardware overhead of spin update circuits scales linearly with spin count, resulting in low spin integration density. To address these challenges, we propose three key innovations: a novel Dynamic-ratio Parallel Annealing (DPA) strategy that dynamically adjusts spin flip ratios during the annealing process while preserving solution quality and significantly improving convergence speed, an approximate computing approach using 8-bit segmented adders that maintains computational accuracy while significantly reducing hardware complexity, and a hardware-optimized hard-σ method that further simplifies probability flip calculations by replacing exponential operations with addition-based approximations. Comprehensive evaluations show Approximate Computing based Ising Machine with Dynamic-ratio Parallel Annealing (ACIM) achieves 10–100× faster convergence than baseline approaches with only 0.54% accuracy loss. The implementation achieves a 40.03% area reduction in the adder circuits used for spin state updates.

Keywords: Ising machine · combinatorial optimization · approximate computing · dynamic annealing

Supported by the research project of National University of Defense Technology 23-ZZCX-JDZ-13.

1 Introduction

Combinatorial optimization problems (COPs), such as VLSI placement and vehicle routing, play a crucial role in modern computing systems [1–4]. However, solving NP-hard COPs using classical von Neumann architectures faces significant challenges as these architectures typically employ exhaustive search methods that incur exponentially increasing computational costs when problem sizes grow beyond 100 variables [5]. This fundamental limitation has driven growing interest in non-von Neumann paradigms like Ising machines, which leverage physical annealing dynamics to bypass the inefficiencies of brute-force search approaches.

Emerging from the study of magnetic systems in physics, the Ising architecture has evolved into a powerful computational paradigm for addressing NP-hard optimization problems [6–9]. While its development from quantum annealers to CMOS-based implementations has demonstrated consistent advantages over von Neumann architectures - particularly in energy efficiency and computational throughput [10–16] - current Ising architectures still confront several interconnected challenges that limit their practical application.

Existing annealing strategies for Ising machines face significant challenges in balancing computational efficiency with solution quality. On one hand, stochastic cellular automata annealing (SCA) [17] and ratio-controlled parallel annealing (RPA) [18] enable rapid parallel updates but often become trapped in local optima. On the other hand, while simulated annealing (SA) theoretically guarantees convergence, its serial single-spin-flip approach requires prohibitively many Monte Carlo steps for large-scale problems.

In addition to these algorithmic limitations, the hardware implementation presents another major bottleneck. The high parallelism requirement forces dedicated update circuits for each spin or spin group, leading to hardware costs that scale linearly with spin count [5,19]. Moreover, the complex arithmetic units needed for spin interaction computations further exacerbate this overhead, demanding more efficient circuit designs.

To address these dual challenges, we propose Dynamic-ratio Parallel Annealing (DPA) to address the limitations of SCA/RPA (local optima trapping) and SA (slow convergence). DPA physically emulates natural annealing by dynamically adjusting spin-flip ratios from 100% parallel updates (global exploration) to single-spin operations (local refinement) through exponential decay. Approximate computing, which has demonstrated significant success in machine learning by reducing hardware costs and power consumption while maintaining accuracy [20–22], is particularly suitable for Ising architectures due to the inherent error tolerance of their stochastic spin update processes and their need for efficient circuit designs. The hard-σ function achieves neural network accuracy comparable to sigmoid functions with substantially lower hardware overhead [23]. We implement this linear approximation in Ising machines to replace the computationally intensive Metropolis and Gibbs (sigmoid-based) probability flip method, completely eliminating both exponential and division operations.

Through the synergistic integration of these innovations, we propose Approximate Computing based Ising Machine with Dynamic-ratio Parallel Annealing (ACIM). The architecture demonstrates three key advantages:

1) Dynamic-ratio Parallel Annealing (DPA): Our implementation achieves 10–100× faster convergence than SA with 8–10% better solution quality than SCA/RPA methods through adaptive parallelism control that optimally balances exploration and exploitation.
2) Approximate Spin Updates: Our 8-bit segmented approximate adders achieves a 40.03% area and 74.73% power consumption in adder circuits used for spin state update, while maintaining 99.46% accuracy in max-cut problems.
3) Hardware-Optimized hard-σ: Our implementation replaces exponential-based probability calculations with addition operations, achieving consistent performance improvements of 0.5% to 1.8% over Gibbs and Metropolis methods across all test scales while completely eliminating complex arithmetic.

2 Background and Related Works

2.1 Ising Model

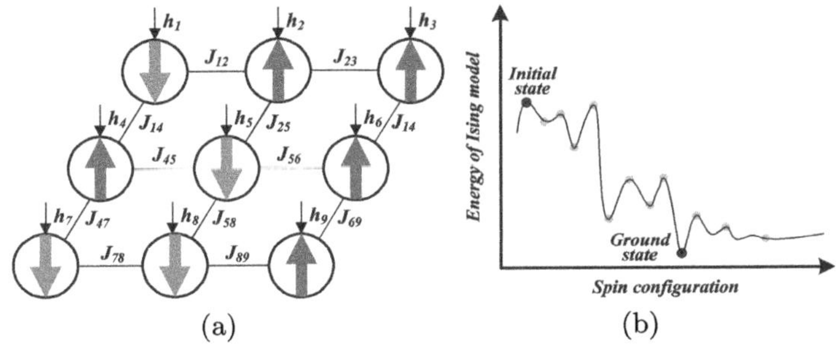

Fig. 1. (a) Ising model. (b) Ising annealing.

The Ising model was originally formulated to describe the behavior of spin particles in magnetic materials [24], as illustrated in Fig. 1(a). In this model, each spin particle is connected to its neighboring spins, with each spin capable of existing in one of two states: spin-up ($\uparrow$, +1) or spin-down ($\downarrow$, −1). The parameter J_{ij} denotes the interaction strength between spins s_i and s_j, while h_i represents the external magnetic field coefficient. The energy of the Ising model is mathematically expressed as follows:

$$E_i(t) = - \sum_{<i,j>}^{N} \left(J_{ij} s_j(t) + h_i \right) s_i(t) \tag{1}$$

In the Ising model, each spin interacts with its neighboring spins through the local field L_i (Eq. 2), which determines spin state updates: when $L_i > 0$, the spin adopts $+1$ (spin-up) state; when $L_i < 0$, it flips to -1 (spin-down). This update mechanism drives the system toward energy minimization, as visualized by the descending trajectory in Fig. 1(b).

$$L_i(t) = \sum_{<ij>} J_{ij}s_j(t) + h_i \tag{2}$$

An efficient approach for local field computation stores the L_i values and avoids redundant calculations through incremental updates (Eq. 3). This strategy significantly reduces computational overhead by only modifying the affected spin interactions when spins are flipped.

$$L_i(t+1) = L_i(t) + 2J_{ij}s_j(t+1) \tag{3}$$

2.2   Annealing Strategies

The Ising architecture initially employed SA to solve COPs. SA is a metaheuristic algorithm inspired by thermodynamic annealing processes. Its core principle involves using a temperature parameter to control the search process: at high temperatures, the algorithm accepts inferior solutions with higher probability to explore the solution space, while gradually converging to low-energy states as the temperature decreases. In SA, each MC step randomly selects only one spin to evaluate its flip probability, executing the flip if the probability condition is met. However, this single-spin update mechanism severely limits SA's annealing speed, making it inefficient for large-scale problems due to its linear time complexity.

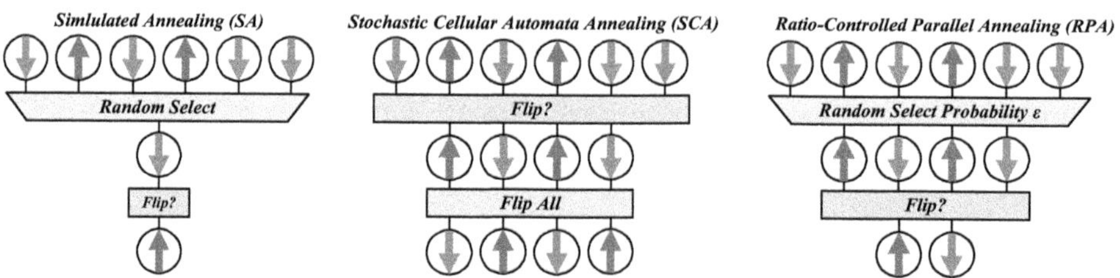

Fig. 2. Existing annealing strategies.

To accelerate the annealing process, researchers developed parallel annealing approaches, with SCA and RPA being the most prominent. SCA enables parallel updates of all spins simultaneously by introducing replica spins to decouple spin dependencies, significantly improving computational speed. RPA further enhances SCA by dynamically controlling the number of spins participating in each update through a predefined ratio parameter ϵ, achieving a balance between convergence and parallelism. While both SCA and RPA accelerate COP solving compared to SA, they tend to fall into local optima more easily. Figure 2 illustrates the working principles of SA, SCA, and RPA annealing strategies.

2.3   Approximate Adder

Approximate computing adders are a class of circuit designs that trade off computational accuracy for reduced hardware overhead and energy consumption [25,26]. Compared to traditional precise adders, approximate adders simplify logic structures or reduce the number of transistors in critical paths, significantly decreasing area and power consumption [27,28]. These adders are particularly suitable for applications where exact precision is not critical, such as image processing and machine learning accelerators [29,30].

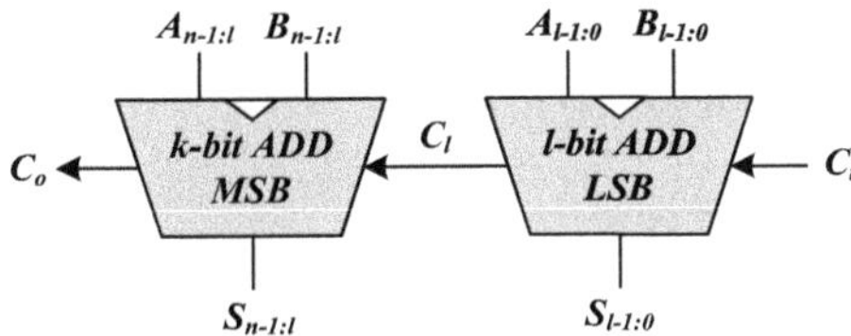

Fig. 3. Segmentation approximate adder.

A segmentation approximate adder is an important variant of approximate adders, composed of a Most Significant Bit (MSB) segment and a Least Significant Bit (LSB) segment, as illustrated in Fig. 3. The MSB segment typically employs precise computation to ensure accuracy in critical bits, while the LSB segment utilizes approximate computation to reduce hardware complexity. This design maintains overall computational accuracy while effectively minimizing energy consumption and area overhead.

In this work, we propose implementing an Ising architecture using a segmentation approximate adder [31]. By strategically allocating bit-widths between the MSB and LSB segments, this approach is expected to significantly reduce hardware costs without compromising the computational precision of the Ising architecture, offering an efficient and feasible solution for large-scale Ising computing systems.

2.4   Probability Flip Strategies

In the Ising model, a probabilistic spin flip strategy is adopted to avoid premature convergence to local optima during optimization. This method gradually reduces the perturbation probability as temperature declines in the annealing process. The diminishing probability allows the system to overcome local energy barriers while ensuring convergence to the ground state with minimal Ising energy. The probabilistic spin flip operation can be mathematically expressed as follows:

$$S_i(t+1) = \begin{cases} -S_i(t), & P(\Delta E_i) > r \\ S_i(t), & P(\Delta E_i) \leq r \end{cases} \tag{4}$$

Here, r is a random number between 0 and 1 ($r \in (0,1)$), and ΔE_i represents the local energy barrier between the two states before and after the spin flip. Given that the current spin state is $S_i(t)$, the spin state after flipping is $-S_i(t)$. Thus, $\Delta E_i = -[-S_i(t)]L_i(t) + S_i(t)L_i(t) = 2S_i(t)L_i(t)$. The common probabilistic flipping methods are as follows:

$$P(\Delta E_i) = \begin{cases} \min\left[1, \exp\left(-\frac{\Delta E_i}{T}\right)\right] & \text{Metropolis} \\ \dfrac{1}{1 + \exp\left(\dfrac{\Delta E_i}{T}\right)} & \text{Gibbs} \end{cases} \tag{5}$$

3   Proposed Ising Architecture

3.1   Architecture Overview

To address the two key challenges of suboptimal annealing strategies, and high hardware costs in spin update circuits, we propose the following solutions: First, we introduce a new DPA strategy that achieves both high parallelism and optimal solutions for combinatorial optimization problems. Second, we implement spin updates using 8-bit approximate adders to significantly reduce hardware overhead. Third, we develop a novel probability flip strategy that transforms complex exponential and division operations into simple addition operations, substantially reducing hardware costs.

These innovations are integrated into our proposed ACIM architecture, as illustrated in Fig. 4. The architecture consists of three main components:

Weight Memory Array: This module consists of multiple memory subarrays that collectively store all spin interaction coefficients J_{ij} required for annealing computations.

Approximate Spin Update Calculator (ASUC): During the annealing process, this module updates local field terms L_i and computes spin flip probabilities $P(\Delta E_i)$ to generate flip flags Δ_i. The implementation uses 8-bit segmented approximate adders to reduce hardware costs while maintaining computational accuracy.

Dynamic-ratio Parallel Annealing Controller (DPAC): This controller processes flip flags Δ_i from the ASUC to implement the Dynamic-ratio Parallel Annealing (DPA) strategy by dynamically adjusting the number of spins to flip at each annealing step through exponential decay.

The annealing process proceeds through iterative computation cycles. During each cycle, the controller first executes probability flip calculations in the ASUC module to generate spin flip flags Δ_i. The DPAC then processes these flags to

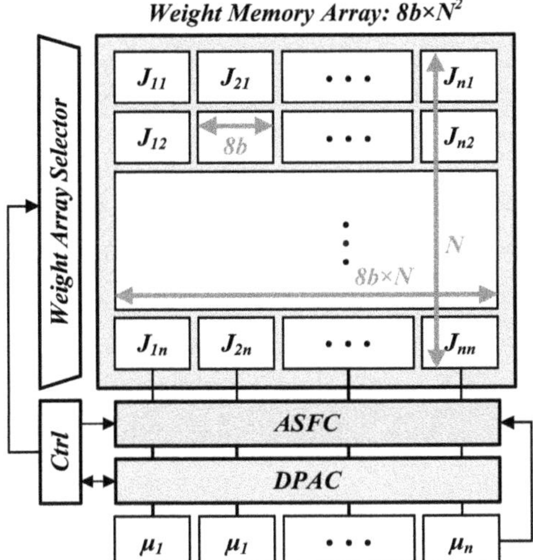

Fig. 4. ACIM architecture.

determine which spins should undergo state transitions. Following spin flips, the system retrieves the corresponding interaction coefficients J_{ij} from memory to update all affected local field terms L_i. This sequence repeats cyclically until the convergence criteria are satisfied, producing the final optimized spin configuration (Fig. 5).

3.2 Dynamic-Ratio Parallel Annealing Controller(DPAC)

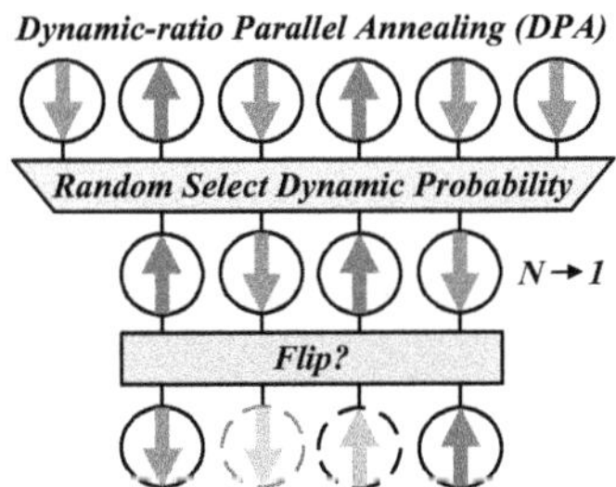

Fig. 5. Dynamic-ratio parallel annealing.

The DPA strategy dynamically adjusts the number of flipped spins during the annealing process through exponential decay (Eq. 6). The decay rate is controlled by coefficient α (empirically set to 1.2–1.5). If this value is higher than 1.5, it will cause rapid attenuation and prematurely terminate the global exploration stage. However, if this value is lower than 1.2, it will weaken the advantage of parallel annealing in convergence speed. The chosen range ensures a progressive transition from fully parallel spin updates ($Flip\_Num = Spin\_Num$) during initial global exploration to single-spin operations ($Flip\_Num = 1$) for final local refinement. Threshold boundaries are automatically generated at power-of-two intervals to maintain hardware efficiency.

$$Flip\_Num = \frac{Spin\_Num}{\alpha^{step}} \tag{6}$$

To implement DPA, we employs a bitmask generator that produces an n-bit mask (where $n = log_2(Spin\_Num)$) corresponding to the current $Flip\_Num$ value. This mask is combined with spin flip flags Δ_i through three key operations: 1) cyclic right-shift using a pseudo-random offset (XorShift16), 2) bitwise AND with the generated mask, and 3) circular left-shift to restore spatial locality. The complete implementation is shown in Fig. 6. The unified architecture maintains compatibility with conventional spin flip strategies (SA, SCA, RPA).

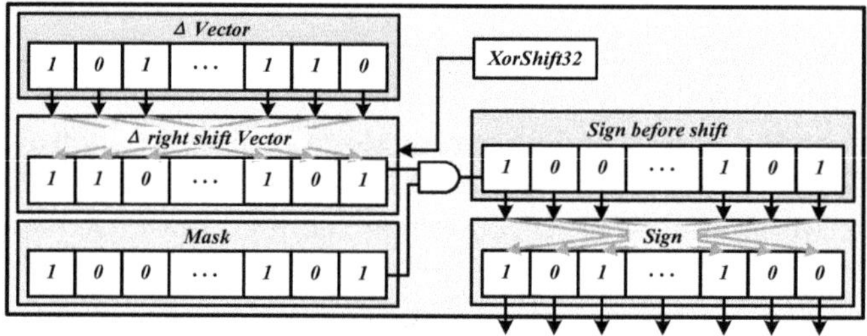

Fig. 6. Dynamic-ratio parallel annealing controller.

The DPA's adaptive ratio balances parallelism and solution quality, addressing the limitations of fixed-ratio strategies like SCA/RPA while avoiding SA's serial bottleneck.

3.3　Approximate Spin Update Calculator (ASUC)

In CMOS-based Ising computing, the stored value "0" represents "+1" in the Ising model, and "1" represents "−1". Letting the stored value be μ, Eq. 3 can be rewritten as:

$$\tilde{L}_i(t + 1) = \tilde{L}_i(t) + (J_{ij} \oplus \{\mu_j(t + 1)\}_n + \mu_j(t + 1)) \tag{7}$$

where $\tilde{L}_i = L_i \gg 1$ (the LSB does not affect spin decisions), and $\{\mu_j\}_n$ is the n-bit expansion of μ_j. The operation $J_{ij} \oplus \{\mu_j\}_n + \mu_j$ toggles between J_{ij} ($\mu_j = 0$) and $-J_{ij}$ ($\mu_j = 1$) via bitwise XOR-addition. After such a transformation, L_i update calculations mainly include XOR calculations and ADD calculations.

Furthermore, we adopt the novel probabilistic flipping formula hard-σ to replace conventional Metropolis/Gibbs sampling, converting complex exponential/division operations into simple XOR and ADD operations (as detailed in Sect. 3.3).

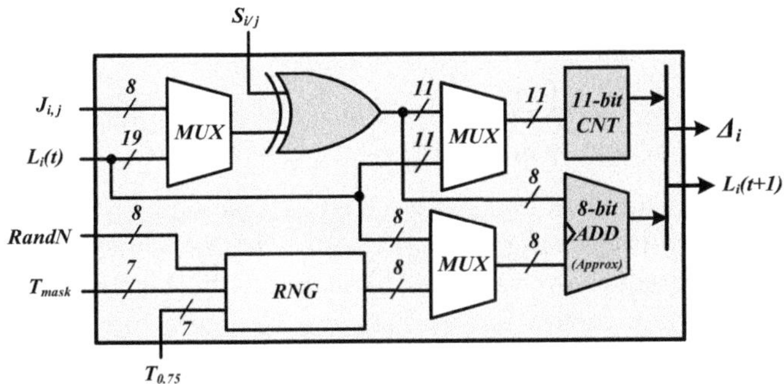

Fig. 7. Approximate spin update calculator.

Consequently, the Ising computation is primarily composed of XOR and ADD operations. This computational simplification enables the Approximate Ising Compute architecture shown in Fig. 7, integrating three core components:

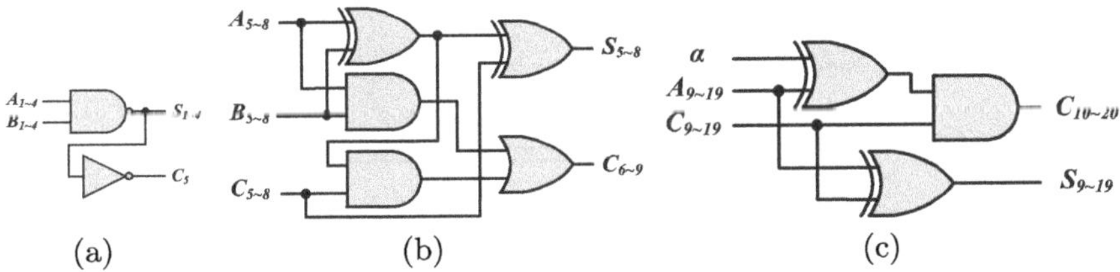

Fig. 8. (a) Approximate adder. (b) Precise adder. (c) Counter.

ADD: In the addition calculation process, our Ising architecture uses an 8-bit adder to complete the Ising calculation. This adder adopts a segmentation approximate adder scheme, where the upper 4 bits perform precise calculation and the lower 4 bits implement approximate calculation. The lower 4 bits use the approximate computing adder [25] in Fig. 8(a), and its circuit expression is as follows.

$$\begin{cases} S_{1-4} = \overline{A_{1\sim4} \cdot B_{1\sim4}} \\ C_5 = A_{1\sim4} \cdot B_{1\sim4} \end{cases} \tag{8}$$

CNT: The architecture implements 2048 spins with 8-bit interaction coefficients J_{ij}, requiring $\tilde{L}_i$ storage of 19 bits (original 20-bit width reduced by discarding the right-shifted LSB). The upper 11-bit update follows four distinct cases: when $(J_{ij} \geq 0, C_8 = 0)$ or $(J_{ij} < 1, C_8 = 1)$, it performs add-0; when $(J_{ij} \geq 0, C_8 = 1)$, add-1; and when $(J_{ij} < 1, C_8 = 0)$, adds binary 11111111111 (equivalent to sub-1 in two's complement). The control parameter α determines the operation mode($\alpha = 0$: add-1, $\alpha = 1$: sub-1). The counter design (shown in Fig. 8(c)) propagates carry signals through chained logic gates until termination, flipping bits via XOR operations during propagation, as shown in Eq. 9.

$$\begin{cases} C_{i+1} = C_i \cdot (\alpha \oplus A_i) \\ S_i = C_i \oplus A_i \end{cases} \tag{9}$$

RNG: The random number generator (RNG), following the approach in [17], produces values within the range $(-num, num)$. It first generates a mask corresponding to the current num value by identifying the highest set bit (position i) in num's binary representation and setting all lower bits in the mask to 1. A 32-bit random number from LFSR32 then undergoes bitwise AND and XOR operations with this mask - the XOR result is selected if the AND result exceeds num, otherwise the AND result is kept. The sign of each output is determined by using bits 31, 23, 15, and 7 as sign bits, ultimately producing four random numbers $T_{0.75}$. The hardware implementation of this RNG is shown in the circuit diagram Fig. 9, where the T_{mask} module handles the masking operations.

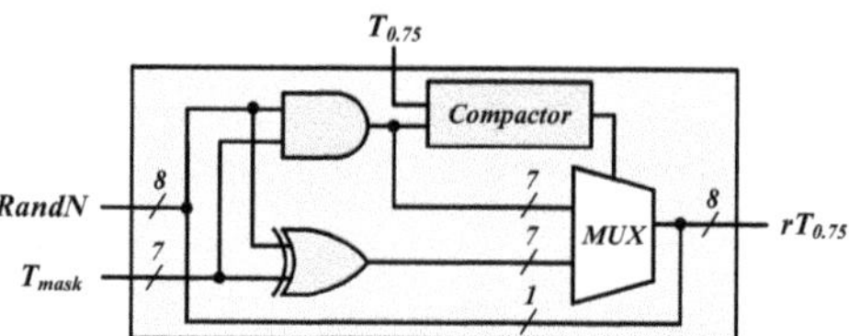

Fig. 9. RNG architecture.

3.4   Hard-σ Spin Flip Probability

Traditional probability flip methods, such as Metropolis and Gibbs (equivalent to the $\sigma(x) = \frac{1}{1+e^{-x}}$), rely on complex exponential operations. To eliminate this overhead, we adopt the hard-σ approximation from deep learning [23], which serves as a lightweight version of $\sigma(x)$. As shown in Fig. 10, $hard$-σ approximates the sigmoid with a piecewise-linear function while preserving similar monotonicity and saturation characteristics:

$$\frac{ReLU6(x+3)}{6} = \frac{\min[\max(0, x+3), 6]}{6} \tag{10}$$

Our Ising model adopts this linear formula, and the probability flip judgment formula can be converted as:

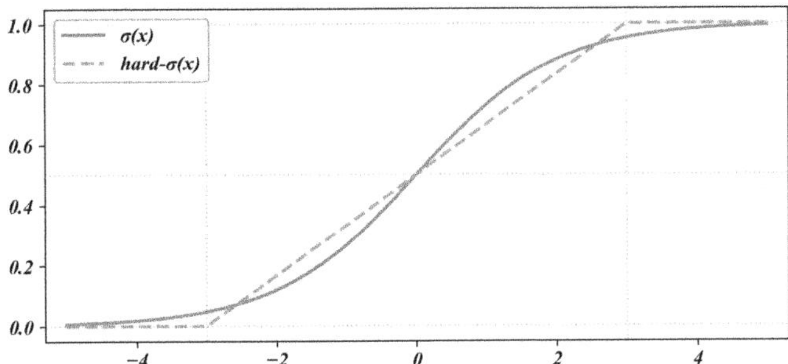

Fig. 10. Comparison of function between $\sigma(x)$ and hard-$\sigma(x)$.

$$S_i(t+1) = \begin{cases} -S_i(t), & \text{if } \Delta E_i + (3-6r)T < 0, \\ S_i(t), & \text{otherwise.} \end{cases} \tag{11}$$

where $r \in (0,1)$ is a random number, $\Delta E_i = 2S_i(t)L_i(t)$. In hardware implementation, to simplify computation, the entire inequality is right-shifted by 2 bits, obtaining $\tilde{L}_i(t) = L_i(t)/2$ and the scaled temperature term. Although this loses precision in the lowest two bits, since the spin flip decision is mainly determined by the sign bit and the probabilistic update mechanism has error tolerance, this approximation has negligible impact on the results. Finally, a hardware-friendly decision condition is obtained:

$$\tilde{L}_i(t) \cdot S_i(t) + \frac{3-6r}{4}T < 0, \tag{12}$$

where $\frac{3-6r}{4}T \in (-0.75T, 0.75T)$. Thus, the formula can be further simplified by substituting $\frac{3-6r}{4}T$ with a random variable $rT_{0.75} \in (-0.75T, 0.75T)$:

$$\mu_i(t+1) = \begin{cases} \overline{\mu_i(t)}, & \tilde{L}_i(t) \oplus \{\mu_i(t)\}_n + rT_{0.75} < 0 \\ \mu_i(t), & \text{otherwise.} \end{cases} \tag{13}$$

This transformation reduces the probability flip calculation to XOR and addition operations, enabling efficient implementation in the ASUC module (Sect. 3.4).

4   Experimental Evaluation

The ACIM architecture was implemented in FinFET technology as a fully-connected 2,048-spin system, with key specifications summarized in Table 1. Our synthesis achieved 800MHz operation within $4.59mm^2$ (8965.24 μm^2 SRAM core and 3351.56 μm^2 logic), integrating 512 weight memory blocks.

Table 1. Experimental Setup.

Configuration	
Technology	FinFET
Memory capacity	4 MB
Frequency	800 MHz
Total area	4.59 mm^2
SRAM core	8965.24 μm^2
Logic core	3351.56 μm^2

The max-cut problem aims to partition vertices (s_i) into two groups ($s_i \in \{-1, 1\}$) to maximize the sum of weights (G_{ij}) on edges connecting vertices in different groups. The objective function is expressed as Eq. 14:

$$H(s)_{\text{maxcut}} = \max \frac{1}{2} \sum_{x<y} G_{xy}(1 - s_x s_y) \tag{14}$$

In this formulation, to maximize the max-cut objective, we aim to minimize its negation. By setting $J_{ij} = -G_{ij}$, $s_i = s_x$, and $h_i = 0$ in Eq. 1, we derive an Ising model that represents the max-cut problem defined by Eq. 14. The Ising model thus solves combinatorial optimization problems through this transformation.

We evaluated the system using a cycle-accurate simulator for max-cut problems at three scales (N = 512, 1024, and 2048). For each size, we generated 5 distinct random graphs with edge weights drawn from U(−128,127) at 50% connection density, executing each instance 10 times with different random seeds. Hardware metrics were extracted from transistor-level designs using SPICE simulations.

4.1   Annealing Strategy Evaluation

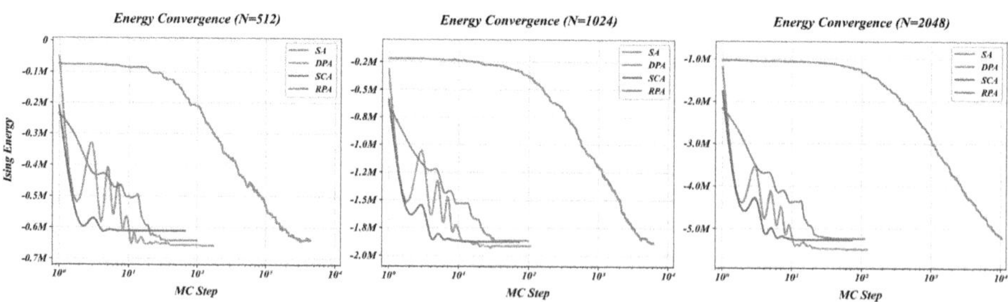

Fig. 11. The convergence of different annealing strategies under various scales.

To verify the convergence properties of DPA, we conducted experiments on the simulator. The energy curves in Fig. 11 clearly demonstrate DPA's superior convergence characteristics. Our quantitative analysis of five randomized trials shows that DPA achieves convergence rates comparable to both SCA and RPA, with all three strategies typically converging within approximately 100 computational steps (Fig. 12). Compared to the SA, DPA exhibits a substantial computational advantage, demonstrating 10–100 times faster convergence (representing a 1–2 order of magnitude improvement).

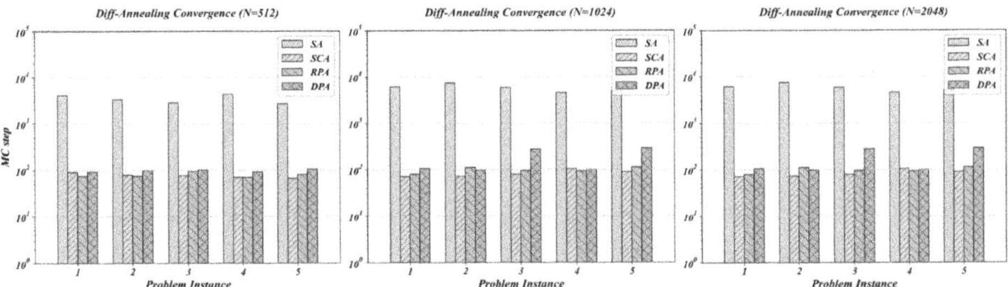

Fig. 12. The steps of different annealing strategies under various scales.

For solution quality evaluation, DPA shows consistent advantages in max-cut problem solving, with the results illustrated in Fig. 13, improving performance by 10% over SCA, 8% over RPA, and 2% over SA. These improvements are particularly pronounced in larger problem instances (N = 2048), where DPA's dynamic adjustment mechanism between parallel and serial updates proves most effective. All experiments were conducted under identical hardware environments and parameter settings to ensure fair comparison.

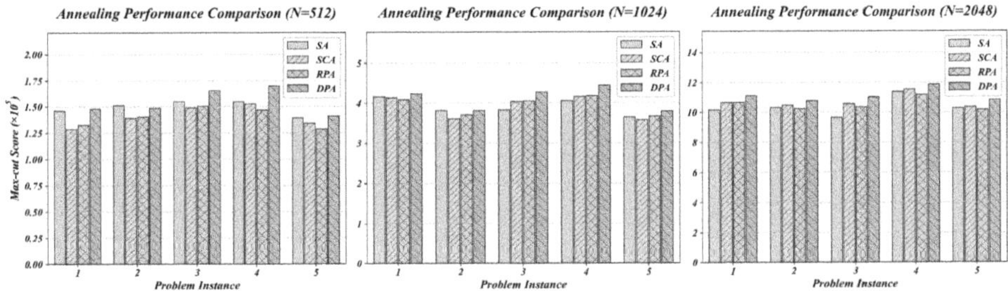

Fig. 13. Max-cut score of different annealing under various scales.

4.2 Approximate Computing Analysis

Table 2. Adder implementation comparison.

Metric	Approx	Precise
Area(μm^2)	7.46	12.44
Static Power(nW)	38.1	151.6
Delya(ps)	109.89	187.84

Our hardware evaluation demonstrates significant benefits from using approximate computing in the Ising architecture. We implemented both the proposed 8-bit segmented approximate adder and a conventional precise adder in FinFET technology, with simulations conducted under typical corner conditions (TT, 0.8 V, 25°C). The approximate adder achieves a 40.03% area reduction (7.46 μm^2 vs 12.44 μm^2), 74.73% static power savings (38.10 nW vs 151.6 nW), and 41.50% faster operation (109.89ps vs 187.84ps latency) compared to the precise version (Table 2).

Compatibility testing with Ising computations through our simulator revealed consistently bounded approximation errors across problem scales: mean errors of 0.06% (N = 512), 0.36% (N = 1024), and 0.54% (N = 2048), as shown in Fig. 14. These results confirm that our approximate computing approach maintains solution quality while delivering substantial improvements in hardware efficiency.

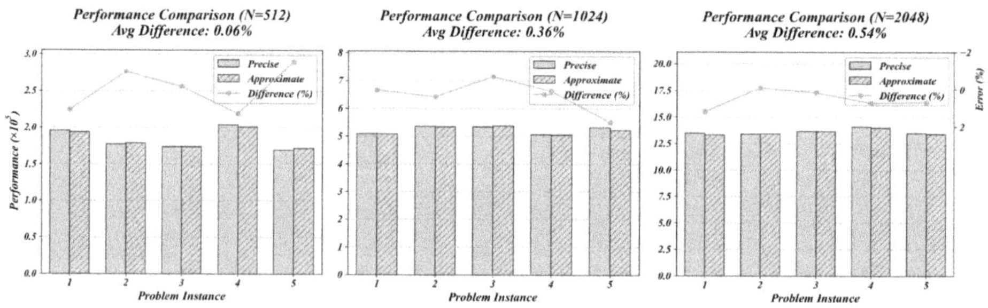

Fig. 14. Max-cut score of different precision under various scales.

4.3 Probability Flip Methods

To validate the effectiveness of the hard-σ probability flip strategy, we conducted systematic experiments on mac-cut problems at different scales, with the results illustrated in Fig. 15. The experimental results demonstrate that the hard-σ method exhibits consistent performance advantages across all three test scales. Specifically, in the N = 512 tests, it achieves computational performance improvements of 1.8% and 1.1% compared to the Gibbs and Metropolis methods, respectively. For N = 1024, the performance advantages are 0.7% and 0.5%, while in large-scale N = 2048 tests, it maintains performance gains of 0.9% and 1.2%.

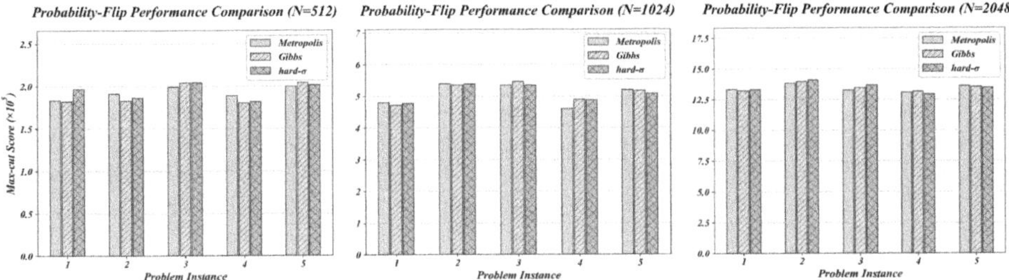

Fig. 15. Max-cut score of different probability flip method under various scales.

From an implementation perspective, the hard-σ method innovatively transforms the time-consuming exponential and division operations of traditional approaches into efficient addition operations (detailed in Sect. 3.4). This computational paradigm optimization not only preserves algorithmic accuracy but, more importantly, significantly enhances hardware implementation efficiency: substantially reducing computational unit resource occupancy while markedly improving operational speed, thereby providing an optimal solution for Ising architecture design.

5   Conclusion

We have presented ACIM, a novel Ising machine architecture that successfully addresses two fundamental challenges in current Ising machines: high hardware overhead in spin update circuits, and suboptimal annealing strategies. Our solution integrates three key innovations: approximate computing circuits, hardware-optimized hard-σ probability flip method, and dynamic-ratio parallel annealing (DPA).

Implemented in FinFET technology, ACIM demonstrates high spin integration density with 800MHz operation for a 2048-spin system consuming 4.59mm^2 area. The DPA strategy dynamically adjusts spin flip ratios from fully parallel to serial operations, achieving 10–100$\times$ faster convergence than simulated annealing while maintaining superior solution quality over SCA/RPA strategies. The 8-bit segmented approximate adder reduces adder module area of spin update circuits by 40.03% and static power by 74.73% compared to precise implementations, while introducing only negligible errors ($\leq$0.54% across problem scales) that are well-tolerated by the stochastic nature of Ising computations. The hard-σ approach replaces exponential-based computations with efficient addition operations, improving performance by 0.5–1.8% while significantly reducing hardware complexity.

References

1. Lucas, A.: ISING formulations of many NP problems. Front. Phys. **2**, 5 (2014)
2. Tanahashi, K., Takayanagi, S., Motohashi, T., Tanaka, S.: Application of ISING machines and a software development for ISING machines. J. Phys. Soc. Jpn. **88**(6), 061010 (2019)
3. Cheng, M.X., Li, Y., Du, D.Z.: Combinatorial optimization in communication networks. Springer (2006). https://doi.org/10.1007/0-387-29026-5_1.pdf
4. Bao, S., Tawada, M., Tanaka, S., Togawa, N.: An ISING-machine-based solver of vehicle routing problem with balanced pick-up. IEEE Trans. Consum. Electron. **70**(1), 445–459 (2023)
5. Su, Y., Kim, T.T.H., Kim, B.: FlexSpin: a CMOS ISING machine with 256 flexible spin processing elements with 8-b coefficients for solving combinatorial optimization problems. IEEE J. Solid-State Circ. **59**(8), 2659–2670 (2024)
6. Yamaoka, M., Yoshimura, C., Hayashi, M., Okuyama, T., Aoki, H., Mizuno, H.: A 20k-spin ISING chip to solve combinatorial optimization problems with CMOS annealing. IEEE J. Solid-State Circuits **51**(1), 303–309 (2015)
7. Takemoto, T., Hayashi, M., Yoshimura, C., Yamaoka, M.: 2.6 a 2× 30k-spin multichip scalable annealing processor based on a processing-in-memory approach for solving large-scale combinatorial optimization problems. In: 2019 IEEE International Solid-State Circuits Conference-(ISSCC), pp. 52–54. IEEE (2019)
8. Su, Y., Kim, H., Kim, B.: CIM-SPIN: a scalable CMOS annealing processor with digital in-memory spin operators and register spins for combinatorial optimization problems. IEEE J. Solid-State Circ. **57**(7), 2263–2273 (2022)
9. Yue, W., et al.: A scalable universal ISING machine based on interaction-centric storage and compute-in-memory. Nature Electronics **7**(10), 904–913 (2024)
10. Bian, Z., Chudak, F., Macready, W.G., Rose, G.: The ISING model: teaching an old problem new tricks. D-wave systems **2**, 1–32 (2010)
11. Johnson, M.W., et al.: Quantum annealing with manufactured spins. Nature **473**(7346), 194–198 (2011)
12. Mu, J., Su, Y., Kim, B.: A 20x28 spins hybrid in-memory annealing computer featuring voltage-mode analog spin operator for solving combinatorial optimization problems. In: 2021 Symposium on VLSI Technology, pp. 1–2. IEEE (2021)
13. Xie, S., Raman, S.R.S., Ni, C., Wang, M., Yang, M., Kulkarni, J.P.: ISING-CIM: a reconfigurable and scalable compute within memory analog ISING accelerator for solving combinatorial optimization problems. IEEE J. Solid-State Circuits **57**(11), 3453–3465 (2022)
14. Honjo, T., et al.: 100,000-spin coherent ISING machine. Sci. Adv. **7**(40), eabh0952 (2021)
15. Okuyama, T., Sonobe, T., Kawarabayashi, K., Yamaoka, M.: Binary optimization by momentum annealing. Phys. Rev. E **100**(1), 012111 (2019)
16. Waidyasooriya, H.M., Hariyama, M.: Highly-parallel FPGA accelerator for simulated quantum annealing. IEEE Trans. Emerg. Top. Comput. **9**(4) (2019)

17. Yamamoto, K., et al.: STATICA: a 512-spin 0.25 m-weight annealing processor with an all-spin-updates-at-once architecture for combinatorial optimization with complete spin–spin interactions. IEEE J. Solid-State Circ. **56**(1), 165–178 (2020)
18. Kawamura, K., et al.: AMORPHICA: 4-replica 512 fully connected spin 336 mhz metamorphic annealer with programmable optimization strategy and compressed-spin-transfer multi-chip extension. In: 2023 IEEE International Solid-State Circuits Conference (ISSCC), pp. 42–44. IEEE (2023)
19. Takemoto, T., et al.: 4.6 a 144kb annealing system composed of $9\times$ 16kb annealing processor chips with scalable chip-to-chip connections for large-scale combinatorial optimization problems. In: 2021 IEEE International Solid-State Circuits Conference (ISSCC), vol. 64, pp. 64–66. IEEE (2021)
20. Gong, J., Saadat, H., Gamaarachchi, H., Javaid, H., Hu, X.S., Parameswaran, S.: ApproxTrain: fast simulation of approximate multipliers for DNN training and inference. IEEE Trans. Comput. Aided Des. Integr. Circuits Syst. **42**(11), 3505–3518 (2023)
21. Chen, C.Y., Choi, J., Gopalakrishnan, K., Srinivasan, V., Venkataramani, S.: Exploiting approximate computing for deep learning acceleration. In: 2018 Design, Automation & Test in Europe Conference & Exhibition (DATE), pp. 821–826. IEEE (2018)
22. Huang, N.C., Chen, S.Y., Wu, K.C.: Sensor-based approximate adder design for accelerating error-tolerant and deep-learning applications. In: 2019 Design, Automation & Test in Europe Conference & Exhibition (DATE), pp. 692–697. IEEE (2019)
23. Howard, A., et al.: Searching forMmobileNetV3. In: Proceedings of the IEEE/CVF International Conference on Computer Vision, pp. 1314–1324 (2019)
24. Brush, S.G.: History of the Lenz-Ising model. Rev. Mod. Phys. **39**(4), 883 (1967)
25. Napoli, E., Zacharelos, E., Strollo, A.G., Di Meo, G.: Approximate full-adders: a comprehensive analysis. IEEE Access (2024)
26. Gupta, V., Mohapatra, D., Park, S.P., Raghunathan, A., Roy, K.: Impact: imprecise adders for low-power approximate computing. In: IEEE/ACM International Symposium on Low Power Electronics and Design, pp. 409–414. IEEE (2011)
27. Gupta, V., Mohapatra, D., Raghunathan, A., Roy, K.: Low-power digital signal processing using approximate adders. IEEE Trans. Comput. Aided Des. Integr. Circuits Syst. **32**(1), 124–137 (2012)
28. Rampeesa, A., Akhila, P., Irfan, M., Rebelli, S., Thoutam, L.R., Ajayan, J.: Design of low power 4-bit Baugh-Wooley multiplier using 1-bit mirror and approximate full adders. In: 2022 2nd Asian Conference on Innovation in Technology (ASIANCON), pp. 1–4. IEEE (2022)
29. Armeniakos, G., Zervakis, G., Soudris, D., Henkel, J.: Hardware approximate techniques for deep neural network accelerators: a survey. ACM Comput. Surv. **55**(4), 1–36 (2022)
30. Rashidi, B.: APPAS: fast and efficient approximate parallel prefix adders and multipliers. J. Supercomput. **80**(16), 24269–24296 (2024)
31. Jiang, H., Han, J., Lombardi, F.: A comparative review and evaluation of approximate adders. In: Proceedings of the 25th Edition on Great Lakes Symposium on VLSI, pp. 343–348 (2015)

Jarm: Automated Remote Memory System for Java Applications

Jiawen Shen, Wenxin Li$^{(\boxtimes)}$, Linxuan Zhong, and Yulong Li

Tianjin Key Laboratory of Advanced Networking, Tianjin University, Tianjin, China
`{shenjiawen,toliwenxin,sylas,toliyulong}@tju.edu.cn`

Abstract. Remote memory systems have attracted increasing attention as they can break the single-machine memory limitation among servers and improve overall memory utilization in data centers. Therefore, mostly implemented in managed languages (i.e., Java), memory-intensive applications can benefit from remote memory systems. However, while page-based remote memory systems support a wide range of applications, they suffer from limited performance due to I/O amplification. Although object-based remote memory systems offer high performance, they are limited to native language applications such as those written in C/C++, and thus cannot support Java applications. Moreover, they typically require manual code modifications, making them less developer-friendly. To achieve high performance and Java-friendliness, we propose Jarm, an object-based remote memory system specialized for Java applications without any manual code modification. Jarm leverages a static-analysis-based automated translation module, which analyzes variable I/O behavior and automatically transforms the original code into an equivalent version that can be directly deployed. Jarm introduces a novel remote object abstraction to enable unified management of memory resources and data lifecycles in Java applications, and leverages a runtime to efficiently control data transmission between local and remote memory. Our evaluation shows that Jarm outperforms prior page-based systems by up to 2.62x normalized throughput with only 0.45% negligible overhead of code translation time.

Keywords: Remote memory system · Java applications · Static analysis

1 Introduction

Memory is the most constrained and least elastic resource [8] attributed to the design of monolithic server in modern data centers, motivating the development of remote memory. It leverages advancements in high-speed networking techniques [11,19,21,22] to provide lower latency and higher throughput than secondary storage, making it increasingly appealing to academia and industry [4,8,12,20].

In modern data centers, large amounts of distributed cloud workloads, including Hadoop [5], Spark [29], Cassandra [1] and SOLR [23] are programmed in managed language [15,27] like Java or Scala. Meanwhile, in the era of big data, with the exponential growth of data volume, the applications mentioned above, as well as other machine learning, data analysis, and key-value storage applications, are progressively shifting towards memory-intensive trends, which require holding a significant amount of data in memory for quick processing, leading to memory becoming a critical resource bottleneck in modern data centers. Remote memory, which allows those applications to utilize both local and remote memory, brings the opportunity to break the limits of standalone resource bottlenecks.

Most existing remote memory system [4,8,27] are page-based, relying on OS-level swapping and file system [27], with the purpose of providing transparency to users and allowing users to deploy their applications on the remote memory platform without the need to make any code modifications. Therefore, they can easily run applications written in various languages without any adaption of code, including Java. However, it will bring data I/O amplification [18] when applications try to access small objects, because a page at least 4KB must be transferred entirely to remote memory. To address the issues with page-based remote memory system and achieve higher performance, some recent efforts [18,30] have attempted to implement fine-grained control at the object level over the swapping of data between local and remote memory. They enable applications to utilize remote memory by means of leveraging their interface abstractions of remote access, but it requires external code modification for developers, and deploying existing legacy applications on these platforms is almost impractical, which can achieve average 41% of code modification(§2.2) with AIFM [18], entailing significant interface replacement costs. It can be time-consuming, labor-intensive, and prone to errors for deployment personnel. Additionally, these works [10,18,25,30] are designed for applications written in native languages (C/C++), hence cannot provide remote memory services for managed languages like Java.

The aforementioned unsatisfied works motivate us that: *Can we design a Java specialized remote memory system that can effectively run legacy Java applications without any code modifications from developers?* To answer this question, we present a user transparent, object-based remote memory system, **Jarm**, for legacy Java applications without any manual code modification. Jarm runs a workflow that automatically analyzes the behavior of variables in Java code before replaces them with custom interfaces, and efficiently manages the data behaviors of remote memory access with a runtime.

First, *how does Jarm analyze and modify a given Java code?* Given the diversity of variable behaviors, the complexity of syntactic structures, the interference from strings and comments, as well as the intricate nature of function calls and parameter passing, simple text-based substitution approaches are inadequate for handling complex application code in a robust and scalable manner. Furthermore, lacking of smart pointer in Java, it is difficult to flexibly manage the lifecycle of objects in an object-based remote memory system, which is one of the reason that AIFM targets to native language applications like C/C++. For this challenge, Jarm leverages a static analysis method, where a variable behavior analyzer conducts an I/O behavior analysis of the variables in the code. Then, the collected analysis information helps Jarm replacer to replace the original I/O interfaces with the interfaces defined by Jarm, while substitute the object types with Jarm-designed remote memory abstraction *RemotableObject*. The abstraction stores references of different objects and manages various aspects of the object's lifecycle, such as allocation, eviction, and retrieval. In this way, Jarm can transform the original Java code into an equivalent version that can be executed on the remote memory platform without any manual efforts.

Second, *how does Jarm efficiently manage Java remote memory access?* With the growing of local worksets, the limited local memory needs to evict overflowing objects to ample remote memory. However, easily evicting objects using a single queue may cause memory thrashing, which could increase unnecessary network round-trips thus slow down the performance. Meanwhile, invalid prefetching strategies fetching lots of unused objects can waste local memory and network bandwidth resources. Additionally, as a Java remote memory system without native RDMA communication primitive support, careful communication designs are necessary to ensure efficient data transfer between local and remote memory. To address the aforementioned challenges, Jarm employs a runtime that efficiently manage the behavior of Java remote memory access, including a multi-priority queue for data eviction, an adaptive strategy for data prefetching, as well as an efficient and reliable communication mechanism. (i) The multi-priority queue adjusts the priority of objects across different queues based on the usage of local memory. To reduce the operation overhead of priority adjustment, Jarm uses a customized queue structure to manage the placement of objects in the queues. (ii) The adaptive data prefetching contains sequence and stride patterns decided dynamically by static program analysis at run-time, where the flexibility can reduce the possibility of wasting local memory and bandwidth resources. (iii) The Jarm communication mechanism transfers data between client and server in a two-side way, where the transmission of messages loading objects will be guaranteed by flag bits and hash tables. Besides, to reduce the frequency of communication between CPU and memory, Jarm buffers the data till the threshold and batches transmission.

We implement Jarm on top of Tai-e [24] with 9K LOC in Java, where Tai-e is a developer-friendly static analysis framework for Java that allows us to build our analyzer, replacer, and an independent runtime that manages the remote memory access. Our experiments demonstrate that Jarm outperforms Fastswap [4] in normalized throughput by up to 2.62x with only 0.45% negligibly overhead of code translation time.

2   Background and Motivation

2.1   Background

There are mainly two types of remote memory systems: page-based and object-based. Page-based remote memory such as Fastswap [4] leverages virtual memory to swap application memory to far RAM instead of a local storage device. Applications use standard language-level pointers to interact with memory objects, while the OS swaps pages between local RAM and far RAM in response to page faults. In contrast, object-based remote memory systems like AIFM [18] are performant due to the fact that object-based swapping allows for more granular control over data management, enabling the system to swap out only the necessary objects rather than entire page. Furthermore, object-based swapping allows systems to better leverage the structure and semantics of the data rather than page-based which is agnostic to applications, leading to more effective caching and retrieval strategies. In consequence, for improved performance, our design space leans toward an object-based system which can better manage memory by combining with the application semantics.

2.2 Motivation

Table 1. The ratio of Java workloads in existing remote memory systems.

Works	Java Workloads	Other Workloads
Infiniswap [8]	VoltDB/GraphX/Spark	Memcached/PowerGraph
LegoOS [20]	Phoenix	Blackscholes/StreamCluster Freqmine/TensorFlow
Fastswap [4]	Spark	Kmeans/Memcached/QuickSort TensorFlow-inception/Linpack
Leap [3]	VoltDB	PowerGraph/Numpy/Memcached
Canvas [28]	Cassandra/Neo4j/Spark MLlib/GraphX	XGBoost/Snappy/Memcached
Hermit [17]	SociaNet/Spark/Cassandra	Memcached/Gdnsd/XGBoost

Java Application Matters in Data Centers. Existing remote memory works evaluate with heterogeneous workloads written by various kinds of languages including native language like C/C++ or managed language like Java, while the latter accounts for a significant proportion. The object-based remote memory systems are invasive, which need to modify the user code, designed for certain language specific, such as AIFM [18] for C/C++ applications. So, we choose some notable non-invasive works to investigate the proportion of Java workloads. Table 1 shows the workloads these remote memory works evaluate with. As shown in the table, Java workloads account for average 39%, which attracts the attention of researchers for their representative of practical production. In addition, Alibaba reports that more than 90% of latency-critical applications are programmed in Java in their datacenters [9]. Therefore, the aforementioned points reveal that the importance of Java workloads during academia and industry, but to our knowledge, there is no efficient object-based remote memory system designed for Java, motivating us to focus on a Java specialized remote memory system.

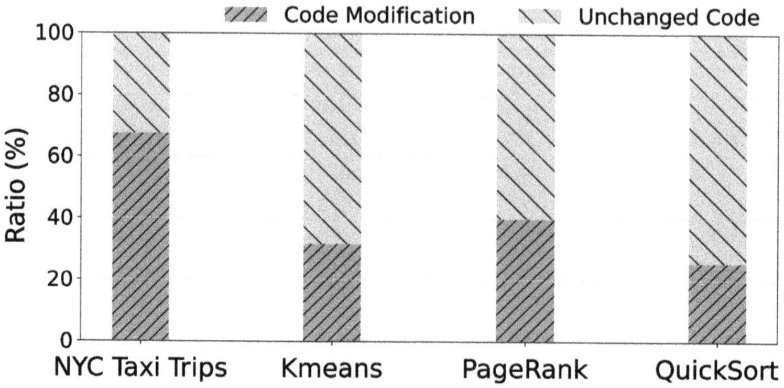

Fig. 1. Costs of porting different legacy Java workloads using AIFM.

Existing Object-Based Remote Memory Systems: User Code Modifications and Nonsupport for Managed Languages. The object-based remote memory systems, while offering performance benefits, require significant modifications to user code, which however can be a barrier to adoption for traditional applications. To reduce the burden of modification for developers, AIFM tries to constrain the program of remote memory parts in a library, but it still requires substantial efforts for developers to port their applications. Figure 1 shows the needed modification ratio of different workloads for running on AIFM. AIFM needs to modify average 41% of the code among the four workloads, where NYC Taxi Trips [18] even needs 67% of the code rewritten. The mean reason lies in the processing of re-architecting the core access patterns of data, including complex far memory access methods and memory management strategies like data evacuation introduced by AIFM. Additionally, for the characteristics of native languages (C/C++) which can utilize pointers to manage remote memory data structures, while managed languages (Java) do not support, it is difficult to directly apply the existing object-based remote memory systems to Java applications. As a result, while these systems may offer performance benefits, the impracticality of adapting existing applications can outweigh the advantages, limiting their applicability in real-world scenarios, which motivates us to try to mitigate the burden of code modification for developers.

3   Jarm Design

Figure 2 shows the overall architecture of Jarm. The goal of Jarm is to provide a transparent object-based remote memory system for legacy applications without users' manual code modification. This section presents the design of the

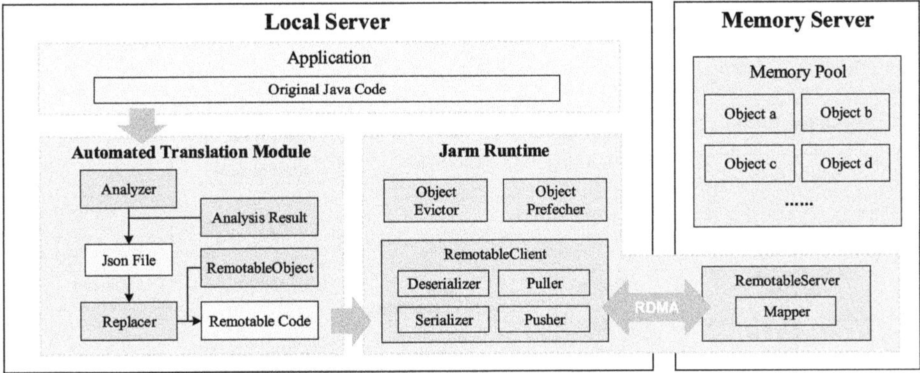

Fig. 2. Jarm Architecture.

Jarm system, detailing the Jarm static analysis workflow with carefully designed abstractions, and the Jarm runtime mechanism.

3.1   Automated Translation Module

The Automatic Translation Module in Jarm provides users with a user transparent middleware that is upward-compatible with their Java code and allows for downward invocation of interfaces for remote memory abstraction, making the entire remote memory system imperceptible to users, as if all data is stored locally and the local memory space is considered significantly large. In other words, the remote memory abstraction (detailed in §3.1.2) is not required to be understood or used by users. This module automatically identifies and analyzes the creation and access behaviors of objects in user code, and replaces these behaviors with the interface of remotable object abstraction.

3.1.1   Automated Analysis and Replacement Workflow

Variable Behavior Analyzer. The first step in implementing the automated translation of Jarm is to identify and analyze the creation and access behavior of objects in the application code. We utilize a Java static analysis tool [24] to generate the corresponding files from the original code, which contain all the IR statements corresponding to the original code, then analyze each individual IR statement.

As shown in the Fig. 3, the analyzer first traverses each IR statement in order and determines if the statement is of type New. Then, taking the New statement as the starting analysis point, it analyzes the IR statements related to the access operations of that object. Specifically, if the identified statement is of type New for object creation, we record the line number at which the corresponding original statement appears in the original code. After that, we further

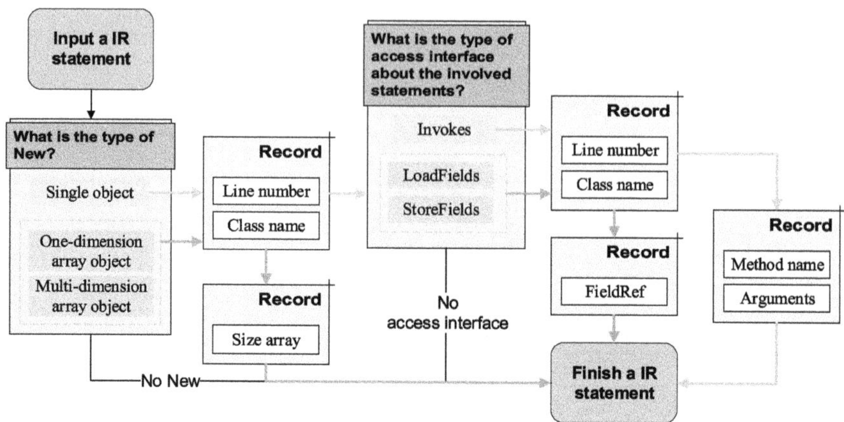

Fig. 3. Jarm Analyzer workflow.

determine the specific allocation object: single object, one-dimensional object array, or multi-dimensional object array. For a single object's new statement, we also record the class name. If the object created by `New` is an object array, we only need to record the class name and the dimension of each dimension of the array. Due to the significant differences in memory allocation and access patterns between one-dimensional and multi-dimensional arrays, where one-dimensional arrays are stored contiguously in memory, while multi-dimensional arrays are typically implemented using nested structures, it is necessary to distinguish between the two. This distinction enables the generation of precise remote memory abstraction interface calls, ensuring correctness of the system. After recording the line number and class name of a single object, we further search for the relevant IR statements that access the type of the `New` object. The object access can be divided into three types: `loadFields` (e.g., `x=v.f`), `storeFields` (e.g., `v.f=x`), and `invokes` (e.g., `v.f()`), which correspond to the three interfaces of `get/set/invoke` in remote memory abstraction. For loadFields and storeFields types, we record LineNumber and FieldRef, while for invokes type, we record LineNumber, MethodName, and Args.

Table 2. Analysis result information in 6 statement types.

Type	LineNumber	ClassName	MethodName	Args	FieldRef
New	✔	✔	✘	✔	✘
NewArray	✔	✔	✘	✔	✘
NewMultiArray	✔	✔	✘	✔	✘
LoadFields	✔	✘	✘	✘	✔
StoreFields	✔	✔	✘	✔	✔
Invokes	✔	✘	✔	✔	✘

```
public void arm_api_test() {        public void arm_api_test(){
  C a1 = new C(10, "hello");          RemotableObject a1 =
  int n = a1.num;                       RemotableObject.Ralloc(C.class, 10, "hello");
  String s = a1.str;                  int n = (int) a1.getField("num");
                                      String s = (String) a1.getField("str");
                           Replace
  a1.setStr("bye");                   a1.invokeFunc("setStr", "bye");
  a1.num = 999;                       a1.setField("num", 999);
}                                   }
```

Fig. 4. Code replacement process.

Analysis Result Representation. Jarm analysis results are stored as a HashMap and passed to the interface replacer in JSON format, which contains all statements related to each actual remote object (i.e., `ObjectRef` in `RemotableObject`) and information (Table 2) about the lines where they appear, the arguments that them needs and so on.

Interface Replacer. Jarm interface replacer leverages the information from the analysis results passed by the analyzer to perform interface replacement on the original code. This replacement transforms all objects into *RemotableObject* types (Fig. 4), replaces all relevant read and write operations with corresponding interface calls, and accurately passes parameter information into the interfaces.

3.1.2 Remotable Object Abstraction

Provided in the form of a Java library, our abstraction allows for flexibility in application deployment, including scenarios where users may prefer not to follow the automated translation process and instead manually specify certain data objects as remote. By utilizing our Java library for remote memory platform, users can achieve a more flexible and adaptable application translation.

Figure 5 shows the API of Jarm remote object. A `RemotableObject` represents an object which can be either placed in local or remote. When each `RemotableObject` is created, it is assigned a unique id(i.e., `rbid`), and their mapping relationships are saved in a global HashMap named `rbidToRoMap`. Therefore, the corresponding `RemotableObject` can be quickly obtained through `rbid`. The instances of `RemotableObject` save the reference to the actual object using the technique of template class. When the actual object is placed on the remote memory (i.e., `isLocal=false`), the value of the reference is set to `null`. All instances of `RemotableObject` share the same sending endpoint instance `remoteClient`, and each pull and push command will call its corresponding sending function.

A RemotableObject object is created as follows:

```
RemotableObject a1 = RemotableObject.Ralloc(C.class, 10, "hello");
```

```java
public class RemotableObject<T> {
  // remote object unique id
  private long rbid;
  // flag indicates whether it's local
  private Boolean isLocal;
  // the reference of actual object, set to null when isLocal=false
  private T objectRef;
  // a endpoint connected to remote server
  private static RemoteClient remoteClient;
  // a map that stores the mapping between rbid and the reference of
     actual object
  private static Map rbidToRoMap;
  ......
  public RemotableObject(Class<?> clazz, Boolean iL, Object ...
      fields);
  public static RemotableObject Ralloc(Class<?> clazz, Object...
      fields);
  public Object getField(String field);
  public void setField(String field, Object newValue);
  public Object invokeFunc(String methodName, Object ... args);
}
```

Fig. 5. Jarm remotable object abstraction.

where, `C.class` returns an object that represents the C class in Java, including information about its methods, fields, inner classes, and other metadata. This is made possible by the Java reflection mechanism, which enables programmers to access class metadata at runtime, such as class name, parent class, constructors, fields, and methods. Combining this with the technique of variable arguments, which allows programmers to declare a method with an ellipsis to indicate that the method can accept any number of same-typed arguments, we can create an instance of the C class within the `Rolloc` and refer to it using an `ObjectRef`, like a string attached to a balloon.

In addition, the remote object API exposes three access interfaces to each `RemotableObject` instance: `getField()`, `setField()`, and `invokeFunc()`, whose encapsulated actual objects can be accessed externally.

- `getField()` represents accessing the specific fields of the actual object, for example, if the `ObjectRef` in `remote_a` points to object `a`, then the operation `remote_a.getField("num")` is equivalent to `a.num`;
- `setField()` represents setting the value of the corresponding field of the actual object, such as `remote_a.setField("num",1)`, which is the same as `a.num=1`;
- `invokeFunc()` represents invoking a certain function of the actual object, for example, `remote_a.invokeFunc("func", 2, "pram1")` is calling the function named `func` in `a`, with the parameter list of (2, "pram1"), i.e., `a.func(2, "pram1")`.

3.2   Jarm Runtime

One of the main reasons for using remote memory is the limited local memory. As the working set increases, the usage of local memory also increases. When the local memory usage reaches a threshold, some data needs to be evicted to remote memory and fetched back when needed. Jarm leverages the runtime module to efficiently manage the data interaction between local and remote memory.

3.2.1   Data Communication Methods In Jarm, communication between local host and remote memory is based on RDMA (Remote Direct Memory Access), which is responsible for the remote client and remote server modules, and initialized when the first instance of the `RemotableObject` is created, establishing a pre-existing connection. Jarm `RemoteClient` has two main communication methods: `Push` and `Pull`.

Push. In the push process, the client sends data packets in the Fig. 6 format, in which the `Flag` is set to `false`, representing a push message type. To minimize network overhead caused by multiple pushes, we store push messages in RDMA data buffer, effectively serving as a cache. When the data buffer reaches its capacity, we employ RDMA write operation to transfer the data to server's RDMA data buffer. Since the receiver remains unaware of one-sided primitives, the sender must additionally transmit a bidirectional primitive (send) message to notify the receiver of incoming data. Furthermore, to ensure that continuous sender writes do not overwhelm the receiver's limited buffer capacity, thereby risking the overwriting of unread data, the sender must wait for the receiver to complete reading dataBuf and respond with an ACK simulated using send primitive before it can resume writing data to the local data buffer. Upon receiving this notification, the receiver first verifies that the Flag bit is set to false, then proceeds to read the data from the data buffer in accordance with the Push Message format, processing each message individually. The receiver evaluates the validity of each message based on the Valid bit, subsequently reading the Rbid and Data information sequentially. These values are then inserted into a maintained HashMap, with Rbid serving as the key and Data as the value.

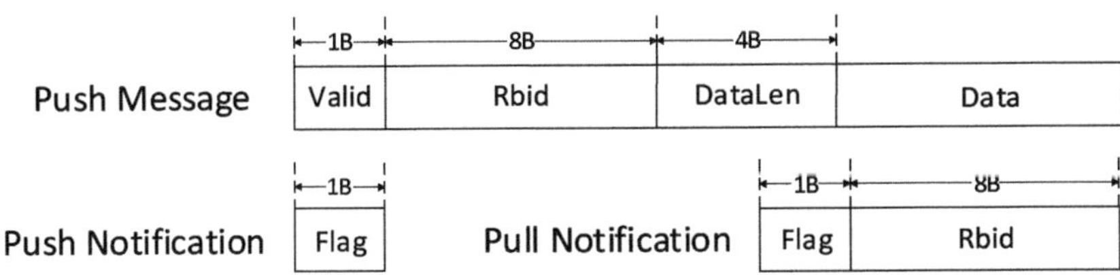

Fig. 6. Communication information format between remote client and server.

Pull. When an application triggers an object miss, it first searches the local data buffer for the object. If the object is indeed present in the data buffer, it directly retrieves the Data field, marking the Valid bit as false to indicate that there is no need to be evicted. Otherwise, it proceeds with a pull operation. The pull operation is implemented using the send primitive, referring to Fig. 6 for the format of the Pull message. Upon receiving the notification, the receiver checks whether the Flag bit is set to true, indicating a pull notification. It then reads the object ID from the remaining bytes and retrieves the corresponding object data from the HashMap, subsequently sending it to the client.

3.2.2   Object Eviction The function of the *Evacuator* thread in Jarm is to monitor the usage of local memory and decide which objects to prioritize for evacuating to remote memory. According to the principle of temporal locality, which states that recently accessed data is likely to be accessed again, we consider using a priority-based approach to assign higher priority to the "hot" objects that are recently and frequently accessed, and lower priority to the "cold" objects that are accessed less frequently. Originally, we considered using a queue to implement priority management. By leveraging its FIFO property, frequently accessed objects would be moved to the end of the queue and assigned higher priority, while infrequently accessed objects would be pushed toward the front and evicted earlier. However, a single queue may lead to excessive priority fluctuations. For example, an object that has not been accessed for a long time but resides near the front of the queue would be moved to the end and granted the highest priority after only a single access, which is clearly suboptimal. To solve this issue, we design a multi-level priority queue with a pipeline-like structure. Specifically, we implemented a multi-priority queue PQ_i, representing priority levels from 1 to N, with lower values indicating higher priority.

We define $Memory_{PQ_i}$ as the memory space occupied by each queue when it is fully loaded. When an application allocates an object through `Ralloc`, the *Evacuator* thread checks the current memory usage of the application. If it exceeds $1 * Memory_{PQ_i}$, which means the PQ1 queue is full, the system enters $State_1$; if it exceeds $2 * Memory_{PQ_i}$, which means the PQ2 queue is full, the system enters $State_2$; if it exceeds $i * Memory_{PQ_i}$, which indicates that the first i queues are full, the system enters $State_i$. In $State_i$, when objects are allocated, the head element of PQ_j (where j is in the range $[1, i]$) will be dequeued and inserted into the tail of PQ_{j+1}, and the newly allocated element will be inserted into PQ_1. When the system reaches $State_N$, all queues are full and the head element of PQ_N needs to be dequeued and evicted to remote memory before the process above repeats until the newly allocated object can be inserted into PQ_1.

To avoid the situation in which all queues are full and each new object allocation needs to wait until PQ_N evacuates an object to remote memory, we set up an auxiliary eviction thread to monitor the local memory constantly and ensure that the usage of PQ_N remains at a tolerable level.

In traditional Java linear storage queues, operations such as enqueue, dequeue, and object removal incur a time complexity of O(N) for removal, which fails to meet the performance requirements of Jarm. To address this, we design a custom data structure, O1Queue, which combines a circular array with a hash table. The array stores the remote object identifiers $rbid$, while the hash table maintains key-value pairs of $\langle rbid, index \rangle$, indicating the position of each $rbid$ in the array. When removing an object, its value is set to -1 to mark it as deleted, and the corresponding entry is removed from the hash table. Both operations have a time complexity of $O(1)$. During enqueue, the object is inserted at the end of the array, and the corresponding mapping is added to the hash table. During dequeue, the value at the head of the array is set to -1, the mapping is removed from the hash table, and the head pointer advances to the next valid entry. Experimental results show that after 200,000 accesses, the total access time using the optimized queue is approximately $1/4$ of that before optimization. When handling one million objects, the additional space overhead of the hash table per queue is around 10 MB, which is also acceptable.

3.2.3   Object Prefetching When fetching the currently requested data from remote memory, the data prefetcher predicts future data demands based on predefined algorithms or policies, and proactively issues prefetch requests to transfer both the predicted and currently needed data into local memory. This design helps reduce the latency of accessing remote data by prefetching it into local memory in advance, thus avoiding transfer delays at the time of actual demand. Based on the principle of spatial locality, data access patterns are typically clustered. For example, array elements are often accessed sequentially, while accesses to two-dimensional arrays typically follow regular stride patterns. Therefore, Jarm adopts two prefetching modes, including sequential access and stride access. The mode is dynamically selected at runtime based on semantic information exposed by static program analysis. When accessing structs or objects, their member fields often exhibit spatial locality. Therefore, Jarm will fetch a member variable of an object from remote memory as well as object's other member variables. Additionally, during the translation phase, Jarm maintains an object reference graph and prefetches all objects within a reference path length of three from the currently fetched object.

4   Implementation

Our core analyzer and runtime libraries are built on top of `Tai-e` [24] with 9K LOC in Java. `Tai-e` is the latest Java static analysis framework available. Compared to other Java static analysis frameworks like `Soot` [26], which classifies all sentences containing the "=" operator as *AssignStmt* type and all data as *Value*, without further distinguishing specific cases such as new, load, and store types in *AssignStmt* and constant, expression and reference in *Value*, or `WALA` [2], which can not obtains information such as the name and type from operands directly, `Tai-e` avoids the issues mentioned above. Furthermore, `Tai-e` provides support for associating each variable with its relevant statement in its IR, which aligns perfectly with our use case. Based on Tai-e, Jarm extends the functionality of Tai-e to enable automated analysis and translation of Java application code. Object creation and access behaviors are seamlessly replaced with remote memory abstraction interfaces, and the Jarm runtime efficiently manages data movement, thereby enabling user transparent, high-performance data transmission of remote memory systems at the application level.

5   Evaluation

5.1   Setup

Testbed. We run experiments on two $m510$ nodes on CloudLab [7] with 8-core Intel Xenon D-1548 v8 CPUs (2.00GHz), 64 GB ECC RAM, and a 10 Gbps Mellanox ConnectX-3 NIC. We use Ubuntu 18.04 (kernel v4.11), the version supports Fastswap. Based on prior work [16], We enable hyper-threads, but disabled CPU C-states, dynamic frequency scaling, transparent huge pages, and kernel mitigations against speculative execution attacks to minimize their interference with experimental results in our system.

Methodology. We compare Jarm with two baselines: (1) local memory system, where applications only use local memory, represents the optimal performance without the latency of remote memory access; (2) Fastswap, the most representative page-based remote memory system, has been widely used in this research domain.

Workloads. We use four common data-intensive workloads provided by CFM [4], including Quicksort, Pgerank, and Kmeans. We set them to be 4 GB, 4 GB, and 5 GB, respectively. The original CFM framework workloads written in C or Python render them incompatible with Java-oriented design targeted by Jarm, we therefore ported these workloads to Java. For the Java runtime configuration, we employ ZGC, one of the state-of-the-art garbage collectors, to guarantee the performance of memory management during application execution. Furthermore, to thoroughly investigate the remote data accessing performance of Jarm, we design a synthesis application Webrequest [18] that simulates typical web service workloads.

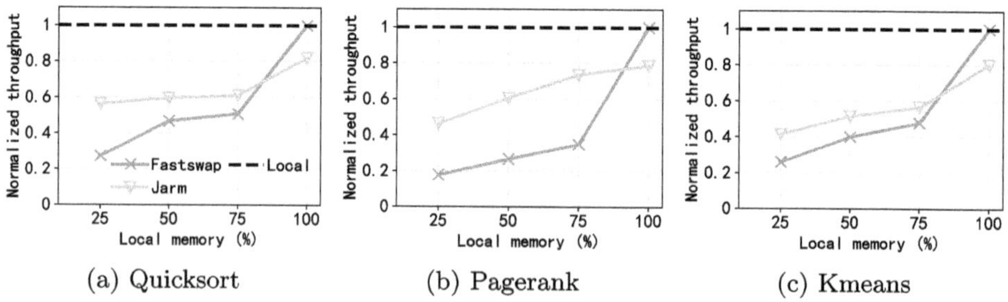

(a) Quicksort   (b) Pagerank   (c) Kmeans

Fig. 7. Comparison of performance between Fastswap and Jarm under different workloads.

5.2 End-to-End Performance

Figure 7 shows the throughput of Jarm compared with Fastswap and local memory in different workloads and the ratio of local memory. Across all workloads, Jarm consistently outperforms Fastswap in terms of overall average performance, with a particularly significant lead when local memory usage is below 75%, which delivers 1.59x higher average performance than Fastswap, with a maximum improvement of up to 2.62x. Specifically, under constrained memory conditions (25% local memory ratio), Jarm consistently deliver average 48.02% (ranging from 41.35% to 56.36%) of the ideal maximum throughput, while Fastswap can only maintain 17.72%-27.39% throughput. Under the QuickSort and PageRank workloads, Jarm achieves an average performance that exceeds Fastswap by 17.06%, even when using only 25% of local memory compared to Fastswap's 75%. Under limited local memory, the high performance of Jarm is attributed to its avoidance of the high overhead caused by page faults in page-granularity systems. By adopting a finer-grained data exchange mechanism, Jarm effectively addresses the issues of excessive software stack overhead, bandwidth waste, and inefficient memory usage inherent in page-granularity designs. In addition, Jarm incorporates application semantics into its data prefetching strategy and reduces resource overhead through operation offloading.

However, when local memory usage reaches 100%, the performance of Jarm falls below Fastswap. In contrast, Fastswap achieves throughput comparable to that of an ideal local system. This is because when all data resides in local memory, application access no longer triggers page faults, and the data exchange paths of page-granularity remote memory systems remain inactive. As a result, the overhead introduced by the granularity and mechanisms of Fastswap disappears, allowing them to reach an ideal optimal performance level. Although Jarm does not require data exchange with remote memory, all objects must be managed through the abstract class *RemoteableObject*, which introduces certain management overhead at the application layer, preventing it from achieving the optimal performance level of a local system.

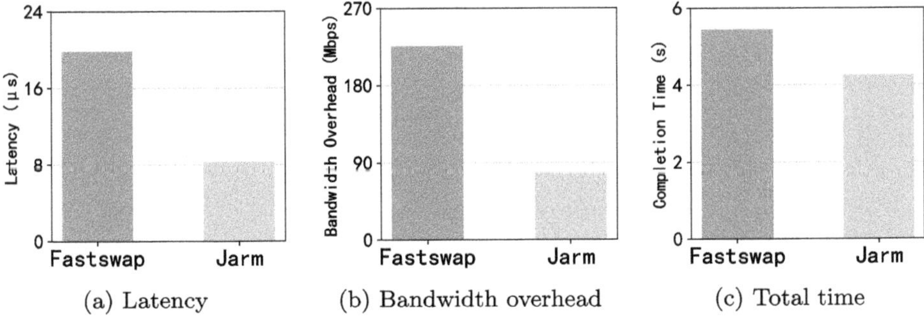

(a) Latency   (b) Bandwidth overhead   (c) Total time

Fig. 8. Comparison of performance between Fastswap and Jarm under Webrequest.

5.3   Data Access Performance

A key advantage that distinguishes Jarm from Fastswap lies in its adoption of a finer-grained data exchange mechanism. Jarm uses the size of individual data objects as the basic unit of data exchange, whereas Fastswap, as a representative of page-granularity remote memory systems, consistently evicts and retrieves data in fixed 4 KB units (i.e., one kernel page). To validate it, we evaluate the data access performance of the two systems using the WebRequest application. We set WebRequest application to perform 10,000 accesses to server-side data. The server consists of a compute node and a memory node, where the compute node has 4 GB of available memory. The total dataset size is 8 GB (i.e., a local memory ratio of 50%) and contains approximately 8.5 million objects, with most object sizes ranging from 64 B to 512 B [6].

As shown in Fig. 8a, Jarm achieves 2.38x lower latency than Fastswap when accessing a single object. This is because data objects are typically 2.3 to 31 times smaller than 4 KB. Fastswap transfers data over the network at a fixed granularity of 4 KB, which not only increases transmission latency but also incurs additional protocol overhead. Due to the 1500-byte Maximum Transmission Unit (MTU) defined by the IEEE 802.3 standard, each 4 KB data block must be split into three segments during transmission. Figure 8b presents the bandwidth overhead of the two systems. Fastswap exchanges data at the page level, often fetching unnecessary data and evicting data that is still needed. This behavior not only increases the frequency of communication with remote memory but also leads to significant network bandwidth waste. The results show that Fastswap consumes an average network bandwidth of 225.6 Mbps over the entire application lifecycle, while Jarm uses only 78.4 Mbps, which represents a 2.88x reduction in bandwidth consumption, thereby improving overall network utilization efficiency. Due to the reasons discussed above, Jarm achieves a 1.28× overall improvement in application performance compared to Fastswap in the WebRequest benchmark (Fig. 8c).

Table 3. The proportion of code translation overhead under different local memory configurations.

Workloads	Code Translation Overhead (%)			
	25%	50%	75%	100%
Quicksort	0.094	0.180	0.198	0.215
Pagerank	0.454	0.592	0.718	0.771
Kmeans	0.160	0.397	0.427	0.573

5.4   Code Translation Overhead

To run conventional Java code on Jarm, the code must be preprocessed by components such as a code analyzer and a code rewriter, which convert the original code into a remotable form recognized by Jarm. This preprocessing introduces a certain amount of overhead. To evaluate the proportion of code translation overhead in the total application runtime, we measure the translation time for four workloads and reported the application completion time under local memory configurations of 25%, 50%, 75%, and 100%. As shown in Table 3, when local memory is configured to 25%, the code translation overhead accounts for less than 0.45% of the total runtime, and the average overhead remains below 0.52% when local memory reaches 100%. Compared to the performance improvement of up to 2.62x achieved by Jarm, this overhead is considered acceptable. Moreover, the same unmodified application requires code translation only once, and subsequent executions can bypass this phase entirely.

6 Related Work

In recent years, remote memory has attracted significant attention from researchers. To build more efficient remote memory systems, a variety of approaches have been proposed, including page-based and object-based designs [4,18,28,30]. However, only few works have focused on achieving a balance between system performance and usability.

Mira [10] proposes a program-behavior-driven far memory system based on MLIR [14], but its design targets native language applications like C/C++. Since C/C++ programs have explicit memory allocation patterns, behavior analysis and code generation can be directly inserted into the compiler pass. In contrast, JVM implicitly controlled memory management in Java programs, therefore MLIR cannot be directly applied to the analysis and instrumentation of Java bytecode. TrackFM [25] is also a compiler-based far memory system. It uses the existing AIFM architecture as the backend and leverages the LLVM [13] compiler to automatically add far memory support to applications. However, relying on AIFM as the backend to ensure system performance, similar to Mira, it is only applicable to native language applications.

In contrast, Jarm targets the Java ecosystem based on Tai-e [24], and provides an automated far memory solution adapted to Java applications, ranging from IR, pointer analysis, to automatic code instrumentation and encapsulation of remote memory APIs.

7 Conclusion

We presented Jarm, a remote memory system designed for Java applications. Jarm makes it easy to run a legacy Java application on remote memory system, as it employs a static analysis workflow to automatically transform Java code into a semantically equivalent version compatible with the remote memory runtime, without requiring any manual modifications. Jarm is also efficient, as it adopts an object-granularity design and introduces a Java-applicable remote object abstraction to manage remote memory data structures, and leverages a runtime to guarantee efficient data transmission between local and remote memory. Our results show that Jarm outperforms page-based approach by up to 2.62× normalized throughput with only 0.45% overhead of code translation time.

Acknowledgments. This work is supported by the National Natural Science Foundation of China under Grant 62432015 and Grant 62202325, the Shandong Provincial Natural Science Foundation Innovation and Development Joint Fund Project under Grant ZR2023LZH010.

References

1. The apache cassandra project. http://cassandra.apache.org/
2. Wala (2018). http://wala.sf.net
3. Al Maruf, H., Chowdhury, M.: Effectively prefetching remote memory with leap. In: 2020 USENIX Annual Technical Conference (USENIX ATC 20), pp. 843–857 (2020)
4. Amaro, E., et al.: Can far memory improve job throughput? In: Proceedings of the Fifteenth European Conference on Computer Systems, pp. 1–16 (2020)
5. Borthakur, D.: The hadoop distributed file system: architecture and design. Hadoop Proj. Website **11**(2007), 21 (2007)
6. Calciu, I., et al.: Rethinking software runtimes for disaggregated memory. In: Proceedings of the 26th ACM International Conference on Architectural Support for Programming Languages and Operating Systems, pp. 79–92 (2021)
7. Duplyakin, D., et al.: The design and operation of {CloudLab}. In: 2019 USENIX Annual Technical Conference (USENIX ATC 19), pp. 1–14 (2019)
8. Gu, J., Lee, Y., Zhang, Y., Chowdhury, M., Shin, K.G.: Efficient memory disaggregation with infiniswap. In: NSDI, pp. 649–667 (2017)
9. Guo, J., et al.: Who limits the resource efficiency of my datacenter: an analysis of alibaba datacenter traces. In: Proceedings of the International Symposium on Quality of Service, pp. 1–10 (2019)
10. Guo, Z., He, Z., Zhang, Y.: Mira: a program-behavior-guided far memory system. In: Proceedings of the 29th Symposium on Operating Systems Principles, pp. 692–708 (2023)
11. Kalia, A., Kaminsky, M., Andersen, D.G.: Using rdma efficiently for key-value services. In: Proceedings of the 2014 ACM Conference on SIGCOMM, pp. 295–306 (2014)
12. Keeton, K.: The machine: an architecture for memory-centric computing. In: Workshop on Runtime and Operating Systems for Supercomputers (ROSS), vol. 10 (2015)
13. Lattner, C., Adve, V.: Llvm: A compilation framework for lifelong program analysis & transformation. In: International Symposium on Code Generation and Optimization, 2004. CGO 2004, pp. 75–86. IEEE (2004)
14. Lattner, C., et al.: Mlir: scaling compiler infrastructure for domain specific computation. In: 2021 IEEE/ACM International Symposium on Code Generation and Optimization (CGO), pp. 2–14. IEEE (2021)
15. Maas, M., Asanović, K., Harris, T., Kubiatowicz, J.: Taurus: a holistic language runtime system for coordinating distributed managed-language applications. ACM SIGPLAN Not. **51**(4), 457–471 (2016)

16. Ousterhout, A., Fried, J., Behrens, J., Belay, A., Balakrishnan, H.: Shenango: achieving high {CPU} efficiency for latency-sensitive datacenter workloads. In: 16th USENIX Symposium on Networked Systems Design and Implementation (NSDI 19), pp. 361–378 (2019)
17. Qiao, Y., et al.: Hermit:{Low-Latency},{High-Throughput}, and transparent remote memory via {Feedback-Directed} asynchrony. In: 20th USENIX Symposium on Networked Systems Design and Implementation (NSDI 23), pp. 181–198 (2023)
18. Ruan, Z., Schwarzkopf, M., Aguilera, M.K., Belay, A.: Aifm: high-performance, application-integrated far memory. In: Proceedings of the 14th USENIX Conference on Operating Systems Design and Implementation, pp. 315–332 (2020)
19. Rumble., S.M.: Infiniband verbs performance (2010). https://ramcloud.atlassian. net/wiki/display/RAM/Infiniband+Verbs+Performance
20. Shan, Y., Huang, Y., Chen, Y., Zhang, Y.: Legoos: a disseminated, distributed {OS} for hardware resource disaggregation. In: 13th {USENIX} Symposium on Operating Systems Design and Implementation ({OSDI} 18), pp. 69–87 (2018)
21. Shrivastav, V., et al.: Shoal: a network architecture for disaggregated racks. In: 16th USENIX Symposium on Networked Systems Design and Implementation (NSDI 19), pp. 255–270 (2019)
22. Sidler, D., Wang, Z., Chiosa, M., Kulkarni, A., Alonso, G.: Strom: smart remote memory. In: Proceedings of the Fifteenth European Conference on Computer Systems, pp. 1–16 (2020)
23. Smiley, D., Pugh, E.: Apache Solr 3 Enterprise Search Server. Packt Publishing (2011)
24. Tan, T., Li, Y.: Tai-e: a developer-friendly static analysis framework for java by harnessing the good designs of classics. In: Proceedings of the 32nd ACM SIGSOFT International Symposium on Software Testing and Analysis, pp. 1093–1105 (2023)
25. Tauro, B.R., Suchy, B., Campanoni, S., Dinda, P., Hale, K.C.: Trackfm: far-out compiler support for a far memory world. In: Proceedings of the 29th ACM International Conference on Architectural Support for Programming Languages and Operating Systems, vol. 1, pp. 401–419 (2024)
26. Vallée-Rai, R. Co, P., Gagnon, E., Hendren, L., Lam, P., Sundaresan, V.: Soot: a java bytecode optimization framework. In: CASCON First Decade High Impact Papers, pp. 214–224 (2010)
27. Wang, C., et al.: Semeru: a {Memory-Disaggregated} managed runtime. In: 14th USENIX Symposium on Operating Systems Design and Implementation (OSDI 20), pp. 261–280 (2020)
28. Wang, C., et al.: Canvas: isolated and adaptive swapping for {Multi-Applications} on remote memory. In: 20th USENIX Symposium on Networked Systems Design and Implementation (NSDI 23), pp. 161–179 (2023)
29. Zaharia, M., Chowdhury, M., Franklin, M.J., Shenker, S., Stoica, I.: Spark: cluster computing with working sets. In: 2nd USENIX Workshop on Hot Topics in Cloud Computing (HotCloud 10) (2010)
30. Zhou, Y., et al.: Carbink:{Fault-Tolerant} far memory. In: 16th USENIX Symposium on Operating Systems Design and Implementation (OSDI 22), pp. 55–71 (2022)

Distributed Energy-Efficient Trajectory Optimization for Multi-UAV Based Agricultural Field Sensing

Puqin Han[1], Zhiqiang Liu[1(✉)], Xu Zhang[2], and Wenjing Li[1]

[1] College of Intelligent Science and Technology, Inner Mongolia University of Technology, Hohhot 010080, China
54388668@qq.com
[2] Inner Mongolia Technical University of Construction, Hohhot 010070, China

Abstract. This paper addresses the energy consumption optimization problem in Multi-UAV collaborative field data collection scenarios. We propose an Energy-Efficient Path Planning Model for Data Collection in Cooperative Multi-UAV Systems (EEPPM-DCCMUS) that jointly optimizes UAV deployment strategy, flight trajectory planning, velocity control, and device transmission power allocation to achieve three objectives: maximizing the minimum transmission rate, minimizing device communication energy consumption, and minimizing total system energy consumption. To solve the EEPPM-DCCMUS problem, we develop an Improved Multi-objective Artificial Hummingbird Algorithm (IMOAHA) that innovatively integrates K-means clustering initialization, Lévy flight-based mutation foraging strategy, and ant colony optimization foraging strategy. Simulation results demonstrate that IMOAHA significantly outperforms the baseline MOAHA algorithm in terms of Pareto front convergence and distribution. Specifically, IMOAHA achieves a minimum transmission rate of 12,754,066.74 bps (comparable to MOAHA), reduces total energy consumption to 12,138.94 J (a 58.9% improvement), decreases total flight distance to 365.05 m (a 53.7% reduction), and lowers energy consumption fluctuation by 81.3%. Compared with existing algorithms including MOAVOA and MOCryStAl, IMOAHA reduces total energy consumption by 25.8%–70.6% and total flight distance by 50.2%–66.3%. This study provides an innovative solution for wide-area coverage and dynamic load balancing of UAV clusters in complex agricultural environments, with significant theoretical and practical implications for promoting large-scale UAV applications in precision agriculture.

Keywords: Multi-UAV cooperation · Energy consumption optimization · Path planning · IMOAHA · Precision agriculture

1 Introduction

Unmanned Aerial Vehicle (UAV) technology has emerged as a pivotal enabler for precision agriculture, particularly demonstrating remarkable advantages in

H. Liu et al. (Eds.): ICA3PP 2025, LNCS 16381, pp. 282–300, 2026.
https://doi.org/10.1007/978-981-95-8399-7_16

crop growth monitoring and yield prediction. Equipped with multispectral and hyperspectral imaging systems, UAV can efficiently capture canopy spectral characteristics, providing critical data for plant health diagnosis and growth status assessment. However, energy consumption during field data acquisition remains a fundamental constraint on operational efficiency and endurance, where path planning rationality directly determines both data collection quality and energy expenditure. This underscores the significant theoretical and practical value of investigating energy optimization methods for collaborative Multi-UAV operations.

The application of UAV technology in the field of agriculture has become a key technical means for the development of precision agriculture. In the research of energy consumption optimization theory and method, scholars have made a series of breakthroughs: the nonlinear energy consumption model proposed by Wu et al. [1] reduces energy consumption by 21% through variable speed strategy; Zhang et al. [2] and Chen et al. [3] used deep reinforcement learning and reinforcement learning algorithms to optimize three-dimensional trajectory planning, respectively. The multi-objective optimization method developed by Zhu et al. [4–6] provides a new idea for system performance balance. The collaborative operation optimization technology of Zeng et al. [7] has significantly improved the energy efficiency of multi-machine collaboration. Xiong et al. [8] provided a new method for energy-saving collection of IoT scenarios by decomposing mixed integer non-convex optimization problems.

The integration of new energy technologies and intelligent management strategies opens up a new way for energy consumption optimization. The innovative solutions proposed by Wu et al. [9] effectively deal with the energy consumption challenges in complex scenarios; the two-layer UAV-MEC model constructed by Tan et al. [10] and Lu et al. [11], the server deployment optimization of Liu et al. [12], and the improved particle swarm optimization algorithm of Wang et al. [13] have promoted the research progress from different dimensions. The current research shows a multi-dimensional development trend: Liu et al. [14] have achieved remarkable results in the field of algorithm optimization; Wu et al. [15] combined self-organizing map with improved PSO algorithm, and Shen et al. [16] improved cluster energy efficiency through joint training and resource allocation mechanism.

The existing research has made significant progress in the optimization of UAV agricultural energy consumption, but there are still the following key problems: First, the existing models focus on single-dimensional optimization, and lack of collaborative modeling of flight dynamics and communication energy consumption; Second, the traditional algorithm has insufficient global search ability and convergence efficiency in complex farmland environment; Finally, most studies do not fully consider the dynamic load balancing requirements specific to farmland scenarios.

Based on this, this paper proposes a Multi-UAV cooperative energy consumption optimization method for field data collection. The main contributions are threefold:(1) We construct an Energy-Efficient Path Planning Model for Data Collection in Cooperative Multi-UAV Systems (EEPPM-DCCMUS) that achieves co-modeling of flight dynamics and communication energy consumption through joint optimization of UAV deployment strategies, flight trajectory planning, velocity control, and device transmission power allocation. (2) To address dynamic load balancing requirements in complex agricultural environments, we design an Improved Multi-objective Artificial Hummingbird Algorithm (IMOAHA) that integrates K-means clustering initialization, Lévy flight mutation, and ant colony optimization strategies to enhance global search capability and convergence efficiency. (3) Simulation results demonstrate that the proposed method significantly reduces total system energy consumption and UAV flight distance while maintaining the minimum transmission rate, providing an effective solution for energy efficiency optimization in agricultural UAV systems.

2   Model to Describe

2.1   System Model

Due to onboard battery capacity limitations, UAV systems face significant challenges in sustaining prolonged communication operations. In Multi-UAV cooperative data collection scenarios, balanced task allocation becomes imperative to prevent both overloading of specific drones with excessive tasks and resource underutilization of others with insufficient assignments, either of which would degrade overall system efficiency. We therefore propose an energy-minimizing path planning framework for Multi-UAV wireless networks. The system architecture, illustrated in Fig. 1, consists of a single base station and multiple UAV endowed with both computational and communication capabilities. These UAV collectively serve a target area that is partitioned into A distinct subregions. The K UAV (represented as set $\mathcal{K} = \{1, 2, ..., K\}$ with individual drones $k_1, k_2, k \in \mathcal{K}$) provide coverage for multiple subregions, while ground devices are denoted as set $\mathcal{R} = \{1, 2, ..., R\}$ ($r \in \mathcal{R}$). The total service duration T is discretized into L equal timeslots t ($t = 1, 2, ..., L$), each with duration $\Delta t = T/L$.

The coordinates of ground devices at timeslot t are denoted as $(x_{r,t}, y_{r,t}, 0)$, while UAV k's coordinates are $(x_{k,t}, y_{k,t}, H)$, where H represents the constant flight altitude (with boundary constraints). The horizontal distance between UAV k and ground device r at timeslot t is:

$$d_{r,k,t}^{H} = \sqrt{(x_{k,t} - x_{r,t})^2 + (y_{k,t} - y_{r,t})^2} \tag{1}$$

At time slot t, the horizontal distance between UAV k_1 and UAV k_2 at time slot t is:

$$d_{k_2,k_1,t}^{H} = \sqrt{\left(x_{k_2,t} - x_{k_1,t}\right)^2 + \left(y_{k_2,t} - y_{k_1,t}\right)^2} \geq d_{\text{safe}} \tag{2}$$

where d_{safe} is the safety distance to avoid UAV collisions.

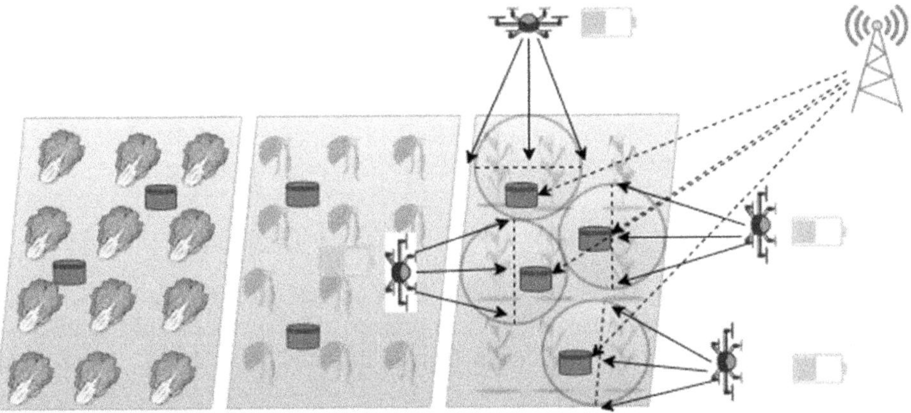

Fig. 1. Multiple UAV work together to collect and transmit farmland crop data based on wireless communication and ground sensor nodes.

The position update formulas of the UAV between time slots, reflecting the continuity of the flight trajectory:

$$x_{k,t} = x_{k,t-1} + v_{k,t-1} \cdot \Delta t \cdot \cos \theta_{k,t-1} \tag{3}$$

$$y_{k,t} = y_{k,t-1} + v_{k,t-1} \cdot \Delta t \cdot \sin \theta_{k,t-1} \tag{4}$$

where $v_{k,t-1}$ is the flight speed at time slot t-1; $\theta_{k,t-1}$ is the flight angle direction, ensuring a smooth trajectory.

2.2 Channel Model

For the scenario of field data collection with multiple UAV collaborating, a distributed trajectory planning strategy is proposed. This strategy assigns a non-overlapping operation area to each UAV and optimizes its flight trajectory to achieve the dual objectives of energy saving and obstacle avoidance. Two channel models are considered in the Multi-UAV network: the Air-to-Air (A2A) channel model and the Air-to-Ground (A2G) channel model. According to 3GPP TR 36.777, the A2A channel between UAV is mainly dominated by the Line-of-Sight (LoS) path, and its Path Loss (PL) can be expressed as:

$$PL_{A2A} = 30.9 + \lfloor 22.25 - 0.5 \cdot \log_{10} |h_{k_2} - h_{k_1}| \rfloor \cdot \log_{10} d_{k_2,k_1,t} + 20 \cdot \log_{10} f_0 \tag{5}$$

where $d_{k_2,k_1,t}$ and $|h_{k_2} - h_{k_1}|$ denote the relative horizontal distance and altitude difference, respectively, between the current UAV (denoted as k_2) and a neighboring UAV (denoted as k_1), f_0 represents the carrier frequency of the A2A link, accounting for the mutual communication capability of the UAV.

The A2G channel employs a Down-Link communication network. Considering that the communication link includes both Line-of-Sight (LoS) and Non-Line-of-Sight (NLoS) paths, the corresponding path losses are given by:

$$PL_{A2G,LoS} = 20\log_{10}(d_{r,k,t}) + 10\log_{10}(f_{c,A2G}) + 20\log_{10}\left(\frac{4\pi}{c}\right) + \alpha_{LoS} \quad (6)$$

$$PL_{A2G,NLoS} = 20\log_{10}(d_{r,k,t}) + 10\log_{10}(f_{c,A2G}) + 20\log_{10}\left(\frac{4\pi}{c}\right) + \alpha_{NLoS} \quad (7)$$

where $d_{r,k,t}$ is the distance between UAV k and ground device r; $f_{c,A2G}$ denotes the system carrier frequency; c represents the speed of light; and α_{LoS}, α_{NLoS} are the additional attenuation factors induced by LoS and NLoS links, respectively.

The probability of a Line-of-Sight (LoS) connection between UAV k and ground device r is expressed as:

$$P_{\text{LoS}} = \frac{1}{1 + b_1 e^{(-b_2(\theta_{k,r,t} - b_3))}} \quad (8)$$

where the elevation angle $\theta_{k,r,t}$ is defined by:

$$\theta_{k,r,t} = \arctan\left(\frac{H}{d_{k,r,t}^H}\right) \quad (9)$$

In these expressions, b_1 and b_2 are environment-dependent constants; $\theta_{k,r,t}$ denotes the elevation angle.

The average path loss between UAV k and ground device r is given by:

$$\overline{PL}_{A2G} = P_{\text{LoS}} \cdot PL_{A2G,LoS} + (1 - P_{\text{LoS}}) \cdot PL_{A2G,NLoS} \quad (10)$$

The data transmission rates, i.e., the rate at which ground device r uploads data to UAV k and the rate at which UAV k uploads data to the base station (BS), can be further designed as:

$$R_{r,k,t} = B\log_2\left(1 + \frac{p_{r,k,t} \cdot h_{k,r,t}}{N_0 \cdot B + I_{r,t}}\right) \quad (11)$$

$$R_{k,BS,t} = B\log_2\left(1 + \frac{p_{k,BS,t} \cdot h_{k,BS,t}}{N_0 \cdot B + I_{k,t}}\right) \quad (12)$$

where $p_{r,k,t} \in [P_{\min}, P_{\max}]$ and $p_{k,BS,t} \in [P_{\min}, P_{\max}]$. Here, $P_{\min}$ denotes the minimum transmission power for devices to upload data, and $P_{\max}$ represents the maximum transmission power for the same. B is the channel bandwidth in hertz (Hz), N_0 signifies the AWGN at the IoT device, and I is the interference term.

For the Air-to-Ground (A2G) scenario, the channel gains between UAV k and device r, as well as between UAV k and the base station, are given by:

$$h_{k,r,t} = \frac{G_0 \cdot d_0^\eta}{d_{k,r,t}^\eta} \quad (13)$$

$$h_{k,BS,t} = \frac{G_0 \cdot d_0^{\eta}}{d_{k,BS,t}^{\eta}} \tag{14}$$

where $d_{k,r,t}$ and $d_{k,BS,t}$ denote the distances between UAV k and ground device r, and between UAV k and the base station, respectively, at time slot t. G_0 represents the channel power gain at a reference distance $d_0 = 1\,\text{m}$, and η is the path loss exponent.

2.3 Energy Consumption Model

(1) Communication Energy Consumption

The transmission energy consumption $E_{r,k}^{DU}(t)$ from the ground device r to the UAV k, and the transmission energy consumption $E_{k,r}^{UB}(t)$ from the UAV relay to the data center, are respectively expressed as:

$$E_{r,k,t}^{\text{comm}} = P_{r,k,t} \cdot \Delta t \cdot \alpha_{r,k,t} \tag{15}$$

$$E_{r,BS,t}^{\text{comm}} = P_{r,BS,t} \cdot \Delta t \cdot \alpha_{r,BS,t} \tag{16}$$

Among them, $\alpha_{r,k,t}$ is used to represent the device access matrix at time slot t. Each element $\alpha_{r,k,t} \in \{0,1\}$ represents a binary indicator. $\alpha_{r,k,t} = 1$ means that UAV k provides service for device r at time slot t; otherwise, $\alpha_{r,k,t} = 0$. Δt represents the time slot interval. $P_{r,k,t}$ is the transmission power for device r to offload tasks to the base station through UAV k, and $P_{k,BS,t}$ is the power for UAV k to transmit data to the base station.

The maximum task amount for the ground device r to relay to the base station through UAV k is:

$$M_{r,k,t}^{\max} = \alpha_{r,k,t} \cdot \Delta t \cdot R_{r,k,t} \tag{17}$$

The communication range of the UAV is limited. If the ground device exceeds the maximum distance, it cannot receive the result of task return. The UAV must promptly process and return the tasks uploaded by the ground device. In addition, considering the limited storage resources of the UAV, when the return delay requirement cannot be met, the task will be offloaded to the base station. Then, the return delay constraint for ground device r is:

$$\sum_{t=1}^{L} \frac{M_{r,k,t}}{f_{k,t} \cdot \alpha_{r,k,t}} \leq D_r^{\max} \tag{18}$$

Among them, $f_{k,t}$ represents the collection frequency allocated by UAV k to device r at time slot t, indicating the number of CPU cycles required to execute a 1 - bit computing task. Therefore, according to Shannon's formula and relevant formulas, the upload energy consumption of UAV k can be rewritten as:

$$E_{k,BS,t}^{\text{comm}} = \frac{\Delta t \cdot P_{k,BS,t}}{B} \cdot \left(2^{\frac{R_{k,BS,t}}{B}} - 1 \right) \tag{19}$$

(2) Flight Energy Consumption

During the execution of computing tasks by the UAV, its own flight energy consumption is a non - negligible item in the overall system. It is defined that the flight energy consumption is only related to the flight speed. If the flight speed is defined as:

$$v_{k,t} = \frac{\sqrt{(x_{k,t} - x_{k,t-1})^2 + (y_{k,t} - y_{k,t-1})^2 + (z_{k,t} - z_{k,t-1})^2}}{\Delta t} \tag{20}$$

Then, the flight energy consumption of UAV k in time slot t is expressed as: The flight energy consumption of UAV k during time slot t is expressed as:

$$E_{k,t}^{\text{fv}} = \frac{1}{2} m \cdot v_{k,t}^2 \cdot \Delta t \tag{21}$$

where m denotes the mass of the UAV.

The total energy consumption of ground device r and UAV k during time slot t are respectively given by:

$$E_{r,t}^{\text{total}} = E_{r,k,t}^{\text{comm}} + \lambda \left(E_{r,k,t}^{\text{comm}} - \overline{E}_{r,t}^{\text{comm}} \right)^2 \tag{22}$$

$$E_{k,t}^{\text{total}} = E_{k,t}^{\text{comm}} + E_{k,t}^{\text{fv}} + \mu \left(E_{k,t}^{\text{fv}} - \overline{E}_{k,t}^{\text{fv}} \right)^2 \tag{23}$$

where λ and μ are penalty coefficients introduced to balance energy consumption variations.

Thus, the total system energy consumption during time slot t is:

$$E_t^{\text{system}} = \omega_1 \sum_{r=1}^{M} E_{r,t}^{\text{total}} + \omega_2 \sum_{k=1}^{K} E_{k,t}^{\text{total}} \tag{24}$$

where ω_1 and $\omega_2 \in [0,1]$ represent the weight factors for local subregion and UAV energy consumption respectively, satisfying:

$$\omega_1 + \omega_2 = 1 \tag{25}$$

2.4   Model Description and Analysis

This study investigates energy-constrained UAV deployment as mobile base stations for sparse-device distributed farmland monitoring systems. Leveraging 3D mobility advantages, UAV can optimize both hovering positions and flight trajectories to minimize transmission distances, thereby enhancing system energy efficiency. Three fundamental challenges emerge: (1)Sparse Coverage in Heterogeneous Fields: The spatial heterogeneity of agricultural monitoring areas with non-uniform device distribution imposes large-scale coverage requirements on UAV. (2) Energy Constraints of Ground Sensors: Strict energy limitations of terrestrial sensor nodes present significant challenges for energy efficiency optimization. (3)Dynamic Energy Consumption of UAV: The time-varying energy consumption patterns during UAV maneuvers critically affect both operational endurance and communication quality-of-service metrics.

To address this, we propose an Energy-Efficient Path Planning Model for Data Collection in Cooperative Multi-UAV Systems ((EEPPM-DCCMUS) that jointly optimizes the following decision variables: UAV deployment strategy $\{X, Y\} = \{x_1, ..., x_M, y_1, ..., y_M\}$, Flight trajectory planning (including visit sequence optimization) $C = \{C_1, ..., C_K\}$, Velocity control $V = \{V_1, ..., V_M\}$, Device transmission power allocation $P = \{P_1, ..., P_I\}$, The model achieves dual optimization objectives: Maximizing the system's minimum transmission rate Minimizing the total energy consumption of both devices and UAV, The solution to EEPPM-DCCMUS can be formally expressed as: $Z=(\{X,Y,C,V,P\},)$ where θ represents the set of optimization parameters.

Max-Min Rate Guarantee: To ensure equitable service quality across all ground devices, we formulate the first objective as maximization of the minimum transmission rate:

$$f_1 = \max \left\{ \min_{r \in \mathcal{M}} R_{r,k,t} \right\} \tag{26}$$

Communication Energy Minimization: For sustainable network operation, the second objective minimizes the total communication energy consumption:

$$f_2 = \min \left\{ \sum_{r \in \mathcal{M}} \sum_{t \in \mathcal{T}} E_{r,k,t}^{\text{comm}} \right\} \tag{27}$$

System-Wide Energy Efficiency: Incorporating both UAV propulsion and communication costs, the third objective minimizes the total system energy expenditure:

$$f_3 = \min \left\{ \sum_{t \in \mathcal{T}} E_t^{\text{system}} \right\} \tag{28}$$

In summary, the EEPPM-DCCMUS framework for UAV-assisted data collection can be formally defined as:

$$\min \mathbf{F} \quad = \{-f_1, f_2, f_3\} \tag{29a}$$

$$s.t. C1: \quad x_{k,t} \in [X_{\min}, X_{\max}], \ y_{k,t} \in [Y_{\min}, Y_{\max}], \ z_{k,t} \in [Z_{\min}, Z_{\max}] \tag{29b}$$

$$C2: \quad P_{r,\min} \leq P_{r,k,t} \leq P_{r,\max} \tag{29c}$$

$$C3: \quad v_{k,t} \leq v_{\max} \tag{29d}$$

$$C4: \quad \alpha_{r,k,t} \in \{0,1\}, \ \sum_{k=1}^{K} \alpha_{r,k,t} = 1 \tag{29e}$$

$$C5: \quad d_{k_1,k_2,t}^{H} \geq d_{\text{safe}} \tag{29f}$$

$$C6: \quad \sum_{l=1}^{L} \frac{M_{r,k,l}}{f_{k,l} \cdot \alpha_{r,k,l}} \leq D_r^{\max} \tag{29g}$$

Where, constraints $C1$ limit the movement range of the UAV; $C2$ limits the transmission power of the ground equipment; $C3$ limits the flight speed of the UAV; $C4$ is the service association constraint between the equipment and the UAV; $C5$ is the safety distance constraint between UAV; $C6$ is the task return delay constraint.

3 Problem Solving

3.1 IMOAHA

The traditional Multi-objective Artificial Hummingbird Algorithm (MOAHA) faces the problems of high energy consumption and long flight distance when solving EEPPM-DCCMUS. Therefore, an Improved Multi-objective Artificial Hummingbird Algorithm (IMOAHA) is proposed to solve the problem. The specific method is to improve the convergence speed and quality of solving EEPPM-DCCMUS by introducing three improved strategies, namely K-means clustering initialization strategy, Levy Flight mutation foraging strategy in guiding foraging process and ant colony optimization (ACO) foraging strategy in regional foraging process. Figure 2 presents the flowchart of IMOAHA.

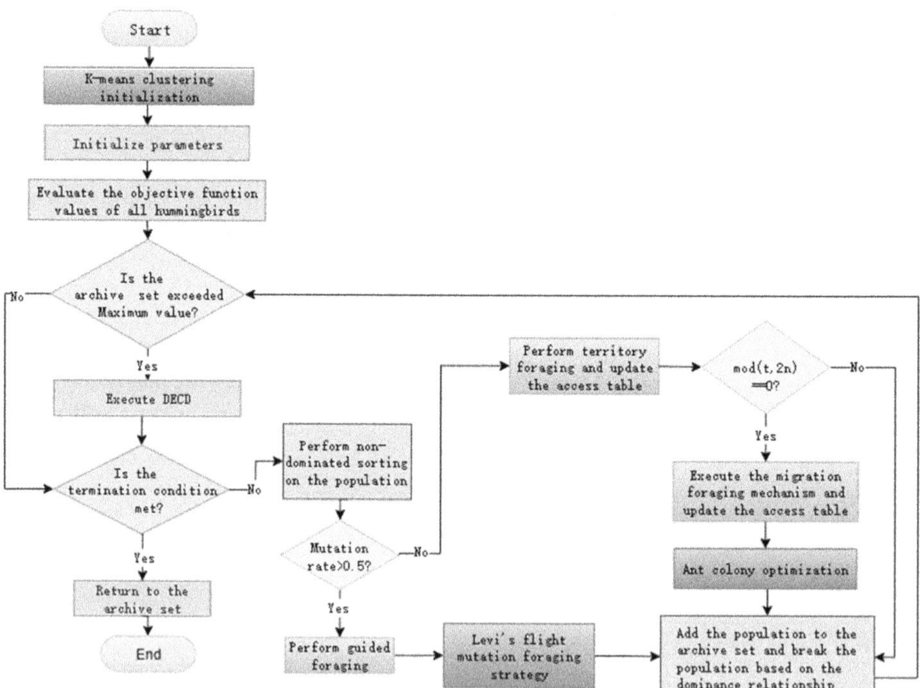

Fig. 2. Multiple UAV work together to collect and transmit farmland crop data based on wireless communication and ground sensor nodes.

(1) K-means Clustering

For the UAV path planning problem, we propose a K-means clustering-based initialization strategy that balances global exploration capability and problem-oriented rationality. This strategy is designed to generate initial trajectories for UAV, with its core concept being to cluster ground devicFor the UAV path planning problem, we propose a K-means clustering-based initialization strategy that balances global exploration capability and problem-oriented rationality. This strategy is designed to generate initial trajectories for UAV, with its core concept being to cluster ground devices and have UAV fly from their initial positions to their respective cluster centers. Below are the mathematical formulation and implementation details of this strategy.es and have UAV fly from their initial positions to their respective cluster centers. Below are the mathematical formulation and implementation details of this strategy.

The objective of K-means is to minimize the within-cluster sum of squares:

$$J = \sum_{k=1}^{K} \sum_{r \in C_k} \|x_r - \mu_k\|^2 \tag{30}$$

where μ_k is the cluster center (the number of UAV), and the cluster center is updated iteratively:

$$\mu_k = \frac{1}{|C_k|} \sum_{r \in C_k} x_r \tag{31}$$

For the k-th UAV, the path from the initial position $p_{k,0} = (x_{k,0}, y_{k,0}, z_{k,0})$ to the target cluster center $\mu_k = (\mu_{k,x}, \mu_{k,y})$ is generated through linear interpolation.

$$p_{k,t} = \left(x_{k,0} + \frac{t}{T}(\mu_{k,x} - x_{k,0}), y_{k,0} + \frac{t}{T}(\mu_{k,y} - y_{k,0}), H \right) \tag{32}$$

(2) Lévy Flight

In the framework of the multi-objective optimization algorithm MOAHA, the Lévy flight mutation mechanism is introduced to enhance the global search capability of the hummingbird algorithm. By combining short-distance movements with occasional long-distance jumps, Lévy flight effectively improves the exploration ability of the algorithm, reducing the risk of falling into local optima. The step distribution of Lévy flight follows a power-law characteristic, with its probability density function expressed as:

$$f(s) \sim \frac{\alpha \Gamma(\alpha)}{s^{\alpha+1}} \quad (s > 0, 1 < \alpha \leq 2) \tag{33}$$

where α is the characteristic exponent of the Lévy stable distribution, and $\Gamma(\cdot)$ is the Gamma function. The step size generation formula is:

$$s = \frac{u}{|v|^{1/\beta}} \tag{34}$$

where $u \sim N(0, \sigma_u^2)$, $v \sim N(0, 1)$ are two normally distributed random variables; β is the Lévy index (typically 1.5). σ_u is calculated by the following formula:

$$\sigma_u = \left(\frac{\Gamma(1 + \beta) \cdot \sin(\pi\beta/2)}{\Gamma((1 + \beta)/2) \cdot \beta \cdot 2^{(\beta-1)/2}} \right)^{1/\beta} \tag{35}$$

The core update formula for Lévy flight mutation is:

$$X_{\text{new}} = X_{\text{best}} + \alpha \cdot s \cdot (X_{\text{best}} - X_{\text{current}}) \tag{36}$$

where X_{new} is the newly generated solution, X_{best} is an excellent individual randomly selected from the archive (Pareto front solution), X_{current} is the current individual, α is the scaling factor, and s is the Lévy flight step vector.

(3) Ant Colony Optimization

During the regional foraging phase of IMOAHA, the Ant Colony Optimization (ACO) mechanism is incorporated to optimize path selection. This mechanism emulates the foraging behavior of ants in nature, utilizing pheromone concentration as a quantitative measure of path quality, which guides ants to preferentially select paths with higher pheromone concentrations, thus achieving self-reinforcing cooperative search. This strategy enhances the selection of superior paths through the accumulation of pheromones, accelerating the algorithm's convergence to high-quality solutions. It also dynamically adjusts the weights of heuristic information and pheromones, balancing historical experience with real-time assessments. Furthermore, it optimizes the UAV's visit order through transition probabilities, reducing communication energy consumption and addressing temporal issues. After each iteration, the algorithm updates the pheromone levels on each path, simulating the pheromones deposited by ants, providing guidance for subsequent searches.The update formula for the pheromone on the path is as follows:

$$\tau_{ij}(n_a, n_a + 1) = (1 - \rho) \cdot \tau_{ij}(n_a, n_a + 1) + \sum_{k=1}^{m_a} \Delta\tau_{ij}^k(n_a, n_a + 1) \tag{37}$$

where the incremental pheromone $\Delta\tau_{ij}^k(n_a, n_a + 1)$ is defined as:

$$\Delta\tau_{ij}^k(n_a, n_a + 1) = \begin{cases} \dfrac{Q}{F_k}, & \text{if ant } k \text{ traverses path } ij \\ 0, & \text{otherwise} \end{cases} \tag{38}$$

In these formulas: Q represents the total amount of pheromone; ρ denotes the pheromone evaporation factor; and F_k is the total objective function value corresponding to the path traversed by the current ant k.

Within the ant colony algorithm framework, the path selection behavior of ant individuals is synergistically regulated by pheromone concentration and heuristic functions. Path quality is positively correlated with pheromone concentration: high-quality paths, with higher pheromone concentrations, are preferentially selected by subsequent ants, forming a positive feedback mechanism. The probability model for ant k transitioning from node i to node j is defined as:

$$p_{ij}^k = \begin{cases} \dfrac{[\tau_{ij}(t)]^\alpha [\eta_{ij}(t)]^\beta}{\sum_{k \in d_k} [\tau_{ij}(t)]^\alpha [\eta_{ij}(t)]^\beta}, & j \in d_k \\ 0, & \text{otherwise} \end{cases} \tag{39}$$

where: $\tau_{ij}(t)$ denotes the pheromone concentration from node i to j at time t; $\eta_{ij}(t)$ denotes the heuristic function from node i to j at time t (typically set as the reciprocal of the Euclidean distance); α is the pheromone heuristic factor, β is the expected heuristic factor, and d_k is the set of selectable nodes for ant k.

3.2 Time Complexity of IMOAHA

The time complexity analysis of IMOAHA focuses on key operations, with the overall complexity dominated by non-dominated sorting, archive maintenance, and crowding distance strategies. The complexities of core steps are as follows: **Non-dominated Sorting**: $\mathcal{O}(T \cdot N^2 \cdot M)$, where T is the maximum number of iterations, N is the population size, and M is the number of objective functions. **Archive Maintenance (Deletion & Update)**: Approximately $\mathcal{O}(T \cdot N^2 \cdot M)$. **Crowding Distance Calculation**: Per-iteration complexity $\mathcal{O}(T \cdot N \cdot M \cdot \log N)$. **Dynamic Elimination Crowding Distance**: Approximately $\mathcal{O}(T \cdot N^2 \cdot M)$. **K-means Initialization**: $\mathcal{O}(N_k \cdot N_r \cdot I)$ (N_k: number of UAV; N_r: number of devices; I: K-means iterations, executed only once during initialization). **Foraging Phase**: Guided foraging (standard MOAHA): $\mathcal{O}(0.5T \cdot N \cdot M)$; Lévy flight: $\mathcal{O}(N \cdot D)$; ant colony optimization: $\mathcal{O}(N \cdot D \cdot Q)$ (D: solution dimension; Q: number of neighborhood solutions); territorial and migratory foraging: $\mathcal{O}(T \cdot N \cdot M)$, $\mathcal{O}(0.5T \cdot N \cdot M)$.

Since the complexities of non-dominated sorting, archive maintenance, and dynamic crowding distance strategies are all $\mathcal{O}(T \cdot N^2 \cdot M)$, and the remaining operations have lower complexities or are executed only once, the overall time complexity of IMOAHA is $\mathcal{O}(T \cdot N^2 \cdot M)$.

4 Simulation Results and Analysis

The simulation is conducted using Matlab 2020a to evaluate the performance of IMOAHA in solving the Multi-UAV trajectory planning and energy-efficient EEPPM-DCCMUS in agriculture. IMOAHA is compared with other multi-objective evolutionary algorithms. The computer used for the simulation is a DELL, with the Windows 10 operating system and an Intel processor i5-7200U.

4.1 Simulation Configuration

The agricultural area is configured as $200\,\text{m} \times 200\,\text{m}$. The default flight altitude of Unmanned Aerial Vehicles (UAV) is set to $50\,\text{m}$, and the mass m of each UAV is $2.0\,\text{kg}$. The speed range of UAV is $[2, 20]\,\text{m/s}$, the communication range is $50\,\text{m}$, and the safety distance is $30\,\text{m}$. The number of UAV is 4, and the number of ground devices is 10. The number of time slots is 24, with a time slot interval $\Delta t = 1\,\text{s}$. The bandwidth B is $1\,\text{MHz}$. The carrier frequency for UAV - to - UAV (A2A) communication is $2.4\,\text{GHz}$, while for UAV - to - ground (A2G) communication, it is $5.8\,\text{GHz}$. The noise power is $-100\,\text{dBm}$, and the transmission power range is $[0, 20]\,\text{dBm}$. To avoid deviations, the maximum number of iterations is set to 200.

To verify the effectiveness of the proposed algorithm, multiple state-of-the-art algorithms are introduced in this study, namely the Multi-objective Artificial Vultures Optimization Algorithm (MOAVOA) [17], Multi-objective Crystal Structure Algorithm (MOCryStAl) [18], Multi-Objective Genetic Algorithm (MOGA) [19], Multi-Objective Particle Swarm Optimization (MOPSO) [20], Non-dominated Sorting Genetic Algorithm II (NSGA-II) [21], Simulated Annealing (SA) [22], Simulated Annealing-Based Multi-objective Optimization Algorithm (AMOSA) [23], and Multi-objective Artificial Hummingbird Algorithm (MOAHA) [24]. To ensure the repeatability of experimental results, all comparison experiments are performed with a fixed number of iterations of 200 simulation parameters and their meanings are shown in Table 1.

4.2 Optimization Result and Analysis

Based on the optimization objectives defined in this study, the Pareto front diagrams of the MOAHA and IMOAHA algorithms are depicted in Fig. 3. The flight route trajectories of UAV generated by the optimal energy - consumption planning strategy are presented in Fig. 4. The time-varying energy consumption of each UAV is illustrated in Fig. 5, and the temporal evolution of total energy consumption is shown in Fig. 6. The total energy consumption and the cumulative flight distance of all UAV across different algorithms are summarized in Table 2. From the comparative experiments, the proposed IMOAHA algorithm demonstrates superior performance in addressing the optimization objectives of this study. Comprehensive Pareto front analysis reveals that IMOAHA exhibits enhanced balance in multi-objective optimization scenarios, making it more suitable for solving EEPPM-DCCMUS.

IMOAHA exhibits superior multi-objective optimization performance compared to MOAHA, as evidenced by its broader Pareto front (Fig. 3) distribution in 3D analysis. The algorithm achieves higher maximum transmission rates and wider energy consumption ranges, demonstrating enhanced exploration of the trade-off space. Notably, IMOAHA forms continuous Pareto surfaces across both high-performance and energy-efficient regions, while MOAHA solutions cluster discontinuously in intermediate ranges. The integration of Lévy flight and ant

Table 1. Simulation Parameters and Their Meanings

Parameter Name	Meaning	Value
K	Number of UAV	4
R	Number of ground devices	10
T	Total service time	24
Δt	Time slot interval (s)	1
H	Default cruising altitude (m)	50
m	UAV mass (kg)	2.0
v	UAV flight speed (m/s)	$[2, 20]$
B	Communication bandwidth (MHz)	1
f_c^{A2A}	UAV-to-UAV communication (GHz)	2.4
f_c^{A2G}	UAV-to-ground device communication (GHz)	5.8
$P_{\max}$	Maximum transmission power (dBm)	20
$P_{\min}$	Minimum transmission power (dBm)	0
Noise	Noise power (dBm)	-100
d_{safe}	Safety distance between UAV (m)	30
comm_range	Communication range (m)	50
α	Characteristic index of Lévy flight	1.5
β	Pheromone heuristic factor	2.0
N_p	Population size	50
N_r	External archive size	100
maxgen	Maximum number of iterations	200
area_size	Area size (m)	$[200, 200]$

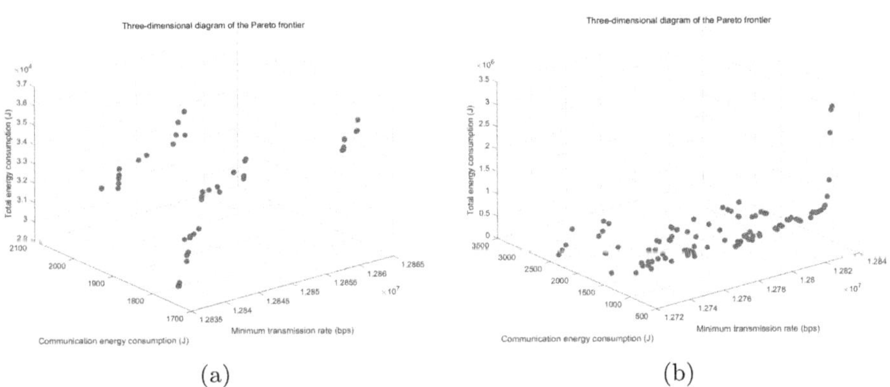

(a)   (b)

Fig. 3. Three-Dimensional Pareto Front Diagrams. (a) Pareto front diagram of MOAHA; (b) Pareto front diagram of IMOAHA.

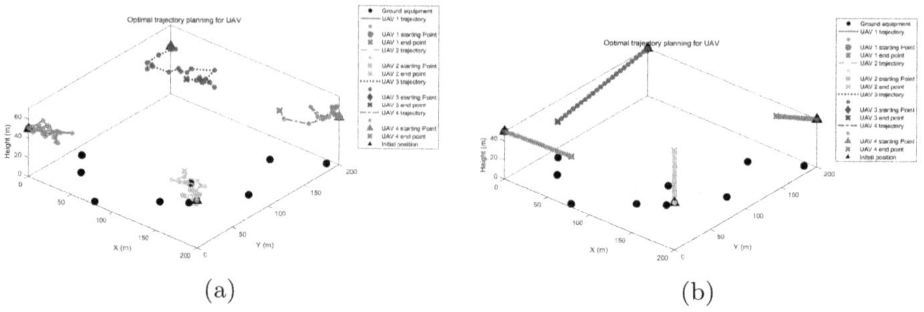

Fig. 4. Comparative Analysis of Optimal UAV Trajectory Planning. (a) MOAHA trajectory showing characteristic; (b) IMOAHA optimized trajectory demonstrating near-linear path.

colony optimization strategies significantly improves solution diversity and quality for UAV cooperative tasks.

IMOAHA demonstrates superior trajectory optimization compared to MOAHA in UAV path planning, as shown in 3D trajectory analyses (Fig. 4). The algorithm generates more efficient flight paths through near-linear trajectories, achieving 53.7% shorter distances and 58.9% lower energy consumption than MOAHA's circuitous routes. By minimizing redundant flight paths (42.99% reduction) and improving coverage accuracy, IMOAHA meets the demanding energy-efficiency requirements for agricultural UAV operations while maintaining task effectiveness.

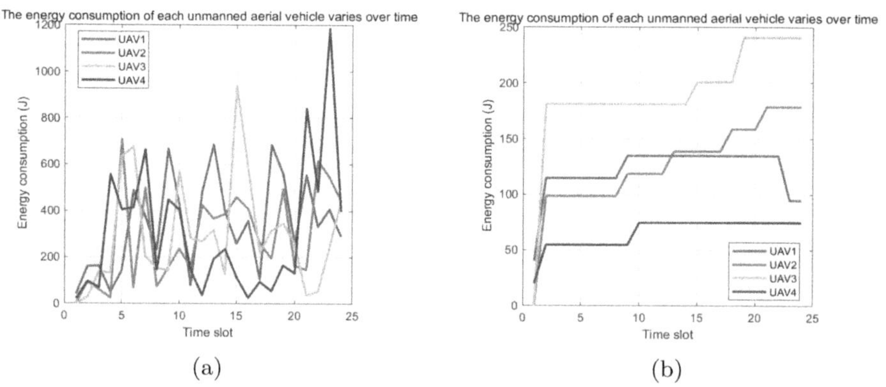

Fig. 5. Time-series variation of UAV energy consumption. (a) MOAHA exhibits higher energy fluctuation; (b) IMOAHA demonstrates stable consumption.

Energy consumption analysis (Fig. 5) demonstrates IMOAHA's superior dynamic regulation capability compared to MOAHA. The algorithm achieves 58.9% smaller energy fluctuations across UAV and reduces peak consumption by 1200J through load balancing, decreasing energy distribution variance by 42.3%. These improvements yield 36.7% lower wasteful energy use, enhancing swarm efficiency for prolonged operations.

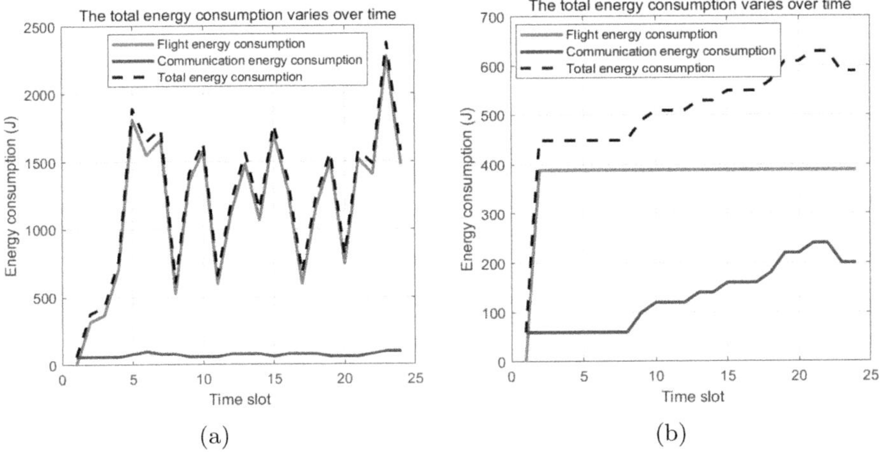

Fig. 6. Time-series variation of total energy consumption. (a) MOAHA shows higher energy volatility; (b) IMOAHA maintains stable consumption profile.

Energy consumption analysis (Fig. 6) reveals IMOAHA's superior energy regulation compared to MOAHA. While MOAHA exhibits unstable flight energy (500–2000 J) with minimal communication costs, IMOAHA maintains stable flight (380J±5%) and balanced communication (200J) through integrated K-means, Lévy flight and ant colony optimization. Experimental results demonstrate that IMOAHA effectively addresses the energy consumption control issues inherent in MOAHA, achieving an 81.3% reduction in flight energy fluctuation amplitude while establishing a superior energy-efficiency optimization framework for Multi-UAV cooperative operations.

As evidenced in Table 2, the proposed IMOAHA algorithm demonstrates superior multi-objective optimization capabilities compared to eight benchmark methods (including MOAHA and MOAVOA). While maintaining comparable communication performance (12.75 Mbps minimum transmission rate, merely 0.64% inferior to the optimal MOAHA), IMOAHA achieves remarkable energy efficiency with total consumption reduced to 12.14 kJ - representing 58.9% and 25.8–70.6% improvements over MOAHA and other algorithms respectively. Simultaneously, it shortens the total flight distance to 365.05 m (53.7% and 50.2–66.3% reductions versus MOAHA and competitors). Notably, IMOAHA establishes new performance benchmarks in both energy consumption and flight distance metrics, validating its exceptional multi-objective optimization capacity and offering an advanced technical framework for agricultural UAV system optimization.

Table 2. Results Obtained by Different Algorithms

Algorithm	Min Transmission Rate (bps)	Total Energy (J)	Total Flight Distance (m)
MOAVOA	1,139,9139.10	41,316.65	1,082.61
MOCryStAl	1,141,4217.45	16,350.60	529.01
MOGA	1,141,0381.80	28,657.19	744.88
MOPSO	1,139,9139.10	41,316.65	1,082.61
NSGAII	1,142,6625.10	38,739.29	1,021.40
SA	1,176,2787.62	16,596.61	372.32
AMOSA	1,143,6797.87	40,253.19	1,019.81
MOAHA	**1,283,6677.32**	29,609.20	788.04
IMOAHA	1,275,4066.74	**12,138.94**	**365.05**

5   Conclusion

Our study introduces an improved multi-objective artificial hummingbird algorithm (IMOAHA) for optimizing energy consumption in Multi-UAV cooperative field data collection, integrating K-means clustering initialization, Lévy flight mutation, and ant colony optimization to enhance global exploration and local search efficiency. Simulations show that IMOAHA outperforms baseline MOAHA, achieving a minimum transmission rate of 12.75 Mbps while reducing total energy consumption by 58.9% (to 12.14 kJ) and flight distance by 53.7% (to 365.05 m), with an 81.3% decrease in flight energy fluctuations and stable communication energy at 200 J. Compared to MOAVOA and MOCryStAl, IMOAHA cuts energy consumption by 25.8%–70.6% and flight distance by 50.2%–66.3%.

The proposed energy-efficient path planning model (EEPPM-DCCMUS) uniquely co-models flight dynamics, communication energy, and task allocation, addressing conventional algorithm limitations in global search and convergence for precision agriculture. Looking forward, further research should focus on validating the algorithm's scalability in larger UAV swarms and more dynamic agricultural settings—incorporating real-world disturbances such as sensor failures and environmental variability—as well as conducting hardware-in-the-loop and field experiments to bridge the gap between simulation and practical deployment.

Acknowledgments. This work was supported in part by the Inner Mongolia Autonomous Region Natural Science Foundation (Grant Nos. 2025MS06003, 2025MS06056), the Inner Mongolia Autonomous Region Science and Technology Planning Project (Grant No. 2021GG0250), the National Natural Science Foundation of China (Grant No. 61962044), and the Fundamental Research Funds for the Central Universities (Grant No. JY20220324).

Disclosure of Interests. The authors declare that there are no competing interests relevant to the content of this article.

References

1. Wu, K., Lu, S., Chen, H., Feng, M., Lu, Z.: An energy-efficient logistic drone routing method considering dynamic drone speed and payload. Sustainability **16**(12), 4995 (2024)
2. Zhang, S., Jin, H., Guo, P.: IRS-assisted energy efficient communication for UAV mobile edge computing. Comput. Netw. **246**, 110387 (2024)
3. Chen, S., Mo, Y., Wu, X., Xiao, J., Liu, Q.: Reinforcement learning-based energy-saving path planning for UAV in turbulent wind. Electronics **13**(16), 3190 (2024)
4. Zhu, X., Zhai, L., Li, N., Li, Y., Yang, F.: Multi-objective deployment optimization of UAV for energy-efficient wireless coverage. IEEE Trans. Commun. **72**(6), 3587–3601 (2024)
5. Lin, L., Wang, Z., Tian, L., Wu, J., Wu, W.: A PSO-based energy-efficient data collection optimization algorithm for UAV mission planning. PLoS ONE **19**(1), e0297066 (2024)
6. Ma, H., et al.: Improved DRL-based energy-efficient UAV control for maximum lifecycle. J. Franklin Inst. **361**(6), 106718 (2024)
7. Zeng, Y., Xia, Y., Chen, S., et al.: Energy-saving offloading strategy for Multi-UAV assisted communication. J. Jilin Univ. (Eng. Technol. Ed.) 1–9 (2024)
8. Xiong, R., Zhang, H., Liang, C., et al.: An energy-efficient data collection method for LoRa networks assisted by Multi-UAV. Chin. J. Comput. **47**(8), 1970–1987 (2024)
9. Wu, L.: Research on energy management of UAV communication payload. Ph.D. dissertation, University of Electronic Science and Technology of China, Chengdu, China (2024)
10. Tan, L., Cao, B., Xia, J., et al.: A two-layer UAV energy consumption optimization method for random computation offloading. Telecommun. Eng. **64**(6) (2024)
11. Lu, W., Zhan, Y., Hua, Q., et al.: Research on energy efficiency optimization method for edge computing system based on UAV wireless energy transmission. J. Electron. Inf. Technol. **44**(3), 899–905 (2022)
12. Liu, J., Yuan, G., Tu, X.: Resource allocation and deployment optimization scheme in UAV-assisted edge computing. Radio Commun. Technol. **51**(2) (2025)
13. Wang, C., Zhou, J., Zhao, Y., He, X., Gu, J., Wang, X.: Optimization of energy consumption of UAV cluster based on K-means improved particle swarm algorithm. In: 2024 IEEE 3rd International Conference on Electrical Engineering, Big Data and Algorithms (EEBDA), Changchun, China, pp. 396–400 (2024)
14. Liu, Y., Liu, S., Liu, X., Liu, Z., Durrani, T.S.: Sensing fairness-based energy efficiency optimization for UAV enabled integrated sensing and communication. IEEE Wirel. Commun. Lett. **12**(10), 1702–1706 (2023)
15. Wu, J., Sun, Y., Wu, Y., Wu, J.: Low energy consumption UAV trajectory optimization algorithm for secure transmission. Comput. Eng. **50**(2), 59–67 (2024)
16. Shen, Y., Qu, Y., Dong, C., Zhou, F., Wu, Q.: Joint training and resource allocation optimization for federated learning in UAV swarm. IEEE Internet Things J. **10**(3), 2272–2284 (2023)
17. Khodadadi, N., Gharehchopogh, F.S., Mirjalili, S.: MOAVOA: a new multi-objective artificial vultures optimization algorithm. Neural Comput. Appl. **34**(23), 20791–20829 (2022)
18. Khodadadi, N., Azizi, M., Talatahari, S., Sareh, P.: Multi-objective crystal structure algorithm (MOCryStAl): introduction and performance evaluation. IEEE Access **9**, 117795–117812 (2021)

19. Murata, T., Ishibuchi, H.: MOGA: multi-objective genetic algorithms. In: Proceedings of 1995 IEEE International Conference on Evolutionary Computation, Perth, WA, Australia, pp. 289–294 (1995)
20. Zhang, X., Xia, S., Li, X., Zhang, T.: Multi-objective particle swarm optimization with multi-mode collaboration based on reinforcement learning for path planning of unmanned air vehicles. Knowl.-Based Syst. **250**, 109075 (2022)
21. Deb, K., Pratap, A., Agarwal, S., Meyarivan, T.: A fast and elitist multiobjective genetic algorithm: NSGA-II. IEEE Trans. Evol. Comput. **6**(2), 182–197 (2002)
22. Kirkpatrick, S., Gelatt, C.D., Vecchi, M.P.: Optimization by simulated annealing. Science **220**(4598), 671–680 (1983)
23. Bandyopadhyay, S., Saha, S., Maulik, U., Deb, K.: A simulated annealing-based multiobjective optimization algorithm: AMOSA. IEEE Trans. Evol. Comput. **12**(3), 269–283 (2008)
24. Khodadadi, N., Mirjalili, S.M., Zhao, W., Zhang, Z., Wang, L., Mirjalili, S.: Multi-objective artificial hummingbird algorithm. In: Biswas, A., Kalayci, C.B., Mirjalili, S. (eds.) Advances in Swarm Intelligence, vol. 1054, pp. 1–12. Springer, Cham (2023)

Chameleon Hash-Based Redactable Blockchains: A Systematic Literature Review

Weilong Lyu[1,2], Shanshan Li[1,2(✉)], He Zhang[1,2], Jingyue Li[3], Zhe Huang[4], and Feng Lin[1,2]

[1] Software Institute, Nanjing University, Nanjing, China
`lvweilong26@126.com`, `{lss,hezhang}@nju.edu.cn`, `211250128@smail.nju.edu.cn`
[2] State Key Laboratory of Novel Software Technology, Nanjing University, Nanjing, China
[3] Department of Computer Science, Norwegian University of Science and Technology, Trondheim, Norway
`jingyue.li@ntnu.no`
[4] School of Software Technology, Zhejiang University, Hangzhou, China
`huangzhe@zju.edu.cn`

Abstract. Blockchain immutability enhances data security but introduces challenges in regulatory compliance, vulnerability remediation, and storage scalability. To address these issues, redactable blockchain technologies have emerged. Among them, chameleon hash-based redactable blockchains (CHRB) are considered the most promising due to their cryptographic soundness, high efficiency, and compatibility with existing architectures. However, research topics in this field are fragmented, and the technical complexity makes it difficult for researchers to identify suitable techniques or design solutions. To systematically understand the CHRB landscape, this study conducts a systematic literature review (SLR) of 40 rigorously selected papers. We identify 10 core research topics, which are categorized into three areas: pre-redaction authorization, redaction process optimization, and post-redaction audit. We also extract 21 key enabling techniques and analyze their associations with the identified topics. Finally, the study summarizes key findings and discusses future research directions. This review provides a structured overview of CHRB's research landscape and technical foundations, aiming to assist researchers and practitioners in navigating this rapidly evolving domain.

Keywords: Chameleon hash · Redactable blockchain · Systematic literature review

1 Introduction

Blockchain has been widely applied in domains such as supply chains [2], IoT [4], and healthcare [21], owing to its decentralization, transparency, and immutability. Its core mechanism lies in a hash-chained structure, where each block contains the hash of its predecessor. Once any data is tampered with, all subsequent

H. Liu et al. (Eds.): ICA3PP 2025, LNCS 16381, pp. 301–321, 2026.
https://doi.org/10.1007/978-981-95-8399-7_17

block hashes become invalid, thereby ensuring the integrity and trustworthiness of the ledger [15]. However, while immutability enhances data security, it also brings practical challenges in terms of regulatory compliance, privacy protection, vulnerability remediation, and storage scalability. For instance, once illegal content is recorded on-chain, it becomes difficult to delete, potentially leading to legal liability [11]; the permanent retention of sensitive user data conflicts with regulations such as the GDPR [6]; the severe consequences arising from vulnerabilities in smart contracts may necessitate costly hard forks for resolution [14]; and the rapidly growing ledger size limits participation by ordinary nodes [16].

To address the inherent conflict between immutability and practical challenges, redactable blockchain technology has emerged. It allows authorized entities to modify or delete on-chain data under specific conditions [17], enabling effective data governance and storage optimization through a controlled redaction mechanism. Currently, there are two main technical approaches: consensus-based and chameleon hash-based solutions [1]. The former typically requires adjusting block structures, executing multiple rounds of consensus, and recalculating the hashes of subsequent blocks, resulting in high costs and latency. In contrast, the latter leverages chameleon hash functions with trapdoor mechanisms, allowing data modification without altering the hash output. Due to high efficiency and compatibility with existing architectures, chameleon hash-based redactable blockchains (CHRB) are widely regarded as the most promising solution [13,19].

In recent years, numerous studies have proposed new solutions by improving chameleon hash algorithms and integrating cryptographic primitives and techniques such as attribute-based encryption [6] and zero-knowledge proofs [16]. These solutions address multiple dimensions, such as enhancing decentralization [3], strengthening security [11], and enabling fine-grained control [14]. However, the fragmentation of research topics and the inherent complexity of cryptographic constructions pose significant challenges for researchers and practitioners in understanding and selecting research directions and approaches. Therefore, conducting a systematic analysis of the CHRB schemes and building a panoramic view to help researchers and practitioners identify key research topics and technical branches holds substantial academic value and practical significance.

Motivated by popularity, thematic dispersion, and technical complexity of the CHRB schemes, we conduct a systematic literature review (SLR) about the research topics and key technologies of CHRB. This research identifies and analyzes 40 CHRB studies, categorizing research into three main aspects: pre-redaction permission control, redaction process optimization, and post-redaction audit. Furthermore, we discuss the commonly adopted algorithms and protocols in CHRB research and employ literature mapping to present the correspondences between technical approaches and research topics.

2 Background and Related Work

2.1 Blockchain Immutability and Block Structure

Blockchain immutability refers to the fundamental characteristic whereby data, once written to the blockchain, cannot be modified or deleted arbitrarily [12]. This property is primarily ensured through two interdependent mechanisms: (i) the hash-linked chain structure that maintains historical consistency, and (ii) collision-resistant hash functions that provide data integrity verification.

Using Bitcoin as a representative example, the block structure of a traditional blockchain consists of two principal components - the block header and block body - as illustrated in Fig. 1. The key fields include the following [7]:

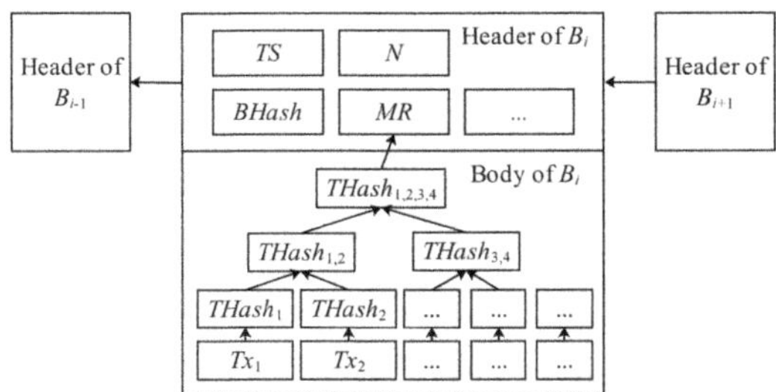

Fig. 1. Block structure of Bitcoin

- **Block header fields**:
  - Parent block hash $BHash$. This field ensures the immutability of preceding blocks.
  - Merkle root MR. The root hash aggregating all transaction hashes in the current block, guaranteeing transaction immutability.
  - Timestamp TS. The field records the UTC timestamp of block creation, determining its chronological position in the chain.
  - Nonce N. A variable parameter in Proof-of-Work (PoW) algorithms that satisfies the target difficulty threshold through hash computation.
- **Block body fields**:
  - A list of transactions $(Tx_1, Tx_2, \ldots, Tx_n)$.
  - Transaction hash list $(THash_1, THash_2, \ldots, THash_n)$. These hashes as leaf nodes are pairwise-hashed layer by layer to generate the Merkle root stored in the header.

2.2 Chameleon Hashing

Chameleon hashing is a specialized cryptographic primitive, first proposed by Krawczyk and Rabin in 2000 [9]. Unlike traditional hash functions, chameleon hashing allows authorized parties who possess the trapdoor to construct valid collisions without altering the hash output. Its formal definition comprises the following five polynomial-time algorithms:

1. $pp \leftarrow$ Setup(1^λ): The Setup algorithm takes a security parameter λ as input and outputs public parameters pp;
2. $(pk, sk) \leftarrow$ KeyGen(pp): The KeyGen algorithm takes the public parameters pp as input and outputs a public key pk and a trapdoor sk. The trapdoor sk serves as the critical secret for forging new variable parameters, while the public key pk is published for hash computation and verification;
3. $CH \leftarrow$ Hash(pk, m, r): The Hash algorithm takes the public key pk, a message m, and a variable parameter r as inputs and outputs a chameleon hash value CH. The variable parameter r ensures identical hash values can be generated for distinct messages;
4. $r' \leftarrow$ Forge(sk, m, r, m'): The Forge algorithm takes the trapdoor sk, original message m, original variable parameter r, and a new message m' as inputs and outputs a new variable parameter r'. This algorithm guarantees that for any $m' \neq m$, a valid r' can be found, such that Hash(pk, m, r) = Hash(pk, m', r') = CH.
5. $\{0, 1\} \leftarrow$ Verify(pk, m, r, CH): The Verify algorithm takes the public key pk, message m, variable parameter r and chameleon hash value CH as inputs and verifies whether the tuple (m, r) produces the given hash value CH. Outputs 1 if verification succeeds, and 0 otherwise.

2.3 Core Workflow of CHRB

The core idea of CHRB is to leverage the trapdoor mechanism of chameleon hashing to enable redactable modifications on traditional blockchain structures. In CHRB, the hash algorithm (e.g., SHA-256) used for computing parent block hash is replaced with the chameleon hash algorithm. When the transactions in block b_i are modified, its block header data changes from m to m'. In this case, it only requires forging a new variable parameter r' to satisfy:

$$hash(m', r') = hash(m, r) = BHash \tag{1}$$

This approach enables modification of block b_i without compromising the integrity verification. Specifically, the core processes of CHRB comprise the following four phases [3]:

1. **Chain Initialization**: A predefined trusted entity invokes the Setup and KeyGen functions to generate public parameters pp along with the chameleon hash's public key pk and trapdoor sk. pp and pk are distributed to all nodes, while sk is securely managed by authorized entities.

2. **Block Generation**: Miners employ the Hash function to perform chameleon hash operations on block content m (including the Merkle root of transactions and other metadata) to construct new blocks. Each block header stores both the chameleon hash value and its corresponding variable parameter r.
3. **Ledger Redaction**: When ledger modification is required, authorized editors possessing sk can invoke the Forge function to compute a new variable parameter r' that matches the modified block content m', thereby enabling transaction content modification without altering the block hash. After performing the modification, the authorized editor broadcasts a message containing modification details to other nodes for verification.
4. **Block Verification**: Upon receiving the block modification message, other nodes call the Verify function to validate the integrity of the modified block. If verification succeeds, they update the block in their local ledger.

2.4   Related Work

To date, only a limited number of review articles on redactable blockchains have been published, as cataloged in Table 1. Politou et al. [12] examined the legal conflict between blockchain immutability and the "right to be forgotten" under GDPR and proposed two technical approaches: "bypassing immutability" and "removing immutability." Vidal et al. [13] investigated revocation mechanisms in blockchain and analyzed the applicability and challenges of eight revocation techniques. Zhang et al. [19] discussed design challenges and evaluation criteria for redactable blockchains and innovatively categorized existing solutions into four classes: consensus-based, chameleon hash-based, meta-transaction-based, and pruning-based approaches. Ye et al. [17] reviewed technical challenges and research opportunities, with particular focus on cryptographic methods for achieving blockchain redactability. Abd et al. [1] classified redactable blockchains into chameleon hash-based and non-chameleon hash-based schemes. They defined redaction implementation challenges and associated security attributes and evaluated each scheme based on these criteria. Fathalla et al. [7] comprehensively summarized fundamental technologies supporting redactable distributed ledgers and classified redactable blockchains into three categories: chameleon hash-based, policy-based, and new block structure-based approaches.

Overall, existing reviews suffer from two significant limitations: (1) **Lack of timeliness**: The most recent reviews were published in 2023 and do not cover the large-scale CHRB solutions that have emerged over the past three years; (2) **Lack of targeted discussion**: Existing reviews treat CHRB as a subcategory of redactable blockchain, and the relatively low proportion of CHRB-related solutions in their included studies suggests that their research questions do not adequately capture the current state of the art in CHRB. To address these gaps, this study conducts **the first SLR** focused exclusively on CHRB, with particular emphasis on its research trends, core topics, and enabling technologies.

Table 1. Redactable Blockchain Reviews.

Literature	Year	Methodology	Total Papers Reviewed	CHRB Coverage
[12]	2019	NLR	11	9.1% (1/11)
[13]	2021	NLR	17	17.6% (3/17)
[19]	2021	NLR	17	52.9% (9/17)
[17]	2023	NLR	13	46.2% (6/13)
[1]	2023	NLR	32	81.3% (26/32)
[7]	2023	NLR	9	55.6% (5/9)
This Paper	2025	SLR	40	100% (40/40)

*NLR: Narrative Literature Review, SLR: Systematic Literature Review

3 Methodology

This study adopts a Systematic Literature Review (SLR) approach, following the guidelines proposed by Kitchenham et al. [8]. SLR is a rigorous method for identifying, evaluating, and synthesizing existing studies to address specific research questions, thereby minimizing bias and enhancing reproducibility. The review team consists of three research students as primary researchers and three supervisors who provide methodological guidance and validate the results.

3.1 Research Questions

This study aims to analyze high-quality peer-reviewed literature on CHRB, with a focus on publication trends, research topics, and adopted technologies. Accordingly, we formulate the following research questions (RQs):

– **RQ1: What are the current research topics in CHRB?** This RQ aims to categorize and summarize the main research topics within CHRB and link them to the corresponding literature.
– **RQ2: What technologies are used in CHRB solutions?** This RQ seeks to identify the core technologies adopted in CHRB studies and explore how these technologies relate to specific research topics.

3.2 Search Strategy

Following the methodologies outlined in [8,20], we developed a search strategy to comprehensively retrieve relevant peer-reviewed publications. Scopus was selected as the primary data source, and an automated search was subsequently conducted. The search string was constructed using the keywords "chameleon hash", "blockchain", and edit-related terms, as shown below:

```
TITLE-ABS-KEY("chameleon hash" AND blockchain AND (redact* OR
rewrit* OR modif* OR edit*))
```

3.3 Study Selection

The search was completed in June 2025. A total of 157 potentially relevant studies were initially retrieved through the automated search. After a rigorous selection process, 40 studies were included for in-depth analysis. The complete list of selected studies is provided in the Appendix. The selection process and related statistics are illustrated in Fig. 2.

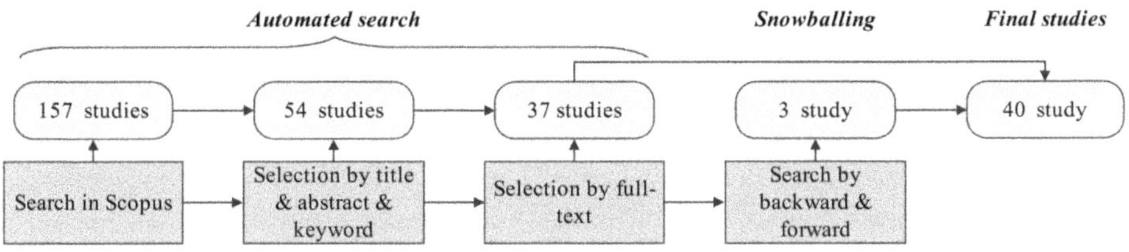

Fig. 2. Study search and selection process

To ensure the scientific rigor and transparency of the selection procedure, we defined one inclusion criterion and four exclusion criteria, as summarized in Table 2. Notably, with the rapid development of blockchain research, the number of related publications has grown explosively across various venues, resulting in significant disparities in research quality. In particular, some emerging workshops or journals with low impact factors may publish studies that lack rigorous peer review or sufficient experimental validation. To ensure the academic reliability and research value of the studies included in this SLR, we established exclusion criterion E3, which employs the CCF (China Computer Federation) Ranking[1] as the quality filter. This criterion helps maintain the quality of the selected literature and minimize subjective bias in the selection process.

3.4 Data Extraction and Synthesis

After finalizing the selected studies, we employed a predefined data extraction form to collect relevant information. The form consists of two categories of data items. The first category includes bibliographic metadata such as title, authors, publication year, publication type and venue. The second category encompasses data items used to answer the research questions, including research background (RQ1), research topics (RQ1), and adopted technologies (RQ2).

---

[1] The CCF Ranking is a widely recognized academic evaluation system in the computer science community in China. It classifies international journals and conferences into three tiers: A (top), B (excellent), and C (good). This classification shows strong alignment with leading international venues such as ACM/IEEE conferences (e.g., SIGCOMM, NDSS) and journals (e.g., TDSC, TIFS), and correlates well with objective metrics like citation count and acceptance rate [10,18].

Table 2. Inclusion and exclusion criteria for this SLR

	Inclusion criteria
I1	Papers that propose a redactable blockchain solution based on chameleon hashing.
	Exclusion criteria
E1	Papers that are not written in English or have no full-text access.
E2	Papers that are not original research.
E3	Papers that are not published in CCF A/B/C-ranked conferences or journals.
E4	Papers that primarily apply existing CHRB techniques rather than proposing or improving such methods.

We applied the thematic analysis method [5] to analyze the second category of data, and followed the six standard steps of thematic analysis: (1) familiarizing with the data, (2) generating initial codes, (3) searching for themes, (4) reviewing themes, (5) defining and naming themes, and (6) producing the final report.

4   Demographics

Figure 3 illustrates the publication trend in the CHRB field. From 2017 to 2020, the field remained in its infancy, with no more than one conference paper published annually. A notable increase occurred in 2021, with the number of conference papers rising to three and journal papers appearing for the first time (four in total). Since then, the number of journal publications has continued to grow, reaching seven in 2023 and increasing further to eleven in 2024. As of June 2025, five journal papers have already been published, while no conference papers have been observed so far. This trend suggests that CHRB research is evolving from an early exploration phase dominated by conferences to a more mature stage characterized by journal-oriented theoretical development.

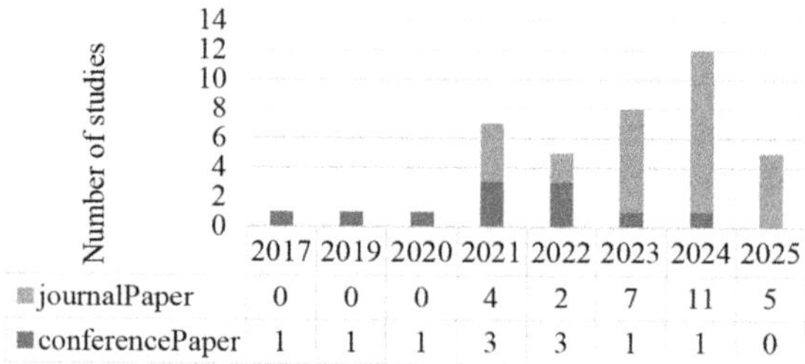

Fig. 3. Annual distribution of reviewed CHRB studies

Figure 4 presents the distribution of publication venues for CHRB research. The data reveals a dual pattern of concentration and diversity: the top-tier security journal *IEEE Transactions on Information Forensics and Security* ranks first with 11 publications, highlighting the journal's focus on CHRB research; *IEEE Transactions on Dependable and Secure Computing* follows with 4 publications. Notably, several studies have also been published in conferences such as *TrustCom* and journals like *IEEE Transactions on Network and Service Management*. In addition, the "Others" category includes 14 venues, each with only one CHRB publication, indicating that CHRB research is additionally disseminated across a broad range of venues. This distribution underscores both the recognition of CHRB research in top-tier journals and the diversity of its research perspectives.

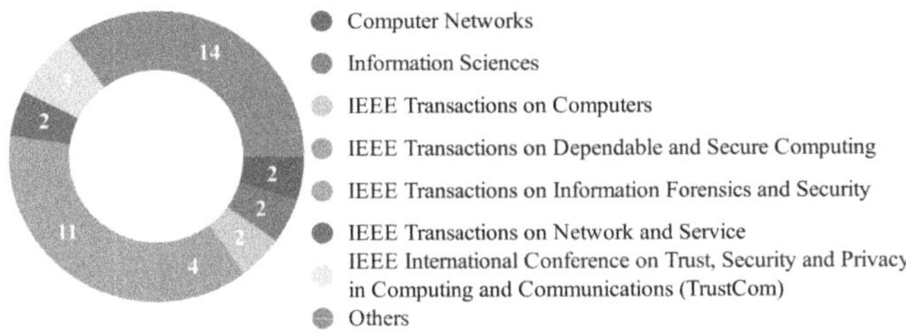

Fig. 4. Distribution of reviewed CHRB studies across publication venues

5   RQ1: Research Topics

By analyzing the research questions, objectives, and contributions of the 40 selected studies, we identified ten secondary topics in CHRB. For clarity, these ten topics were clustered into three primary research topics—**Pre-redaction Permission Control**, **Redaction Process Optimization**, and **Post-redaction Audit**—as illustrated in Fig. 5. Several secondary topics were further refined into tertiary ones where additional granularity was needed.

5.1   T1 Pre-redaction Permission Control

Pre-redaction permission control provides access-control mechanisms that determine *who*, *where*, *how many*, and *when* redactions are permitted.

T1.1 Authorization-based Redaction (*WHO* can edit): The basic CHRB model allows any trapdoor holder to edit ledger data, posing risks of malicious tampering. This topic enforces identity-based policies, granting redaction rights only to trapdoor holders who meet specific authorization criteria. For instance, Derler et al. [S2] proposed Policy-based Chameleon Hashing (PCH), which integrates Attribute-Based Encryption (ABE) with chameleon hashing to enforce policy-driven redaction control. Building on PCH, Tian et al. [S21] introduced a tree-based revocation mechanism using the KUNodes algorithm, enabling dynamic updating and revocation of redaction rights.

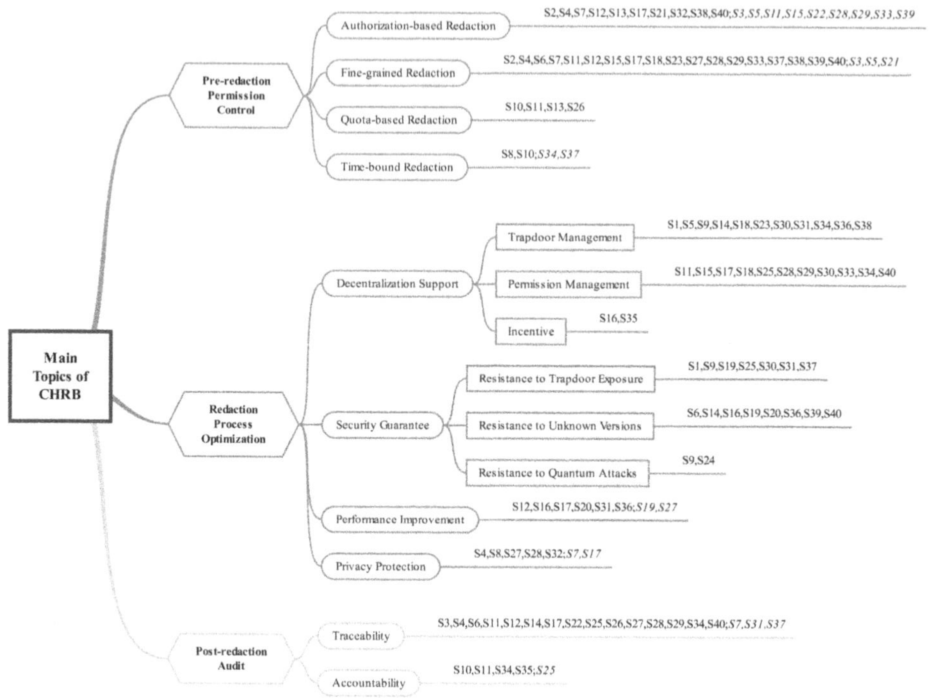

Fig. 5. Current research topics in CHRB. Each secondary or tertiary topic is annotated with the IDs of corresponding studies. IDs in *italic* indicate studies that do not explicitly state the topic as a research focus but address it implicitly in their proposed solutions.

T1.2 Fine-grained Redaction (*WHERE* to edit): In the basic CHRB model, a trapdoor holder can arbitrarily modify an entire block, increasing the risk of malicious or unintended alterations to unrelated data. This topic mitigates such risks by restricting redactions to specific data units (e.g., individual transactions or fields). For example, Zhang et al. [S5] employed Chameleon Hashing with Ephemeral Trapdoors (CHET) to enable transaction-level redaction using ephemeral trapdoors derived from a master trapdoor. Jia et al. [S6] applied chameleon hashing to the integrity protection of specific fields, enabling only authorized modifications under corresponding trapdoors.

T1.3 Quota-based Redaction (*HOW MANY* edits): The basic CHRB model lacks constraints on the number of redactions, which allows unlimited modifications of the same data by the trapdoor holder. This may lead to abuse and compromise system consistency. This topic introduces verifiable quota mechanisms to ensure compliance while preserving blockchain's core values. For instance, Li et al. [S13] proposed a k-time controllable redaction scheme based on vector commitments. Duan et al. [S26] introduced t-time Chameleon Hashing (t-CH), which enforces a strict redaction cap on edits through a trapdoor self-destruction mechanism.

T1.4 Time-bound Redaction (*WHEN* to edit): In the basic CHRB model, trapdoor holders can edit historical data at any time, creating risks of abuse due to the absence of temporal restrictions. This topic enforces time-bound redaction policies to ensure that edits are permitted only within predefined time windows. For example, Li et al. [S37] introduced a time-lock mechanism to restrict transaction rewriting. Huang et al. [S8] proposed the Time-Updatable Chameleon Hash (TUCH), where the randomness for collision generation expires periodically, rendering the trapdoor valid only within a time window Δt.

5.2   T2 Redaction Process Optimization

Redaction process optimization tackles centralisation risks, attack surfaces, performance overhead, and privacy concerns introduced by redaction functionality.

T2.1 Decentralization Support: In the basic CHRB model, both trapdoor and permission management rely on a trusted entity, which contradicts the trustless design principle of blockchain. This topic leverages distributed technologies to eliminate single points of trust, and comprises three tertiary topics:

T2.1.1 Decentralized Trapdoor Management: This topic aims to enable distributed collaboration for trapdoor generation and storage. For example, Li [S18] proposed Non-Interactive Threshold Chameleon Hashing (NITCH), where trapdoor shares are independently generated by committee members using the Distributed Key Generation (KDG) protocol. Ateniese et al. [S1] applied secret sharing techniques to split the trapdoor and store the shares across multiple nodes in a distributed manner.

T2.1.2 Decentralized Permission Management: This topic focuses on decentralizing access control decisions across multiple entities. For instance, Ma et al. [S15] employed Multi-Authority Attribute-Based Encryption (MA-ABE) to enable cross-domain permission management without a central coordinator. Tian et al. [S29] used threshold secret sharing to distribute attribute keys among committee members, who must collaborate to reconstruct the key and grant access.

T2.1.3 Incentive: This topic addresses the lack of motivation for nodes to participate in redaction and verification tasks in decentralized settings. For example, Xu et al. [S16] embedded the version verification of the redacted blocks into the consensus rules, mandating miners to update modification records, with noncompliance resulting in the loss of mining rewards. Zhao et al. [S35] developed an incentive model combining reputation mechanisms and game-theoretic analysis to economically incentivize compliant redaction behavior.

T2.2 Security Guarantee: The integration of chameleon hash functions and ledger redaction mechanisms, while enhancing the compliance of CHRB systems, also introduces new security threats. This topic focuses on CHRB-specific security challenges and addresses potential risks through innovative supplementary mechanisms. It consists of three tertiary topics:

T2.2.1 Resistance to Trapdoor Exposure: As the core secret tied to redaction authority, the confidentiality of the chameleon hash trapdoor is critical. This topic focuses on mitigating exposure risks during transmission and computation. For instance, in CHRB systems based on CHET (Chameleon Hash with

Ephemeral Trapdoors) [S38], ephemeral trapdoors are generated for each transaction to limit the impact of potential trapdoor leakage. Wu et al. [S30] adopted multi-party computation (MPC) protocols to enable collaborative computation without revealing trapdoors. Dai et al. [S25] proposed a design where trapdoors are made public, relying on consensus verification to ensure secure redaction.

T2.2.2 Resistance to Unknown Versions: Repeated redactions on the same block can lead to version conflicts and compromise global ledger consistency. This topic emphasizes efficient version control to synchronize the latest state across the network. For example, Jia et al. [S14] employed RSA accumulators to aggregate the latest information of all redacted blocks. Tian et al. [S20] constructed a blockchain authentication tree using aggregatable vector commitments, ensuring state consistency through dynamic root commitment updates.

T2.2.3 Resistance to Quantum Attacks: Most existing chameleon hash constructions rely on assumptions such as the Discrete Logarithm Problem (DLP) or RSA, making them susceptible to future quantum attacks. This topic explores post-quantum secure designs by building chameleon hash functions based on hard problems like Short Integer Solution (SIS) and Learning With Errors (LWE). For instance, Li et al. [S24] proposed a tagged chameleon hash (tCH) scheme based on lattice-based cryptography and used it to design a post-quantum secure redactable blockchain.

T2.3 Performance Improvement: Existing CHRB solutions often suffer from significant computational and communication overhead due to the integration of cryptographic primitives and distributed protocols. This topic aims to improve execution and verification efficiency by optimizing the core redaction process. For example, Guo et al. [S17] employed off-chain computation techniques to outsource costly operations, such as bilinear pairings, to external cloud servers. Shen et al. [S19] combined universal accumulators with non-interactive succinct proofs to achieve efficient version control and block verification.

T2.4 Privacy Protection: The explicit linkage between redaction actions and participant identities may result in privacy breaches and targeted attacks. This topic focuses on anonymity-preserving techniques that prevent redaction activities from being traced back to specific entities. For example, Panwar et al. [S4] proposed an anonymous redaction framework combining dynamic group signatures with zero-knowledge proofs. Huang et al. [S8] introduced a linkable-and-redactable ring signature scheme, allowing miners to perform compliant redactions while preserving anonymity within a dynamic public key ring.

5.3   T3 Post-redaction Audit

Post-redaction audit aims to establish a comprehensive regulatory framework for CHRB systems, ensuring the transparency, traceability, and accountability of redaction operations.

T3.1 Traceability: In the basic CHRB model, only the final redacted result is retained, resulting in the loss of critical audit information generated during the redaction process. This topic utilizes logging and version control mechanisms to achieve end-to-end tracking. For example, in the scheme proposed by Zhang et al. [S34], editors are required to submit signed redaction proposals prior to modification, enabling non-repudiation of operations. Huang et al. [S28] introduced a traceable ring signature (TRS) scheme that preserves editor privacy under normal conditions while allowing authorized parties to reveal identities when necessary, thus balancing privacy protection and regulatory compliance.

T3.2 Accountability: The absence of punitive mechanisms in the basic CHRB model allows malicious nodes to carry out unauthorized redaction operations at negligible cost. This topic focuses on constructing mechanisms for responsibility attribution and punishment, enabling penalty mechanisms such as monetary fines or permission revocation against non-compliant participants. For instance, Xu et al. [S10] proposed a deposit-based mechanism in which smart contracts automatically confiscate the violator's funds. Zhao et al. [S35] designed a behavior evaluation framework for CHRB that dynamically links redaction actions with node reputation, automatically triggering permission downgrades or node disqualification when violations are detected.

Key finding of RQ1: Early studies (before 2022) focused on foundational functionalities such as permission control and core process design, while recent research (since 2023) has shifted toward decentralized architectures, security enhancement, and auxiliary mechanisms (e.g., incentives and accountability), indicating a transition from functional verification to system-level optimization.

6   RQ2: Adopted Techniques

By analyzing the contributions, preliminaries, and schemes of the included studies, we identify 21 major techniques in CHRB research, as shown in Table 3. These techniques are introduced from two perspectives: cryptographic tools and mechanisms, and distributed protocols and mechanisms.

6.1   Cryptographic Tools and Mechanisms

Cryptographic tools and mechanisms primarily focus on leveraging the core security properties (such as confidentiality, non-repudiation, etc.) provided by cryptographic primitives to directly achieve specific security objectives or functionalities in CHRB. These technologies typically serve as foundational elements for secure protocols, with their security grounded in standard computational hardness assumptions. This category comprises the following 10 techniques.

A) Digital Signature (DS): A signature tool based on asymmetric cryptography used to ensure message authenticity and non-repudiation. In CHRB, DS is applied in *T1.1 Authorization-based Redaction* [S7, S13] for verifying editor identity, and in *T3.1 Traceability* [S3, S25] for tracking malicious modifications. Notably, several variants of DS have also been applied in existing studies: Ring Signatures (RS) [S8, S17, S28] and Group Signatures [S4] provide anonymity; Policy-based Sanitizable Signature (P3S) [S7] supports dynamic binding and verification of field-level access policies; Identity-Based Signature (IBS) [S22] simplifies key management by allowing direct use of identity identifiers for authoriza-

Table 3. Major Techniques in CHRB.

Categories	Techniques	Topics and Studies
Cryptographic Tools and Mechanisms	Digital Signature	T3.1(S3,S4,S6,S7,S10,S11,S14,S15,S17, S22,S25,S26,S31,S34,S40); T1.1(S7,S13); T2.4(S4,S8,S17,S28); T2.1.2(S18,S30)
	Attribute-Based Encryption	T1.1(S2,S3,S4,S5,S11,S12,S15,S17,S21, S22,S28,S29,S33,S38,S39); T3.1(S3,S29); T2.1.2(S5,S15,S28,S33)
	Ephemeral Trapdoor	T1.2(S2,S3,S4,S5,S6,S11,S12,S15,S17, S21,S23,S28,S29,S33,S38,S39); T2.2.1(S1,S19,S31,S38)
	Accountable Assertion	T1.2 & T3.1(S37)
	Zero-Knowledge Proof	T2.2.1(S1,S13,S30,S37,S38); T2.4(S4,S7)
	Traitor Tracing	T3.1(S12,S22,S28)
	Vector Commitment	T1.3(S13); T2.2.2 & T2.3(S20)
	Accumulator	T2.2.2(S14,S19,S36,S39)
	Merkle Hash Tree	T2.2.2(S16)
	Succinct Proof	T2.3(S19)
Distributed Protocols and Mechanisms	Distributed Key Generation	T2.1.1(S9,S14,S18); T2.1.2(S40)
	Secret Sharing	T2.1.1(S1,S14,S23,S30,S31,S34,S36,S38); T2.1.2(S17,S29); T2.4(S32)
	Multi-party Computation	T2.1.1(S1,S9); T2.2.1(S30)
	Committee	T2.1.2(S11,S18,S25,S29,S34,S40)
	Consistent Hash Ring	T2.1.2(S11)
	Distributed Random Beacon	T2.1.2(S18)
	Reputation/Activity	T2.1.2 & T3.2(S11,S25,S34,S35); T2.1.3(S35)
	Deposit	T3.2(S10)
	Privileged Token	T1.3(S10,S11); T1.4(S10)
	Time Lock	T1.4(S10,S37)
	Off-chain Computation	T2.3(S12,S17,S27); T2.4(S27)

tion; Threshold Signature (TS) [S18, S30] enables distributed threshold control, ensuring that redactions require collaboration among multiple parties.

B) Attribute-Based Encryption (ABE): An attribute-based access control encryption tool that allows decryption rights to be dynamically defined based on user attributes. In CHRB, ABE is primarily used to encrypt trapdoors, thereby supporting the dynamic assignment of redaction rights in *T1.1 Authorization-based Redaction* [S2, S4]. Notably, several variants of ABE have also been applied in existing studies: ABE with Traceability (ABET) [S3, S29] enables malicious users tracing; Multi-Authority ABE (MA-ABE) [S5, S15] and Decentralized ABE (D-ABE) [S28, S32] provide federated or decentralized permission management; Matchmaking ABE (MABE) [S12] supports dynamic permission binding and bidirectional policy enforcement; Revocable ABE (RABE) [S21] supports dynamic revocation of permissions.

C) Ephemeral Trapdoor (ET): A trapdoor protection mechanism in chameleon hash schemes that allows collision generation using temporary trapdoors derived from a master trapdoor without exposing the master itself. In CHRB, ET is primarily applied to enforce fine-grained access control in *T1.2*

Table 4. List of included primary studies

ID	Title
S1	Redactable blockchain-or-rewriting history in bitcoin and friends
S2	Fine-grained and controlled rewriting in blockchains: chameleon-hashing gone attribute-based
S3	Policy-based chameleon hash for blockchain rewriting with black-box accountability
S4	ReTRACe: Revocable and traceable blockchain rewrites using attribute-based cryptosystems
S5	Redactable transactions in consortium blockchain: controlled by multi-authority CP-ABE
S6	Redactable blockchain supporting supervision and self-management
S7	Fine-grained and controllably redactable blockchain with harmful data forced removal
S8	Scalable and redactable blockchain with update and anonymity
S9	Quantum resistant key-exposure free chameleon hash and applications in redactable blockchain
S10	K-time modifiable and epoch-based redactable blockchain
S11	CDEdit: Redactable blockchain with cross-audit and diversity editing
S12	One-time rewritable blockchain with traitor tracing and bilateral access control
S13	Redactable blockchain with k-time controllable editing
S14	Redactable blockchain from decentralized chameleon hash functions
S15	Redactable blockchain in decentralized setting
S16	SEREDACT: Secure and efficient redactable blockchain with verifiable modification
S17	Online/offline rewritable blockchain with auditable outsourced computation
S18	Wolverine: A scalable and transaction-consistent redactable permissionless blockchain
S19	Verifiable and redactable blockchains with fully editing operations
S20	VRBC: A verifiable redactable blockchain with efficient query and integrity auditing
S21	Revocable policy-based chameleon hash for blockchain rewriting
S22	Accountable and fine-grained controllable rewriting in blockchains
S23	Fine-grained redactable blockchain using trapdoor-hash
S24	Tagged chameleon hash from lattices and application to redactable blockchain
S25	PRBFPT: A practical redactable blockchain framework with a public trapdoor
S26	Controlled redactable blockchain based on T-times chameleon hash and signature
S27	Redact4Trace: A solution for auditing the data and tracing the users in the redactable blockchain
S28	DARB: Decentralized, accountable and redactable blockchain for data management
S29	Accountable fine-grained blockchain rewriting in the permissionless setting
S30	Redactable consortium blockchain based on verifiable distributed chameleon hash functions
S31	Redactable blockchain-based secure and accountable data management
S32	NANO: Cryptographic enforcement of readability and editability governance in blockchain databases
S33	Secure redactable blockchain with dynamic support
S34	Dynamic trust-based redactable blockchain supporting update and traceability
S35	Concordit: A credit-based incentive mechanism for permissioned redactable blockchain
S36	Redactable blockchain from decentralized chameleon hash functions, revisited
S37	Message control for blockchain rewriting
S38	Redactable blockchain supporting rewriting authorization without trapdoor exposure
S39	Resilient and redactable blockchain with two-level rewriting and version detection
S40	PRBC: A practical redactable blockchain incorporating chameleon hash functions with attribute control

Fine-grained Redaction [S6, S21], and to enhance *T2.2.1 Resistance to Trapdoor Exposure* by limiting the impact scope of any leakage [S1, S31].

D) Accountable Assertion (AA): A cryptographic primitive that ensures both confidentiality and extractability. It allows entities to generate verifiable, concealed commitments to assertions and supports extraction of keys or identity information in case of contradictory claims for accountability. In CHRB, AA is used to bind redaction rights to specific transactions or fields, thereby supporting *T1.2 Fine-grained Redaction*, and provides non-interactive traceability capabilities for *T3.2 Accountability* [S37].

E) Zero-Knowledge Proof (ZKP): A verification tool that allows a prover to convince a verifier of a statement's truth without revealing any additional information. In CHRB, ZKP is primarily used in *T2.2.1 Resistance to Trapdoor Exposure* to conceal trapdoor information during collision computation [S30, S37], and in *T2.4 Privacy Protection* to protect the identity of participants involved in redaction [S4, S7].

F) Traitor Tracing (TT): A mechanism for identifying and tracking insider malicious users who illegally distribute keys or content. In CHRB, TT is primarily applied in *T3.1 Traceability* to identify the responsible party behind unauthorized redaction actions [S12, S22].

G) Vector Commitment (VC): A cryptographic tool that allows a short commitment to a fixed-length vector. The committer can bind the entire vector and generate verifiable proofs for specific elements. In CHRB, VC is mainly used for quota management in *T1.3 Quota-based Redaction* [S13], version control in *T2.2.2 Resistance to Unknown Versions* [S20], and verification acceleration in *T2.3 Performance Improvement* [S20].

H) Accumulator: A cryptographic primitive that compresses a set of elements into a fixed-size accumulator value. It supports membership proofs for each element, without the need to store the entire original set. In CHRB, Accumulator is mainly used in *T2.2.2 Resistance to Unknown Versions* to verify whether a block is the most recently version [S14, S19].

I) Merkle Hash Tree (MHT): A tree-structured hash verification tool that organizes data elements as leaf nodes and computes parent hashes up to a unique root. It allows efficient verification of data existence and integrity. Similar to accumulators, MHTs are mainly used for latest version verification [S16].

J) Succinct Proof (SP): A verification tool that compresses the verification process of complex computations into a fixed-size proof, enabling verifiers to confirm correctness in a very short time without re-executing the original computation. In CHRB, SP is primarily used in *T2.3 Performance Improvement* to optimize the efficiency of querying and verifying the latest version [S19].

6.2 Distributed Protocols and Mechanisms

Distributed protocols and mechanisms primarily focus on how to enable collaboration among participating nodes through the design of specific interaction rules, processes, and organizational structures to address coordination, trust, efficiency, and security challenges in distributed environments. These technologies depend on message passing and consensus formation among nodes, with their security and effectiveness often built upon cryptographic tools while also considering node behaviors (honest or Byzantine) and network models (synchronous or asynchronous). This category includes the following 11 techniques.

A) Distributed Key Generation (DKG): A distributed protocol where multiple parties collaboratively generate a key without any single party having complete control. In CHRB, DKG is primarily used in *T2.1.1 Decentralized Trapdoor Management* and *T2.1.2 Decentralized Permission Management* to enable the distributed generation of trapdoors [S9, S14] or attribute keys [S40].

B) Secret Sharing (SS): A distributed protocol that splits a secret into multiple shares and distributes them among participants, such that the original secret can only be reconstructed when a predefined threshold is met. In CHRB, SS is primarily used in *T2.1.1 Decentralized Trapdoor Management* and *T2.1.2 Decentralized Permission Management* for the shared storage and fault-tolerant protection of trapdoors [S1, S36] or attribute keys [S17, S29].

C) Multi-party Computation (MPC): A distributed computing protocol that enables multiple parties to jointly compute a function over private inputs without revealing them. In CHRB, MPC is primarily used in *T2.1.1 Decentralized Trapdoor Management* [S1, S9] to securely and decentrally perform trapdoor-related computations, and in *T2.2.1 Resistance to Trapdoor Exposure* [S30] to prevent leakage of trapdoor information.

D) Committee: A distributed governance mechanism in which a group of nodes is elected to represent the network in executing critical tasks, thereby improving efficiency and reducing the cost of full-node participation. Committee is used in *T2.1.2 Decentralized Permission Management* for collective voting on redaction permissions and authorization of redaction operations [S25, S40].

E) Consistent Hash Ring (CHR): A distributed protocol that uses virtual ring mapping to achieve dynamic load balancing, minimizing data migration when nodes join or leave. In CHRB research, CHR is mainly applied in *T2.1.2 Decentralized Permission Management* for dynamic management of committee membership [S11].

F) Distributed Random Beacon (DRB): A protocol for generating unpredictable and publicly verifiable random numbers in a distributed manner. In CHRB, DRB is primarily used in *T2.1.2 Decentralized Permission Management* to ensure fair selection of committee members and resist collusion attacks [S18].

G) Reputation/Activity: A dynamic scoring mechanism that evaluates participant reliability based on their historical behavior. In CHRB, Reputation/Activity is primarily used in *T2.1.2 Decentralized Permission Management* to select reliable committee members [S25, S34], as well as in *T2.1.3 Incentive* and *T3.2 Accountability* to provide internal incentives and punishments [S35].

H) Deposit: An economic guarantee mechanism in which participants stake assets as collateral, which may be slashed upon misconduct. In CHRB, Deposit is used in *T3.2 Accountability* to penalize malicious redaction behavior [S10].

I) Privileged Token (PT): A digital credential or security token used to grant holders specific access rights or operational privileges in decentralized systems. In CHRB, PT is primarily employed in *T1.3 Quota-based Redaction* and *T1.4 Time-bound Redaction* to regulate the number [S11] and validity period [S10] of redaction actions.

J) Time Lock (TL): A smart contract mechanism that delays the execution of certain operations by enforcing a mandatory waiting period. In CHRB, TL is used in *T1.4 Time-bound Redaction* to constrain the valid time window for redaction operations [S10,S37].

K) Off-chain Computation (OC): A technique that offloads computations to off-chain systems while only submitting final results or essential proofs to the blockchain. In CHRB, OC is primarily employed in *T2.3 Performance Improvement* to efficiently generate collisions and redaction proofs [S12, S17], as well as in *T2.4 Privacy Protection* to safeguard users' private data [S27].

Key finding of RQ2: (1) Existing CHRB schemes typically integrate multiple cryptographic tools and distributed protocols to enhance system security and functional completeness; (2) However, the combined use of diverse algorithms and protocols introduces significant computational and communication overhead, which becomes a major performance bottleneck in CHRB systems.

7 Discussion and Validity Analysis

7.1 Implications to Academia

This study identifies several key research topics and technical directions in CHRB and summarizes the core design features of existing solutions. However, notable challenges remain in state consistency, scalability, and incentive mechanisms.

First, enabling ledger redaction increases the risk of inconsistent states across nodes. While accumulators and vector commitments support for version control, frequent or asynchronous redactions can still undermine system reliability. Future research could explore embedded version metadata, synchronization strategies aligned with consensus, and lightweight broadcasting to enhance consistency.

Second, the integration of various cryptographic tools and distributed protocols introduces substantial overhead, limiting practical deployment. To address this, modular architectures, succinct proof systems, and hardware acceleration could be explored to balance functionality and performance.

Finally, existing works offer limited enforceable penalties against malicious redactions, and redaction tasks impose non-trivial costs on participating nodes. Without proper incentives, rational nodes may escape responsibility. Although deposit and reputation mechanisms have been introduced, their economic and game-theoretic designs remain preliminary. Future work should develop robust incentive and accountability frameworks that integrate behavior evaluation, economic rewards, and automated enforcement.

7.2 Threats to Validity

During the search phase, this study used Scopus as the sole database rather than incorporating multiple digital libraries. However, Scopus already includes most major sources recommended by the guidelines of Kitchenham et al. [8], and only three additional eligible studies were identified through backward and forward snowballing, suggesting that the search coverage was relatively comprehensive. Regarding the selection criteria, to ensure the quality of the included studies while balancing workload, we limited the scope to journals and conferences listed in the CCF ranking. Although this restriction may have led to the omission of some relevant works from non-CCF venues, it effectively improved the reliability of our analysis. In future work, we plan to remove this restriction and further expand the literature sources to enhance the generalizability of our conclusions. To minimize researcher bias, we adopted the following measures: (1) Each paper was independently reviewed by at least two researchers; (2) a predefined data extraction form was used to standardize the analysis process; and (3) regular group meetings were held to resolve disagreements by consensus. These measures significantly enhanced the objectivity of our classification and conclusions.

8 Conclusion

This paper presents a comprehensive analysis of CHRB through a systematic literature review from two dimensions: research topics and core technologies. By reviewing 40 high-quality primary studies, we identify three main research topics: pre-redaction permission control, redaction process optimization, and post-redaction audit. Additionally, we summarize 21 key technologies, including cryptographic tools (e.g., attribute-based encryption, zero-knowledge proofs) and distributed protocols (e.g., secret sharing, multi-party computation). Based on these findings, we construct a CHRB technical framework that illustrates the relationships among research topics, adopted techniques, and representative studies, offering a structured reference for future research and development.

Future work will include (1) expanding the scope to cover all solution approaches in redactable blockchains; (2) broadening literature sources to incorporate non-CCF-ranked publications and relevant gray literature; (3) explicitly linking threat models, security properties, and techniques to support scheme selection; and (4) discussing specific optimization measures in the implementation of CHRB to address real-world deployment challenges.

Acknowledgments. This work is supported by the National Natural Science Foundation of China (No.62302210, No.62572237), the Jiangsu Provincial Key Research and Development Program (No.BE2021002-2), the Natural Science Foundation of Jiangsu Province (No.BK20241195), and the Innovation Project and Overseas Open Project of State Key Laboratory for Novel Software Technology (Nanjing University) (ZZKT2025A12, ZZKT2025B18, ZZKT2025B20, ZZKT2025B22, KFKT2025A17, KFKT2025A19, KFKT2025A20, KFKT2024A02, KFKT2024A13, KFKT2024A14, KFKT2023A09, KFKT2023A10).

[Primary studies selected for this SLR.]

(See Table 4).

References

1. Abd Ali, S.M., Yusoff, M.N., Hasan, H.F.: Redactable blockchain: comprehensive review, mechanisms, challenges, open issues and future research directions. Future Internet **15**(1), 35 (2023)
2. Ahamed, N.N., Vignesh, R., Alam, T.: Tracking and tracing the halal food supply chain management using blockchain, rfid, and qr code. Multimedia Tools Appl. **83**(16), 48987–49012 (2024)
3. Ateniese, G., Magri, B., Venturi, D., Andrade, E.: Redactable blockchain-or-rewriting history in bitcoin and friends. In: 2017 IEEE European Symposium on Security and Privacy (EuroS&P), pp. 111–126. IEEE (2017)
4. Cai, T., Chen, W., Zhang, J., Zheng, Z.: Smartchain: a dynamic and self-adaptive sharding framework for iot blockchain. IEEE Trans. Serv. Comput. (2024)
5. Cruzes, D.S., Dyba, T.: Recommended steps for thematic synthesis in software engineering. In: 2011 International Symposium on Empirical Software Engineering and Measurement, pp. 275–284. IEEE (2011)
6. Derler, D., Samelin, K., Slamanig, D., Striecks, C.: Fine-grained and controlled rewriting in blockchains: chameleon-hashing gone attribute-based. In: 26th Annual Network and Distributed System Security Symposium, NDSS 2019 (2019)
7. Fathalla, E., Wang, C., Li, X., Gazda, R., Wu, H.: Redactable distributed ledgers: a survey. Distrib. Ledger Technol. Res. Pract. **2**(3), 1–26 (2023)
8. Kitchenham, B.A., Charters, S.: Guidelines for performing systematic literature reviews in software engineering. Technical report. Technical report, ver. 2.3 ebse technical report. ebse (2007)
9. Krawczyk, H.M., Rabin, T.D.: Chameleon hashing and signatures. US Patent 6,108,783 (2000)
10. Li, X., Rong, W., Shi, H., Tang, J., Xiong, Z.: The impact of conference ranking systems in computer science: a comparative regression analysis. Scientometrics **116**, 879–907 (2018)
11. Li, Y., Liu, S.: Tagged chameleon hash from lattices and application to redactable blockchain. In: IACR International Conference on Public-Key Cryptography, pp. 288–320. Springer, Heidelberg (2024)
12. Politou, E., Casino, F., Alepis, E., Patsakis, C.: Blockchain mutability: challenges and proposed solutions. IEEE Trans. Emerg. Top. Comput. **9**(4), 1972–1986 (2019)
13. Vidal, F.R., Ivaki, N., Laranjeiro, N.: Revocation mechanisms for blockchain applications: a review. In: 2021 10th Latin-American Symposium on Dependable Computing (LADC), pp. 01–10. IEEE (2021)
14. Wang, W., Peng, H., Duan, J., Wang, L., Hu, X., Zhao, Z.: Resilient and redactable blockchain with two-level rewriting and version detection. IEEE Trans. Inf. Forensics Secur. **20**, 1163–1175 (2025)
15. Wu, H., Tang, Y., Shen, Z., Tao, J., Lin, C., Peng, Z.: Telex: two-level learned index for rich queries on enclave-based blockchain systems. IEEE Trans. Knowl. Data Eng. (2025)
16. Wu, X., Du, X., Yang, Q., Wang, N., Wang, W.: Redactable consortium blockchain based on verifiable distributed chameleon hash functions. J. Parallel Distrib. Comput. **183**, 104777 (2024)
17. Ye, T., Luo, M., Yang, Y., Choo, K.K.R., He, D.: A survey on redactable blockchain: challenges and opportunities. IEEE Trans. Netw. Sci. Eng. **10**(3), 1669–1683 (2023)

18. Yu, D., Xiang, B.: Discovering topics and trends in the field of artificial intelligence: using LDA topic modeling. Expert Syst. Appl. **225**, 120114 (2023)
19. Zhang, D., Le, J., Lei, X., Xiang, T., Liao, X.: Exploring the redaction mechanisms of mutable blockchains: a comprehensive survey. Int. J. Intell. Syst. **36**(9), 5051–5084 (2021)
20. Zhang, H., Babar, M.A., Tell, P.: Identifying relevant studies in software engineering. Inf. Softw. Technol. **53**(6), 625–637 (2011)
21. Zhou, X., et al.: Federated distillation and blockchain empowered secure knowledge sharing for internet of medical things. Inf. Sci. **662**, 120217 (2024)

GP-BFT: A Low-Latency BFT Protocol Under Grouped Multi-committee Parallel Consensus

Changsong Yang[1,2], Shukang Wei[1,2], Yujue Wang[3(✉)], Hai Liang[1,2], and Donglin Yao[1,2]

[1] Guangxi Key Laboratory of Cryptography and Information Security, Guilin University of Electronic Technology, Guilin, China
[2] Guangxi Engineering Research Center of Industrial Internet Security and Blockchain, Guilin University of Electronic Technology, Guilin, China
[3] Hangzhou Innovation Institute of Beihang University, Hangzhou, China
`wyujue@buaa.edu.cn`

Abstract. The Byzantine Fault Tolerance State Machine replication (BFT-SMR) protocol ensures the correct operation of distributed systems in the presence of faulty or malicious replicas. Traditional Byzantine protocols typically employ a single-leader model to drive consensus, whereas this model exhibits significant performance bottleneck: (1) The single-leader model requires all replicas to participate in the consensus, resulting in increased communication overhead among replicas and higher consensus latency; (2) The single-leader model incurs a communication cost of $O(n^2)$ during the timeout phase. To address these issues, this paper proposes a BFT-SMR protocol, GP-BFT, based on multi-committee parallel consensus. GP-BFT initially divides online replicas randomly into multiple committees, forming a committee set, and follows a 2-chain commit mechanism to achieve parallel consensus among committees. By transitioning from full-replica consensus to intra-committee consensus, GP-BFT reduces commit latency while ensuring safety and liveness. Experimental results demonstrate that in scenarios with large-scale replicas and transactions, GP-BFT maintains high throughput with lower latency.

Keywords: BFT-SMR · Multi-Committee · Parallelism · Large-scale replicas · Low-latency

1 Introduction

Since the emergence of Bitcoin [19], blockchain technology has garnered widespre-ad attention from both academia and industry. Blockchain is a distributed ledger characterized by decentralization, transparency, security, anonymity, and immutability [21,22]. The rise of blockchain has driven research into the efficiency and stability of its underlying Byzantine Fault Tolerance

H. Liu et al. (Eds.): ICA3PP 2025, LNCS 16381, pp. 322–338, 2026.
https://doi.org/10.1007/978-981-95-8399-7_18

(BFT) [6] protocols. BFT protocols ensure the security of distributed systems by tolerating malicious replicas or Byzantine faults [16]. In infrastructure such as the financial industry, which requires rapid response, low-latency BFT protocols are fundamental to guaranteeing high reliability and real-time performance.

However, traditional BFT protocols frequently adopt a single-leader model [1,4,20]. In the stable state, the leader is responsible for proposing, and the protocol progresses through verification and negotiation involving all replicas. As the scale of the network increases, the communication overhead among replicas experiences a marked increase, leading to a significant rise in the latency of BFT protocols. In the event of timeout handling being triggered, the failure of replicas to process messages within the message delivery delay, denoted by the variable Δ, results in the leader switching rounds via a timeout mechanism to ensure liveness. In large-scale networks, this further degrades the performance of BFT protocols.

Therefore, the development of a low-latency BFT protocol that is suitable for large-scale scenarios is imperative. Nevertheless, the reduction of protocol latency gives rise to the following challenges:

- **Bottleneck of Replica Count.** Consensus latency depends on communication delay among replicas. BFT protocols require $2f + 1$ replicas to exchange messages like blocks and proposals for consensus. Consequently, the protocol's latency increases as the number of replicas in the network grows.
- **Estimation of Message Delay** Δ. BFT protocols require replicas to deliver messages within time Δ. Failure triggers a timeout to ensure liveness. However, estimating Δ is challenging: if too large, replicas wait excessively; if too small, frequent timeouts increase overhead. Consequently, inaccurate estimation increases both latency and communication costs.

1.1   Our Contributions

Single-leader consensus protocols have insignificant performance bottleneck in small-scale networks due to the small number of replicas and low commit delays. Although timeouts are difficult to accurately estimate, the impact is limited due to fast network transmission. However, in large-scale networks, these problems become prominent. For this reason, this paper proposes GP-BFT, a low-latency BFT protocol based on multi-committee parallel consensus, with the following main contributions.

1) This paper proposes a multi-committee parallel proposal scheme, which divides all replicas into multiple committees for parallel consensus, reducing the number of replicas in each committee, lowering communication overhead among replicas, decreasing overall protocol latency, and improving system performance.
2) This paper proposes a fast synchronization scheme for timeout replicas, traded replicas. By leveraging the characteristics of multi-committee parallel consensus, it utilizes committees that have completed proposals to enable replicas in timeout committees to quickly complete block commitments, avoiding

the overhead of view switching in timeout committees and reducing system latency.

3) Successfully designed and implemented GP-BFT, with experiments demonstrating that it is a low-latency, high-throughput BFT protocol suitable for large-scale network environments.

1.2 Organization

The remainder of this paper is organized as follows. Section 2 introduces related work. Section 3 presents the system model and basic theory related to this paper. Section 4 describes the design ideas of the protocol in detail and analyzes its liveness and safety. Section 5 reports the experimental evaluation results. Finally, this paper concludes in Sect. 6.

2 Related Works

Synchronous BFT protocol assumes an upper bound on message latency of Δ and requires the replica to wait for the leader proposal within Δ. PiLi [9] and Streamlet [8] optimized the latency by simplifying the locking mechanism and introducing additional communication phases to avoid complex view changes and to ensure security and activation. Sync HotStuff [1] optimizes the latency by adopting an asynchronous locking mechanism, with a stable state latency is $2\Delta + \delta$ (δ is an upper bound on the actual latency), a design that effectively balances performance and fault tolerance in synchronous environments. Shrestha et al. [2] proposed a rotating leader protocol with an optimal latency of $\Delta + \delta$, but it does not support fault tolerance, which restricts its applicability in real distributed systems.

In addition, Hybrid-BFT [18] optimizes the decision latency through optimistic responsiveness, with an average latency of 3δ in optimistic conditions and $1.5\delta + O(\delta)$ in the worst case. Hamster [10] protocol is the first to introduce coding into a synchronous BFT protocol, with throughput growing linearly with the number of replicas. The throughput grows linearly with the number of replicas, which is better than Sync HotStuff [1] and suitable for high load scenarios. Guo et al. [14] proposed a weak synchronization model, which laid the foundation for the subsequent research. Based on this, Huang et al. [15] designed a leaderless weakly synchronous BFT protocol, which adopts the leaderless model to balance the load between replicas and improve robustness and scalability.

Partial synchronization model assumes that the network may be temporarily out of synchronization but resumes synchronization after an unknown global stability time (GST), which is closer to the actual network but requires more communication rounds. PBFT [7], as a benchmark for BFT protocols, has a normal communication complexity of $O(n^3)$ and anomalous time up to $O(n^4)$ through three-step consensus. Subsequent studies such as Tendermint [5], FBFT [17], AAR [3], Wendy [13] optimized the consensus structure to reduce the communication overhead. Mainstream protocol HotStuff [24] used parallel pipeline and linear view change to reduce communication cost. Jolteon [12] improved throughput and reduces latency by optimizing 3-chain proposal mechanism.

Additionally, Gao et al. [11] used a randomly selected master replica mechanism to dramatically reduce network communication and introduced a large number compression algorithm to improve security and encryption efficiency when message types are large. Yao et al. proposed a leaderless BFT protocol, DyBFT [23], which effectively mitigates the performance degradation caused by the increase in the number of replicas and the transaction volume by maintaining a collection of local committees and using parallel path proposals and local submission of blocks without additional communication overheads.

These solutions can significantly reduce communication complexity and latency by simplifying processes, optimizing view transitions, enhancing parallelism, and introducing new mechanisms. They made BFT protocols more efficient and adaptable to actual network requirements, providing more robust solutions for distributed systems. However, they do not adequately address latency issues in large-scale scenarios.

3   System Model and Definitions

3.1   Models

Consider a distributed system consisting of $n = 4f$ replicas, f represents the number of Byzantine replicas. In this system, Byzantine replicas can behave arbitrarily. Each replica has a unique identifier p_i, for $0 \leq i < n$. There is an adversary in the system who can control all the Byzantine replicas and coordinate their actions. In this model, every pair of replicas can communicate with each other. The adversary can arbitrarily delay message delivery, but it is assumed that the message delivery between honest replicas is reliable, and honest replicas are able to exchange state information with each other, ensuring that messages between honest replicas will eventually reach their destinations.

In the key-distribution module, assume the existence of a public-key infrastructure (PKI) and standard digital signatures. A replica i signature on a message m is denoted by $(m)_i$ and is called a signature share. Additionally, assume a t-threshold quorum signature scheme, when any t distinct signature shares $(m)_i$ on the same message m are collected, they can be combined into a single quorum signature of the same length. Furthermore, assume a collision-resistant hash function $H(x)$ that maps messages of arbitrary length to outputs of a fixed length. It is also assumed that the adversary's computational resources are limited, preventing it from breaking any of the above assumptions. Finally, assume the PKI is trusted and securely distributes key pairs to all replicas.

State Machine Replication (SMR) has been identified as a significant technique for resolving the replica consistency problem. In this model, each replica is implemented as a Deterministic Finite-State Machine (DFSM), thereby ensuring that the processing results for the same input remain consistent. The majority of BFT protocols are based on the concept of Byzantine Fault-Tolerant State

Machine Replication (BFT-SMR), which is a process that aims to achieve consensus. In the BFT-SMR model, client requests are submitted in the form of logs, and honest replicas are responsible for storing and replicating these logs. In order to guarantee the reliability of the system, BFT-SMR must satisfy the following two essential properties:

- **Safety**: Each honest replica can record only one transaction at the same log position. If any two non-faulty replicas submit transactions d and d' at the same log position, then it must hold that $d = d'$.
- **Liveness:** Every transaction proposed by a client will eventually be submitted and recorded in the logs by all honest replicas.

Furthermore, a BFT protocol is required to guarantee that the submitted transactions satisfy external validity requirements, that is, the transactions submitted by the replicas can be verified through specific predicates. In order to facilitate implementation, GP-BFT performs a pre-validation of the validity of transactions when replicas process messages.

3.2 Definitions

In order to facilitate the description of the protocol, specific terminology is employed throughout the execution process. Furthermore, authentic replicas will automatically discard messages that do not conform to the standard format.

- *Replicas, Committees, and Committee Sets*: During initialization, a fixed number of replicas are generated. Each replica is a triple [*id, pk, address*] containing unique *ID*, public key, and network address. Every 4 replicas form a committee, with all committees constituting the committee set $\mathbb{G} = \{g_1, \ldots, g_k\}$. Here, g_i denotes a committee formed by randomly partitioning replicas, and $|\mathbb{G}| = k$.
- *Level-1 Leader, Level-2 Leader*: During the protocol implementation process, leaders are divided into two categories, where level-1 leaders exist in each committee and are primarily responsible for driving consensus within the committee. level-2 leaders are elected in each round from a collection of all level-1 leaders.
- *Local timeout, Global Timeout*: Local timeout refers to a local timeout when a timeout occurs for a single committee in the committee collection. A global timeout occurs when the number of committees in which a timeout occurs exceeds half of the total number of committees in the committee set.
- *Round Number*: The protocol is advanced using a round iteration mechanism. When the system is initialized, the initial round is set to $r_{\text{init}} = 0$. Each replica in the committee maintains its local current round r_{cur}. Upon completion of the internal consensus, the replica increments its local round by 1. In the current round r, the committee g_i assigns its level-1 leader $L_{g_i}^r$.
- *Block Format*: Block $B = [ID,\ qc,\ round,\ txn]$. Here, ID is the hash $H(qc,\ round,\ txn,\ author)$ with *author* being the randomly elected level-2 leader, *round* indicates current round, *qc* is the quorum certificate of the

predecessor block, and *txn* represents batched transactions from the mempool.

- *Quorum Certificate (QC)*: In QC contains [*round, digest, signs*], where $round = B.round$ and $digest = B.ID$. The *signs* field summarizes the share of valid signatures from $2f + 1$ replicas. The latest generated QC is denoted as qc_{last}.
- *Timeout Certificate (TC)*: Aggregated from $2f + 1$ timeout message shares with structure [$qc_{last}, author, Vec_{timeout}$]. Here, qc_{last} is the latest local QC, *author* denotes the current round leader, and $Vec_{timeout}$ represents the timeout message set.
- *Request Synchronous Massage (ReqSyncMSG)*: Broadcast by the level-1 leader of the current timeout committee, containing a TC and the current timeout round. Recipients are level-1 leader of all committees.
- *Re-Commit Massage (ReCommitMSG)*: The *ReCommitMSG* will carry a block, and the replica of the *ReqSyncMSG* received will broadcast the block in the committee to resubmit the block and advance the round.

4   GP-BFT Protocol

4.1   Protocol Design

To overcome performance bottleneck in single-leader models, GP-BFT restructures the consensus architecture into a multi-committee grouping framework. The initialization phase creates n replicas, which are randomly partitioned into k committees that form $\mathbb{G}$. Under stable state, replicas execute three core operations: Block Generation, Block Vote, Block Commit. Algorithm 1 formalizes the stable state execution flow of GP-BFT.

Randomness generation employs a coin-threshold mechanism for two scenarios: intra-committee level-1 leader election and cross-committee level-2 leader election.

- **Intra-election**: committees maintain local random numbers, set coin thresholds. Public random numbers are generated when a replica receives $2f + 1$ valid shares.
- **Cross-election**: The leader set maintains global randomness with threshold $|\mathbb{G}|/2 + 1$. Exchanging shares during block submission generates randomness for level-2 leader election.

During block generation, level-1 leaders create blocks with author set to the current round's level-2 leader and broadcast them intra-committee.

During the block voting phase, the block is broadcast among all replicas and must receive votes from at least $2f + 1$ replicas to be considered valid. To ensure the security and consistency of the voting process, GP-BFT establishes two safety rules during this phase. First, the round of the block must be greater than the last voting round of the replica, preventing duplicate submissions of the same block. Second, the round of the QC contained in the block must be consecutive with the current block's round, i.e., satisfying $B.qc.round + 1 = block.round$, ensuring that the block carries its parent block's QC and adheres to the 2-chain commit rule, thereby enhancing block security. After satisfying these two rules, replicas will vote for the block in the current round r and send their votes to the level-1 leader of round $r + 1$ in their committee for aggregation, thus completing the voting process.

During the block commit phase, when processing block B in round r, it is first necessary to determine the two nearest ancestor blocks B_0 and B_1 of the block before voting. If B_0, B_1 and B are three consecutive blocks, a commit operation for block B_0 is triggered and stored locally. Subsequently, the share of signatures for block B_0 is broadcast. If more than half of the level-1 leaders recognize the commit request, the block commit is executed; otherwise, the commit operation is stopped and waits for the local timeout to be triggered.

Algorithm 1. GP-BFT protocol in stable state.

Local variables:

$\quad r_{cur} \leftarrow 1$, $qc_{last} \leftarrow genesisBlock.qc$, $commit_{last} \leftarrow 0$

1: **function** ADVANCEROUND(round)
2: $\quad$ *reset Timer* $\qquad\qquad\qquad\qquad\qquad\qquad\qquad$ ▷ Reset the timer for replica
3: $\quad r_{cur} \leftarrow round + 1$
4: **end function**
5: **procedure** QC(vote)
6: $\quad qc_{last} \leftarrow wait\ for\ 2f + 1\ valid\ votes$
7: $\quad$ **if** p_i *is next leader* **then** $\qquad\qquad\qquad\qquad$ ▷ p_i is the current replica
8: $\quad\quad Blcok \leftarrow GenerateBlock \qquad\qquad\qquad\qquad\qquad$ ▷ Embed qc_{last}
9: $\quad\quad BroadCast(Block)$
10: $\quad$ **end if**
11: $\quad$ AdvanceRound($qc_{last}.round$) $\qquad\qquad\qquad$ ▷ Advance the round with qc
12: **end procedure**
13: **procedure** VOTEANDSTORE(Block)
14: $\quad$ *Store Block* $\qquad\qquad\qquad\qquad\qquad$ ▷ Store block into the memory pool
15: $\quad$ **if** *find Block's ancestors* B_0, B_1 **then**
16: $\quad\quad$ *broadcast the* $(B_0)_{sig}$ *to other level-1 leaders*
17: $\quad$ **end if**
18: $\quad$ Vote(Block)
19: **end procedure**
20: **procedure** COMMIT(Block)
21: $\quad$ **if** *wait for* $\frac{G}{2} + 1$ *valid signature* AND $commit_{last} < block.round$ **then**
22: $\quad\quad Queue \leftarrow block.parent \qquad\qquad\qquad\qquad$ ▷ Election level-2 leader
23: $\quad$ **end if**
24: $\quad commit_{last} \leftarrow block.round$
25: $\quad$ *commit block from Queue*
26: **end procedure**

Figure 1 shows the complete process of GP-BFT in stable state. Following the consensus process, the black circle denotes the leader $L_{g_i}^r$ of the i group of committees at round r. In round $r + 1$, the committees g_i vote within the group on block B_r. The leader $L_{g_i}^r$ aggregates the votes to form the QC and collects more than half of the partial signatures by broadcast to complete the round. During this time, the primary leader of round $r + 1$ generates B_{r+1} and sets $B_{r+1}.author$ as the secondary leader for that round based on a random number after validation is complete and B_{r-1} is submitted. In round $r+2$, B_r is certified and the committees submit B_r, vote B_{r+1} and generate B_{r+2}, and finally submit B_r.

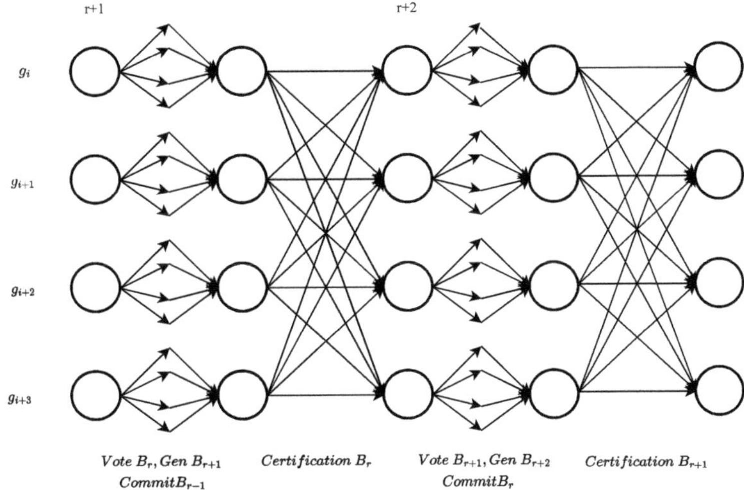

Fig. 1. Steady-State GP-BFT with 4 replicas per group, Featuring two-round block submission: voting on round r block, generation of round $r + 1$ block, commitment of round $r - 1$ block, and certification of round r block (Aggregated Vote).

During the timeout handling, network fluctuations may inevitably trigger timeouts when replica timers expire. Under the multi committee parallel consensus model, while multiple committees process requests concurrently, network instability can cause partial committees to experience local timeouts, whereas global timeout are statistically less probable. To resolve timeouts efficiently, committees that have completed round r consensus assist timeout committees in round r for rapid block submission. In the event of global timeout, the round advancement mechanism is activated immediately. Algorithm 2 illustrates the core pseudocode for GP-BFT in the case of timeouts.

In local timeout scenarios where a committee set triggers the timeout mechanism, the level-1 leader $L_{g_i}^r$ of the affected committee aggregates broadcast *timeout* messages into a TC, then broadcasts a *ReqSyncMSG* to other committees' current-round level-1 leaders. Recipients validate the attached QC, and upon successful verification, they send *ReCommitMSG* to restart block submission in the timeout committee. During global timeout, $L_{g_t}^r$ directly advances rounds using TC and updates $qc_{last} = TC.qc$ upon receiving $\frac{\mathbb{G}}{2} + 1$ *ReqSyncMSG*.

Algorithm 2. GP-BFT in timeout state.

1: **upon** *timer expired*
2:     *BroadCast(timeout)*   ▷ Intra-group broadcast
3: **procedure** HANDLETC(timeout)
4:     $TC \leftarrow$ *wait for* $2f + 1$ *timeout*
5:     **if** p_i *is current level-1 leader* **then**
6:         *BroadCast ReqSyncMSG(TC)*   ▷ broadcast to other level-1 leader
7:     **end if**
8: **end procedure**
9: **procedure** HANDLEREQSYNCMSG(*ReqSyncMSG*)
10:     *ReqSyncMSG Set* $\leftarrow$ *collect ReqSyncMSG*
11:     **if** *ReqSyncMSG Set.size* $= \frac{|\mathbb{G}|}{2} + 1$ **then**
12:         $UpdateQC(TC.qc_{last})$
13:         $AdvanceRound(TC.qc_{last}.round)$
14:     **else**
15:         $Block \leftarrow Verify(QC)$ ▷ from Mempool
16:         *Send ReCommitMSG(Block)*   ▷ send to $TC.author$
17:     **end if**
18: **end procedure**
19: **procedure** RECOMMIT(Block)
20:     **upon** *reciver ReCommitMSG*
21:         *BroadCast(Block)*
22:         *commit Blocks and its ancestors*
23: **end procedure**

Figure 2(a) shows that the committee g_{i+3} has a timeout in round $r + 1$, at this point in the validation of the B_r stage other level-1 leaders have collected more than half of the signatures, at this point will be in the $r + 2$ round of the submission of the block, for g_{i+3} the committee in the $r + 2$ round will be broadcasting a *timeout* message, by $L_{g_{i+3}}^{r+2}$ leaders aggregate to form a TC. During the verification phase of the block in round $r + 2$ the other level-1 leaders broadcast the block signature normally, at this time the timeout committee level-1 leader will broadcast *ReqSyncMSG*, at the same time the level-1 leader of the timeout committee will collect more than half of the signatures of the B_{r+1}, and receive the signatures of the other level-1 leaders in round $r + 3$. *ReCommitMSG* sent by the other leader will submit B_r and B_{r+1}. Figure 2(b) illustrates the process where a plenary timeout occurs, at which point less than half of the block signatures will be collected when the block is verified. In the round $r + 2$ each committee group will trigger a timeout aggregation TC so that the level-1 leaders will broadcast *ReqSyncMSG*, this process will collect more than half of the *ReqSyncMSG*, and thus the entire committee group will advance in the round.

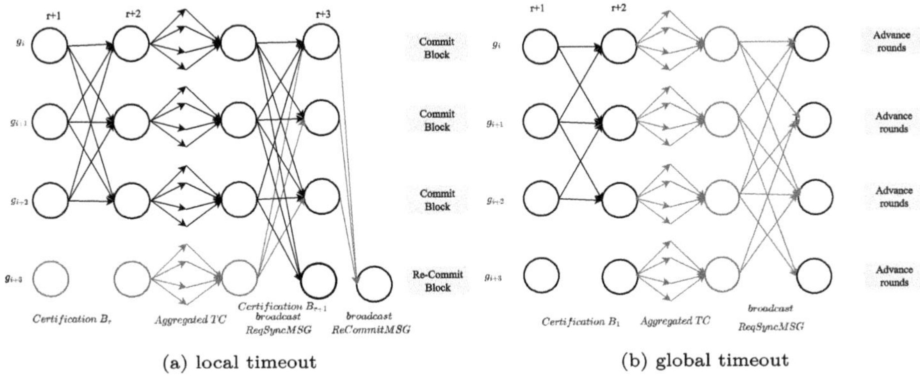

(a) local timeout   (b) global timeout

Fig. 2. Timeout handling mechanisms in GP-BFT: Local and global state. In the figure, red parts indicate timeout states, while black parts indicate stable states. (a) Local timeout state: The local timeout committee aggregates TC, and the stable state committee assists in block submission. (b) Global timeout state: After collecting more than half of *ReqSyncMSG* (including TC), the committee group advances the round. (Color figure online)

4.2 Correctness Analysis

Lemma 1. *If a global timeout occurs in round r, the set of committees must be able to proceed to round $r + 1$.*

Proof. The occurrence of a global timeout in round r means that a majority of the committees have suffered a local timeout. At this point, the remaining few committees, when attempting to submit a block, fail to do so because they are unable to collect a sufficiently large number of block signatures, and will eventually trigger a timeout as well. During the timeout processing phase, the Department will collect $\frac{|\mathbb{G}|}{2} + 1$ *ReqSyncMSG* and proceed directly to round advancement.

Lemma 2. *If no more than half of the committees in the round r experience a local timeout, then, surely, the block can be submitted.*

Proof. Let there be k committees in the round r that experience a local timeout $(k < |\mathbb{G}|/2)$. At this point, the remaining $|\mathbb{G}| - k$ committees (more than half of them) will broadcast their block signatures during the commit phase. With the majority of non-timed-out committees, these remaining committees are able to collect a sufficient number of signature shares and successfully commit the block. At the same time, the partially timed-out committees will quickly synchronize the submitted block by broadcasting *ReqSyncMSG*, and eventually all committees advance together to the round $r + 1$.

Lemma 3. *If the set of committees triggers a timeout in round r, it must advance to round $r + 1$.*

Proof. From Lemma 1 and Lemma 2, it is clear that when there are more than half of the timeout committees, the system advances directly to the round. When there are no more than half of the timeout committees, the system advances to the round after submitting the block. Therefore, regardless of the number of timeout committees, the system always advances to the round $r + 1$.

Lemma 4. *If level-1 leader L_1 (committee g_1) proposes block B_1 for transaction TX and level-1 leader L_2 (committee g_2) proposes block B_2 for TX ($B_1 \neq B_2$), they will not be submitted simultaneously.*

Proof. During the block submission phase, each level-1 leader broadcasts the signatures of its blocks to be submitted. Depending on the distribution of block support:

- half of each. Only half of the signatures can be collected by all level-1 leaders. At this point the committee group triggers a global timeout (by Lemma 1) and the system advances directly to the round $r + 1$.
- B_1 is in the majority. Then its collected signatures fulfill the majority condition and are successfully submitted. A committee supporting B_2 that is unable to collect enough signatures triggers a local timeout and advances to the round $r + 1$ after synchronizing the submitted B_1 by broadcasting a *ReqSyncMSG* (by Lemma 2).
- B_2 is in the majority. Then submit B_2 and advance to a new round after synchronizing the other committees.

Lemma 5. *If the adversary does not control more than half of the committee, the consensus cannot be disturbed.*

Proof. Consider the case where the number of groups is odd or even:

- If the final number of groupings is odd, assume that $|\mathbb{G}| = 2k + 1$, the adversary controls at most k committees, and the honest party holds $k + 1$ to collect enough signatures to submit a legitimate block.
- If the final number of groupings is even, assume that $|\mathbb{G}| = 2k$, the adversary controls at most k committees, the honest side also holds k, and neither side can reach $k + 1$ signatures, triggering a global timeout and advancing the round.

In summary, the adversary can neither submit a malicious block nor prevent the system from advancing normally.

Lemma 6. *For $n = 4f$ replicas divided into k groups, where the model with 4 replicas in each group of committees does not control more than half of them.*

Proof. In committee, each group contains 4 replicas when the size of the committee set is $|\mathbb{G}| = \frac{n}{4}$. Assuming that there are 2 Byzantine replicas in each group, the number of Byzantine replicas is $f = 2k$ when the final number of groups is odd, and $f = 2k + 1$ when the final number of groups is even.

- $f = 2k$, then the Byzantine replicas form exactly k committees (2 Byzantine replicas per committee), the number of legitimate committees is also k, and the proportion of Byzantine committees is exactly half of the total, and according to Lemma 5, there is no control over the consensus process.
- $f = 2k + 1$, then the Byzantine replicas form k committees (2 Byzantine replicas per committee), the remaining 1 Byzantine replica can not form a full committee, then the number of legitimate committees is $k+1$, the proportion of Byzantine committees is less than half.

If each committee consists of more than 2 Byzantine replicas, the above shows that the number of legal committees is still in the majority, and this is proven.

Theorem 1. *(Safety) If two committees submit blocks B'_r and B_r in the round $r + 2$, then $B'_r = B_r$.*

Proof. Within the committee, according to the 2-chain submission rule, at round $r + 2$ the block from round r of will be mentioned, at which point B_r extends its ancestor block, i.e., $QC_r \rightarrow B_{r-1} \rightarrow QC_{r-1} \rightarrow \cdots \rightarrow B_1 \rightarrow QC_{genesis} \rightarrow B_0$, whereby block B_r is created with a unique prefix defined, so that all replicas within the committee submit the same block at each position. By Lemma 5 and Lemma 6, Byzantine replicas cannot control the entire protocol. Within the committee group, according to Lemma 4, it is possible to derive blocks that do not create conflicts. At this point, the two committees submit blocks B'_r and B_r, respectively, then $B'_r = B_r$.

Theorem 2. *(Liveness) In the protocol, if a transaction is proposed, then honest replicas must be able to commit to that transaction.*

Proof. Assume that a block has been committed in the round r, and other honest replicas will also commit these blocks. When $r - 1$, the assumption is valid. Assume that when $r = k$, the assumption holds. Then, for $r = k + 1$, since transactions can be committed before the round k and the blocks are the same, for the round $k + 1$, if the system is stable, transactions can definitely be committed. Lemma 3 indicates that in the timeout case, transactions may be committed in the round $k + 1$, or all may time out and be committed in the round $k + 2$. Therefore, for honest replicas, transactions proposed by the client will always be committed.

Theorem 3. *Dividing n replicas into groups of 4 will be the most efficient.*

Proof. Due to the limitations of the Byzantine model, the number of replicas n must satisfy $n = 3f + 1$ to withstand byzantine attacks. The number of groups is k, with m replicas per group, thus $k = \frac{n}{m}$. Global timeout necessitates advancing the round, which incurs the highest overhead. Therefore, we consider the probability of global timeout to assess the impact of group count on consensus efficiency. Assuming network fluctuations affect each replica equally, let the probability of timeout per group be p. The probability of a global timeout occurring is $P \geq p^k$. Thus, a larger number of groups results in a smaller probability of global timeout. At this point, the minimum number of replicas required by the Byzantine model is $n = 4$, where P is minimized.

In summary, GP-BFT ensures both liveness and safety, and grouping replicas into sets of 4 during the partitioning process proves to be the most efficient approach.

5   Evaluation

The GP-BFT protocol is implemented in Rust, using Tokio for asynchronous network communication, ed25519-dalek4 for signing, and RocksDB for data persistence. It uses TCP for replica communication. We have improved the Rust code and test scripts of Jolteon. This solution primarily focuses on transitioning from a single-leader consensus to a multi-committee consensus strategy and adjusting the timeout policy. To ensure consistency in experimental conditions, Δ will remain unchanged.

During the experiment, each replica deploys clients to send fixed data transactions to the replicas. Transactions eventually reach the memory pool and the backend management system enables fast propagation. The transaction size is 512 bytes and the memory pool batch is 50 KB. Latency and throughput of the test protocol consensus: throughput is the number of submitted transactions per second and latency is the average time from reception to submission. Experiments were conducted using 24 vCPUs (Intel Xeon Gold 6128@3.40GHz), 64GB of memory, and Ubuntu Server 23.10 system. The 12 and 24 replica scenarios were tested and each set of experiments was repeated 3 times for 1 minute each to minimize the error.

To simulate an actual network environment, the client transaction input rate starts at 10k tx/s and increases by 10k each time, with the Δ value set to 1000 ms for all protocols. Figure 3 shows the results of the no-failure replica experiment without any manual intervention. As the number of replicas and the transaction input rate increase, the throughput of Jolteon and GP-BFT continues to rise, with GP-BFT eventually outperforming HotStuff and Jolteon. In terms of latency, Jolteon and GP-BFT perform similarly in small-scale replica environments, but as the number of replicas increases and the input rate rises, GP-BFT achieves lower latency, demonstrating its advantage in large-scale replica environments. Figure 4 shows the results for 2 and 4 failed replicas out of 12 and 24 replicas, respectively. Initially, the non-failed replicas are grouped, thereby minimizing the impact of failed replicas on GP-BFT, and the performance is similar to that in Fig. 3.

No-Failure Replicas: Figure 3 shows the performance of the protocol in a no-failure replica scenario. With 12 and 24 replicas, GP-BFT has an advantage over Jolteon in terms of latency for large transaction inputs as the transaction input rate increases. Specifically, HotStuff's throughput reaches its peak at a transaction rate of $40k\ tx/s$, primarily because HotStuff requires three phases to complete a commit. As the transaction input rate increases, the time required to fill its queue pipeline also increases, placing higher demands on the memory

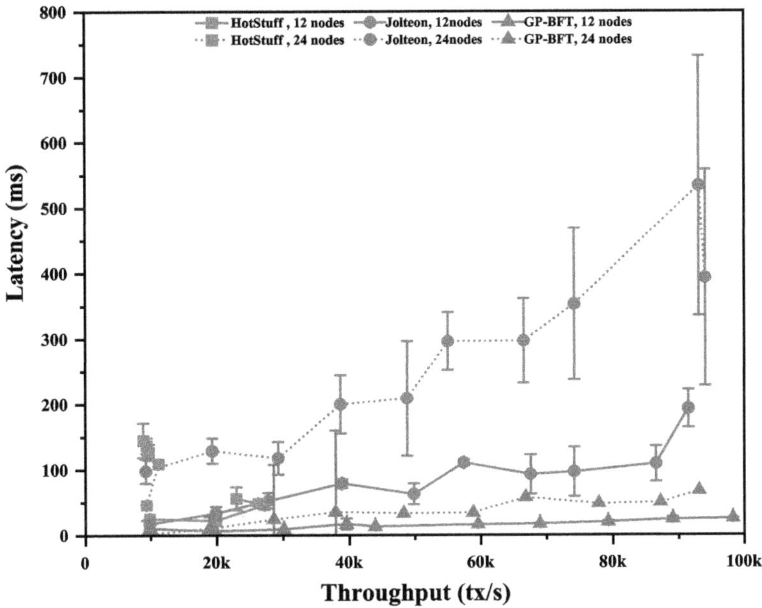

Fig. 3. Throughput-latency performance comparison of HotStuff, Jolteon, and GP-BFT in stable state with 12 or 24 replicas.

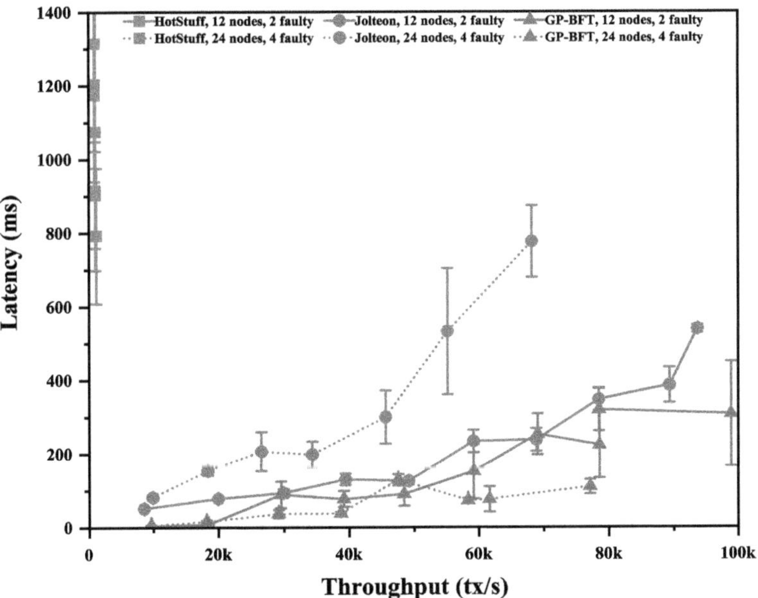

Fig. 4. Comparison of throughput-latency performance between HotStuff, Jolteon, and GP-BFT in scenarios with 12 or 24 replicas, under conditions where replica failures occur.

pool's performance. However, under the same memory pool model, the performance difference in throughput between Jolteon and GP-BFT is relatively small. Thanks to GP-BFT's group parallel commit feature, which reduces communication latency between replicas, GP-BFT outperforms other protocols in overall latency. This indicates that GP-BFT can maintain superior performance even under high loads.

Faulty Replicas: Figure 4 shows the performance of each protocol in the presence of faulty replicas. HotStuff experiences a significant performance degradation when faulty replicas are introduced, while Jolteon and GP-BFT also exhibit increased latency. In Jolteon, when a failed replica becomes the leader, the timeout mechanism is triggered due to the expiration of the waiting timer, leading to frequent round advances and further increasing latency. In contrast, GP-BFT's design effectively groups surviving replicas during the protocol initialization grouping phase to facilitate consensus. When comparing the performance of 24 replicas with 12 replicas, we find that 24 replicas perform better than 12 replicas. This is because, with fewer replicas (e.g., 12 replicas), the committee may trigger timeouts in the same round, thereby increasing latency. In the case of 24 replicas, however, due to the larger number of committee groups, the probability of simultaneous timeouts is relatively low. Timeout committees only need to perform rapid synchronization with non-timeout committees, significantly reducing latency.

In summary, GP-BFT performs well in adaptive network environments and in the presence of replica failures. Compared with HotStuff and Jolteon, GP-BFT achieves significant improvements in latency, primarily due to its structural advantages in group consensus. Experimental results show that in scenarios with small replica scales and transaction volumes, GP-BFT performs similarly to other protocols. However, in large-scale environments, the grouping advantage of GP-BFT is significantly reflected in lower latency. Therefore, GP-BFT is a BFT protocol with high throughput and low latency that is suitable for large-scale scenarios.

6   Conclusion

This paper investigated the limitations of BFT protocol in the single-leader model, such as, the overhead of recreating blocks after a leader switch increases consensus latency, and, all replicas participate in consensus, and the communication overhead between them affects the performance of the consensus protocol. To address these issues, this paper proposed GP-BFT, a multi-committee parallel BFT consensus protocol. By controlling the number of replicas in each committee group and enabling parallel consensus among these groups, GP-BFT reduces communication overhead between replicas. Additionally, during timeout periods, GP-BFT leverages committees that have already reached consensus to quickly commit blocks, thereby adapting to environments with a large number of replicas and high transaction volumes. Experimental results demonstrated that GP-BFT maintains high throughput and low latency while ensuring system liveness and safety.

Acknowledgement. This work is supported by the Guangxi Natural Science Foundation (2024GXNSFAA010453, 2025GXNSFGA069004), the National Natural Science Foundation of China (62562021, 62562019, 62172119), and the Zhejiang Provincial Natural Science Foundation of China (LZ23F020012).

References

1. Abraham, I., Malkhi, D., Nayak, K., Ren, L., Yin, M.: Sync hotstuff: simple and practical synchronous state machine replication. In: 2020 IEEE Symposium on Security and Privacy (SP), pp. 106–118 (2020). https://doi.org/10.1109/SP40000.2020.00044
2. Abraham, I., Nayak, K., Shrestha, N.: Optimal good-case latency for rotating leader synchronous BFT. In: Bramas, Q., Gramoli, V., Milani, A. (eds.) 25th International Conference on Principles of Distributed Systems (OPODIS 2021). Leibniz International Proceedings in Informatics (LIPIcs), vol. 217, pp. 27:1–27:19. Schloss Dagstuhl - Leibniz-Zentrum für Informatik, Dagstuhl (2022). https://doi.org/10.4230/LIPIcs.OPODIS.2021.27
3. Abspoel, M., Attema, T., Rambaud, M.: Brief announcement: malicious security comes for free in consensus with leaders. In: Proceedings of the 2021 ACM Symposium on Principles of Distributed Computing, PODC'21, pp. 195–198. Association for Computing Machinery, New York (2021). https://doi.org/10.1145/3465084.3467953
4. Amir, Y., Coan, B., Kirsch, J., Lane, J.: Prime: byzantine replication under attack. IEEE Trans. Dependable Secure Comput. **8**(4), 564–577 (2011). https://doi.org/10.1109/TDSC.2010.70
5. Buchman, E.: Tendermint: byzantine fault tolerance in the age of blockchains (2016). http://hdl.handle.net/10214/9769
6. Castro, M., Liskov, B.: Practical byzantine fault tolerance. In: Proceedings of the Third Symposium on Operating Systems Design and Implementation, OSDI '99, pp. 173–186. USENIX Association (1999). https://dl.acm.org/doi/10.5555/296806.296824
7. Castro, M., Liskov, B.: Practical byzantine fault tolerance. In: Proceedings of the Third Symposium on Operating Systems Design and Implementation, OSDI '99, pp. 173–186. USENIX Association (1999). https://dl.acm.org/doi/proceedings/10.5555/296806
8. Chan, B.Y., Shi, E.: Streamlet: textbook streamlined blockchains. In: Proceedings of the 2nd ACM Conference on Advances in Financial Technologies, AFT '20, pp. 1–11. Association for Computing Machinery, New York (2020). https://doi.org/10.1145/3419614.3423256
9. Chan, T.H.H., Pass, R., Shi, E.: PiLi: an extremely simple synchronous blockchain. Cryptology ePrint Archive, Paper 2018/980 (2018). https://eprint.iacr.org/2018/980
10. Fu, X., Li, M., Zeng, Q., Li, T., Yang, S., Guan, Y., Liu, C.: Hamster: a fast synchronous byzantine fault tolerant protocol. IEEE Trans. Inf. Forensics Secur. **20**, 2664–2676 (2025). https://doi.org/10.1109/TIFS.2025.3544034

11. Gao, M., Lu, G., Wang, Z., Gao, Y.: Wpbft: an improved consensus algorithm based on the hotstuff algorithm. In: 2023 2nd International Conference on Artificial Intelligence and Blockchain Technology (AIBT), pp. 56–59 (2023). https://doi.org/10.1109/AIBT57480.2023.00018

12. Gelashvili, R., Kokoris-Kogias, L., Sonnino, A., Spiegelman, A., Xiang, Z.: Jolteon and ditto: network-adaptive efficient consensus with asynchronous fallback. In: Eyal, I., Garay, J. (eds.) Financial Cryptography and Data Security, pp. 296–315. Springer, Cham (2022). https://doi.org/10.1007/978-3-031-18283-9_14

13. Giridharan, N., Howard, H., Abraham, I., Crooks, N., Tomescu, A.: No-commit proofs: defeating livelock in BFT. Cryptology ePrint Archive, Paper 2021/1308 (2021). https://eprint.iacr.org/2021/1308

14. Guo, Y., Pass, R., Shi, E.: Synchronous, with a chance of partition tolerance. In: Boldyreva, A., Micciancio, D. (eds.) Advances in Cryptology - CRYPTO 2019, pp. 499–529. Springer, Cham (2019). https://doi.org/10.1007/978-3-030-26948-7_18

15. Huang, K., Hou, R., Zeng, Y.: Lwsbft: leaderless weakly synchronous bft protocol. Comput. Netw. **219**, 109419 (2022). https://doi.org/10.1016/j.comnet.2022.109419

16. Lamport, L., Shostak, R., Pease, M.: The Byzantine generals problem, pp. 203–226. Association for Computing Machinery, New York (2019). https://doi.org/10.1145/3335772.3335936

17. Malkhi, D., Nayak, K., Ren, L.: Flexible byzantine fault tolerance. In: Proceedings of the 2019 ACM SIGSAC Conference on Computer and Communications Security, CCS '19, pp. 1041–1053. Association for Computing Machinery, New York (2019). https://doi.org/10.1145/3319535.3354225

18. Momose, A., Cruz, J.P., Kaji, Y.: Hybrid-BFT: Optimistically responsive synchronous consensus with optimal latency or resilience. Cryptology ePrint Archive, Paper 2020/406 (2020). https://eprint.iacr.org/2020/406

19. Nakamoto, S.: Bitcoin: a peer-to-peer electronic cash system (2008). https://bitcoin.org/bitcoin.pdf

20. Sui, X., Duan, S., Zhang, H.: Marlin: two-phase bft with linearity. In: 2022 52nd Annual IEEE/IFIP International Conference on Dependable Systems and Networks (DSN), pp. 54–66 (2022). https://doi.org/10.1109/DSN53405.2022.00018

21. Wang, X., Duan, S., Clavin, J., Zhang, H.: Bft in blockchains: from protocols to use cases. ACM Comput. Surv. **54**(10s), 1–37 (2022). https://doi.org/10.1145/3503042

22. Wen, B., Wang, Y., Ding, Y., Zheng, H., Qin, B., Yang, C.: Security and privacy protection technologies in securing blockchain applications. Inf. Sci. **645**, 119322 (2023). https://doi.org/10.1016/j.ins.2023.119322

23. Yao, D., et al.: Dybft: leaderless bft protocol based on locally adjustable valid committee set mechanism. World Wide Web **28**(1), 17 (2025). https://doi.org/10.1007/s11280-024-01324-w

24. Yin, M., Malkhi, D., Reiter, M.K., Gueta, G.G., Abraham, I.: Hotstuff: bft consensus with linearity and responsiveness. In: Proceedings of the 2019 ACM Symposium on Principles of Distributed Computing, PODC '19, pp. 347–356. Association for Computing Machinery, New York (2019). https://doi.org/10.1145/3293611.3331591

EncloPC: Power Control for Server Enclosures with Heterogeneous Workloads at Power Peaks

Wenli Zheng[1(✉)] and Xiaorui Wang[2]

[1] School of Computer Science, Shanghai Jiao Tong University, Shanghai, China
`zheng-wl@cs.sjtu.edu.cn`
[2] Department of Electrical and Computer Engineering, The Ohio State University, Columbus, USA

Abstract. A server enclosure hosts multiple servers together with cooling fans and needs to control their total power consumption under a given budget. Existing methods for computer system power control either only manipulate CPU power states or assume all the workloads have homogeneous optimization goals, leaving workload-specific optimization with different hardware actuators an unsolved problem, especially at the enclosure level. In this paper, we propose EncloPC, a hierarchical framework for enclosure-level power control. During a power peak, EncloPC dynamically determines the power allocation between the servers running interactive and batch workloads, respectively. EncloPC features two power controllers for the two groups of servers, designed rigorously based on feedback control theory to guarantee the power control accuracy even with modeling errors. Both controllers jointly adapt CPU and memory frequencies for maximized power efficiency and profits. The evaluation results show that compared to state-of-the-art solutions, EncloPC can achieve up to 70% reduction in processing delay for interactive workloads, and an additional annual profit of up to $6.7 Million for a data center hosting 50,000 servers.

Keywords: Power control · Server enclosure · Interactive and batch workloads

1 Introduction

Server enclosures are widely used in data center modularization as they are self-contained and space-efficient. Server enclosures are typically equipped with oversubscribed power supply units (PSUs), i.e., the PSU would be overloaded if all the contained servers achieve their power peaks simultaneously. Hence, power capping [22] is important for power safety by monitoring and capping the total power consumption of all the enclosure components, mainly including the servers and cooling fans, in real time. Usually, the ideal case is to make the total power consumption just equal to the power budget, for both power safety and optimized system performance. In addition, enclosure operators can use server- or enclosure-level uninterruptible power supply (UPS) batteries to increase the budget, which raises the significance of accurately

Wenli Zheng's work is sponsored by the National Natural Science Foundation of China under grant number 62172270.

controlling the power consumption if a specific time duration of UPS discharge needs to be guaranteed.

However, existing power capping and control solutions usually neglect the heterogeneity in either hardware actuators or workload-specific optimization goals, while a server enclosure can run different workloads with different hardware components. Many of those solutions only use CPU dynamic voltage and frequency scaling (DVFS) to manipulate hardware power states [13,21], and some others focus on GPUs [36] or purely rely on software knobs [15,24]. A few studies jointly manage CPU DVFS and memory DVFS [1,5], but assume all the workloads have the same optimization goal like energy minimization. Moreover, most of those solutions are designed for server clusters, i.e., the given power budget is only used by servers, while in a server enclosure the cooling fans share the power budget with the servers. Such a difference leads to a dynamic sub-budget for the servers, because the variation of server power consumption causes a varying fan speed and thus varying fan power consumption. Therefore, enclosure power control is still an unsolved problem when different hardware components and optimization goals are both taken into consideration.

In this paper, we propose EncloPC, a control framework for enclosure-level power capping. In contrast to existing work, EncloPC maximizes the total profits generated on a server enclosure based on workload-specific optimization goals, by coordinating the CPU and memory DVFS and considering the power variation of cooling fans. EncloPC features a power allocator and two power controllers. The power allocator dynamically determines the total server power budget, and divides it into two groups of servers respectively running interactive and batch workloads with linear programming, whose result maximizes the total profits of running the two types of workloads. The two power controllers are rigorously designed based on multi-input-multi-output (MIMO) feedback control theory for guaranteed control accuracy and system stability, dynamically optimizing the CPU and memory frequencies according to the delay sensitivity and resource affinity of workloads. Under the power constraint, the servers running interactive workloads minimize their processing delays, and the servers running batch workloads optimize their energy efficiency. We compare EncloPC with five state-of-the-art solutions in experiments, including a feedback controller with the adaptation of both CPU and memory [2], an open-loop controller that optimizes the IPS performance [5], and a UPS-based power capping scheme [12]. The results show that EncloPC can achieve up to 70% better system performance for interactive workloads than existing methods, and up to $6.7 Million additional annual profits for a data center hosting 50,000 servers.

Specifically, we make the following major contributions:

- We propose to do server enclosure power capping through workload-specific feedback control with coordinated CPU and memory DVFS.
- We propose a systematic framework EncloPC that coordinates two server power controllers given their total power budget dynamically impacted by the cooling fans of the enclosure.
- We design novel MIMO power control algorithms for power-efficient processing of interactive and batch workloads, rigorously based on optimal control theory.

The controllers coordinate heterogeneous hardware actuators for workload-specific optimization goals.

The rest of the paper is organized as follows. Section 2 discusses the related work. Sections 3 and 4 present the design and its critical details. Sections 5 and 6 present the evaluation and analysis. Section 7 concludes the paper.

2   Related Work

Power peaks of computer systems (e.g., server clusters and data centers) caused by heavy workload can overload the power infrastructure that is oversubscribed. To avoid such overload for safety, a variety of solutions have been proposed. The software-only solutions typically reduce the power consumption by temporary workload reduction, with the knobs like throughput throttling [15, 16] and workload suspension [24]. Some solutions even encourage users to actively reduce their workload [10]. Another mainstream type of solutions use the hardware knobs, of which most are CPU DVFS [13, 21] and a few can be memory DVFS [1, 5], to throttle the power of devices. Since the aforementioned two types of power capping solutions can sacrifice the workload processing performance and thus the profits, discharging an energy storage device (ESD) [12, 37] to increase the power supply capacity becomes a popular knob. Although the discharge capacity of an ESD is limited, such a method works based on the assumption that a power peak does not last too long and can be covered by the discharge duration. Similarly, the studies on computational sprinting [6, 11] make use of the time duration that the power infrastructure can tolerate overloading, to avoid performance degradation during power peaks. It is worth noting that most power capping solutions rely on power models to map power manipulation to performance variation, which can introduce non-trivial errors due to the limitations of modeling techniques or environmental dynamics.

Power control is a research topic closely related with power capping, but additionally addressing how to guarantee the gap between the power budget (power cap) and the actual power consumption is negligible, even if the power models are inaccurate. Traditional methods [30, 32] typically adopt feedback control algorithms for theoretically guaranteed control accuracy and stability of power consumption, along with the optimized computing performance determined by CPU frequencies. Some later work [2] jointly controls the power of both CPU and main memory, for more power saving and shorter execution time. Recent work like SprintCon [39] further integrates power control with computational sprinting for enhanced system performance during workload bursts, though still only manipulating CPU power for batch workloads. In summary, prior studies on power control neglect the differences in optimization goals between different types of workloads, and most of them regard CPU DVFS as the unique hardware actuators. Since the per-watt profits are different for different workloads, dynamically adjusting the division of total power budget to each type of workload is necessary to optimally making use of the power budget. EncloPC improves power allocation between interactive and batch, CPU-intensive and memory-intensive workloads in addition to power control, along with taking the power consumption of cooling fans into consideration since they share the power budget with servers, and thus is better at maximizing the profits of a server enclosure with heterogeneous workloads.

3   System Design

In this section, we present the system design of the power control framework EncloPC, and the techniques for server hardware adaptation.

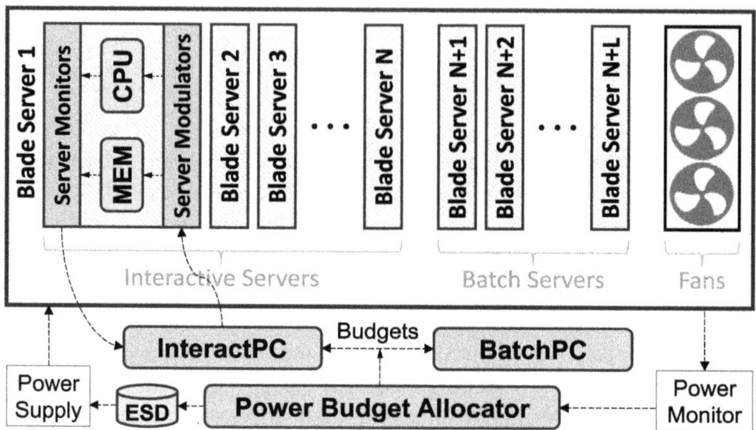

Fig. 1. The hierarchy and control loops of EncloPC, with the server enclosure and power system architecture (only the interaction between 1 controller and 1 server is plotted here for clarity).

3.1   System Design

We assume the server enclosure can colocate two types of workloads, i.e., interactive and batch (classified based on QoS requirements), on the same server for utilization benefits. However, during a power peak, some servers may have only batch workloads to run (e.g., due to data skew of interactive workloads), and the other servers defer batch workloads such that they run only interactive workloads without severe interferences [34]. As the result, each server runs either batch or interactive workload during the power peak. We do not assume deferring all the batch workloads because the deadlines of some batch workloads may be approaching, and missing the deadlines can cause too much revenue loss. Therefore, the servers are divided into two groups depending on the types of workloads running on them.

We briefly illustrate how EncloPC works in Fig. 1. The Power Budget Allocator dynamically allocates the power budgets between them based on their different demands, before sending the two budgets to their respective power controllers, InteractPC and BatchPC, with respect to the two server groups. In order to control the server power consumption while optimizing the processing of interactive and batch workloads, the two controllers need to manipulate the power states of hardware components (i.e., the frequencies of CPU and memory in this paper, and can be extended to other components like GPU without loss of generality) based on the workload types and processing statistics. Therefore, for both the two controllers, after the power allocator determines the UPS power discharge rate and the two server group power budgets, a feedback control loop is executed periodically as follows:

1. The power monitors, as the first feedback provider, report the power consumption to the controllers.
2. The server monitors, as the second feedback provider, report the workload type, utilization of each CPU core, and number of memory accesses to the controllers.
3. Based on the collected feedback data and the power budgets, the controller determines the new frequencies of the CPU cores and memories in the next period for the servers under its control.
4. The server modulators adjust the CPU and memory frequencies according to the controller outputs.

EncloPC can make use of UPS batteries (if available) as the ESDs to provide additional power during power peaks. The desired ESD discharge rate P_s is determined based on the ESD capacity E and the desired discharge duration T_s (i.e., how long the ESD should be discharged). For simple calculation, $P_s = E/T_s$. The data center operator can determine T_s based on the prediction of power peaks, e.g., how much workload need to be handled and how long the peak will last. Then EncloPC controls the discharge rate to P_s to guarantee T_s, and optimizes the profits of co-processing interactive and batch workloads at the same time. If a power peak lasts longer than predicted, EncloPC can work with the total power budget equal to the enclosure power capacity.

3.2 Hardware Power State Adaptation

EncloPC manipulates the power states of two major power-consuming components in each server for power control, CPU and main memory. For CPU, we use DVFS to manipulate the frequency and thus power state of each core, because DVFS introduces minimal overheads with negligible response time. We assume all the workloads do single-thread processing for concise presentation focusing on the power control related issues, though EncloPC can be integrated with existing solutions [18] to further allocate the power budget of a workload to a set of cores running its threads.

For main memory, we adopt frequency scaling as well, as proposed in prior studies [1,5]. Compared to the solutions that reschedule memory access time for DRAM power-down or deactivate part of the DRAM chips to reduce memory power, frequency scaling allows memory accesses on all the DRAM chips at any time, and requires no particular changes of the workload or DRAM microarchitecture. Therefore, frequency scaling can be a more appropriate approach for memory power adaptation, though not all the existing servers are designed to support changing the memory frequency at runtime. For those servers, EncloPC can still adjust the CPU frequency for power control.

4 Algorithm Design

In this section, we discuss the power modeling and power allocation methods, along with the detailed power control algorithms used in EncloPC.

4.1 Power Modeling

Previous research has shown that the relationship between processor power and frequency is approximately linear in a certain range [9, 14] under stable utilization. Therefore, we adopt their linear model to calculate the CPU power of server i in time slot k based on the frequency $f_{i,j}^c$ of its core j:

$$p_i^c(k) = \sum_j K_{i,j}^c f_{i,j}^c(k) + C_i^c \tag{1}$$

where $K_{i,j}^c$ is the slope coefficient and C_i^c is the frequency-independent part of CPU power. With the coefficient matrix $\mathbf{K}_i^c = [K_{i,1}^c, K_{i,2}^c, ..., K_{i,J}^c]$ and frequency matrix $\mathbf{F}_i^c = [f_{i,1}^c, f_{i,2}^c, ..., f_{i,J}^c]^T$ for a server with J cores, we have the matrix form: $p_i^c(k) = \mathbf{K}_i^c \mathbf{F}_i^c(k) + C_i^c$.

We model the memory power based on the power model of DDR3 SDRAM in [26], which shows that the DRAM power can be divided into two parts: the first part is proportional to the DRAM frequency, and the second part is independent of the frequency. Hence the memory power can be modeled as a function of the frequency f_i^m:

$$p_i^m(k) = K_i^m f_i^m(k) + C_i^m \tag{2}$$

where K_i^m is the slope parameter and C_i^m is the frequency-independent part of memory power. The power consumed by the other memory components other than the DRAM chips (e.g., I/O buses and buffers) is also counted in C_i^m. For generality, we can derive the server power model as:

$$p_i(k) = \mathbf{K}_i \mathbf{V}_i(k) + C_i \tag{3}$$

where $\mathbf{V}_i = [f_{i,1}^c, f_{i,2}^c, ..., f_{i,J}^c, f_i^m(k)]^T$, and $\mathbf{K}_i = [K_{i,1}^c, K_{i,2}^c, ..., K_{i,J}^c, K_i^m(k)]$.

It is worth noting that although some studies [20, 23] provide sophisticated power models regarding CPU utilization, memory accesses and other factors as variables, our simple models fit power control better for two reasons. First, the key reason for employing power models in power control is to predict the power under a new frequency, but changes of frequencies can lead to unpredictable changes of utilization/accesses, disabling those sophisticated models. Second, simple models speed up solving the optimization problems of control algorithms in each control period, which is important for online operation. In addition, EncloPC can tolerate some system modeling errors with its feedback control design, weakening the advantage of sophisticated models. If the actual relationship is highly non-linear, we can use a piecewise-linear model instead.

While we regard the CPU power and memory power as the major objects in enclosure power control, in reality there are other server components and enclosure components that can impact the enclosure power variation, e.g., the enclosure cooling fans that are shared by the servers. The fan speed is usually self-adaptive to the server temperature. Therefore, the fan power consumption can increase when the server power increases with generating a greater amount of heat, resulting in system errors. Such errors are also handled by the feedback control of EncloPC.

4.2 Power Allocator

The power budget allocator is designed to handle two tasks: 1) determine the ESD power discharge rate, and 2) assign the group power budgets to the two power controllers. It operates with a period longer than the settling time of the power controllers, such that the power budget allocation will not change before the power consumption converges to the power budget, and thus the whole system is stable.

EncloPC controls the ESD discharge rate by controlling the enclosure power consumption, and the power budget allocator divides the power load to the ESD and the enclosure power supply unit (PSU) by operating a configurable switch. For example, if the power budget is 2500 W and the enclosure power capacity is 2000 W, then EncloPC controls the enclosure power within 2500 W, and the power budget allocator sources 2000 W from the enclosure PSU and sources the rest of power demand from the ESD. The configurable switch can be operated with duty cycling [8], as shown in Fig. 2(a), to allocate power load between the two power supply sources. Figure 2(b) illustrates the situation when the real power consumption is lower than the budgeted value, e.g., 2250 W, for which the duty ratio is adjusted to save UPS energy.

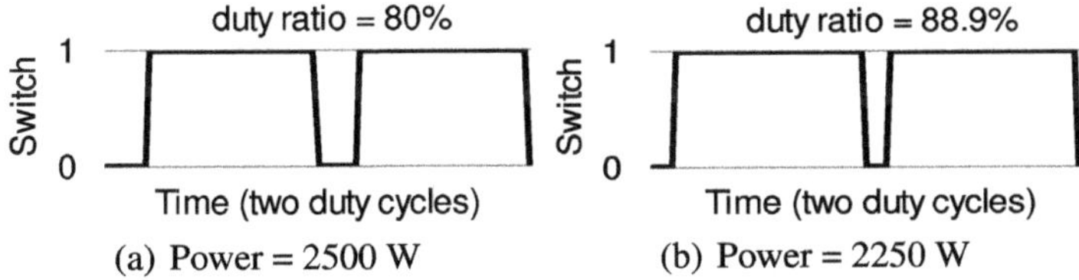

(a) Power = 2500 W (b) Power = 2250 W

Fig. 2. Duty cycling control of the configurable switch.

The power budget allocator subtracts the fan power from the total amount of available power to obtain the total server power budget, based on the assumption that the fan speed changes much slower than the server power does, as the change of temperature takes time. Then, the allocator divides the total server power budget to the two server groups in the way that maximizes the total profits, which include the revenue and energy cost of processing workloads. We model the relationship between the revenue of interactive workloads and their system performance (processing delay) based on the annual revenue of eBay data centers ($120,000 per server with 20% utilization on average [25]) and the studies about the revenue loss caused by processing delay [7]. We use RL to denote the per-watt revenue loss caused by reducing PB_I from PB_I^O to $P_I(k-1)$, and RL is proportional to the average server utilization. PB_I is the power budget for the servers running interactive workloads, PB_I^O is the optimal value of PB_I that optimizes the system performance (typically the power of running all interactive workloads at peak frequencies), and $P_I(k-1)$ is the power consumption of these servers measured in the last control period. In total, the revenue loss due to power capping is $(PB_I^O - PB_I(k))RL$. To address the energy cost of interactive workloads saved by reducing power budget, we have $RL' = RL - EP$, where EP is the electricity price.

For the energy cost model of batch workloads, the energy consumption is the product of power consumption and execution time, and the execution time is inversely proportional to the processing speed. We derive the processing speed based on a CPI model [5], which considers both the CPU and memory frequencies. We use EC to denote the per-watt energy cost caused by reducing PB_B from PB_B^O to $P_B(k-1)$. PB_B is the power budget for the servers running batch workloads, PB_B^O is the optimal value of PB_B that minimizes the energy consumption, and $P_B(k-1)$ is the total power consumption of respective servers in the last control period. In addition, EC is proportional to the electricity price EP. Hence the additional energy cost due to power capping is $(PB_B^O - PB_B(k))EC$.

In order to maximize the total profits generated on the enclosure, we formulate the power budget allocation problem as a linear optimization problem with the models of revenues and costs. The power budget allocator can solve this problem with linear programming, which is a commonly used solution to linear optimization problems (e.g., [27]). Specifically, the problem is: given the per-watt revenues and costs (of interactive and batch workloads) achieved in the last period, how to divide the total server power budget of the enclosure PB to the two server groups, such that the total profit loss PL due to power capping is minimized. The objective function to be minimized is:

$$PL(k) = (PB_I^O - PB_I(k))RL' + (PB_B^O - PB_B(k))EC \qquad (4)$$

subject to the constraint that $PB_I(k) + PB_B(k) \leq PB(k)$, i.e., the sum of two server group power budgets is no more than the total server power budget of the enclosure. Since the problem has only two manipulated variables (PB_I and PB_B) and can be solved by linear programming, the computational overhead of executing power budget allocation is negligible.

The power budget allocator does not involve the revenue of batch workloads because we assume all the batch workloads must be processed before deadlines, and thus their revenue is regarded as a constant. If their deadlines are missed, the data center will have respective revenue loss. The revenue can be calculated as $\sum R_j b_j$, where b_j is the amount of jobs in batch class j finished before deadlines, and R_j is the per-job revenue of the class, according to the revenue model in [17]. In addition, if a sudden shift in workload mix occurs just after an allocation decision, the total profit during this allocation period might be affected, but the two power controllers can still guarantee respective maximized profits with power safety.

4.3 Control Algorithm of InteractPC

We design the controllers according to the *Model Predictive Control* (MPC) [19] theory, in order to make use of its advantages on handling multi-input-multi-output feedback control problems. For example, the system stability guaranteed by MPC can ensure the power consumption of the server enclosure converges to a pre-determined value, under the constraint that the system inputs are bounded in a reasonable range.

The controller InteractPC traces a reference trajectory to close in on its control target, i.e., the group power budget P_g (equaling to PB_I for InteractPC), to smooth the converging process and avoid severe overshoot. We set the reference trajectory ref like some prior studies [39] also using MPC in computer system power control as follows:

$$ref\left(k+x|k\right) = P_g - e^{-\frac{\tau}{\tau_{ref}}x}\left(P_g - P_t\left(k\right)\right) \tag{5}$$

where P_t is the controlled variable. For InteractPC, it is the total power consumed by all the servers running interactive workloads. The time constant τ_{ref} can be tuned to balance the overshoot and convergence time.

In each control period, the controller minimizes a cost function as follows, using the vector V_{max} to indicate the maximum frequencies of CPU and memory.

$$W\left(k\right) = \sum_{n=1}^{H_p} \|p_t\left(k+n|k\right) - ref\left(k+n|k\right)\|_{Q(n)}^2 +$$
$$\sum_{n=0}^{H_c-1} \|\Delta\mathbf{V}\left(k+n|k\right) + \mathbf{V}\left(k+n|k\right) - \mathbf{V}_{max}\|_{\mathbf{R}(n)}^2 \tag{6}$$

The first term is the tracking error, which measures the gap between p_t and ref in a series of period with the length H_p (called the prediction horizon). It means the controller predicts the variation of the controlled variables in the next H_p periods, given the assumption that the current manipulation will keep the same. For instance, a current manipulation that increases the CPU frequency by 50 MHz can be assumed to be repeated in the next H_p periods. The second term in the cost function is the control penalty, which measures the gaps between the manipulated CPU/memory frequencies and their respective maximum values in a series of period with the length H_c (called the control horizon). Generally, the smaller such gaps, the better computational performance the servers achieve. Therefore, the entire control algorithm of InteractPC is designed to not only control the server power consumption, but also optimize the server performance at the same time. Note that the controlled variable $P_t\left(k\right)$ in Equation (5) is reported by the power monitor in every control period as the feedback, which provides the reference for the closed-loop controller to eliminate inevitable system errors like fan power variation and power modeling errors. Hence the controller re-solves the optimization problem that minimizes the cost function in every control period.

To optimally allocate power budgets to different servers and server components, InteractPC manipulates the value of $\mathbf{R}$, the weight of control penalty in Eq. (6). $\mathbf{R}$ consists of the weights for CPU and memory frequencies, i.e., $\mathbf{R} = [\mathbf{R^c}, \mathbf{R^m} \times \mathbf{D}]$. Given N_s servers running interactive workloads, $\mathbf{R^c}$ is a vector of R_i^c for $i = 1, 2, ..., N_s$, and R_i^c is the ratio of Server i's CPU utilization over the sum of all the servers' CPU utilizations. Similarly, R_i^m is the ratio of Server i's memory accesses per second over the total amount of per-second memory accesses of all the servers. We multiply $\mathbf{R^m}$ by a factor $\mathbf{D}$ as in [2], where $\mathbf{D} = \mathrm{diag}[d_1, d_2, ..., d_{N_s}]$, in order to allocate power between CPU and memory based on the workload characteristics (e.g., CPU intensive or memory intensive). For Server i, $d_i = u_i^m f_i^c / f_i^m$, where u_i^m is the utilization of memory cycles, f_i^c is the CPU frequency and f_i^m is the memory frequency. When the workload is more memory intensive, d_i has a larger value and motivates the controller to allocate more power to memory than CPU. In addition, $\mathbf{R}$ and Q (the weight for tracking error) can be used to respectively tune the relative significance of each period in the control and prediction horizons if needed.

4.4 Differences for BatchPC

The controller design introduced above is for InteractPC that manages the servers running interactive workloads, and hence the control penalty is related to system performance. For BatchPC, we change the control penalty to be the energy consumption of the servers running batch workloads, involving both the power consumption and execution time. We can estimate the execution time with short-term profiling, which takes milliseconds to count the CPU cycles and cache misses, assuming the same execution pattern repeats itself throughout the execution of the whole batch workload. Then we use a progress model [5] to calculate how the frequency scaling of CPU and memory impacts the execution time, and multiply the execution time by the power consumption to derive the energy consumption. As lower power consumption usually leads to longer execution time, there is a most energy-efficient combination of CPU and memory frequencies for each server to minimize its energy consumption. Hence for BatchPC we replace V_{max} in Eq. (6) with the most energy-efficient frequencies of the CPU and memory, to avoid running the servers at unneeded high frequencies. Accordingly, we set the upper bound of the power budget for BatchPC in the power budget allocator, such that we can avoid a waste of power budget. BatchPC also applies additional lower bounds for the manipulated CPU and memory frequencies in order to make sure the batch workloads can be completed before the deadlines. The lower bounds are dynamically adjusted based on the progress of execution and remaining time before deadlines.

We determine the values of $\mathbf{R}$ for BatchPC in a similar way as InteractPC, i.e., in proportional to the resource usage, because a higher utilization indicates changing the frequency can have a greater impact on energy consumption. $\mathbf{R}^{m}$ are also multiplied by $\mathbf{D}$ to enlarge the control penalty on decreasing the memory frequency when running memory-intensive batch workloads, because this can significantly extend the execution time with relatively small power saving. Similarly, when $\mathbf{D}$ has smaller values for CPU-intensive workloads, the control penalty on decreasing the CPU frequency becomes larger to prevent a great impact on the execution time.

4.5 Discussions

While we derive the optimal power budget based on assuming the predictability of upcoming workloads or power peaks here, in practice this can be another problem. It has been discussed in [38], which proposes multiple heuristic strategies to dynamically adjust the ESD power budget made according to prediction. Such approaches are orthogonal to EncloPC, and can be integrated together to develop more efficient, thorough, and sophisticated solutions for enclosure-level power control. In addition, consecutive power peaks can make it hard to recharge used ESDs in time, requiring particular strategies in real-world deployment for optimized power budgeting.

The power budget allocator and power controllers can run on one of the blade servers or an external computer. We implement the algorithms as a Matlab script on a laptop, which consumes 0.26 s for computation with 4.16 J energy in each control period. Although the overheads can increase with more servers, an efficient parameter space exploration strategy [28] can avoid unnecessary quadratic program problem solving, significantly reducing the runtime overheads.

5   Evaluation Methodology

In this section, we introduce the evaluation methodologies including experiment setup and baseline solutions.

5.1   Experiment Setup

We simulate an enclosure of blade servers based on power models, whose parameters are determined by profiling tests on physical servers and cooling fans. The fan power is simulated by a cubic model fitting the profiling data collected from an IBM BladeCenter H enclosure. In the profiling tests, the fan power is determined by a PID controller that controls the peak server temperature to its temperature limit, like the typical solution of current practice. The parameters of the PID controller are tuned with the Ziegler-Nichols method. As the result, the fitted cubic model can output a value of fan power given a value of server power. Since the parameters in the fan power model can have different values when the layout of servers and fans changes, in the simulation we use the same layout as the IBM BladeCenter H enclosure, which hosts 14 blade servers and 2 fans, and the power capacity of the enclosure is 2600 W. We assume the UPS discharges at 400 W during power peaks, and thus the servers and fans can use 3000 W power in total. Each of the 14 servers is equipped with a 4-core CPU and 4 GB DDR3 memory. The CPU frequency ranges from 0.8 to 1.9 GHz like AMD Opteron 6168 processors, and the memory frequency ranges from 200 to 800 MHz as in [5].

To consider the impact of modeling error, we use more accurate models for power measurement in the simulation than the simple power models used in the controller design. For CPU power we use a model taking both frequency and utilization [9] as inputs. For memory power we use the model from a MICRON technical note [26] like [5], taking both the frequency and the number of memory accesses as inputs. To set the idle power of each server, we refer to the study in [33], which shows that half of the servers have idle power more than 30% of the peak server power, though the most energy-efficient ones can achieve 20%. Hence we set the idle power to be 60 W and equal to 18.6% of the peak server power, considering the development of energy proportionality.

After the two controllers set the processor and memory frequencies, we calculate the average processing delay of interactive workloads and the processing progress of batch workloads. We apply four of the interactive workload traces, collected from a Wikipedia data center [29], on the four cores of each server, and make different servers run different traces. The processing delay is calculated according to the testing results in [35], which shows the impact of frequencies on the throughput of running SPEC jbb2005 benchmark. We select SPEC jbb2005 as the reference here because this benchmark emulates a client/server system that processes real-world applications like modern data centers [4]. The batch workload traces are collected by running SPEC CPU2006 benchmarks on a physical server with the same processor and memory settings as the simulated server. We use the system statistics in the traces, including used CPU cycles, cache misses, and the hit latencies on cache and memory, to calculate the processing progress of batch workloads [5].

Data centers usually process batch workload when the interactive workload is relatively low, which means only a subset of the servers in an enclosure need to handle the interactive workloads (called *interactive servers* for short), and the remaining servers can run the batch workloads (called *batch servers* for short). In the simulation, we set up three scenarios for different numbers of interactive servers: 12, 8 and 4, and thus the numbers of batch servers are 2, 6 and 10, respectively, since we have 14 servers in total. If all the batch workloads on a certain server are completed before the end of power peak, the batch server can be shutdown for power saving.

5.2 Baseline Solutions

To evaluate EncloPC, we compare it with five state-of-the-art baselines, including two single-server controllers *CoScale* [5] and *Single* [2], and three multi-server controllers *HPL* [3], *CPUPC* [31] and *OptPerf*.

The *CoScale* solution subtracts the fan power from the total amount of available power, and evenly divides the remaining part to each server. Similar to the CoScale scheme proposed in [5], it finds the combination of CPU and memory frequencies of each server that optimizes the performance of this server within the given power budget. While *CoScale* optimizes the performance of each server locally, it cannot coordinate the power states of different servers based on their different workloads for global optimization.

Single is designed based on [2], which uses an MPC controller to coordinate the CPU frequency and the number of active memory banks, for single-server power control and performance optimization. In our evaluation, we replace the adaptation of active memory banks by the adaptation of memory frequency for the consistency with the other solutions, and evenly allocate the total server power budget to each server. Similar to *CoScale*, *Single* cannot shift power across servers.

The *HPL* solution controls the power consumption of all the blade servers in an enclosure as described in [3]. It allocates the power budget to the servers in proportional to CPU utilizations, and determines the CPU frequency of each server with a PID controller. *HPL* excludes memory adaptation and maintains the memory frequency at its maximum value. Different from the other baselines and EncloPC, *HPL* has no optimization of system performance.

CPUPC adopts the MPC algorithm to control the total enclosure power and optimize the system performance, by manipulating the CPU frequencies. Similar to *HPL*, *CPUPC* has no memory adaptation, no coordination of server power and fan power, and no particular strategy for the batch workloads that may not desire the optimized performance.

Since none of the above four baselines handle batch workloads in a different way from interactive workloads, to highlight the significance of this issue, we design a new baseline solution *OptPerf*. It controls the enclosure power consumption with the MPC algorithm, and jointly adapts all the CPUs and memories to optimize the overall system performance.

6   Results

In this section, we present the results of comparing EncloPC with the baseline solutions. First of all, we validate that all the tested solutions can realize power capping for the server enclosure, such that we can fairly compare them in the other aspects. Here we assume 8 interactive servers and 6 batch servers (3 for CPU intensive and 3 for memory intensive). Figure 3 shows that with different power caps, EncloPC and the baselines can successfully realize power capping, though some solutions may underutilize the available power when the power cap is relatively high (e.g., 3000 W). For EncloPC, this is because the batch servers are run at non-peak frequencies to minimize their energy consumption. For *CoScale* and *Single*, this is because they can assign a higher power budget to an interactive server than it demands, and thus the power budget is underutilized.

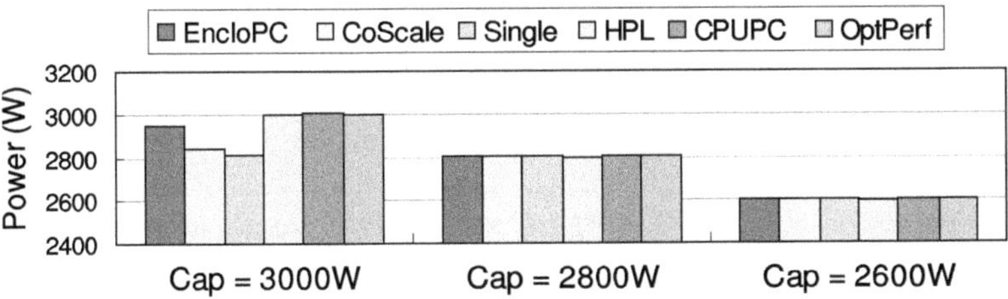

Fig. 3. Average power consumption of the whole enclosure at the steady state with different power budgets.

In the following subsections, we first show the benefits of integrating power control with ESD discharge in Sects. 6.1 and 6.2. Then we compare EncloPC and the baselines in terms of the delay of processing interactive workloads (Sect. 6.3), the energy consumed by batch workloads (Sect. 6.4), and the total profit loss (Sect. 6.5).

6.1   Impact of ESD Discharge Power on Profits

Although the enclosure power consumption can be controlled to meet some lower power caps without safety problems, it is likely to impact the profits generated on the enclosure, mainly due to the performance degradation of interactive servers. This is the reason we integrate power control with ESD discharge to raise the power cap. With UPS batteries as the most widely used ESD, here we test different UPS power discharge rates to evaluate the impact of using stored energy on improving the total profit. We test 200 W UPS power for the scenario that the UPS energy capacity deployed for the enclosure is only half of the original value, and test 0 W UPS power for no UPS energy is used.

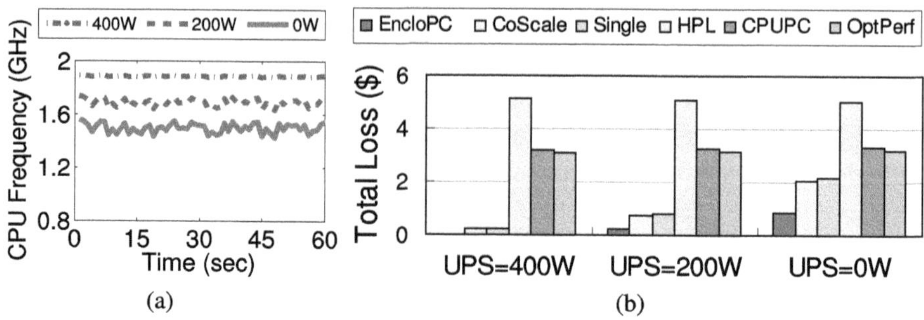

Fig. 4. (a) Average CPU frequency of interactive servers for EncloPC, and (b) total loss of the profit generated on the enclosure during the power peak, with different UPS discharge rates.

With a lower UPS discharge rate, EncloPC and the baselines need to degrade the power states of servers. For example, as shown in Fig. 4(a), EncloPC can run the processors of interactive servers close to the peak frequency (1.9 GHz) with 400 W UPS power, but need to slow them down to around 1.7 or 1.5 GHz on average with 200 W or zero UPS power, respectively, while the frequency is dynamically adjusted due to the workload fluctuation in the Wikipedia traces. Similarly, the two single-server baselines *CoScale* and *Single* also slow down the interactive servers, because their power budgets of interactive servers are proportionally reduced. Consequently, as shown in Fig. 4(b), their profit loss is significantly increased with lower UPS discharge rates. The three multi-server baselines *HPL*, *CPUPC* and *OptPerf* are not affected much because they run interactive servers at very low frequencies even with 400 W UPS power, and thus lower UPS power does not lead to much additional profit loss. Although EncloPC can adjust the power allocation and get less affected than the single-server baselines, discharging 400 W UPS power can still reduce the loss by 96% and 99% compared to 200 W UPS power and no UPS power.

6.2   Impact of Power Control on ESD Discharge

In the experiments above, we make the five baselines control the UPS power discharge rate to the configured values like EncloPC by adding the ability of controlling power discharge rate to them, while their original methods allows greedy usage of stored energy. In order to show the impact of power control on ESD discharge, here we make the five baselines to greedily discharge UPS instead of controlling the power, i.e., allowing the enclosure to consume as much power as demanded when the UPS energy is still available for power capping. Clearly, this time they will run out of UPS energy quickly, and for the remaining time they resume their respective power control schemes to control the enclosure power consumption within 2600 W. Then we compare the five baselines with greedy UPS usage to EncloPC.

The comparison in Fig. 5(a) between *CoScale* with greedy UPS usage and EncloPC clearly shows the difference: while EncloPC can keep ESD discharging in the 12 min, *CoScale* has to finish discharging in less than 7 min because of the high power consumption (the power varies with the workload fluctuation in the Wikipedia traces [29]).

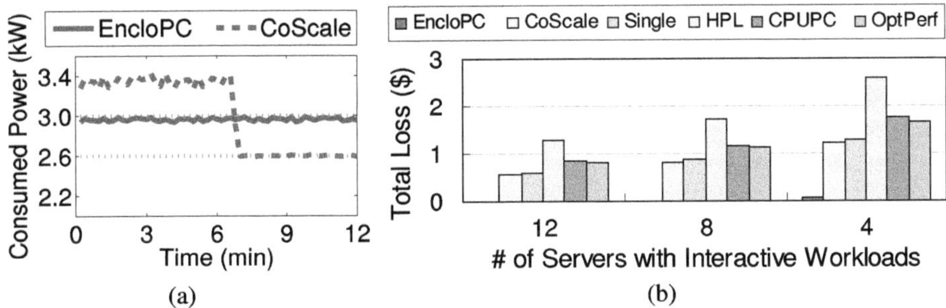

Fig. 5. (a) Enclosure power consumption with 2.6 kW enclosure power capacity, and (b) total profit loss, with greedy UPS usage (baselines) against controlled UPS discharge rate (400 W for EncloPC, so 3 kW in total).

The other single-server baseline *Single* has a similar problem as *CoScale*. In Fig. 5(b), we can see the greedy UPS usage leads to significantly more profit loss for *CoScale* and *Single* due to the early termination of discharging. The three multi-server baselines have more profit loss as well, despite that their interactive servers can use much more power before running out of UPS energy, because after that the limited power cap still leads to high profit loss for them. In total, EncloPC can reduce the profit loss for 92–99% by controlling the UPS power discharge rate in power capping.

6.3 Delay for Interactive Workload

EncloPC realizes power control with optimal power allocation between interactive and batch servers. In Fig. 6, we can see that the two single-server solutions can achieve good system performance for interactive workloads, but the three multi-server baselines lead to significantly increased processing delay with fewer interactive servers and more batch servers. The reason is that the single-server solutions determine the per-server power budget independently of the workloads, i.e., the per-server power budget is always 1/14 of the total server power budget. Such power budget is sufficient to run the interactive servers at the peak CPU and memory frequencies most of the time, because they are usually underutilized (16.5% utilization on average according to the traces from Wikipedia [29]). On the other hand, those multi-server baselines are designed to give higher power budgets to the servers with higher utilizations, which means the batch servers are given more power since they are usually fully utilized (when processing SPEC CPU2006 benchmarks). Hence the power budgets allocated to interactive servers are much less, and the processing delay of interactive workloads become much longer. EncloPC keeps good system performance like *CoScale* and *Single*, because the power budget allocator tends to assign more power to the interactive servers for maximized total profits. In total, EncloPC can reduce the processing delay by up to 70% (when compared to *HPL*).

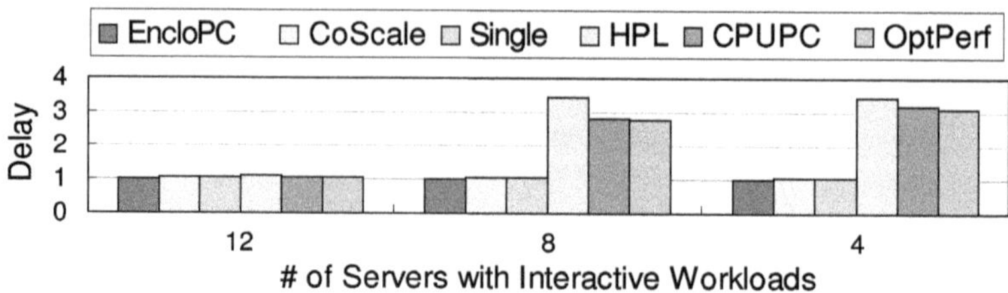

Fig. 6. Comparison on normalized system performance (processing delay) of the servers running interactive workloads.

6.4 Energy Costs of Batch Workload

Compared to the revenue of interactive workloads, the revenue of batch workloads is not affected by power capping as they are completed before the deadlines, as mentioned in Sect. 4.2. In this set of experiments, we compare the costs of extra energy consumption. Based on the results presented in Fig. 7, we can see that the two single-server solutions cause relatively high energy consumption in all the three scenarios, and the reason is also that they evenly allocate the per-server power budget independently of the workloads. Such power budget is inadequate for the batch servers because they are usually fully utilized. Consequently, the execution time of batch workloads is significantly extended, leading to high energy consumption. On the other hand, the three multi-server solutions consume less energy, because they allocate more power to the batch servers and thus shorten the execution time of batch workloads. However, since they are designed to improve the system performance of batch servers rather than energy saving, their energy consumption is still not minimized. In contrast, EncloPC explicitly minimizes the total energy consumption of batch servers, and therefore consumes the least energy among all the tested solutions. In total, EncloPC can reduce the energy consumption by up to 13% (when compared to *Single*).

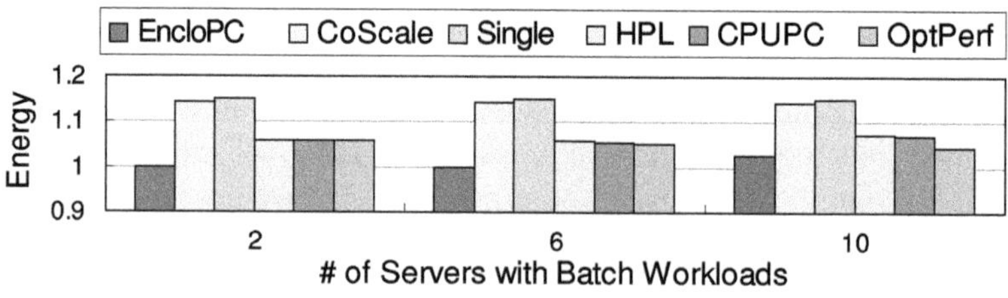

Fig. 7. Comparison on normalized energy consumption of the servers running batch workloads.

6.5 Total Profits

The total profit loss during the power peak is equal to the sum of revenue loss and extra energy cost. Figure 8(a) shows the impact of processing delay on the revenue loss during the power peak. We can see that the revenue loss is almost zero when the normalized

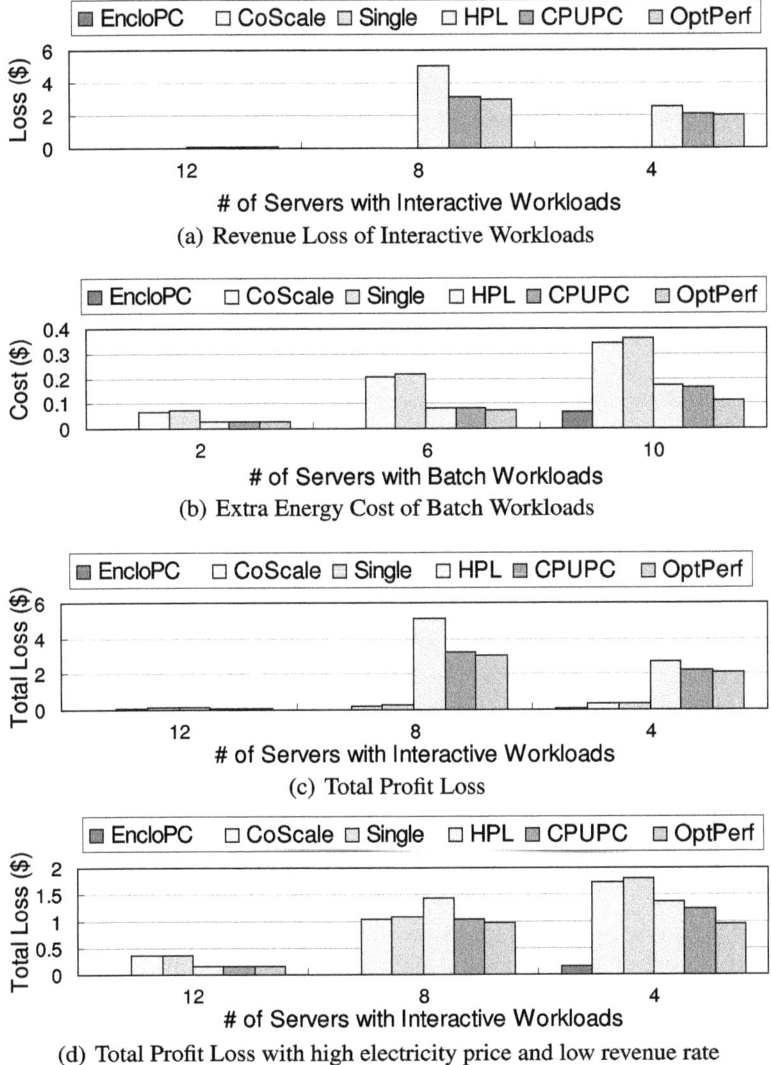

(a) Revenue Loss of Interactive Workloads

(b) Extra Energy Cost of Batch Workloads

(c) Total Profit Loss

(d) Total Profit Loss with high electricity price and low revenue rate

Fig. 8. The revenue loss (a), extra energy cost (b), and their sum, the total profit loss during the power peak (c). For (d), we increase the electricity price by 5 times and decrease the revenue rate by 5 times.

delay is close to 1 (for EncloPC and the two single-server baselines), but increases to 2–5 dollars with longer delays (for the three multi-server baselines). Figure 8(b) shows the impact of energy consumption on the electricity cost, with the electricity price at $0.05/kWh [8]. As expected, the extra energy cost rises up with the increase of batch servers and energy consumption. The extra cost for EncloPC to process each batch task is still less than $0.1 on average even for the 10 servers running batch workloads, while for the single-server solutions the extra cost can be up to \$0.37 per task. As shown in Fig. 8(c), EncloPC achieves the least profit loss. The advantage of EncloPC over *OptPerf* demonstrates the significance of handling interactive and batch workloads separately in power control. Since EncloPC is an enclosure-level method, each enclosure in a data center is controlled independently. Therefore, for a large data center hosting 50,000 servers as introduced in [25], if the batch workloads are processed once a day, EncloPC can save the total profit by up to \$6.7 Million annually (when compared to *HPL*).

While in Fig. 8(c) the revenue loss of interactive workloads is the dominant factor, the balance between it and the energy cost of batch workloads can vary from one data center to another and depend on the electricity price. Therefore, we set up another scenario by increasing the electricity price by 5 times and decreasing the revenue rate by 5 times, such that the energy cost can also be an important factor that affects the total profit loss. We present the results in Fig. 8(d), in which we can see *CoScale* and *Single* cause more profit loss than *CPUPC* and *OptPerf* for they have more energy costs. Despite of the changes of scenario settings, EncloPC still achieves significantly less profit loss than any baseline.

7　Conclusion

Existing work on computer system power control are point solutions that address only one or two of the challenges faced by server enclosure power control, which requires joint consideration of heterogeneous workload optimization goals and hardware actuation. In this paper, we have proposed EncloPC, power control for server enclosure with workload-specific optimization. In sharp contrast to existing work, EncloPC is a systematic framework that actively controls the power consumption by adapting both CPU and memory frequencies. Within the desired power budget, EncloPC optimizes the processing delay for interactive workloads and minimizes the energy consumed by batch workloads. EncloPC is rigorously designed based on MIMO MPC control theory for guaranteed control accuracy and system stability. The evaluation results show that EncloPC achieves up to 70% better system performance for interactive workloads and 13% more energy savings for batch workloads.

References

1. Chen, J., Manivannan, M., Goel, B., Pericàs, M.: Joss: joint exploration of CPU-memory DVFS and task scheduling for energy efficiency. In: International Conference on Parallel Processing (ICPP), pp. 828–838 (2023)
2. Chen, M., Wang, X., Li, X.: Coordinating processor and main memory for efficient server power control. In: International Conference on Supercomputing (ICS), pp. 130–140 (2011)
3. Company, H.P.D.: HP power capping and HP dynamic power capping for ProLiant servers (2011). http://h20565.www2.hp.com/hpsc/doc/public/display?docId=emr_na-c01549455
4. Standard Performance Evaluation Corporation: Specjbb 2005. https://www.spec.org/jbb2005/

5. Deng, Q., Meisner, D., Bhattacharjee, A., Wenisch, T., Bianchini, R.: CoScale: coordinating CPU and memory system DVFS in server systems. In: Annual IEEE/ACM International Symposium on Microarchitecture (MICRO), pp. 143–154 (2012)
6. Fan, S., Zahedi, S.M., Lee, B.C.: The computational sprinting game. In: International Conference on Architectural Support for Programming Languages and Operating Systems (ASPLOS), pp. 561–575 (2016)
7. Forrest, B.: Bing and google agree: slow pages lose users. http://radar.oreilly.com/2009/06/bing-and-google-agree-slow-pag.html
8. Govindan, S., Sivasubramaniam, A., Urgaonkar, B.: Benefits and limitations of tapping into stored energy for datacenters. In: Annual International Symposium on Computer Architecture (ISCA), pp. 341–352 (2011)
9. Horvath, T., Skadron, K.: Multi-mode energy management for multi-tier server clusters. In: International Conference on Parallel Architectures and Compilation Techniques (PACT), pp. 270–279 (2008)
10. Hossen, M.R., Ahmed, K., Islam, M.A.: Market mechanism-based user-in-the-loop scalable power oversubscription for HPC systems. In: International Symposium on High-Performance Computer Architecture (HPCA), pp. 485–498 (2023)
11. Huang, Z., Joao, J.A., Rico, A., Hilton, A.D., Lee, B.C.: DynaSprint: microarchitectural sprints with dynamic utility and thermal management. In: Annual IEEE/ACM International Symposium on Microarchitecture (MICRO), pp. 426–439 (2019)
12. Kontorinisy, V., et al.: Managing distributed ups energy for effective power capping in data centers. In: ISCA (2012)
13. Kumbhare, A.G., et al.: Prediction-based power oversubscription in cloud platforms. In: USENIX Annual Technical Conference (USENIX ATC), pp. 473–487 (2021)
14. Lefurgy, C., Wang, X., Ware, M.: Power capping: a prelude to power shifting. Springer Cluster Comput. J. Netw. Softw. Tools Appl. 11(2), 183–195 (2008)
15. Li, S., et al.: Thunderbolt: throughput-optimized, quality-of-service-aware power capping at scale. In: Symposium on Operating Systems Design and Implementation (OSDI), pp. 1241–1255 (2020)
16. Liu, F., Zhou, Z., Jin, H., Li, B., Li, B., Jiang, H.: On arbitrating the power-performance tradeoff in SaaS clouds. IEEE Trans. Parallel Distrib. Syst. 25(10), 2648–2658 (2014)
17. Liu, Z., et al.: Renewable and cooling aware workload management for sustainable data centers. In: SIGMETRICS (2012)
18. Ma, K., Wang, X., Wang, Y.: DPPC: dynamic power partitioning and control for improved chip multiprocessor performance. IEEE Trans. Comput. 63(7), 1736–1750 (2014)
19. Maciejowski, J.M.: Predictive Control with Constraints. Prentice Hall (2002)
20. Maseda, T., Enes, J., Expósito, R.R., Touriño, J.: Automated approach for accurate CPU power modelling. In: International Conference on Cluster Computing (CLUSTER), pp. 97–107 (2024)
21. Piga, L., et al.: Expanding datacenter capacity with DVFS boosting: a safe and scalable deployment experience. In: International Conference on Architectural Support for Programming Languages and Operating Systems (ASPLOS), vol. 1, pp. 150–165 (2024)
22. Raghavendra, R., Ranganathan, P., Talwar, V., Wang, Z., Zhu, X.: No power struggles: coordinated multi-level power management for the data center. In: International Conference on Architectural Support for Programming Languages and Operating Systems (ASPLOS), pp. 48–59 (2008)
23. Rohbani, N., Darabi, S., Sarbazi-Azad, H.: PF-DRAM: a precharge-free DRAM structure. In: Annual International Symposium on Computer Architecture (ISCA), pp. 126–138 (2021)
24. Sakalkar, V., et al.: Data center power oversubscription with a medium voltage power plane and priority-aware capping. In: International Conference on Architectural Support for Programming Languages and Operating Systems (ASPLOS), pp. 497–511 (2020)

25. Sverdlik, Y.: An eBay server's worth. DatacenterDynamics (2013)
26. Micron Technology: Calculating memory system power for DDR3 (2007). http://www.micron.com/~/media/Documents/Products/Technical%20Note/DRAM/TN41_01DDR3_Power.Pdf
27. Teodorescu, R., Torrellas, J.: Variation-aware application scheduling and power management for chip multiprocessors. In: International Symposium on Computer Architecture (ISCA), pp. 363–374 (2008)
28. Tondel, P., Johansen, T., Bemporad, A.: An algorithm for multi-parametric quadratic programming and explicit MPC solutions. In: Conference on Decision and Control (CDC), vol. 2, pp. 1199–1204 (2001)
29. Urdaneta, G., Pierre, G., Steen, M.: Wikipedia workload analysis for decentralized hosting. Comput. Netw. **53**(11), 1830–1845 (2009)
30. Wang, X., Chen, M., Lefurgy, C., Keller, T.: SHIP: scalable hierarchical power control for large-scale data centers. In: International Conference on Parallel Architectures and Compilation Techniques (PACT), pp. 91–100 (2009)
31. Wang, X., Chen, M.: Cluster-level feedback power control for performance optimization. In: International Symposium on High Performance Computer Architecture (HPCA), pp. 101–110 (2008)
32. Wang, X., Chen, M., Fu, X.: MIMO power control for high-density servers in an enclosure. IEEE Trans. Parallel Distrib. Syst. **21**(10), 1412–1426 (2010)
33. Wong, D., Annavaram, M.: KnightShift: scaling the energy proportionality wall through server-level heterogeneity. In: Annual IEEE/ACM International Symposium on Microarchitecture (MICRO), pp. 119–130 (2012)
34. Yang, H., Breslow, A., Mars, J., Tang, L.: Bubble-flux: precise online QoS management for increased utilization in warehouse scale computers. In: Annual International Symposium on Computer Architecture (ISCA), pp. 607–618 (2013)
35. Zhang, X., Shen, K., Dwarkadas, S., Zhong, R.: An evaluation of per-chip nonuniform frequency scaling on multicores. In: USENIX Conference on USENIX Annual Technical Conference (USENIX ATC), p. 19 (2010)
36. Zhao, D., et al.: Sustainable supercomputing for AI: GPU power capping at HPC scale. In: Symposium on Cloud Computing (SoCC), pp. 588–596 (2023)
37. Zheng, W., Ma, K., Wang, X.: Hybrid energy storage with supercapacitor for cost-efficient data center power shaving and capping. IEEE Trans. Parallel Distrib. Syst. **28**(4), 1105–1118 (2017)
38. Zheng, W., Wang, X.: Data center sprinting: enabling computational sprinting at the data center level. In: International Conference on Distributed Computing Systems (ICDCS), pp. 175–184 (2015)
39. Zheng, W., et al.: SprintCon: controllable and efficient computational sprinting for data center servers. In: International Parallel and Distributed Processing Symposium (IPDPS), pp. 815–824 (2019)

Updatable Verifiable Credential Sharing with Selective Disclosure

Jui-Yung Lin[1], Xingfu Yan[2(✉)], Wing W. Y. Ng[1], Gong Zheng[2], and Ying Gao[1]

[1] South China University of Technology, Guangzhou, China
`3956ray@gmail.com, {wingng,gaoying}@scut.edu.cn`
[2] South China Normal University, Guangzhou, China
`xfyan78@163.com, gongzheng@scnu.edu`

Abstract. Credential sharing is a critical mechanism for building trust, protecting privacy and improving collaboration efficiency in modern information systems. However, conventional credential management systems often rely on centralized management, which results in single point of failure and lack of user control over credentials. Moreover, traditional credential sharing mechanisms can easily lead to excessive disclosure of user privacy while ensuring credibility. To address these challenges, this paper proposes SelDUVCS, an adaptable system for verifiable credential sharing with selective disclosure, which minimizes sensitive information exposure through selective disclosure and enhances the system's privacy protection capability. By integrating vector commitment techniques with blockchain technology, SelDUVCS ensures verifiability of disclosed credentials while significantly reducing privacy exposure and eliminating reliance on centralized entities. Our experimental evaluation demonstrates that SelDUVCS achieves substantial improvements in privacy protection and credential verification efficiency, offering a viable alternative to traditional credential sharing frameworks.

Keywords: Credential Sharing System · Blockchain · Vector commitment · Selective Disclosure

1 Introduction

In the context of the evolving information society, a wide range of applications, including online education, government services, healthcare, and finance, rely on secure and trustworthy user authentication. This authentication has become a foundational component of modern digital infrastructure. Digital credentials (e.g., academic certificates, employment records, and digital IDs) have become the standard means of asserting identity, qualifications, and permissions in digital environments. As a key component of digital credentials, credential sharing has become increasingly essential-particularly as digital interactions across organizational and domain boundaries require trusted, cross-domain collaboration. Despite the growing significance of credential sharing, current credential sharing

H. Liu et al. (Eds.): ICA3PP 2025, LNCS 16381, pp. 359–378, 2026.
https://doi.org/10.1007/978-981-95-8399-7_20

systems face several challenges, including centralized management, high maintenance burden on issuing authorities, and limited user control over personal credentials. These challenges raise concerns [1–3].

Centralized management (i.e., all data is stored and processed in a single data center or storage node) is prone to a single point of failure (SPOF). If this center fails or is compromised, it can severely affect the availability of the entire system [4,5]. For example, in 2008, Netflix experienced a critical database storage failure[1], resulting in a three-day service outage that disrupted customer operations. A SPOF not only leads to system downtime, but also increases the risk of data breaches and unauthorized access, potentially resulting in data loss and business interruptions. According to Forrester's research, approximately 30% of companies that use centralized storage have experienced at least one major failure[2], leading to prolonged data loss or substantial disruption of services. As a distributed ledger (or database) technology, blockchain has the characteristics of decentralization, transparency, immutability, and traceability in practical applications [6,7], which is naturally suitable for solving the issues faced by centralized management in digital credential systems [8,9]. These characteristics make blockchain an important candidate for building a more secure and reliable digital credential system, which can enhance the controllability and privacy protection of user data while improving the robustness of the system.

A significant challenge in credential sharing is balancing user privacy with secure verification. Traditional methods often lead to excessive data disclosure, i.e., sharing more information than necessary during the verification process. This practice not only contravenes the data minimization principle but also exposes users to privacy risks such as identity theft or targeted attacks. Especially in blockchain-based systems, data transparency amplifies the impact of any disclosure [10]. To mitigate this, secure selective disclosure mechanisms have been developed to allow users to prove the validity of their credentials while revealing only the minimal subset of information required. Among existing solutions, redactable signatures [11] have emerged as an intuitive cryptographic primitive for achieving selective disclosure. They enable authorized redactors to redact certain parts of a signed message without invalidating the signature. Building on this, several selective disclosure schemes have been proposed [8,9]. However, these approaches often incur significant computational and communication overhead due to their reliance on complex computation operations, making them less suitable for real-time or large-scale applications. More critically, these approaches typically focus on static credentials and offer limited or no support for updates when credential attributes change over time. This lack of updatability presents a fundamental limitation in real-world scenarios where user information evolves dynamically.

In practical applications, credential attributes such as employment status, academic degrees, licenses, or health records often change, requiring updates without reissuing the entire credential. However, most verifiable credential

---

[1] Netflix: https://about.netflix.com/news/completing-the-netflix-cloud-migration.
[2] Forrester: https://www.forrester.com/report/.

schemes do not support efficient updates and require revalidation and reissuance upon any change, which limits flexibility and sustainability. Thus, there is a need for a verifiable credential sharing mechanism that supports efficient updates while maintaining selective disclosure.

We propose SelDUVCS, an updatable blockchain-based credential sharing scheme using vector commitments. SelDUVCS enables selective disclosure and fine-grained control over data sharing. The credential owner commits to the attributes and publishes public parameters on-chain for integrity and verifiability. When a request is received, the owner generates commitments for the requested attributes. SelDUVCS supports efficient updates to credentials and commitments when attributes change. The main contributions of this paper are as follows:

- SelDUVCS is a blockchain-based verifiable credential sharing scheme that supports both selective disclosure and efficient credential updates. Unlike traditional approaches that require full reissuance upon data changes, our scheme allows users to efficiently update credential attributes without revealing unrelated information, improving flexibility in dynamic application scenarios.
- SelDUVCS adopts a distributed credential sharing architecture by integrating vector commitments with blockchain. By storing public parameters and decentralized identifiers (DIDs) on-chain, the system ensures transparency, tamper-resistance, and eliminates the single point of failure common in centralized credential infrastructures.
- We provide a formal security analysis of SelDUVCS, proving its unforgeability and data integrity under the Computational Diffie-Hellman (CDH) assumption. We also formally verify the correctness of selective disclosure operations. Extensive performance evaluations demonstrate that SelDUVCS achieves low verification overhead and is suitable for real-world deployment in dynamic and decentralized environments.

2   Related Work

In this section, we briefly review the credential sharing and vector commitments.

2.1   Credential Sharing

In distributed systems and blockchain applications, credential sharing plays a fundamental role, allowing users to prove their attributes or identity while ensuring privacy protection. In 2001, Camenisch et al. [12] introduced the anonymous credential system, which ensures the unlinkability of the identity of users through randomized signatures. However, the size of their signature grows linearly with the number of attributes in the signature. To address this issue, Camenisch et al. [13] proposed an idemix anonymous credential system based on blind signatures and zero-knowledge proofs. This approach enables the sharing of credentials without disclosing the attributes of the credentials, thus maintaining privacy. Building on these foundations, the LOSSW scheme [14] improves upon the Waters [15] signature scheme by supporting sequential aggregation of signatures. Despite its advantages, this approach incurs additional communication costs as it requires interaction between the signatories. Subsequently, Nguyen et al. [16] proposed Gradubique which utilizes the identity management, authorization mechanisms, and channel architecture of Hyperledger Fabric [17] to enforce fine-grained access control and ensure privacy protection in credential sharing. Additionally, Fan et al. [18] integrated linear secret sharing and zero-knowledge proofs in traffic networks, enabling verifiable credential sharing while preserving identity privacy and data integrity in decentralized systems.

Despite these advances, the issue of excessive data disclosure remains a concern during the credential sharing and usage process. To mitigate this, several selective disclosure techniques have been proposed in alignment with the principle of data minimization. The Coconut scheme [19] achieves selective disclosure of credentials by combining blind signatures and threshold signature techniques, which ensures the privacy of the user and the security of the credential through the controlled disclosure of attributes. In 2020, W3C team introduced the concept of verifiable credentials, a standardized digital credential format that supports data minimization[3], selective disclosure, and protection against correlation. This framework allows users to disclose only the essential information, thereby safeguarding privacy. Belchior et al. [20] introduce SSIBAC, which integrates DIDs and verifiable credentials, compiles access policies into verifiable presentation (VP) requests, and uses VPs with zero knowledge predicate proofs to authorize with selective disclosure while preserving unlinkability across organizations. Xu et al. [21] proposed a blockchain-based identity management and authentication scheme in mobile network, achieving selective disclosure through redactable signatures. Building on this, Mukta [9] introduced CredChain, a redactable signature-based self-sovereign identity (SSI) framework, enhancing privacy protection and verification, and serving as a reference for credential-sharing applications. Later, Liu et al. [8] used ring and redactable signatures for selective data disclosure, enabling controlled identity verification while ensuring privacy and security during data sharing.

2.2 Vector Commitment

Vector commitment is a cryptographic technique that allows one to commit to a sequence of values. It enables the verification of specific elements within the sequence without revealing or recomputing the entire commitment. The concept of vector commitments can be traced back to the introduction of Merkle trees in 1988 [22]. Merkle trees, a fundamental data structure, are primarily used to verify subsets of data, thereby reducing the computational overhead associated with data verification. The formalization of vector commitment schemes was proposed by Libert et al. [23], followed by Catalano et al. [24]. Inspired by Merkle trees, these schemes allow verification of arbitrary elements in a dataset without

---

[3] W3C: https://www.w3.org/TR/vc-data-model/.

recalculating the entire vector commitment. Subsequently, Papamanthou et al. [25] proposed a data-streaming vector commitment, which further improved verification efficiency. However, none of these approaches introduced methods for aggregating multiple vector commitments.

Afterwards, Gorbunov et al. introduced Pointproof [26], a method utilizing polynomial commitments to aggregate multiple vector commitments. In the same year, Tomescu [27] proposed an alternative approach within the polynomial commitment framework, incorporating a Merkle tree structure to create the AMT vector commitment scheme. This addressed the issue of partial verification inherent in traditional Merkle trees. Building on these advancements, Hyperproofs [28] designed a framework that can aggregate and maintain multiple vector commitment proofs. This approach targets efficient verification of multiple vector commitments, reducing storage and computational overhead in scenarios requiring validation of several elements.

3 Preliminaries

In this section, we briefly introduce bilinear pairing, CDH assumption, and vector commitments.

3.1 Bilinear Pairing

Let G_1, G_2 and G_T be three cyclic groups of prime order p. Denote the generators of G_1 and G_2 as g_1 and g_2, respectively. A bilinear pairing is a map $e : G_1 \times G_2 \to G_T$ that satisfies the following properties:

1. **Bilinearity:** For all $u \in G_1$, $v \in G_2$, and $a, b \in \mathbb{Z}_p$, it holds that $e(u^a, v^b) = e(u, v)^{ab}$.
2. **Non-degeneracy:** The pairing satisfies $e(g_1, g_2) \neq 1$. This ensures that the pairing is meaningful and non-trivial.
3. **Computability:** There exists an efficient algorithm to compute $e(u, v)$ for all $u \in G_1$ and $v \in G_2$.

3.2 CDH Assumption

Let G be a cyclic group with a prime order q, where g is a generator of G. Given a tuple (g, g^a, g^b), where $a, b \in \mathbb{Z}_q^*$, it is computationally infeasible for any probabilistic polynomial-time (PPT) adversary $\mathcal{A}$ to compute g^{ab} with non-negligible advantage.

3.3 Vector Commitment [24]

Vector commitment is designed to commit to an sequence of values, allowing a specific position in the sequence to be opened at a later time. Vector commitments are non-interactive primitives and can be described through the following algorithms:

- **VC.KeyGen**$(1^k, q) \rightarrow pp$: Given a security parameter k and a vector size q (where $q = poly(k)$), this algorithm produces public parameters pp.
- **VC.Comp**$_{pp}(m_1, \ldots, m_q) \rightarrow C$: Given pp and q messages $m_1, \ldots, m_q \in \mathcal{M}$, this algorithm generates a commitment C and auxiliary information aux, which describes the committed message contents.
- **VC.Open**$_{pp}(m, i, aux) \rightarrow \Lambda_i$: This algorithm generates a proof Λ_i which is used for verifying whether the message m is the i-th message in the commitment. If the commitment has been updated, aux may include corresponding update details.
- **VC.Ver**$_{pp}(C, m, i, \Lambda_i) \rightarrow 0/1$: The verification algorithm checks the validity of the commitment using C, m, i, and Λ_i. It returns 1 if and only if Λ_i is valid, C is derived from $m_1, \ldots, m_q$, and $m = m_i$.
- **VC.Update**$_{pp}(C, m, m', i) \rightarrow (C', U)$: The committer who produced C wants to change the i-th message to m' to update C. The algorithm takes the old message m, the new message m' and the position i as input and outputs a new commitment C' together with an update information U.
- **VC.ProofUpdate**$_{pp}(C, \Lambda_j, m', U) \rightarrow \Lambda'_j$: A client who owns a proof Λ_j, that is valid w.r.t. to C for some message at position j, can runs this algorithm to compute an updated proof Λ'_j, where Λ'_j is valid w.r.t. C' which contains m' as the new message at position i.

It is later possible to open the commitment w.r.t. specific positions (e.g., to prove that m_i is the i-th committed message). More precisely, vector commitments are required to satisfy what we call position binding. Position binding states that an adversary should not be able to open a commitment to two different values at the same position.

4   System Model and Security Model

4.1   System Model

As illustrated in Fig. 1, we propose an overall architecture of SelDUVCS. The system comprises four entities: Credential Authority(CA), the Data Owner (DO), the Data User (DU), and the Blockchain. The roles of these entities in the system are defined as follows:

- Credential Authority (CA): The CA is assumed to be a trusted entity responsible for issuing credentials and their commitments, generating public parameters, and publishing them to the blockchain platform. Each credential consists of multiple attributes supporting selective disclosure, which allows data owners (DO) to share specific attributes at their discretion. In addition, the CA publishes the public parameters to IPFS.
- Data Owner (DO): A data owner is registered on the blockchain platform and holds credentials. When sharing credentials, the DO aims to selectively conceal sensitive information intended for non-shared targets.
- Data User (DU): DUs access the data shared by DOs during the data-sharing process.

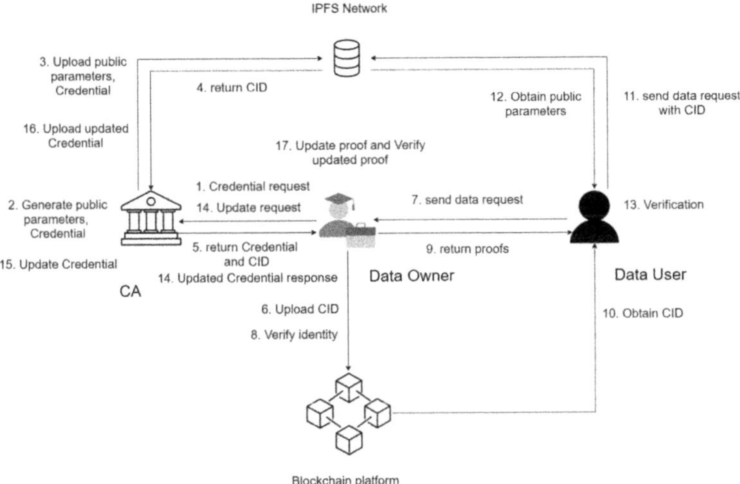

Fig. 1. The system model of the SelDUVCS

- Blockchain platform: An immutable entity that ensures the privacy of transactions. On the blockchain, DO and DU can maintain anonymity during private data-sharing transactions.
- InterPlanetary File System (IPFS): IPFS is responsible for storing public parameters and credentials, and generates a Content Identifier (CID) based on the data file, allowing users to retrieve the corresponding data via the CID.

4.2   Threat Model

Following the schemes [8,9], we assume that the CA is trusted and the blockchain platform acts as a neutral third party to register DIDs, neither of which are not susceptible to corruption. Based on this assumption, we outline the following potential threats.

- **Malicious DOs:** Some DOs are malicious. In practice, malicious DOs could forge the credentials to respond to the DU's request or just make impersonation attacks. However we assume DOs, even malicious DOs, could process the request honestly.
- **Semi-honest DUs:** DUs are honest-but-curious and could be curious about DOs' identity information. So they perform the credential verification honestly but they may learn some knowledge during this process.
- **Forgery Attack:** A malicious DO may forge credentials that do not correspond to their certain attributes or true identity, thereby impersonating other users or creating fabricated attributes. This type of deception can lead to severe security breaches, particularly in authentication systems, where

an attacker could exploit forged credentials to gain unauthorized access to restricted resources or manipulate/steal sensitive data, etc.
- **Unauthorized Data Access:** A DU may attempt to access sensitive information that the DO has not authorized for disclosure. For example, in academic credential verification scenarios, DUs may infer additional non-shared targets information, such as student ID numbers or home addresses, during the process, violating privacy and increasing the risk of data leakage.

4.3   Designed Goals

To address the potential attacks mentioned above, the security objectives of SelDUVCS are defined as follows:

- **Selective Disclosure:** DOs should be able to selectively disclose specific attributes of a credential in response to a request from a DU while concealing non-essential information. This ensures minimizing the disclosure of information, thereby preventing the excessive exposure of sensitive data.
- **Unforgeability:** Malicious DOs cannot forge credentials, that is, a DO forges the credentials or an attribute that does not exist in its credential, the verification will fail.
- **Efficiency of Credential Verification:** When selectively disclosing multiple attributes, the efficiency of verifying credential should be ensured. Even though the number of disclosed attributes or credentials increases, the system should minimize both computational and communication overhead.
- **Practicality:** Beyond privacy protection, the proposed scheme must be practical. This implies that both the computational complexity and communication overhead should be feasible and reasonable for real-world applications.

5   The Designed Scheme

In this section, we firstly proposed a vector commitment based on CDH with using blockchain platform, IPFS network and smart contract, then based on this scheme we construct our SelDUVCS scheme.

5.1   Workflow of SelDUVCS

Figure 2 presents the comprehensive system flowchart. During the system initialization phase, essential system parameters are generated. CA, DOs and DUs register themselves on the blockchain platform to obtain their own decentralized identifiers (DID). Then, CA sets some parameters for generating credentials.

Next in credential generation phase, a DO sends a credential request to CA which generates a credential $M = \{m_1, \cdots, m_q\}$, the corresponding commitment C for the DO and the public parameters pp. CA uploads, credential M, commitment C and parameters pp to the IPFS network. The IPFS network generates the corresponding CID for the public parameters and commitment C.

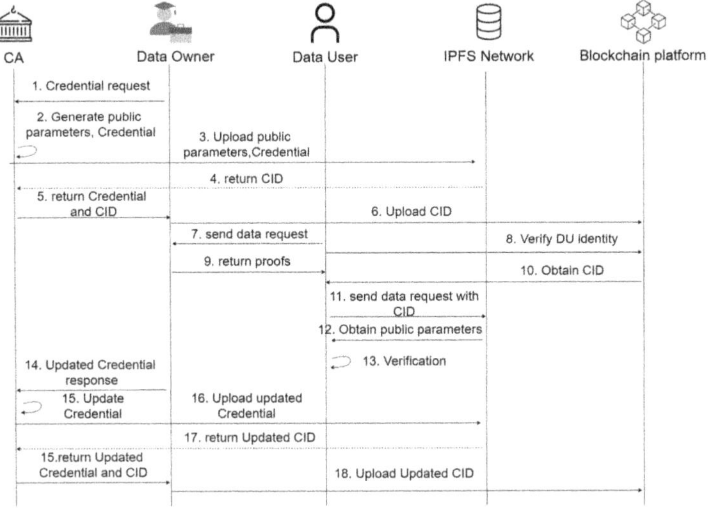

Fig. 2. The workflow of the SelDUVCS

In credential sharing phase, DU submits a credential usage request to the DO who computes the proofs of each attribute to be disclosed. Then, DO sends the proofs to DU. In the credential verification phase, DU retrieves the required CID from blockchain platform, the DU acquires the public parameters from the IPFS network utilizing the CID. After DU gets the required data, Once the DU has obtained the necessary data, it employs the public parameters to verify the proofs. This verification process ensures that the disclosed information, denoted as m_i is trustworthy and originates from a CA.

5.2 Designed Smart Contract

Our SelDUVCS system involves the smart contract to manage data, including register, update and get. Upon receiving the credential issued by the CA and the CID generated by IPFS, the DO invokes the register function in the smart contract to register the CID on the blockchain and enforce access control. Only DUs who have passed DID-based verification are authorized to access the CID. Moreover, the DO can invoke the smart contract to update the CID, ensuring that the updated data can be correctly accessed and utilized. Our designed smart contract mainly includes four algorithms, namely, Register, Update and GetInfo.(shown in Algorithm 1)

5.3 SelDUVCS Details

- *System Initialization*: Given the security parameter k, the public parameters $pp_1 = \{\mathbb{Z}_p, \mathbb{G}, \mathbb{G}_T, g, e\}$ is generated by the system, where $\mathbb{G}$ is a cycle group with prime order p, e is a bilinear map $e : \mathbb{G} \times \mathbb{G} \to \mathbb{G}_T$, and $g \in \mathbb{G}$ is a

---

Algorithm 1. CID Management Contract

Require: Function name, invoked parameters
Ensure: CID registration and update functions established

```
 1: struct CIDInfo {
 2:     address owner;                                    ▷ Content owner address
 3:     uint fileSize;                                        ▷ File size in KB
 4:     uint timestamp;                                  ▷ Last update timestamp
 5: }
 6: mapping(string ⇒ CIDInfo) public cidData;
 7: function REGISTER(string memory cid, uint256 fileSize)
 8:     public                                         ▷ Invoked by DU for new CID
 9:        require(msg.sender ∈ DIDRegistry)      ▷ Only the DO can register the CID
10:        require(fileSize > 0)                         ▷ Non-empty file constraint
11:        cidData[cid].owner ← msg.sender
12:        cidData[cid].fileSize ← fileSize
13:        cidData[cid].timestamp ← block.timestamp
14: function UPDATE(string memory targetCID, string memory newCID,
    uint256 newFileSize)
15:     public                                          ▷ Owner-initiated CID update
16:        require(cidData[targetCID].owner == msg.sender)  ▷ Ownership verifi-
        cation
17:        require(bytes(newCID).length > 0)              ▷ Non-empty CID constraint
18:        require(newFileSize > 0)
19:     if keccak256(abi.encodePacked(targetCID)) ≠
        keccak256(abi.encodePacked(newCID)) then
20:        address originalOwner ← cidData[targetCID].owner  ▷ Preserve owner-
        ship
21:        cidData[newCID].owner ← originalOwner
22:        cidData[newCID].fileSize ← newFileSize              ▷ Size update
23:        cidData[newCID].timestamp ← block.timestamp
24:        emit CIDUpdated(targetCID, newCID, originalOwner, newFileSize)
25: function GETINFO(string memory cid)
26:     public  view  returns  (address owner,  uint256 fileSize,  uint256
        timestamp)                              ▷ CID information retrieval[3,7](@ref)
27:        CIDInfo memory info ← cidData[cid]
28:        return (info.owner, info.fileSize, info.timestamp)
```

---

generator. CA, DOs and DUs register themselves on the blockchain platform to obtain their own decentralized identifiers (DID)[4].

- *Setup*: CA randomly selects a set of q numbers $z_1, \ldots, z_q \leftarrow \mathbb{Z}_p$ and computes $h_i = g^{z_i}$ (for $i = 1, \ldots, q$) and $h_{i,j} = g^{z_i z_j}$ (for $i, j = 1, \ldots, q, i \neq j$). CA uploads the $pp_2 = \{\{h_i\}_{i \in [q]}, \{h_{i,j}\}_{i,j \in [q], i \neq j}\}$ to the IPFS network. The IPFS network returns the CID corresponding to pp_2.

---

[4] Each DID is globally unique and associated with a public key that is included in a DID Document. This document allows others to verify the identity of the holder by ensuring that the public key matches the DID holder's private key.

- *Credential Generation*: A DO may provide his/her credentials to a DU for some purpose. A DO's credentials may include identification cards, degree certificates, and other information, which are issued by different CAs. For simplicity, we limit the discussion on one CA situation.
  - DO prepares their own information and then sends the requirement for a credential to the CA.
  - The CA checks the requirement and forms a credential $M = \{m_1, \cdots, m_q\}$, where m_i is the i-th attribute of the credential. Then CA computes the commitment C of the credential M as follows:

$$C = h_1^{m_1} h_2^{m_2} \cdots h_q^{m_q}. \tag{1}$$

  - Then, the CA sends the $\{M, C, CID\}$ to the DO. The DO deploy a smart contract to register the CID on the blockchain platform.
- *Credential Sharing*: DU sends a request to use the credential and submits its DID to DO for identity verification.
  - The DO verifies DU's identity. Upon receiving the DID from the DU, the DO initiates an identity authentication process by interacting with the blockchain. Specifically, the DO queries the blockchain to retrieve the corresponding CID, and return to DU.
  - After pass the DID verification, DO choose to disclose certain attributes $m_i \in M$ that satisfy the request of DU while concealing rest attributes. For each attribute m_i, DO computes the proof P_i as follows:

$$P_i = \prod_{j=1, j \neq i}^{q} h_{i,j}^{m_j} \tag{2}$$

  to prove that m_i is a attribute in the credential M.
  - Assume that attributes satisfying DU is $m_i \subset M$. DO compute the proofs and sends $\{m_i, P_i\}$ to DU.
- *Credential Verification*:
  - After receiving $\{m_i, P_i\}$ from DO, DU obtains the required CID from the blockchain platform, and downloads the required public parameters pp_2 from IPFS network by using the CID.
  - Then, DU verifies each proof P_i by using the VC.Ver algorithm as follows:

$$e(C/h_i^{m_i}, h_i) \overset{?}{=} e(P_i, g) \tag{3}$$

  m_i is the data at i-th position of $M = (m_1, \ldots, m_q)$ by computing.
- *Credential Update*: DO needs to update the credential attributes, which involves the following steps:
  - DO sends a credential update request to the CA and provide the updated information with CID.
  - Based on the received information, the CA forms a updated credential $M' = \{m_1, \cdots, m', \cdots, m_q\}$, where m_i is the i-th attribute of the credential, m' represent the message that the DO intends to update. Then CA computes the updated commitment C' of the updated credential M' as follows:

$$C' = C \cdot h_i^{m'-m} \tag{4}$$

- After updating the credential, the CA uploads $\{M', C'\}$ to the IPFS network and returns $\{M', C', CID'\}$ to the DO. Then, DO performs an update of the CID' on the blockchain through a smart contract invocation.
- *Proof Update:* After receiving the updated credential and commitment, the DO performs the following operations to update the proof:
  - Update all proofs corresponding to attributes except for m', using the following formula:

$$P'_j = P_j \cdot \left(h_i^{m'-m} \right)^{z_j} = P_j \cdot h_{j,i}^{m'-m} \tag{5}$$

  j is the position where the attribute m' is updated, and i corresponds to the other attributes in M'.
  - After computing P'_j, verify the correctness of the updated proof using Eq. (3).

Note. *The proof P_i can be computed offline. Since a credential can be used repeatedly, once DO receives $\{M, C\}$ from CA, it can calculate proofs for each attribute m_i in advance during an offline phase.*

6   Security Analysis

Theorem 1. *If the CDH problem is difficult, our SelDUVCS satisfies unforgeability.*

Proof. If we want to prove that SelDUVCS satisfies unforgeability, we have to prove that the vector commitment has the position binding property, i.e., an adversary cannot open $M_I = \{m_1, \cdots, m_I\}$ and $M'_I = \{m'_1, \cdots, m'_I\}$ at same positions.

Assume there exists a PPT adversary $\mathcal{A}$ that wins the Cert-Unforgeability Game in a non-negligible advantage. We construct a simulator $\mathcal{B}$ that can solve the CDH problem by running $\mathcal{A}$. The game process can be described as follows:

Setup:

1. $\mathcal{B}$ receives a Square-CDH instance (g, g^a), and aims to compute g^{a^2}
2. $\mathcal{B}$ randomly select an index $i \in \{1, \ldots, q\}$, which is the index at which the adversary $\mathcal{A}$ attempts to break the Cred-Unforgeability.
3. For all $j \in [q]$ and $j \neq i$, $\mathcal{B}$ selects $z_j \in \mathbb{Z}_p$ and sets:

$$h_i = g^a, \quad h_{i,j} = (g^a)^{z_j}, \quad h_j = g^{z_j}.$$

For position i, set $h_i = g^a$, The public parameters $pp = \{g, \{h_i\}_{i \in q}, \{h_j\}_{j \in q, j \neq i}\}$ are given to $\mathbb{A}$

Adversary Execution:

1. **Adversary's Forgery:** Assume that adversary $\mathcal{A}$, after receiving the public parameters pp, successfully forges two distinct data-commitment and proof pairs (m, P_i) and (m', P'_i) for position i, where $m \neq m'$.

2. **Compute g^{a^2}**: Algorithm $\mathcal{B}$ computes g^{a^2} as follows:

$$g^{a^2} = \left(\frac{P_i}{P_i'}\right)^{(m-m')^{-1}}$$

where P_i and P_i' are the proofs output by adversary $\mathcal{A}$.

3. **Verification:** Algorithm $\mathcal{B}$ verifies the proofs forged by adversary $\mathcal{A}$ using the following equation:

$$e\left(C, h_i\right) = e\left(h_i^m, h_i\right) e\left(P_i, g\right) = e\left(h_i^{m'}, h_i\right) e\left(P_i', g\right).$$

From this, it follows:

$$e\left(h_i, h_i\right)^{m-m'} = e\left(\frac{P_i'}{P_i}, g\right).$$

From the above algorithms, we can derive the following conclusion: If the adversary $\mathcal{A}$ successfully forges two valid commitment-proof pairs with a probability of ϵ, then algorithm $\mathcal{B}$ can compute g^{a^2} with a probability of ϵ/q. However, this contradicts the CDH assumption. Therefore, under the CDH assumption, the proposed scheme ensures the unforgeability of credentials as well as the integrity and security of the system.

6.1   Other Security Goals

- **Selective Disclosure:** In SelDUVCS, a DO can selectively disclose specific attributes from the credential based on the verification requirements. In response to the *req*, the data owner reveals a subset of credential attributes. The revealed subset is the minimum needed to satisfy the verifier. Due to the computational intractability of the CDH problem, it is mathematically infeasible for the data user to infer undisclosed data from the returned results. This ensures that non-disclosed content remains invisible, enabling selective disclosure of the credential.
- **Identity Hiding:** In our SelDUVCS scheme, the identity information of the DO is kept secret. In the SelDUVCS, credential attributes are computed and uploaded in such a way that they remain indistinguishable to others without the corresponding verification messages. These attributes are computed through vector commitments and subsequently uploaded to the blockchain platform, ensuring tamper resistance. During the verification process, besides the required messages, unnecessary information is concealed within the vector commitment. For the identity message C, it is implicitly derived from a subset of the credentials.
- **Resistance to Forgery Attacks**: Forgery could involve the manipulation of credential attributes. Since the credential attributes are concealed within the vector commitment, if a malicious Data Owner (DO) were able to forge a

certificate lacking a specific attribute, they could exploit their ability to solve the CDH puzzle, as demonstrated in Theorem 2, with the corresponding proof provided.

– **Resistance of impersonation attack:** For a malicious data user, impersonating others in our SelDUVCS scheme would be a challenging task. One potential approach to impersonation involves credential forgery. During the registration process, we obtain the DID from the blockchain platform. When credentials are uploaded via the CA, they are associated with this DID. Consequently, the data user is unable to gain unauthorized access to the data owner's credentials.

7   Performance Evaluation

In this section, we present a detailed analysis of the performance of the proposed scheme. The experiments are conducted on a server running Ubuntu 6.5.0-35-generic with an Intel(R) Xeon(R) Silver 4210 CPU @ 2.20GHz. The implementations developed in Python (version 3.10.12). For bilinear pairing operations, we employ the pypbc library, utilizing a Type A pairing curve. The pairing parameters are configured with a 128-bit prime field and an 80-bit group order.

Each experiment was repeated 100 times, and the average execution time is recorded. We evaluate the runtime performance of each algorithmic component. To evaluate the practical performance in SelDUVCS, we compare it with existing work PERCE [8], focusing on three key stages: credential generation, credential editing for selective disclosure, and credential verification. These stages directly impact the efficiency and usability of credential sharing.

We design two evaluation scenarios to assess the efficiency of the proposed scheme and conduct an additional experiment to measure the gas cost on the blockchain. In the first scenario, the number of credential attributes is fixed while the disclosure level is varied. In the second scenario, the disclosure level is fixed and the number of attributes is varied. The final experiment evaluates the gas cost incurred by executing the SelDUVCS smart contracts on the blockchain.

7.1   Performance Under Fixed Attribute Number

To evaluate the impact of different disclosure levels on system performance, we fixed the number of shared credential attributes at $\gamma = 100$ and $\gamma = 100$, and compared the efficiency of our SelDUVCS with that of PERCE [8] under these conditions.

As shown in Fig. 3a, Fig. 3c, The performance trends for both 100 and 500 attributes in SelDUVCS exhibit similar patterns. As the disclosure percentage increases from 10% to 50%, the execution times of both the CredGen and Verify phases grow linearly, with the overall cost increasing by nearly fivefold. At the highest disclosure level (50%), the CredGen time remains below 0.032 s for 100 attributes and below 0.018 s for 500 attributes, while the Verify time remains below 0.015 s for 100 attributes and below 0.8 s for 500 attributes.

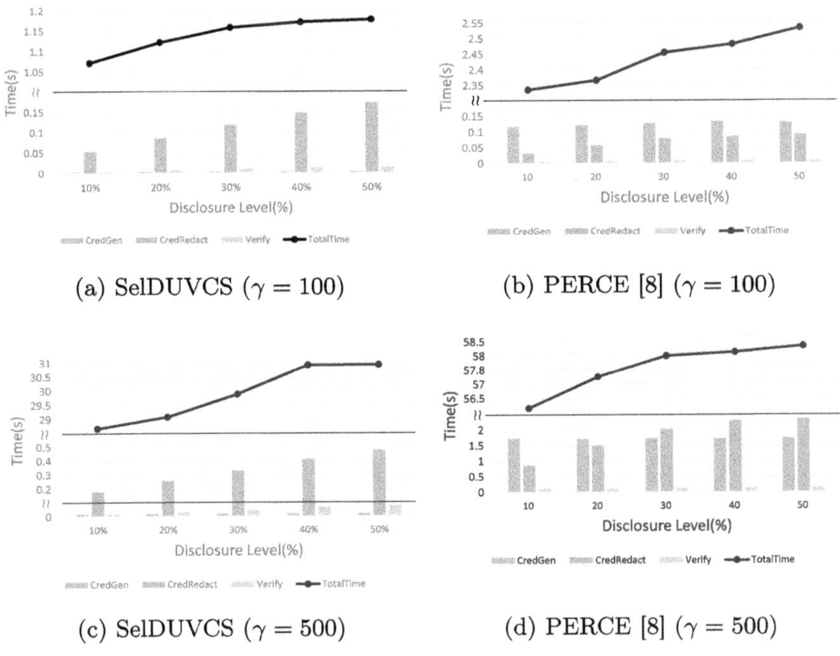

(a) SelDUVCS ($\gamma = 100$)   (b) PERCE [8] ($\gamma = 100$)

(c) SelDUVCS ($\gamma = 500$)   (d) PERCE [8] ($\gamma = 500$)

Fig. 3. Performance Comparison Under Varying Disclosure Levels

Compared to CredGen and Verify, CredRedact contributes a larger portion to the total execution time, making it one of the primary sources of performance overhead. The execution time of CredRedact increases by nearly fivefold between the lowest and highest disclosure levels. This higher cost primarily stems from the use of exponential operations with a time complexity of $O(n^2)$.

In contrast, for PERCE [8], as shown in Fig. 3b and Fig. 3d, the credential generation times remain nearly constant, with execution times reaching 0.13 s for 100 attributes and 1.7 s for 500 attributes at the highest disclosure level (50%). The time cost of CredRedact increases with the disclosure percentage, showing a threefold difference between 10% and 50%. This is because in the PERCE scheme, the redaction process involves removing undisclosed data when computing the redactable signature. As a result, the computational overhead peaks at higher disclosure levels due to the increased number of required exponential operations.

At the highest disclosure level (50%), the Verify time remains efficient, staying below 0.009 s for 100 attributes and below 0.03 s for 500 attributes. This is attributed to the design of the PERCE scheme, where only a single redactable signature needs to be verified regardless of the number of disclosed attributes.

Clearly, although the PERCE scheme exhibits lower growth in Verify time compared to SelDUVCS, its performance in terms of CredGen, CredRedact, and overall execution time increases more significantly. This indicates that our proposed scheme demonstrates better stability with respect to varying disclosure levels when the number of credential attributes is fixed.

7.2 Performance Under Fixed Disclosure Level

In this section, we fix the disclosure percentage and evaluate how increasing the number of credential attributes affects the execution time of each algorithm. Table 1 shows that at a 20% disclosure level, the execution times of CredGen and Verify in SelDUVCS increase linearly with the number of attributes. For the smallest dataset (50 attributes), CredGen and Verify take 0.002 and 0.0031 s, respectively, growing to 0.079 and 0.158 s for the largest dataset (2500 attributes).

Table 1. Execution Time Comparison at 20% Disclosure Level (where TotalTime includes Keygen)(**Unit: s**)

Execution	Schemes	50	100	500	2500
CredGen	PERCE [8]	0.053	0.122	1.702	37.056
	SelDUVCS	0.002	0.003	0.017	0.080
CredRedact	PERCE [8]	0.019	0.057	1.497	34.949
	SelDUVCS	0.027	0.086	1.964	46.931
Verify	PERCE [8]	0.005	0.006	0.023	0.111
	SelDUVCS	0.003	0.006	0.032	0.159
TotalTime	PERCE [8]	0.637	2.336	59.508	2398.194
	SelDUVCS	0.301	1.121	29.063	715.897

In contrast, CredRedact exhibits exponential growth, with execution time rising from 0.0267 s (50 attributes) to 46.93 s (2500 attributes). As shown in Eq. 2, the cost is directly related to the number of attributes, with a time complexity of $O(n^2)$, causing total execution time to increase exponentially from 0.3 to 715 s.

In PERCE, as shown in Table 1, CredGen and CredRedact exhibit exponential growth with increasing attributes. For the smallest dataset (50 attributes), CredGen takes 0.07 s and CredRedact 0.02 s, growing to 37.1 and 35 s for the largest dataset (2500 attributes). In contrast, the Verify phase scales linearly, with execution time increasing from 0.005 s to 0.12 s.

Table 2 shows that at the 50% disclosure level, SelDUVCS exhibits similar performance trends as at 20%. For the smallest dataset (50 attributes), CredGen takes 0.002 s and Verify 0.0078 s. For the largest dataset (2500 attributes), these values increase to 0.081 s and 0.41 s, respectively. The CredRedact phase also sees a substantial increase, from 0.049 s (50 attributes) to 46.93 s (2500 attributes).

Table 2. Execution Time Comparison at 50% Disclosure Level (where TotalTime includes Keygen)(**Unit: s**)

Execution	Schemes	50	100	500	2500
CredGen	PERCE [8]	0.054	0.129	1.721	37.428
	SelDUVCS	0.002	0.003	0.018	0.082
CredRedact	PERCE [8]	0.024	0.091	2.353	55.414
	SelDUVCS	0.049	0.173	4.695	118.911
Verify	PERCE [8]	0.006	0.009	0.028	0.138
	SelDUVCS	0.008	0.014	0.079	0.411
TotalTime	PERCE [8]	0.666	2.463	56.251	2467.57
	SelDUVCS	0.321	1.167	31.683	802.137

In PERCE, the performance trends at the 50% disclosure level are also consistent with those observed at the 20% level. For the smallest dataset (50 attributes), the execution times are 0.06 s for CredGen and 0.023 s for CredRedact. For the largest dataset (2500 attributes), the times increase to 37.4 s and 55 s, respectively. The Verify phase continues to grow linearly, with execution time increasing from 0.005 s to 0.12 s as the number of attributes increases.

7.3 Gas Cost

To evaluate the gas cost of SelDUVCS, we use the Remix IDE[5] on Ethereum [29]. Compilation uses Solidity 0.4.23 with compiler-default EVM settings, and experiments run on the Remix VM (Mainnet fork).

After compiling and deploying our smart contract into the above configured Remix we analyze the gas cost of each algorithm for which is one of main concern cost factors in Ethereum. The results of gas costs are shown in Fig. 4, where we set Gas Limit as 3,000,000 gas and each gas is worth 3GWei(corresponding to 0.000009USDT for an exchange rate 3000 Eth/USDT). One can find that the deployment spends the most gas of 9.422USDT(which includes the 8.326USDT). All other algorithms namely Register and Update spend 1.063USDT and 1.136USDT respectively. It is worthing noted that the deployment and register are only execution once, but Update algorithm will be triggered multiple times when DO needs to update their data.

---

[5] https://remix.ethereum.org/.

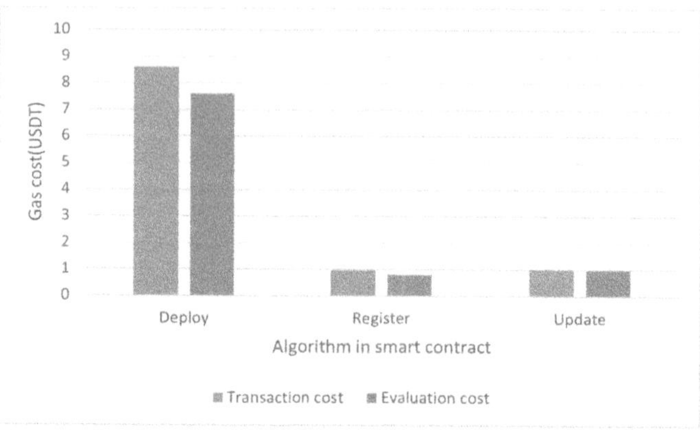

Fig. 4. Gas cost evaluation

8   Conclusion

In this work, we present SelDUVCS, a blockchain-based updatable credential-sharing framework that uses vector commitments to enable selective disclosure. SelDUVCS addresses the critical issue of excessive privacy disclosure in credential sharing by enabling fine-grained disclosure with reduced computational overhead. Our evaluation demonstrates key advantages. First, for fixed attributes, SelDUVCS outperforms PERCE in execution time for CredGen and Verify phases, with linear growth compared to PERCE's exponential growth. Though CredRedact has an exponential time complexity $O(n^2)$, SelDUVCS remains suitable for large-scale verification because the CredGen and Verify stages of SelDUVCS still maintain linear growth compared to the PERCE scheme. The design integrates IPFS with blockchain-anchored content identifiers (CIDs), ensuring immutability via smart contracts while mitigating on-chain storage costs. Deployment on Ethereum demonstrates manageable gas fees (one-time deployment and per-update)

However, the SelDUVCS scheme still encounters limitations in computational overhead. The $O(n^2)$complexity for redaction remains a bottleneck at scale. Future work will focus on optimizing CredRedact and Verify (e.g., batch verification, cross-commitment aggregation). Additionally, we plan to develop a multi-chain gas adaptation mechanism for deployment in resource-constrained environments such as L2 Rollups and IoT edge networks. Furthermore, potential linkability concerns in repeated credential presentations will be explored, with solutions potentially involving credential re-randomization.

Acknowledgments. We thank the reviewers that provided suggestions to improve this paper. This work was supported in part by the National Natural Science Foundation of China (Grant No. 62302173) and the Guangdong Basic and Applied Basic Research Foundation (Grant No. 2024A1515010143).

References

1. Brands, S.: A technical overview of digital credentials (2002)
2. Mishra, R.A., Kalla, A., Braeken, A., Liyanage, M.: Privacy protected blockchain based architecture and implementation for sharing of students' credentials. Inf. Process. Manag. **58**(3), 102512 (2021)
3. Bonomi, L., Huang, Y., Ohno-Machado, L.: Privacy challenges and research opportunities for genomic data sharing. Nat. Genet. **52**(7), 646–654 (2020)
4. Zhou, Q., Huang, H., Zheng, Z., Bian, J.: Solutions to scalability of blockchain: a survey. IEEE Access **8**, 16440–16455 (2020)
5. Huang, J., Kong, L., Chen, G., Min-You, W., Liu, X., Zeng, P.: Towards secure industrial IoT: blockchain system with credit-based consensus mechanism. IEEE Trans. Ind. Inf. **15**(6), 3680–3689 (2019)
6. Islam, M.R., et al.: A review on blockchain security issues and challenges. In: 2021 IEEE 12th Control and System Graduate Research Colloquium, pp. 227–232 (2021)
7. Nakamoto, S., Bitcoin, A.: A peer-to-peer electronic cash system. Bitcoin **4**(2), 15 (2008). https://bitcoin.org/bitcoin.pdf
8. Liu, Y., He, D., Feng, Q., Luo, M., Choo, K.K.R.: Perce: a permissioned redactable credentials scheme for a period of membership. IEEE Trans. Inf. Forensics Secur. **18**, 3132–3142 (2023)
9. Mukta, R., Martens, J., Paik, H.Y., Lu, Q., Kanhere, S.S.: Blockchain-based verifiable credential sharing with selective disclosure. In: 2020 IEEE 19th International Conference on Trust, Security and Privacy in Computing and Communications (TrustCom), pp. 959–966. IEEE (2020)
10. Li, T., Wang, H., He, D., Jia, Yu.: Blockchain-based privacy-preserving and rewarding private data sharing for IoT. IEEE Internet Things J. **9**(16), 15138–15149 (2022)
11. Sanders, O.: Efficient redactable signature and application to anonymous credentials. In: IACR International Conference on Public-Key Cryptography, pp. 628–656. Springer, Heidelberg (2020)
12. Camenisch, J., Lysyanskaya, A.: An efficient system for non-transferable anonymous credentials with optional anonymity revocation. In: International Conference on the Theory and Applications of Cryptographic Techniques, pp. 93–118. Springer, Heidelberg (2001)
13. Camenisch, J., Van Herreweghen, E.: Design and implementation of the idemix anonymous credential system. In: Proceedings of the 9th ACM Conference on Computer and Communications Security, pp. 21–30 (2002)
14. Steve, L., Ostrovsky, R., Sahai, A., Shacham, H., Waters, B.: Sequential aggregate signatures, multisignatures, and verifiably encrypted signatures without random oracles. J. Cryptol. **26**(2), 340–373 (2013)
15. Waters, B.: Efficient identity-based encryption without random oracles. In: Annual International Conference on the Theory and Applications of Cryptographic Techniques, pp. 114–127. Springer, Heidelberg (2005)
16. Nguyen, T.: Gradubique: an academic transcript database using blockchain architecture (2018)
17. Androulaki, E., et al.: Hyperledger fabric: a distributed operating system for permissioned blockchains. In: Proceedings of the Thirteenth EuroSys Conference, pp. 1–15 (2018)
18. Fan, K., et al.: A secure and verifiable data sharing scheme based on blockchain in vehicular social networks. IEEE Trans. Veh. Technol. **69**(6), 5826–5835 (2020)

19. Sonnino, A., Al-Bassam, M., Bano, S., Meiklejohn, S., Danezis, G.: Coconut: threshold issuance selective disclosure credentials with applications to distributed ledgers. arXiv preprint arXiv:1802.07344 (2018)
20. Belchior, R., Putz, B., Pernul, G., Correia, M., Vasconcelos, A., Guerreiro, S.: Ssibac: self-sovereign identity based access control. In: 2020 IEEE 19th International Conference on Trust, Security and Privacy in Computing and Communications (TrustCom), pp. 1935–1943. IEEE (2020)
21. Xu, J., Xue, K., Tian, H., Hong, J., Wei, D.S.L., Hong, P.: An identity management and authentication scheme based on redactable blockchain for mobile networks. IEEE Trans. Veh. Technol. **69**(6), 6688–6698 (2020)
22. Merkle, R.C.: A digital signature based on a conventional encryption function. In: Conference on the Theory and Application of Cryptographic Techniques, pp. 369–378. Springer, Heidelberg (1987)
23. Libert, B., Yung, M.: Concise mercurial vector commitments and independent zero-knowledge sets with short proofs. In: Theory of Cryptography Conference, pp. 499–517. Springer, Heidelberg (2010)
24. Catalano, D., Fiore, D.: Vector commitments and their applications. In: Public-Key Cryptography-PKC 2013, pp. 55–72 (2013)
25. Papamanthou, C., Shi, E., Tamassia, R., Yi, K.: Streaming authenticated data structures. In: Advances in Cryptology-EUROCRYPT 2013, pp. 353–370 (2013)
26. Gorbunov, S., Reyzin, L., Wee, H., Zhang, Z.: Pointproofs: aggregating proofs for multiple vector commitments. In: Proceedings of the 2020 ACM SIGSAC Conference on Computer and Communications Security, pp. 2007–2023 (2020)
27. Nicolescu, I.A.T.: How to keep a secret and share a public key (using polynomial commitments). PhD thesis, Massachusetts Institute of Technology (2020)
28. Srinivasan, S., Chepurnoy, A., Papamanthou, C., Tomescu, A., Zhang, Y.: Hyperproofs: aggregating and maintaining proofs in vector commitments. In: 31st USENIX Security Symposium (USENIX Security 22), pp. 3001–3018 (2022)
29. Wood, G., et al.: Ethereum: a secure decentralised generalised transaction ledger. Ethereum Proj. Yellow Pap. **151**(2014), 1–32 (2014)

sensitive data. For example, the human rights group network for human rights documentation in Burma (ND-Burma) [10] activists collect a lot of data on human rights violations on mobile devices (these sensitive data are encrypted). When activists cross the border, the border inspectors check the stored data in mobile devices. Once the existence of the encrypted data is found by the border inspectors, they coerce activists to disclose the decryption keys. Another example is when a videographer smuggled the evidence of human rights violations out of Syria [22]. Without the high-efficiency data protection mechanisms, the videographer had to hide a MicroSD card in a wound on his arm, for protecting the encrypted sensitive data. These compelling examples illustrate the critical need for advanced data protection mechanisms that go beyond traditional encryption.

In order to defend against powerful adversaries, plausibly deniable (PD) technology is proposed to hide the existence of sensitive data. The concept of PD is that sensitive data are encrypted into a ciphertext, and the ciphertext can be decrypted into a different innocuous plain text with a decoy key. Only using the true key, the ciphertext can be decrypted into the original sensitive data. Therefore, facing the threat attacks of a powerful adversary, the device owner can disclose the decoy key only, protecting sensitive information. The theoretical foundation of plausible deniability was established by Canetti et al. [5], who introduced the concept of deniable encryption schemes that allow senders to generate "fake random choices" that make ciphertexts appear to encrypt different plaintexts than they actually do. This groundbreaking work laid the foundation for numerous practical implementations, including hidden volume systems like TrueCrypt and VeraCrypt, which have become essential tools for protecting sensitive data in high-risk environments.

A variety of PD approaches have been proposed over the past few decades, which can be generally categorized into two types, according to storage medium: HDD-based PD approaches [15, 21], and SSD-based PD approaches [6, 11, 17, 26]. HDD-based approaches are proposed for HDD that supports in-place updates. These approaches achieve the plausible deniability property with the steganographic technique or the hidden volume technique, defending against coercive adversaries. The SSD-based PD approaches are proposed for flash memory. These approaches achieve the plausible deniability property with the specific flash file system, the specific flash translation layer, or the specific flash voltage encoding to cooperate with flash characteristics. The evolution of plausible deniability techniques has been closely tied to the underlying storage technologies and their unique characteristics. Early implementations focused primarily on hard disk drives, leveraging the in-place update capabilities and the relatively straightforward block-level access patterns. As storage technologies have evolved, so too have the requirements and challenges for implementing effective deniable storage systems.

However, achieving deniability in emerging IMR-based hard disks is much more challenging because: First, applying existing SSD-based PD approaches to IMR storage medium is inapplicable. In SSD, flash memory cells are programmed with out-of-place updates, whereas IMR supports in-place updates.

That is, flash memory cell can be programmed only when it has never been written to or has been erased, whereas in IMR program operation can be done at any time. Hence, SSD-based PD approaches become inadaptable for IMR due to these different storage characteristics. Second, existing HDD-based PD approaches are also inapplicable when directly applied to IMR storage medium. When the existing hidden volume technique is used in IMR directly, IMR will suffer from deniability compromises (the security analysis is explained later). On the other hand, the steganographic technique can be applied to IMR disks with a large amount of redundant data. But it leads to the inefficient use of disk space and increased I/O operations, which makes it unacceptable for IMR-based devices. The unique architectural characteristics of IMR technology introduce several novel challenges for implementing plausible deniability. The interlaced track structure, where bottom tracks are partially overlapped by adjacent top tracks, creates complex dependencies between data blocks that must be carefully managed to maintain both performance and security properties. Traditional hidden volume approaches, which rely on simple partitioning of storage space, fail to account for these interdependencies and can inadvertently expose the existence of hidden data through observable patterns in the IMR zone structures. Furthermore, the read-modify-write (RMW) operations required for updating bottom tracks in IMR systems create additional security vulnerabilities that are not present in conventional storage systems. These operations can leave forensic traces that may be detectable by sophisticated adversaries, potentially compromising the deniability properties of the storage system. The challenge lies in developing data management strategies that can effectively mask these operational patterns while maintaining the performance benefits that make IMR technology attractive for large-scale storage deployments.

In this paper, we introduce the first practical PD IMR system, SecureIMR, that achieves security against adversaries in IMR. SecureIMR supports two volumes, i.e., a public volume and a hidden volume, to write public data and sensitive data into IMR disks, respectively. Under the public volume, an authorized user stores public non-sensitive data. When operating in the hidden volume, the authorized user stores sensitive data where the existence of sensitive data needs to be denied. The public volume allocates free blocks from the head of the free zone pool, whereas the hidden volume allocates free blocks from the tail of the free zone pool, trying to avoid over-writing sensitive data by public data. With the enhanced data allocation strategies and programm strategies, SecureIMR stealthily isolates the two volumes. *To the best of our knowledge, our work is the first design that incorporates practical plausible deniability into IMR.* Our research represents a significant advancement in the field of secure storage systems, addressing a critical gap in the literature regarding plausible deniability for emerging storage technologies. The SecureIMR system introduces several novel concepts and techniques that extend beyond traditional approaches to deniable storage, specifically addressing the unique challenges posed by IMR technology.

- We introduce the first concrete attacks on a naive hidden volume-based PD system for IMR. Our attacks can successfully compromise deniability provided by the naive hidden volume-based PD system.
- We proposed the first practical PD IMR system, SecureIMR, that achieves security against coercive adversaries on IMR.
- We provide the security analysis and the implementation of SecureIMR using the open-source IMR simulator, IMRSim. The results show that SecureIMR incurs a negligible overhead, when compared with conventional non-PD IMR storage.
- Additionally, our work contributes to the broader understanding of security-performance trade-offs in modern storage systems. Through experimental evaluation, we demonstrate that it is possible to achieve security guarantees without significantly compromising the performance benefits that make IMR technology attractive for enterprise and cloud storage applications.

In the rest of this paper, Sect. 2 presents background and motivation. Section 3 elaborates on the details of the proposed SecureIMR scheme. We present security analysis and discussion in Sect. 4 and Sect. 5. The experimental results are analyzed in Sect. 6. And we conclude this paper in Sect. 7.

2 Background and Motivation

In this section, we give a brief background on interlaced magnetic recording (IMR), hidden volume technique used for deniable encryption. Then, we give the attack scenarios for attacking a naive hidden volume-based PD System for IMR.

2.1 Interlaced Magnetic Recording

Interlaced Magnetic Recording (IMR) represents a paradigm shift in magnetic storage technology, emerging as a response to the fundamental physical limitations that have constrained the continued scaling of conventional magnetic recording (CMR) systems. As the storage industry approaches the superparamagnetic limit, where thermal energy begins to randomly flip magnetic bits, innovative approaches like IMR have become essential for maintaining the trajectory of capacity growth that has characterized the storage industry for decades.

In hard disk drive (HDD), a sector (or called block) is the basic I/O unit, and multiple blocks constitute a track. The tracks in IMR are recorded in an interlaced manner, where two types of tracks exist, namely the double-sided squeezed track (or called bottom track) and the non-squeezed track (or called top track). Bottom tracks are overlapped partially by two adjacent top tracks. And, top tracks can be written with a narrower width than bottom tracks, and thus slightly lower capacity than that of bottom tracks (e.g., 80% that of bottom tracks). In IMR, bottom tracks are written first, and then top tracks are written.

Furthermore, aggregating multiple interlaced tracks as zones[2] (also called bands, or track groups) can efficiently manage read and write operations [2,23].

In IMR, top tracks can be rewritten as many times as needed. If the usage is greater than the total capacity of the bottom tracks, overwriting a bottom track will destroy the valid data stored in the two adjacent top tracks. Hence, the read-modify-write (RMW) strategy is used to overwrite a bottom track. That is, if the two adjacent top tracks have valid data, RMW first reads the two adjacent top tracks to a backup region. And then, RMW updates the bottom track with the received updated data. At last, after the bottom track is updated, RMW rewrites the adjacent top tracks with the backed-up data in the backup region.

2.2 Hidden Volume Technique

The field of plausible deniability has evolved significantly since its theoretical foundations were established in the late 1990s. Early work [5] introduced the fundamental concepts of deniable encryption, demonstrating that it is theoretically possible to construct encryption schemes where the same ciphertext can plausibly decrypt to different plaintexts depending on the key used. This seminal work laid the groundwork for numerous practical implementations and theoretical extensions.

The practical implementations of deniable storage systems has involved numerous engineering challenges and design trade-offs. Early practical systems like StegFS [15] employed steganographic techniques to hide data within seemingly innocuous file structures. However, these approaches often suffered from limited storage efficiency and vulnerability to statistical analysis attacks that could detect the presence of hidden data through anomalous patterns in file system metadata.

Hidden volume systems, e.g., TrueCrypt [1], Rubberhose [24], MobiFlage [20], DEFTL [11], MDEFTL [12], have become the most widely deployed practical implementations of plausible deniability. The hidden volume technique initially fills the entire disk with noise-like random data, and then creates two volumes, namely a public volume (or called an outer volume) and a hidden volume (or called a PD volume). The public volume is placed across the entire disk, where public non-sensitive data are encrypted using a decoy key. The key insight is that the hidden volume appears as random data within the free space of the outer volume, making its detection extremely difficult without knowledge of the encryption key. The hidden volume is placed in a secret offset within the disk, e.g., towards the end of the disk, where sensitive data are encrypted using a true key. Therefore, regular public data are stored in the public volume, whereas sensitive data will be stored in the hidden volume.

The deniability of the hidden volume technique comes from the fact that the user reveals the decoy key when facing a powerful adversary, and claims that the

---

[2] If IMR does not aggregate multiple tracks as zones, we take it as IMR with single zone. As zones improve I/O performance, IMR with multi-zone is used as default, unless otherwise stated.

remaining space only contains the initialized random data. The adversary can only decrypt public data with the decoy key, and he/she cannot differentiate the encrypted sensitive data from the initialized noise-like random data.

The hidden volume technique can ensure deniability when an adversary can access the disk only once (i.e., a single snapshot adversary). Note that, when an adversary can periodically access the disk (i.e., a multi-snapshot adversary), the hidden volume technique suffers from deniability compromises. This is because, the multi-snapshot adversary can detect the data changes in the free space of the public volume, by comparing the different snapshots.

2.3   Attack Scenarios

We provide concrete attack scenarios, in which the adversary is able to compromise deniability provided by the prior hidden volume-based PD system for IMR.

Adversarial Model: We assume that there is a powerful adversary who captures a device equipped with IMR. The attacker can directly gain physical access to the IMR disk through the known IMR interface commands, while bypassing the upper file system layer and the block drive layer. Only after the IMR disk is unmounted or powered off, the adversary can access the IMR disk. Hence, he/she cannot capture the running state in the IMR storage system. This is because, with the running state, it is virtually impossible to provide strong plausible deniability for all the PD approaches [4]. Furthermore, the adversary is only a single snapshot adversary, i.e., the adversary can have access to the disk only once.

The adversary coerces the device owner to reveal encryption keys. But, if the adversary does not observe any evidence of hidden data, the adversary gives up coercing the device owner further. The adversary is knowledgeable about the PD design. That is, the existence of the PD system is not the evidence of hiding data information. Only when the adversary finds the existence of sensitive data, he/she can coerce the owner to reveal encryption keys further (except for the decoy keys given first).

Attacking a Naive Hidden Volume-Based PD System for IMR: First, we show how to apply a naive hidden volume technique to IMR.

A naive hidden volume technique is applied as follows. Initially, we first fill the entire bottom tracks (in free state) with random data, and we then fill the entire top tracks (in free state) with random data. The two volumes, a public volume and a hidden volume, are created in IMR. In the public volume, we select free zones from the head of the free zone pool (FZP) to allocate zones for public data writes. After allocating the incoming data in one block within the selected zone, the block state changes to valid. In the hidden volume, we select the free zones from the tail of FZP to allocate zones for sensitive data writes. Under the two volumes, we use the classic two-phase data allocation strategy[3] to allocate data to a specific location within the selected zone. When update operations are performed to bottom tracks, RMW backs up and restores the data of the adjacent top tracks, once the two adjacent top tracks have valid data.

When an adversary tries to decrypt data in this naive hidden volume-based IMR, the adversary can identify four types of IMR zones, as shown in Fig. 2.

- *Type I:* Zones are filled with random data.
- *Type II:* Zones are filled with random data, corrupted data, and meaningful public data.
- *Type III:* Zones are filled with both random data, and corrupted data.
- *Type IIII:* Zones are filled with public data (or maybe including random data).

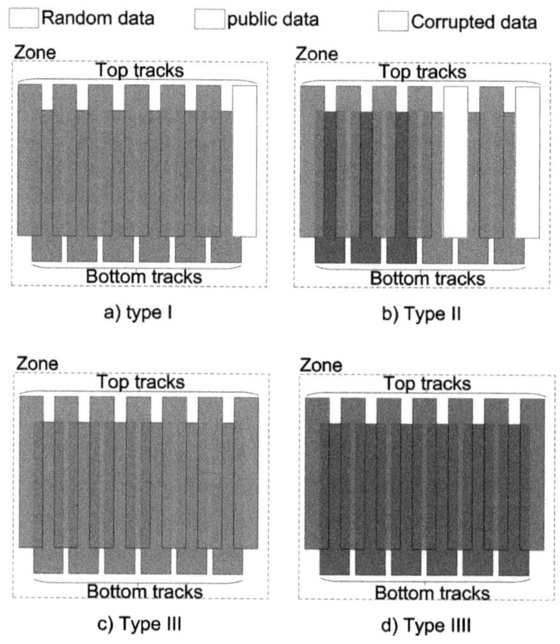

Fig. 2. zone types in IMR.

When all the data in a zone are decrypted as random data, this is a *type I* zone which is only filled with the initialized random data, as shown in Fig. 2(a). When the adversary tries to decrypt all the data stored in the *type IIII* zones using the decoy key, he/she should successfully decrypt data as meaningful non-sensitive public data, as shown in Fig. 2(d).

---

[3] In the first phase, if the usage is less than the total capacity of the bottom tracks, data are written to the bottom tracks sequentially. Then, in the second phase, space will be allocated from top tracks sequentially.

When all the data in a zone are decrypted as random data, corrupted data, and public data, the adversary thinks this is a *type II* zone (as shown in Fig. 2(b)) that has been used to store some public data in this zone. Note that the corrupted data are due to the special nature of IMR. Since we use the classic two-phase data allocation in IMR, public data are orderly stored the bottom tracks first, followed by the top tracks. In writing bottom tracks if the usage is less than the total capacity of the bottom tracks, programming can be done without any performance penalty in conventional IMR disks, i.e., programming data with in-place update. However, due to the initialized random data on the two adjacent top tracks, the data integrity of the adjacent top tracks will be substantially corrupted [9]. Although there are corrupted data except for random data and public data in IMR, we can plausibly interpret corrupted data due to the special nature of IMR.

When the adversary tries to decrypt data stored in one *type III* zone (as shown in Fig. 2(c)), he/she decrypts data as random data and corrupted data. Since one IMR disk without the hidden volume can contain only three types of IMR zones at most (i.e., *type I*, *type II*, and *type IIII*), the adversary knows the existence of the encrypted sensitive data. This is because *type III* zones are only written with sensitive data under the hidden volume (but with the decoy keys, these data can only be decrypted as random data). As programming sensitive data to the bottom tracks, the adjacent top tracks will be substantially corrupted (i.e., the corrupted data in some top tracks). Therefore, when the adversary finds out a *type III* zone, it leads to deniability compromise. This observation motivates us to design a practical PD system for IMR disks to eliminate deniability compromises.

Challenges in IMR-Based Deniable Storage: Therefore, the implementation of plausible deniability in IMR-based storage systems presents several unique challenges that distinguish it from conventional storage technologies. These challenges arise from the fundamental architectural differences of IMR technology and the complex interdependencies between different track types within the interlaced recording structure.

One of the primary challenges is the management of read-modify-write operations required for updating bottom tracks. When data in a bottom track needs to be modified, the system must first read the affected data from the two adjacent top tracks, perform the modification to the bottom track, and then rewrite the top track data. This process creates observable patterns that could potentially be exploited by adversaries to detect the presence of hidden data or to infer information about data access patterns.

Another significant challenge is the zone-based organization of IMR storage, where tracks are grouped into zones with specific characteristics and constraints. The allocation and management of these zones must be carefully orchestrated to maintain plausible deniability while ensuring efficient storage utilization and performance. Traditional hidden volume approaches that rely on simple partitioning schemes are inadequate for addressing the complex zone management requirements of IMR systems.

3   SecureIMR Design

The existing hidden volume-based PD systems used for IMR cannot defend against the adversary due to the special nature of IMR. To achieve deniability, we propose a practical PD IMR system for IMR-based devices, SecureIMR. Note that, in this paper, we elaborate our design with the classic two-phase data allocation, while the proposed scheme is applicable to IMR with other data allocation schemes (e.g., three-phase data allocation). The adaptability of SecureIMR to different data allocation schemes is a crucial design feature that ensures the system applicability across various IMR implementations. While we focus on the two-phase allocation strategy for clarity and simplicity, the underlying principles and security mechanisms can be extended to more complex allocation schemes that may be employed in future IMR systems. When the hidden volume technique is adopted into one IMR disk to achieve plausible deniability, two important aspects should be taken into consideration:

- *How to fill the entire IMR disk with random data initially?* The filling order can affect the data integrity due to the interlaced track layout in the IMR disk. The initialization process must carefully account for the track dependencies and RMW operations that occur during the random data filling process. Improper initialization can create detectable patterns thatcompromise deniability.
- *Where is the hidden volume placed in the IMR disk?* The hidden volume is commonly placed towards the end of the disk. Since the separation granularity for the public volume and hidden volume can be zone-level, track-level, or block-level, the placement location for the hidden volume can be different. The choice of granularity affects both security properties and performance characteristics, requiring careful optimization based on the specific threat model and performance requirements.

To address these two design issues, a novel SecureIMR scheme is proposed in IMR-based storage system to achieve deniability.

3.1   SecureIMR Overview

SecureIMR works in two volumes: the public volume and the hidden volume. SecureIMR hides sensitive data into the hidden volume. SecureIMR first fills the entire IMR disk with random data, and then each block state is changed to "initial" (a new block state in SecureIMR). This initialization process is carefully orchestrated to avoid creating detectable patterns that could reveal the presence of hidden data. The "initial" state serves as a crucial abstraction that enables the system to maintain consistent behavior across both public and hidden volumes.

To assist data allocation and prevent sensitive data from being overwritten by pubic data, SecureIMR maintains one free zone pool (FZP). FZP keeps track of the physic zone IDs in ascending order, which is available to be allocated to the new data. Note that FZP is stored in the reserved blocks[4]. SecureIMR allocates the zones from the head of FZP under the public volume, and selects the zones from the tail of FZP under the hidden volume. With this no overlap zone policy, SecureIMR prevents sensitive data from being overwritten by public data. FZP represents a critical component of the SecureIMR architecture, serving as the central coordination mechanism for zone allocation across both volumes. FZP maintains not only the availability status of zones but also tracks their allocation history and usage patterns to ensure that zone allocation decisions maintain plausible deniability properties.

Upon being coerced, the user should be able to convince the adversary that the embedding sensitive data are nothing but just normal initialized random data. To achieve this goal, under the hidden volume, the programming command to the bottom track is performed with RMW (both for first write commands and update commands). The consistent use of RMW operations for all bottom track writes in the hidden volume is a key innovation that eliminates the Type III zone vulnerability identified in naive implementations. By ensuring that all hidden volume operations follow the same patterns as normal IMR operations, SecureIMR maintains statistical indistinguishability between zones containing hidden data and those containing only public data.

3.2 Logical Storage Layout

The logical storage layout of SecureIMR is designed to provide a clear separation between public and hidden volumes while maintaining the appearance of a conventional IMR storage system. Figure 3 illustrates the overall organization of the SecureIMR storage layout.

The storage layout demonstrates the fundamental principle of SecureIMR: the public volume grows from the beginning of the storage space, while the hidden volume grows from the end. This bidirectional allocation strategy creates a natural buffer zone between the two volumes, reducing the risk of accidental overwrites while maintaining plausible allocation patterns.

The reserved area at the beginning of the disk contains critical system metadata, including the free zone pool (FZP) and other control structures necessary for SecureIMR operation. This area is carefully protected and encrypted to prevent information leakage that could compromise the security of either volume.

3.3 Public/Hidden Volume Management

Conventional IMR maintains a block translation layer that has a block-level mapping table (MT) to translate the logical block address (LBA) of data to

---

[4] The reserved blocks are only used to store regular metadata. The reserved blocks are at the beginning of the IMR disk.

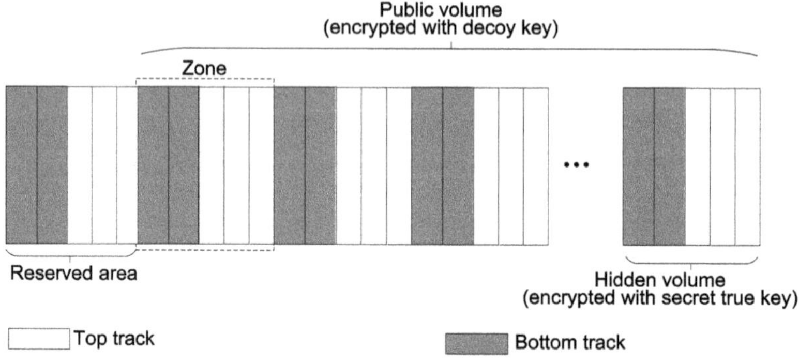

Fig. 3. The logical storage layout of SecureIMR.

the physical block address (PBA) [7]. In SecureIMR, a mapping table is used for each volume. As one user can access the IMR disk either under the public volume or the hidden volume, we can map each volume to a series of blocks with the corresponding MT.

Initializing: During initialization, SecureIMR fills the entire device medium with random data. SecureIMR first fills the entire bottom tracks with random data, from the outer diameter (OD) tracks to the inner diameter (ID) tracks. And then, SecureIMR fills the entire top tracks with random data in the opposite direction, avoiding the data loss of these random data. These random data will be used to deny the existence of the hidden sensitive data. After initialization, random data are stored in all the blocks. SecureIMR adds a new state, called "initial", to exhibit this characteristic, that is all the blocks are changed to the "initial" state. Then, SecureIMR initializes the public volume across the entire IMR disk (except for the reserved blocks) to store public data, as shown in Fig. 3. The public data will be encrypted with the decoy key. The corresponding mapping table is empty, and we can initially fill it with "0". To avoid the potential booting-time attack, the user should first initialize the public volume, and then initialize the hidden volume. When the user needs to initialize the hidden volume, SecureIMR uses the zone from the tail of FZP to store sensitive data, as shown in Fig. 3. The sensitive data will be encrypted with the true key. Similarly, the corresponding mapping table is empty initially. Without the true key, anyone is unable to differentiate sensitive data from the initialized random data. Hence, he/she is unable to identify the existence of the hidden volume. Under each volume, there is only one active zone for serving the user write requests, and the write request data will be allocated to one candidate block within the active zone under a given data allocation strategy, e.g., the classic two-phase data allocation. After all the free blocks within the active zone are allocated, this zone becomes deactivated, and another zone in FZP will be chosen as the new active zone.

Public/Hidden Volume Mounting: The public volume (resp. hidden volume) is encrypted by the decoy key (resp. true key). SecureIMR derives the key from a password with a key derivation function, e.g., HKDF [16]. If the user wants to boot the public volume (resp. hidden volume), the user enters the decoy password (resp. true password). With the correct decoy password (resp. true password), the user derives the decoy key (resp. true key) and decrypts the public (resp. hidden) mapping table. And then, the blocks used for the public volume (resp. hidden volume) are localized, and the public volume (resp. hidden volume) is decrypted and mounted. If the entered password is not the correct one, mounting the public/hidden volume is aborted.

3.4   Data Allocation and Block State Management

The data allocation strategy in SecureIMR is designed to maintain plausible deniability while optimizing performance. The system implements sophisticated block state management that tracks the lifecycle of data blocks across both volumes.

Figure 4 illustrates the block state transitions in both conventional IMR and SecureIMR systems. The introduction of the "initial" state is a key innovation that enables SecureIMR to maintain consistent behavior across different operational scenarios. The "initial" state serves multiple purposes in SecureIMR: it provides a consistent representation for blocks containing random initialization data, it enables the system to distinguish between truly free blocks and blocks that have been initialized with random data, and it supports the plausible deniability properties by ensuring that hidden data appears indistinguishable from initialized random data.

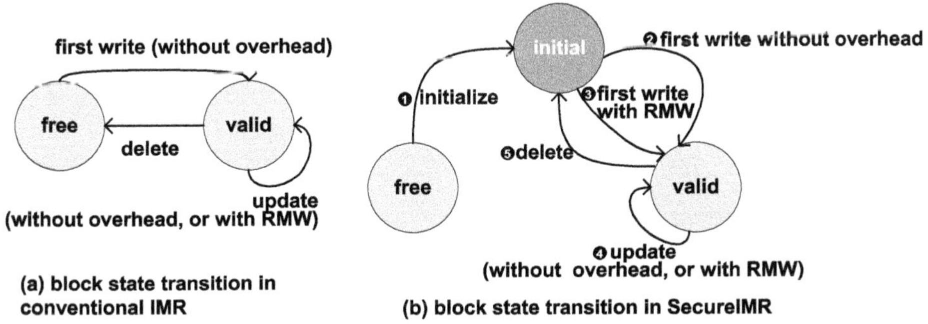

Fig. 4. Block state transitions.

Data Allocating in the Public Volume: After initialization, all the "free" blocks become "initial" blocks (❶ in Fig. 4(b)). Then, under the public volume, SecureIMR allocates blocks for first writes to the next write point of the active zone (❷). When the active zone cannot assign a new "initial" block, another new zone from the head of FZP is selected as the new active zone. When the deletion commands are processed to securely erase a deleted file, the state of the block returns to the "initial" state (❺). SecureIMR performs other regular operations (e.g., read operations, update operations that are ❹) as usual. As no design changes are needed to achieve deniability, we neglect these regular operations in the public volume. Next, we will show several key operations in the hidden volume in the following.

Data Allocating in the Hidden Volume: Under the hidden volume, SecureIMR allocates the blocks for the first writes to the next write point of the active zone (❸ if writing to bottom tracks, ❷ if writing to top tracks). When performing the first write operation to a bottom block, conventional in-place write (without RMW overhead) will destroy the initialized random data stored in the adjacent top blocks. Therefore, SecureIMR performs RMW, if we need to execute first write to bottom tracks (❸). When performing update operations, RMW is used to overwrite a bottom block as usual (❹). When the active zone cannot assign a new "initial" block, another new zone from the tail of FZP is selected as the active zone. When the deletion commands are processed to securely erase a deleted file, the state of the block also returns to the "initial" state (❺). Hence, by avoiding *type III* zones, SecureIMR accommodates the special nature of IMR to achieve plausible deniability.

The consistent use of RMW operations for all bottom track writes in the hidden volume is crucial for maintaining deniability. This approach ensures that hidden volume operations create the same observable patterns as normal IMR operations, preventing adversaries from detecting the presence of hidden data through analysis of zone types or operation patterns.

4   Security Analysis

In this section, we show that SecureIMR can provide deniability in IMR.

Theorem 1. *The deniability provided by SecureIMR in an IMR disk is comparable to the hidden volume technique, when facing a single-snapshot adversary.*

Proof. When a powerful adversary coerces a victim user to reveal encryption keys, the victim user can disclose the decoy key. Then, the adversary tries to compromise deniability with the decoy key in two attack scenes.

Scene 1: When the adversary decrypts all the public data with the decoy key, the adversary performs forensic analysis on IMR. First, the adversary can identify up to three block states of all the blocks: "free", "initial", and "valid". Without the true key, the sensitive data stored in the hidden volume only can be decrypted as random data. Thereby the block states of these blocks are "initial". As the adversary is knowledgeable about the design of the PD system, the three states cannot provide any "special" indications for the attack. Second, the adversary can identify up to three types of IMR zones: *Type I*, *Type II* and *Type IIII* (in Fig. 2). Since SecureIMR avoids the occurrence of *Type III* zone, only other three types of zones cannot lead to deniability compromises, as explained in Sect. 2.

Scene 2: When the adversary decrypts all the public data with the decoy key, the adversary performs additional I/O operations (write operations, read operations, and delete operations). In this case, the adversary tries to find a sighting of disk state changes, in order to compromise deniability. Firstly, performing reads does not positively correlate with disk state changes. And secondly, writes and deletes lead to the disk state changes. However, as hidden data can only be decrypted as random data under the public volume, the state changes due to writes and/or deletes, cannot provide any "special" indications for suspecting the existence of hidden data.

To summarize, Theorem 1 shows that, when facing a single-snapshot adversary, SecureIMR can provide deniability in IMR.

5   Discussion

Encryption. SecureIMR can integrate with encryption technology efficiently. The employed encryption algorithm makes ciphertext indistinguishable from random data, e.g., AES-XTS.

Weak Password Attacks. The security of all existing PD systems rely on the confidentiality of the true keys. There are several weak password (WP) attacks, e.g., online/offline brute force attacks. So, SecureIMR forces users to set high-entropy passwords that make IMR more resilient to WP attacks.

Denial of Service (DOS) Attacks. Denial of service (DOS) attack is inherent to all existing PD systems [6]. So, SecureIMR can not protect against DOS attack. Specifically, the attacker can simply sustainedly perform delete/write in disk to destroy all the data, including the potential hidden data. However, this DOS attack can be usually addressed with the important data backup on other backup devices [15].

Facing a Multiple-Snapshot Adversary. When facing a single-snapshot adversary, SecureIMR can provide deniability in IMR. However, at present, SecureIMR is not able to defend against a multiple-snapshot adversary, who can periodically access the disk. By comparing multiple snapshots, the adversary can detect unaccountable changes in the IMR storage state. Defending against a multiple-snapshot adversary is much more challenging, which will be pursued in our future work.

6   Performance Evaluation

In this section, we present the experimental methodology and the setting details, and then analyze the I/O results with comparison to the conventional IMR disk that cannot provide any deniability, showing the effectiveness of SecureIMR.

6.1   Experimental Methodology

We implemented SecureIMR storage system in an open-source IMR simulator, IMRSim [25]. To incorporate SecureIMR, IMRSim is modified to support the public volume and the hidden volume, where each volume has the corresponding mapping table. The free zone pool is added into the reserved area of the simulated IMR disk. We simulate a 128 GB IMR disk [19], and the parameters are listed in Table 1. For performance testing, we use the well-known benchmark, fio [3] with the non-buffered I/O operations to evaluate workloads. To implement the SecureIMR storage system, we first execute the initialization phase, by filling IMR with random data.

Table 1. The Main IMR Specification

Parameter	Value
capacity	128 GB
zone number	32
zone size	256 MB
block number per bottom track	568
block number per top track	456
block size	4096 B
data transfer rate	600 MB/s
rotational speed	5400 rpm
cache buffer size	64 MB
data allocation strategy	two-phase data allocation

In the experiments, we implemented and compared the following schemes: 1) `Baseline`: This is the default setting for conventional IMR disk, which cannot provide any deniability. 2) `SecureIMR Public`: This is the public volume used in the proposed SecureIMR system. 3) `SecureIMR Hidden`: This is the hidden volume used in SecureIMR.

6.2   Experimental Results

Read Performance: Figure 5(a) plots the normalized read response time comparison under different disk space usages. The results are normalized to `Baseline`. We observed that the read performance of both the public volume and the hidden volume of SecureIMR are almost the same as `Baseline`. This is because, SecureIMR does not need to modify read command to achieve deniability.

Write Performance: Next, we compare the normalized write response time for different schemes in Fig. 5(b). The results are also normalized to `Baseline`. We make two key observations from Fig. 5(b). First, the write operation in the

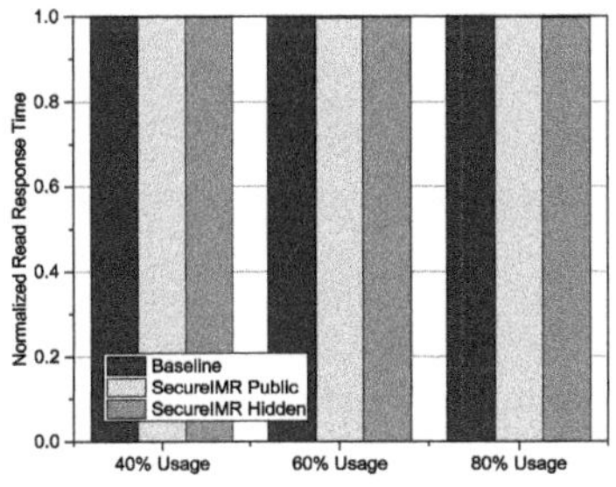 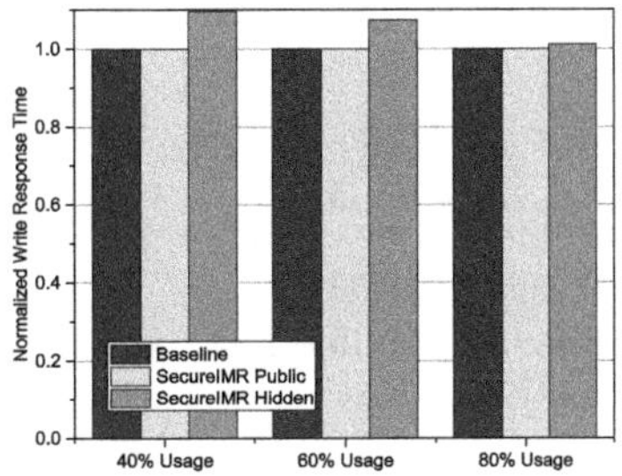

(a) The read latency comparison   (b) The write latency comparison

Fig. 5. The performance results under different disk space usages.

public volume of SecureIMR is comparable to `Baseline`, leading to similar write performance. Second, the write performance in the hidden volume of SecureIMR slightly degrades compared to `Baseline`. However, when the disk space usage is substantially increasing, the performance difference between `SecureIMR Hidden` and `Baseline` becomes weaker. This is because, under the higher disk space usage in `Baseline`, more update operations are performed with RMW.

7   Conclusion

This paper presents SecureIMR, the first practical plausibly deniable storage system specifically designed for Interlaced Magnetic Recording (IMR) technology. Our work addresses a critical gap in the literature by developing novel techniques for implementing deniable storage on emerging magnetic recording technologies that present unique challenges not addressed by existing approaches. The key contributions of this work include the identification and analysis of specific vulnerabilities in naive applications of existing deniable storage techniques to IMR technology, the development of sophisticated zone management and data allocation strategies that maintain plausible deniability while optimizing performance, and the implementation and evaluation of a complete deniable storage system that demonstrates practical feasibility. The analyses and evaluations show that, SecureIMR achieves plausible deniability, while incurring a negligible overhead.

Acknowledgement. This work is supported in part by the National Natural Science Foundation of China (grant number 61902136 and 61932010), and in part by Open Project Program of State Key Lab of Processors, Institute of Computing Technology, CAS under Grant No. CLQ202503, and by Fundamental Research Funds for the Central Universities (grant number 2019kfyXJJS090).

References

1. Truecrypt. http://truecrypt.sourceforge.net/
2. Amer, A., Holliday, J., Long, D.D., Miller, E.L., Pâris, J.F., Schwarz, T.: Data management and layout for shingled magnetic recording. IEEE Trans. Magn. **47**(10), 3691–3697 (2011)
3. Axboe, J.: Flexible I/O tester. https://github.com/axboe/fio. Accessed 2016
4. Blass, E.O., Mayberry, T., Noubir, G., Onarlioglu, K.: Toward robust hidden volumes using write-only oblivious ram. In: Proceedings of the 2014 ACM SIGSAC Conference on Computer and Communications Security, pp. 203–214 (2014)
5. Canetti, R., Dwork, C., Naor, M., Ostrovsky, R.: Deniable encryption. In: Annual International Cryptology Conference, pp. 90–104. Springer (1997)
6. Chen, C., Chakraborti, A., Sion, R.: PEARL: Plausibly Deniable Flash Translation Layer using WOM coding. arXiv, pp. 1109–1126 (2020). http://arxiv.org/abs/2009.02011
7. Hajkazemi, M.H., Kulkarni, A.N., Desnoyers, P., Feldman, T.R.: Track-based translation layers for interlaced magnetic recording. In: USENIX Annual Technical Conference, pp. 821–832 (2019)
8. Hwang, E., Park, J., Rauschmayer, R., Wilson, B.: Interlaced magnetic recording. IEEE Trans. Magn. **53**(4), 1–7 (2017). https://doi.org/10.1109/TMAG.2016.2638809
9. Hwang, E., Yoon, S.: Generalized interlaced magnetic recording with flexible inter-track interference cancellation. IEEE Trans. Magn. **54**(11), 1–5 (2018). https://doi.org/10.1109/TMAG.2018.2836361
10. Ingold, J.: Password case reframes fifth amendment rights in context of digital world. http://www.denverpost.com/news/ci_19669803. Accessed 01 Mar 2012
11. Jia, S., Xia, L., Chen, B., Liu, P.: DEFTL: implementing plausibly deniable encryption in flash translation layer. In: ACM Conference on Computer and Communications Security, pp. 2217–2229 (2017). https://doi.org/10.1145/3133956.3134011
12. Jia, S., Zhang, Q., Xia, L., Jing, J., Liu, P.: MDEFTL: incorporating multi-snapshot plausible deniability into flash translation layer. IEEE Trans. Dependable Secure Comput. **14**(8) (2021). https://doi.org/10.1109/TDSC.2021.3100897
13. Khati, L., Mouha, N., Vergnaud, D.: Full disk encryption: bridging theory and practice. In: Cryptographers' Track at the RSA Conference, pp. 241–257. Springer (2017)
14. Marchon, B., Pitchford, T., Hsia, Y.T., Gangopadhyay, S.: The head-disk interface roadmap to an areal density of Tbit/in2. Adv. Tribol. **2013** (2013)
15. Pang, H., Tan, K.L., Zhou, X.: StegFS: a steganographic file system. In: International Conference on Data Engineering, vol. i, pp. 657–667 (2003)
16. Percival, C., Josefsson, S.: The scrypt password-based key derivation function. Technical report (2016)
17. Peters, T.M., Gondree, M.A., Peterson, Z.N.J.: DEFY: a deniable, encrypted file system for log-structured storage. In: Annual Network and Distributed System Security Symposium (2015). https://doi.org/10.14722/ndss.2015.23078
18. Salo, M., et al.: The structure of shingled magnetic recording tracks. IEEE Trans. Magn. **50**(3), 18–23 (2014)
19. Seagate: Desktop HDD product manual standard models st1000dm003 st500dm002. https://www.seagate.com/www-content/product-content/desktop-hdd-fam/en-us/docs/100768625b.pdf. Accessed 2015

20. Skillen, A., Mannan, M.: Mobiflage: deniable storage encryptionfor mobile devices. IEEE Trans. Dependable Secure Comput. **11**(3), 224–237 (2013)
21. TrueCrypt: Free open source on-the-fly disk encryption software. http://www.truecrypt.org/
22. Westhead, R.: How a Syrian refugee risked his life to bear witness to atrocities. https://www.thestar.com/news/world/2012/03/14/how_a_syrian_refugee_risked_his_life_to_bear_witness_to_atrocities.html. Accessed 14 Mar 2012
23. Wu, F., et al.: Tracklace: data management for interlaced magnetic recording. IEEE Trans. Comput. **70**(3), 347–358 (2020)
24. Yu, X., Chen, B., Wang, Z., Chang, B., Zhu, W.T., Jing, J.: Mobihydra: pragmatic and multi-level plausibly deniable encryption storage for mobile devices. In: International Conference on Information Security, pp. 555–567. Springer (2014)
25. Zeng, Z., Chen, X., Yang, L.T., Cui, J.: Imrsim: a disk simulator for interlaced magnetic recording technology. NPC (2022)
26. Zuck, A., Li, Y., Bruck, J., Porter, D.E., Tsafrir, D.: Stash in a flash. In: USENIX Conference File Storage Technologies, pp. 169–185 (2018)

Heterogeneous Parallel Optimization Research of Plasma Guiding Center Orbit Simulation

Xianqian Meng[1,2], Tao Liu[1,2(✉)], Baofeng Gao[1,2], Ying Guo[1,2], Jingshan Pan[1,2], Dawei Zhao[1,2], and Xiaoming Wu[1,2]

[1] Key Laboratory of Computing Power Network and Information Security, Ministry of Education, Shandong Computer Science Center (National Supercomputer Center in Jinan), Qilu University of Technology (Shandong Academy of Sciences), Jinan, China
[2] Shandong Provincial Key Laboratory of Computing Power Internet and Service Computing, Shandong Fundamental Research Center for Computer Science, Jinan, China
liutao@sdas.org

Abstract. Due to the complex and highly nonlinear physical process of magnetic confinement fusion, the traditional numerical simulation methods have gradually failed to meet the high requirements of computational accuracy and speed in the research. Therefore, the adoption of high-performance computing technology has become the key to improve the simulation efficiency and accuracy. Plasma simulation plays a key role in optimizing the design of magnetic confinement fusion devices. In this paper, we propose a general heterogeneous parallel optimization strategy for the orbital simulation program of plasma guiding center (ORBIT), which covers the triple optimization methods of heterogeneous computing porting, dynamic load balancing and I/O storage optimization. In order to evaluate the effectiveness of the optimization strategy, this paper conducts comprehensive tests for different data sizes, and the test results show that the triple parallel optimization program achieves speedup ratios ranging from 16.45 to 25.6 times under a single process. In addition, when scaling up to 256 processes, the program performs very well in terms of weak scalability, up to 87%. This result shows that the optimization scheme proposed in this paper can effectively improve the computational efficiency of large-scale plasma simulation and maintain a significant performance improvement on heterogeneous platform, and also provides a reference value for more high-performance computing applications in the future.

Keywords: High-performance computing · Heterogeneous computing platform · Guiding center orbit simulation · Parallel optimization

1 Introduction

Magnetic confinement fusion, as an important path to realize controllable fusion energy, has the ultimate goal of constructing stable and efficient fusion reac-

H. Liu et al. (Eds.): ICA3PP 2025, LNCS 16381, pp. 398–418, 2026.
https://doi.org/10.1007/978-981-95-8399-7_22

tors to meet the future demand for clean energy [1]. Among various types of magnetic confinement devices, tokamak [2] has become one of the core directions of global fusion research due to its relatively mature technology path and good experimental results. However, accurate simulation and in-depth analysis of the motion behavior of charged particles in the plasma are urgently needed to achieve efficient energy confinement and reduce energy loss in devices such as tokamaks.

Charged particles in plasma exhibit nonlinear, high-dimensional dynamics in a complex three-dimensional electromagnetic field, and their microscopic behavior directly affects the macroscopic transport properties and plasma stability. Therefore, the guiding center orbit model, as a simplified modeling method of particle orbits based on neglecting the high-speed cyclotron motion, has become an important tool for understanding and predicting the particle transport behavior [3]. The ORBIT program, as a representative numerical implementation of this model, has been widely used in particle orbit simulation and transport studies in magnetic confinement fusion.

Although the capabilities of the ORBIT program have improved over the years, the code is still a serial application, which is now a barrier to guiding center orbit simulation. Therefore, the introduction of high-performance computing, especially heterogeneous parallel computing architecture [4], has become a key path to improve the performance of ORBIT. Notably, the natural independence between particle orbits in the guiding center orbit model provides ideal conditions for mass parallelization.

The main contribution of this paper is to propose a general heterogeneous parallel optimization strategy, including heterogeneous computing porting, dynamic load balancing and I/O storage optimization methods. In this paper, the ORBIT program is ported and optimized to a CPU-GPU heterogeneous environment, enriching the ecology of heterogeneous applications for large-scale particle simulation.

The paper is structured as follows. Section 2 presents the theoretical basis and analysis of guiding center orbit simulation. Section 3 describes the heterogeneous computating porting for guiding center orbit simulation. Section 4 describes the dynamic load balancing for guiding center orbit simulation. Section 5 describes the I/O storage optimization for guiding center orbit simulation. Section 6 gives the experimental results and performance analysis. Section 7 describes the related work. Finally, Sect. 8 gives the conclusion.

2 Theoretical Basis and Analysis of Guiding Center Orbit Simulation

In magnetic confinement fusion research, simulation of charged particle trajectories in complex magnetic fields is key to understanding plasma confinement properties, energetic particle transport, and magnetohydrodynamic instabilities. However, the numerical solution of this problem faces two central challenges:

multi-scale physical coupling [5] and the need for long-term evolutionary predictions. To address these challenges, this section will focus on the theory of guiding centers and its numerical integration methods, as well as on the analysis of hotspot functions.

2.1 Guiding Center Theory and Numerical Integration Methods

In the magnetic confinement fusion device, the motion of charged particles is mainly controlled by the Lorentz force [6]. The Lorentz force equation is expressed as Eq. (1).

$$\frac{d\mathbf{v}}{dt} = \frac{q}{m}(\mathbf{E} + \mathbf{v} \times \mathbf{B}) \tag{1}$$

where q, m, and $\mathbf{v}$ denote the particle charge, mass, and velocity vector respectively, $\mathbf{E}$ represents the electric field vector, and $\mathbf{B}$ the magnetic flux density vector.

However, the difference in spatial and temporal scales between the fast gyration [7] and the slow-variable drift [8] of the particle trajectories due to the magnetic field gradient and curvature makes it computationally expensive to solve the full orbital equations directly. For this reason, the guiding center equation neglects the details of the cyclotron phase through adiabatic conditions, and retains only the evolution equations for the guiding center position $\mathbf{X}$ and the parallel velocity $v_\parallel$ which drastically simplifies the calculation. The guiding center equation can be expressed as Eq. (2).

$$\frac{d\mathbf{X}}{dt} = v_\parallel \mathbf{b} + \mathbf{V}_d, \quad \frac{dv_\parallel}{dt} = -\frac{\mu}{m}\mathbf{b} \cdot \nabla B + \frac{q}{m}E_\parallel \tag{2}$$

In Eq. 2, $\mathbf{b}$ is the unit vector in the direction of the magnetic field; $\mathbf{V}_d$ is the drift velocity of the guide core, caused by the gradient and curvature of the magnetic field; μ is the magnetic moment, which represents the ratio of the perpendicular kinetic energy to the magnetic field; and $E_\parallel$ is the parallel electric field component.

The guiding center drift velocity $\mathbf{V}_d$ consists of a variety of physical effects, mainly including curvature drift, gradient drift, and electric field drift [9]. The specific representation of the guiding center drift velocity $\mathbf{V}_d$ is in the form of Eq. (3).

$$\mathbf{V}_d = \frac{mv_\parallel^2}{qB^2}\mathbf{b} \times (\mathbf{b} \cdot \nabla)\mathbf{b} + \frac{\mu}{qB}\mathbf{b} \times \nabla B + \frac{\mathbf{E} \times \mathbf{B}}{B^2} \tag{3}$$

Three of these terms correspond to curvature drift, gradient drift, and electric field-induced $\mathbf{E} \times \mathbf{B}$ drift, respectively. The expression for $\mathbf{V}_d$ shows that all of its components depend on the magnetic field through a fork-product form. These terms are ultimately related only to the magnitude of the magnetic field B and not to the specific components of the magnetic field. The magnetic field can be expressed as Eq. (4).

$$\mathbf{B} = \nabla \zeta \times \nabla \psi + F(\psi)\nabla \zeta \tag{4}$$

where ψ is the poleward flux, ζ is the toroidal angular coordinate, and $F(\psi)$ is the toroidal field function.

In guiding center orbit simulation, numerical integration methods are used to solve differential equations, especially when the equations cannot be solved by analytical methods. The Runge-Kutta 4th order (RK4) method [10] is ideal for solving the equations of motion of the guiding center (Eq. (2)) because of its high accuracy and numerical stability.

The basic idea of the RK4 method is to gradually approximate the exact solution by performing multiple calculations of the differential equation within each time step. Specifically, the method obtains a more accurate solution by computing four different slopes within each step (i.e., four estimates of the differential at the current point), and then weighting and averaging these slopes. By iteratively updating the guiding center position $\mathbf{X}$ and the parallel velocity $v_{\parallel}$, the RK4 method is able to accurately model the state of the guiding center at each moment in time, thus providing a reliable numerical solution for orbital simulations.

RK4 approximates the solution by a four-step slope-weighted average (show Eq. (5)):

$$y_{n+1} = y_n + \frac{h}{6}(k_1 + 2k_2 + 2k_3 + k_4) \tag{5}$$

where the parameter h is the time step, which determines the integration accuracy and computational efficiency; k_1, k_2, k_3, and k_4 are the intermediate slopes, which are obtained by weighting the values of the functions at different positions.

2.2   Profiling of Hotspot Functions

Based on the theoretical model and numerical integration method described above, the guiding center orbit simulation program faces high computational cost during long time evolution or multi-particle simulation. In order to locate the performance bottleneck, this section uses the gprof tool to analyze the performance of the program, focusing on the computational time consumption of the core functions, so as to propose targeted optimization schemes. According to the analysis results (see Fig. 1), the hotspots of the program are mainly concentrated in the field function, ptrba function and onestep function. These three functions constitute the core computational link of the guiding center orbit simulation.

The field function interpolates the magnetic field vector, magnetic field gradient tensor, and other magnetic field-related quantities (e.g., curvature) from the discrete magnetic field grid according to the current position of the particle. The ptrba function computes all drift terms (e.g., curvature drift, gradient drift, etc.) and handles possible perturbed fields (e.g., magnetic perturbations) based on the output of the field function. The onestep function integrates the field and ptrba functions, completes the orbital advancement for a full time step, and updates the particle state using the RK4 numerical method.

With the increasing number of simulated particles and spatial resolution, the ORBIT program has gradually exposed computational bottlenecks under the traditional CPU serial or OpenMP multi-threaded framework, especially when dealing with large-scale data, it faces the problems of excessively long computation time and low resource utilization, which makes it difficult to satisfy the stringent requirements of the future fusion simulation on the coexistence of real-time and high-precision. Therefore, in the subsequent part of this paper, we will present the improvements in three areas, namely heterogeneous computing porting, dynamic load balancing, and I/O storage optimization for the guiding center orbit simulation program, in order to achieve a significant improvement in the efficiency of the ORBIT program.

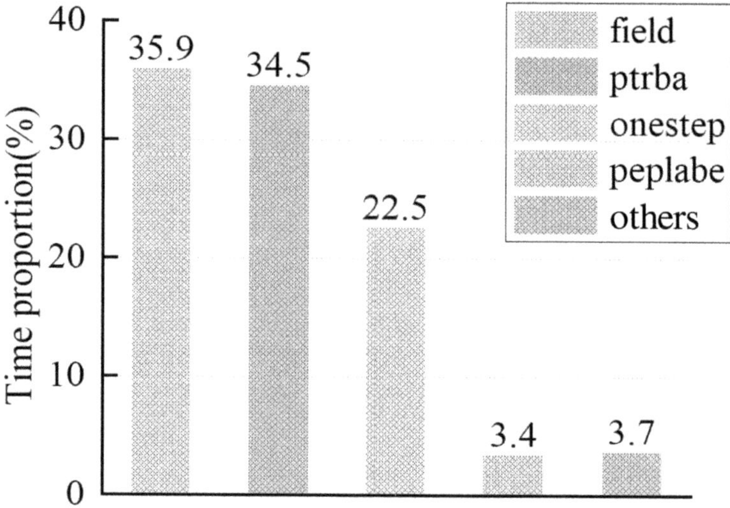

Fig. 1. The hotspots of the ORBIT program.

3 Heterogeneous Computing Porting for Guiding Center Orbit Simulation

3.1 Computing Kernel Optimization

This section uses the RK4 algorithm as an example to introduce computational optimization. As the core numerical integrator of the ORBIT program, the performance of the RK4 algorithm directly impacts the overall simulation efficiency. However, the serial computation of the RK4 algorithm faces significant performance bottlenecks, especially when calculating slopes on a per-particle basis. Since each RK4 step requires access to global magnetic field data, memory bandwidth becomes a bottleneck, preventing the full utilization of the CPU's multi-core parallel computing capabilities. As the number of particles increases, computation time grows exponentially, thereby affecting the overall simulation efficiency.

The independent computation of each particle's trajectory in the Algorithm 1 makes it a natural fit for the SIMT model for GPUs. To achieve efficient parallelization of the Algorithm 1, we assign particles to independent CUDA threads, and each thread block is responsible for processing a set of particles which are parallelized by thread-level parallelism and instruction-level parallelism. Then, by using the Nsight Compute analysis tool, we found that the memory bandwidth of the algorithm is one of the performance bottlenecks. To address this issue, we cached particle states in registers and optimized the GPU's memory hierarchy (e.g., shared memory) to reduce accesses to global memory, resulting in a 3.2% increase in bandwidth utilization. These strategies maximize the computational power of heterogeneous platforms, accelerate data access, and alleviate memory bandwidth bottlenecks.

Algorithm 1: RK4 Algorithm Porting Optimization

Data: $y_0 \in \mathbb{R}^n$, $[t_0, t_{\text{end}}]$, h, b

Result: y_{end}

1 **Initialize:**

   – Allocate memory for y_d, $k_{1,d}$, $k_{2,d}$, $k_{3,d}$, $k_{4,d}$, $y_{\text{temp},d}$

   – $y_d \leftarrow y_0$, $t \leftarrow t_0$

   – Compute $N_b \leftarrow \lceil n/b \rceil$

while $t < t_{\text{end}}$ **do**

    Stage 1:

    Kernel f: $\langle N_b, b \rangle \rhd (t, y_d) \rightarrow k_{1,d}$

    Stage 2:

    $y_{\text{temp},d} \leftarrow y_d + \frac{h}{2} k_{1,d}$

    Kernel f: $\langle N_b, b \rangle \rhd (t + \frac{h}{2}, y_{\text{temp},d}) \rightarrow k_{2,d}$

    Stage 3:

    $y_{\text{temp},d} \leftarrow y_d + \frac{h}{2} k_{2,d}$

    Kernel f: $\langle N_b, b \rangle \rhd (t + \frac{h}{2}, y_{\text{temp},d}) \rightarrow k_{3,d}$

    Stage 4:

    $y_{\text{temp},d} \leftarrow y_d + h k_{3,d}$

    Kernel f: $\langle N_b, b \rangle \rhd (t + h, y_{\text{temp},d}) \rightarrow k_{4,d}$

    State Update:

    Kernel Update: $\langle N_b, b \rangle \rhd y_d \leftarrow y_d + \frac{h}{6}(k_{1,d} + 2k_{2,d} + 2k_{3,d} + k_{4,d})$ $t \leftarrow t + h$

Return: $y_{\text{end}} \leftarrow y_d$

3.2 Cross-Device Communication Optimization

After initially migrating the hotspot code to the GPU, we found that frequent CPU-GPU data transfers had become a performance bottleneck. Specifically, this manifested as follows: before each call to the GPU kernel, input data needed to be transferred from host memory to device memory; after computation, the results must be transferred back from the GPU to the CPU, increasing communication overhead. Due to the synchronous nature of data transfers, GPU compute units often remain idle while waiting for data transfers, and PCIe bandwidth limitations further reduce computational efficiency. Many intermediate computation results are repeatedly transferred between the CPU and GPU, yet these data are only used internally within the GPU, and transferring them back to the CPU serves no practical purpose, resulting in unnecessary data copying.

To reduce communication overhead, we use a device memory persistence strategy, the core idea of which is to extend the lifetime of data in GPU memory, avoid unnecessary CPU-GPU data transfers, and thus minimize the number of communications. This is achieved by storing intermediate variables in different types of memory (such as global memory) in device variable files, eliminating all intermediate data transfers, and only transferring data during program initialization and final result output.

4   Dynamic Load Balancing for Guiding Center Orbit Simulation

In the heterogeneous computing porting of ORBIT programs, the traditional task allocation strategy leads to a serious CPU-GPU load imbalance. Take the core function field as an example, the GPU side is overloaded while the CPU is idle, mainly due to the particle computation division that does not consider the architectural differences and the lack of dynamic adjustment mechanism.

To address these issues, this section proposes a dynamic load adjustment algorithm that balances the load through adaptive task scheduling and adjustment of scheduling parameters, thereby maximizing the utilization of computational resources.

4.1   Dynamic Load Adjustment Algorithm

This dynamic load adjustment algorithm can dynamically adjust the amount of computing tasks for both CPU and GPU according to the size of their load capacity, so that the heterogeneous platform tasks loaded to CPU and GPU can be completed in the same time period. The specific process is as follows:

Let cpu_time and gpu_time be the computation time of CPU and GPU respectively, B_{cpu} and B_{gpu} be the number of data blocks allocated by CPU and GPU respectively, the scheduling parameter ∂ ($\partial > 0$) be the degree of load inequality of the program, ∂_1 be the parameter of termination condition, local_time be the computation time of the task at this point in time, and min_time be the optimal time for task scheduling. The smaller the scheduling parameter ∂ is the smaller the program load inequality, ∂ can be expressed as Eq. (6).

$$\partial = \left| 1 - \frac{\text{cpu\_time}}{\text{gpu\_time}} \right| \tag{6}$$

In task scheduling, the CPU and GPU are each assigned a corresponding number of data blocks, with B_{cpu} initially set to 1 and B_{gpu} set to the remaining number of data blocks. Task scheduling calculates the time for each, and uses Eq. (6) to calculate the scheduling parameter ∂. In order to avoid the frequent task migration overhead caused by over-scheduling, and at the same time ensure that the system strikes a balance between computational efficiency and resource utilization, this algorithm sets the load variance threshold $\partial = 0.2$. The selection of this threshold is based on the fact that when the threshold value is too small, it causes unnecessary scheduling overheads and when the threshold value is too large, both loads deviate too much.

First, the basic growth component of task scheduling is logarithmic growth. This is because logarithmic growth rapidly increases task block allocation in the early stages and then levels off in the later stages, avoiding the unreasonable overloading of the CPU caused by exponential growth. In order to better achieve load balancing between the CPU and GPU, error weights need to be added. Error weights dynamically adjust the task allocation ratio based on the current execution time error between the CPU and GPU, balancing the load between the two.

Combining the basic growth component and the error weight, the change in the number of data blocks in the early stages of the CPU is expressed as Eq. (7).

$$B_{\mathrm{cpu}} = B_{\mathrm{cpu}} + \left[a \cdot \log(b \cdot t + 1) + c \cdot \left| \frac{\mathrm{gpu\_time} - \mathrm{cpu\_time}}{\mathrm{gpu\_time} + \mathrm{cpu\_time}} \right|^d \right] \tag{7}$$

Among them, the first and second parts are the basic growth part and the error weight, respectively. Parameter a controls the growth rate, b adjusts the impact of t on growth, t represents the number of scheduling times, c is the weight of the load adjustment error impact, and d represents the sensitivity of the exponential adjustment error. When $d > 1$, it is more sensitive to larger errors.

To avoid rapid load adjustments causing computational instability or inaccurate results, when the load balancing condition is met, the algorithm appropriately controls the rate of change of B_{cpu}. In this case, load distribution is performed according to $B_{\mathrm{cpu}} = B_{\mathrm{cpu}} + 1$, and the maximum values of cpu_time and gpu_time at this time are recorded as local_time, and then compare it to min_time.

If the current local_time is less than min_time, it indicates that the current task scheduling performance is better than the previous optimal state. In this case, local_time is assigned to mintime, and B_{cpu} and B_{gpu} are recorded. However, if local_time is not less than mintime, meaning that the current task execution time has not been reduced to the optimal time range, the algorithm will continue to execute the predefined scheduling cycle until the specified termination conditions are met.

The termination condition parameter ∂_1 is expressed as Eq. (8).

$$\partial_1 = 1 - \frac{\mathrm{cpu\_time}}{\mathrm{gpu\_time}} \tag{8}$$

When the value of ∂_1 exceeds 0.2, it indicates that in the current task scheduling, the CPU is undertaking a greater share of computational tasks or heavier workload compared to the GPU, resulting in a significant imbalance in load distribution between the two. Therefore, this state should be treated as a termination condition, and the scheduling loop should be halted. As shown in Algorithm 2.

Algorithm 2: Adaptive CPU-GPU Load Balancing System

---

1   **Main: if** *history matches* $n_{particles}$ **then**
2   Reuse distribution
3   **goto** Final
4   **Dynamic Tuning:** $b \leftarrow 128$, $n_{cycles} \leftarrow \lceil n_{particles}/b \rceil$
5   **for** $k \leftarrow 1$ **to** n_{cycles} **do**
6   GPU: Launch k blocks
7   Sync
8   Measure t_{GPU}
9   CPU: Process $k{\times}b$ particles
10   Measure t_{CPU}
11   **if** $t_{GPU} > 1.2 t_{CPU}$ **then**
12       $k \leftarrow k + \lceil 4\log(2k+1) + 4\frac{|t_{GPU}-t_{CPU}|}{t_{GPU}+t_{CPU}} \rceil$
13   **else if** $|t_{GPU} - t_{CPU}| \leq 0.2\max(t_{GPU}, t_{CPU})$ **then**
14   Update Y_{opt}
15   **else**
16   Save to history_data.txt
17   Terminate
18   **Final:** $n_{CPU} \leftarrow Y_{opt}{\times}b$; $n_{GPU} \leftarrow n_{particles}-n_{CPU}$

4.2   Persistent Caching Strategy

In the aforementioned dynamic load balancing algorithm, one issue that arises is that repeatedly executing task scheduling for duplicate input data results in excessive time overhead. To address this issue, a two-dimensional array structure cache table can be established to store historical data. Each row of this cache table stores the key parameters of a task scheduling instance, specifically comprising the following two columns: the first column represents the number of CPU data blocks, and the second column represents the input data volume N. Combining text files with arrays enables persistent storage to preserve historical data across program execution cycles. The specific steps are as follows:

First, the algorithm creates a text file during runtime to store historical data, serving as the "data persistence layer". Upon each startup, the algorithm parses the file and loads the historical records into the two-dimensional array to restore the previous scheduling history. Subsequently, during task scheduling, the algorithm iterates through the second column of the array to search for matching records. If a match is found, the historical results are directly returned to avoid redundant calculations; if no match is found, a new scheduling task is executed, and the results are updated in both the array and the text file for future use.

In terms of cache query and update strategies, this algorithm adopts a first-in, first-out (FIFO) strategy to manage the cache table. First, a variable position is introduced to record the position of the next update. The write pointer is updated cyclically through modulo operations (i.e., position = (position+1) mod MAX_SIZE), avoiding the branch prediction overhead caused by conditional judgments. By updating the cache table and text file data based on position, this FIFO strategy ensures efficient utilization of cache space and cyclic updates of historical data. The task scheduling cache process is shown in Fig. 2.

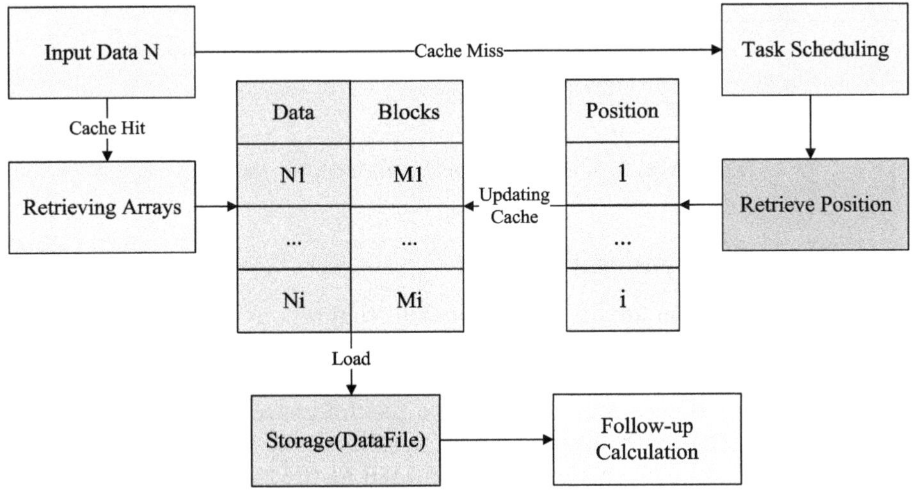

Fig. 2. Task scheduling cache processing process.

5 I/O Storage Optimization for Guiding Center Orbit Simulation

With the completion of heterogeneous computational porting (Sect. 3) and dynamic load balancing (Sect. 4), the bottleneck of the ORBIT simulation shifts to data storage. In the 400,000-particle ORBIT simulation, the I/O time share is as high as 43.69% and there is a redundancy problem in the text storage. Existing solutions such as HDF5 can reduce the storage space, but sacrifice the readability of data; while the traditional text format can not balance the efficiency and accuracy. For this reason, this paper proposes to optimize the text storage efficiency by adaptive grouping strategy, using GPU-accelerated index base extraction and tail normalization, and constructing two-level indexes (see Fig. 3).

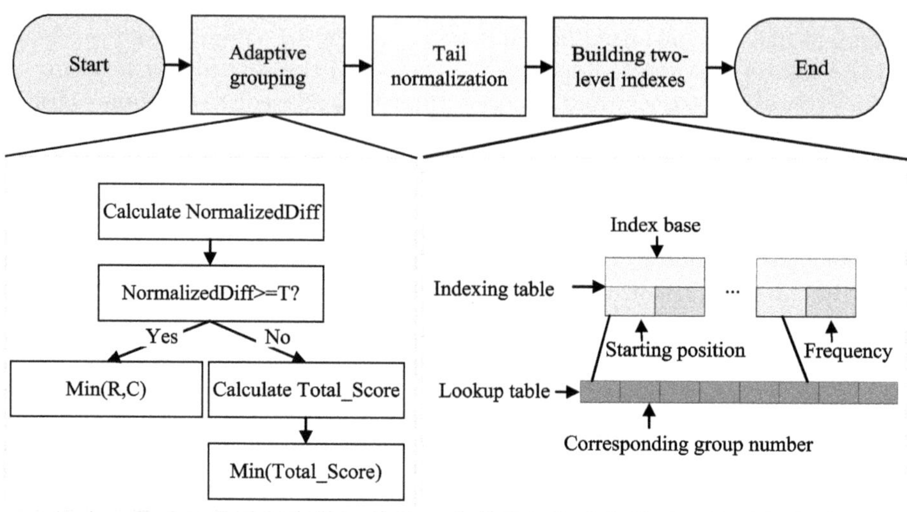

Fig. 3. I/O storage optimization process.

5.1   Adaptive Grouping Policy

In storage optimization for guiding center orbit simulation, the adaptive grouping strategy aims to dynamically adjust the grouping method based on the actual structure of the data, thereby reducing redundant storage and improving storage efficiency. First, it is necessary to determine the difference in the number of rows and columns in the text output data. If the difference in the number of rows and columns is large, the smaller side is selected as the basis for grouping; if the difference is small, a multidimensional analysis is performed to determine the optimal grouping strategy. The main process of the adaptive grouping strategy is as follows:

Let R represent the number of data rows (Rows), C the number of data columns (Columns), and T the threshold for row-column differences. The normalized difference NormalizedDiff for calculating the row-column ratio is expressed as Eq. (9).

$$\text{NormalizedDiff} = \frac{|R - C|}{R + C} \tag{9}$$

The range of NormalizedDiff results is $[0, 1)$. Values closer to 1 indicate greater differences between rows and columns, while values closer to 0 indicate that the number of rows and columns is similar.

If NormalizedDiff is not less than T, it indicates that the difference in the number of rows and columns is large, and the side with the smaller number of rows and columns is selected as the grouping; if NormalizedDiff is less than T, it indicates that the difference in the number of rows and columns is small, and multidimensional analysis is performed.

Multi-dimensional analysis selects the optimal grouping direction based on two indicators through weighted comprehensive scoring. The two indicators include index variance analysis and spatial continuity assessment. Among them, index variance analysis is used to select the direction with higher index consistency within the group, which dominates the grouping decision; spatial continuity assessment is used to select the direction with stronger local continuity within the group, which assists in performance optimization.

The variance of the exponent Var is expressed as Eq. (10).

$$\text{Var} = \frac{1}{N} \sum_{i=1}^{N} (\text{Exp}_i - \mu_{\text{Exp}})^2 \tag{10}$$

Spatial continuity assessment Cont is expressed as Eq. (11).

$$\text{Cont} = \frac{1}{N-1} \sum_{i=1}^{N-1} \text{Corr}(\text{Exp}_i, \text{Exp}_{i+1}) \tag{11}$$

Among them, N represents the total number of data points in the current group, Exp_i represents the exponential part of the i-th data point, μ_{Exp} represents the mean of the exponentials in the current group, and $\text{Corr}(\text{Exp}_i, \text{Exp}_{i+1})$ uses the Pearson correlation coefficient to determine the correlation between adjacent exponent.

The weighted comprehensive score Total_Score of the multi-dimensional analysis is expressed as Eq. (12).

$$\text{Total\_Score} = 0.7 \times \text{Var} + 0.3 \times \text{Cont} \tag{12}$$

By comparing the comprehensive scores from multi-dimensional analysis and selecting the smaller group as the basis for grouping, it is possible to effectively solve the problem of difficulty in grouping when the number of rows and columns is similar. Multi-dimensional analysis not only avoids the limitations of relying solely on the ratio of rows and columns, but also combines the actual distribution characteristics of the data to more accurately select the optimal grouping method.

5.2   Index Base Extraction and Tail Normalization

In the storage process of floating-point numbers, the exponent part usually occupies more storage space, while the mantissa part has a more significant impact on computational accuracy. By extracting the exponent part and normalizing the mantissa, storage space can be significantly reduced while maintaining computational accuracy.

First, the program needs to calculate the exponent part of each floating-point number, analyze the data within the group, and extract the exponent of each floating-point number. Based on these exponents, the largest exponent within the group is selected as the common exponent base. By utilizing the parallel processing capabilities of the GPU, this operation can efficiently determine the common exponent base and store it in a global memory array.

Once the common index base is determined, the program normalizes the floating-point numbers within the group. Specifically, the program adjusts all floating-point numbers according to the common index base, retaining only the mantissa and removing the exponent part, thereby significantly reducing storage space. This method not only reduces storage requirements but also ensures that numerical accuracy is retained and information loss is avoided. As shown in Algorithm 3.

Algorithm 3: Parallel Data Formatting using Maximum Exponent Normalization

Data: Device arrays: `ppd[]`, `ddum[]`, `xpp[]`, `zpp[]`, `rdum[]`, `pcdm[]`
Number of particles: `nprt`

1 **Phase 1: Compute Maximum Exponents**
2 **for** *each thread idx = 1* **to** *nprt in parallel* **do**
3   Load $\texttt{val1} \leftarrow \texttt{ppd}[idx]$, ..., $\texttt{val6} \leftarrow \texttt{pcdm}[idx]$
4   **if** *all values non-zero* **then**
5   **for** $k \in \{1,\dots,6\}$ **do**
6           $\texttt{exp}_k \leftarrow \lfloor \log_{10}(|\texttt{val}_k|) \rfloor$
7           $\texttt{atomicMax}(\texttt{d\_max\_exp}[k], \texttt{exp}_k)$

8 **Phase 2: Normalize Data**
9 **for** *each thread idx = 1* **to** *nprt in parallel* **do**
10   Load $\texttt{val1},\dots,\texttt{val6}$
11   **if** *all values non-zero* **then**
12   **for** $k \in \{1,\dots,6\}$ **do**
13           $\texttt{rel\_exp}_k \leftarrow \texttt{exp}_k - \texttt{d\_max\_exp}[k]$
14           $\texttt{mantissa}_k \leftarrow \texttt{val}_k/10^{\texttt{d\_max\_exp}[k]}$
15   Update $\texttt{array}_k[idx] \leftarrow \texttt{mantissa}_k$

16 **Output:** Normalized mantissas in `ppd[]`, ..., `pcdm[]`;
17 Maximum exponents in `d_max_exp[1..6]`.

5.3 Build Two-Level Index

Due to the excessive number of groups, duplicate public index bases may occur. To address this issue, this study designed an efficient two-level index structure aimed at solving the redundant storage problem caused by duplicate public index bases. This architecture effectively reduces the storage space of redundant data by constructing a two-level index structure, counting and mapping public index bases.

The first-level data structure is an index table entry, which includes the common index base, the starting position of the corresponding lookup table, and the number of elements. The index bases in the index table are arranged in ascending order to facilitate subsequent lookups. The second-level data structure is the main structure, which includes the index table, lookup table, actual index value array, total number of elements, and number of unique index bases. The lookup table is used to store the group numbers corresponding to each index base, while the actual index value array stores the values of the common index bases. The total number of elements records the total number of index bases, and the number of unique index bases indicates the number of index bases that appear uniquely.

The program reads the index bases from the index table in sequence, iterates through the actual index value array, and sequentially stores the corresponding group numbers for each index base into the lookup table. Whenever an index base and its group number are stored in the lookup table, the program records the current storage position to calculate the starting position of the next index base in the lookup table and stores it in the first-level data structure. Through the mapping relationship between the index table and the lookup table, the program can quickly look up the group number corresponding to the current index base, ensuring that each identical index base is output only once, thereby effectively reducing redundant data.

6 Experimental Results and Performance Analysis

6.1 Experimental Environment Configuration

This evaluation was based on a CPU-GPU heterogeneous computing environment with a hardware configuration of eight NVIDIA A100 GPUs. The specific software and hardware configurations are shown in Table 1. To test the program's versatility, all tests used small-scale, medium-scale, and large-scale input data. The data scale settings were designed to fully utilize the computing power of heterogeneous resources while demonstrating the impact of different data scales on optimization results. All results are averages of three rounds of testing.

Table 1. The hardware and software used in the experimental tests.

Hardware	CPU Model	Intel Xeon Platinum 8358
	Number of CPUs	2
	GPU Model	NVIDIA A100
	Number of GPUs	8
	GPU Memory	8*40 GB
Software	System	Linux 4.18.0
	Compiler	mpif90 and nvfortran

6.2 Performance Analysis

To test the effectiveness of program porting and optimization, we set the number of simulated particles to 100,000, 200,000, and 400,000 in single-process optimization, representing small, medium, and large data volumes, respectively. Figure 4 shows the acceleration achieved by different optimizations compared to the original serial ORBIT program. In the figure, the horizontal axis represents different data scales, labeled as small, medium, and large, respectively. The vertical axis represents the acceleration ratio, indicating the acceleration effects of different optimization methods at different data scales. Table 2 demonstrates the generalization of the three optimization methods.

Table 2. Optimization levels and strategies.

Optimization level	Included strategies
Level1 optimization	Heterogeneous computing porting
Level2 optimization	Heterogeneous computing porting + Dynamic load balancing
Level3 optimization	Heterogeneous computing porting + Dynamic load balancing + I/O storage optimization

During the heterogeneous computing porting process, we initially ported the program to a heterogeneous environment through hotspot function analysis, computing kernel optimization, and cross-device communication optimization. Test results show that, compared with the original serial ORBIT program, heterogeneous computing porting optimization improved execution speed by 12.34 to 18.42 times under different data scales.

Based on heterogeneous computing porting, we further propose dynamic load balancing optimization. We design task scheduling to reasonably allocate data volumes between CPUs and GPUs, thereby utilizing CPU idle time to further accelerate program execution time. Test results show that compared to the original serial ORBIT program, heterogeneous computing porting and load balancing optimization improve speed by 13.58 to 19.71 times across different data scales.

I/O storage optimization based on heterogeneous computing porting and dynamic load balancing optimization can reduce the character count of floating-point data output and file storage size, thereby improving the overall efficiency of the program. Therefore, the triple parallel optimization method for guiding center orbit simulation can significantly improve program efficiency. Compared with the original ORBIT program, the triple parallel optimization method achieves acceleration improvements of 16.45 to 25.6 times across different data scales.

The speedup from optimization becomes more significant as the data size increases. This is because larger data sizes make computational operations more intensive, and fixed overheads such as transfers are spread over more computational operations, reducing their proportion of overall program elapsed time.

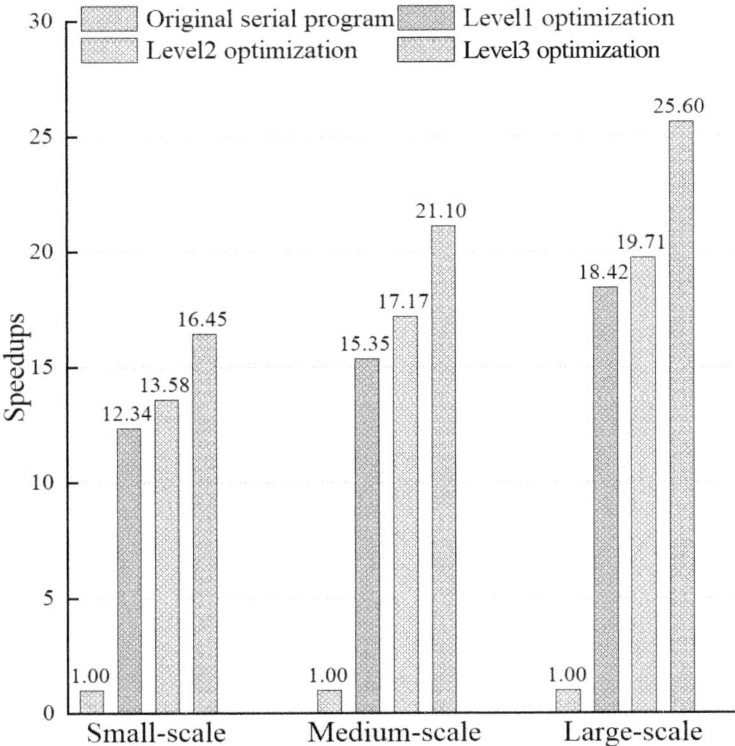

Fig. 4. Acceleration ratio test for different scales of single nodes.

Compared with the original program, the optimized program shows a significant decrease in the time spent in the field function, ptrba function, and onestep function (see Fig. 5), from the original 92.9% to 28.08%. This optimization significantly improves the execution efficiency of the program, indicating a significant improvement in the performance of these three key functions.

6.3   Scalability

In the strong scalability test, three particle numbers with different scales are set, namely 3,200,000, 4,800,000, and 7,200,000 particles, corresponding to small, medium, and large input data sets, respectively. In this section, the execution time of the program with two processes was used as the baseline time for calculating the parallel efficiency of the program with other numbers of processes. The program used was the triple parallel optimized program.

In Fig. 6, a, b, and c represent the parallel efficiency of the program after parallel optimization at different data scales. Taking Fig. 6 as an example, the horizontal axis represents the number of processes in the test, and the vertical axis represents the parallel efficiency of the program after triple parallel optimization. As shown in Fig. 6, the parallel efficiency of the program decreases when

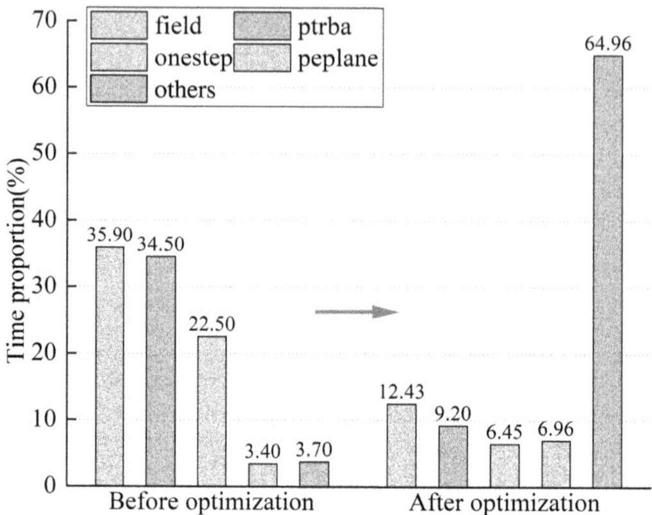

Fig. 5. Changes in hotspot functions before and after optimization.

the number of processes increases to 128 or more, mainly due to the fact that MPI communication takes more time and a large number of processes accessing shared resources leads to contention and communication delays. This leads to a decrease in strong scalability as the number of processes increases.

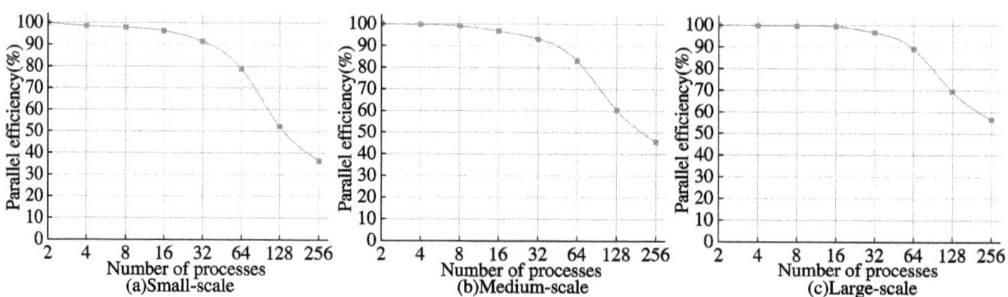

Fig. 6. Strong scalability with different scales.

In the weak scalability test, three particle numbers with different scales are set. To keep the computational load consistent across each process, the specific particle counts were set to 200,000 * N, 400,000 * N, and 800,000 * N, where N represents the number of processes used, corresponding to small-scale, medium-scale, and large-scale input data, respectively. The execution time of the program with two processes was used as the baseline test time to calculate the parallel efficiency of the program compared to other numbers of processes (see Fig. 7). Similarly, the program used was the triple parallel optimized version.

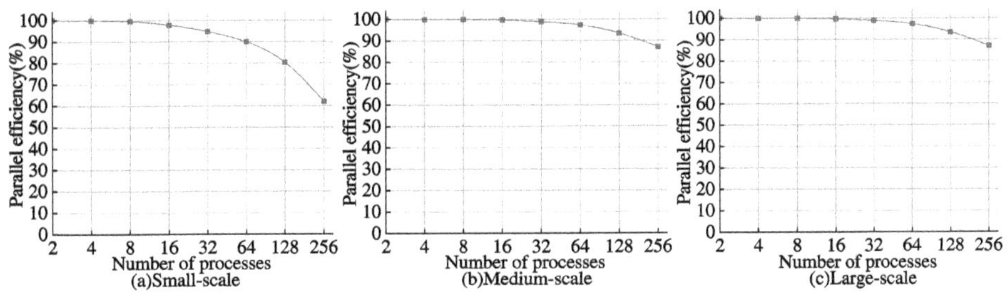

Fig. 7. Weak scalability with different scales.

6.4 Verification of Results

In this section, we summarize the results obtained in various optimization scenarios. Table 3 compares the test results of the original program and the optimized program in a single-process environment.

Table 3. Comparison of test results for single processes.

	Small-scale	Medium-scale	Large-scale
Original serial program	5.99463E-06	8.41195E-06	8.05811E-06
Level1 optimization	5.99463E-06	8.41195E-06	8.05811E-06
Level2 optimization	5.99463E-06	8.41195E-06	8.05811E-06
Level3 optimization	5.99463E-06	8.41195E-06	8.05811E-06

Table 4 compares the test results of the original program and the optimized program in multi-process environments. Since the number of simulated particles increases with the number of processes in the test, while the number of particles simulated by each process remains constant, the total computational workload varies when using different numbers of processes, leading to differences in computational results between processes.

7 Related Work

In magnetic confinement fusion plasma simulations, guiding center orbit simulations and GPU acceleration are two key research directions. The numerical implementation of guiding center orbit simulations has undergone significant development. [11] established a numerical solution framework for the guiding center motion equations, employing a fourth-order Runge-Kutta method, which serves as the computational foundation for programs such as ORBIT. [12] developed a three-dimensional high-precision magnetic field calculation method, significantly improving the accuracy and numerical stability of the drift velocity term in the guiding center equations. [13] proposed a numerical method

Table 4. Comparison of test results for different processes.

	Number of process	Small-scale	Medium-scale	Large-scale
Original parallel program	16	8.57051E-06	9.91182E-06	1.02108E-05
	32	8.93871E-06	1.01639E-05	1.03645E-05
	64	9.91588E-06	1.04776E-05	1.04537E-05
	128	9.91165E-06	1.03996E-05	1.05276E-05
	256	9.87957E-06	1.02655E-05	1.03838E-05
Parallel optimization program	16	8.57051E-06	9.91182E-06	1.02108E-05
	32	8.93871E-06	1.01639E-05	1.03645E-05
	64	9.91588E-06	1.04776E-05	1.04537E-05
	128	9.91165E-06	1.03996E-05	1.05276E-05
	256	9.87957E-06	1.02655E-05	1.03838E-05

to address the reverse semi-Lagrangian method in guiding center models. [14] implemented guiding center simulations in the PLUTO code. [15] used Lie transformation theory to systematically renormalize the guiding center equations of motion. [16] resolved the structural preservation problem faced by traditional numerical methods in guiding center orbit simulation. [17] explored the symplectic integral method for guiding center motion in non-regular coordinate systems. [18] simplified the calculation of the drift term in the guiding center equations.

Significant achievements have also been made in GPU acceleration for plasma simulations. [19] proposed a GPU-accelerated PIC algorithm. [20] introduced an efficient GPU parallelization strategy for the binary collision module in particle-in-cell simulations. [21] developed a 2D GPU code named PM2D for kinetic simulations of laser-plasma instabilities in large-scale plasmas.

In summary, significant progress has been made in guiding center orbit simulation and GPU acceleration technology in magnetic confinement fusion plasma simulation. However, although existing research has provided important computational support for plasma simulation, existing CPU-GPU heterogeneous computing still faces challenges such as uneven utilization of computing resources and limited parallel computing efficiency. Therefore, to optimize the ORBIT program in a CPU-GPU heterogeneous environment, it is imperative to propose new optimization strategies to overcome existing computational bottlenecks. Given this, this paper is dedicated to the heterogeneous parallelization optimization of the ORBIT program, which holds significant practical significance and provides a potential solution to enhance the computational efficiency and stability of large-scale plasma simulations.

8   Conclusion

This study proposes an efficient parallel implementation of the plasma-guided center-of-mass orbit simulation program ORBIT on a CPU-GPU heterogeneous platform based on NVIDIA's CUDA architecture platform. Through triple parallel optimization—heterogeneous computing porting, dynamic load balancing, and I/O storage optimization—the program's performance bottlenecks are significantly reduced. Finally, the program's runtime is tested across three different scales to validate the effectiveness of heterogeneous parallelism. Experimental results show that under a single-process, single-GPU configuration, the optimized program achieves acceleration of up to 16.45 to 25.6 times across three different scales. When scaled to 256 processes, the program exhibits weak scalability of 62.02 to 87% across the three scales.

Future work includes researching more complex parallel patterns, expanding to multi-node platforms, and optimizing cross-GPU communication mechanisms; combining efficient and accurate machine learning models to improve the efficiency and accuracy of load balancing algorithms; designing hierarchical caching or prefetching strategies to optimize memory access efficiency and further improve the performance limits of heterogeneous computing platforms.

Acknowledgement. This work is supported by Shandong Provincial Natural Science Foundation under Grant ZR2024LZH009, National Key Research and Development Program under Grant 2024YFE0210800, Natural Science Foundation under Grant 62495062 and Beijing Natural Science Foundation under Grant L242017.

References

1. Aksakal, II., et al.: Plasma confinement and nuclear fusion (2025)
2. Dal Molin, A., et al.: Measurement of the gamma-ray-to-neutron branching ratio for the deuterium-tritium reaction in magnetic confinement fusion plasmas. Phys. Rev. Lett. **133**(5), 055102 (2024)
3. Trent, T., et al.: Covariant guiding center equations for charged particle motions in general relativistic spacetimes. Astrophys. J. **987**(1), 101 (2025)
4. Silvano, C., et al.: A survey on deep learning hardware accelerators for heterogeneous HPC platforms. ACM Comput. Surv. **57**(11), 1–39 (2025)
5. Fang, H., et al.: A multiscale material-structure-hydroelasticity coupled analytical model for floating sandwich structures with hierarchical cores. Mar. Struct. **79**, 103055 (2021)
6. Michaud, A.: Demystifying the lorentz force equation. J. Mod. Phys. **13**(5), 776–838 (2022)
7. Kolesnichenko, Y.I., et al.: Ion cyclotron emission in maxwellian plasmas. Phys. Plasmas **31**(4) (2024)
8. Burby, J.W.: Guiding center dynamics as motion on a formal slow manifold in loop space. J. Math. Phys. **61**(1) (2020)
9. Salahshoor, M.: Electron mixing performance of a magnetron sputtering cathode. J. Appl. Phys. **137**(6) (2025)
10. Aida, M.: Fourth-order Runge-Kutta method for solving applications of system of first-order ordinary differential equations. Enhanced Knowl. Sci. Technol. **2**(1), 517–526 (2022)

11. Gorelenkov, N.N., et al.: Energetic particle physics in fusion research in preparation for burning plasma experiments. Nucl. Fusion **54**(12), 125001 (2014)
12. Sánchez, E., et al.: A quasi-isodynamic configuration with good confinement of fast ions at low plasma β. Nucl. Fusion **63**(6), 066037 (2023)
13. Piao, X., et al.: An efficient trajectory tracking algorithm for the backward semi-lagrangian method of solving the guiding center problems. J. Comput. Phys. **418**, 109664 (2020)
14. Mignone, A., et al.: A guiding center implementation for relativistic particle dynamics in the pluto code. Comput. Phys. Commun. **285**, 108625 (2023)
15. Cary, J.R., et al.: Hamiltonian theory of guiding-center motion. Rev. Mod. Phys. **81**(2), 693–738 (2009)
16. Ellison, C.L., et al.: Degenerate variational integrators for magnetic field line flow and guiding center trajectories. Phys. Plasmas **25**(5) (2018)
17. Albert, C.G., et al.: Symplectic integration with non-canonical quadrature for guiding-center orbits in magnetic confinement devices. J. Comput. Phys. **403**, 109065 (2020)
18. Pavone, A., et al.: Machine learning and Bayesian inference in nuclear fusion research: an overview. Plasma Phys. Controlled Fusion **65**(5), 053001 (2023)
19. Xiong, Q.: Application of GPU-accelerated particle-in-cell simulations in magnetic reconnection associated with energy conversion between field and particles. J. Suppl. Ser. **264**, 3 (2024)
20. Lehe, R., et al.: An efficient GPU parallelization strategy for binary collisions in particle-in-cell plasma simulations. In: Proceedings of the Platform for Advanced Scientific Computing Conference, pp. 1–8 (2025)
21. Ma, H., et al.: PM2D: a parallel GPU-based code for the kinetic simulation of laser plasma instabilities at large scales. Comput. Phys. Commun. **304**, 109295 (2024)

Facess: A Fast Access Method for Shared Data Among Multi-cores in Parallel Programs

Junhui Wang[1,2], Jierong Tang[1,2(✉)], Shiyuan Wang[1,2], Yijing Peng[1,2], Hongwei Zhou[1,2], Quanyou Feng[1,2], Libo Huang[1,2], and Yongwen Wang[1,2]

[1] College of Computer Science and Technology, National University of Defense Technology, Changsha, China
tangjierong@nudt.edu.cn
[2] Key Laboratory of Advanced Microprocessor Chips and Systems, Changsha, China

Abstract. With the increasing core count in modern multi-core processors, directory-based cache coherence protocols face significant challenges related to the "three-hop problem," which introduces substantial latency in shared data access. This paper proposes a Fast Access Method for Shared Data (Facess) to address this issue. The Facess approach introduces a directory cache between each processor's private cache controller and the network interface, modifies cache block status tag bits, and redesigns bus, directory, and processor state machines. Experimental results using the GEM5 simulation platform with PARSEC benchmarks show that Facess achieves remarkable latency improvements: approximately 84% for instruction-related data, 80% for data load operations, and 78% overall latency reduction.

Keywords: Multi-Core Processors · Cache Coherence · Directory Protocol · Latency Optimization · Facess Method

1 Introduction

In 2001, following the launch of the first commercial multicore processor, the IBM Power4 [1], companies such as Intel and AMD subsequently introduced several processors with 2–4 cores [2,3]. In 2007, MIT and Tilera Corporation collaborated to develop the Tile64 processor [4], marking the first instance of interconnecting 64 processor cores via an on-chip two-dimensional network. The success of Tile64 signified the transition of processor design into the multicore era, with Chip Multi-Processors (CMP) gradually superseding single-core processors as the new direction in processor development. In recent years, companies such as Intel, AMD, Marvel, Amazon, Alibaba, and Sophoh have successively launched processors with 64 cores [5,6].

In these processors, the architectural designs can broadly be categorized into two types: cluster-based hierarchical architectures and flat architectures.

H. Liu et al. (Eds.): ICA3PP 2025, LNCS 16381, pp. 419–437, 2026.
https://doi.org/10.1007/978-981-95-8399-7_23

In cluster-based hierarchical architectures, multicore processors initially organize 2–8 cores into a cluster, followed by connecting multiple clusters using an on-chip network. For instance, ThunderX3 [7] employs a ring network structure. In flat architectures, all processor cores are interconnected directly through an on-chip network. For example, Amazon utilizes a mesh network structure for connectivity. This demonstrates that on-chip networks have become the default interconnection method in multicore processors.

In processors utilizing on-chip networks, performance enhancement is commonly achieved through a three-level cache strategy. Typically, the L1 and L2 caches are private to the processor core, whereas the L3 cache is shared within a cluster or among all cores. Maintaining cache coherence among multiple cores thus emerges as a pivotal challenge. Presently, two predominant cache coherence protocols are employed in CMP (Chip Multi-Processor) systems: snooping-based and directory-based protocols [8–10]. Directory-based protocols keep a separate directory associated with main memory that stores the state of each block of main memory. Each entry in this centralized directory may contain several fields depending on the protocol, for example, a dirty bit, a bit indicating whether or not the block is cached, pointers to the caches that contain the block, etc. The snooping-based cache coherence protocol is generally applied in CMP systems with bus or ring interconnect structures, facilitating cache coherence transactions through broadcasting. In contrast, the directory-based cache coherence protocol employs a directory to store details on the state of shared cache data and the distribution of its copies, with messages transmitted in a point-to-point manner via the directory.

However, as the number of processor cores increases, broadcasting methods can lead to network congestion, thereby necessitating the shift towards directory-based approaches. Compared to snooping-based methods, directory-based approaches enable targeted message sending, thus reducing the volume of unnecessary communications. Consequently, given the extensive number of nodes requiring monitoring, high-performance processors often resort to directory-based cache coherence protocols due to their efficiency and scalability in handling the complexities of modern multicore environments [11]. For example, Unlike the prior generation of Intel Xeon processors that supported four different snoop modes (no-snoop, early snoop, home snoop, and directory), the Intel Xeon processor Scalable family of processors only supports the directory mode [12].

However, within the context of directory-based cache coherence protocols, there persists an issue known as the "three-hop problem". Indirect misses do incur the extra latency of sending a message on the on-chip interconnection network to the specified private cores [13]. Such messages are sent in parallel, and responses are typically sent directly to the original requester, resulting in a "three-hop protocol". The indirect miss latency for coherence is from 1.5 to two times larger than the latency of a non-coherent miss.

This issue significantly impacts the speed at which processor cores can access data. Studies have demonstrated that reducing the latency of data retrieval to 1 could potentially increase performance by multiple folds. Additionally, numerous articles on prefetching indirectly address this problem, highlighting the importance of optimizing data access times to enhance overall processor efficiency. The discussions around prefetching strategies underscore the critical nature of addressing the latency challenge to improve the speed and performance of multicore processors in handling complex computations and tasks.

Therefore, we propose a Fast access (Facess) method for share data. The main idea is that we set a directory cache in each network interface. Once a new request is issued, the cache is searched. If there is a hit, the request will be split into two messages - one data request message is sent directly to the owner node; one coherence request message is sent to the corresponding directory controller. Then, the owner node and the controller send back data and acknowledgement separately. This will can reduce the latency of request sharply.

The rest of the paper is organized as follows. A brief review of the relevant work is given in Sect. 2. The scenarios of potential shared data are presented in Sect. 3. We describe our Facess method in detail in Sect. 4. The experimental results are presented in Sect. 5. Finally, our conclusion is presented in Sect. 6.

2   Related Work

To accelerate shared data access, research efforts have primarily focused on two approaches: data prefetching and direct data access.

Data prefetching typically utilizes relevant algorithms to predict and precompute data that the processor core pipeline may need, fetching it from memory or other nodes and storing it in the processor core. This allows subsequent data requests from the pipeline to be fulfilled directly from the local private cache. For example, Prodigy [14] introduces a "Data Indirection Graph" (DIG) to encode data structures and their indirect relationships, combined with a dedicated SRAM prefetcher for intelligent prefetching, resulting in significant data access latency and performance improvement. On the other side, the core idea of DMP (Differential Matching Prefetching) [15] leverages differential matching technology to detect correlation between index arrays and data arrays, predicting future access addresses and preloading data accordingly. Compared to existing methods, DMP supports various complex indirect access patterns including single-level, multi-level, multi-path, and range accesses, while dynamically adjusting prefetch distance based on runtime conditions to maintain high prefetch coverage and accuracy across different application scenarios. Although these methods can more accurately match data access patterns, they still face challenges with large network transmission delays for accessed data, which may impact prefetching effectiveness.

In the domain of direct data access, related work generally accelerates access by caching shared data information. Li et al. [16] observed that cache-missed target data has a high probability of residing in the cache of a neighboring node. They proposed a hardware architecture based on the Cluster Cache Monitor (CCM), where the processor is divided into multiple clusters, each consisting of multi cores and a CCM. The CCM contains tag information and cache state information. When a cache miss occurs in one core, the CCM detects whether the requested data can be served by another cache within the same cluster. Upon a hit, data is directly transferred from the neighboring core, avoiding long-distance access. However, if the data is not within the cluster, the traditional access path is followed, resulting in long-distance access even when data resides in a neighboring cluster. This paper addresses this limitation. Similarly, Arm introduced the concept of Snoop Filter [17] in cluster-based network architectures. The Snoop Filter tracks cache lines present in the cores of a cluster, with tags for all cached shared memory stored in a directory within the interconnect. All shared accesses query this snoop filter, which provides two possible responses: HIT (data is on-chip with a vector indicating the cluster containing the data) or MISS (fetch from external memory). This approach reduces snoop traffic by favoring directed snoops over broadcast snoops when possible, substantially decreasing snoop response traffic. Christian et al. [18] proposed Proximity Coherence, which establishes dedicated links between nearby caches. Load misses are forwarded to nearby caches via these dedicated links. Their paper explores Proximity Coherence, a novel scheme where L1 load misses are optimistically forwarded to nearby caches through new dedicated links, demonstrating improved communication efficiency. Additionally, the Locality-aware Adaptive Cache Coherence Protocol [19] aims to enhance cache performance and energy efficiency in multi-core processors. Its core idea is to determine whether to retain or replicate data blocks in the cache based on their spatio-temporal locality. Specifically, the protocol tracks each cache line's usage through low-overhead hardware-level runtime analysis and only privately caches or replicates data blocks with high locality, thereby reducing unnecessary cache pollution and network communication.

3   Analysis of Data Sharing Scenarios

In multicore processors, data requested by one processor core may already be cached in another processor core. As shown in Fig. 1, processor core $C00$ wants to obtain a copy of data at address X, which is already stored in the L1 or L2 cache of $C01$. Assuming the processor uses a directory-based coherence protocol and a Mesh on-chip network with XY routing, the directory information for address X is stored in directory $D63$. Next, we will explain the entire data request process in detail.

First, processor core $C00$ obtains the directory number based on address X and sends a data request packet to directory node $D63$. After receiving the request, the directory node queries the directory to determine the data cache node $C01$ and further forwards the request to the target node. After processor core $C01$ sends the data to $C00$, it synchronously returns coherence information to $D63$ to update the directory information. As shown in the figure, although the requesting node and target node are physically adjacent, the entire data access requires 62 hops to complete. Assuming no other packets compete for the link in the network and each hop takes only 2 clock cycles, the entire access process requires 124 clock cycles.

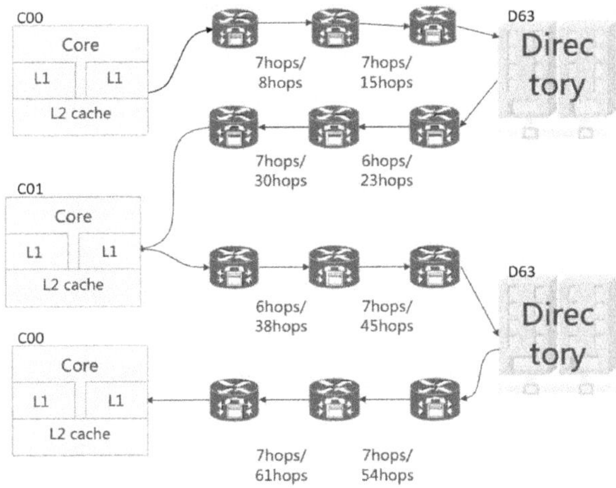

Fig. 1. Example of Shared Data Access

It should be noted that the above process assumes only one request in the network. Experiments show that when running the blackscholes application from the PARSEC benchmark suite [22], there are numerous cases where data at a single address is shared and accessed by all processor cores. One processor core first fetches the data from memory, and then other processor cores quickly request copies of this data in sequence. After the processor core with modification rights completes data modification and updates, other nodes re-acquire new copies. This pattern occurs hundreds of times even for small-scale data. Assuming the directory for this data is $D8$, the average latency for a single request would be 136 clock cycles even without network congestion. Calculations show that if we can bypass the directory node to obtain data directly, the latency can be reduced by approximately half.

Table 1 presents the proportion of shared accesses among all accesses when running the PARSEC benchmark suite on a 16-core processor. The results indicate that shared data accesses account for approximately 60% of all accesses. Particularly for critical load-type instructions, shared data accesses account for approximately 68% of the total. Reducing the access latency for shared data would significantly improve system performance.

4   The Facess Method

In this section, we propose Facess to bypass directory nodes and enable direct access to shared data. First, we will introduce the overview of our Facess method. Next, we will present the specific details of the overall architecture. In general, packets cause state changes, which in turn trigger new packets and other changes. Therefore, we first introduce the packet types and directory cache entry states that will be used. Then, we present the main state machines. After that, to

Table 1. Proportion of Shared Data Accesses

Applications	Fetch	Load	Store	total
splash2.fft	33.68	69.4	53.12	61.66
parsec.blackscholes	39.01	71.15	47.82	61.02
splash2x.water_nsquaered	39.43	68.65	46.45	59.61
splash2.lu_cb	34.46	69.24	45.95	59.12
splash2.water_spatial	37.67	66.64	45.76	58.06
splash2.lu_ncb	37.03	67.21	57.76	61.83
parsec.swaptions	41.13	70.71	55.65	63.81
splash2.cholesky	39.23	67.05	46.27	58.93
splash2x.lu_ncb	32.09	66.65	49.18	58.08
average	37.08	68.52	49.77	60.24

illustrate specific implementation details, we describe the microarchitecture of the interface modules and directory controller. Finally, we use concrete examples to clearly demonstrate the entire data flow process.

4.1   Overview of the Facess Method

Before formally introduce our work, we first elaborate on the communication patterns of packets in network-on-chip (NoC) based and directory-based coherence protocols. As shown in Fig. 4, assume each processor core has private L1 and L2 caches. For shared data, the two levels of private caches are typically queried first. Only when both caches miss, packets are sent into the network. Unlike bus architectures, requests in NoCs are forwarded based on destination nodes rather than physical addresses. Therefore, the L2 cache controller maintains a global System Address Map, which stores the correspondence between physical addresses and nodes. For example, physical addresses 0-128GB typically correspond to Flash address space; physical memory space for shared data is interleaved across different directory nodes based on different bits of the physical address. When both caches miss, the request queries the System Address Map using the physical address to obtain the directory node number before generating the corresponding packet. Taking the CHI protocol as an example, a request packet includes destination ID, source ID, transaction ID, opcode, physical address, data size, and memory attributes. The destination ID is obtained by querying the mapping table with the physical address.

To bypass directory nodes and enable direct access to shared data, we need to modify the network interconnection interface modules and directory controllers. The interconnection interface module is further divided into request interface module and receive interface module. In the request interface module, we add a directory cache to retain recently used directory information. Each entry contains a physical address field and a corresponding node ID field, indicating the current owner of the data at that address. When the miss of both caches in the processor core occurs, the request queries both the address space mapping table and the directory cache simultaneously. If the directory cache misses, the request is encapsulated as a normal request packet. Otherwise, the request is split into two requests: a coherence request packet and a data request packet. The coherence request packet is sent to the directory node as usual, while the data request is sent directly to the processor core node where the data resides. The directory node then returns an ack packet to confirm the correctness of the directory cache content; the receive interface module sends the data directly to the request interface module along with additional information to maintain coherence (Fig. 2).

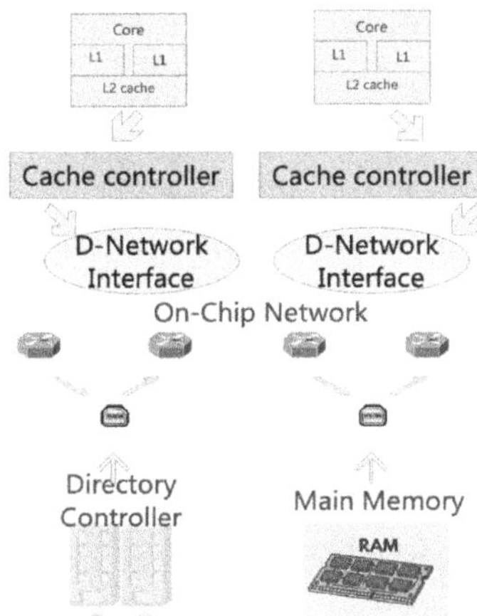

Fig. 2. Overall Architecture Diagram

4.2   Communication Protocol Packets

In directory-based coherence protocols, each directory maintains a vector table indicating which processor cores hold data for corresponding addresses. To elaborate on the communication process, we present the main packet types. Here, we use the hammer-like protocol from the GEM5 simulator [21] as the base protocol, but our approach is applicable to other protocol types. We categorize and describe the new packet types according to different channels.

Channel from Request Interface Module to Directory Controller. In this channel, all packets originate from the request interface module. The specific packet types are as follows:

– GETS type. Requests shared access to a cache block.

- GETX type. Requests exclusive access to modify a cache block.
- PUTX type. Requests write-back of a cache block.
- GETC type. Requests coherence information for a cache block.

Channel from Request Interface Module to Request Receive Module.
In this channel, all packets originate from the request interface module. The specific packet type is as follows:

- GETD type. Requests direct access to a cache block.

Channel from Directory Controller to Request Receive Module. In this channel, all packets originate from the directory controller. The specific packet types are as follows:

- INV type. Requests invalidation of a cache block.
- F-GETS. Forwarded cache block share request.
- F-GETX. Forwarded cache block modify request.

Channel from Directory Controller to Request Send Module. In this channel, all packets originate from the directory controller. The specific packet types are as follows:

- ACK-C. Acknowledges cache block coherence information.
- M-DATA. Data from memory.

Channel from Request Receive Module to Request Send Module. In this channel, all packets originate from the request receive module. The specific packet types are as follows:

- ACK. Acknowledgment of cache block.
- DATA. Cache block data.

Channel from Request Receive Module to Directory Controller. In this channel, all packets originate from the request receive module. The specific packet type is as follows:

- UNBLOCK. Non-blocking request for cache block.

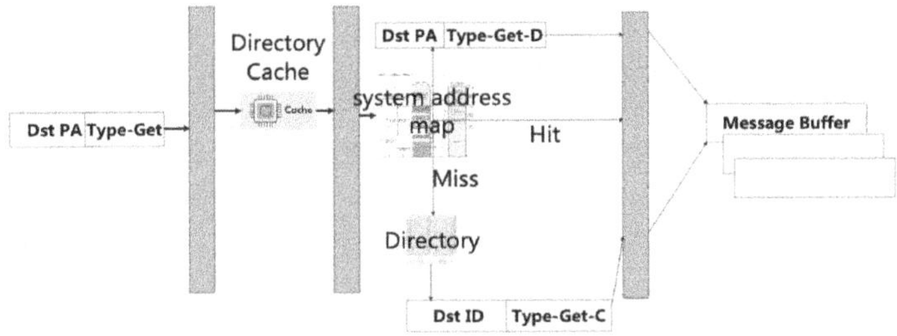

Fig. 3. Block Diagram of Interface Send Module

4.3   The Microarchitecture of the Interface Modules and Directory Controller

Interface Send Module. As shown in Fig. 3, there are two stages before packets are sent to the network. For Get-D type requests, we first query the directory cache. If the directory cache hits, it means the desired data exists in another processor core's cache. The request is then split into two packets: a get-c type packet and a Get-D type packet. The get-c packet notifies the directory controller to update the directory vector table; the Get-D type packet retrieves the data.

The structure of the directory cache is shown in Fig. 4. Similar to ordinary data caches, the directory cache uses a set-associative organization. It contains a tag array and a data array. The tag array stores partial physical addresses, while the data array contains the node ID of the data owner. When querying the directory cache, part of the physical address is used as an index, and the remainder is used to match the tag. If the directory cache hits, it returns the node ID where the data resides.

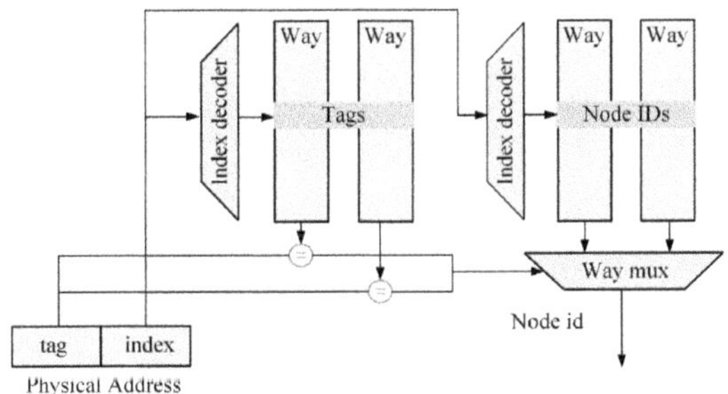

Fig. 4. Schematic Diagram of Directory Cache

For get-c type packets, we use the destination physical address to query the address mapping table. The node ID obtained from the address mapping table serves as the packet's destination ID, while the node ID from the directory cache is included as payload. When the directory controller receives the packet, it verifies that the node ID from the directory cache matches the one in the directory vector table. For get-d type packets, the node ID from the directory cache is used as the packet's destination ID. Both packets are buffered in the network interface queue and sent to the network sequentially according to the physical layer protocol's handshake mechanism.

Interface Receive Module. If we receive a get-d message, the flow is as follows. We check if the data of the physical address is in the cache according to the snoop filter. If the data is there, send the data back to the request node (Type-SendBack-D). Here, we should notice that in case the data is modified later. The state of the data should change to read-only state. If the original state is E or M, then it change to O state. Later, if the directory sends an invalidation message, the node invalidate the corresponding cache entry.

As shown in Fig. 5, when the interface receive module receives an M-type message, it indicates that valid data has been received. Then, we only need to wait for an ack-C type message to merge the data message and coherence message into a single request completion message.

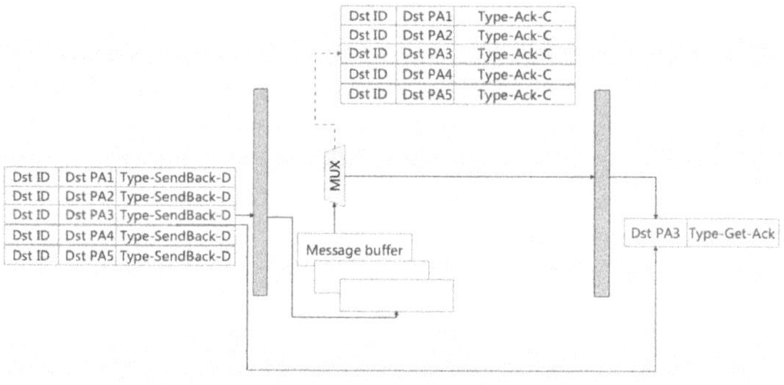

Fig. 5. The diagram of NI

For other types of requests, the processing method is consistent with the normal flow.

Directory Controller. For each directory, except the common messages, it can receive a new type message - get-c. The message tells that the network interface has sent out a get-d message to the data node directly.

In addition to other normal types of packets, the directory controller receives get-c type packets. This packet notifies the directory controller that the sender has already sent a get-d type packet to another processor core node to directly obtain the relevant data. If the node ID obtained by querying the directory for the physical address in the get-c packet matches the node ID in the packet, the controller sends an ack-C type packet to the requesting node. Furthermore, for getx-c type packets, it also sends an invalidation packet to the original destination data node. Finally, the directory controller updates the directory vector table with the latest data sharing status.

Besides, the basic maintainment policy of the directory cache is as follows: If the directory controller finds that the vector table of a directory entry contains more than 2 nodes, it sends a message to update the directory cache to all processor cores. Processor cores can use LRU or RRIP policies to update the directory cache. Additionally, if the owner of a cache block changes or the valid state of the cache block changes, the directory controller will send a message to all nodes to update the directory cache.

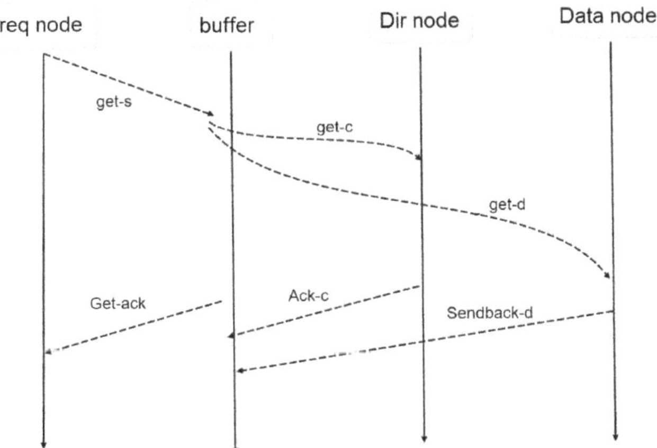

Fig. 6. Schematic Diagram of Scenario 1 Packet Interaction

4.4   Case Study of Communication Scenarios in Facess Method

Scenario 1: Acquiring Share of Data. As shown in Fig. 6, for get-s type requests, the requesting node first splits it into a get-c packet and a get-d packet. After the directory node and data node send back the ack-c packet and d packet, the requesting node combines the information from both packets to confirm that the get-s request has been completed. The process is the same even if two nodes request data from the same address simultaneously.

Scenario 2: Acquiring Ownership of Shared Data. For get-x type requests, the processing flow is similar to the get-s type. However, as shown in Fig. 7, it should be noted that inv-d type packets may arrive at the requesting data node before get-d type packets. Here, we introduce a new state - IW. The IW state means the data is waiting for a get-d request. Once the get-d type packet arrives, the cache state is changed to I state.

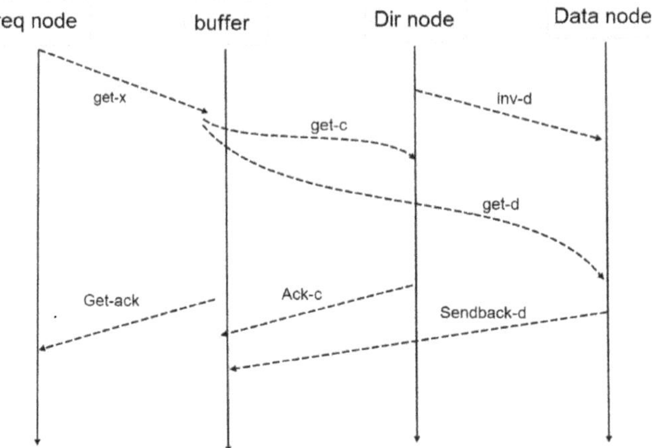

Fig. 7. Schematic Diagram of Scenario 2 Packet Interaction

Complex Scenario. In this section, we will discuss the scenario where two nodes send get-x requests almost simultaneously.

As shown in Fig. 8, Node 1 and Node 2 send get-x requests for the same address. Then, the get-c packet from Node 1 arrives at the directory node first. However, the get-d packet from Node 2 arrives at the data node first. To maintain coherence, the directory node will first buffer the packet from Node 2 and then send an inv-d message to Node 1. It should be noted that the inv-d message needs to include the requesting node ID. This way, after the data node sends the data back to the requesting node, it will invalidate the corresponding data. After sending the ack-c packet, the directory node continues processing the request from Node 2. At this point, the directory node will discover that the data node ID carried in the request from Node 2 is incorrect. It will handle this request as a normal request packet: send a get-x request to the current owner node, and then forward the data to Node 2.

It can be seen that node 1 and node 2 send get-x requests for the same address. Then, the get-c message from node 1 arrives at directory node first. However, the get-d message from node 2 arrives at data node first. To hold coherence, the directory node will stall the get-c message from node 2 and send the inv-d message for the get-c message from node 1. It should be noticed that the inv-d message will contains the request node id. Thus, the data node will invalidate the data after sending the data back to the requesting node. After sending the ack-c message, the directory node continue to deal the request from

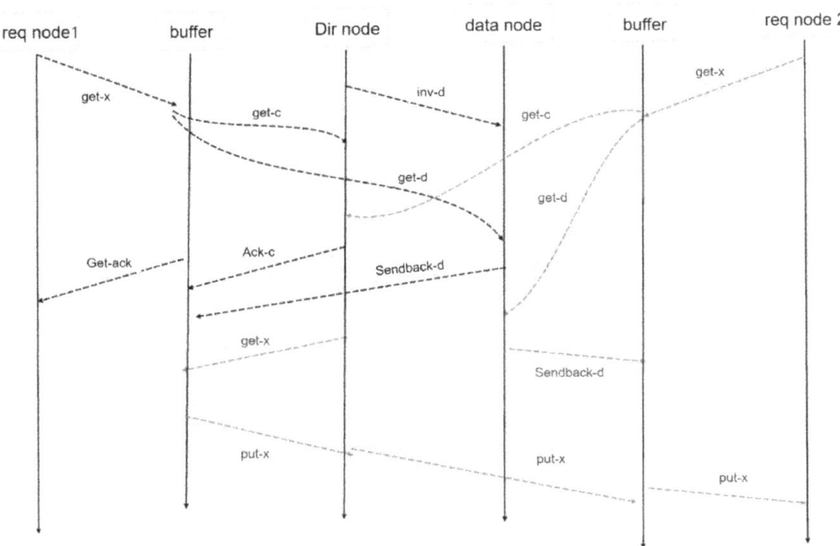

Fig. 8. Schematic Diagram of Complex Scenario Packet Interaction

node 2. This time, the directory node will find that the request carry a false data node id. It will deal this request as common case: send get-x request to current owner node and forward data to node 2 later.

5   Experiments

5.1   Experimental Setup

Regarding the simulated multi-core processor, this experiment builds a simulation platform based on the GEM5 full-system simulator [21], using an X86 architecture out-of-order multi-core processor model. Each processor core is equipped with two levels of private caches: L1 instruction and data caches (32KB, 2-way set associative) with a latency of 1 clock cycle; the private L2 cache (2MB, 8-way set associative) has an access latency of 10 clock cycles. The on-chip interconnection network uses a 4×4 two-dimensional Mesh topology, implemented through the Garnet 2.0 interconnection network, with link transmission latency configured to 1 clock cycle and supporting 10 virtual channels to avoid protocol-level deadlocks. Network nodes use 4-cycle pipeline structure, supporting XY routing, using Round-Robin for port arbitration, with each virtual network supporting 4 virtual channels. The cache coherence protocol uses MOESI_Hammer implementation, which maintains states through a distributed directory structure, supporting five states across M, O, E, S, and I, effectively reducing communication overhead in Non-Uniform Memory Access (NUMA) scenarios. For the method proposed in this paper, at the cache and directory controller, we add new message types and new states as described in the above section. The directory cache is emulated by a normal cache.

In terms of software, we selected the Linux kernel version 4.9.186 and Ubuntu LTS 18.04. After the system starts, we use the PARSEC benchmark suite for relevant tests. PARSEC [22] is a standardized benchmark suite for multi-core and many-core architectures, designed to reflect typical parallel workloads in modern shared-memory systems. It includes compute-intensive applications such as fluidanimate and blackscholes, and memory-intensive applications such as canneal and streamcluster. The former has a high thread-level parallelism ratio, making it suitable for cache coherence stress testing; the latter has a high last-level cache miss rate, which can effectively evaluate interconnection network congestion. To run the PARSEC benchmark suite on the GEM5 platform, we modified the operating system image file and placed the compiled PARSEC benchmark suite into the image file. The specific experimental testing method is: first set the path variable to the directory where the operating system and Linux kernel are located, then start GEM5 in full-system mode (Table 2).

Table 2. The GEM5 Configuration Parameters.

Component	Configuration
Processor	X86 cores
L1 Caches	Split private I&D, 32 KB 2-way set associative, 64-byte block size
L2 Caches	private,2MB 2-way set associative, 64-byte block size
Directory Caches	different size and ways, 64-byte block size
Network	Garent, 8*8 2D Mesh, 1 cycle on-chip link latency, 10 virtual network
Router	4 cycle router pipeline, XY routing, 4VC per virtual network, Round-Robin queuing
Operating System	Linux Ubuntu LTS 18.04, kernel 4.9.186

5.2   Experimental Results

To compare and analyze the effects brought by the Facess method, we statistically analyzed the relative time required to access data when using Facess versus the traditional method for different types of requests. Additionally, to explore the impact of different directory sizes, we statistically analyzed the hit rate and resulting reduction in access time for different cache sizes.

The Improvement of Access Latency. For the PARSEC benchmark suite, we statistically analyzed the reduction in access time brought by the Facess method for different programs in the PARSEC benchmark suite. The cache size was set to 4 entries with a 2-way set associative structure. For operations such as Fetch/Load/Store, the latency ratios were recorded for directory cache hit cases, where the numerator is the latency of directly obtaining data from other caches after a cache hit, and the denominator is the total latency of the traditional

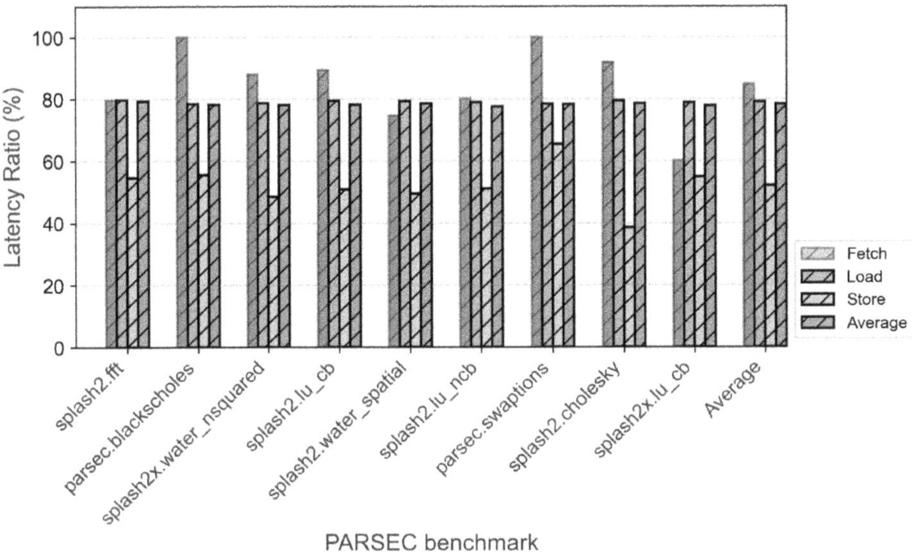

Fig. 9. Schematic Diagram of Latency Improvement

path: first accessing the directory node to query the cache location node, then obtaining data from the corresponding node's cache. The average latency is the ratio of the sum of the former to the sum of the latter.

As can be seen from Fig. 9, for all applications, instruction-related data achieves a latency improvement of approximately 84%, while data load-related operations achieve a latency improvement of around 78%. Data store and load-related operations show significant improvement, with latency reduced by half. Overall, data loads encounter the most cache misses, thus requiring the highest proportion of directory cache queries. Therefore, for all types of data accesses, the improved total latency is approximately 78% of the original. Through analysis, it can be found that the proportion of latency improvement is basically consistent across all programs. This indicates that the Facess method can achieve basically consistent results for all programs, demonstrating strong universality.

Furthermore, the instruction data access latency of one application achieved the lowest value of 40%, significantly below the average, indicating that the average distance between the requester, directory, and data owner nodes is the longest, thus enabling the Facess method to achieve the best results. Furthermore, for two of the applications, the instruction-type access latency ratio is 1. This is because no directory cache hits were observed in the measurements, resulting in no significant improvement.

Directory Cache Hit Ratios. First, we explored the variation of cache hit ratio with cache capacity when the directory cache associativity is fixed at 2-way. For capacities below 1KB, directory cache can still be implemented using

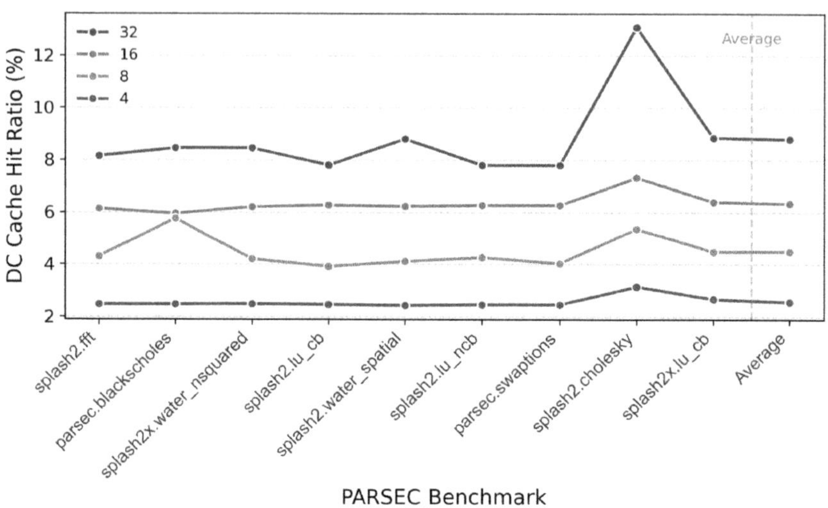

Fig. 10. Schematic Diagram of Directory Cache Hit Rates for Different Programs

registers. Since each directory entry size is 16 bytes, we conducted comparative analysis for directory cache total capacities ranging from 4 to 32 entries.

As shown in Fig. 10, the cache hit ratio generally exhibits an increasing trend with larger cache sizes across all applications, with the 32-entry configuration consistently achieving the highest hit ratios (average: 8.62%), followed by 16-entry (6.34%), 8-entry (4.53%), and 4entry (2.59%). This confirms the positive correlation between cache capacity and hit ratio performance. All applications demonstrate consistent performance ranking across cache configurations, indicating stable memory access patterns.

Furthermore, for the single test program Blackholes, we further explored the potential benefits of increasing the directory cache size. Therefore, we set the cache associativity to 2-way and the total number of entries ranging from 32 to 16384.

Figure 11 shows, when the number of cache entries increases from 32 to 16384, the hit rate shows a significant upward trend (7.66% to 20.76%), with a cumulative increase of 13.1% points, indicating that increasing cache capacity can effectively reduce cache misses.

In addition, Fig. 12 shows the variation trend of directory cache hit ratio under different cache associativity configurations when the cache size is 4096. The experiment contains 8 different associativity configurations (2, 4, 8, 16, 32, 64, 128, and 256-way), measuring and recording the corresponding cache hit rates. The chart uses a logarithmic coordinate to display associativity changes, with point graphs presenting discrete data and line graphs showing the overall variation trend.

As can be seen from Fig. 12, the cache hit rate shows an slight upward trend with increasing associativity, rising from 15.71% in the 2-way configuration to

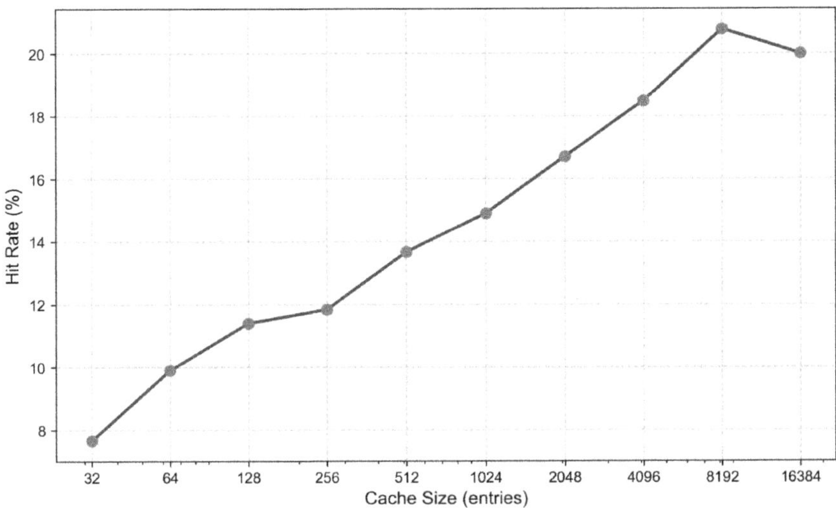

Fig. 11. Schematic Diagram of Hit Rate Variation Trend with Different Directory Cache Sizes

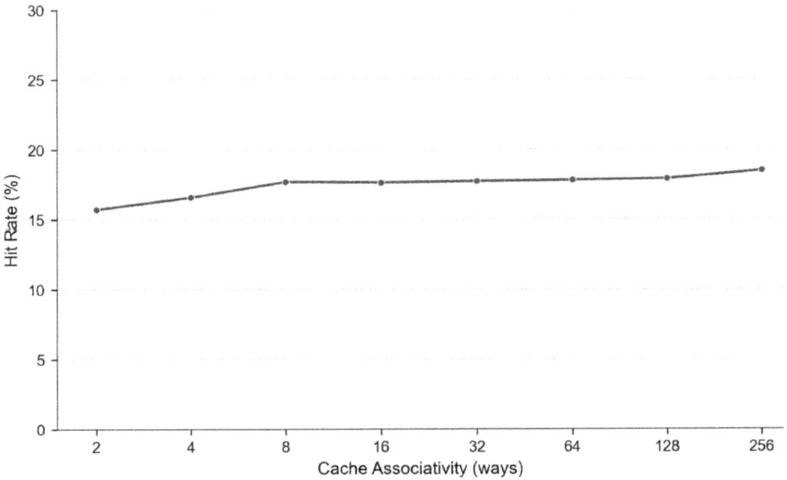

Fig. 12. Schematic Diagram of the Impact of Different Associativities on Hit Rate

18.47% in the 256-way configuration. This indicates that increasing cache associativity can effectively reduce conflict misses and improve system performance. The 8-way configuration is an important inflection point for performance gains, achieving a hit rate of 17.68%, which is an 18.7% improvement compared to the 2-way configuration, while further increasing to 128-way only provides an additional 0.21% improvement. From a cost-performance trade-off perspective, the 8-way configuration can be considered the optimal choice for practical applications.

6   Conclusion

This paper presents Facess, an optimized directory-based cache coherence protocol that effectively mitigates the three-hop problem in multi-core processors. By introducing a directory cache at the network interface level and optimizing cache block state management, Facess significantly reduces shared data access latency across various application scenarios. Experimental validation demonstrates consistent performance improvements across different cache configurations, with the 8-way associative cache emerging as the optimal design point. The proposed method shows strong generalizability, achieving average latency reductions of 78% across benchmark applications. Future work will explore adaptive cache management policies and integration with heterogeneous computing architectures to further enhance performance scalability.

References

1. Tendler, J.M., Dodson, J.S., Fields, J.S., Le, H.Q., Sinharoy, B.: POWER4 system microarchitecture. IBM J. Res. Dev. **46**, 5–26 (2002)
2. Asanovic, K., et al.: A view of the parallel computing landscape. Commun. ACM **52**(10), 56–67 (2009)
3. Stackhouse, B., et al.: A 65 nm 2-billion transistor quad-core Itanium processor. IEEE J. Solid-State Circuits **44**(1), 18–31 (2009)
4. Bell, S., et al.: TILE64 - processor: a 64-core SoC with mesh interconnect. In: Proceedings of IEEE International Solid-State Circuits Conference (ISSCC), pp. 88–598 (2008)
5. Zhang, C.: Mars: a 64-core ARMv8 processor. In: 2015 IEEE Hot Chips 27 Symposium (HCS), Cupertino, CA, USA, pp. 1–23 (2015)
6. Brown, N., Jamieson, M., Lee, J., Wang, P.: Is RISC-V ready for HPC prime-time: evaluating the 64-core Sophon SG2042 RISC-V CPU. In: Proceedings of the SC '23 Workshops of the International Conference on High Performance Computing, Network, Storage, and Analysis, pp. 1566–1574 (2023)
7. Sugumar, R., Shah, M., Ramirez, R.: Marvell ThunderX3: next-generation arm-based server processor. IEEE Micro **41**(2), 15–21 (2021)
8. Sorin, D., Hill, M., Wood, D.: A Primer on Memory Consistency and Cache Coherence. Synthesis Lectures on Computer Architecture (2011)
9. Yang, Q., Thangadurai, G., Bhuyan, L.N.: Design of an adaptive cache coherence protocol for large scale multiprocessors. IEEE Trans. Parallel Distrib. Syst. **3**(3), 281–293 (1992)
10. Pong, F., Dubois, M.: Verification techniques for cache coherence protocols. ACM Comput. Surv. **29**(1), 82–126 (1997)
11. Patel, V., Biswas, S., Chaudhuri, M.: Leveraging cache coherence to detect and repair false sharing on-the-fly. In: 2024 57th IEEE/ACM International Symposium on Microarchitecture (MICRO), Austin, TX, USA, pp. 823–839 (2024)

12. Arafa, M., Fahim, B., Kottapalli, S., et al.: Cascade lake: next generation Intel Xeon scalable processor. IEEE Micro **39**(2), 29–36 (2019)
13. Derebasoglu, E., Kadayif, I., Ozturk, O.: Coherency traffic reduction in manycore systems. In: 2022 25th Euromicro Conference on Digital System Design (DSD), Maspalomas, Spain, pp. 262–267 (2022)
14. Talati, N., et al.: Prodigy: improving the memory latency of data-indirect irregular workloads using hardware-software co-design. In: Proceedings of IEEE International Symposium on High Performance Computer Architecture (HPCA), pp. 654–667 (2021)
15. Fu, G., Xia, T., Luo, Z., Chen, R., Zhao, W., Ren, P.: Differential-matching prefetcher for indirect memory access. In: Proceedings of IEEE International Symposium on High Performance Computer Architecture (HPCA), pp. 439–453 (2024)
16. Li, G., Temam, O., Liu, Z., Guo, S., Wang, D.: Cluster cache monitor: leveraging the proximity data in CMP. Int. J. Parallel Prog. **43**(6), 1054–1077 (2015)
17. Pellegrini, A., et al.: The Arm Neoverse N1 platform: building blocks for the next-gen cloud-to-edge infrastructure SoC. IEEE Micro **40**(2), 53–62 (2020)
18. Fensch, C., Barrow-Williams, N., Mullins, R.D., Moore, S.: Designing a physical locality aware coherence protocol for chip-multiprocessors. IEEE Trans. Comput. **62**(5), 914–928 (2013)
19. Kurian, G., Khan, O., Devadas, S.: The locality-aware adaptive cache coherence protocol. In: Proceedings of International Symposium on Computer Architecture (ISCA), Tel-Aviv, Israel, pp. 523–534 (2013)
20. Jiang, Q., Lee, Y.C., Zomaya, A.Y.: The power of ARM64 in public clouds. In: 2020 20th IEEE/ACM International Symposium on Cluster, Cloud and Internet Computing (CCGRID), Melbourne, VIC, Australia, pp. 459–468 (2020)
21. Binkert, N., et al.: The gem5 simulator. SIGARCH Comput. Archit. News **39**(2), 1–7 (2011)
22. Bienia, C., Kumar, S., Singh, J.P., Li, K.: The PARSEC benchmark suite: characterization and architectural implications. In: Proceedings of the International Conference on Parallel Architectures and Compilation Techniques (PACT), Toronto, Ontario, Canada, pp. 72–81 (2008)

A Mapping Strategy Optimization Framework for Systolic Array Accelerators

Xin Yuan, Chenyang Wang, Qi Wu, Hengshan Yue, Haixiao Xu[✉],
and Xiaohui Wei

College of Computer Science and Technology, Jilin University, Changchun, China
{yuanxin23,chenyang23}@mails.jlu.edu.cn,
{wuqi,yuehs,haixiao,weixh}@jlu.edu.cn

Abstract. Systolic array accelerators have emerged as a mainstream architecture for deep neural networks (DNNs) owing to their efficient dataflow and high parallelism. Mapping strategies determine how computations and data transfers are orchestrated, and thus have a critical impact on accelerator performance. Comprehensively representing feasible mapping strategies (i.e., the mapping space) and efficiently evaluating their performance are essential for fully exploiting accelerator capabilities. However, existing methods inadequately model the data path from off-chip memory to processing elements (PEs) in systolic arrays, leading to suboptimal mapping strategies and degraded performance. Additionally, under constrained buffer capacities, conventional loop-nest scheduling provides insufficient temporal locality for effective on-chip data reuse. Moreover, previous performance models struggle to balance simulation accuracy with efficiency. To overcome these limitations, we propose a novel mapping strategy optimization framework for systolic array accelerators, introducing three key innovations: (1) fine-grained tiling and 6D loop-nest scheduling strategy to capture data movement from off-chip memory to the systolic array, enabling full data path modeling and better utilization of on-chip resources; (2) loop reversal optimization to prioritize computations on buffered data across iterations, enhancing temporal locality and on-chip data reuse; and (3) an event-driven performance model advancing simulations only at critical scheduling and dependency events, achieving high accuracy with significant efficiency gains. Experimental results demonstrate that our framework reduces latency by up to 37.63% and lowers energy consumption by up to 19.97% compared to the state-of-the-art FAMS framework. In addition, our performance model achieves an average 38.3× speedup in mapping strategy evaluation.

Keywords: Deep Neural Network Accelerator · Systolic Array · Mapping Space Exploration · Dataflow

1 Introduction

In recent years, DNNs have achieved breakthrough progress in artificial intelligence domains such as computer vision [1,2] and natural language pro-

© The Author(s), under exclusive license to Springer Nature Singapore Pte Ltd. 2026
H. Liu et al. (Eds.): ICA3PP 2025, LNCS 16381, pp. 438–457, 2026.
https://doi.org/10.1007/978-981-95-8399-7_24

cessing [3,4], with model parameter scales continuously expanding. Large-scale DNNs are characterized by intensive computational and memory requirements. These demands far exceed the capabilities of traditional general-purpose CPU/GPU architectures [5]. To overcome this bottleneck, domain-specific accelerators specialized for DNN tasks have been developed [6–8]. Among various neural network accelerator architectures, systolic arrays have attracted considerable attention for their massively parallel processing capabilities. The systolic array is composed of processing elements (PEs) arranged in a two-dimensional grid. It demonstrates exceptional performance in processing large-scale matrix multiplication operations that are fundamental to DNNs, and has been successfully validated and deployed at scale in commercial products such as Google TPU [8] and Tesla FSD [9].

Systolic array accelerators support multiple approaches for partitioning computational workloads into tiles and scheduling them in a specific order, with each referred to as a mapping strategy. Differences in hardware resource utilization and data reuse efficiency across mappings can lead to substantial performance variations [10]. Experimental results show that, under identical workload and hardware configurations, the energy efficiency gap between the best and worst mapping strategies can reach up to 15.2 ×.

To fully leverage the potential of systolic array accelerators for DNN workloads, it is essential to comprehensively represent the mapping space and efficiently evaluate mapping strategies [11]. However, the mapping space is typically large and complex due to the intricate data movements across a multi-level memory hierarchy. Furthermore, feasible mappings are constrained by the hardware architecture and workload characteristics. Therefore, accurately representing the mapping space and efficiently identifying optimal mappings remains a challenging task.

Many studies have proposed notations to represent mapping strategies on systolic array accelerators [10–15]. Representative examples include the compute-centric notation Timeloop [10], the data-centric notation MAESTRO [12,13], the relation-centric notation TENET [14], and the state-of-the-art memory-centric notation FAMS [11]. These notations enhance the expressiveness of mapping strategies and facilitate accelerator performance optimization.

Despite these advances, several limitations remain:

First, existing approaches fail to comprehensively capture the full data movement path. Timeloop lacks accurate modeling of inter-PE dataflow. MAESTRO and TENET offer limited support for off-chip data transfers. The recently proposed FAMS framework pays insufficient attention to data movement between on-chip buffers and the PE array. Such incomplete representations of the mapping space hinder full exploitation of accelerator performance.

Second, the tile scheduling order defined by loop nests, as adopted by FAMS and Timeloop, inherently exhibits poor temporal locality when buffer capacity is limited, resulting in ineffective on-chip data reuse. Specifically, many data tiles are evicted before they can be reused, leading to frequent off-chip memory accesses.

Third, existing performance models struggle to balance accuracy and efficiency. Timeloop, MAESTRO, and TENET adopt analytical models for fast evaluation, but their simplifications reduce accuracy. In contrast, FAMS achieves higher accuracy through cycle-accurate simulation, but incurs substantial simulation overhead.

To address these limitations, this paper proposes a novel mapping framework for systolic array accelerators, featuring three key innovations:

1. **Fine-Grained Tiling and Scheduling Strategy**: We propose a fine-grained task tiling scheme combined with a 6D loop-nest scheduling strategy to uniformly define data transfers from off-chip storage to the systolic array. This enables comprehensive modeling of the full data path. Experimental results show up to 31.78% reduction in latency and an average improvement of 13.10% in energy efficiency compared to the state-of-the-art notation.
2. **Dynamic Loop Reversal Optimization**: We introduce a dynamic loop reversal technique that adaptively reorders tile execution to prioritize those depending on data retained in on-chip buffers. This enhances temporal locality and improves on-chip data reuse. When combined with the first optimization, it yields an additional 11.63% average improvement in data reuse.
3. **Event-Driven Performance Model**: We develop an event-driven performance model that advances simulation only at critical execution events and skips predictable phases. Compared to state-of-the-art performance model, it achieves a 38.3 $\times$ speedup in performance evaluation while maintaining high accuracy.

2   Background

2.1   Systolic Array Accelerator

Figure 1 illustrates the architecture and dataflow of a systolic array accelerator. The systolic array adopts a two-dimensional grid topology composed of numerous PEs. Each PE integrates a multiply-accumulate (MAC) unit and local registers, and connects to its neighboring PEs for direct data forwarding. Edge PEs interface with on-chip global buffers through dedicated interconnects. During execution, data is loaded from array edges and propagated in a systolic manner across the array. This design enables deep overlap between computation and data movement, improves data reuse, and reduces memory access overhead [16]. As a result, systolic array accelerators deliver high performance in large-scale matrix multiplication, a fundamental operation in DNNs.

2.2   Mapping

A mapping strategy defines how to partition DNN computations and orchestrate the transfer of associated data across multi-level memory hierarchies [17]. In systolic array accelerators, mapping involves data transfers across three memory levels: off-chip memory, on-chip buffers, and the local registers within PEs,

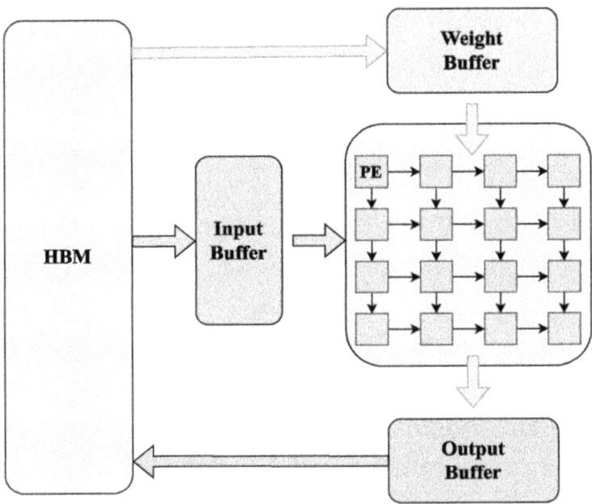

Fig. 1. Detailed architecture and dataflow of a systolic array accelerator

as illustrated in Fig. 1. The choice of mapping strategy critically impacts performance, as different partitioning and scheduling decisions lead to varying resource utilization, compute-transfer parallelism, and on-chip data reuse.

Identifying the optimal mapping strategy is crucial for accelerator hardware design, as system performance depends jointly on the DNN workload, the hardware architecture, and the mapping itself [10]. For a given hardware and workload, multiple valid mappings may exist; thus, selecting the optimal one is essential for fairly evaluating and comparing different accelerator designs.

However, identifying optimal mappings remains a challenging task. The large set of mapping strategies forms a high-dimensional and complex mapping space. This space is constrained by both hardware resource limitations and the characteristics of the computational task. Traditional experience-based approaches often fall short of achieving optimal performance. Therefore, systematic mapping space exploration has become essential: by comprehensively representing the mapping space and employing structured search methods, truly optimal mapping strategies can be identified through exhaustive exploration.

3   Related Work and Motivation

3.1   Related Work

Extensive research has focused on constructing comprehensive mapping space and developing performance models for evaluating mapping strategies [10–15].

In terms of representing the mapping space, early work such as Timeloop proposed a computation-centric notation that uses multidimensional loop nests to describe the mapping space across multi-level memory hierarchies, as illustrated in Fig. 2(b). However, it lacks the ability to model inter-PE dataflows within the array. MAESTRO introduced a data-centric notation focusing on inter-PE communication. TENET further proposed a relation-centric notation that uses relational expressions to uniformly describe PE interconnects and data distribution strategies, offering stronger expressiveness and flexibility. Although MAESTRO and TENET improve the understanding of on-chip dataflows and extend the mapping space, they do not sufficiently consider task tiling strategies constrained by limited on-chip memory. Therefore, they are unable to accurately model off-chip data movement. FAMS proposes a memory-centric mapping notation. It represents the tiling and scheduling strategy along the data path from off-chip memory to the global buffers using a 3D loop nest, enabling precise characterization of off-chip data movement. However, it overlooks fine-grained data transfers from the global buffers to the systolic array. Its tiles are typically large enough to support multiple computation rounds on the array. Using such coarse-grained tiles as the basic unit for scheduling and resource management limits compute-transfer parallelism and hinders effective utilization of on-chip resources.

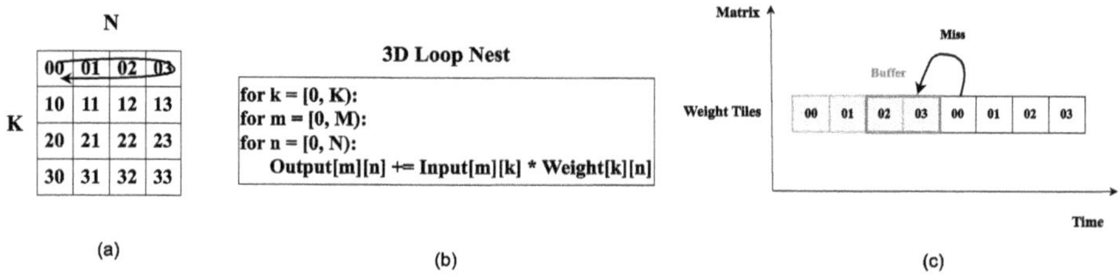

Fig. 2. An example of matrix multiplication tiling and scheduling: (a) a tiled weight matrix; (b) a 3D loop nest; (c) access sequence of weight tiles under loop-nest scheduling, where the red box indicates the on-chip buffer contents when m changes. (Color figure online)

For mapping strategy evaluation, Timeloop, MAESTRO, and TENET employ analytical models based on abstracted DNN execution behaviors. While these models enable fast evaluation, they fail to capture resource contention and data dependencies, limiting their accuracy. In contrast, FAMS incorporates a cycle-accurate simulator that models accelerator behavior with high fidelity. However, its high computational cost makes it impractical for large-scale design space exploration [18].

3.2 Motivation

Although existing studies have made significant progress in representing the mapping space and mapping strategy evaluation for systolic array accelerators, our analysis reveals three key limitations that constrain their effectiveness:

Incomplete Data Path Modeling. Timeloop fails to model dataflows between PEs within the systolic array. MAESTRO and TENET lack support for off-chip data transfers. FAMS overlooks data movement between the global buffers and the systolic array. Collectively, such modeling limitations prevent existing methods from fully exploiting the performance potential of systolic array accelerators for DNN workloads.

Data Reuse Distance Bottleneck Under Loop-Nest Scheduling. Under the tile scheduling order defined by loop nests, all tiles of the same matrix share the same data reuse distance—that is, the number of unique data element accesses between two accesses to the same tile (e.g., as illustrated in Fig. 2(a) and (b), the reuse distance of each tile is four tiles). When this reuse distance exceeds the buffer capacity, tiles may be evicted before being reused, as shown in Fig. 2(c). Under a First-In, First-Out (FIFO) buffer management policy, weight tiles cannot be reused on-chip. However, this critical issue remains largely unaddressed in existing work.

Accuracy–Efficiency Imbalance in Performance Evaluation. Analytical models simplify DNN execution to enable fast mapping strategy evaluation, but at the cost of accuracy. Cycle-accurate models provide high-fidelity hardware modeling, but incur significant simulation overhead. Existing evaluation approaches struggle to balance accuracy and efficiency.

These limitations underscore the need for a comprehensive mapping optimization framework with three key capabilities: (1) modeling the full data path from off-chip memory to PEs to enable global mapping optimization; (2) optimizing loop-nest scheduling to enhance on-chip data reuse under constrained buffer capacity; and (3) efficiently and accurately evaluating the performance of mapping strategies. Addressing these challenges is essential to fully unlocking the performance potential of systolic array accelerators for modern DNN workloads.

4   Methodology

4.1   Mapping Space Representation Overview

Figure 3 illustrates our mapping space representation of DNN tasks on systolic array accelerators, highlighting the mapping choices made at each stage of the process. The mapping process begins with task transformation: for convolutional operations, an im2col transformation restructures them into matrix multiplications, ensuring compatibility with the systolic array computation paradigm.

To respect on-chip memory capacity constraints, we partition the full matrix multiplication task into tiles, each representing the amount of data the systolic array can process in a single computation round. To fully exploit data reuse, we further group tiles to enhance data locality, with the execution order defined by a 6D loop nest. We also introduce a loop reversal strategy that dynamically adjusts scheduling to reduce the reuse distance of critical data, thereby improving on-chip data reuse. Finally, the specific mapping of tiles onto the systolic array is determined by dataflow patterns. For clarity, the complete representation of loop reversal is omitted in Fig. 3 due to its complexity.

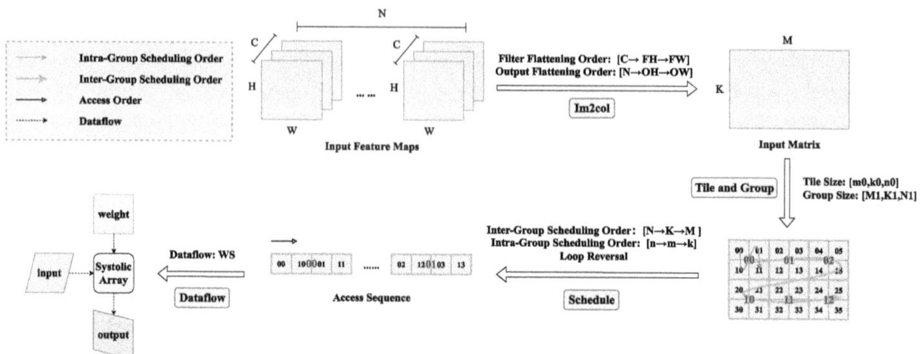

Fig. 3. Mapping space representation overview

Overall, the proposed framework captures the constraints of systolic array capacity through fine-grained tiling, while implicitly modeling on-chip buffer constraints through tile grouping. It uniformly represents data transfer from off-chip storage to global buffers and subsequently from global buffers to the systolic array using a 6D loop nest. Furthermore, it employs dataflow patterns to represent inter-PE data movement, thereby achieving comprehensive, end-to-end modeling of the entire data path.

It is worth noting that the im2col transformation adopts fixed implementation choices, such as row-major or column-major ordering, while dataflow patterns similarly follow predefined modes, such as Weight Stationary (WS). As these are well-established techniques, this paper does not elaborate on their internal mechanisms. Instead, the following sections focus on our three key techniques: fine-grained tiling and scheduling, dynamic loop reversal optimization, and the event-driven performance model.

4.2 Fine-Grained Tiling and Scheduling

In this section, we first introduce a fine-grained tiling strategy that partitions the workload into tiles, each executable in a single computation round on the systolic array, with the goal of improving pipeline efficiency and on-chip resource utilization. We then propose a 6D loop-nest scheduling strategy to enhance data locality and maximize on-chip data reuse. Our method is illustrated using a representative matrix multiplication workload with dimensions $[M, K, N]$, where M, K, and N denote the sizes of the operand matrices.

Fine-Grained Tiling Strategy. Given the capacity constraints of multi-level memory hierarchies, workload data must be partitioned into smaller tiles and transferred to compute units in a defined order. To support this process, we employ a fine-grained tiling strategy that divides tensor operations into tiles of size $[m_0, k_0, n_0]$, where m_0, k_0, and n_0 denote the tile sizes along the corresponding dimensions. Each tile can be executed in a single round on the systolic array.

This tile granularity is constrained by two key factors. First, to ensure that each tile is mapped only once onto the array, its dimensions must conform to the structural limitations imposed by the dataflow pattern and the array shape. For example, under the WS dataflow, the tile size must satisfy the following constraint:

$$k_0 \leq R, \quad n_0 \leq C \tag{1}$$

where R and C denote the number of rows and columns in the systolic array, respectively. Second, to enable pipelined parallelism between computation and data transfer, the tile size must also satisfy the following constraints to ensure that the system can prefetch the next tile while computing the current one:

$$\begin{cases} m_0 \cdot k_0 \leq \dfrac{B_{\text{input}}}{2} \\[2ex] k_0 \cdot n_0 \leq \dfrac{B_{\text{weight}}}{2} \\[2ex] m_0 \cdot n_0 \leq \dfrac{B_{\text{output}}}{2} \end{cases} \tag{2}$$

Here, B_{input}, B_{weight}, and B_{output} denote the capacities of the global buffers for the input, weight, and output matrices, respectively.

This tile serves as the basic scheduling unit for coordinating data transfer and computation. Specifically, when a matrix multiplication tile is scheduled, the system prefetches the corresponding operand tiles from off-chip memory. Once these tiles are ready in the buffer, they can be loaded into the array for computation. After computation, if no subsequent operations depend on the tiles, they can be evicted from the buffer. This fine-grained scheduling mechanism unifies the granularity of data transfers and execution order across the data path—from off-chip memory to on-chip buffers, and from buffers to the systolic array. By coordinating these stages at tile granularity, our method enables fine-grained pipelined parallelism between data transfer and computation, and improves utilization of on-chip resources.

6D Loop-Nest Scheduling Strategy. For scheduling order, a common approach is to use a 3D loop nest $[M \rightarrow K \rightarrow N]$, as shown in Fig. 4(a), where the relative order of M, K, and N specifies the loop nesting hierarchy from outermost to innermost. However, this scheduling strategy results in large reuse distances of tiles. For example, the order in Fig. 4(a) leads to a reuse distance of four for weight tiles. If the buffer can accommodate all four tiles, the weight data remains on-chip and can be reused across multiple m-loop iterations. Otherwise, data eviction occurs at every iteration. Under a FIFO buffer management policy, all weight tiles must be reloaded from off-chip memory at each m-loop iteration. Consequently, the total off-chip traffic for weights increases from $K \times N$ to $M_0 \times K \times N$, where M_0 denotes the number of m-loop iterations. This phenomenon explains why FAMS may yield optimal mappings with insufficient on-chip data reuse in scenarios with constrained buffer capacity or large-scale workloads.

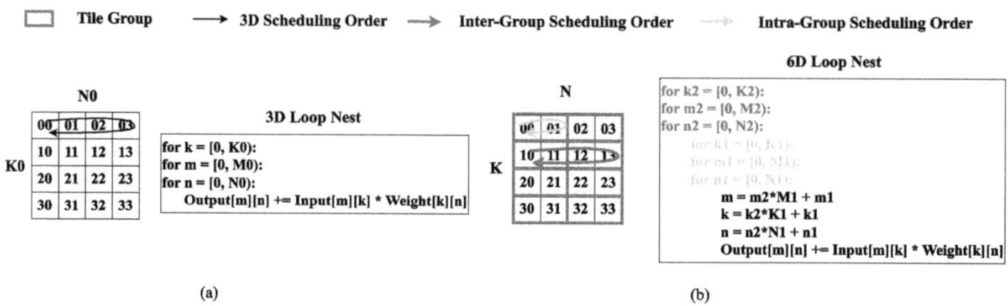

Fig. 4. Comparison between 3D loop nest (a) and 6D loop nest (b)

To further improve data reuse, we extend the original 3D loop structure to a 6D loop nest $[M \to K \to N] \to [m \to k \to n]$, as illustrated in Fig. 4(b). The tiles defined by the inner 3D loops are referred to as a tile group. Here, $[M \to K \to N]$ specifies the inter-group scheduling order, while $[m \to k \to n]$ defines the intra-group scheduling order. The parameters $[M_1, K_1, N_1]$ denote the number of tiles in each dimension within a tile group. Local grouping and scheduling of tiles enhance data locality. Each tile group is considered a computational unit capable of fully reusing buffered data during execution.

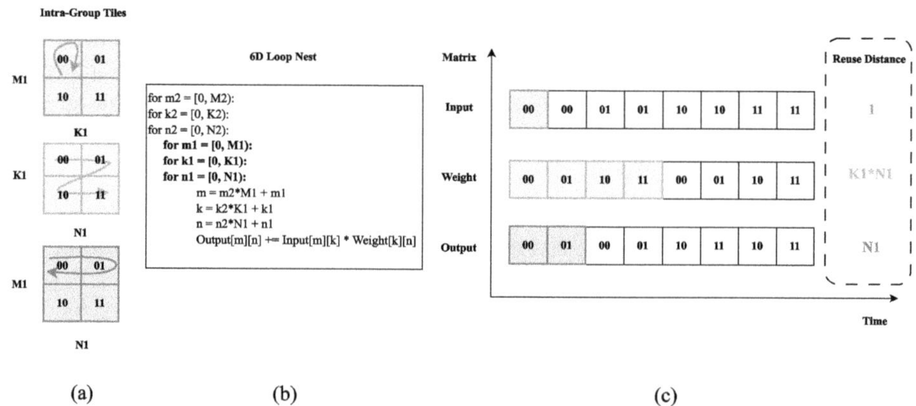

Fig. 5. Reuse distances of operand tiles during intra-group computation: (a) intra-group tile layouts of the input, weight, and output matrices; (b) corresponding 6D loop-nest structure; (c) access sequences of operand tiles.

To ensure such reuse, the tile group must satisfy specific size constraints. As illustrated in Fig. 5, under a given intra-group scheduling order, tiles from the three operand matrices exhibit different reuse distances. If the buffer capacity is smaller than the reuse distance, the corresponding data may be evicted before its next access, reducing on-chip data reuse. To prevent this, we impose the following constraints on tile group size:

$$\begin{cases} m_0 \cdot k_0 \cdot d_{\text{input}} \leq B_{\text{input}} \\ k_0 \cdot n_0 \cdot d_{\text{weight}} \leq B_{\text{weight}} \\ m_0 \cdot n_0 \cdot d_{\text{output}} \leq B_{\text{output}} \end{cases} \tag{3}$$

Here, d_{input}, d_{weight}, and d_{output} represent the tile reuse distances of the input, weight, and output tiles, respectively. In the example shown in Fig. 4, weight tiles are grouped in pairs, allowing each tile to be reused M_1 times during intra-group computation. A tiling and scheduling configuration is considered valid only if all the above constraints are satisfied.

4.3 Dynamic Loop Reversal Optimization

In the previous section, we discussed tile reuse distances within tile groups and enforced full on-chip data reuse during intra-group computations by applying constraints on tiling and scheduling. Since both intra-group and inter-group scheduling are structured as 3D loop nest, they exhibit similar data reuse patterns.

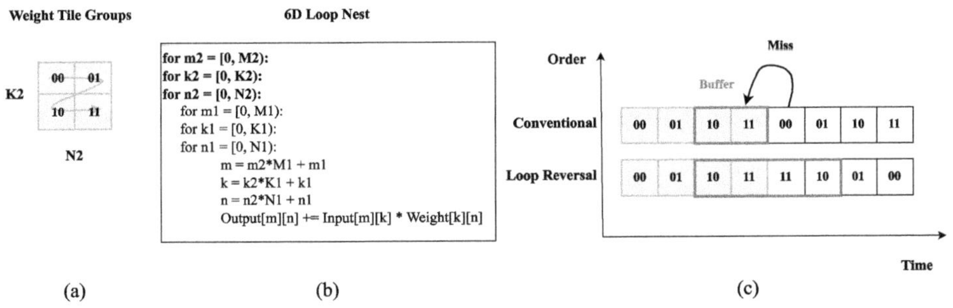

Fig. 6. Illustration of inter-group loop reversal optimization: (a) layout of weight tile groups; (b) corresponding 6D loop-nest structure; (c) comparison of access sequences for weight tile groups with and without loop reversal, where red boxes indicate the on-chip buffer contents when m_2 changes. (Color figure online)

However, group-level reuse cannot be reliably achieved solely through tiling or scheduling adjustments. For example, reusing a weight tile group across two adjacent scheduling instances requires satisfying the following constraint:

$$k_0 \cdot n_0 \cdot K_1 \cdot N_1 \cdot d_{\text{weight\_group}} \leq B_{\text{weight}} \tag{4}$$

Here, $d_{\text{weight\_group}}$ denotes the reuse distance of a weight tile group. This constraint is not solely determined by scheduling and tiling, but also depends on the full dimensions of the matrix. For example, under the scheduling order shown in Fig. 6(b), satisfying Fig. 4 would require the buffer to hold the entire weight matrix. If this condition is not satisfied, weight tiles will be evicted before they can be reused, resulting in the loss of on-chip reuse opportunities.

To preserve limited on-chip data reuse under such conditions, we propose a loop reversal optimization strategy. This approach reverses the loop iteration order after each group-level iteration, allowing computations involving data that remain in the buffer to be prioritized in the subsequent iteration, thereby enhancing temporal locality. As shown in Fig. 6 (c), under the conventional scheduling order, a buffer miss occurs at the beginning of each new m_2-loop iteration, causing previously loaded data to be evicted. After applying loop reversal, computations involving buffered data (i.e., *tile 10* and *tile 11*) are prioritized, enhancing on-chip reuse. Additionally, although the next access to *tile 00* and *tile 01* is delayed, their data reuse distances remain unchanged. This indicates that loop reversal does not compromise data reuse opportunities.

Algorithm 1: 6D loop-nest scheduling after loop reversal optimization

---

Input: $A \in \mathbb{R}^{M \times K}$, $B \in \mathbb{R}^{K \times N}$
outer extents $(M_2, K_2, N_2) \rightarrow (g_m, g_k, g_n)$
inner extents $(M_1, K_1, N_1) \rightarrow (t_m, t_k, t_n)$
Output: $C = AB \in \mathbb{R}^{M \times N}$

```
1  for gm ← 0 to M2−1 do
2      dir1 ← (−1)^gm                                    // flip gk direction every gm
3      for gk ∈ [0, K2−1]step dir1 do
4          dir2 ← (−1)^(gm+gk)                           // flip gn direction every gk
5          for gn ∈ [0, N2−1]step dir2 do
6              dir3 ← (−1)^(gm+gk+gn)      // flip tm, tk, tn directions every gn
7              for tm ∈ [0, M1−1]step dir3 do
8                  for tk ∈ [0, K1−1]step dir3 do
9                      for tn ∈ [0, N1−1]step dir3 do
10                         m ← gm M1 + tm
11                         k ← gk K1 + tk
12                         n ← gn N1 + tn
13                         Cm,n ← Cm,n + Am,k Bk,n
```

Algorithm 1 illustrates an example of the final scheduling order after loop reversal optimization. Notably, the intra-group scheduling order is also reversed after each tile group is processed. This design is necessary because group size is not directly constrained by buffer capacity; instead, only the tile reuse distance is restricted to fit within the buffer. Consequently, the total size of a tile group may still exceed the buffer capacity.

4.4 Event-Driven Performance Model

In mapping space exploration, an accurate and efficient performance model is important for identifying high-quality strategies. Analytical models enable fast evaluation but rely on idealized assumptions, which limit their fidelity [10,12–14]. In contrast, recent work adopts cycle-accurate simulation to achieve higher precision [11], but the associated computational overhead hinders large-scale exploration.

We observe that the primary performance bottleneck of cycle-accurate simulators stems from cycle-by-cycle modeling of hardware behavior and states. While this approach helps understand software mapping and hardware behavior, it incurs significant computational cost. Notably, under a given mapping strategy, dataflow patterns are typically regular and predictable, offering opportunities to optimize simulation performance.

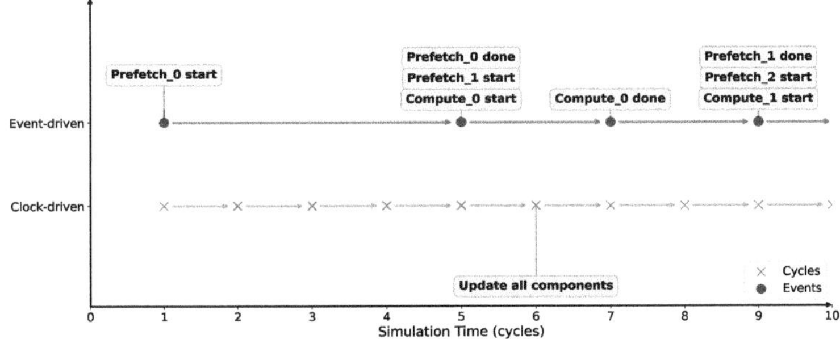

Fig. 7. Comparison of clock-driven and event-driven simulation paradigms

Based on this observation, we propose an event-driven simulator that avoids both idealized assumptions and the cycle-by-cycle tracking of hardware states. As illustrated in Fig. 7, the event-driven simulation paradigm advances the global state only when critical execution events occur—such as the start and end of tile transfers or the initiation and completion of tile computations—while skipping the simulation of predictable behaviors. This approach efficiently captures key scheduling events, data dependencies, and resource conflicts, thereby significantly improving simulation efficiency without compromising evaluation accuracy.

Our simulator is implemented using the SimPy framework, which is a powerful discrete-event simulation framework in Python for building process-interaction models. Energy estimation follows the modeling principles introduced in Eyeriss [19]. As shown in Fig. 8, the simulator takes as input the neural network workload, mapping strategy, and hardware configuration, and outputs key performance metrics, including total latency, energy consumption, PE utilization, and data reuse rates.

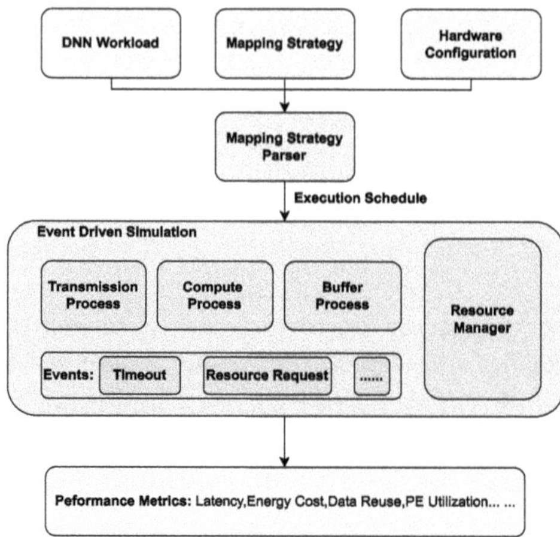

Fig. 8. Overview of the proposed event-driven performance model

5 Experimental Results

5.1 Experimental Setup

To systematically evaluate our mapping framework, we develop an automated methodology that explores the design space of practical systolic array accelerators. Specifically, we target five representative array sizes—32×32, 64×64, 128×128, 256×256, and 512×512 PEs—spanning a broad spectrum of deployment scenarios.

We use square matrix multiplication as the benchmark workload, with dimensions scaling from $R \times R \times R$ to $16R \times 16R \times 16R$ in powers of two, where R denotes the systolic array dimension. To examine diverse resource conditions, we vary on-chip buffer capacity from $2R^2$ Bytes to the full workload size (measured as the size of a single global buffer, consistent throughout this section), and adjust DRAM bandwidth from $0.25R$ to R Bytes per cycle.

All data are represented as 8-bit integers. Tile size exploration is performed at the granularity of the array dimension to balance search efficiency and coverage of the mapping space. For each configuration, we exhaustively evaluate all valid mapping strategies. To ensure a fair comparison, we adopt the memory-centric notation proposed by FAMS [11] as our baseline.

All experiments are conducted on a system equipped with an Intel® Core™ i5-12600KF processor (10 cores, 4.9 GHz) and 32 GB of RAM. This configuration serves as the default evaluation setup; scenario-specific variations are detailed in the corresponding experimental sections.

5.2 Performance Model Evaluation

In this section, we evaluate both the accuracy and efficiency of the proposed performance model. For accuracy, we compare our model against open-source cycle-accurate systolic array simulator SCALE-Sim [20, 21]. The comparison uses a fixed 64×64 array and workloads ranging from $64 \times 64 \times 64$ to $2048 \times 2048 \times 2048$, with both simulators applying identical mapping strategies. For efficiency, we explore the mapping space on systolic arrays ranging from 32×32 to 256×256 and compare simulation time with FAMS's cycle-accurate simulator.

Accuracy Validation. Figure 9 presents the absolute and relative errors in latency estimation by our simulator, with SCALE-Sim as the reference baseline. The average relative error is 0.799%, primarily due to unmodeled factors such as memory access latency. These deviations are considered acceptable for large-scale mapping space exploration.

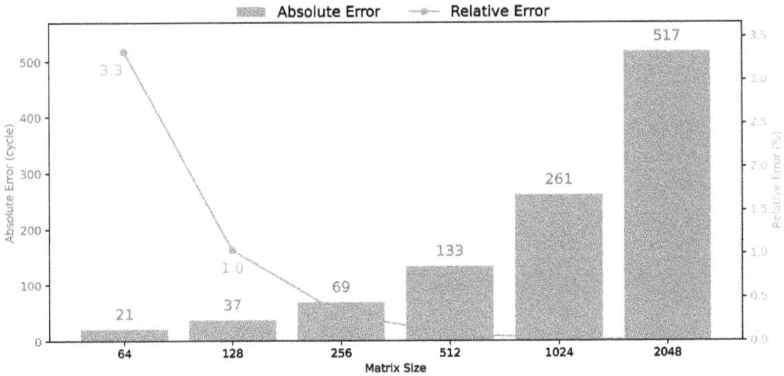

Fig. 9. Latency estimation error compared to SCALE-Sim across matrix sizes

Simulation Efficiency. Experimental results show that our event-driven performance model achieves an average simulation time of only 0.0277 s, compared to 1.0629 s for FAMS's cycle-accurate simulator, yielding a 38.30× speedup, as shown in Table 1. This substantial improvement in simulation efficiency enables practical exploration of large-scale mapping spaces, which would otherwise be infeasible with cycle-accurate simulation.

Scalability Enabled by the Event-Driven Paradigm. As shown in Fig. 10, we evaluate the average simulation time across configurations with varying systolic array sizes and workload scales (see Sect. 5.1 for details). The cycle-accurate simulator exhibits a sharp increase in simulation time as workload size grows, reaching 19.89 s in large-scale scenarios. In contrast, our event-driven model maintains consistently low simulation times with minimal sensitivity to scaling. This difference stems from their fundamental simulation paradigms: cycle-accurate simulators incur overhead proportional to the total number of execution cycles, which grows rapidly with workload complexity. In comparison, the

Table 1. Comparison of simulation time between the proposed model and FAMS

Simulation Time Metric	FAMS (s)	Proposed Model (s)
Maximum	19.899	0.3174
Minimum	0.1391	5.97×10^{-5}
Median	0.2383	0.0017
Average	1.0629	0.0277
Average Speedup		38.30×

event-driven model's overhead is primarily determined by the number of simulation events, which depends on the relative scale between workload and array size rather than the absolute computational intensity. This decoupling enables robust and scalable performance, particularly for large arrays and complex workloads.

5.3 Impact of Fine-Grained Tiling and Scheduling

This section evaluates the impact of fine-grained tiling and scheduling on system performance. We compare our mapping strategies with FAMS across a range of configurations, focusing on four key metrics: execution latency, energy consumption, energy-delay product (EDP), and DRAM traffic.

The results consistently show improvements across all metrics. For latency-optimal mapping strategies, our method achieves an average latency reduction of 7.76%. For energy-optimal mappings, it reduces energy consumption by 3.64% and lowers DRAM traffic by 17.81% on average. Most notably, under EDP-optimal mapping strategies, our approach achieves an average 11.59% reduction in EDP. These results demonstrate the effectiveness of fine-grained tiling and scheduling in enhancing on-chip resource utilization and data reuse.

In the following discussion, we analyze the experimental results shown in Fig. 11 to demonstrate how our fine-grained tiling and scheduling contribute to performance improvements over FAMS. The baseline hardware configuration features a 128×128 systolic array, 1024KB buffer size, and 128 Bytes/cycle off-

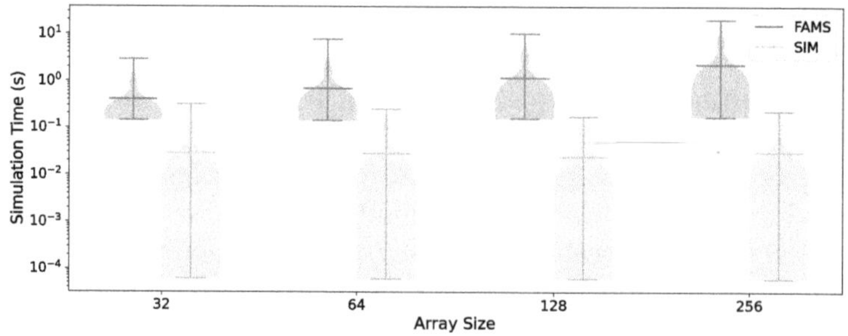

Fig. 10. Average simulation time distribution across array sizes

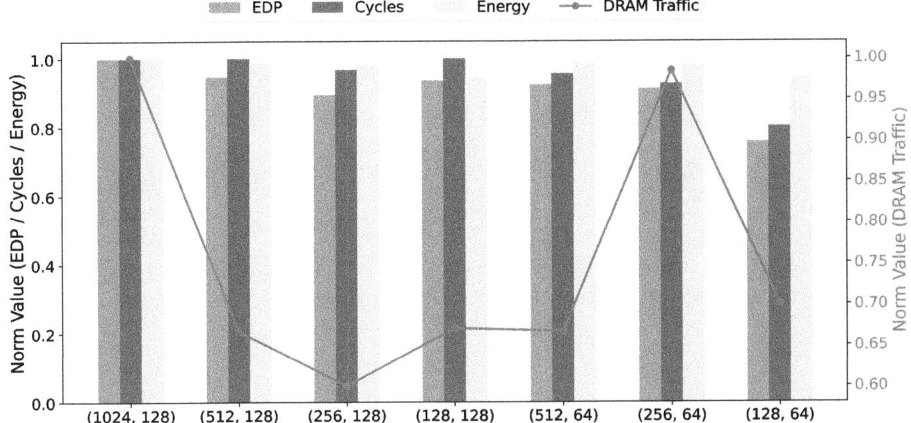

Fig. 11. Comparison of our EDP-optimal mapping strategies and FAMS's across diverse hardware configurations. The x-axis (a, b) represents the hardware configuration, where a denotes the global buffer capacity (KB) and b specifies the off-chip bandwidth (Bytes/cycle). All results are normalized to those obtained under FAMS's optimal mappings.

chip communication bandwidth, with a fixed workload of $1024 \times 1024 \times 1024$ matrix multiplication. Buffer capacity and communication bandwidth are varied based on this baseline configuration.

When buffer capacity is sufficient, as in configuration $(1024, 128)$ of Fig. 11, complete on-chip data reuse is guaranteed. FAMS's optimal mapping adopts the same tile size as ours, which supports single-round execution on the array and minimizes pipeline latency. As a result, both FAMS and our method achieve identical optimal performance.

In configurations $(256, 128)$, $(128, 128)$, and $(128, 64)$, due to limited buffer capacity, FAMS's optimal mappings still adopt tile sizes that match the systolic array dimensions. These cases correspond to scenario (a) in Fig. 4. In such settings, the 3D loop-nest scheduling used by FAMS is overly coarse-grained and results in poor data locality. In contrast, our 6D loop-nest scheduling leverages tile grouping to significantly improve on-chip data reuse, thereby enhancing overall performance.

In the remaining configurations—$(512, 128)$, $(512, 64)$, and $(256, 64)$—the optimal mappings identified by FAMS adopt tile sizes that require multiple computation rounds on the systolic array, allowing data reuse across rounds. However, these tile sizes are limited to no more than half of the buffer capacity to accommodate prefetching. This restricts the full utilization of on-chip buffers. In contrast, our approach supports prefetching and releasing at fine-grained tile granularity, decoupling group size from direct buffer capacity constraints. As a result, our optimal mappings feature larger tile groups and more efficient buffer utilization. Since full on-chip data reuse during intra-group computations is guaranteed (as illustrated in Sect. 4.2), our method achieves higher data reuse

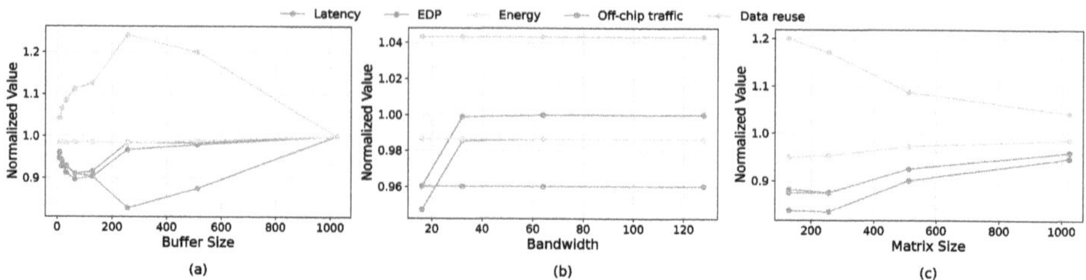

Fig. 12. Impact of loop reversal optimization under varying hardware and workload conditions: (a) on-chip buffer size; (b) off-chip communication bandwidth; (c) matrix dimensions.

and reduces off-chip memory traffic. Furthermore, our latency benefits from the pipelining enabled by fine-grained tiling. Under bandwidth-constrained conditions, the reduced off-chip communication further lowers latency, contributing to overall performance improvements.

In summary, experimental results confirm the effectiveness of our tiling and scheduling strategies. Fine-grained tiling enables efficient overlap between data transfer and computation, thereby reducing pipeline latency. Coordinated task scheduling and resource management at this granularity significantly improve on-chip resource utilization. Additionally, the 6D loop-nest scheduling enhances data locality and maximizes on-chip data reuse.

5.4   Impact of Dynamic Loop Reversal

Dynamic loop reversal optimization reduces off-chip memory traffic by reducing data reuse distances across loop iterations. Based on our tiling and scheduling strategy, comprehensive experiments show that loop reversal improves on-chip data reuse by 11.63% on average, reduces off-chip memory accesses by 8.30%, and lowers the EDP by 4.71%.

To evaluate how the benefits of loop reversal vary with different hardware configurations and workloads, we conduct a detailed analysis shown in Fig. 12. The baseline setup consists of a 64×64 systolic array, 8KB global buffer, 16 Bytes/cycle off-chip bandwidth, and a $1024 \times 1024 \times 1024$ matrix multiplication workload. Based on this setup, we independently vary buffer size, off-chip bandwidth, and matrix dimensions to assess the impact of each factor. All metrics are normalized to the proposed fine-grained tiling and scheduling baseline.

As shown in Fig. 12(a), when buffer capacity is the primary performance bottleneck, increasing the buffer size enables more intermediate tiles to be retained across iterations, thereby amplifying the data reuse benefits of loop reversal. The optimization is most effective when the buffer can hold one-fourth of the matrix. However, as buffer capacity continues to grow, it naturally alleviates data reuse distance bottlenecks and reduces off-chip memory traffic. As this effect becomes dominant, the performance gains from loop reversal gradually diminish.

Table 2. Performance metrics normalized to FAMS across different array sizes

Array Size	EDP	Latency	Energy	DRAM Traffic	Data Reuse
32 × 32	0.7978	0.8946	0.9020	0.7702	1.3963
64 × 64	0.8287	0.8966	0.9312	0.7579	1.4290
128 × 128	0.8526	0.8950	0.9566	0.7498	1.4547
256 × 256	0.8699	0.8951	0.9737	0.7523	1.4481
512 × 512	0.8803	0.8952	0.9844	0.7523	1.4481
Average	**0.8458**	**0.8953**	**0.9495**	**0.7565**	**1.4352**

As shown in Fig. 12(b), when off-chip bandwidth is the performance bottleneck, reducing off-chip memory traffic through loop reversal shortens communication time and enhances overall performance. However, once bandwidth is no longer a limiting factor, the performance gains begin to plateau. Figure 12(c) demonstrates that as workload size increases, the fixed off-chip memory access savings per iteration are amortized over a larger computational volume, leading to diminishing relative benefits.

In summary, loop reversal optimization significantly enhances performance in scenarios with limited on-chip hardware resources. It reduces energy consumption by lowering off-chip memory accesses. Under bandwidth-constrained conditions, it further reduces communication latency, contributing to overall performance improvements.

5.5 Impact of Joint Optimization

While the previous sections evaluated the individual effects of fine-grained tiling with scheduling and loop reversal, we now assess their combined impact. Compared to FAMS, the joint optimization reduces EDP by up to 50.08%, with an average reduction of 15.43% across all configurations. It also achieves average reductions of 24.35% in off-chip memory traffic and 10.48% in latency. Figure 2 summarizes the normalized performance metrics across different systolic array configurations.

6 Conclusion

This paper presents a mapping optimization framework for systolic array accelerators, introducing three key innovations: (1) A fine-grained tiling and scheduling strategy that provides a unified representation of data transfers from off-chip memory to the systolic array, thereby enhancing pipeline parallelism and on-chip resource utilization; (2) Dynamic loop reversal optimization, which reduces data reuse distances to improve on-chip buffer efficiency under resource-constrained conditions; (3) An event-driven performance model, which advances simulation based on critical execution events, striking a balance between accuracy and efficiency.

Compared to the state-of-the-art baseline, our approach leads to significant performance improvements, including average reductions of 15.43% in EDP, 10.48% in latency, and 5.05% in energy consumption. The proposed performance model delivers a 38.3 × average simulation speedup while maintaining high evaluation accuracy.

In summary, our mapping framework enables efficient exploration of large-scale mapping spaces and consistently delivers superior mapping performance across diverse hardware configurations and workloads. It offers a practical and scalable solution for mapping space exploration in systolic array accelerators.

Acknowledgments. This work was supported by the National Natural Science Foundation of China under Grant Nos. 62272190 and 62302190.

References

1. Krizhevsky, A., Sutskever, I., Hinton, G.E.: Imagenet classification with deep convolutional neural networks. Commun. ACM **60**(6), 84–90 (2017)
2. Dosovitskiy, A., et al.: An image is worth 16x16 words: transformers for image recognition at scale. arXiv preprint arXiv:2010.11929 (2020)
3. Vaswani, A., et al.: Attention is all you need. In: Advances in Neural Information Processing Systems, vol. 30 (2017)
4. Devlin, J., Chang, M.W., Lee, K., Toutanova, K.: Bert: pre-training of deep bidirectional transformers for language understanding. In: Proceedings of NAACL-HLT, pp. 4171–4186 (2019)
5. Alwani, M., Chen, H., Ferdman, M., Milder, P.: Fused-layer CNN accelerators. In: 2016 49th Annual IEEE/ACM International Symposium on Microarchitecture (MICRO), pp. 1–12. IEEE (2016)
6. Parashar, A., Rhu, M., Mukkara, A., Puglielli, A., Venkatesan, R., Khailany, B., et al.: SCNN: an accelerator for compressed-sparse convolutional neural networks. ACM SIGARCH Comput. Archit. News **45**(2), 27–40 (2017)
7. Zhang, C., Li, P., Sun, G., Guan, Y., Xiao, B., Cong, J.: Optimizing FPGA-based accelerator design for deep convolutional neural networks. In: Proceedings of the 2015 ACM/SIGDA International Symposium on Field-Programmable Gate Arrays, pp. 161–170 (2015)
8. Jouppi, N.P., et al.: In-datacenter performance analysis of a tensor processing unit. In: Proceedings of the 44th Annual International Symposium on Computer Architecture, pp. 1–12 (2017)
9. Talpes, E., et al.: Compute solution for tesla's full self-driving computer. IEEE Micro **40**(2), 25–35 (2020)
10. Parashar, A., et al.: Timeloop: a systematic approach to DNN accelerator evaluation. In: Proceedings of ISPASS, pp. 304–315 (2019)
11. Sun, H., et al.: FAMS: a framework of memory-centric mapping for DNNs on systolic array accelerators. IEEE Trans. Very Large Scale Integr. (VLSI) Syst. 1–14 (2025)
12. Kwon, H., Chatarasi, P., Sarkar, V., Krishna, T., Pellauer, M., Parashar, A.: MAESTRO: a data-centric approach to understand reuse, performance, and hardware cost of DNN mappings. IEEE Micro **40**(3), 20–29 (2020)

13. Kwon, H., Chatarasi, P., Pellauer, M., Parashar, A., Sarkar, V., Krishna, T.: Understanding reuse, performance, and hardware cost of DNN dataflow: a data-centric approach. In: Proceedings of the 52nd Annual IEEE/ACM International Symposium on Microarchitecture, pp. 754–768 (2019)
14. Lu, L., et al.: TENET: a framework for modeling tensor dataflow based on relation-centric notation. In: 2021 ACM/IEEE 48th Annual International Symposium on Computer Architecture (ISCA), pp. 720–733 (2021)
15. Yang, X., et al.: Interstellar: using halide's scheduling language to analyze DNN accelerators. In: Proceedings of the Twenty-Fifth International Conference on Architectural Support for Programming Languages and Operating Systems, pp. 369–383 (2020)
16. Xu, R., Ma, S., Guo, Y., Li, D.: A survey of design and optimization for systolic array-based DNN accelerators. ACM Comput. Surv. **56**(1), 1–37 (2023)
17. Chen, Y.H., Emer, J., Sze, V.: Using dataflow to optimize energy efficiency of deep neural network accelerators. IEEE Micro **37**(3), 12–21 (2017)
18. Sun, H., Shen, J., Zhang, C., Liu, H.: A coarse-and fine-grained co-exploration approach for optimizing DNN spatial accelerators: improving speed and performance. Electronics **14**(3), 511 (2025)
19. Chen, Y.H., Krishna, T., Emer, J.S., Sze, V.: Eyeriss: an energy-efficient reconfigurable accelerator for deep convolutional neural networks. IEEE J. Solid-State Circuits **52**(1), 127–138 (2016)
20. Samajdar, A., Zhu, Y., Whatmough, P., Mattina, M., Krishna, T.: SCALE-Sim: Systolic CNN Accelerator Simulator (2019)
21. Samajdar, A., Joseph, J.M., Zhu, Y., Whatmough, P., Mattina, M., Krishna, T.: A systematic methodology for characterizing scalability of DNN accelerators using scale-sim. In: 2020 IEEE International Symposium on Performance Analysis of Systems and Software (ISPASS), pp. 58–68. IEEE (2020)

LFBC: A Lifecycle-Managed False Bubble Flow Control Scheme for Torus Networks

Haofei Zhang[ID] and Youmeng Li[(✉)][ID]

College of Intelligence and Computing, Tianjin University, Tianjin 300072, China
{zhanghaofei,liyoumeng}@tju.edu.cn

Abstract. This paper proposes a Lifecycle-managed False Bubble Flow Control (LFBC) scheme to address the starvation issue in torus networks caused by the introduction of critical bubbles. Traditional Flit Bubble Flow Control (FBFC) avoids deadlock by reserving idle buffers as critical bubbles, but it often leads to starvation under high load or during long-packet transmissions, degrading network performance. While the existing Bubble Dateline Flow Control (BDFC) alleviates some of these issues through virtual channel partitioning, it still faces performance bottlenecks when handling large volumes of cross-dimensional or long messages. The LFBC scheme dynamically monitors the load status of routers within the current dimension by introducing false bubbles and a dimension counter. False bubbles are temporarily marked from idle buffers and can promptly replace critical bubbles when starvation occurs, thereby reducing its frequency. The dimension counter globally tracks the distribution of false bubbles and coordinates the replacement of critical bubbles to ensure efficient resource utilization. Experimental results demonstrate that, compared to FBFC and BDFC, LFBC achieves up to 40% and 30.8% reductions in latency under uniform random, transpose, shuffle, and neighbor traffic patterns, respectively, while significantly improving throughput in non-uniform traffic modes. Additionally, LFBC exhibits better adaptability and stability in long-packet transmissions and large-scale network expansions, maintaining high buffer utilization.

Keywords: Torus On-chip Network · Bubble Flow Control · Lifecycle · False Bubble

1 Introduction

Since the introduction of multi-core chip research in the late 1990s, Network-on-Chip (NoC) [11] has emerged as a critical and rapidly evolving field of study. With the increasing number of on-chip computing cores, the efficient transmission of inter-core data has created an urgent demand for greater communication bandwidth.

Hemani et al. [12] proposed NoC as a new paradigm for interconnect communication. First, NoC can provide highly scalable bandwidth with relatively small

H. Liu et al. (Eds.): ICA3PP 2025, LNCS 16381, pp. 458–478, 2026.
https://doi.org/10.1007/978-981-95-8399-7_25

area overhead, delivering near-linear bandwidth growth as the number of nodes increases. Second, NoC adopts the layered design philosophy of off-chip communication protocols, enabling more comprehensive power management from the physical layer to the application layer. Additionally, NoCs with regular topologies feature fixed-length local interconnects, facilitating reusable design—any type of processor core can connect to the NoC router via a network interface. This enhances the reusability of communication modules, reducing both design and verification costs. As a result, NoC has become the most attractive solution for on-chip communication, garnering widespread attention from researchers and engineering designers.

Torus network [6] stands out due to its structural symmetry, which effectively reduces router hop count and lowers network latency. This characteristic helps avoid congestion in central regions, a common issue in Mesh networks during data transmission. Owing to these advantages, the Torus topology has been adopted in numerous industrial-grade products, such as SGI's BlueMountain and IBM's BlueGene/Q systems [1].

2D-Torus networks [6] are a structurally symmetric topology that can effectively reduce router hops and network latency. These characteristics can effectively prevent data packets from being congested in the middle area like a mesh network during transmission. Due to its structural advantages, many products in industry adopt torus topology.

The solutions to address the aforementioned issues can be primarily categorized into three approaches:

First is the Dateline scheme proposed by Dally [7]. This solution requires manual configuration of Dateline and at least two prioritized virtual channels. In each dimensional ring network, only high-priority channels are permitted to cross the Dateline, and must switch to low-priority virtual channels when doing so. The proposed scheme achieves deadlock avoidance by breaking cyclic dependencies, but at the cost of significantly degraded buffer utilization. In this architecture, high-priority channel buffers remain virtually idle unless the destination node happens to be the current router. This design leads to substantial resource wastage in NoC implementations. With the issue becoming particularly pronounced as network scale increases, routers at topological boundaries exhibit especially severe virtual channel underutilization.

The second is the escape virtual channel proposed by Duato [8]. Duato theory proves that cyclic dependency graph is a sufficient and unnecessary condition for network deadlock free. Even if there are loops in the cyclic dependency graph, as long as there are acyclic sub parts, the cyclic dependency can be removed to avoid deadlock. However, in torus network, because there are natural rings in each dimension, its acyclic sub parts still need to use special methods to avoid deadlock. Moreover, it will introduce more virtual channels and make the design more complex in the torus topology which already has a circular dependency graph, so few researchers and enterprises use the escape virtual channel with the torus topology.

The third is the bubble flow control scheme [4]. By reserving an idle buffer in the ring as a bubble, and controlling that the newly injected and dimension transferred messages outside the ring cannot occupy the reserved space, it ensures that deadlock in the ring is avoided. The latest bubble flow control scheme has reduced the bubble size to the flit size [16]. Although, this scheme solves the deadlock problem in the dimension and improves the buffer utilization, the starvation phenomenon of its cross dimensional transmission still has a significant impact on the network performance. In the cross dimensional transmission of bubble flow control, the message transferred to dimension cannot occupy the critical bubble, and the ordinary idle buffer of the downstream router must be able to accommodate the size of the entire message before message transmission. This restriction has a more serious impact on the transmission of long messages, because long messages may also face the situation that short messages occupy the downstream buffer. In addition, because the bubble flow control mechanism only uses one virtual channel, when the message is blocked, the subsequent transmissible messages will also be blocked, which will lead to the queue head blocking problem, further limiting the performance of the bubble flow control mechanism.

To address the starvation issue in torus networks caused by the introduction of bubble flow control, this paper proposes a lifecycle-managed false bubble flow control scheme(LFBC). By leveraging false bubbles and dimension counters to implicitly perceive the load status of routers within the current dimension, the scheme can promptly replace critical bubbles with those in lightly loaded routers when starvation requests occur, thereby reducing the frequency of starvation. Since the lifecycle-based false bubble flow control does not partition virtual channels, it maintains strong performance even during long packet transmissions. Additionally, this scheme does not employ a Dateline strategy, thus avoiding performance degradation caused by a large number of packets crossing the Dateline in the network.

2 Related Work

The flow control mechanism manages the allocation of network buffers and links. Traditional flow control mechanisms mainly include store and forward [7], virtual direct [13], virtual channel and wormhole flow control [6]. Now, more and more scholars focus on the unique possibilities and limitations of network on chip to study the flow control mechanism.

Balasubramanian and Rajeev [2] realized that different links can have different bandwidth and transmission speed by using controllable upper metal wiring layer and heterogeneous interconnection size in chip manufacturing. Kumar and Amit [15] proposed that flit bypass the pipeline of router through bypass, which can reduce the transmission delay, dynamic power consumption and buffer turnover time of router. Krishna and Tushar [14] proposed smart flow control mechanism, which allows data to be transmitted to multi hop router in one clock cycle.

Hosseinfarrokhbakht [10] proposed a novel flow control called fastflow, which improves the level of packets by time division multiplexing and predefines non overlapping channels for these high priority packets to transmit them without buffer. This flow control method not only provides high throughput, but also solves protocol level and network level deadlock. Fallin and Chris et al. [9] Proposed a buffer free flow control technology, which can avoid the buffer area and power consumption overhead. In addition, some scholars have also studied the overhead of fault-tolerant mechanisms caused by various flow control strategies [19].

The Bubble Flow Control is a crucial technique for preventing deadlock. This method reserves idle buffers as bubbles within the ring and ensures that newly injected packets or those transitioning from other dimensions cannot occupy the reserved space, thereby guaranteeing deadlock avoidance in the ring. There are two classic implementations of the bubble flow control mechanism: the Localized Bubble Scheme (LBS) [18] and the Critical Bubble Scheme (CBS) [5]. Both LBS and CBS were proposed for virtual cut-through switching and can efficiently handle fixed-length packets. However, when the network carries a significant number of variable-length packets, LBS and CBS may lead to deadlock [1] because the transmission of short packets can fragment the bubbles reserved for long packets. To avoid deadlock, LBS and CBS treat all packets as long packets in networks with variable-length packet transmission, which reduces buffer utilization and consequently degrades network performance.

To address the design limitations of LBS and CBS, MA [16] proposed the Flit Bubble Flow Control (FBFC) based on wormhole switching [3,8]. Unlike LBS and CBS, FBFC prevents deadlock by maintaining a single flit-sized idle buffer unit in the ring, eliminating the need to treat all packets as long packets. As a result, in networks with variable-length packet transmission, FBFC not only avoids deadlock but also improves buffer utilization. Besides, Aniruddh and ramrakhyani [20] proposed a static bubble scheme, which can be applied to the underlying network at the time of design, and expand the subset of routers by adding buffers so that any dependency chain has at least one static bubble to ensure that the network is deadlock free.

In 2024, we [21] proposed the Bubble Dateline Flow Control (BDFC) mechanism to address the performance bottlenecks inherent in both Dateline and bubble flow control schemes. Without additional hardware costs, BDFC simultaneously alleviates starvation in bubble flow control and improves buffer utilization of the dateline technique, ultimately optimizing overall network performance.

3 Motivation

3.1 Starvation Condition in FBFC

The fundamental idea of the FBFC scheme [16] is shown in Fig. 1. Within a certain dimension of the network, there are four routers: R0, R1, R2, and R3. The gray squares represent critical bubbles, which can only be utilized by packets within the ring and are inaccessible to packets outside the ring. When a critical bubble is occupied by an upstream packet flit, the ordinary idle bubble released upstream will be marked as a critical bubble. Critical bubbles ensure the orderly transmission of packets within the ring, preventing mutual blocking and thereby avoiding deadlock. The white squares denote ordinary idle buffers, which can be used by packets both within and outside the dimension. The blue squares indicate that these buffers in the virtual channels are occupied by packet flits. The green squares represent packets outside the ring, with the arrow direction indicating the data transmission direction. Dimension-changing packets are not permitted to use critical bubbles.

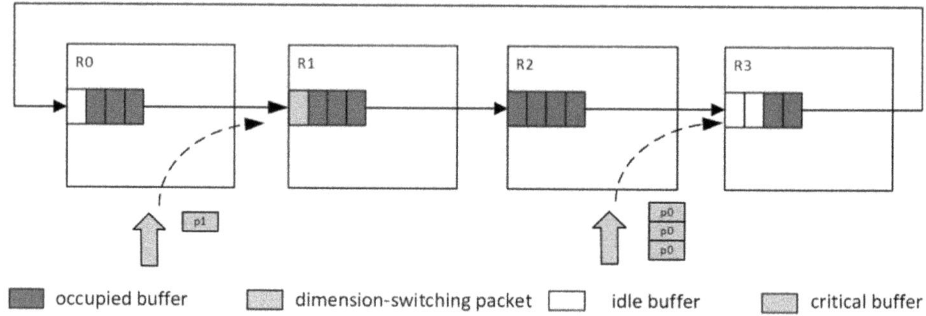

Fig. 1. Starvation in FBFC under High Task-injection Rates.

In FBFC, when the network injection rate is high, starvation issues may arise. This occurs because the buffer resources required by injected or dimension-changing packets exceed those needed for packets transmitting within the ring. In Fig. 1, dimension-external packets P0 and P1 cannot complete their dimension changes. Although P1 has a packet length of only 1, its downstream router R1 has only one idle buffer, which is a critical bubble. Since critical bubbles cannot be occupied by packets outside the ring, P1 does not meet the dimension-changing condition. Meanwhile, P0 has a packet length of 3, but its downstream router R3 has only two idle buffers, so it also fails to satisfy the dimension-changing requirement. To address this starvation issue, the solution is to freeze packet injection and dimension-changing operations in all routers within the ring except R1. This allows packets within the ring to continue advancing toward their destination routers or transitioning to other dimensional rings, thereby freeing up sufficient idle buffers in the current dimension to meet the dimension-changing transmission of P1. Once P1 successfully changes dimensions, the injection and dimension-changing operations in other routers are resumed. The same logic applies to P0; however, P0 also faces the issue of short packets from router R2 preempting buffers in R3, making it even more difficult for the long packet P0 to meet the dimension-changing condition.

Furthermore, starvation issues may also occur when the network injection rate is low. As shown in Fig. 2, critical bubbles in the network may reside in router R3 for an extended period, preventing packet P0 from meeting the dimension-changing condition.

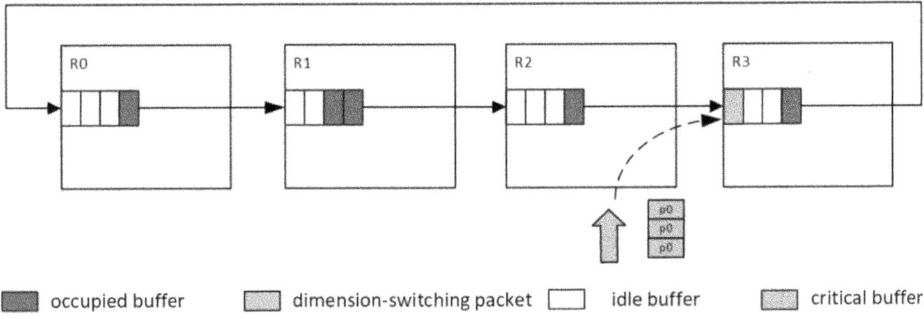

Fig. 2. Starvation in FBFC under High Task-injection Rates

Although researchers have investigated starvation issues and proposed solutions such as using ejection queues [17] as intermediate buffers, these approaches still suffer from head-of-line blocking. Moreover, when the injection rate is high and the network becomes saturated, the utilization of ejection queues increases significantly, thereby diminishing their effectiveness in alleviating starvation during dimension-changing operations.

3.2 Starvation Condition in BDFC

BDFC [21] shown in Fig. 3 integrates both bubble and Dateline schemes. It divides a channel into two virtual channels, one of which does not need to inject critical bubbles, so it does not need the downstream router to have an idle buffer that can receive the entire message during injection and dimension transfer, reducing the frequency of starvation. Besides, the scheme introduces Dateline flow control idea, so that the packets crossing the date change line can select the virtual channel vc0/vc1 according to the degree of network congestion without deadlock.

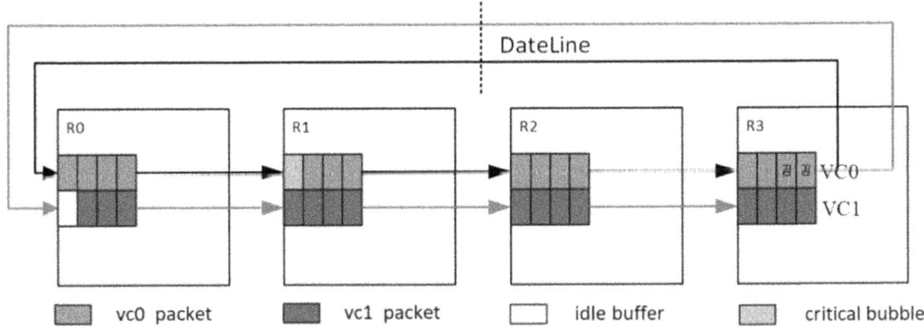

Fig. 3. Bubble Dateline Flow Control.

Although the BDFC flow control mechanism effectively reduces the starvation of FBFC and improves its buffer utilization, it will lead to a significant decline in network performance in the following two scenarios:

(1) When the network contains a large number of dimension transfer messages that need to cross the date line, the BDFC flow control mechanism can not effectively use the introduced virtual channel to solve the starvation condition, but also waste the buffer overhead.

As shown in Fig. 4, it is assumed that messages P1, P2 and P3 are messages that need to cross the date change line when transferring dimensions. In order to avoid deadlock, these three messages can only be transmitted through virtual channel vc0, not virtual channel vc1. Even though virtual channel vc0 is already quite congested at this time, virtual channel vc1 has a lot of idle buffer. When a large number of these packets appear in the network, the performance of BDFC flow control mechanism will be greatly reduced. And when the network size is larger, the number of messages that need to cross the date change line will be more, and BDFC will be more restricted by this.

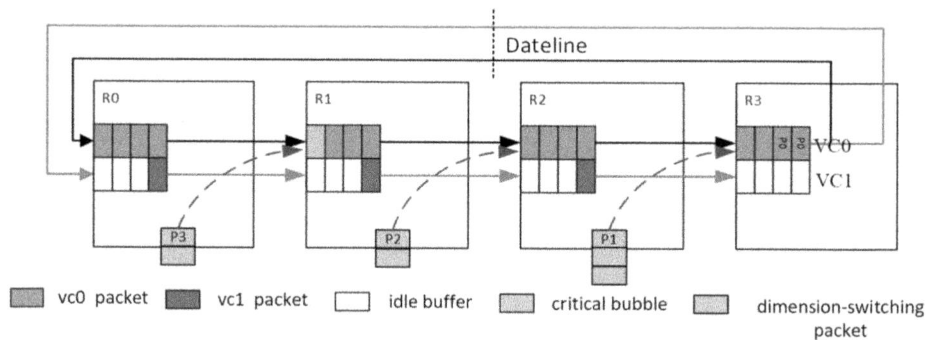

Fig. 4. BDFC Contains a Large Number of Cross Dateline Messages

(2) When the length of the message transmitted in the network is large, it is difficult for the virtual channel vc0 to meet its dimension transfer or injection request, resulting in the blockage of the network at a node, which greatly reduces the performance of the whole network.

Because BDFC divides an input channel into two virtual channels, the depth of each virtual channel queue is relatively shallow. As shown in Fig. 5, although vc0 of R3 router only uses one idle buffer, message P1 cannot be transferred successfully because message P1 needs to contain four ordinary idle buffer downstream. Message P2 can also not be converted to dimension successfully. Although all buffer of virtual channel vc0 of R2 router are not used by flit, it contains critical bubbles, and messages that cannot be converted to dimension cannot be used. Therefore, when the network contains a large number of long packets, its performance will also be significantly reduced, even lower than the FBFC flow control mechanism.

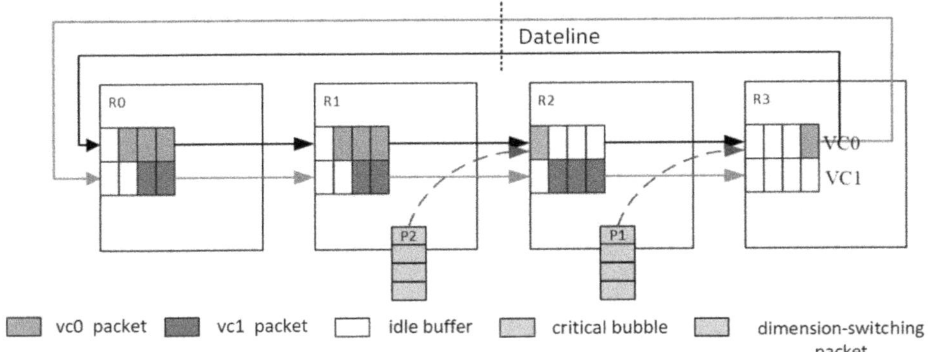

Fig. 5. BDFC Contains a Large Number of Long Message.

Through the above analysis, it is revealed that both FBFC and BDFC may still encounter starvation under certain specific conditions, even when the aggregate idle buffers across all routers in the current dimension exceed the size of injected or dimension-changing packets. Consequently, coordinated utilization of idle buffer resources among all intra-dimensional routers can more effectively alleviate starvation occurrences.

4 Lifecycle-Managed False Bubble Flow Control

4.1 False Bubble

False bubbles is proposed in this paper, which to dynamically monitor the transmission load within the current dimension, thereby effectively addressing starvation issues inherent in the critical bubble scheme. A false bubble represents an idle buffer in the router that can be temporarily designated as a false bubble during a specific lifecycle period.

False bubbles may originate from idle buffers located at routers that are upstream of, but non-adjacent to, the router containing the critical bubble. When starvation occurs, the false bubble can be converted into critical bubble, while the original critical bubble reverts to an idle buffer. This mechanism enables leapfrog advancement of critical bubbles, effectively eliminating starvation induced by critical bubble stagnation.

The generation conditions for false bubbles require that:

1. No adjacent downstream router within the current dimension contain critical bubbles;

2. The number of credit tokens maintained persistently exceeds a predefined clock cycle threshold.

The generation of false bubbles must occur at upstream routers because the marking information of critical bubbles is carried and transmitted via credit tokens, and is marked at routers upstream of those containing critical bubbles. The generation of false bubbles must occur at upstream routers because the marking information of critical bubbles is carried and transmitted via credit tokens, and is marked at routers upstream of those containing critical bubbles. As shown in Fig. 6, at cycle 0, the critical bubble is located at router R2 and marked by router R1. Since R1 currently holds only one credit from router R2, its Critical Bubble marker (CB) register is set to true. Consequently, router R1 can only transmit one flit to R2, which will occupy the critical bubble of R2. At cycle 1, a flit of packet P1 is transmitted to R2. When router R1 ejects the flit of packet P1, it generates a credit and sends it to its upstream router R0, carrying the critical bubble flag information. Upon receiving this credit, the credit counting module in R0 records the credit and detects that it also contains the critical bubble flag. Therefore, router R0 sets its CB marker register to true, indicating that router R1 possesses a critical bubble.

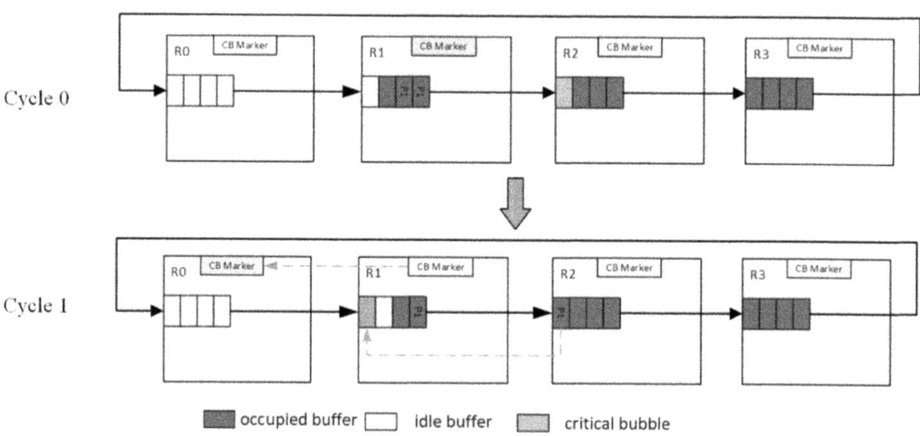

Fig. 6. LFBC Credit Transmission.

Each output port of the router integrates three key components: (1) a timer for false bubbles (FB timer), (2) a marker register (FB) to flag the presence of false bubbles in adjacent downstream routers, and (3) a marker register (CB) to indicate the existence of critical bubbles in adjacent downstream routers. The counter initiates its timing operation when two conditions are simultaneously satisfied: the current router contains neither critical bubbles nor false bubble markers, and the number of held downstream credit tokens reaches or exceeds a predetermined threshold(FB timer threshold). Upon reaching the configured threshold, the counter triggers two concurrent actions: it dispatches a false bubble marker request to the dimension counter and sets the local False Bubble marker register (FB) to logic '1', thereby establishing the presence of a false bubble in the adjacent downstream router.

As shown in Fig. 7, the preset value of credit count in the current network configuration is 3, and both R0 and R1 contain three idle buffers. The fake bubble timer (FB timer) of their adjacent upstream routers meets the condition to start counting. When the timer reaches the preset threshold, the router sets the FB marker to true, indicating that its downstream router contains a fake bubble, while simultaneously sending a marker request to the dimension counter.

When a fake bubble is occupied by normally transmitted packets, the router sends a cancellation request to the dimension counter and resets its local FB marker register to 0.

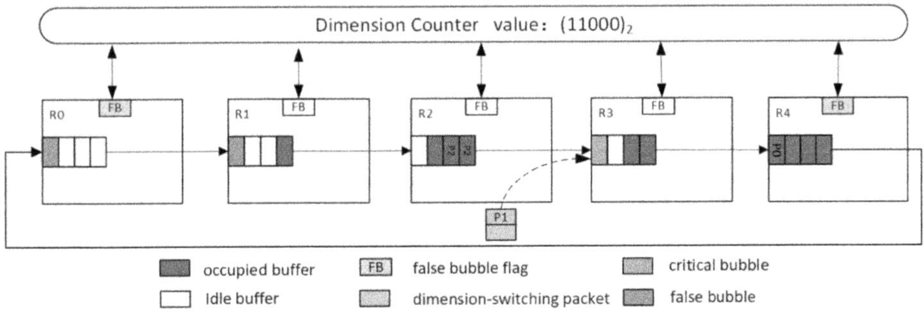

Fig. 7. LFBC Theoretical Schematic

4.2   Dimension Counter

Each dimensional ring is equipped with a dimension counter, which tracks the fake bubble status within the current dimension. The dimension counter allocates a single-bit flag to each router in the dimension to indicate whether the router contains a fake bubble. When an upstream router sends a fake bubble marker request, it sets the corresponding downstream router's flag bit in the dimension counter to 1. Conversely, upon receiving a fake bubble cancellation request, the downstream router's flag bit is reset to 0. By introducing the dimension counter, global awareness of buffer load conditions within the dimensional ring is achieved. If a router in the dimension experiences starvation-induced congestion, the dimension counter can detect the load status of other routers in the same dimension. Consequently, it facilitates the redirection of critical bubbles to less congested routers, thereby mitigating network congestion caused by starvation.

In Fig. 7, the torus network employs a 5×5 configuration, thus the dimension counter utilizes a 5-bit width where each bit corresponds to a specific router's fake bubble (FB) status. In the current scenario, since both R0 and R1 have their FB markers set to 1, the dimension counter value is "11000", indicating that routers R0 and R1 contain fake bubbles.

4.3 Transition from False Bubble to Critical Bubbles

When newly injected packets or dimension-switching packets cause starvation in the network due to critical bubbles, the router sends a starvation status notification to the dimension counter. Based on the current fake bubble markers in the network, the dimension counter resets the fake bubble marker bit (setting it to 0) for the fake bubble located farthest from the critical bubble. Subsequently, it sends a fake bubble reset signal to the corresponding router containing this fake bubble, while simultaneously transmitting a starvation acknowledgment signal back to the router that reported the starvation condition.

Upon receiving the fake bubble reset signal from the dimension counter, the router with the fake bubble marker clears its FB marker register (setting it to 0) and sets its CB marker register to 1. This indicates that the fake bubble in this router has been converted into a critical bubble.

When the router that initiated the starvation request receives the acknowledgment signal from the dimension counter, it resets its CB marker register to 0, signifying that the critical bubble has been released and its buffer space has become available again. If the router fails to receive an acknowledgment from the dimension counter while still experiencing starvation caused by critical bubbles, it will continue to send starvation status notifications to the dimension counter until the starvation condition is resolved.

In Fig. 7, router R2 needs to forward packet P1 to downstream router R3. Since P1 is a dimension-switching packet with a length of 2 flits, it currently cannot meet the dimension-switching condition. This is because router R3 has only 2 flits of available buffer space, one of which is occupied by a critical bubble.

Under these conditions, the starvation counter within router R2 initiates a waiting count for packet P1. When the predefined time threshold is reached and P1 still fails to complete dimension-switching (indicating persistent starvation), R2 issues a starvation request to the dimension counter. The dimension counter then searches for fake bubbles within the current dimension. If present, it selects the fake bubble farthest from the critical bubble. The dimension counter simultaneously sends an acknowledgment to R2 and a set marker command to R4. Upon receiving the acknowledgment, R2 clears its downstream critical bubble marker (setting it to 0) and reinitiates a downstream channel request, enabling successful dimension-switching for P1. When R4 receives the set marker command, it performs three operations: (1) resets the fake bubble marker to 0, (2) sets the CB marker register to 1, and (3) clears the FB marker register to 0, thereby completing the replacement of a fake bubble with a critical bubble.

If during the counting period the critical bubble migrates to another router due to dimension-internal packet transmission (e.g., where packet P2's transmission to R3 might relocate the critical bubble to R2), the starvation counter pauses but does not reset. The counter remains suspended until P1 successfully completes dimension-switching. Should the critical bubble return to R3 while P1 remains unswitched, the starvation counter resumes accumulation from its previous value. This mechanism accelerates the triggering of subsequent starvation requests, facilitating the redistribution of critical bubbles to less congested routers within the dimension. This approach significantly enhances the probability of successful dimension-switching for P1, effectively alleviating packet congestion and improving overall network throughput.

5   Router Microarchitecture

Figure 8 illustrates the microarchitecture of the LFBC router, which consists of the following key components: input unit, route computation (RC) unit, starvation generation (SG) unit, false bubble (FB) marker register, critical Bubble (CB) marker register, credit counter (credit_cnt), false bubble signal generation logic, crossbar allocation unit and crossbar transmission unit.

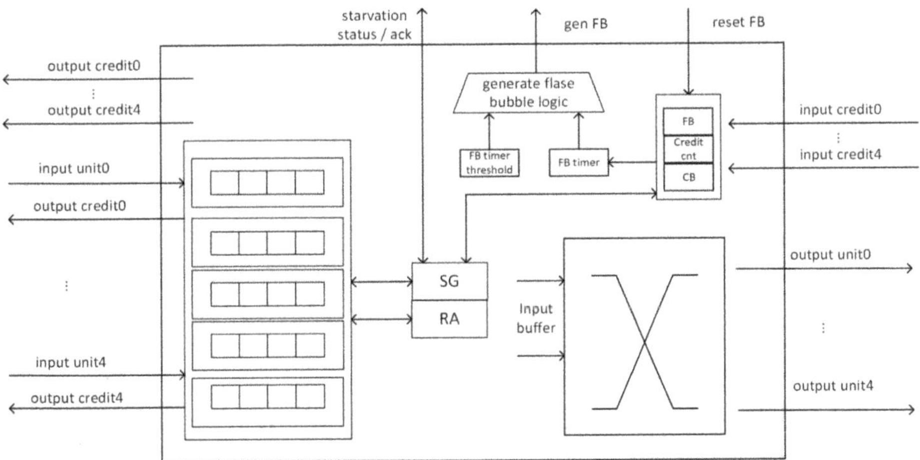

Fig. 8. LFBC Router Microarchitecture

Each input port will send a credit to the upstream based on the remaining number in its buffer, and the upstream will only send a flit to the downstream when holding the letter of credit. At each output port, downstream letters of credit are received and counted, and it also contains information about critical bubbles. If the downstream channel contains critical bubbles, they will be recorded in the CB unit. When there are critical bubbles downstream, the VA module will determine whether the message is a dimension transfer or injection message. If it is a dimension transfer or injection message, the downstream channel application needs to hold a number of ordinary letters of credit equal to or greater than the length of the message to complete the application.

FB timer is the module that generates false bubble counting. When both FB and CB units are 0, the FB timer counter will only perform an addition count when the number of letters of credit held by the FB timer counter is greater than or equal to the generated flag value. When it is less than the set value, it will perform a subtraction count until the counter value is 0 or greater than the false bubble generation threshold. When the value of the FB timer counter is greater than the threshold, a tag request will be sent to the dimension counter for tagging and the counter will be reset to zero.

If a false bubble has already been generated, when FB is equal to 1, FB timer will perform an addition count when the number of letters of credit held is less than or equal to the cancelNum value, and a subtraction count when it is greater than this value, until the counter value is 0 or equal to the threshold. When the counter value is greater than the threshold, a cancelNum request will be generated. If the number of letters of credit held is 0, there is no need to wait until the FB timer reaches the threshold to send a unmark request, as the false bubble has already been used and FB is directly set to 0. In addition, when the critical bubble reaches the router containing false bubbles due to transmission within the ring, a false bubble cancellation request will also be sent to the dimension counter, and the FB register will be set to 0.

The credit_cnt in Fig. 8 is the letter of credit statistics module. When a flit of buffer is available downstream, it will receive a letter of credit carrying critical bubble marker information. After receiving the letter of credit, credit_cnt will first accumulate and then check if its critical bubble marker field is 1. If it is 1, set the CB register to 1 and FB to 0, and send a false bubble cancellation request to the dimension counter.

6   Experiment

6.1   Experiment Setup

Booksim2 is an open-source, clock-cycle accurate NoC software simulator that modularizes various components of the NoC, supporting multiple topologies, routing algorithms, router microstructures, and arbiter algorithms. It also features high scalability and detailed behavioral analysis. We modified Booksim2 to implement BDFC, FBFC-C (Flit Bubble Flow Control-Critical, hereinafter referred to as FBFC), and LFBC for performance comparison.

The specific experimental configuration parameters are shown in Table 1. The number of virtual channels for LFBC and FBFC is set to 1, while the number of virtual channels for BDFC scheme is 2, so the number of buffers for each virtual channel in BDFC is half of the total buffer. The proportion of single slice messages transmitted in the experimental setup network is 80%, and the proportion of messages with three slice sizes is 20%.

In order to evaluate the performance of LFBC under different network sizes, experiments were conducted on three topology sizes: 4x4, 8x8, and 16x16. Four traffic modes, including uniform random, neighbor, tanspose, and shuffle, were selected for comparison. The traffic in both uniform random and neighbor modes

Table 1. Experimental Parameter.

Parameters	Values
Network Topology	4x4torus; 8x8torus; 16x16torus
Traffic Patterns	uniform random, transpose, shuffle, neighbor
Routing Algorithm	XY
VC	BDFC(2VCs), LFBC(1VC), FBFC-C(1VC)
Buffers/VC	6flits
Warm-up time	3000cycles
Run time	7000cycles

is relatively evenly distributed in the network, and the performance in uniform random mode is generally used as a benchmark in simulations. In tanspose mode, there will be relatively more dimension conversion messages, which is more likely to trigger the starvation condition of the three flow control mechanisms. Before testing, set 3000 clock cycles to preheat and ensure the network reaches a stable state, reducing statistical errors. The remaining 7000 clock cycles are used to measure performance under this traffic mode.

6.2   Performance

Figure 9 shows the delay comparison of three flow control mechanisms, LFBC, BDFC, and FBFC, under four traffic modes of uniform random, neighbor, transport, and shuffle in an 8x8 network. Among them, uniform random and neighbor represent their latency performance in uniform traffic mode, while transport and shuffle demonstrate the latency comparison of three mechanisms under non-uniform traffic.

In Fig. 9(a), the relationship between delay and injection rate for three flow controls in uniform flow mode is shown, where LFBC improved by 40% and 3.7% compared to FBFC and BDFC, respectively. The reason why BDFC performance is lower than LFBC is that in uniform random mode, some packets of BDFC are determined to cross the date line during injection, which limits its performance even in network congestion by only selecting the virtual channel VC0 for transmission. The reason why LFBC does not show much performance improvement compared to BDFC is that the false bubbles it generates only work for a small period of time when the network becomes congested. In uniform mode, packets are transmitted relatively evenly in the network, and the congestion status of each router is not significantly different. When the injection rate is low, multiple false bubbles will be generated in each dimension. As the injection rate increases, the transformed packets begin to experience starvation and request the replacement of critical bubbles. At this time, some routers are still relatively idle, so the replacement of critical bubbles can be carried out. But not long after replacement, other routers will also enter a relatively congested state, at which point

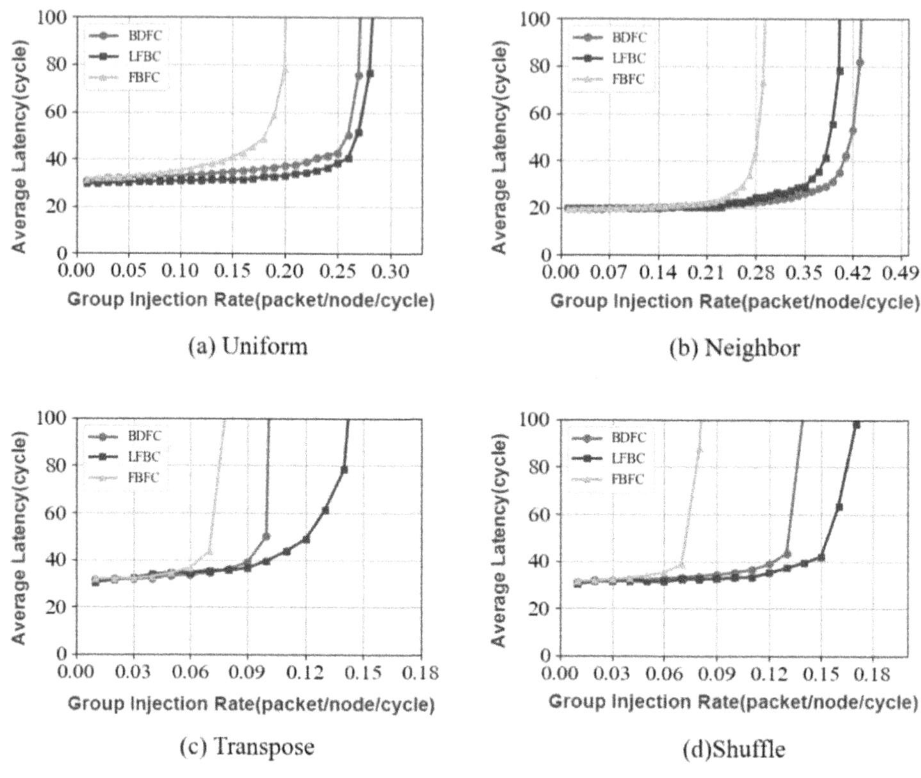

Fig. 9. Performance in uniform random, transpose, shuffle and neighbor traffic mode.

most routers cannot generate false bubbles to reduce the starvation condition in their dimension, thereby limiting the performance improvement of LFBC.

In Fig. 9(b), the reason why the performance of BDFC is 7.5% higher than that of LFBC is that in neighbor mode, the distance between the source node and the destination node is shorter, resulting in lower events of message dimension conversion. This allows the advantages of BDFC's virtual channel to be more fully utilized. However, LFBC still improved by 37.9% compared to FBFC, mainly because when starvation occurs, FBFC freezes the injection and dimension conversion of other routers in the dimension ring until the hungry packets are successfully converted, which greatly affects the transmission of other packets in the network. LFBC does not restrict the transmission of other packets, but transfers critical bubbles to routers containing false bubbles, critical bubbles are transferred to routers with relatively idle dimensions, thereby improving the overall network performance.

Figure 9(c) and (d) demonstrate the performance of three flow control mechanisms in non-uniform flow mode. In the transport traffic mode, the maximum injection rate of LFBC relative to BDFC and FBFC has increased by 40% and 100%, respectively. The reason why LFBC has such a significant improvement in this mode is mainly because the source and destination nodes of the message are symmetrical in this mode, which leads to a large number of messages that need to be transformed in the network, making it easier to trigger starvation condition. Although the second virtual channel of BDFC does not need to satisfy atomicity, which improves its performance in this mode, as more and more messages need to cross the date line, even if there is still space for the other virtual channel to continue transmission, the transmission delay of messages containing critical bubble channels has exceeded the maximum transmission delay that the network can accept, limiting its further performance improvement.

However, the overall injection rate of the actual network has not reached a very high level, and LFBC is more likely to generate false bubbles at this time. When packets are congested due to critical bubbles during dimension conversion, the false bubbles generated by LFBC can quickly respond to starvation requests, greatly reducing congestion and improving the overall performance of the network. In shuffle flow mode, the maximum injection rate of LFBC compared to BDFC and FBFC flow control mechanisms increased by 30.8% and 125%, respectively. In shuffle traffic mode, message transmission is relatively random and traffic distribution is uneven, which can easily lead to local network congestion, while other areas have overlapping idle scenarios. So when packets are transferred in congested areas, LFBC is more likely to transfer them to idle locations farther away from the congested area, because LFBC will prioritize the use of false bubbles that are farthest from critical bubbles, resulting in fewer critical bubbles in each congested area and improving the overall performance of the network.

Figure 10 shows the saturation throughput of three flow control mechanisms, LFBC, BDFC, and FBFC, under the 8x8 network uniformrandom and transport flow control mechanisms. The XY dimensional sequential path and bubble scheme ensure that messages already injected into the network can continue to be forwarded downwards, but booksim2 will exit the simulation when the delay exceeds 500 clock cycles. In order to obtain accurate saturation throughput, we set a larger exit simulation delay so that it can have a higher packet injection rate without exiting the simulation.

From Fig. 10, it can be seen that in uniform random mode, LFBC has improved by 7.5% and 48.3% compared to BDFC and FBFC, respectively. In transport mode, LFBC has increased by 31.3% and 75% compared to BDFC and FBFC, respectively. LFBC has a greater throughput improvement in non-uniform traffic mode compared to BDFC and FBFC in uniform traffic mode. This is mainly because hotspots are more likely to occur in non-uniform traffic mode, making it easier for routers in idle areas within the dimension to generate false bubbles. When the congested area of the network suffers from starvation condition due to critical bubbles, it can effectively transfer the critical bubbles to areas with relatively idle transmission, allowing more flips to reach the destination router in the same time range and increasing the saturation throughput of the LFBC flow control mechanism. In uniform flow mode, the false bubbles in LFBC only play an important role in the period before network congestion, so

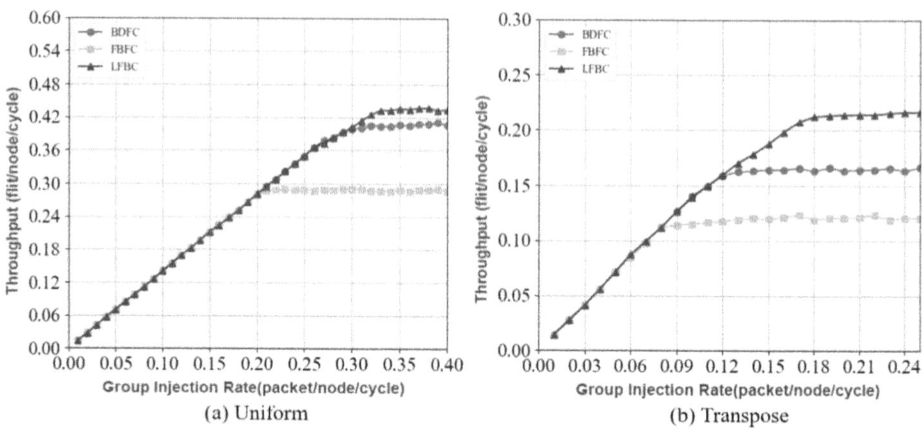

Fig. 10. Throughput in uniform random, transpose traffic mode.

the improvement compared to BDFC is relatively limited. Regardless of the flow control mechanism, FBF eliminates starvation condition by injecting and transforming dimensions into routers through freezing, which limits the improvement of FBFC throughput.

6.3   Sensitivity to Packet Size and Scalability

The essence of the starvation condition is that the ordinary idle buffer downstream cannot meet the space allocation conditions required for transmitting messages during injection and dimension conversion. Therefore, different message lengths will affect the frequency of congestion occurrence, thereby having a significant impact on the saturated throughput of the network. Figure 11(a) shows the comparison of saturated throughput of three flow control mechanisms under different packet sizes in the 8x8 network uniform random traffic mode. From the graph, it can be seen that FBFC performs the best when the packet size is 6, but there is not much difference compared to packet sizes of 3 and 5. This is mainly because in the current setting, the buffer size is 12, which is less than half the buffer length. If there are no critical bubbles in the channel, the input port can accommodate at least two messages at the same time, so the frequency of congestion has not increased, and the difference in saturation throughput is not significant.

When the message size of BDFC is 3, the saturation throughput is the highest. When the message size exceeds 3, the saturation throughput begins to decrease. When the message size is 6, the saturation throughput is the lowest. The reason for this phenomenon is that BDFC has two virtual channels, each with a size of 6 buffer units. Therefore, when the message size is greater than 3, the converted and injected messages can only receive one message, reducing the buffer utilization and increasing the frequency of congestion. When the packet

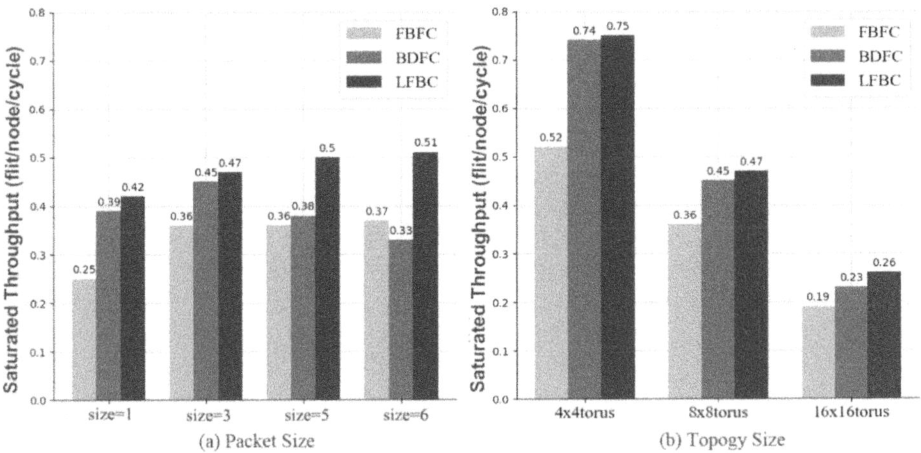

Fig. 11. Comparison of saturated throughput under different packet sizes and different topologies.

size is 6, the downstream normal idle buffer for injection and dimension conversion must be completely idle and free of critical bubbles, which greatly increases the probability of congestion. In this scenario, its saturation throughput is even lower than FBFC.

LFBC, similar to FBFC, has only one virtual channel and has not exceeded half of its buffer capacity even when the packet size is 6, so its saturation throughput has not decreased. Through testing in scenarios with different message sizes, the drawbacks of BDFC in long message mode were exposed, namely that when the message length is large, if BDFC performance needs to be maintained, a larger capacity of buffer is required. However, LFBC does not have the disadvantage of BDFC, as it can transmit longer packets relative to BDFC without causing a decrease in network performance.

Figure 11(b) evaluates the scalability of LFBC under uniform random traffic mode. As the network size increases, the dimensionality also increases, and there will be more and more packets converted to dimensionality, making it easier for packets to trigger starvation during transmission. The saturation throughput of the three flow control mechanisms also decreases accordingly. LFBC improved by 1.4%, 4.4%, and 13.1% compared to BDFC at 4x4 torque, 8x8 torque, and 16x16 torque, respectively. This is mainly because as the network size expands, the dimension counter of the LFBC flow control mechanism can record more false bubbles. When a certain area is blocked due to critical bubbles, the dimension counter can replace false bubbles farther away from the congested area with critical bubbles, thereby making the congested area in the network farther away from the critical bubbles, reducing the frequency of starvation condition, and making its scalability better compared to the BDFC mechanism.

6.4   False Bubble Timer Threshold and Buffer Utilization

Figure 12(a) shows a comparison of the saturation throughput of the LFBC flow control mechanism generating false bubbles at different thresholds (FB timer threshold). It can be seen that it has the highest saturation throughput at a threshold of 17, while its performance is the worst at a threshold of 2. This is mainly because when the threshold is relatively low, the router will quickly send false bubble tagging requests to the dimension counter, but it may take another two clock cycles for the credit it holds to be used, resulting in a false idle tag, and may even replace critical bubbles when the router is about to become congested, causing the network to fall into new starvation condition. When the threshold is 17 cycles, it indicates that the downstream buffer has been relatively idle for 17 consecutive clock cycles, indicating that the packet transmission in the current area is relatively idle. The generated false bubbles can vaguely represent the load situation of downstream routers. And when the threshold is set to a higher value, its saturation throughput tends to decrease. This is mainly because when the threshold is set too high, there will be fewer false bubbles in the dimension, or it will be difficult to generate false bubbles because when the threshold is lower than the set value, subtraction counting will be performed. Therefore, if the threshold is set too high, even if the load in a certain area is small, no false bubbles will be generated. Therefore, when a router experiences starvation due to dimensionality conversion, it requests bubble replacement from the dimension counter without any available false bubble scenarios. Therefore, when the threshold is large, its saturation throughput tends to decrease.

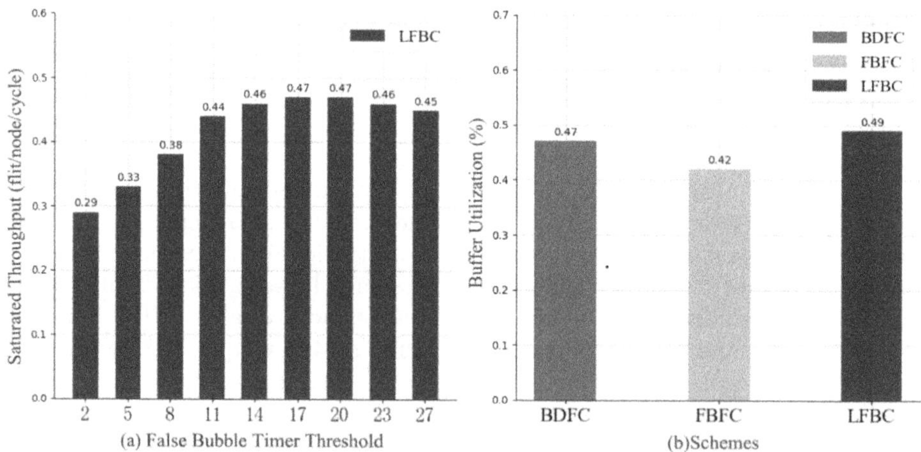

Fig. 12. Comparison of LFBC's saturation throughput at different thresholds and buffer utilization with other schemes.

Figure 12(b) shows the average buffer utilization of three flow control mechanisms, BDFC, FBFC, and LFBC, under stable network operation. They are

47%, 42%, and 49% respectively, with the FBFC mechanism having the lowest average buffer utilization rate because it freezes the injection and transfer of packets from other routers in the dimension when hungry occurs, until the hungry packets are successfully transmitted downwards, which directly leads to a decrease in its average buffer utilization. When starvation occurs, the LFBC mechanism will move critical bubbles to areas with lighter message loads within the dimension, reducing the frequency of starvation occurrence and not freezing the transmission of other injected and converted messages within the dimension. Therefore, its average buffer utilization is relatively high. The buffer utilization of BDFC is only 2% lower than that of LFBC, mainly because it contains two virtual channels, reducing the frequency of packet head blocking. In addition, in the BDFC mechanism, one virtual channel can continue to transmit the entire packet without using the buffer of downstream routers.

7   Conclusion

This paper proposes an LFBC mechanism to address the starvation condition in FBFC and the bottleneck of transmitting long messages in BDFC. The LFBC mechanism introduces the concepts of false bubbles and dimension counters. False bubbles, like regular idle buffer, can be used by any message, representing the status of downstream routers. The dimension counter is used to receive false bubble tagging requests, critical bubble cancellation requests, and false bubble usage requests. When the downstream buffer of a router is relatively idle and reaches the set threshold, it will send a flag to the dimension counter for recording. At this time, the dimension counter can vaguely perceive the congestion of message transmission within the dimension. When an injection or dimension transfer message experiences starvation due to a critical bubble, a replacement request will be sent to the dimension counter. Based on the recorded situation, the dimension counter will replace the farthest false bubble from the congested area with the critical bubble, thereby alleviating the congestion at that location. Because LFBC does not divide one channel into two virtual channels, it does not need additional buffer when the network needs to transmit long messages.

References

1   Adiga, N.R., et al.: Blue gene/l torus interconnection network. IBM J. Res. Dev. **49**(2.3), 265–276 (2005)
2. Balasubramonian, R., Muralimanohar, N., Ramani, K., Cheng, L., Carter, J.B.: Leveraging wire properties at the microarchitecture level. IEEE Micro **26**(6), 40–52 (2006)
3. Chen, L., Pinkston, T.M.: Worm-bubble flow control. IEEE (2013)
4. Chen, L., Wang, R., Pinkston, T.M.: Critical bubble scheme: an efficient implementation of globally aware network flow control. In: 2011 IEEE International Parallel & Distributed Processing Symposium, pp. 592–603. IEEE (2011)

5. Chen, L., Wang, R., Pinkston, T.M.: Critical bubble scheme: an efficient implementation of globally aware network flow control. IEEE (2011)
6. Dally, W.J., Seitz, C.L.: The torus routing chip. Distrib. Comput. **1**, 187–196 (1986)
7. Dally, W.J., Towles, B.P.: Principles and practices of interconnection networks. Elsevier (2004)
8. Duato, J.: A new theory of deadlock-free adaptive routing in wormhole networks. IEEE Trans. Parallel Distrib. Syst. **4**(12), 1320–1331 (1993)
9. Fallin, C., Craik, C., Mutlu, O.: Chipper: a low-complexity bufferless deflection router. In: 2011 IEEE 17th International Symposium on High Performance Computer Architecture, pp. 144–155. IEEE (2011)
10. Farrokhbakht, H., Gratz, P.V., Krishna, T., San Miguel, J., Jerger, N.E.: Stay in your lane: a NOC with low-overhead multi-packet bypassing. In: 2022 IEEE International Symposium on High-Performance Computer Architecture (HPCA), pp. 957–970. IEEE (2022)
11. Guerrier, P., Greiner, A.: A generic architecture for on-chip packet-switched interconnections. In: Proceedings Design, Automation and Test in Europe Conference and Exhibition 2000 (Cat. No. PR00537), pp. 250–256 (2000)
12. Hemani, A., et al.: Network on chip: an architecture for billion transistor era. In: Proceeding of the IEEE NorChip Conference (2000)
13. Kleinrock, L.: A new computer communication switching technique. Comput. Netw. **3** (1979)
14. Krishna, T., Chen, C.H.O., Kwon, W.C., Peh, L.S.: Breaking the on-chip latency barrier using smart. In: 2013 IEEE 19th International Symposium on High Performance Computer Architecture (HPCA), pp. 378–389. IEEE (2013)
15. Kumar, A., Peh, L.S., Jha, N.K.: Token flow control. In: 2008 41st IEEE/ACM International Symposium on Microarchitecture, pp. 342–353. IEEE (2008)
16. Ma, S., Wang, Z., Liu, Z., Jerger, N.E.: Leaving one slot empty: flit bubble flow control for torus cache-coherent NoCs. IEEE Trans. Comput. **64**(3), 763–777 (2013)
17. Ouyang, Y., Sun, C., Li, R., Wang, Q., Li, J.: Transit ring: bubble flow control for eliminating inter-ring communication congestion. J. Supercomput. **79**(2), 1161–1181 (2023)
18. Puente, V., Izu, C., Beivide, R., Gregorio, J.A., Vallejo, F., Prellezo, J.M.: The adaptive bubble router. J. Parallel Distrib. Comput. **61**(9), 1180–1208 (2001)
19. Pullini, A., Angiolini, F., Bertozzi, D., Benini, L.: Fault tolerance overhead in network-on-chip flow control schemes. In: Proceedings of the 18th Annual Symposium on Integrated Circuits and System Design, pp. 224–229 (2005)
20. Ramrakhyani, A., Krishna, T.: Static bubble: a framework for deadlock-free irregular on-chip topologies. In: 2017 IEEE International Symposium on High Performance Computer Architecture (HPCA), pp. 253–264. IEEE (2017)
21. Zhang, H., Li, Y., Deng, Y., Fang, Q.: BDFC: a new flow control mechanism fortorus networks. In: International Conference on Algorithms and Architectures for Parallel Processing (2024)

Z-STREAM: A Transparent Reordering Engine for High-Performance Zoned Namespace SSDs

Canghao Wen, Jinhua Cui[(✉)], Xuanxuan Fu, and Laurence Tianruo Yang

Huazhong University of Science and Technology, Wuhan 430074, China
`{m202474002,jhcui,fuxuanxuan,ltyang}@hust.edu.cn`

Abstract. Zoned Namespace (ZNS) SSDs offer compelling advantages over conventional SSDs by exposing a zoned storage interface that helps mitigate write amplification and improve performance. However, their strict sequential-write constraint within each zone poses a significant challenge for many real-world applications that naturally generate out-of-order write requests. Unaligned or out-of-order writes in ZNS can trigger costly read-modify-write operations or command rejections, leading to severe performance degradation.

To address this issue, this paper proposes Z-STREAM, a novel, two-level collaborative scheduler. The first level is a lightweight host-side batch sorter (HBS), implemented in the NVMe driver path, which intercepts and reorders I/O requests within small batches based on their logical block addresses. The second level is the device-side compensating write set engine (CWS-Engine), a transparent reordering engine within the ZNS firmware. The CWS-Engine utilizes a per-zone software buffer to absorb out-of-order writes. When writes are evicted from this buffer, the engine generates persistent metadata for a novel structure we term the compensating write set (CWS). CWS metadata, efficiently managed by a dedicated binary search tree (BST), ensures correct data retrieval during read operations while minimizing performance overhead.

We have implemented and integrated this entire collaborative architecture into the FEMU simulator. Our evaluation demonstrates that the proposed system potently and transparently transforms chaotic write streams into ZNS-compliant sequential writes, reducing write amplification to a near-ideal 1.003 and improving throughput by up to 3.82x on representative workloads.

Keywords: Zoned Namespace SSDs · I/O Scheduling · Host-Device Co-design

1 Introduction

The explosive growth of data-intensive applications, from large-scale databases to artificial intelligence workloads, has fueled an insatiable demand for high-performance, cost-effective storage solutions. Flash-based solid state drives

H. Liu et al. (Eds.): ICA3PP 2025, LNCS 16381, pp. 479–499, 2026.
https://doi.org/10.1007/978-981-95-8399-7_26

(SSDs) have become a cornerstone in modern data centers due to their superior performance over traditional hard disk drives. However, conventional SSDs employ an internal flash translation layer (FTL) that, while providing a standard block device interface, introduces considerable overheads such as garbage collection, which can lead to unpredictable performance and high write amplification (WA) [1].

To overcome these limitations, the storage industry has introduced Zoned Namespace (ZNS) SSDs as a new interface paradigm [2]. ZNS devices divide their storage space into fixed-size regions called zones and then expose this zoned architecture to the host. ZNS devices enforce a strict sequential write constraint within each zone, requiring that data be written sequentially by advancing a zone write pointer. The zoned architecture grants the host greater control, leading to lower WA, reduced on-device DRAM requirements, and more predictable I/O performance [3].

Despite the potential benefits of ZNS SSDs, their sequential write constraint presents a marked hurdle for widespread adoption. Many existing file systems, databases, and key-value stores are not designed for this constraint and frequently issue small, random, or out-of-order write requests. Upon arrival at a ZNS device, such requests face two undesirable outcomes: the device either rejects the commands outright or triggers costly internal read-modify-write operations. Both scenarios severely degrade performance and increase write amplification, which in turn diminishes the lifespan of the SSD. To avoid these issues, the most direct workaround is to enforce a queue depth of one ($QD = 1$) per zone. However, this simple method, while guaranteeing sequentiality, creates a profound architectural mismatch. By serializing all I/O operations, it strands the rich internal parallelism of the SSD and fundamentally negates the architectural benefits of the multi-queue NVMe interface. This mismatch is the primary obstacle preventing ZNS from achieving its full performance potential and necessitates an intelligent layer to bridge the gap between application I/O patterns and the device's physical constraints.

Existing solutions often focus on a single layer of the stack. Host-side solutions attempt to address the out-of-order write problem by buffering and reordering write requests in the operating system before sending the commands to the device. However, host-side schedulers lack crucial, real-time information about the internal state of the device (e.g., buffer utilization, exact write pointer positions). In contrast, device-side solutions use internal buffers within the SSD firmware to absorb and reorder out-of-order writes. Although powerful, such device-side solutions miss the opportunity to perform low-cost, early-stage reordering before requests consume valuable on-device resources and bandwidth.

In this paper, we propose a novel, two-level collaborative scheduling architecture that synergizes the strengths of both the host and the device to transparently and efficiently manage out-of-order writes. Our system comprises two main components. The first is the host-side batch sorter (HBS). Implemented in the NVMe driver path of the host, the HBS acts as a lightweight first-pass filter that intercepts ZNS write commands, collects the commands in a small batch, and performs a quick sort based on the starting logical block address (SLBA) before dispatching the commands. Such a sorting process resolves easily reorderable write patterns with minimal overhead. The second component is the compensating write set engine (CWS-Engine). For write streams that remain out-of-order after the initial pass of HBS, a more powerful, state-aware engine within the device firmware takes over. CWS-Engine utilizes a per-zone software buffer to absorb out-of-order writes. To handle buffer eviction, the engine can write data out-of-order to the flash media while ensuring data integrity. To track such non-sequential placements, CWS-Engine generates and manages persistent metadata via its core data structure, the compensating write set (CWS). CWS efficiently records the mapping of logical addresses to corresponding physical locations, with the CWS metadata managed by a per-zone interval tree for fast lookups during read operations.

We have implemented the Z-STREAM architecture within the FEMU simulator, modifying both host-side I/O logic and the device-side ZNS FTL. Our primary contributions are:

1. The design of a novel two-level, collaborative host-device architecture for transparently handling out-of-order writes on ZNS SSDs.
2. The implementation of HBS, a lightweight, batch-based sorter in the host I/O path that pre-processes write requests.
3. The design of CWS-Engine, a device-side firmware module featuring in-storage buffering and a novel metadata scheme (CWS) for managing out-of-order data placement.
4. A comprehensive evaluation demonstrating that the collaborative architecture significantly reduces write amplification and improves throughput compared to baseline.

The remainder of this paper is structured as follows. Section 2 reviews the background of ZNS technology. Section 3 details the design and implementation of the collaborative scheduling architecture. Section 4 describes the experimental setup, and Sect. 5 presents and analyzes the evaluation results. Section 6 discusses related work, and Sect. 7 concludes the paper.

2   Background

This section briefly covers the foundational principles of Zoned Namespace (ZNS) technology and the architectural origins of write disorder in modern I/O stacks.

2.1   Zoned Namespace (ZNS) Fundamentals

ZNS is a storage interface that exposes an SSD's storage as a collection of contiguous, fixed-size regions called zones, reducing the semantic gap between the host and the device [2]. The most critical principle of ZNS is the sequential write constraint [4]. This constraint dictates that within any active zone, data must be written sequentially to a write pointer, as illustrated in Fig. 1. A write to any

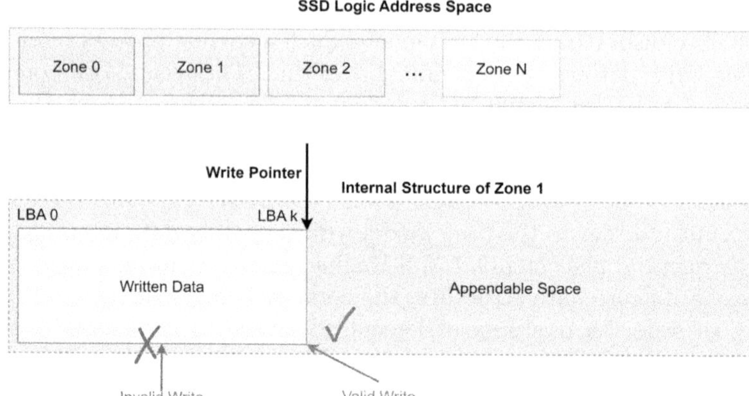

Fig. 1. An illustration of the sequential write constraint within a single ZNS zone.

other location is rejected. Data cannot be modified in-place; to reclaim space, the host must issue a command to reset the entire zone.

The lifecycle of a zone is governed by a state machine (Fig. 2), transitioning from empty to open, and finally to a read-only full state. By giving the host direct control over data placement and garbage collection, the ZNS model promises to significantly reduce write amplification (the ratio of data physically written to NAND versus data requested by the host), lower on-device DRAM costs, and deliver more predictable I/O performance.

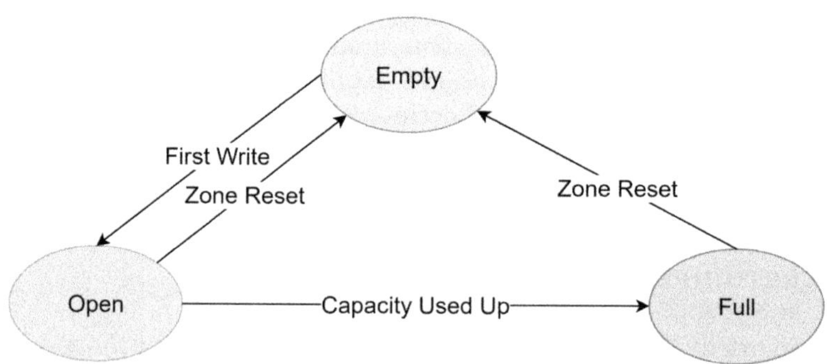

Fig. 2. The lifecycle of a ZNS zone.

2.2 The Origin of Write Disorder in the I/O Stack

Logically sequential writes from applications often arrive at the SSD out of order. This disorder stems from the parallelism in modern I/O stacks, where multi-core

CPUs dispatch I/O requests to different NVMe Submission Queues (SQs) [5,6]. This reordering prevents host software from being able to guarantee sequential arrival at the device, creating a fundamental conflict with the ZNS model.

3   Motivation

The background analysis reveals a fundamental conflict: ZNS hardware demands strict sequentiality, yet the host I/O stack inherently produces disorder. The current section deconstructs the fundamental conflict by quantifying the severe performance penalty imposed by current workarounds and analyzing the profound inadequacies of existing solutions, thereby establishing the critical motivation for our work.

3.1   The Performance Penalty of the Standard Write Model

To guarantee data integrity, the standard ZNS write model enforces a pessimistic yet necessary workaround: each zone is restricted to a queue depth of one ($QD = 1$). Although this approach ensures correctness, it serializes all I/O to a zone. The serialization creates a critical bottleneck, preventing ZNS SSDs from leveraging their rich internal parallelism.

To quantify this performance penalty, we present a preview of our key evaluation results from Sect. 5, comparing the standard $QD = 1$ model against our proposed high-performance Z-STREAM architecture. The results, summarized in Fig. 3, demonstrate that the $QD = 1$ limitation causes the severe bottleneck in application-level workloads. In an OLTP workload, which simulates the small random writes common in databases, our Z-STREAM architecture achieves a 3.49x throughput speedup over the standard model. The penalty persists even

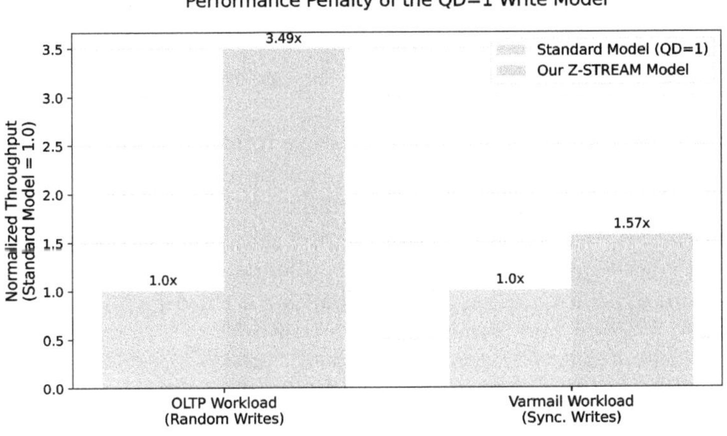

Fig. 3. Normalized throughput comparison.

in metadata-heavy synchronous workloads like Varmail, where Z-STREAM still provides a 1.57x performance improvement. The data in Fig. 3 reveals that the $QD = 1$ restriction is a severe rather than minor performance issue in the software ecosystem for ZNS SSDs.

3.2 The Limitations of Existing Solution Approaches

To overcome the performance bottleneck of the $QD = 1$ model, two primary solution approaches can be considered: resource-intensive buffering and semantic changes to the host-device contract.

A conceptually simple approach is to use a large in-device DRAM buffer [7] to absorb and reorder out-of-order writes before committing them sequentially to flash. However, the naive buffering approach conflicts with the core ZNS design principle of minimizing on-device DRAM. ZNS SSDs are designed to be low-cost and resource-efficient by minimizing on-device DRAM. A buffer large enough to handle realistic concurrent workloads would require a prohibitive amount of expensive and power-safe DRAM. Such a requirement reintroduces the very cost and complexity that ZNS technology aims to eliminate.

An industry-prevalent approach to mitigating the performance bottleneck of the $QD = 1$ model is the zone append model [8]. The append model achieves high queue depth performance by shifting the task of assigning the final LBA from the host to the device. However, this name-after-write approach introduces a significant trade-off, as it is fundamentally incompatible with the name-before-write assumption on which the existing software ecosystem heavily relies [9]. Adopting the zone append model has severe consequences, beginning with the loss of deterministic data placement. The host can no longer place data at specific logical addresses, preventing crucial optimizations like guaranteeing file contiguity and leading to fragmentation. Furthermore, this breaks fundamental system-level services, such as software RAID, which requires pre-calculating parity locations—a process that clashes with the append model's post-write address assignment. Finally, integrating this new paradigm demands a costly re-architecting of the software stack, requiring deep modifications to file systems and databases. This high barrier to entry has severely limited the append model's adoption in practice.

The analysis of these approaches reveals an intractable dilemma for developers, forcing a difficult compromise between two flawed alternatives. On one hand, the zone append model sacrifices compatibility with the vast existing software ecosystem; on the other hand, resource-intensive buffering sacrifices the core cost-efficiency principle of ZNS itself. This dilemma has significantly hampered ZNS adoption, exposing a clear research gap for a third path, a solution that is both resource efficient and semantically compatible.

3.3 Design Goals

The analysis of the existing solution approaches reveals the need for a third path: a solution that resolves the conflict between performance and compatibility without compromise. The work presented in this paper is driven by the goal of forging this path. We aim to design an architecture that transparently maximizes ZNS hardware performance for both legacy and future applications. To unleash such performance, the Z-STREAM architecture is engineered to meet three fundamental design goals.

First, the solution must provide application-agnostic transparency. It must operate entirely within the storage stack (host driver and device firmware), remaining invisible to applications, file systems, and the operating system kernel. Second, the architecture must deliver unconstrained performance by eliminating the $QD = 1$ bottleneck. The resulting write throughput should be competitive with the theoretical optimum of the append model. Finally, and most critically, the solution must ensure preservation of host-centric semantics. The established name-before-write model must be preserved, guaranteeing that the host retains full and deterministic control over the logical address space.

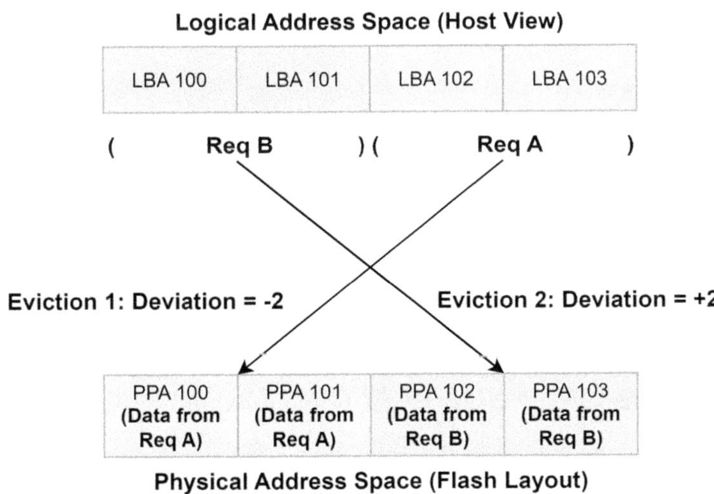

Fig. 4. Schematic diagram of CWS formation mechanism.

Our proposed Z-STREAM architecture meets these goals through its collaborative host-device design, which combines a lightweight host-side sorter (HBS) with an efficient, metadata-driven in-storage engine (CWS-Engine). The key to preserving host-centric semantics lies in the novel metadata mechanism of CWS-Engine, the compensating write set (CWS). As illustrated in Fig. 4, the engine allows out-of-order writes to be physically placed in different locations on the flash media. A CWS tracks the deviation between the intended logical address of a write and its actual physical address. When a set of writes have swapped locations in a way that their cumulative deviation sums to zero, they form a CWS. This mechanism allows the device to maintain a perfectly sequential logical view for the host while handling writes out-of-order internally, achieving both high performance and full compatibility with minimal metadata overhead.

Table 1 summarizes this analysis, highlighting how the Z-STREAM architecture is uniquely positioned to satisfy all three design goals simultaneously, unlike the existing standard and append models.

Table 1. Comparison of ZNS Solution Approaches Against Key Design Goals

Design Goal	Standard Model (QD = 1)	Append Model	Our Approach (Z-STREAM)
App. Transparency	Yes	No	Yes
Performance	No	Yes	Yes
Semantics	Yes	No	Yes

Note: App. Transparency = Application-Agnostic Transparency; Performance = Unconstrained Performance; Semantics = Preservation of Host-centric Semantics.

4   System Design and Implementation

Z-STREAM implements a two-level, collaborative architecture to transparently handle out-of-order writes for ZNS SSDs. The core strategy of Z-STREAM is to partition the reordering problem across the host and the device. A lightweight and stateless pre-processing stage on the host resolves simple, localized write disorder. In parallel, a powerful and state-aware engine on the device manages complex, long-range dependencies. Z-STREAM aims to maximize ZNS SSD performance by transparently reordering I/O requests and optimizing data placement.

4.1   Overall Architecture

Figure 5 illustrates the path of a write request through the system architecture. Out-of-order writes are categorized into two classes: local disorder and global disorder. Local disorder, where sequential requests are shuffled within a small window, is resolved by the Host-side Batch Sorter (HBS). Global disorder involves more complex dependencies, which is managed by the CWS-Engine within the device firmware. The dual-layer approach synergizes the ability of the host to handle simple cases with the capacity of the device for managing complex patterns. The entire process remains transparent to the application.

4.2   The Host-Side Batch Sorter (HBS)

HBS constitutes the first stage of the Z-STREAM collaborative architecture, operating as a lightweight pre-processing filter within the host's NVMe driver

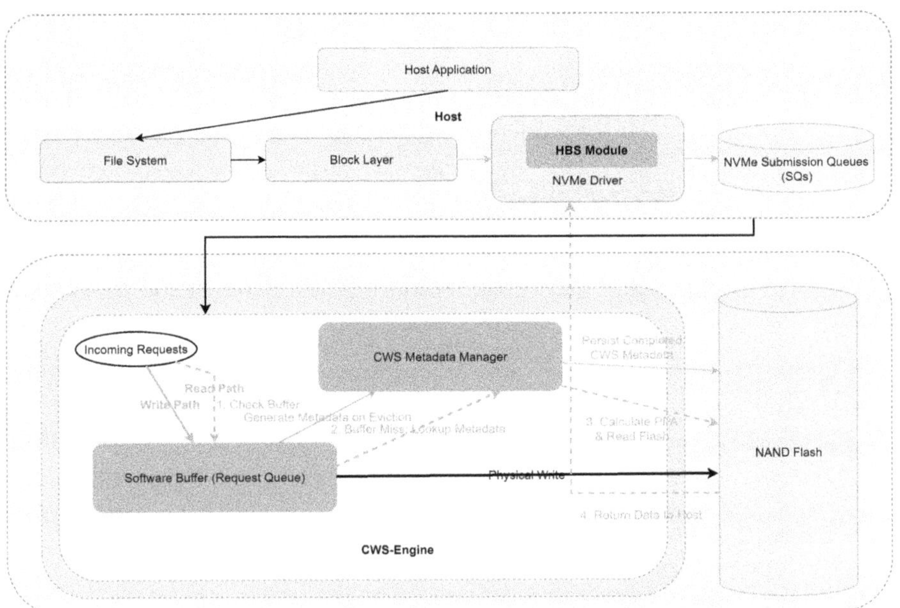

Fig. 5. Overview of the Z-STREAM collaborative architecture. The components and data flows highlighted in green represent the novel contributions introduced in this work. (Color figure online)

path. The primary design goal of HBS is to resolve localized write disorder at a minimal computational cost before I/O requests are transmitted to the ZNS device. This goal is achieved through the three-step process detailed in Algorithm 1. First, HBS intercepts ZNS write commands from a submission queue. Secondly, the commands are then collected into a temporary batch to undergo an in-memory sort. Finally, the sorted commands are dispatched in a newly ordered sequence.

The rationale for placing an initial reordering stage on the host originates directly from the architectural origins of write disorder. As described in Sect. 2.2, modern multi-core systems can dispatch I/O requests to different hardware submission queues in parallel, causing logically sequential writes to arrive at the device out of order. HBS is explicitly designed to counteract such out-of-order write patterns. The stateless design of HBS avoids complex state management and the need for persistent data buffering on the host, ensuring a negligible impact on CPU resources. Such a lightweight implementation allows HBS to effectively handle simple reordering opportunities and, in turn, reduces the workload on the more resource-intensive, state-aware CWS-Engine on the device.

A critical parameter in the HBS implementation is the batch size, which dictates how many requests are collected before a sort is performed. The batch size presents a fundamental trade-off between reordering effectiveness and I/O latency. A larger batch increases the likelihood of capturing a longer sequence of logically adjacent writes, thus maximizing the reordering potential. However, a larger batch also increases the dispatch latency for requests held in the batch. Conversely, a smaller batch minimizes latency, but is less effective for highly chaotic write streams. In our reference implementation, the batch capacity is configured to 128 requests. This value was determined to be sufficiently large to capture and resolve the typical I/O disorder observed in multi-queue host environments, while ensuring the introduced batching latency remains an order of magnitude smaller than the baseline end-to-end I/O completion latency. Thus, it strikes a favorable balance between reordering effectiveness and latency overhead.

Algorithm 1. Host-side Batch Sorter (HBS) Logic

```
 1: procedure PROCESSSUBMISSIONQUEUE(SQ)
 2:     Batch ← NEWARRAY(MAX_ZNS_BATCH_SIZE)          ▷ Initialize a batch
 3:     count ← 0
 4:     while SQ is not empty and count < MAX_ZNS_BATCH_SIZE do
 5:         Cmd ← SQ.POP_FRONT
 6:         if Cmd is a ZNS write command then
 7:             Batch[count] ← Cmd                      ▷ Add ZNS write to batch
 8:             count ← count + 1
 9:         else
10:             DISPATCHIMMEDIATE(Cmd)      ▷ Dispatch non-ZNS cmds immediately
11:         end if
12:     end while
13:     if count > 0 then
14:         SORT(Batch, key=SLBA)                        ▷ Sort batch by SLBA
15:         for i ← 0 to count − 1 do
16:             DISPATCHTODEVICE(Batch[i])     ▷ Dispatch sorted batch to device
17:         end for
18:     end if
19: end procedure
```

The sorting mechanism within HBS is direct and effective. Once the batch is filled, the collected requests are sorted based on their Starting Logical Block Address (SLBA). The sorting process transforms a potentially disordered stream into a locally sequential one. By delivering a more orderly request stream to the SSD, HBS significantly alleviates the pressure on the internal buffers of the device and on CWS-Engine. The pre-processing performed by HBS enables an effective division of labor. HBS handles simple, local disorder, which allows CWS-Engine to dedicate its advanced mechanisms to resolving complex, long-range global disorder. Such collaboration between the host and the device is the core principle of the Z-STREAM architecture.

4.3 The Device-Side CWS-Engine

The design of the CWS-Engine is informed by existing research on optimized data placement strategies for ZNS SSDs [10]. CWS-Engine functions as the core component of the Z-STREAM architecture and operates within the ZNS firmware. The primary function of CWS-Engine is to virtualize the zone write pointer, a process that decouples the host-visible logical write pointer from the true physical write pointer on the NAND media. As a result, writes appear sequentially from the perspective of the host. CWS-Engine implements the virtualization of the zone write pointer by staging out-of-order writes in an internal software buffer. A novel metadata structure, CWS, is then utilized to track the non-sequential data placement. The combination of in-storage buffering and CWS metadata allows the device to service writes out-of-order physically while preserving the strict sequential abstraction required by the host.

CWS-Engine maintains a per-zone state structure composed of several key components. A central element is the in-memory write buffer, a request-oriented buffer for caching out-of-order writes. The capacity of this buffer is configured to 128 KiB per zone in the reference implementation, representing a practical trade-off between the reordering capability and the footprint of DRAM on the device. Another key component is the CWS metadata management module, which manages the life cycle of CWS metadata. CWS metadata management module includes a builder to generate records after buffer eviction and uses a search-optimized Binary Search Tree (BST) for fast retrieval. Finally, contextual state pointers link the zone state to global device structures to enable context-aware resource management.

The write path of CWS-Engine, detailed in Algorithm 2, follows a multi-stage logic that prioritizes sequentiality while gracefully handling disorder. The logic first checks for a fast path that allows perfectly sequential writes to bypass buffering entirely. If the fast path cannot be taken, the engine attempts an opportunistic flush to write any newly sequential requests from the buffer. Any write that remains out-of-order is staged in the buffer. The oldest request is evicted to flash when the buffer is full, an action that in turn triggers CWS metadata generation and finalization checks.

Algorithm 2. Z-STREAM Engine Write Path Logic

```
 1: procedure PROCESSBACKENDWRITE(ZS, Req)
 2:    if Req.SLBA = ZS.LWP and ZS.Buffer is empty then
 3:        WRITETOFLASH(Req, Req.SLBA)                       ▷ Stage 1: Fast path
 4:        return
 5:    end if
 6:    OPPORTUNISTICFLUSH(ZS)                                ▷ Stage 2
 7:    if Req.SLBA = ZS.LWP then
 8:        WRITETOFLASH(Req, Req.SLBA)
 9:        OPPORTUNISTICFLUSH(ZS)                     ▷ Re-check after sequential flush
10:        return
11:    end if
12:                                              ▷ Stage 3: Buffering and Eviction
13:    while ZS.Buffer.size + Req.size > CAPACITY do
14:        OldestReq ← ZS.BUFFER.POP_FRONT
15:        PhysicalAddr ← ZS.PWP
16:        WRITETOFLASH(OldestReq, PhysicalAddr)
17:        UPDATECWS(OldestReq, PhysicalAddr)
18:        if ISCWSCOMPLETE(ZS) then                   ▷ Check after updating CWS
19:            FINALIZECWSMETADATA(ZS)            ▷ Consolidate and store metadata
20:        end if
21:    end while
22:    ZS.BUFFER.PUSH_BACK(Req)
23: end procedure
```

The CWS metadata system is the key to the transparency of the Z-STREAM architecture. The core principle of the system is that a set of out-of-order writes forms a compensating set when the sum of page-level deviations for all writes in the set equals zero. The deviation for each write is defined as $D = P_{physical} - P_{logical}$.

Once the cumulative deviation for a set of writes totals zero ($\sum D_i = 0$), the current CWS is considered complete. Upon completion, the collected displacement events are consolidated using run-length encoding into a compact metadata record. The record is then inserted into a per-zone Binary Search Tree (BST), indexed by the starting logical page number of the CWS. A standard BST is used because the inherently random nature of CWS formation makes performance-degrading insertion orders highly unlikely. This choice prioritizes implementation simplicity and low firmware overhead, as the read path requires efficient point lookups, for which a BST is ideally suited, rather than more complex range queries that would necessitate a structure like an interval tree. The amortized construction cost for a single CWS is O(P), where P represents the number of pages within the set. A subsequent lookup for the record is efficient O(log N), where N is the number of CWS records stored in the tree. With an efficient O(P) construction and O(log N) lookup, the resulting per-write overhead is negligible.

To illustrate, consider a simple example with two single-page write requests, A (for logical page 100) and B (for logical page 101), arriving out of order. Assume the zone's physical write pointer (PWP) is at physical page 100. First, request B arrives and is evicted, writing its data to physical page 100. Its deviation is calculated as $D_B = P_{\text{physical}}(100) - P_{\text{logical}}(101) = -1$. The PWP advances to 101. Next, request A is evicted and written to physical page 101, yielding a deviation of $D_A = P_{\text{physical}}(101) - P_{\text{logical}}(100) = +1$. The engine checks the cumulative deviation, $\sum D = D_B + D_A = -1 + 1 = 0$. With the sum at zero, a CWS is formed, and a single, compact metadata record is generated for this two-page swap.

4.4   Read Path, State Management, and Robustness

The Z-STREAM architecture is designed for correctness and reliability. The read path is augmented to first check the in-memory write buffer for requested data. If the data are not present in the buffer, the path proceeds to query the CWS metadata BST to find the true physical address of a given logical page. Any potential latency increase from the query is negligible. The reason is that a single, predictable in-memory tree lookup is orders of magnitude faster than a physical NAND read operation.

To maintain state consistency, a zone reset command triggers a routine that cleans up all associated Z-STREAM resources.

Ensuring data consistency across power failures is a critical design requirement. CWS-Engine achieves power failure safety through a dedicated recovery scan mechanism. Consistent with standard device buffer designs, data residing in the volatile per-zone write buffer is not protected across power loss; host-side timeouts will trigger retransmissions for these unacknowledged writes. For all persisted data, integrity is guaranteed. Upon system restart, the recovery scan reads Logical Page Numbers (LPNs) stored in the spare area of each physical flash page—a standard practice for robust flash management [1]. This process reconstructs the CWS metadata tree from first principles, ensuring that all on-media data, including pages belonging to partially completed sets, is correctly indexed and accessible after recovery. A successful reconstruction guarantees data integrity and allows normal operation to resume.

To handle edge cases where a CWS remains incomplete, the system implements a fail-safe mechanism called CWS compaction. Triggered by a timeout or a resource threshold, the compaction process consolidates all data belonging to the incomplete CWS and rewrites the data sequentially. Such a process incurs a temporary write amplification hit as a deliberate trade-off to guarantee long-term system stability and prevent resource leakage.

5   Evaluation

This section presents a comprehensive evaluation of the proposed Z-STREAM architecture. The evaluation aims to validate the performance, efficiency and transparency of the architecture. The analysis progresses from a low-level micro-architectural analysis to high-level application benchmarks to achieve this goal.

5.1   Experimental Setup

All experiments are conducted using the FEMU NVMe SSD emulator [11]. The detailed experimental platform configuration is summarized in Table 2. Key parameters include a host system with an Intel Core CPU and 64 GB of DRAM, along with an emulated ZNS SSD featuring 64 GB capacity, 256 MB zones, and TLC NAND flash.

The evaluation focuses on the performance delta between the proposed system and a real-world baseline. The following four schemes are compared:

- ZNS-Standard (Baseline): A standard ZNS SSD model that uses the `mq-deadline` I/O scheduler to enforce a strict queue depth of one (QD = 1) for each zone.
- HBS-only: An ablation configuration where only the host-side batch sorter is enabled, sending reordered requests to a standard ZNS device model without CWS-Engine.
- CWS-Engine-only: An ablation configuration where only the device-side CWS-Engine is enabled, directly processing the original, unordered I/O stream from the host.
- Z-STREAM (Proposed System): The complete, two-level collaborative architecture, where both HBS and CWS-Engine are enabled.

Table 2. Experimental Platform Configuration

Component	Specification
Host System	Intel Core CPU, 64 GB DRAM
Operating System	Linux Kernel 6.8
File System	F2FS on top of ZNS device
Emulator	FEMU
Emulated ZNS SSD	
Capacity	64 GB
Zone Size	256 MB
NAND Type	TLC
Page Read Latency	50 µs
Page Program Latency	500 µs
Block Erase Latency	5 ms

A carefully selected set of workloads from Filebench [12], Sysbench [13], and FIO [14] is used to stress different aspects of the system. The chosen profiles cover a spectrum of I/O patterns, from the metadata-intensive nature of *varmail*, to the random-write stress of *oltp*, and the mixed I/O of *fileserver*. This selection ensures a holistic evaluation.

5.2   Micro-Benchmark Analysis

To isolate the raw I/O capabilities and scalability of the proposed architecture, a pure random write stress test is conducted using FIO at varying QD. The results, shown in Fig. 6, starkly illustrate the performance differences. The random write IOPS at varying queue depths (Fig. 6a) demonstrate the raw I/O capabilities of each scheme, while the QoS latency under high concurrency (Fig. 6b) highlights their scalability and stability.

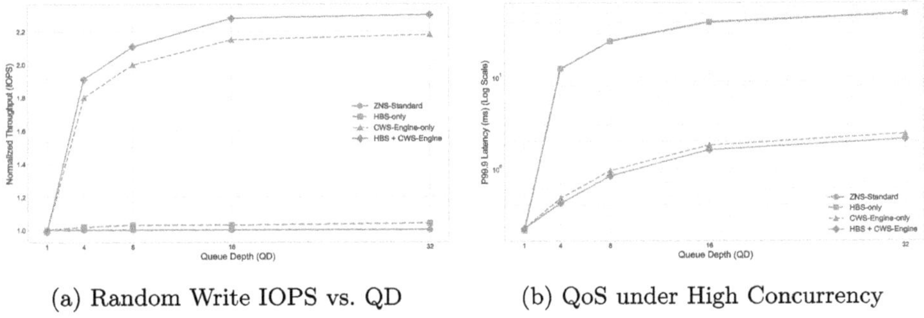

(a) Random Write IOPS vs. QD   (b) QoS under High Concurrency

Fig. 6. Performance ablation study. (a) shows the write throughput, and (b) shows the QoS latency.

The ZNS-Standard and HBS-only configurations both show no throughput scalability, with performance remaining flat regardless of the application-level queue depth. This outcome is expected, as both configurations are ultimately bottlenecked by the ZNS sequential write constraint at the device level, which necessitates I/O serialization at the host.

For the ZNS-Standard scheme, this serialization is handled directly by the `mq-deadline` I/O scheduler, which enforces a strict queue depth of one for each zone. As a result, any potential for parallelism from a high application-level queue depth is eliminated.

The HBS-only configuration faces the same fundamental limitation. While HBS pre-sorts requests on the host, it cannot alter the ZNS device's intrinsic intolerance for out-of-order writes. Without the device-side buffering capabilities of the CWS-Engine, the system must still rely on the I/O scheduler to serialize writes to each zone to guarantee correctness. Therefore, although HBS improves the logical ordering of the request stream, HBS-only cannot overcome the serialization bottleneck by itself, and the performance of HBS-only also fails to scale.

In stark contrast, architectures equipped with the CWS-Engine demonstrate excellent scalability. The root of this scalability lies in the engine's ability to effectively virtualize the zone write pointer. By buffering out-of-order requests, the CWS-Engine decouples the host's command submission process from the physical NAND programming sequence. This decoupling allows the host to maintain a high queue depth and enables the device to fully leverage its internal parallelism to service the accumulated requests. The full Z-STREAM system achieves the highest throughput, peaking at a 2.30x speedup at QD = 32, validating this architectural approach.

The impact on Quality of Service (QoS) is even more dramatic. Tail latency (P99.9) for the baseline and HBS-only configurations explodes at high queue depths, reaching a catastrophic 50 ms. In contrast, the full Z-STREAM system maintains excellent QoS, keeping tail latency to a mere 2.1 ms at QD = 32. The reduction in tail latency compared to the baseline is greater than 23x, transforming the user experience from unpredictable and laggy to consistently responsive for latency-sensitive applications.

5.3   Application-Level Performance Analysis

The real-world impact of the proposed architecture is evaluated under a suite of realistic application workloads covering diverse I/O patterns. The results presented in Fig. 7, confirm the significant benefits of Z-STREAM. The full Z-STREAM system consistently outperforms the baseline and individual component configurations, demonstrating the effectiveness of the two-level architecture.

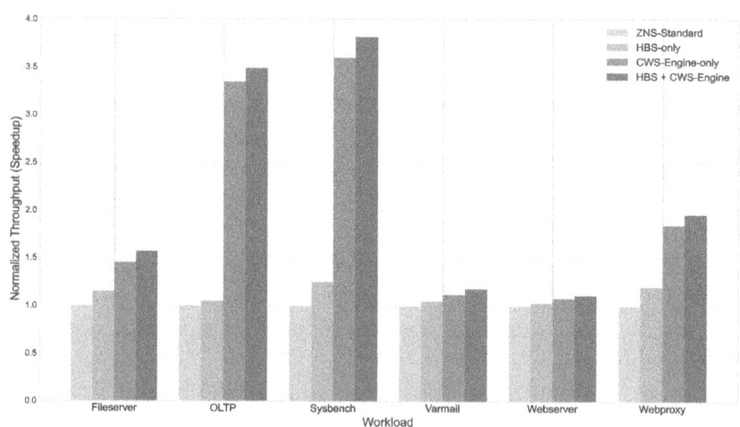

Fig. 7. Normalized Throughput across Application-Level Benchmarks.

Specifically, in write-intensive workloads that are notoriously difficult for ZNS devices, the Z-STREAM architecture delivered remarkable speedups. A speedup of 3.49x was achieved in the *oltp* profile, which simulates the small random writes typical of key-value stores, and a 3.82x speedup was recorded in the mixed-I/O *fileserver* profile. Performance improvements stem from the ability of the Z-STREAM architecture to absorb and reorder chaotic writes. The capability of Z-STREAM not only accelerates the write path but also reduces internal device contention, boosting concurrent read performance.

Even in scenarios dominated by synchronous operations, such as the 'fsync'-heavy *varmail* workload, the Z-STREAM system achieved a 1.57x speedup by reducing the critical path latency by approximately 50%. In read-centric workloads like *webserver* and *webproxy*, the system yielded modest gains of 1.18x and 1.11x, respectively. The results demonstrate that the write-focused optimizations introduce no read penalty. In fact, the optimizations improve overall system responsiveness by minimizing I/O interference.

Independent validation with the fileio test from Sysbench corroborated the previous results. The test showed a 1.96x improvement in the write rate and a profound 62% reduction in the 95th percentile latency, from 18.61 ms to 7.04 ms. Such results highlight the ability of Z-STREAM to deliver the stable and predictable performance crucial for database-like applications.

The consistent performance advantage of the full Z-STREAM system over the CWS-Engine-only configuration highlights the synergistic effect of the two-level collaborative design. HBS acts as an efficient low-cost filter, resolving localized disorder on the host. This pre-processing delivers a more orderly I/O stream to the device, significantly boosting the CWS-Engine's efficiency. Specifically, the smoother stream increases the hit rate of the sequential "fast path" (bypassing the buffer entirely) and improves the logical locality of requests that do enter the buffer. This alleviates pressure on the limited device-side resources, allowing the CWS-Engine to focus its capabilities on resolving more complex, global out-of-order patterns, ultimately achieving higher overall throughput.

5.4   System Overheads

A critical aspect of the Z-STREAM architecture is storage efficiency. The efficiency is evaluated by tracking the overhead of the CWS metadata system and the resulting Write Amplification (WA) under the demanding *oltp* workload.

To ensure clarity, the Write Amplification is calculated using the standard formula presented in Sect. 2. The numerator comprehensively accounts for all bytes written to the physical media, including user data and persistent CWS metadata. Even with this strict accounting, the measured WA remained exceptionally low, typically ranging from 1.001 to 1.003.

The near-ideal WA is achieved through two key techniques: compact, run-length-encoded metadata and an efficient storage strategy that co-locates metadata and data within the same zone. Because the metadata is atomically erased with the zone upon reset, Z-STREAM avoids the garbage collection overhead common in other designs. The resulting storage overhead is minimal with a measured metadata write ratio of only 0.015% in the worst-case workload.

6   Related Work

The shift towards zoned storage is an outstanding trend in the storage community, with many researchers arguing for the adoption of ZNS to overcome the limitations of conventional SSDs [3]. The Z-STREAM architecture builds on prior research efforts but is distinguished by a two-level, host-device collaborative approach. Related work is subsequently categorized into three main areas.

6.1   Application and File System Adaptation

A distinguished body of work focuses on making upper-layer software ZNS-aware. The process involves re-engineering data-intensive applications to natively generate sequential write patterns. For log-structured file systems like F2FS [15], such an adaptation is relatively natural. For more complex applications like LSM-tree-based key-value stores, a major research thrust has been to modify the compaction and flushing mechanisms of the applications.

For example, ZenFS explores how to manage the SSTable placement in RocksDB on ZNS devices [16]. Other research has proposed compaction-aware zone allocation strategies to co-design the internal data flow of the LSM-tree with the zone layout of the device [17]. Although such approaches can achieve high efficiency, a primary drawback is the lack of transparency, requiring bespoke engineering effort for each application. In contrast, Z-STREAM provides a transparent solution.

6.2   Host-Side I/O Scheduling

Another line of research investigates solutions purely at the host level. Host-level schedulers aim to buffer and reorder requests before the requests reach the device. For example, Fair-ZNS [18] proposes a self-balancing I/O scheduling scheme to address fairness issues. However, all host-only solutions suffer from a fundamental limitation: a lack of device-state awareness. The host is typically blind to real-time buffer utilization or the exact physical write pointer location within the SSD. Such information asymmetry can lead to suboptimal decisions. Z-STREAM overcomes the limitation of host-side unawareness by complementing the host-side HBS with the device-side CWS-Engine, an engine that possesses complete and accurate knowledge of the internal state of the device.

6.3   In-Storage Intelligence and System Integration

Embedding more intelligence within the device firmware is a powerful approach. Previous related research has explored various techniques. Some works have focused on optimizing internal device operations, such as proposing new LSM-style garbage collection schemes that are better suited for the ZNS model [19]. Beyond single-device optimizations, system-level research like RAIZN [20] has explored how to build reliable RAID arrays using ZNS devices, highlighting new challenges that zoned storage brings to system-level data protection.

Concurrent research, such as Land of Oz [10], has also proposed an in-storage indirection layer to resolve out-of-order writes by tracking the deviation between logical and physical page addresses. Z-STREAM is distinguished from such work in two fundamental ways. First is its novel Compensating Write Set (CWS) metadata scheme, which does not track individual page deviations but rather generates a single, compact record only when a set of deviations cumulatively sums to zero, offering a potentially more efficient metadata footprint. Second, and more critically, is Z-STREAM's two-level collaborative architecture. The presence of the host-side HBS, which pre-processes the I/O stream to reduce the burden on the device, is a foundational design difference not present in purely device-centric solutions like Land of Oz.

Other device-level research, such as eZNS [21], focuses on a complementary problem: providing an elastic zoned namespace to manage inter-zone resource allocation and quality of service. Such work is orthogonal to our own, as Z-STREAM addresses the fundamental challenge of handling intra-zone write sequence constraints, making the basic ZNS write model transparently high-performance for all applications.

Although device-centric approaches are powerful, such solutions can still be overwhelmed by highly random workloads. The ultimate novelty of Z-STREAM lies in the synergy between the host and the device. HBS acts as a first-pass filter, smoothing the I/O stream and reducing the pressure on the more resource-intensive CWS-Engine. Such collaborative offloading is the key differentiator of Z-STREAM, combining the strengths of both host-side and device-side approaches to provide a complete, transparent and high-performance solution.

7   Conclusion

This paper addresses the fundamental conflict between the strict sequential write constraint of Zoned Namespace (ZNS) SSDs and the prevalence of out-of-order I/O patterns in modern applications. The paper presents the design, implementation, and evaluation of Z-STREAM, a novel, two-level collaborative architecture that successfully virtualizes a random-write interface over sequential-write ZNS hardware. The proposed solution bridges the semantic gap by intelligently dividing the responsibility of write reordering between the host and the device. The architecture combines a lightweight Host-side Batch Sorter, acting as a low-overhead pre-processor, with a powerful, state-aware CWS-Engine operating within the device firmware. The result is a fully transparent system that enhances the performance and usability of ZNS technology without requiring any application-level modifications.

Acknowledgement. This work is supported in part by the National Natural Science Foundation of China (grant number 61902136 and 61932010), and in part by Open Project Program of State Key Lab of Processors, Institute of Computing Technology, CAS under Grant No. CLQ202503, and by Fundamental Research Funds for the Central Universities (grant number 2019kfyXJJS090).

References

1. Cai, Y., Ghose, S., Haratsch, E.F., Luo, Y., Mutlu, O.: Error characterization, mitigation, and recovery in flash-memory-based solid-state drives. Proc. IEEE **105**(9), 1666–1704 (2017)
2. NVM Express: NVMe Zoned Namespaces (ZNS) Command Set Specification. NVM Express (2025)
3. Stavrinos, T., Berger, D.S., Katz-Bassett, E., Lloyd, W.: Don't be a blockhead: zoned namespaces make work on conventional SSDs Obsolete. In: Proceedings of the 18th Workshop on Hot Topics in Operating Systems (HotOS 2021) (2021)
4. Tehrany, N., Trivedi, A.: Understanding NVMe Zoned Namespace (ZNS) flash SSD storage devices. arXiv preprint arXiv:2206.01547 (2022)
5. Chen, Z., Guo, Z., Xia, Y., Wu, J.: MQ-ZNS: a multi-queue-aware I/O scheduler for zoned namespace SSDs. In: 2021 International Conference on Field-Programmable Technology (FPT), pp. 222–230 (2021)
6. Park, S., Kim, D., Lee, Y.: A queue-aware scheduling algorithm for multi-tenant NVMe SSDs. In: 2022 IEEE International Conference on Networking, Architecture and Storage (NAS), pp. 1–8 (2022)
7. Yang, J., Lee, S., Ahn, S.: Selective power-loss-protection method for write buffer in ZNS SSDs. Electronics **11**(7), 1086 (2022). https://doi.org/10.3390/electronics11071086, https://www.mdpi.com/2079-9292/11/7/1086
8. Bjørling, M.: Zone append: a new way of writing to zoned storage. In: USENIX Vault 2020 (2020)
9. Ha, J.Y., Yeom, H.Y.: zCeph: achieving high performance on storage system using small zoned ZNS SSD. In: Proceedings of the 38th ACM/SIGAPP Symposium on Applied Computing, pp. 1342–1351 (2023)
10. Wang, Y., Zhou, Y., Wu, F., Zhang, J., Yang, M.C.: Land of OZ: resolving orderless writes in zoned namespace SSDs. IEEE Trans. Comput. **73**(11), 2520–2533 (2024). https://doi.org/10.1109/TC.2024.3441866
11. Li, H., Hao, M., Tong, M.H., Sundararaman, S., Bjørling, M., Gunawi, H.S.: The CASE of FEMU: cheap, accurate, scalable and extensible flash emulator. In: 16th USENIX Conference on File and Storage Technologies (FAST 2018), pp. 83–90 (2018)
12. FileBench Community: FileBench: a file system and storage benchmark. GitHub Repository (2024). https://github.com/filebench/filebench
13. Kopytov, A.: SysBench: a system performance benchmark. GitHub Repository (2024). https://github.com/akopytov/sysbench
14. Axboe, J.: FIO: flexible I/O tester. https://github.com/axboe/fio (2024)
15. Lee, C., Sim, D., Hwang, J.J., Cho, S.W.: F2FS: a new file system for flash storage. In: 13th USENIX Conference on File and Storage Technologies (FAST 2015), pp. 273–286 (2015)
16. Oh, M., Yoo, S., Choi, J., Park, J., Choi, C.E.: ZenFS+: nurturing performance and isolation to ZenFS. IEEE Access **11**, 26344–26357 (2023)
17. Lee, H.R., Lee, C.G., Lee, S.W., Kim, Y.K.: Compaction-aware zone allocation for LSM-tree based key-value store on ZNS SSDs. In: Proceedings of the 14th USENIX Workshop on Hot Topics in Storage and File Systems (HotStorage 2022) (2022)
18. Liu, R., Tan, Z., Shen, Y., Long, L., Liu, D.: Fair-ZNS: enhancing fairness in ZNS SSDs through self-balancing I/O scheduling. IEEE Trans. Comput. Aided Des. Integr. Circuits Syst. **41**(11), 3721–3734 (2022)

19. Choi, G., Lee, K., Oh, M., Choi, J., Jhin, J., Oh, Y.: A new LSM-style garbage collection scheme for ZNS SSDs. In: Proceedings of the 12th USENIX Workshop on Hot Topics in Storage and File Systems (HotStorage 2020) (2020)
20. Kim, T., et al.: RAIZN: redundant array of independent zoned namespaces. In: Proceedings of the 28th ACM International Conference on Architectural Support for Programming Languages and Operating Systems (ASPLOS 2023), pp. 660–673 (2023)
21. Min, J., Zhao, C., Liu, M., Krishnamurthy, A.: eZNS: an elastic zoned namespace for commodity ZNS SSDs. In: 17th USENIX Symposium on Operating Systems Design and Implementation (OSDI 2023) (2023)

Weight-and-Load-Driven Approximation Methodology for Energy-Efficient Neural Network Accelerators

Zhiqiang Wen, Jianmin Zhang, Yan Sun[✉], and Shangshang Yao

College of Computer Science and Technology, National University of Defense Technology, Changsha 410073, China
yansun@nudt.edu.cn

Abstract. Neural networks in embedded systems confront computational efficiency and energy consumption challenges arising from neural network inference. Although existing approximate computing techniques can reduce energy consumption, they lack runtime-reconfigurable approximation levels, making it difficult to meet the precision requirements of diverse scenarios. In this paper, we propose a time-efficient methodology that maps runtime-reconfigurable approximate multipliers to weight intervals while respecting tight accuracy-loss thresholds. It jointly considers per-layer computational volume and weight-interval fault-tolerance in 8-bit quantized networks, addressing prior neglect of volume disparities and misjudged weight significance. Experiments on CIFAR-10, CIFAR100 and GTSRB show that our method achieves average energy savings of 21.82% (up to 24.7%) on ResNet and outperforms state-of-the-art methods by >20% on VGG. Configuration completes in 0.6 h on GPU, yielding a 7.5x speedup over prior works.

Keywords: Approximate multiplier · Deep neural networks · Design space exploration · Low power

1 Introduction

With the progression of artificial intelligence technologies, machine learning is increasingly permeating the core domains of general-purpose computing and embedded systems. In embedded systems with limited resources, the deployment of neural networks confronts challenges of both computational efficiency and energy consumption, which has rendered hardware acceleration for precision-power trade-offs a hotspot.

Given the error-tolerant characteristics and low-power computing requirements of applications such as the Internet of Things (IoT) [3,25] and artificial intelligence [24,27], approximate computing has emerged as an effective solution to address resource overhead issues by trading off precision and energy

---

First Author and Second Author contribute equally to this work.

H. Liu et al. (Eds.): ICA3PP 2025, LNCS 16381, pp. 500–518, 2026.
https://doi.org/10.1007/978-981-95-8399-7_27

efficiency. A primary characteristic of these applications is that their computations are mainly composed of intensive matrix multiplications. Previous research mainly focused on reducing the cost of multipliers [11], for that multiplication operations consume significantly more energy. For example, hardware accelerators have integrated thousands or even more MAC units [9] to keep up with the required throughput. For instance, Google's Tensor Processing Units (TPUs) integrate 4K MAC units [2]. However, such a large number of MAC units results in a significant increase in energy consumption.

Multipliers are the main components of MAC. Traditional multipliers have been implemented through netlist pruning or some approximate components [5,20,22,23]. Although automation frameworks [12] for designing approximate circuits have been proposed to expedite the process, these methods also employ fixed approximation levels, and the circuits lack runtime reconfigurability. This limitation restricts their ability to dynamically adapt to different precision requirements across diverse operational scenarios, given that different applications often have varying degrees of fault tolerance. Designing customized approximate circuits for each specific application presents a considerable challenge. For instance, in neural networks, different layers have different error tolerance capabilities. Some layers have very poor error tolerance, and slight approximation may lead to significant precision loss.

In contrast, runtime-reconfigurable approximation techniques [13] offer a more flexible solution. They can dynamically adjust precision levels based on the specific requirements of different applications. This adaptability enables better energy efficiency without significantly compromising overall performance and accuracy. Therefore, runtime-reconfigurable approximation techniques hold great potential for addressing the increasingly diverse and evolving demands of modern computing applications.

In this study, we propose a novel approximation method for energy-efficient neural network inference based on runtime-reconfigurable multipliers, taking both computational volume and the weight fault-tolerance of quantized neural networks into consideration. Our main contributions are as follows:

1. We propose an efficient approximation strategy that considers both the computational volume differences between layers and the varying fault tolerance of weight intervals.
2. We propose a time-efficient methodology that can complete the configuration process of approximate multipliers with fewer inference attempts.
3. Based on our strategy, we modify the previous hardware design to support more flexible runtime configuration.
4. Experimental results show that compared with other state-of-the-art works, our method has advantages in energy saving and time consumption.

2   Related Work

In the rapidly evolving landscape of energy-efficient computing, approximate multipliers have become a focal point of research. As the cornerstone arithmetic units, modern multiplier designs mainly pursue two paths: crafting approximate compressors that deliberately trade accuracy for power, or aggressively pruning netlists to exploit structural redundancy. Among these, genetic-programming-driven approximate multipliers have recently gained traction for their ability to automatically discover non-intuitive yet highly efficient architectures. Capitalizing on this trend, ALWANN [13] introduces layer-wise multi-objective optimization, assigning distinct approximate multipliers to individual layers and thus navigating the Pareto frontier between energy efficiency and model accuracy.

Significant advances have also been made in MAC-level mapping strategies. One approach interleaves approximate multipliers with complementary error polarities, positive and negative, to cancel cumulative errors, broadening the applicability of approximate computing while slashing energy [14]. Building on this insight, both genetic algorithms and particle-swarm optimization have been harnessed to place these error-balanced multipliers within MAC arrays, further boosting operational efficiency [4].

However, the aforementioned schemes rely on fixed-error approximate multipliers and lack runtime reconfigurability, limiting their suitability for dynamic contexts such as neural-network inference, where certain layers or data streams exhibit extremely low approximation tolerance. To overcome this rigidity, a dual-quality approximate compressor [1] has been introduced that dynamically toggles between exact and approximate modes, greatly enhancing multiplier flexibility and energy efficiency across diverse workloads. Complementarily, Jain [8] isolates power-hungry yet accuracy-trivial logic blocks, enabling runtime-reconfigurable circuits via logic isolation and further improving power–accuracy adaptability.

Building on reconfigurable circuits, Tasoulas [21] presents LVRM, a runtime-approximation framework tailored for neural-network inference. LVRM couples runtime-reconfigurable approximate multipliers with weight-oriented mapping to slash energy without network retraining. Spantidi [18] further employs Probabilistic Signal Temporal Logic (PSTL) to formally encode precision constraints and derive optimal weight-to-multiplier mappings, minimizing energy while guaranteeing accuracy. By removing manual tuning, retraining, or weight modification, both approaches dramatically cut configuration complexity and runtime overhead.

3 Motivation

3.1 Computational Volume Disparities

Neural networks employ a deep multi-layer architecture, with state-of-the-art networks sometimes comprising hundreds of convolutional layers. Previous weight-oriented mapping strategies, such as LVRM [21], rarely take into account the pronounced differences in computational volume across layers. Indeed, computational volume can grow or shrink by orders of magnitude within a single network. Furthermore, existing methods tend to liberally assign approximate multipliers to layers that exhibit high fault tolerance. This leads to a paradoxical situation where high-tolerance yet low-computational-volume layers receive an abundance of approximation, thereby squandering precious precision, while high-computational-volume layers are left with only minimal approximation and thus fail to capture the full energy-savings potential that approximation offers. Figure 1 illustrates these stark computational-volume disparities across layers in three representative networks: ResNet20 [6], VGG13 [15] and MobileNetV2 [7].

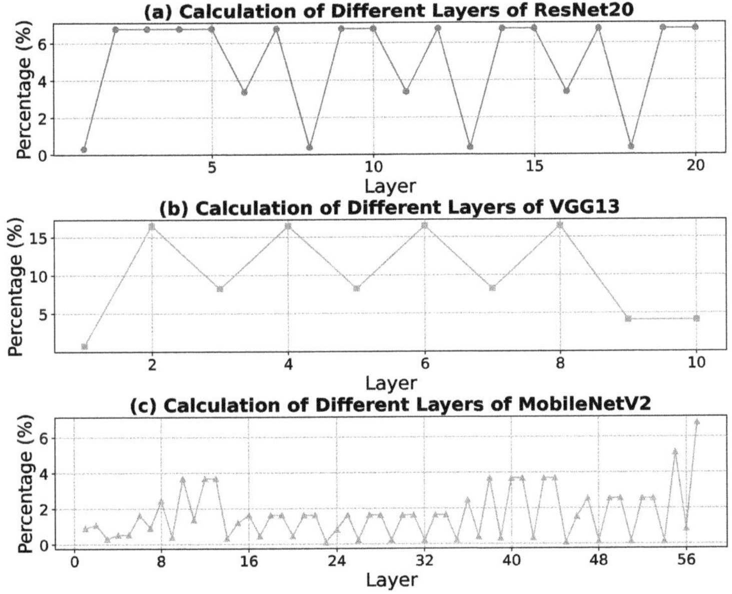

Fig. 1. Computational volume across layers of various neural networks (a) Resnet20; (b) VGG13; (c) MobilenetV2.

It is evident that in well-structured networks, only a small number of layers exhibit clearly abnormal computational volumes—either substantially higher or substantially lower than the average. For example, in ResNet20, the computational volume of individual layers can be classified into three groups: low, medium, and high. The highest computational volume is more than sixty times greater than the lowest, and it is approximately twice that of the second-highest.

In contrast, less regular architectures such as VGG and MobileNetV2, especially MobileNetV2, demonstrate much more significant differences in computational volume among layers. Applying a milder approximation level to layers with higher computational volume can often yield greater energy savings than applying a higher approximation level to layers with lower computational volume, because the absolute number of operations in high-volume layers is substantially larger. Therefore, while the fault tolerance of a layer provides important guidance for determining where approximation can be safely applied, the computational volume of each layer also plays a crucial role in maximizing overall energy savings.

3.2   Difference in Weight Interval Importance

Recent studies adopt weight-oriented methods [18,21] that assign distinct approximate multipliers to fine-grained weight intervals within each layer. Earlier work often assumes a normal-like weight distribution centered near 0 or 128, yet this assumption does not always hold, especially when weights deviate substantially from 128. In such cases, only a small fraction of the layer is effectively covered. Figure 2 illustrates the weight distributions of two networks: most layers follow a near-normal pattern, while several layers clearly depart from it.

Why adopt a finer-grained, weight-oriented multiplier allocation? If every weight carried equal importance, such a strategy would be unnecessary. The essence of weight-oriented mapping is to maximize approximation in tolerant intervals while preserving exact computation for sensitive ones.

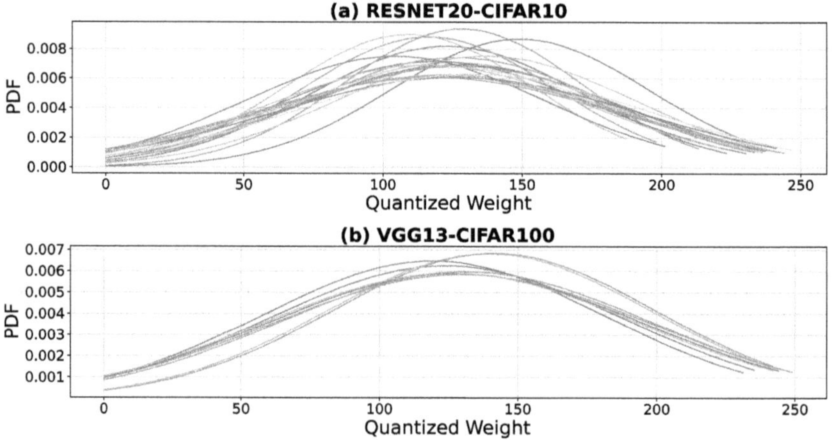

Fig. 2. The distribution of weights in each layer of the neural networks (a) Resnet20-Cifar10; (b) VGG13-Cifar100.

By examining the 8-bit quantized weights of the first ten convolutional layers in ResNet20, we evenly partition the weights of each layer into four intervals (Q1-Q4) and the number of weights in the four intervals is roughly the same. For instance, if a convolutional layer has 40,000 weight parameters, we devide them into four weight intervals of approximately equal size, with each weight interval containing about 10,000 weight parameters. Given the characteristics of the convolutional layer, we believe that the computational volumes obtained by four weight intervals are roughly equivalent. Then we assign a dedicated approximate multiplier to each weight interval. The resulting accuracy curves in Fig. 3 reveal clear differences in interval importance within the same layer, providing direct guidance for our weight-oriented approximation strategy.

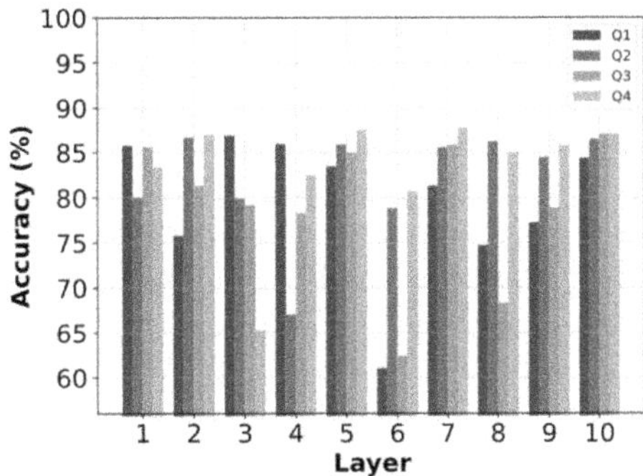

Fig. 3. The accuracy of the first 10 layers of ResNet20 after replacing the exact multiplier with approximate multipliers.

3.3 Time Overhead

Previous exploration approaches mainly rely on genetic algorithms or exhaustive layer-by-layer search to decide which weight intervals should be replaced. The former often takes a long time to converge, while the latter incurs a cost that grows linearly with the number of layers because each layer is explored at fine granularity.

To overcome these drawbacks, we reduce the total number of exploration attempts. Since the replacement proceeds layer-by-layer and interval-by-interval, binary search emerges as a more efficient way to pinpoint the optimal weight ranges without the overhead of exhaustive enumeration.

4 Proposed Methodology

As is illustrated in Fig. 4, our workflow firstly presents AXMUL, a runtime-reconfigurable approximate multiplier built on the 4:2 compressors. Next, we quantitatively profile the computational volume (MAC count) and weight-interval fault-tolerance across every layer of a pre-trained network. Finally, leveraging these computational volume and weight-interval fault-tolerance, we propose an allocation strategy that uniformly quantizes the DNN to 8 bits and maps three AXMUL modes to distinct weight intervals within each layer.

4.1 Multiplier Design

As is mentioned, multiple reconfigurable approximate multiplier designs have been investigated. In this work, AXMUL is built on a 4:2 compressor proposed by [1] which employs power gating. It comprises three sections: the exact portion, the approximate portion, and the shared portion; the exact and approximate portions are mutually exclusive. The operating principle is shown in Fig. 5.

As shown in Fig. 6, building on a basic Wallace multiplier and the approximate compressor mentioned before, we establish three accuracy levels: AXMUL0

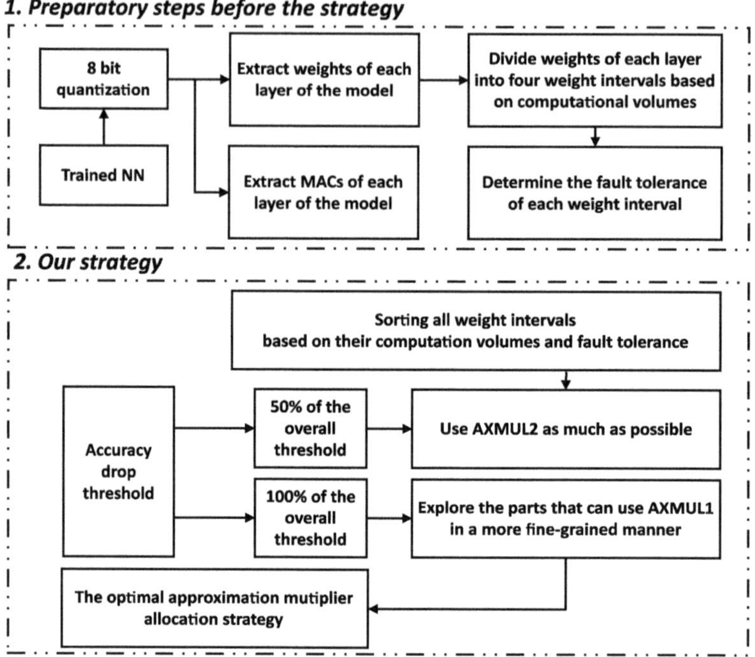

Fig. 4. The overview of the proposed methodology.

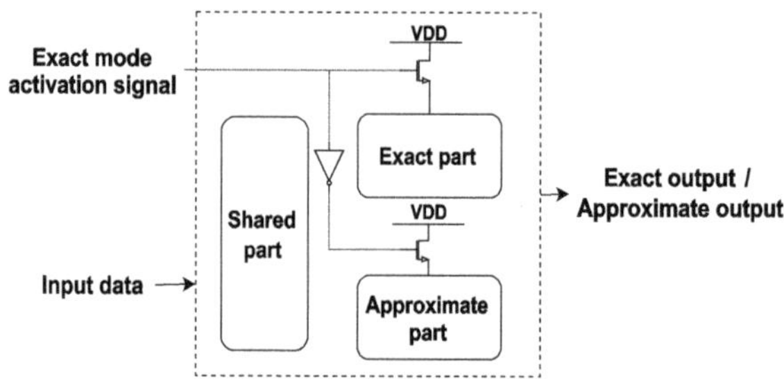

Fig. 5. The operating principle for reconfigurable compressors.

(exact), AXMUL1 (using approximate compressors after Mode1) and AXMUL2 (using approximate compressors after Mode2).

4.2 Measurement of Computational Volume Disparities

Neural networks are multilayered structures with significant differences in computational volume across layers. To measure these differences, we utilize ptflops [16] under PyTorch, as TensorFlow lacks a convenient tool for this purpose. This step only requires deriving the computational volume of each layer based on the model, thus, it consumes almost no time. The differences in com-

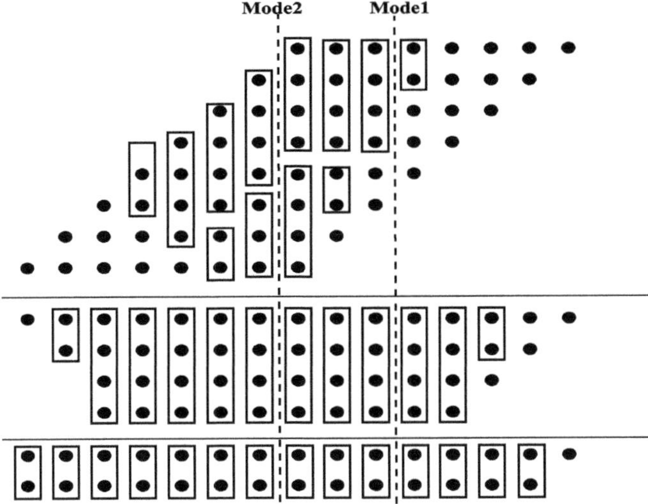

Fig. 6. The multiplier structure used in this paper and the method for partitioning different approximation modes.

putational volumes across different layers in Fig. 1 are obtained from ptflops. Taking ResNet20 as an example, the computational load of the second convolutional layer accounts for about 6.8% of the total, while the eighth layer is close to 0%, indicating significant variations.

4.3   Variations in Weight Importance

Previous methods assumed a symmetric, normal-like weight distribution and concentrated on the high-density region around the center. Suppose we extract all weights from a convolutional layer, yielding 40,000 weights in total. We split these weights into four frequency-balanced intervals (Q1 – Q4) covering the 0–255 range, each containing roughly 10,000 weights, and then assess the sensitivity of each interval. By perturbing one interval at a time while keeping the remaining three exact, we generate an importance ranking for every layer. This experiment reveals pronounced differences in interval sensitivity within the same layer, validating our weight-oriented allocation strategy.

In this step we only replace a small subset of multipliers in each layer. Because the affected portion is tiny (only part of one layer), accuracy remains largely intact, allowing us to retain the approximation across most layers. When we test the AXMUL2 multiplier which is our highest-error variant, the resulting accuracy drops are too small to clearly separate robust from fragile intervals. To expose these differences we temporarily employ even higher-error multipliers. Crucially, these exaggerated-error units are used solely for diagnosis; they are not deployed in the final design.

We therefore conduct the evaluation with four multipliers spanning distinct error levels. As depicted in Fig. 7, while the ranking of well-performing intervals fluctuates, poor-performing intervals are consistently identifiable across all precision settings. Detecting these under-performing regions is more important than obtaining an exact ranking, since our goal is to apply approximation only to fault-tolerant intervals and to protect sensitive ones. Hence, we leverage higher-error multipliers solely to probe the fault tolerance of each interval.

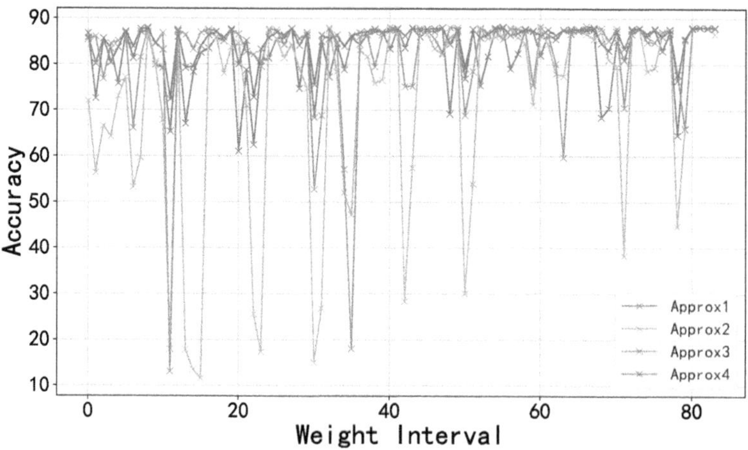

Fig. 7. Use different approximate multipliers to explore the significant of different intervals.

4.4 Weight and Load-Based Mapping

Step 1 Data Filtering: In the preliminary steps, we use the AXMUL multiplier to estimate the fault-tolerance of different weight intervals within a layer. This approach, while not precisely ranking the fau lt tolerance of different intervals, effectively identifies those with poor fault tolerance. However, we cannot completely discard these data. When there is redundancy in precision, these intervals with poor fault tolerance can still be utilized. Moderate approximations to them may work better than aggressive approximations to intervals with better fault tolerance. Instead, we assign them a lower priority. Therefore, the original data are initially divided into two parts: high tolerance and low tolerance, as in shown in Fig. 8.

To filter out the unfavorable components of approximation as much as possible, we distinguish low-tolerant data based on two criteria: one part is the weight intervals unsuitable for approximation according to their numerical values and the other part is the weight intervals selected based on a proportion. The former ensures that weight intervals with absolutely low fault tolerance are avoided for approximation, while the latter, using a proportional method, ensures that

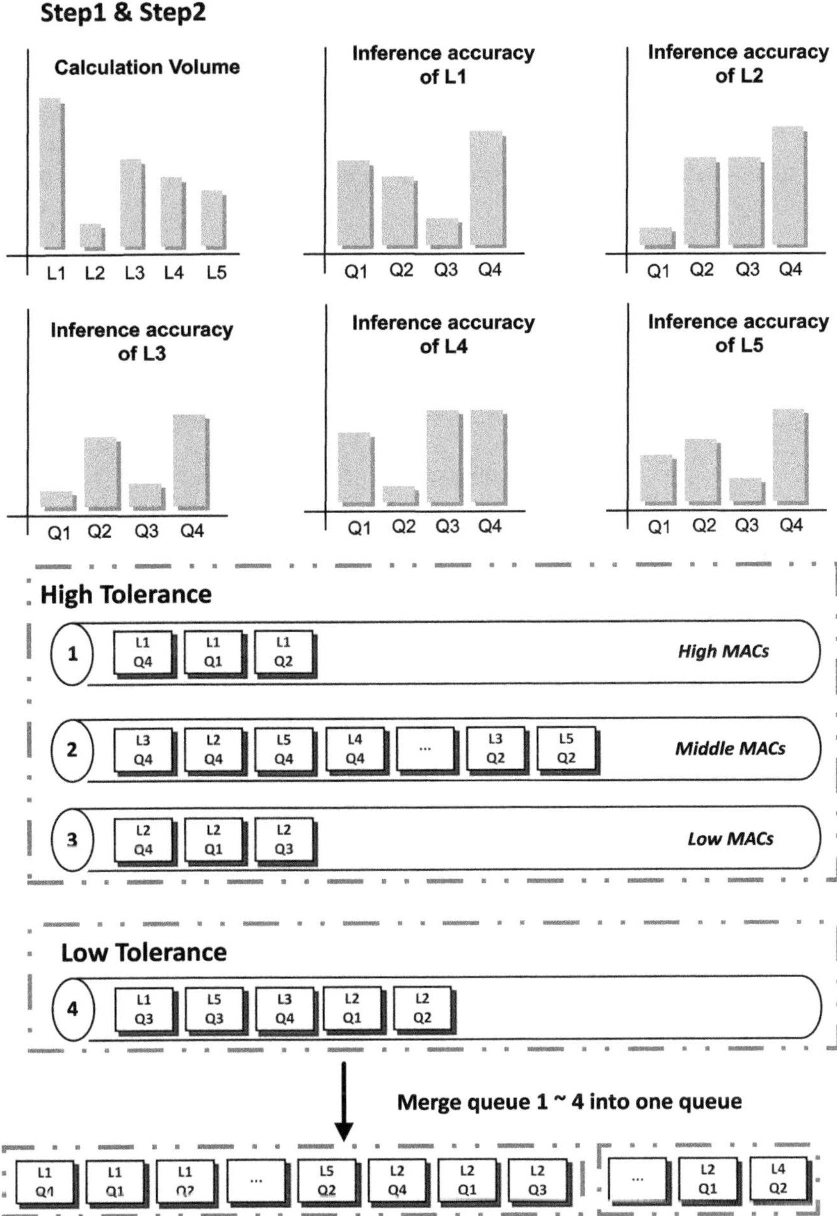

Fig. 8. Illustration of Step1 and Step2 for sorting the weight intervals according to computational volumes and fault tolerance.

some relatively poorly performing weight intervals are excluded. We combine these two methods and finally take the union of their results. For the former, we specify that weight intervals with a precision drop of over 20% (relative to the exact result) after replacement would be selected, and for the latter, we selected the bottom 10% of all intervals after sorting them by precision. These selected weight intervals were then combined to form the first part of the data unsuitable for approximation. At this stage, we do not consider the computational volume, as approximating these intervals might have a significant impact on precision.

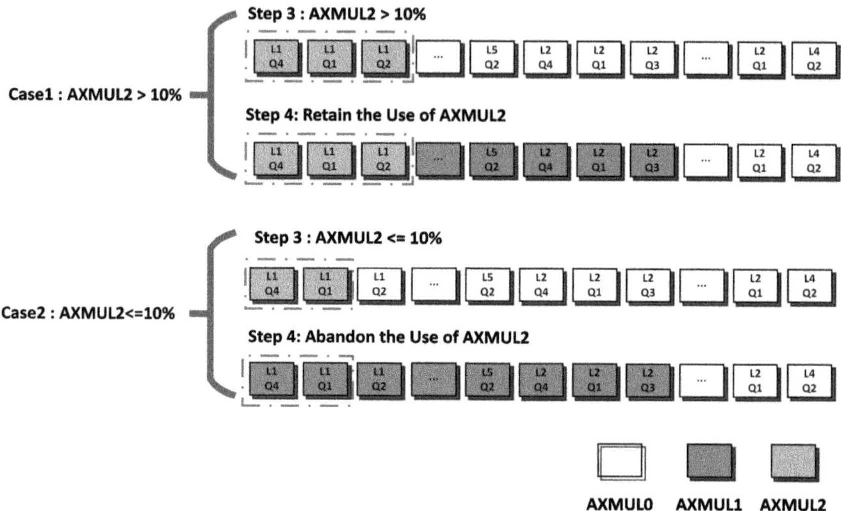

Fig. 9. Illustrations of Step 3 and Step 4 (Case 1: AXMUL2 Usage > 10% in Step 3; Case 2: AXMUL2 Usage 10% in Step 3).

Step 2 Weight Interval Comprehensive Ranking: In Step 1, we incorporate the importance of computational volume. For the high tolerant part, we further divide it into three groups: High MACs, Middle MACs and Low MACs. The basis for division is the average computational volumes of all layers. We consider layers more than twice the average computational volume as High MACs, those computational volumes less than a quarter of the average as Low MACs and the rest as Middle MACs. We assume that the computational volumes of weight intervals within the same group were comparable, and within each group, the intervals were sorted by fault tolerance in descending order. Then, the three groups are merged in the order of High MACs, Middle MACs and Low MACs into one queue. Finally, the low tolerance part is appended to the end of the group. As previously mentioned, we did not ignore this part; when there is precision redundancy, we would consider approximating these parts as well.

As shown in Fig. 7, among all layers, Layer 1 has the highest computational load, more than twice the average computational load, entering the High MAC category. However, Layer 1 has a low tolerance for error in the Q3 range, thus it falls into the Low Tolerance category. Additionally, although Layer 2 has a strong error tolerance in the Q4 range, its computational load is too low, so it still falls into the Low MACs category. After sorting each column by computational load and assigning weight intervals, the columns are internally sorted from high to low error tolerance.

Step 3 AXMUL2 Allocation: We start allocating approximate multipliers from this step. As approximation can lead to accuracy loss, we will set a accuracy drop threshold, Limit. Since AXMUL1 has a relatively small precision loss, we aimed to achieve a significant scale effect through AXMUL1. Therefore, we set a threshold for the precision drop in this stage at LMT1 = Limit/2. For instance, if the acceptable accuracy drop is 2%, then in this stage, the maximum allowable accuracy drop for AXMUL2 is 1% and another 1% for AXMUL1 in Step4. We used a binary search to explore the optimal position. Initially, we apply AXMUL2 to the first half of the queue. If the precision drop fails to reach the threshold LMT1, we expand the boundary to apply AXMUL2 to the first 3/4 of the data; conversely, we applied AXMUL2 to the first 1/4 of the data. This process was repeated until we found the position where as many intervals as possible could use the AXMUL2 multiplier without exceeding the precision drop threshold.

Step 4 AXMUL1 Allocation: The method for this step is similar to Step3, but the overall threshold was set to LMT2 = Limit. Again, we employ a binary search approach from the end of AXMUL2 to determine the maximum replacement boundary, ensuring that the precision drop did not exceed the specified limit LMT2 while maximizing the use of AXMUL1 multipliers. As shown in Fig. 9, if the usage of AXMUL2 exceeds 10% in Step 3, we consider this proportion to be not insignificant and acceptable; however, if the usage of AXMUL2 is less than 10%, we believe that the use of AXMUL2 is insufficient. Using AXMUL1, which has lower error, offers more advantages in terms of power consumption, and therefore, we discard the AXMUL2 mapped in Step 3 and instead use AXMUL1 in the same manner.

4.5   Run-Time Reconfiguration

The circuit design for run-time reconfiguration in this paper follows the circuit design in LVRM [21]. Since the overall design of the MAC in the LVRM has been preliminarily evaluated, we only elaborate on our method without further discussion (Fig. 10).

According to the approach outlined in the previous section, we need to store a 32-bit LayerID in each convolutional layer. Within this LayerID, three 8-bit data segments, L1, L2 and L3 are used to divide the weights of each layer into four intervals (Q1 to Q4). Additionally, an 8-bit Conf data segment indicates the multiplier selected for each of the four intervals. We define a 2-bit control signal ctrl with the following encoding: 2'b00 for AXMUL0, 2'b01 for AXMUL1, and 2'b11 for AXMUL2. For instance, if Conf = 8'b11000111, it signifies that

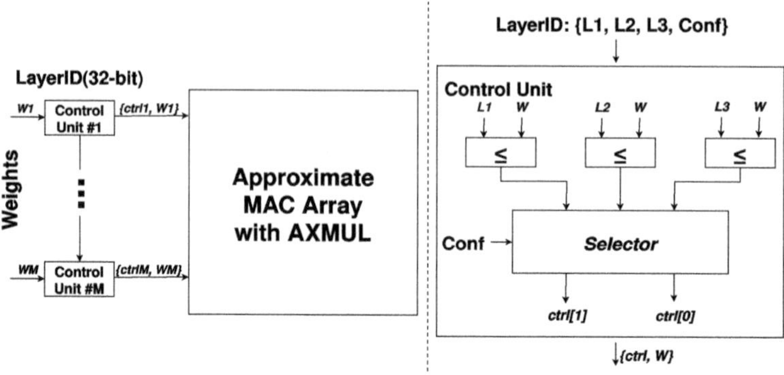

Fig. 10. Circuit diagram for generating control signals for reconfigurable approximate multipliers.

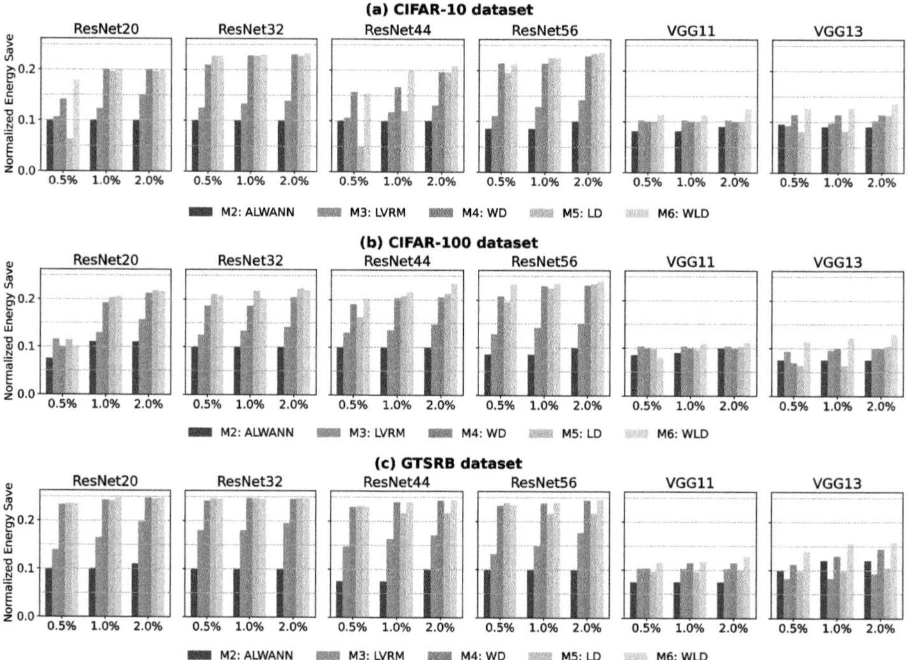

Fig. 11. Normalized energy reduction on different datasets (a) CIFAR10; (b) CIFAR100; (c) GTSRB.

Q1 to Q4 utilize AXMUL2 (2'b11), AXMUL0 (2'b00), AXMUL1 (2'b01) and AXMUL2 (2'b11), respectively.

Three 8-bit comparators generate a 3-bit signal, with all possible values being 3'b000, 3'b100, 3'b110 and 3'b111, corresponding to the weight ranges (Q1 - Q4), respectively. This 3-bit signal, along with Conf, is fed into the Selector to retrieve

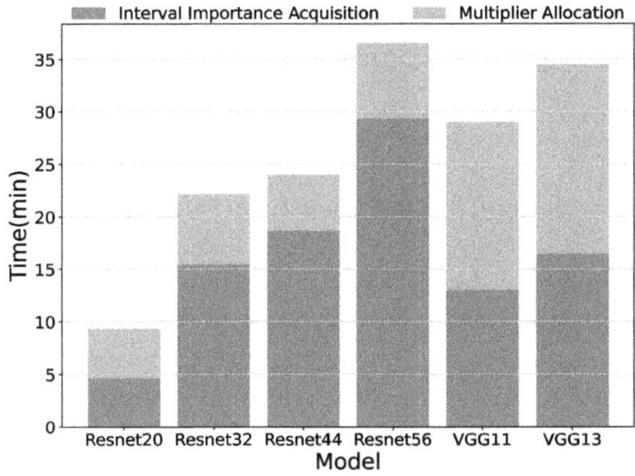

Fig. 12. Execution time of our approach on CIFAR100.

the multiplier configuration stored in the convolutional layer. For example, if the comparator outputs 3'b110, indicating the weight lays in Q3, the Selector chooses the multiplier configuration of Q3, which is 2'b01 and then outputs ctrl alongside the weight W.

5 Experiment and Evaluation

5.1 Experimental Setup

The experiments are conducted on a system equipped with a Xeon(R) Gold 6430 with 16 cores and a NVIDIA GeForce RTX 2080Ti with 11 GB GPU memory, which is utilized for training and inference of deep neural networks.

Experiments are conducted using Tensorflow, with our modifications made to the tf-approximate code [13]. Energy consumption is calculated based on the MACs obtained from ptflops [16] in Pytorch, combined with the power consumption of AXMUL. Multipliers are described in Verilog HDL and synthesized with the compile_ultra command at the critical path delay of the exact multiplier using Design Compiler.

5.2 Benchmarks

Models evaluated includes ResNet20, ResNet32, ResNet44, ResNet56, VGG11, VGG13. Three datasets, CIFAR10 [10], CIFAR100 [10] and GTSRB [19] are employed. The primary metrics measured are the energy savings and inference time for each individual model.

Before diving into the experimental results, we first provide a brief overview of the main approximation methods used in our experiments. The following approximation methods are evaluated and compared in detail:

(M1) Exact Multiplier: All layers of the neural network utilize the exact multiplier without any approximation.

(M2) ALWANN: This method employs multiple fixed-error approximate multipliers. Using multi-objective optimization, different multipliers are allocated to different layers.

(M3) LVRM: This method utilizes the proposed LVRM approximate multiplier and applies error correction through bias updates. It follows a four-step approach for weight-mapped multipliers: first ranking all layers by fault tolerance, then mapping all layers to LVRM2, followed by partial mapping to LVRM2, and finally partial mapping to LVRM1.

(M4) LD, Considering only computational volumes: To examine the impact of computational complexity on the results, we only consider the differences in computational volumes between layers, and sort them according to their fault tolerance.

(M5) WD, Considering only the importance of weights: To examine the impact of the importance of weights, we simply sort all weight intervals according to their fault tolerance.

(M6) WLD, Considering computational volumes and fault tolerance: This is the approximate multiplier allocation strategy proposed in this paper, which takes into account both the computational volumes of each layer in the neural network and the fault tolerance of the weight intervals.

5.3   Energy Evaluation

We evaluate using the power consumption of the exact multiplier as the baseline, focusing on the power reduction achieved compared to this exact case. The formula for calculating relative energy reduction is as follows:

$$P_{dec} = 1 - \frac{\sum P_{AXMUL_i} \times Calc_{AXMUL_i}}{P_{exact} \times Calc_{all}} \tag{1}$$

P_{AXMUL_i} represents the power consumption of AXMUL_i (where $i = 0, 1, 2$), $\text{Calc}_{\text{AXMUL}_i}$ represents the computational volume of the interval using AXMUL_i, P_{exact} represents the power consumption of the exact multiplier, $Calc_{all}$ represents the computational volumes of the intervals and P_{dec} represents the relative power reduction ratio.

5.4   Experimental Results

Figure 11 demonstrate the experimental results of energy savings after applying the approximate multiplier to six types of deep neural network models. To facilitate comparison and maintain consistency with previous work, we selected the same three accuracy drop thresholds: 0.5%, 1.0%, 2.0%. We compared our energy savings with other methods using the same thresholds.

Figure 11(a) shows the results of different methods on the CIFAR10 dataset. Our proposed method (M6) outperforms others in most scenarios and accuracy loss thresholds. At the 0.5% accuracy threshold, our method achieves average power reductions of 20.11%, 23.39% and 23.6% on Resnet20, Resnet32 and

Resnet56 respectively. However, when it comes to Resnet44, it delivers a moderate reduction of 17.86% but still outperforms other approaches. ALWANN (M2) employs optimization with various fixed-precision multipliers to find optimal solutions. However, due to time limits and lack of fine-grained weight exploration, it finally achieves only about 9% average energy reduction. Similar to our work, LVRM (M3) employs a multipler with several approximate modes to accommodate varying accuracy thresholds. However, LVRM initial extensive use of the LVRM2 multiplier which has the maximum error, coupled with its focus on approximating only a subset of central-axis weights, restricts its weight exploration. Across the three thresholds, LVRM attains average power reductions of 12.88%, 13.4% and 14.95%, with a peak power reduction up to 15.59% (Resnet56, 2%). Apart from Resnet, in the case of VGG networks, the energy reduction is markedly lower than that of Resnet, attributable to VGG's fewer convolutional layers and reduced fault tolerance for approximation. Our analysis of the results reveals that, in most instances, even for the largest 2% threshold, less than 10% of weights intervals can utilize AXMUL2. Consequently, in line with the constraint established in Step 4 of Sect. 4.4, our method predominantly defaults to using solely AXMUL1 for approximation. Under the VGG networks with a accuracy threshold of 2%, the maximum power consumption is 15.82%. Our method's improvement is no more than 6% higher than those of other methods.

Figure 11(b) presents the CIFAR100 dataset results. It is a more complex dataset and shows slightly inferior performance compared to CIFAR10. Yet, it still delivers good results in most scenarios and thresholds. Our approach achieves average power reductions of 15%, 23% and 21.5% on Resnet20, Resnet32 and Resnet44 respectively. Similarly, the performance on VGG mirrors that on CIFAR10. As is mentioned in Sect. 4.3, we use the approximate multiplier with larger errors for error tolerance evaluation which can sometimes yield inaccurate weight interval importance data. This issue isn't evident in CIFAR10, likely due to its simplicity and high fault tolerance. But on more complex datasets, for instance, under a 0.5% accuracy threshold, while our method works well on CIFAR10, it shows some inconsistent results on CIFAR100. For Resnet20 with 0.5% threshold, the power reductions of M4-M6 are only 9.99%, 11.38% and 9.87%, which are 1%–2% lower than LVRM. However, our approach still performs better in most scenarios.

Figure 11(c) presents results of GTSRB which also follow a similar pattern to the above results. GTSRB is a relatively simple dataset. Even the less fault-tolerant VGG networks also show great energy reduction. The average energy reductions for the four ResNet models are 28.39%, 29.13%, 27.8% and 27.3% and those for VGGs are 14.1% and 14.05%.

Across the three datasets, we observe that energy gains tend to rise with the increasing layers of neural networks, especially Resnet56. This aligns with findings in Spantidi's work [17]. Our method has a key edge over previous ones in efficiently exploring the design space of deep neural networks. In the subsequent section, we will proceed to delve into this matter in greater depth. Moreover, we propose considering computational volumes for replacement and set up a control group using only computational volumes (M5) for ranking. Results show that relying solely on computational volumes doesn't perform well under low thresholds but achieves the highest power reduction under a 2% threshold in some cases (Resnt32-CIFAR100). It successfully identifies the weight ranges that are computationally expensive but with relatively poor performance. As M4-M6 all use weight intervals, their overall results are quite close. However, M6 outperforms M4 in most cases across the three thresholds and six neural networks, showing better stability. In contrast, M5, which depends purely on fault tolerance, is less stable and prone to early stopping during exploration, though it delivers good results under loose accuracy conditions. Overall, after considering the computational volumes and the importance of weights, our approach outperforms methods that only account for a single factor in most cases.

5.5 Time Consumption

In this section, we evaluate the execution time of our proposed method. As mentioned in Sect. 3, previous approaches have exhibited suboptimal performance in terms of time consumption. To conduct the evaluation, we utilized the 2080Ti GPU with 22GB VRAM and implemented certain modifications based on the code of tf-approximate [13]. Since tf-approximate employs a look-up table (LUT) approach for handling approximate multipliers, the inference speed of a convolutional layer may experience some degradation after approximation. Therefore, to minimize time overhead, we aim to reduce the number of approximate substitutions as much as possible, replacing only the essential components.

Our method involves two primary steps: (a) Interval Importance Acquisition: This component incurs significant overhead. Since we only need to determine relative magnitude relationships, we use 25% of the test data for evaluation. This step corresponds to Sect. 4.3. (b) Multiplier Allocation: We have transformed all intervals into a one array according to our method. This step corresponds to Sect. 4.4.

The main time cost of the method lies in the step of Interval Importance Acquisition. Take Resnet56 as a typical example; nearly 80% of the time is consumed in this step, while Multiplier Allocation takes up less than 20%. As the model depth increases, the advantage of our proposed method become more prominent. When compared with other state-of-art works, our method achieves significant improvements. For example, ALWANN [13] requires 7.5 days for Resnet50. RETSINA [26] and LVRM [21] take 4.5 h and 2 h respectively for Resnet56, while our method only takes 0.6 h, achieving 7.5x and 3.4x speedups respectively (Fig. 12).

6 Conclusion

In this paper, we propose an approximate methodology driven by weight-interval fault-tolerance and computational volumes for energy-efficient neural network inference. By analyzing the variations in computational volumes and weight-interval fault-tolerance across different layers of neural networks, our method devises an efficient and time-saving mapping strategy for approximate multipliers. Experimental results demonstrate that this approach achieves substantial average energy reduction across several common neural network models while effectively controlling precision loss.

Acknowledgments. This work is supported by the National Key Research and Development Program of China (No. 2022YFB2803405).

References

1. Akbari, O., Kamal, M., Afzali-Kusha, A., Pedram, M.: Dual-quality 4:2 compressors for utilizing in dynamic accuracy configurable multipliers. IEEE Trans. Very Large Scale Integr. (VLSI) Syst. **25**(4), 1352–1361 (2017)
2. Cass, S.: Taking ai to the edge: Google's TPU now comes in a maker-friendly package. IEEE Spectr. **56**(5), 16–17 (2019)
3. Choudhary, P., Bhargava, L., Suhag, A.K.: Designing of energy-efficient approximate multiplier circuit for processing unit of IoT devices. SN Comput. Sci. **4**(5), 506 (2023)
4. Devi, D.N., Kumar, G.A., Gowda, B.G., Rao, M.: PSO optimized design of error balanced weight stationary systolic array architecture for CNN. In: 2024 25th International Symposium on Quality Electronic Design (ISQED), pp. 1–8. IEEE (2024)
5. Esposito, D., Strollo, A.G.M., Napoli, E., De Caro, D., Petra, N.: Approximate multipliers based on new approximate compressors. IEEE Trans. Circuits Syst. I Regul. Pap. **65**(12), 4169–4182 (2018)
6. He, K., Zhang, X., Ren, S., Sun, J.: Deep residual learning for image recognition. In: Proceedings of the IEEE Conference on Computer Vision and Pattern Recognition, pp. 770–778 (2016)
7. Howard, A., Zhmoginov, A., Chen, L.C., Sandler, M., Zhu, M.: Inverted residuals and linear bottlenecks: mobile networks for classification, detection and segmentation. In: Proceedings of CVPR, pp. 4510–4520 (2018)
8. Jain, S., Venkataramani, S., Raghunathan, A.: Approximation through logic isolation for the design of quality configurable circuits. In: 2016 Design, Automation & Test in Europe Conference & Exhibition (DATE), pp. 612–617. IEEE (2016)
9. Jouppi, N.P., et al.: In-datacenter performance analysis of a tensor processing unit. In: Proceedings of the 44th Annual International Symposium on Computer Architecture, pp. 1–12 (2017)
10. Krizhevsky, A., Hinton, G., et al.: Learning multiple layers of features from tiny images (2009)
11. Momeni, A., Han, J., Montuschi, P., Lombardi, F.: Design and analysis of approximate compressors for multiplication. IEEE Trans. Comput. **64**(4), 984–994 (2014)
12. Mrazek, V., Hanif, M.A., Vasicek, Z., Sekanina, L., Shafique, M.: autoax: an automatic design space exploration and circuit building methodology utilizing libraries of approximate components. In: Proceedings of the 56th Annual Design Automation Conference 2019, pp. 1–6 (2019)
13. Mrazek, V., Vasicek, Z., Sekanina, L.: Design of quality-configurable approximate multipliers suitable for dynamic environment. In: 2018 NASA/ESA Conference on Adaptive Hardware and Systems (AHS), pp. 264–271. IEEE (2018)

14. Park, G., Kung, J., Lee, Y.: Design and analysis of approximate compressors for balanced error accumulation in mac operator. IEEE Trans. Circuits Syst. I Regul. Pap. **68**(7), 2950–2961 (2021)
15. Simonyan, K., Zisserman, A.: Very deep convolutional networks for large-scale image recognition. arXiv preprint arXiv:1409.1556 (2014)
16. Sovrasov, V.: Flops counting tool for neural networks in pytorch framework (2018)
17. Spantidi, O., Zervakis, G., Anagnostopoulos, I., Amrouch, H., Henkel, J.: Positive/negative approximate multipliers for DNN accelerators. In: 2021 IEEE/ACM International Conference on Computer Aided Design (ICCAD), pp. 1–9. IEEE (2021)
18. Spantidi, O., Zervakis, G., Anagnostopoulos, I., Henkel, J.: Energy-efficient DNN inference on approximate accelerators through formal property exploration. IEEE Trans. Comput. Aided Des. Integr. Circuits Syst. **41**(11), 3838–3849 (2022)
19. Stallkamp, J., Schlipsing, M., Salmen, J., Igel, C.: Man vs. computer: benchmarking machine learning algorithms for traffic sign recognition. Neural Netw. **32**, 323–332 (2012)
20. Taheri, M., et al.: Adam: adaptive fault-tolerant approximate multiplier for edge DNN accelerators. arXiv preprint arXiv:2403.02936 (2024)
21. Tasoulas, Z.G., Zervakis, G., Anagnostopoulos, I., Amrouch, H., Henkel, J.: Weight-oriented approximation for energy-efficient neural network inference accelerators. IEEE Trans. Circuits Syst. I Regul. Pap. **67**(12), 4670–4683 (2020)
22. Vakili, B., Akbari, O., Ebrahimi, B.: Efficient approximate multipliers utilizing compact and low-power compressors for error-resilient applications. AEU-Int. J. Electron. Commun. **174**, 155039 (2024)
23. Venkatachalam, S., Ko, S.B.: Design of power and area efficient approximate multipliers. IEEE Trans. Very Large Scale Integr. (VLSI) Syst. **25**(5), 1782–1786 (2017)
24. Xu, R., Ma, S., Wang, Y., Guo, Y., Li, D., Qiao, Y.: Heterogeneous systolic array architecture for compact CNNs hardware accelerators. IEEE Trans. Parallel Distrib. Syst. **33**(11), 2860–2871 (2021)
25. Yao, S., Zhang, L.: FHAM: FPGA-based high-efficiency approximate multipliers via LUT encoding. In: 2022 IEEE 40th International Conference on Computer Design (ICCD), pp. 487–490. IEEE (2022)
26. Zervakis, G., Amrouch, H., Henkel, J.: Design automation of approximate circuits with runtime reconfigurable accuracy. IEEE Access **8**, 53522–53538 (2020)
27. Zhao, Y., Lu, J., Chen, X.: A dynamically reconfigurable accelerator design using a sparse-winograd decomposition algorithm for CNNs. Comput. Mater. Continua **66**(1) (2021)

HT-DLC: A Data Layout and Computing Design for Logic-in-Memory Convolution with High Throughput

Ziyi Bo[1,2], Nuo Xu[1,2(✉)], Kang Liu[1,2], Anning Zhao[1,2], Chenghao Tan[1,2], and Libo Huang[1,2(✉)]

[1] College of Computer, National University of Defense Technology, Changsha, China
`{bo_ziyi,liukang,zhaoanning,tanchenghao24}@nudt.edu.cn`
[2] Key Laboratory of Advanced Microprocessor Chips and Systems, Changsha, China
`{xunuo,libohuang}@nudt.edu.cn`

Abstract. In-memory computing performs calculations in place where data is stored, eliminating the bandwidth limitations of the traditional von Neumann architecture. Existing methods have applied it to convolution calculations, but they fail to fully exploit the parallelism of the array, resulting in low throughput. We propose HT-DLC, a data layout and computing design for logic-in-memory convolution. We consider the data-dependent characteristics of the multiply-accumulate (MAC) operations that constitute the convolutional layer. By storing the data required by a MAC in a single row of the array to reduce inter-row replication, we maximize the parallelism of the array. And we consider the case of storing multiple MACs in a single array row and design a method for selecting input data of MAC to improve area utilization. When performing convolution on grayscale images using convolutional kernels of different widths, compared to stateful logic image processing accelerator based on the state-of-the-art synthesis flow STAR, our implementation of image convolution achieves an average throughput improvement of $15.11\times$ and an average reduction of 93.11% in the computational delay of a single MAC, while the area utilization is reduced by an average of 7.28%. The results show that using HT-DLC for convolution can fully exploit the parallelism of the array, resulting in higher execution efficiency and lower average delay per MAC.

Keywords: In-memory Computing · Convolution · Stateful Logic · Parallel Computing

1 Introduction

With the advancement of technology, there has been an explosive growth in data volume and an increase in the complexity of data processing. The bottleneck of the von Neumann architecture, namely the communication between memory and processors, is becoming more and more evident in its impact on the improvement of computer performance. The separation of computation and storage in computers leads to frequent reading and writing of data from memory during operation. The read and write speed of memory is far behind the processing speed of the processor, causing the processor to always wait for data and thus restricting performance. In-memory computing, which completes data processing directly in memory, is expected to be a solution to the bottleneck of the von Neumann architecture [24,28]. In-memory computing [11] relies on memristors, which have a simple structure and possess both memory and resistive properties. On one hand, data is stored by distinguishing between the high and low resistance states of the memristor to represent logic '0' and logic '1'. Moreover, these devices can maintain their current state after power loss, exhibiting non-volatility. On the other hand, their physical structure changes based on the amount of charge or the direction of current passing through them, such as ion migration or the formation/breakage of conductive filaments, which is macroscopically manifested as a change in resistance. By controlling the magnitude and direction of voltage or current, we can precisely adjust the resistance of the memristor, facilitating the definition and implementation of logic operations, such as IMP, NOR, NOT, and so on. By selecting appropriate logic gates to form a complete logic family, we can combine logic gates to perform complex computations.

Early in-memory computing was primarily based on analog computing. To address issues such as low precision and poor scalability, digital in-memory computing has gradually evolved. The process nodes have continuously advanced from 28nm, 12nm to 5nm, with a peak energy efficiency as high as 258.5TOPS/W and an area efficiency as high as 221 TOPS/mm^2 [4,10,23]. Any complex computation that can be decomposed into a combination of logic gates can be completed in digital in-memory computing without theoretical loss of precision.

Existing work has attempted to use in-memory computing to achieve high parallelism and energy efficiency for various algorithms, such as accelerating the number theoretic transform [17], neural networks inference [3,9,13–15,19], and training [8,22], Fast Fourier Transform [12], database processing [18], graph processing [1], encryption algorithms [16], etc. Research has fully utilized the characteristics of different devices, such as ReRAM as a new type of non-volatile memory, which has the characteristics of high density, low power consumption, and reconfigurability [5,21]; the voltage-controlled magnetic anisotropy (VCMA) of spin-orbit torque magnetic random access memory (SOT-MRAM) improves the computational efficiency of each storage cycle [9,22]; FeFET uses ferroelectric polarization direction to modulate channel conductivity, achieving non-volatile storage and logic control with low latency and low power consumption, supporting high-frequency computation [14]; SRAM, based on transistor storage, has high precision and reliability, with mature processes [19].

For the task of accelerating convolution on digital in-memory computing arrays, existing work has optimized convolution execution from three dimensions: computation methods, algorithms, and data characteristics. Some studies use combinations of pure logic gates to perform complex computations: In binary neural networks, convolution degenerates into XNOR (exclusive NOR) and bit counting(popcount). When convolution is extended to INT8, it involves bit-wise computation through combinations of logic gates followed by accumulation [5,21]. Some studies integrate standard digital units, such as adders, shifters, and LUTs, next to SRAM arrays, configuring computational logic adjacent to storage units [2,4,10,20,23,25,27]. For convolution algorithm optimization, Hao Wu et al. reduced the number of energy-intensive multiplication operations using the Winograd algorithm to achieve acceleration [23]. Additionally, leveraging the sparsity of data in neural networks, different optimization methods are applied for different types of sparsity, skipping invalid computations to accelerate convolution and improve energy efficiency [4,23].

To balance precision and energy efficiency, current digital in-memory computing generally optimizes through the synergy of quantization and approximate computing. It adopts configurable quantization with INT4/8/16 (e.g., supporting dynamic switching between different bit widths ranging from 4b to 16b [10]), and introduces MB-XNOR encoding to mitigate the precision loss of multi-bit activation. Meanwhile, by using approximate compressors and adders to reduce the area by 55%, and combining approximate-aware training, the accuracy on CIFAR-10 was brought back from 25.2% to 86.9%, ultimately achieving a 10× improvement in energy efficiency through the synergy of low-bit width and approximation [20].

Current research mostly executes two-dimensional bit-plane hierarchical mapping on input image tensors: first, pixels are spatially divided into tiles of size $H_{tile} \times W_{tile} \times C_{tile}$, and then these tiles are unfolded into SRAM/MRAM arrays in row-major or column-major order. The k-bit quantized data of each pixel is further split into k bit-planes, with MSB to LSB mapped layer by layer to different rows or columns to ensure channel continuity and pixel locality. Decoders or crossbars support row column interchange to match the array aspect ratio and maximize parallel utilization. Existing work has not yet considered the spatial characteristics of convolution kernels and image data to optimize the execution of convolution on digital in-memory computing arrays.

We found that when all array rows activate the same column devices to complete the same logic gate, the array always works with the maximum parallelism and the highest throughput. We define the array row utilization as the sum of activated array rows per clock cycle $Rows_t$ divided by the delay T and then divided by the total number of array rows $Rows_{total}$ during the execution process, to measure the degree of parallelism utilization in the computation process, that is, the proportion of clock cycles in which array rows participate in the computation. After statistical analysis, we found that when performing convolution on grayscale images using a 3×3 1-bit convolution kernel on a 512×512 array, the array row utilization of IMAGING [5] is 63.96%, and that of STAR [21] is 31.64%.

$$array\ row\ utilization = \frac{\sum_{t=1}^{t=T} Rows_t}{Rows_{total} * T} \tag{1}$$

The issues with current methods are: 1) The storage mode of image data based on pixel positions limits the parallelism of array rows, resulting in low array row utilization; 2) Solving the data dependency between MACs determined by the characteristics of the convolution kernel through redundant storage of image rows or batch-by-batch computation leads to low throughput of the array.

Therefore, targeting the efficient execution of convolutional layers in neural networks on digital in-memory computing arrays, this paper explores the storage of data based on the distribution characteristics of convolution kernels and considers the storage and arrangement of input data when multiple MACs constituting the convolutional layer are stored in the same row of the array. We optimize the stateful logic synthesis process for convolution calculations to accelerate the execution of convolutional layers. Our contributions are as follows:

1. We redesigned the storage and execution methods of MACs for image convolution based on the characteristics of memristor crossbar arrays, further considering the scenario where a single array row can accommodate multiple MACs.
2. Through stateful logic synthesis and processing, possible logic gates are combined into composite gates to optimize the execution sequence and reduce execution delay.
3. Compared with existing methods, when executing convolutional layers using our method, the array throughput is increased by an average of 15.11 times, the computation delay of a single MAC is reduced by an average of 93.11%, and the area utilization is decreased by an average of 7.28%.

The organization of this paper is as follows. Starting from Sect. 2, the structure and working principle of the in-memory stateful logic array and the characteristics of convolution are introduced. Section 3 describes the design of HT-DLC, including the storage method of image data and the execution method of logic sequences. Section 4 compares this method with STAR and IMAGING for evaluation. Section 5 concludes the paper and summarizes the article.

2   Background

2.1   Digital In-Memory Stateful Logic Array

Stateful Logic Gates Based on Memristive Arrays. The digital memristor array consists of horizontal wordlines, vertical bitlines, memristors located at the crosspoints, and peripheral circuits used to drive the bitlines and wordlines, as shown in Fig. 1. By applying voltage or grounding to the selected bitlines and wordlines, the configuration of stateful logic gates within the memory array can be accomplished. As indicated by the blue dashed box in Fig. 1(a), under the applied voltage, devices A, B, and C form a stateful logic gate. The initial resistance state of device C is either high or low. The voltage drop across device C is determined by the resistances of devices A and B, as well as the series resistance R_S. When the resistance states of A and B are in certain combinations, the voltage drop across device C will exceed/fall below its set/reset voltage, causing the formation/breakage of conductive filaments within device C and a change in its resistance state. If the high resistance state is defined as logic '0' and the low resistance state as logic '1', a logical relationship can be established between the transformed resistance state of device C and the initial resistance states of devices A and B. By adjusting the magnitude of the applied control voltage, different conditional transformation operations can be realized, thereby implementing stateful logic gates with various logical functions. Figure 1 (b) illustrates the implementation of an AND gate. Figure 2 presents the SPICE simulation results of various stateful logic gates based on the Stanford memristor device model [7].

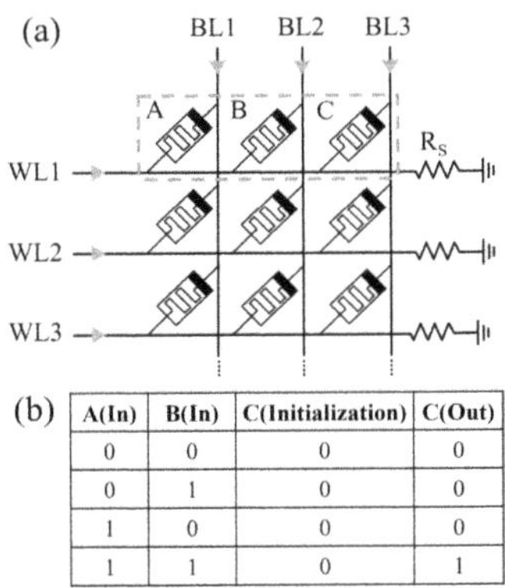

A(In)	B(In)	C(Initialization)	C(Out)
0	0	0	0
0	1	0	0
1	0	0	0
1	1	0	1

Fig. 1. The implementation of AND gates on digital memristive array.

The Cascading and Parallelism of Logic Gates. The design of cascaded logic gates based on memristors is the core of in-memory computing. After decomposing complex computations into combinations of simple logic gates, the input and output devices of the logic gates are configured in the same array row/column. The output of each logic gate is stored in the device as the input of the next logic gate. By applying different voltage combinations in a time-sequenced manner, complex computations can be completed. This process is known as cascading. As shown in the Fig. 3, this is an example of implementing a one-bit full adder through the cascading of in-memory stateful logic gates. Here, a and b are the addends, cin_0 is the carry-in from the lower bit, C is the carry-out, S is the sum, and the number of execution steps is 7.

When multiple array rows are activated, applying a fixed voltage to the bitlines of the array can simultaneously trigger logic gates in each activated row. This means that logic gates of the same type and at the same position in different rows can be executed in parallel within a single clock cycle, as shown in Fig. 4. Therefore, digital processing-in-memory (PIM) supports large-scale parallelism, significantly enhancing computational throughput.

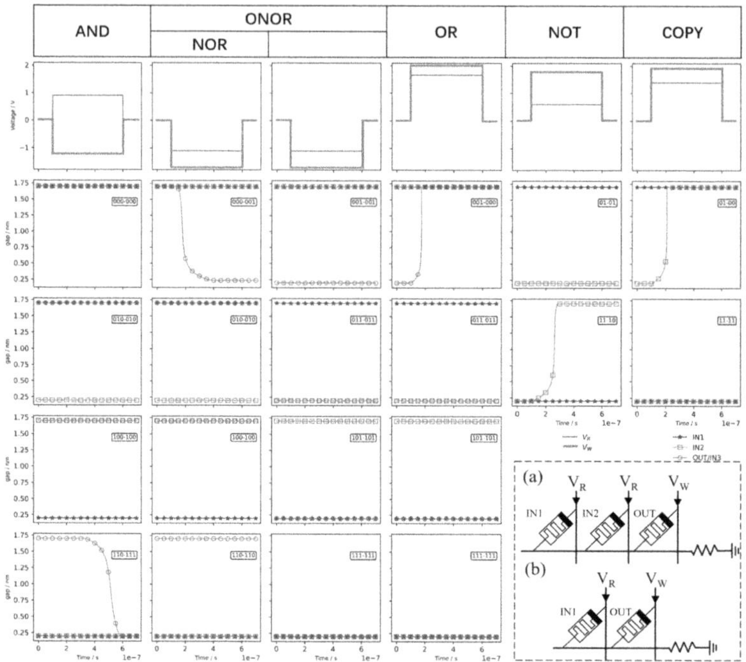

Fig. 2. SPICE verification results of various stateful logic gates. The first four columns represent PMR-two-3AND, PMR-two-3NOR, PMR-three-3ONOR, and PMR-two-3OR, corresponding to (a); the last two columns represent PMR-one-2NOT and PMR-one-2COPY, corresponding to (b). The first row indicates the applied voltages V_R and V_W, and the subsequent rows show the resistance changes corresponding to different resistance input combinations (represented by gap length for resistance).

2.2 Convolution

The essence of convolution is to achieve the weighted superposition operation of two functions through integration or summation. This operation dynamically extracts local features through a sliding window mechanism. In traditional fields, convolution is widely applied. With the development of deep learning, convolution, as the basic operator of Convolutional Neural Networks (CNN), realizes the hierarchical extraction and expression of features through the cascading of multiple convolution kernels. In Binary Neural Networks, where the weights and activations are either 0 or 1, the original multiply-accumulate operations in the convolutional and fully connected layers can be realized by logic operations (XNOR) and bit counting (popcount) operations.

The convolutional layer extracts features through local connectivity and weight sharing, transforming the input data into lower-dimensional features while preserving the spatial structure of the input data. In contrast, the fully connected layer performs classification or regression operations on the input features, where each input neuron is connected to each output neuron, disregarding

cycle	Gate	a	b	cin_0	N6	N7	N8	N9	C	N10	S
0		a	b	cin_0	N6	N7	N8	N9	C	N10	S
1	AND	IN	IN			OUT					
2	NOR	IN	IN		OUT						
3	NOR				IN	IN	OUT				
4	AND			IN			IN	OUT			
5	OR					IN		IN	OUT		
6	NOR			IN			IN			OUT	
7	NOR							IN		IN	OUT

Fig. 3. Implementation of a 1-bit full adder using Logic-in-memory.

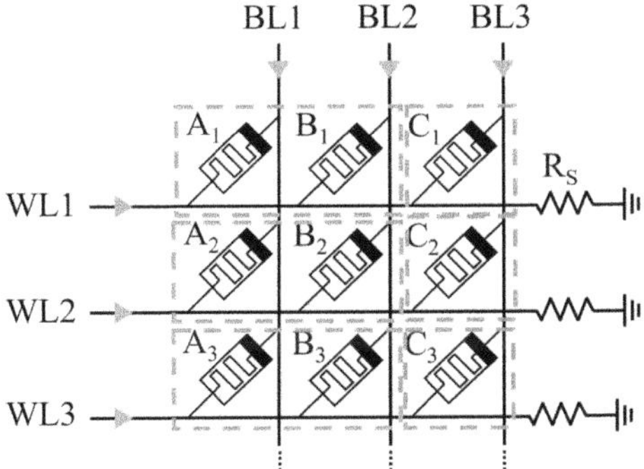

Fig. 4. Parallel execution of the logical operations $C_i = AND(A_i, B_i)$, i=1,2,3.

the spatial structure of the original data. When implementing both convolutional and fully connected layers within an in-memory stateful logic array, the fully connected layer requires only that the image data and weights be stored in the same row or column of the array in a one-to-one correspondence to meet its computational needs. However, during the sliding of the convolutional window, the input data participates in multiple computations. Thus, the storage of data and execution for the convolutional layer are more complex than those for the fully connected layer.

The IMAGING method stores image data in the array according to the position of the pixels, as shown in Fig. 5. By using redundant storage to eliminate data dependencies, IMAGING enables parallel computation of column-wise MACs. In the second step, partial MAC results are obtained and then copied to the first row corresponding to the MAC input data to obtain the final MAC results. When performing convolution with a convolution kernel of size 3*3 and stride 1, 6 rows of image data execute 4 MAC operations, actually occupying 12 rows of the array.

The STAR method, like IMAGING, stores image data in the array according to the position of the pixels. To improve the area utilization of the array, STAR

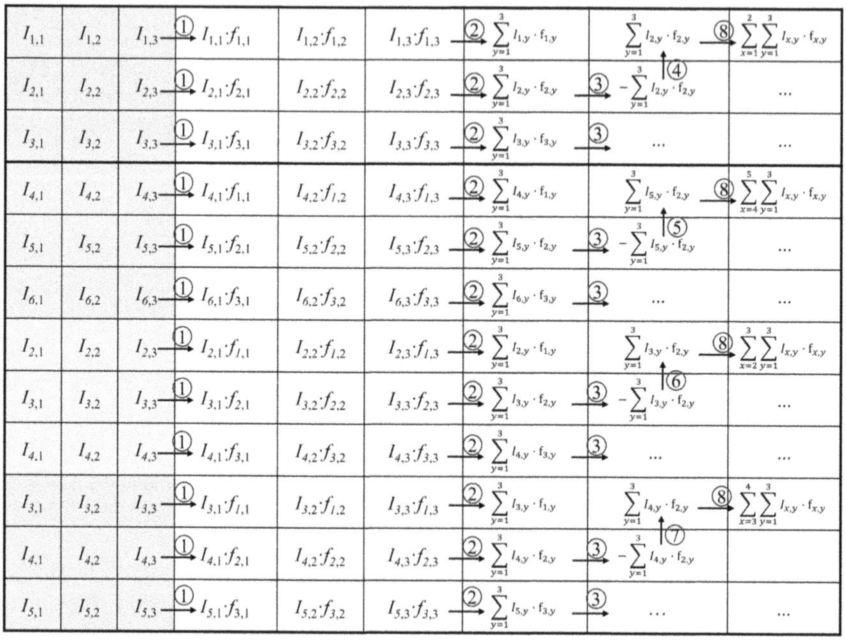

$I_{1,1}$	$I_{1,2}$	$I_{1,3}$ —①→ $I_{1,1}{\cdot}f_{1,1}$	$I_{1,2}{\cdot}f_{1,2}$	$I_{1,3}{\cdot}f_{1,3}$ —②→ $\sum_{y=1}^{3} I_{1,y}\cdot f_{1,y}$	$\sum_{y=1}^{3} I_{2,y}\cdot f_{2,y}$ —⑧→	$\sum_{x=1}^{2}\sum_{y=1}^{3} I_{x,y}\cdot f_{x,y}$
$I_{2,1}$	$I_{2,2}$	$I_{2,3}$ —①→ $I_{2,1}{\cdot}f_{2,1}$	$I_{2,2}{\cdot}f_{2,2}$	$I_{2,3}{\cdot}f_{2,3}$ —②→ $\sum_{y=1}^{3} I_{2,y}\cdot f_{2,y}$	—③→ $-\sum_{y=1}^{3} I_{2,y}\cdot f_{2,y}$ ↑④	...
$I_{3,1}$	$I_{3,2}$	$I_{3,3}$ —①→ $I_{3,1}{\cdot}f_{3,1}$	$I_{3,2}{\cdot}f_{3,2}$	$I_{3,3}{\cdot}f_{3,3}$ —②→ $\sum_{y=1}^{3} I_{3,y}\cdot f_{3,y}$	—③→ ...	...
$I_{4,1}$	$I_{4,2}$	$I_{4,3}$ —①→ $I_{4,1}{\cdot}f_{1,1}$	$I_{4,2}{\cdot}f_{1,2}$	$I_{4,3}{\cdot}f_{1,3}$ —②→ $\sum_{y=1}^{3} I_{4,y}\cdot f_{1,y}$	$\sum_{y=1}^{3} I_{5,y}\cdot f_{2,y}$ —⑧→	$\sum_{x=4}^{5}\sum_{y=1}^{3} I_{x,y}\cdot f_{x,y}$
$I_{5,1}$	$I_{5,2}$	$I_{5,3}$ —①→ $I_{5,1}{\cdot}f_{2,1}$	$I_{5,2}{\cdot}f_{2,2}$	$I_{5,3}{\cdot}f_{2,3}$ —②→ $\sum_{y=1}^{3} I_{5,y}\cdot f_{2,y}$	—③→ $-\sum_{y=1}^{3} I_{5,y}\cdot f_{2,y}$ ↑⑤	...
$I_{6,1}$	$I_{6,2}$	$I_{6,3}$ —①→ $I_{6,1}{\cdot}f_{3,1}$	$I_{6,2}{\cdot}f_{3,2}$	$I_{6,3}{\cdot}f_{3,3}$ —②→ $\sum_{y=1}^{3} I_{6,y}\cdot f_{3,y}$	—③→ ...	...
$I_{2,1}$	$I_{2,2}$	$I_{2,3}$ —①→ $I_{2,1}{\cdot}f_{1,1}$	$I_{2,2}{\cdot}f_{1,2}$	$I_{2,3}{\cdot}f_{1,3}$ —②→ $\sum_{y=1}^{3} I_{2,y}\cdot f_{1,y}$	$\sum_{y=1}^{3} I_{3,y}\cdot f_{2,y}$ —⑧→	$\sum_{x=2}^{3}\sum_{y=1}^{3} I_{x,y}\cdot f_{x,y}$
$I_{3,1}$	$I_{3,2}$	$I_{3,3}$ —①→ $I_{3,1}{\cdot}f_{2,1}$	$I_{3,2}{\cdot}f_{2,2}$	$I_{3,3}{\cdot}f_{2,3}$ —②→ $\sum_{y=1}^{3} I_{3,y}\cdot f_{2,y}$	—③→ $-\sum_{y=1}^{3} I_{3,y}\cdot f_{2,y}$ ↑⑥	...
$I_{4,1}$	$I_{4,2}$	$I_{4,3}$ —①→ $I_{4,1}{\cdot}f_{3,1}$	$I_{4,2}{\cdot}f_{3,2}$	$I_{4,3}{\cdot}f_{3,3}$ —②→ $\sum_{y=1}^{3} I_{4,y}\cdot f_{3,y}$	—③→ ...	...
$I_{3,1}$	$I_{3,2}$	$I_{3,3}$ —①→ $I_{3,1}{\cdot}f_{1,1}$	$I_{3,2}{\cdot}f_{1,2}$	$I_{3,3}{\cdot}f_{1,3}$ —②→ $\sum_{y=1}^{3} I_{3,y}\cdot f_{1,y}$	$\sum_{y=1}^{3} I_{4,y}\cdot f_{2,y}$ —⑧→	$\sum_{x=3}^{4}\sum_{y=1}^{3} I_{x,y}\cdot f_{x,y}$
$I_{4,1}$	$I_{4,2}$	$I_{4,3}$ —①→ $I_{4,1}{\cdot}f_{2,1}$	$I_{4,2}{\cdot}f_{2,2}$	$I_{4,3}{\cdot}f_{2,3}$ —②→ $\sum_{y=1}^{3} I_{4,y}\cdot f_{2,y}$	—③→ $-\sum_{y=1}^{3} I_{4,y}\cdot f_{2,y}$ ↑⑦	...
$I_{5,1}$	$I_{5,2}$	$I_{5,3}$ —①→ $I_{5,1}{\cdot}f_{3,1}$	$I_{5,2}{\cdot}f_{3,2}$	$I_{5,3}{\cdot}f_{3,3}$ —②→ $\sum_{y=1}^{3} I_{5,y}\cdot f_{3,y}$	—③→ ...	...

Fig. 5. The storage and execution methods of IMAGING.

does not use redundant storage for image data. Instead, it needs to select MAC operations for parallel computation that have no data dependencies each time, as illustrated by the two MACs in Fig. 6. The convolution kernel is loaded onto the array rows as control information, so it does not require additional area in the array rows. In the second step, partial MAC results are obtained. These partial results are then copied to the first row corresponding to the MAC input data to continue the computation, thereby obtaining the results of the first round of MAC operations. Subsequently, the process of selecting MAC operations continues between rows until all MAC operations are completed, yielding the final convolution results.

3 Method

In this chapter, we elaborate on the entire process of HT-DLC for designing the initial layout and efficient execution of convolution in the stateful logic array. The flowchart is shown in the Fig. 7. Firstly, based on the characteristics of the convolution kernel, we designed the storage and layout methods for the individual MAC that constitute the convolution. Then, we performed synthesis and mapping of a single convolution computation, namely Multiply-Accumulate(MAC), to obtain the optimal execution sequence. In this way, we obtained an efficient initial data layout and execution scheme for a MAC within a single array row.

$I_{1,1}$	$I_{1,2}$	$I_{1,3}$ —①→ $I_{1,1}{\cdot}f_{1,1}$	$I_{1,2}{\cdot}f_{1,2}$	$I_{1,3}{\cdot}f_{1,3}$ —②→ $\sum_{y=1}^{3} I_{1,y}\cdot f_{1,y}$	$\sum_{y=1}^{3} I_{2,y}\cdot f_{2,y}$ —⑥→ $\sum_{x=1}^{2}\sum_{y=1}^{3} I_{x,y}\cdot f_{x,y}$
$I_{2,1}$	$I_{2,2}$	$I_{2,3}$ —①→ $I_{2,1}{\cdot}f_{2,1}$	$I_{2,2}{\cdot}f_{2,2}$	$I_{2,3}{\cdot}f_{2,3}$ —②→ $\sum_{y=1}^{3} I_{2,y}\cdot f_{2,y}$ —③→ $-\sum_{y=1}^{3} I_{1,y}\cdot f_{1,y}$ ④	...
$I_{3,1}$	$I_{3,2}$	$I_{3,3}$ —①→ $I_{3,1}{\cdot}f_{3,1}$	$I_{3,2}{\cdot}f_{3,2}$	$I_{3,3}{\cdot}f_{3,3}$ —②→ $\sum_{y=1}^{3} I_{3,y}\cdot f_{3,y}$ —③→	...	...
$I_{4,1}$	$I_{4,2}$	$I_{4,3}$ —①→ $I_{4,1}{\cdot}f_{1,1}$	$I_{4,2}{\cdot}f_{1,2}$	$I_{4,3}{\cdot}f_{1,3}$ —②→ $\sum_{y=1}^{3} I_{4,y}\cdot f_{1,y}$	$\sum_{y=1}^{3} I_{5,y}\cdot f_{2,y}$ —⑥→ $\sum_{x=1}^{2}\sum_{y=1}^{3} I_{x,y}\cdot f_{x,y}$
$I_{5,1}$	$I_{5,2}$	$I_{5,3}$ —①→ $I_{5,1}{\cdot}f_{2,1}$	$I_{5,2}{\cdot}f_{2,2}$	$I_{5,3}{\cdot}f_{2,3}$ —②→ $\sum_{y=1}^{3} I_{5,y}\cdot f_{2,y}$ —③→ $-\sum_{y=1}^{3} I_{5,y}\cdot f_{2,y}$ ⑤	...
$I_{6,1}$	$I_{6,2}$	$I_{6,3}$ —①→ $I_{6,1}{\cdot}f_{3,1}$	$I_{6,2}{\cdot}f_{3,2}$	$I_{6,3}{\cdot}f_{3,3}$ —②→ $\sum_{y=1}^{3} I_{6,y}\cdot f_{3,y}$ —③→	...	...

Fig. 6. The storage and execution methods of STAR.

Finally, we extended the method to the layout and data storage of multiple MACs within a single array row. The specific methods are as follows.

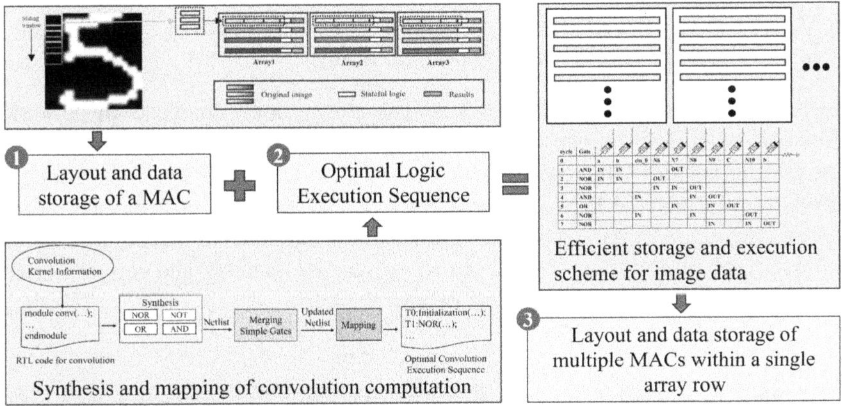

Fig. 7. An Overview of HT-DLC.

3.1 Layout and Data Storage of a MAC

For each MAC operation that constitutes the convolutional layer, existing implementation methods vary. One approach involves performing multiplication between image elements and the corresponding elements of the convolution kernel, followed by addition. Tommaso Zanotti et al. designed a half-adder chain for the execution of addition and further improved it by proposing a binary adder tree architecture based on SIMPLY. Compared to the half-adder chain, the performance of the adder tree is enhanced, but the overhead of executing a single addition increases with the height of the tree [26]. Moreover, since all data are always involved in the computation, there is significant computational redundancy. Another approach involves storing data in the array according to pixel positions and dividing the MAC process into several partial MAC and a final addition based on the rows where the data are located. Specifically, the process is first divided into H partial MAC according to the height H of the filter, which are executed in parallel between array rows. Then, these partial results are moved to an idle position on the array row, copied row by row to the first row belonging to that MAC, and added one by one to obtain the final MAC result [5].

Regarding the data reuse issue in convolution calculations between rows of image data, IMAGING addresses it by redundantly storing the rows that are reused by the convolution kernel on the array rows. In contrast, the STAR method, aiming to save device area, stores the image data in the array only once and completes the convolutional layer by performing MAC operations that do not involve data reuse each time. The performance of array execution is limited by two aspects:

1. Inter-row copying cannot be performed in parallel and must be done row by row, which has a significant impact on computational delay.
2. In the second stage of execution, the parallelism of array rows is not fully utilized, affecting the throughput of the array.

Therefore, we concatenate the rows of the data block selected by the sliding window end-to-end and store them in a single row of the array, as shown in Fig. 8. As the sliding window moves, there is data dependency between the image data required by adjacent MACs, and this part of the data is redundantly stored in the array. However, during the execution of a single MAC, there is no need for data movement or copying. The array rows execute exactly the same operations, which can maximize the parallelism of the array.

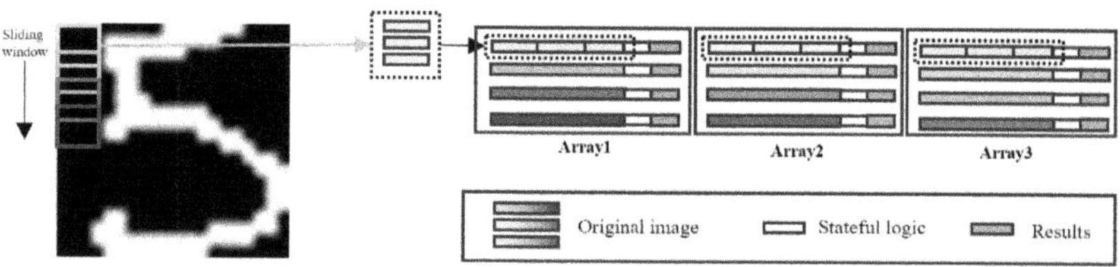

Fig. 8. The storage of image data in the array.

3.2   Synthesis and Mapping of Convolution Computation

This section describes our method for optimizing the logic execution sequence. The STAR synthesis flow, in order to save space for data storage, actively inserts erasing operations between execution sequences on the one hand to reduce the number of devices used for stateful logic. On the other hand, it prioritizes the execution of logic operations corresponding to nodes that are easier to erase, thereby achieving lower devices for stateful logic and higher array area utilization. As described in Sect. 3.1, when we store the sub-operation of convolution, namely the MAC, in a single row of the array, we can consider the MAC computation as a whole and use stateful logic synthesis and mapping to obtain an efficient stateful logic execution sequence.

The synthesis and mapping process is shown in Fig. 9. First, a complete logic family is selected, such as the logic family composed of PMR-two-3NOR, PMR-two-3OR, PMR-one-2NOT, and PMR-two-3AND. A CMOS-oriented stateful logic synthesis tool is used to transform the RTL code expressing complex computations, such as the code for performing convolution operations, into a gate-level netlist composed of simple logic gates. Subsequently, the gate-level netlist is optimized by synthesizing simple gates that meet certain conditions into composite gates [6], in order to reduce execution delay. For example, both NOR, OR and NOR, NOR may potentially be combined into a PMR-three-3ONOR. The simple gates that are replaced will not be executed, and their results cannot serve as inputs to other logic gates. At the same time, inputs in a composite gate that are covered by the output cannot serve as covered inputs for other composite gates. Finally, the logic sequence is mapped to a single row of the crossbar array, saving area through device reuse to obtain the stateful logic execution sequence. In the convolutional layer, each convolution kernel is applied to the entire image data. When the MAC calculations executed on different array rows use the same convolution kernel, the computations between array rows are fully executed in parallel. Therefore, we incorporate the information of the convolution kernel into the RTL code to customize the logic synthesis sequence for different convolution kernels. On the one hand, this can reduce the complexity of the logic execution sequence. On the other hand, it lowers the storage requirements for convolutional computations.

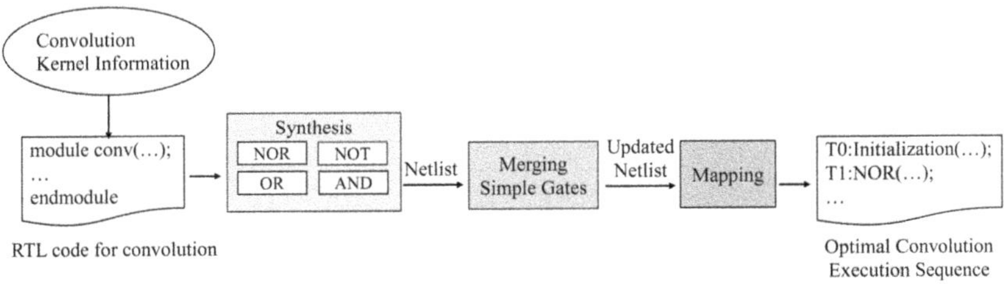

Fig. 9. The synthesis and mapping flow for a single MAC operation.

3.3   Layout and Data Storage of Multiple MACs

Affected by the stride of the convolution kernel, the data used in MAC computations may be reused. Therefore, when selecting multiple MACs to be executed on a single array row, their spatial relationships on the image are taken into consideration. Taking the storage and computation of four MACs as an example, the spatial relationships may be random, consecutive, semi-dense, and dense, as shown in Table 1. The last method, under the condition of storing and computing the same four MACs, occupies the fewest devices and has the highest data reuse ratio. So, when storing data on a single array row, reusing data as much as possible can accommodate the largest number of MACs. To improve the area utilization of the array, we aim to reduce the multiple storage of image data caused by data dependencies between MACs and choose dense arrangement as much as possible to save device area.

Table 1. The scheme for storing and computation four MACs on a single array row.

Arrangement	The shape of the data	The distribution of MACs	Area
Random	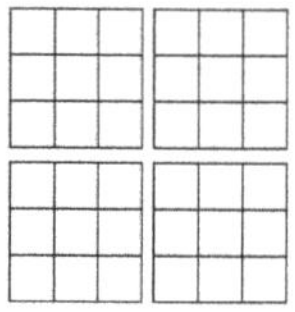	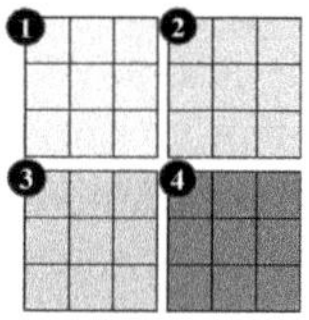	36
Consecutive	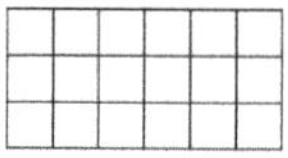	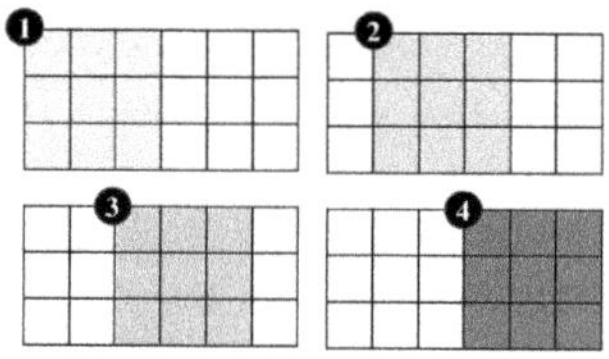	18
Semi-dense	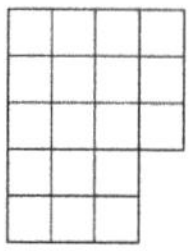	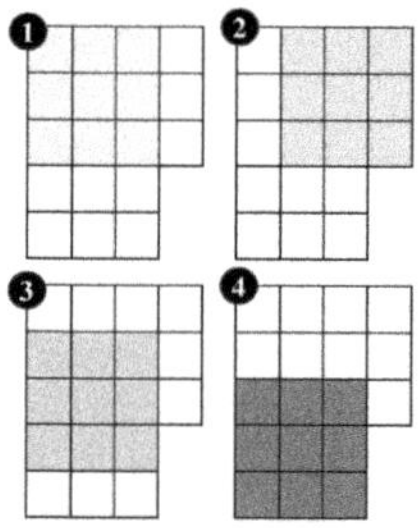	18
Dense	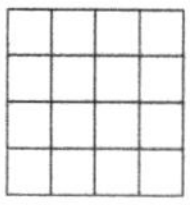	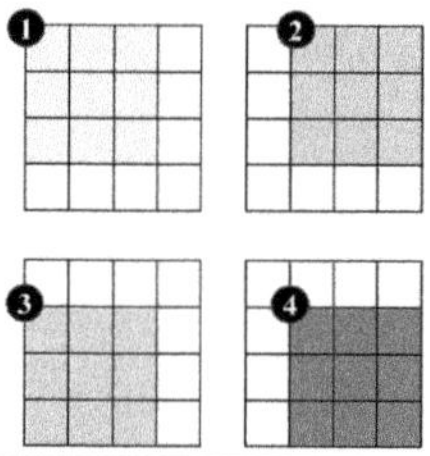	16

We first consider the case where the number of MACs accommodated by the array row is N. There are two optimal MAC arrangement scenarios, as shown in Fig. 10. Considering that the data interchange between i and m has no effect on the result, an additional constraint $i \geq m$ is added. Assuming the convolution kernel size is 3×3 and the stride is 1, the area of the MAC group in Fig. 10(a) is $(i + 2) \times (m + 2) + n + 2$, which is $N + 2 \times (i + m) + 6$, and the area of the MAC group in Fig. 10 (b) is $(i + 2) \times (m + 2)$, which is $N + 2 \times (i + m) + 4$. The problem is then transformed into finding the minimum value of $i + m$. By iterating through $i = [1, \ldots, \lceil \sqrt{N} \rceil]$, the corresponding area is calculated, and the i and m with the smallest area are recorded. In this way, we obtain the area S required for the dense arrangement of any number of MACs.

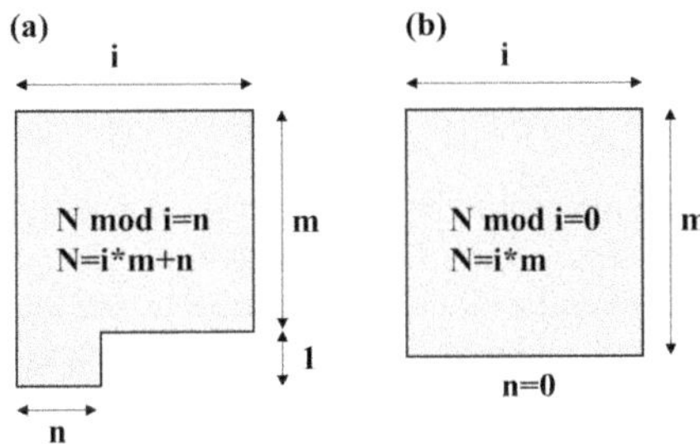

Fig. 10. The MAC arrangement scenario, where N is the number of MACs that the array row can accommodate. i represents the number of MACs arranged horizontally.

Apart from the devices required for storing intermediate results and outputs for each MAC, it is assumed that the remaining space in the array row can store X bits of image data. We iterate through the possible number of MACs j that a single array row can accommodate and compare S_j with X, that is, the number of devices required for the dense arrangement of j MACs with the actual storage space available, until we obtain the storage solution with the fewest required devices for the maximum number of MAC that can be accommodated in a single row of the array.

Next, we need to place the data on the array according to the storage solution. To enable full parallelism of the array rows, we concatenate the rows of data blocks with the same shape end-to-end and place them on different rows of the same array for execution, as shown in Fig. 11. Figure 11 (a) shows the MAC matrix formed by the movement of the sliding window of the convolution kernel over the image data. There is data reuse between adjacent MACs. Five MACs form a MAC group, which is stored in a single row of the array. Although the pink and blue data blocks appear to have the same shape, they are not entirely equivalent when computed on the array because the convolution kernel matrix is not necessarily centrally symmetric. Therefore, data blocks of different shapes need to be stored on different arrays. At any moment during MAC execution, the data activated by the column selection in all array rows are in completely equivalent relative positions within their respective data blocks. This ensures that the array operates at full load with parallel computation.

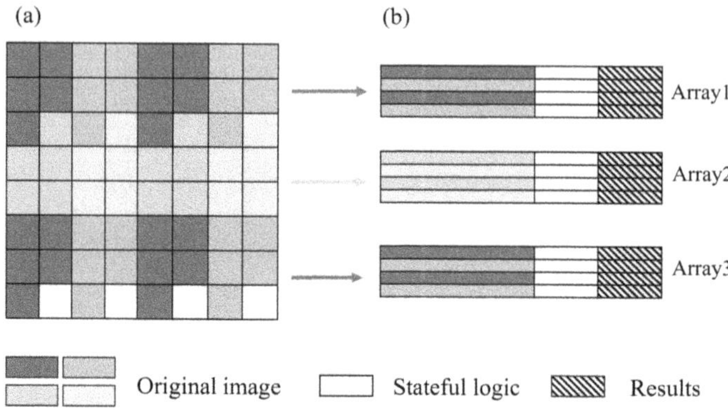

Fig. 11. The position distribution of the five MACs on the image and their storage on the array. In (a), each block represents the data required for a MAC, and there is data dependency between adjacent data blocks. According to the shape of the MAC clusters, the data of MAC groups with the same arrangement are concatenated end-to-end by rows and stored for execution on the same array in (b).

4   Experiments and Evaluation

In this chapter, we compare the execution of convolution in memristor-based crossbar array between HT-DLC and two image processing accelerators based on MAGIC, namely STAR and IMAGING. They both use the MAGIC stateful logic family, which is composed of NOR and NOT logic operations. However, STAR optimizes the synthesis flow based on IMAGING and employs various techniques to improve area utilization. In contrast, the logic family we use consists of AND, NOR, NOT, and OR operations. We reduce delay by optimizing storage and synthesizing composite gates. To obtain the logic execution sequences for the three methods, we decompose the RTL code of the convolution operation into logic gates using a CMOS-oriented logic synthesis tool, then process it with different synthesis flows, and finally map it to array rows. We have verified the impact of array size and convolution complexity on the acceleration of convolution by the three methods.

4.1   Performance Comparison of Binary Convolution Execution on Arrays of Different Sizes

Binary convolution, where the pixel bit-width of the image data is 1, and the bit-width of each element in the convolution kernel is also 1, allows the multiplication to be simplified to a logical XNOR operation. The STAR and IMAGING methods divide the execution of a MAC into three parts: multiplication (XNOR), addition 1, and addition 2. In contrast, our proposed method treats the MAC as a single unit. Table 2 lists the relevant parameters for executing a MAC using the three methods. M_{STAR} and $M_{IMAGING}$ denote the number of MACs filled in the column direction for the two methods, while N_{STAR} and $N_{IMAGING}$ represent the number of MACs filled in the row direction. Our method uses the approach described in Sect. 3.1 to store the data required by the MAC in a single array row, eliminating the need for replication operations.

Table 2. Steps of a single MAC and the area used for stateful logic in binary convolution with a 3×3 convolution kernel.

	Calculation type	Area for Stateful logic	Steps for computation	Steps for intra-row replication	Steps for inter-row replication
	Xnor (MUL)	7	39		
STAR	Add1	8	23	$2*N_{STAR}$	$2*M_{STAR}*N_{STAR}$
	Add2	8	28		
	Xnor (MUL)	27	28		
IMAGING	Add1	18	19	$2*N_{IMAGING}$	$2*M_{IMAGING}$
	Add2	22	23		
Ours	Conv	18	57	0	0

We conducted binary convolution using a 3*3 convolution kernel with a stride of 1 on arrays of various sizes, ranging from 32×32 to 1024×1024. We compared the three methods from three different aspects:

Area Utilization. We revised the definition of area utilization for STAR [21] to the actual proportion of data stored in a single array, excluding any redundantly stored or replicated data. As shown in Fig. 12(a), for each array size, the throughput is normalized to the result of STAR. When the array is small, the area utilization of HT-DLC method lags behind that of the STAR method because the array is too small to provide enough devices for the complete MAC to execute stateful logic. As the array size increases, the gap in area utilization between the two methods gradually narrows. When the array size is 1024×1024, the difference in area utilization between the two methods is 3.51%. The IMAGING method, which does not use an erasing operation to save the number of devices required for stateful logic and intermediate result storage, and redundantly stores image data, has a lower area utilization.

Throughput. Throughput is measured as the number of MAC operations executed per cycle. In Fig. 12(b), the throughput for each array size is normalized to the result of our proposed method. The throughput of our method is labeled above the bars. As the array size increases, the throughput of all three methods increases, but the growth rates of IMAGING and STAR are much slower than that of HT-DLC.

Average Delay per MAC. As shown in Fig. 12(c), the average delay per MAC execution decreases as the array size increases. The average delay of the HT-DLC method is significantly lower than that of the other two methods. STAR improves area utilization by reducing the area for redundant inputs. However, due to data dependencies between adjacent MACs in convolution calculations, STAR requires multiple rounds of computation to complete the convolution corresponding to the data stored in the array, whereas both IMAGING and our proposed method require only one round of computation.

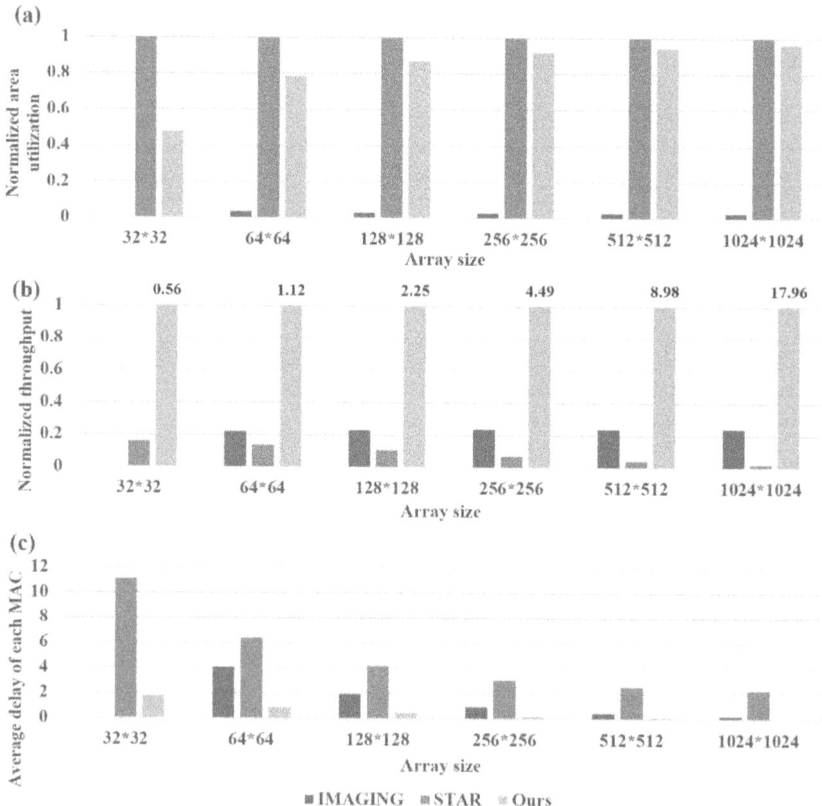

Fig. 12. Comparison of performance between STAR, IMAGING, and Ours.

The results indicate that our proposed method, compared to the optimal method STAR, achieves higher throughput and lower execution delay with only a minor increase in area utilization, especially pronounced on larger arrays.

4.2 Performance Comparison of Arrays Under Different Convolution Kernel Bit-Widths

This section investigates the performance of arrays executing convolution operations with varying complexities. We selected grayscale images with 8-bit pixels and conducted experiments on convolutional layers with filters of 1-bit, 2-bit, 3-bit, and 4-bit. For array sizes of 512×512 and 1024×1024, we measured the throughput, average delay for a MAC, and area utilization of the three methods. When the array size was 512×512, the IMAGING method could not complete data storage and execution for filters of 3-bit and 4-bit. We compared the three methods from three aspects:

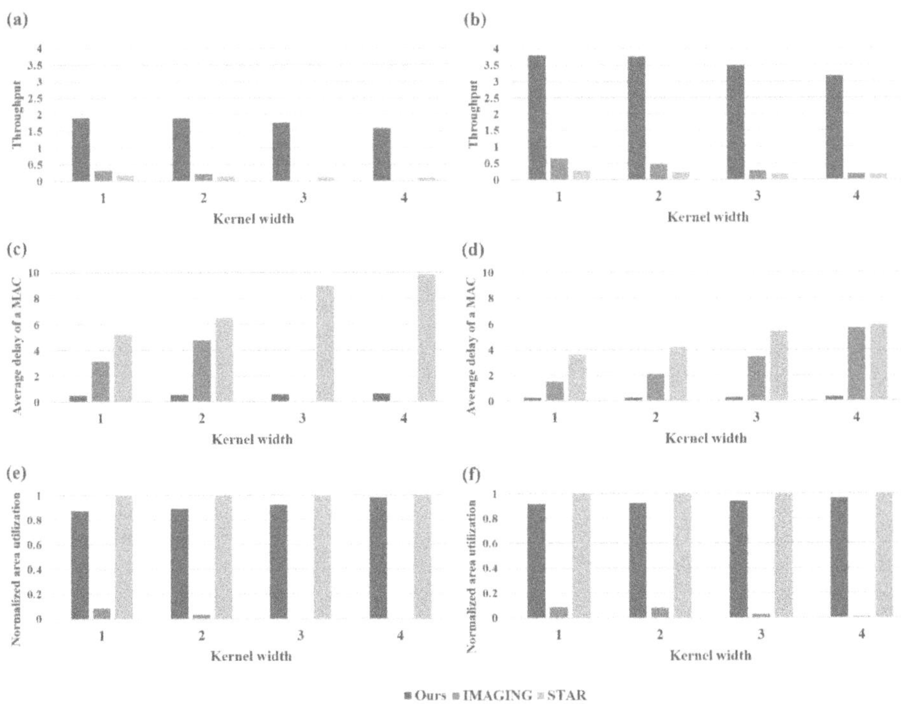

Fig. 13. Comparison of performance between STAR, IMAGING, and Ours, where (a), (c) and (e) are executed with array size of 512*512 and (b), (d) and (f) are executed with array size of 1024*1024.

Throughput. As shown in Fig. 13(a) and (b), the throughput decreases as the bit-width of the convolution kernel elements increases. Larger array sizes yield higher throughput. On average, the throughput of our proposed method is 15.11 times that of the STAR method.

Average Delay per MAC. As shown in Fig. 13(c) and (d), we compare the average delay per MAC execution for the three methods during convolution operations. As can be seen from the figure, the average delay of our proposed method is very low, with a reduction of 93.11% compared to that of STAR.

Area Utilization. As shown in Fig. 13(e) and (f), we normalized the data to the results of STAR and compared the area utilization of the three methods. Relatively speaking, the more complex the computation, the closer the area utilization of our proposed method is to that of the STAR method. According to statistics, the average loss in area utilization is 7.28%.

5 Conclusion

Existing logic-in-memory implementations for neural network convolution layer have not fully utilized the parallelism of array rows, wasting computational resources of the array. This paper proposes HT-DLC, a design for convolution acceleration based on logic-in-memory. HT-DLC designs a new image data storage method, which is extended from the storage of a single MAC to the selection and arrangement of multiple MACs. This approach fully leverages the parallelism of the array while maximizing the area utilization of image data. Additionally, HT-DLC optimizes the logic execution sequence by synthesizing simple logic gates into composite gates, reducing latency and thereby enhancing the throughput of arrays. Compared to the existing image convolution acceleration method with STAR, HT-DLC fully utilizes the computational power and parallelism of array rows, significantly enhances the array throughput and reduces the computational delay of a single MAC with only a minor reduction in area utilization.

Acknowledgments. This work is supported by Key Laboratory of Advanced Microprocessor Chips and Systems of NUDT. This work is supported by National Natural Science Foundation of China under Grant 62202483, and the research foundation from NUDT under Grant 22-ZZCX-046-02.

References

1. Ahn, J., Hong, S., Yoo, S., Mutlu, O., Choi, K.: A scalable processing-in-memory accelerator for parallel graph processing. In: Proceedings of the 42nd Annual International Symposium on Computer Architecture, pp. 105–117. ISCA'15, ACM (2015)
2. Angizi, S., He, Z., Parveen, F., Fan, D.: Imce: Energy-efficient bit-wise in-memory convolution engine for deep neural network. In: 2018 23rd Asia and South Pacific Design Automation Conference (ASP-DAC), pp. 111–116. IEEE (2018)
3. Castaneda, O., Bobbett, M., Gallyas-Sanhueza, A., Studer, C.: Ppac: a versatile in-memory accelerator for matrix-vector-product-like operations. In: 2019 IEEE 30th International Conference on Application-specific Systems, Architectures and Processors (ASAP), pp. 149–156. IEEE (2019)

4. Fujiwara, H., et al.: A 5-nm 254-tops/w 221-tops/mm2 fully-digital computing-in-memory macro supporting wide-range dynamic-voltage-frequency scaling and simultaneous mac and write operations. In: 2022 IEEE International Solid- State Circuits Conference (ISSCC), IEEE (2022)
5. Haj-Ali, A., Ben-Hur, R., Wald, N., Ronen, R., Kvatinsky, S.: Imaging: In-memory algorithms for image processing. IEEE Transactions on Circuits & Systems. Part I: Regular Papers pp. 4258–4271 (2018)
6. Hu, Y., Xu, N., Feng, C., Tong, W., Liu, K., Fang, L.: Losss-logic synthesis based on several stateful logic gates for high time-efficient computing. In: 2024 29th Asia and South Pacific Design Automation Conference (ASP-DAC), pp. 805–811. IEEE (2024)
7. Jiang, Z., et al.: A compact model for metal-oxide resistive random access memory with experiment verification. IEEE Trans. Electr. Devices pp. 1884–1892 (2016)
8. Jin, H., et al.: Rehy: A reram-based digital/analog hybrid pim architecture for accelerating cnn training. IEEE Trans. Parallel Distributed Syst. 1–1 (2021)
9. Kim, T., Jang, Y., Kang, M.G., Park, B.G., Lee, K.J., Park, J.: Sot-mram digital pim architecture with extended parallelism in matrix multiplication. IEEE Trans. Comput. **71**(11), 2816–2828 (2022)
10. Lee, C.F., et al.: A 12nm 121-tops/w 41.6-tops/mm2 all digital full precision sram-based compute-in-memory with configurable bit-width for ai edge applications. In: 2022 IEEE Symposium on VLSI Technology and Circuits (VLSI Technology and Circuits), IEEE (2022)
11. Lehtonen, E.e., Poikonen, J.H., Laiho, M.: Two memristors suffice to compute all boolean functions. Electr. Lett. 230–231 (2010)
12. Leitersdorf, O., Boneh, Y., Gazit, G., Ronen, R., Kvatinsky, S.: Fourierpim: high-throughput in-memory fast fourier transform and polynomial multiplication. Memories Mater. Devices Circuits Syst. **4**, 100034 (2023)
13. Leitersdorf, O., Ronen, R., Kvatinsky, S.: Convpim: evaluating digital processing-in-memory through convolutional neural network acceleration. arXiv (2023)
14. Long, Y., et al.: A ferroelectric fet-based processing-in-memory architecture for dnn acceleration. IEEE J. Exploratory Solid-State Comput. Devices Circuits **5**(2), 113–122 (2019)
15. Manglik, A., et al.: Neon: Enabling efficient support for nonlinear operations in resistive ram-based neural network accelerators. arXiv p. 11 (2022)
16. Oved, B., Leitersdorf, O., Ronen, R., Kvatinsky, S.: Hashpim: high-throughput sha-3 via memristive digital processing-in-memory. In: 2022 11th International Conference on Modern Circuits and Systems Technologies (MOCAST), pp. 1–4. IEEE (2022)
17. Pakala, A., Chen, Z., Yang, K.: Mbsntt: a highly parallel digital in-memory bit-serial number theoretic transform accelerator. IEEE Trans. Very Large Scale Integr. (VLSI) Syst. **33**(2), 537–545 (2025)
18. Perach, B., Ronen, R., Kimelfeld, B., Kvatinsky, S.: Understanding bulk-bitwise processing in-memory through database analytics. IEEE Trans. Emerg. Top. Comput. **12**(1), 7–22 (2024)
19. Tu, F., et al.: Sdp: Co-designing algorithm, dataflow, and architecture for in-sram sparse nn acceleration. IEEE Trans. Comput. Aided Des. Integr. Circuits Syst. **42**(1), 109–121 (2023)
20. Wang, D., Lin, C.T., Chen, G.K., Knag, P., Krishnamurthy, R.K., Seok, M.: Dimc: 2219tops/w 2569f2/b digital in-memory computing macro in 28nm based on approximate arithmetic hardware. In: 2022 IEEE International Solid- State Circuits Conference (ISSCC), pp. 266–268. IEEE (2022)

21. Wang, F., et al.: Star: synthesis of stateful logic in rram targeting high area utilization. IEEE Trans. Comput. Aided Des. Integr. Circuits Syst. **40**(5), 864–877 (2021)
22. Wang, H., Zhao, Y., Li, C., Wang, Y., Lin, Y.: A new mram-based process in-memory accelerator for efficient neural network training with floating point precision. In: 2020 IEEE International Symposium on Circuits and Systems (ISCAS) (2020)
23. Wu, H., et al.: A 28-nm 19.9-to-258.5-tops/w 8b digital computing-in-memory processor with two-cycle macro featuring winograd-domain convolution and macro-level parallel dual-side sparsity. IEEE J. Solid-State Circuits **60**(1), 347–361 (2025)
24. Xu, N., Park, T., Yoon, K.J., Hwang, C.S.: In-memory stateful logic computing using memristors: gate, calculation, and application. Phys. Status Solidi-Rapid Res. Lett. 1–22 (2021)
25. Yan, B., et al.: A 1.041-mb/mm2 27.38-tops/w signed-int8 dynamic-logic-based adc-less sram compute-in-memory macro in 28nm with reconfigurable bitwise operation for ai and embedded applications. In: 2022 IEEE International Solid- State Circuits Conference (ISSCC), IEEE (2022)
26. Zanotti, T., Pavan, P., Puglisi, F.M.: Study of RRAM-Based Binarized Neural Networks Inference Accelerators Using an RRAM Physics-Based Compact Model. IntechOpen (2023)
27. Zhang, Y., et al.: Time-domain computing in memory using spintronics for energy-efficient convolutional neural network. IEEE Trans. Circuits Syst. I Regul. Pap. **68**(3), 1193–1205 (2021)
28. Zou, X., Xu, S., Chen, X., Yan, L., Han, Y.: Breaking the von neumann bottleneck: architecture-level processing-in-memory technology. Sci. China (Inf. Sci.) 60–69 (2021)

Software Systems and Programming Models

GRIS: A Graph Neural Network-Based Reinforcement Learning Approach for Instruction Scheduling

Qijun Zheng and Youmeng Li[✉]

College of Intelligence and Computing, Tianjin University, Tianjin 300072, China
{zqjzqj,liyoumeng}@tju.edu.cn

Abstract. To enhance the instruction-level parallelism and pipeline utilization of the high-performance vector accelerator FT-Matrix, this paper proposes an automatic instruction scheduling method for assembly instructions based on Graph Neural Networks (GNN) and Reinforcement Learning (RL). The method models the instructions and their dependencies within a basic block as a Directed Acyclic Graph (DAG). It leverages GNN to compress the instruction state space, extract dependency features, and generate a low-dimensional global embedding representation. Building on this foundation, a Deep Q-Network (DQN)-driven reinforcement learning framework is designed to dynamically learn the optimal scheduling policy through a multi-objective reward function that includes the number of parallel instructions, critical path optimization, execution cycle, type balance, and task completion incentives.

Experiments conducted on a set of operators (covering compute-intensive, data transfer, and control-flow operations) of the FT-Matrix3000 accelerator demonstrate the effectiveness of the proposed method. Compared to the manually optimized baseline, the instruction execution cycle is reduced by up to 9.6%, and the pipeline utilization for control-flow operators is increased from 36.2% to 41.8%. This method overcomes the limitations of traditional heuristic algorithms, which are prone to local optima and high computational overhead, and provides an end-to-end adaptive solution for efficient instruction scheduling in AI-accelerated hardware.

Keywords: Instruction Scheduling · Graph Neural Network · Reinforcement Learning · Vector Accelerator

1 Introduction

Artificial Intelligence emerged in the middle of the 20th century and has witnessed unprecedented development over the past decade or so. In particular, the rise of deep learning technology in recent years, especially the wide application of the Transformer architecture [7] [6], has become a key factor driving the rapid development of models. Model performance has a log-linear relationship with the

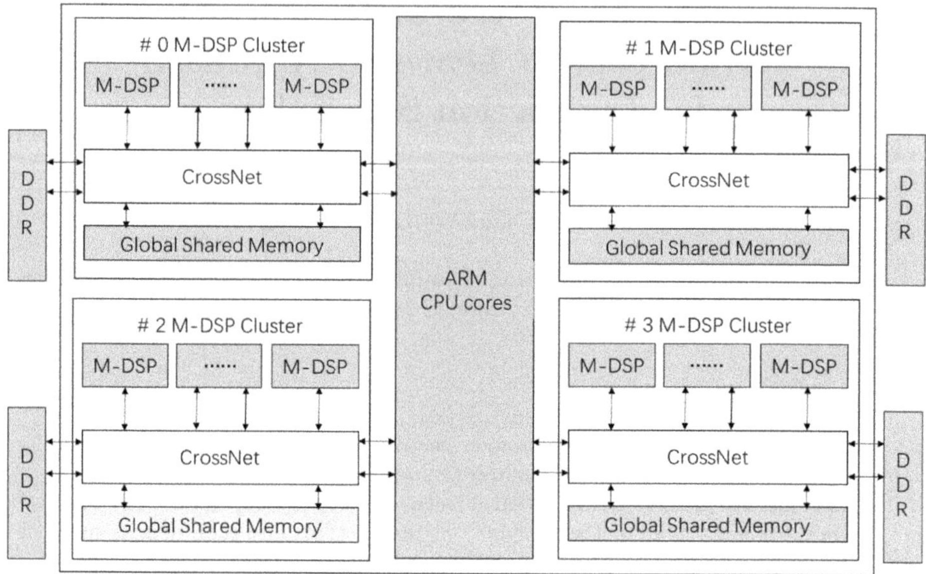

Fig. 1. FT-Matrix Heterogeneous Accelerator.

number of parameters, the dataset size, and the total amount of computational power [5]. Computational power has become a key factor in improving model performance. In order to meet the growing demand for computational power, AI acceleration technology has undergone profound changes at both the hardware and algorithm levels [14].

At the algorithmic level, AI acceleration technology comprehensively improves the efficiency and performance of models through systematic optimization methods, from inference, model optimization to distributed computing and other dimensions [2]. At the model level, techniques such as quantization, sparsification, and knowledge distillation further optimize model efficiency by reducing model storage requirements and computational complexity [4,11]. At the hardware level, various innovative architectures continue to emerge to meet the growing demand for computational power. Among them, the FT-Matrix [1], shown in Fig. 1 is a heterogeneous accelerator for high-performance computing, which is characterized by multiple M-DSP cores, support for single-instruction multiple-data (SIMD) streams and very long instruction words (VLIW), and a multilevel storage structure. Its powerful parallel computing capability has great potential in areas such as artificial intelligence.

Assembly instruction optimization plays a critical role in AI acceleration, especially for high-performance heterogeneous vector accelerators like the FT-Matrix. It consists of 8 CPU cores and 96 M-DSP cores, where the structure of a single M-DSP core is shown in Fig. 2. FT-Matrix significantly improves computational efficiency with its multi-core architecture, SIMD and VLIW. These features allow a single instruction to process multiple data elements

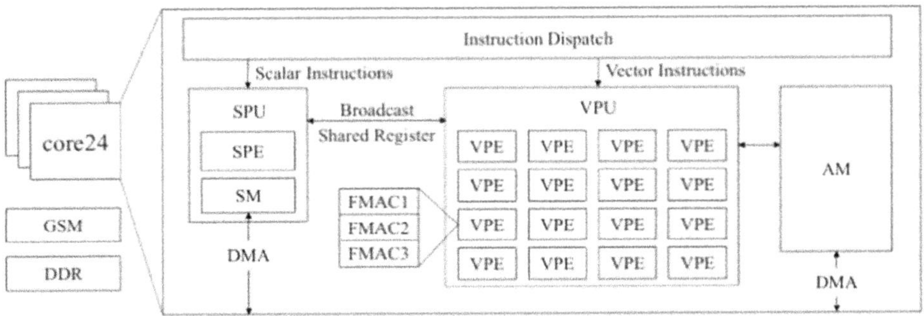

Fig. 2. M-DSP Architecture.

simultaneously, thereby dramatically reducing the number of instructions and increasing computational throughput when processing large-scale parallel data. For example, in the matrix operations of deep learning, SIMD can process the computation of multiple elements simultaneously, significantly speeding up the computation. VLIW, on the other hand, further improves the execution efficiency of instructions by executing multiple operations in a single instruction cycle. This optimization enables the hardware to utilize the limited computational resources more efficiently, thus significantly improving the inference speed and training efficiency of the model without increasing the hardware cost. However, as model complexity increases and computational demand continues to rise, the assembly instruction scheduling problem becomes more prominent. In particular, efficiently scheduling multiple instructions within a single time cycle to maximize the utilization of computational resources in each cycle has become a key challenge. This challenge involves not only the optimization of hardware design but also the implementation of fine-grained instruction scheduling strategies at the software level to ensure that instructions in each time cycle are executed efficiently, thereby avoiding resource waste and performance bottlenecks.

2   Related Work

The current development of operators for AI acceleration hardware, such as FT-Matrix, still relies heavily on manual optimization by engineers for instruction scheduling. This approach is inefficient, empirically dependent, and poorly generalized, making it difficult to meet the demand for rapid model iteration. In search of automation, researchers have tried traditional heuristic algorithms (e.g., greedy, genetic, simulated annealing, ant colony algorithms). However, these methods generally suffer from inherent defects such as susceptibility to local optima, huge computational overhead, parameter sensitivity, or inefficiency, which make it difficult to meet the stringent requirements of AI acceleration hardware on high-performance and high-efficiency instruction scheduling [16]. Therefore, it is crucial to develop new automated instruction scheduling methods that are more efficient and adaptable to AI acceleration scenarios to significantly improve computational efficiency and hardware resource utilization.

Traditional instruction optimization techniques in compilers still exhibit limitations in optimizing complex code structures and system constraints. To further overcome these challenges, recent years have seen the introduction of machine learning techniques in the field of compiler optimization [10]. Google Research proposed the MLGO [15] (Machine Learning Guided Compiler Optimizations) framework, which integrates machine learning models into LLVM to replace traditional heuristic decision-making methods. Another significant advancement is CompilerGym [3], developed by Research. CompilerGym provides a reinforcement learning environment that enables researchers to explore and optimize compiler decision-making processes through machine learning models. It supports various optimization tasks, such as instruction scheduling and loop optimization, and offers a standardized interface to facilitate rapid experimentation and deployment of new optimization strategies.

In the field of scheduling optimization, methods combining reinforcement learning and graph neural networks have achieved significant progress, offering novel approaches to solving complex scheduling problems. The synergistic advantages of reinforcement learning and graph neural networks have been extensively validated across multiple domains. In path planning, Kool et al. [8] proposed an RL framework based on GNNs and attention mechanisms, modeling urban nodes as a graph structure and dynamically generating path sequences via policy gradient algorithms. This method not only achieved near-optimal solutions on multiple benchmark datasets but also significantly reduced computational time. In neural architecture search (NAS), Zoph et al. [17] employed RL to autonomously explore network architecture spaces, while subsequent studies (e.g., You et al., 2020) further incorporated GNNs for graph representation learning of candidate architectures, accelerating the discovery of high-performance designs. Song et al. [12] utilized GNNs and deep RL to address complex job-machine scheduling problems. Additionally, Lee et al. [9] applied deep RL to solve DAG task scheduling, enabling parallel execution of compilation pipeline tasks across multiprocessors with a 3% performance improvement over other methods. These studies demonstrate that the integration of RL and GNNs not only resolves challenging combinatorial optimization problems but also finds practical applications in real-world engineering scenarios.

In the context of assembly instruction scheduling, Fabian et al. proposed a method combining graph neural networks and reinforcement learning [13], aiming to prevent register spilling by reducing register pressure during instruction scheduling. This approach achieved performance gains across multiple architectures compared to traditional compilers. Huanting Wang introduced SuperSonic, a meta-optmizer based on reinforcement learning and multi-task learning techniques, which automatically identifies and adjusts suitable RL architectures from training benchmarks. The tuned RL models can then be deployed to optimize new programs.

In summary, compiler instruction optimization techniques still face numerous challenges, which are primarily reflected in the following three aspects:

(1) Traditional optimization techniques rely on manually designed rules and heuristic methods, making it difficult to achieve optimal solutions for complex programs and diverse hardware architectures. The order of optimization steps significantly impacts the final results, yet traditional approaches struggle to accurately predict the interactions between different optimization techniques, often requiring extensive manual intervention and iterative adjustments.
(2) Different software structures and hardware architectures impose varied demands on optimization techniques. Most current optimization methods are manually designed and fine-tuned by experts, making them slow to adapt to new hardware architectures or software requirements, with high development and maintenance costs. As hardware complexity increases, compilers need a deeper understanding of hardware characteristics to fully exploit parallel computing capabilities and resources. However, traditional compilers have limited hardware-aware optimization capabilities, hindering their ability to maximize the performance potential of modern hardware.
(3) Although advancements in machine learning have introduced new opportunities for compiler optimization, current applications remain in their early stages. Furthermore, applying machine learning to compiler optimization faces challenges such as data collection, model design, and real-time performance, particularly in complex hardware architectures and large-scale programs.

3   Overview

During the DSP program compilation process, the assembly code can be partitioned into several basic blocks through flow control instructions. For the problem of instruction scheduling within a basic block, the instructions and their dependencies within a basic block can be modeled as DAG, where nodes represent instructions and edges represent dependencies between instructions. This graph structure makes GNNs an ideal choice for modeling instruction dependencies. GNNs are capable of accurately capturing the dependencies between instructions and generating compact and informative graph representations that provide detailed contextual information for RL. Compared to traditional methods, GNNs excel in processing graph-structured data and can effectively handle complex interactions between nodes, giving them a unique advantage in modeling instruction dependencies within basic blocks. Meanwhile, RL dynamically adjusts scheduling strategies based on this information to improve overall performance. This combined approach not only effectively addresses the complexity and dynamics of intra-basic block instruction scheduling but also automatically optimizes the scheduling strategy through continuous learning, thereby enhancing scheduling efficiency and performance.

To address the assembly instruction scheduling problem within a basic block, this paper proposes GRIS, a novel instruction scheduling approach that combines graph neural networks with reinforcement learning to optimize instruction sequences.

3.1 Assembly Instruction Sequence Modeling Design

The assembly instructions within the basic block are modeled by extracting instruction features and analyzing data dependencies, transforming the assembly instruction sequence into a DAG. In the DAG, nodes represent instruction features, and edges represent dependencies between instructions. This modeling process provides a clear representation of instruction features and dependencies for subsequent scheduling optimization.

3.2 GNN-Based State Compression

Based on the modeling of assembly instruction sequences, GNNs are utilized to process graph-structured data. GNNs efficiently represent DAGs by learning the features of nodes and edges and capturing complex dependencies. GNNs aggregate information from neighboring nodes and update node features to reflect their positions and roles within the graph. This representation not only compresses the state space and reduces computational complexity but also provides rich contextual information for reinforcement learning.

3.3 Optimization of Instruction Scheduling Policies Driven by Reinforcement Learning

Based on the graph representation generated by the GNN, the instruction scheduling policy is optimized using a deep Q-network (DQN). The DQN selects the optimal action for each state by learning the optimal action-value function (Q-function). The DQN learns the optimal scheduling policy stably through experience replay and target network update mechanisms. During training, the DQN selects actions based on the current state and updates the Q-function based on the reward signals fed back from the environment, dynamically adjusting the scheduling policy to adapt to different dependencies and execution environments.

In summary, the method proposed in this study achieves efficient and dynamic instruction scheduling through the modeling and mechanism design of assembly instruction sequences, GNN-based state space compression, and RL-driven scheduling strategy optimization. This method can adapt to the complexity and dynamics of the instruction scheduling problem, automatically optimize the scheduling strategy, and improve scheduling efficiency and performance. Specific implementation details will be elaborated in subsequent sections.

Moreover, since the FT-Matrix architecture employs a proprietary instruction set, existing assembly instruction optimization algorithms designed for conventional instruction sets cannot be directly applied. Therefore, the GRIS can serve as an efficient solution to rapidly develop compiler optimizations for this architecture, enabling full utilization of its hardware resources and significantly improving code execution efficiency.

4   Methodology

4.1   Problem Modeling

A Markov Decision Process (MDP) is a mathematical framework for solving sequential decision-making problems and is widely used in reinforcement learning tasks. In an MDP, an agent observes the environment in a given state and selects an action based on the current state. The environment then transitions to a new state based on the agent's choice and provides a corresponding reward. The agent's goal is to find an optimal decision path through continuous trial and learning, which maximizes the long-term cumulative reward. In the assembly instruction optimization for FT-Matrix, since the instruction scheduling decision has continuity and long-term effects, it can be modeled as a Markov decision process whose goal is to maximize the pipeline utilization of the assembly instruction sequence. The four main components—state space, action space, state transition, and reward function—are described in detail below.

State Space S. In the reinforcement learning framework, the state space S represents all possible system states, and the state $s \in S$ represents the system state at a certain moment. To more clearly express the feature information of instructions and the dependencies between them, this method employs a Graph-based Representation to model the assembly instructions in a Basic Block. Each state s can be represented as a Directed Graph G, that is:

$$s = G(X, A) \tag{1}$$

Here, X denotes the feature matrix, which stores the feature information of each instruction. A denotes the adjacency matrix, which describes the dependencies between instructions.

Action Space A. In the reinforcement learning framework, the action space A defines the set of all possible actions that an agent can execute at each time step. In the FT-Matrix assembly instruction scheduling problem, the action $a \in A$ represents the scheduling action chosen by the agent in the current time period, that is selecting a set of executable instructions from the *ReadyQueue* and scheduling them to the current time period. The *ReadyQueue* stores instructions that either do not depend on preceding instructions or whose preceding instructions have already been executed. Assuming that the basic block contains n assembly instructions, the action at each time step can be represented as a set of instructions:

$$a = \{i_1, i_2 \ldots i_k\}, \quad j \in ReadyQueue, \quad k \leq C \tag{2}$$

Here, *ReadyQueue* denotes the set of instructions whose data dependencies have been satisfied and are ready for immediate execution, meaning they do not depend on any instruction that has not yet been executed. C denotes the maximum number of instructions that can be scheduled in parallel within the current time cycle, which is determined by the hardware resource limitations of the FT-Matrix.

State Transfer. In the reinforcement learning framework, state transition describes how the environment shifts from the current state s to the next state s' after the agent performs an action. In the FT-Matrix assembly instruction scheduling task, state changes are reflected in the updates to the feature matrix X, the adjacency matrix A, and the ready queue *ReadyQueue*. Whenever the agent selects and executes an action a, a subset of instructions is scheduled to the current time cycle, and the state consequently changes to a new state $s' = G(X', A')$.

Reward Function R. In the reinforcement learning framework, the reward function R guides the agent to learn the optimal strategy step by step by quantifying the quality of its decisions. Each time the agent selects and executes an action a from state s, the environment transitions from state s to state s' and provides a reward value r, which measures the quality of the current scheduling action.

4.2   Overall Framework of the Methodology

The overall flow of this method is shown in the Fig. 3. It mainly consists of three core phases: assembly instruction sequence modeling, GNN-based state compression, and RL-driven instruction scheduling. The input to the system is the original assembly instruction sequence, and the output is the assembly instruction sequence after scheduling is completed.

The task of the assembly instruction sequence modeling phase is to convert the assembly instruction code into structured data suitable for processing by graph neural networks and reinforcement learning. This phase also involves implementing the action generation function, state transition mechanism, and reward function. Next is the GNN-based state compression phase, which aims to transform the graph-structured data represented by the feature matrix X and the adjacency matrix A into a global vector representation, thereby significantly reducing the state space. This is achieved through a GNN-based encoder that learns a compact representation of the graph-structured data. Finally, there is the RL-driven instruction scheduling phase. In this phase, based on the global vector representation learned by the GNN network in the previous phase, the DQN network learns the optimal instruction scheduling strategy, thereby selecting the optimal action $a*$ under the current state s to form the instruction scheduling scheme and achieve efficient instruction scheduling optimization.

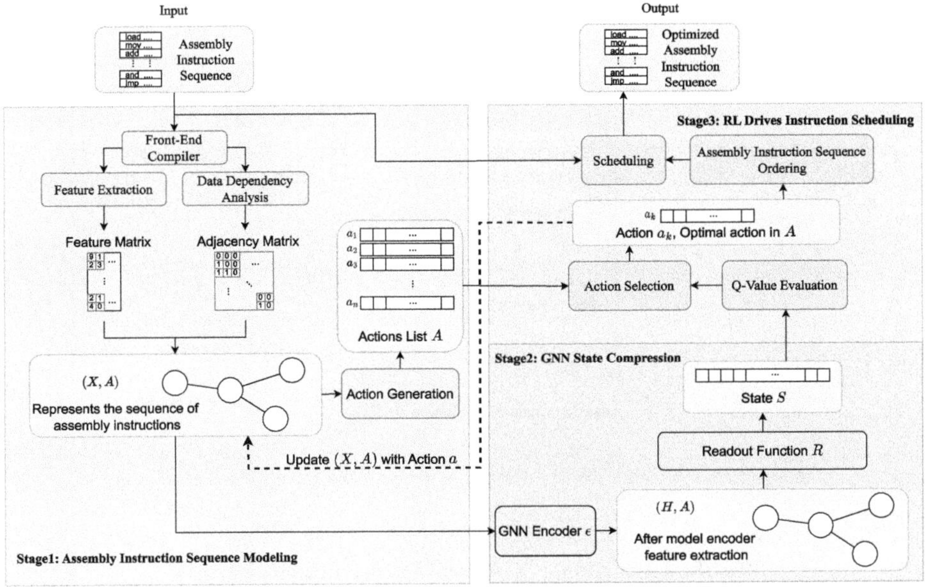

Fig. 3. Overall Framework of the Methodology.

4.3 Assembly Instruction Sequence Modeling

To enable the GNN and RL models to process the assembly instruction scheduling task, the sequence of assembly instructions needs to be converted into structured data representations, namely the feature matrix X and the adjacency matrix A. The feature matrix X stores the feature information of each assembly instruction, while the adjacency matrix A describes the data dependencies between instructions. The Fig. 4 illustrates how a sequence of assembly instructions is converted into a feature matrix and an adjacency matrix.

Construction of the Feature Matrix X. Assuming that the basic block contains n instructions, the dimension of the feature matrix X is $n \times d$, where d represents the feature dimension of the instructions. For each row x_i of matrix X, it corresponds to the feature information of instruction i. The feature information can be divided into five parts: assembly instruction type, instruction execution cycle, number of operands contained in the instruction, hardware resources required, and whether it is on the longest path, as shown in the table. The value of d is set to 16, meaning that each instruction can be encoded as a feature vector of length 16, with different features occupying different numbers of positions. These feature vectors are stacked in rows to form the feature matrix X (Table 1).

Construction of the Adjacency Matrix A. After the data dependency analysis, if instruction j can only be executed after instruction i has been executed, then $A_{ij} = 1$ in the adjacency matrix; otherwise, $A_{ij} = 0$. Given the directed

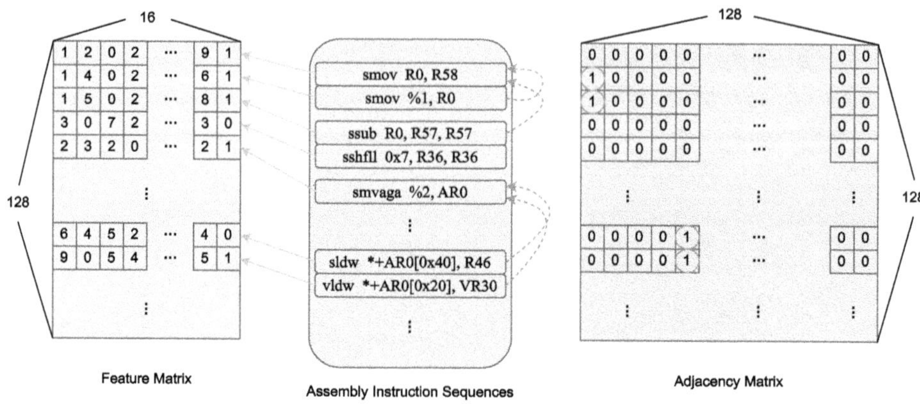

Fig. 4. Construction of Feature Matrix and Adjacency Matrix.

Table 1. Encoding of Assembly Instruction Features

Feature Name	Feature Description
Instruction Execution Cycle	The time required for an instruction to complete execution on the processor, measured in clock cycles. It can be obtained through assembly instruction semantic analysis and hardware manuals, reflecting the duration of instruction execution.
Number of Operands in Instruction	The number of data or address information involved in the instruction, including registers, memory addresses, and immediate numbers. It focuses on the number of operands that affect data dependencies, obtained through assembly instruction syntax analysis, reflecting the complexity of the instruction.
Longest Path	Whether the instruction is located on the longest execution path from the program's entry to exit. Judged by static analysis tools, marked as 1 if on the longest path, otherwise 0, related to program performance bottlenecks.
Assembly Instruction Type	Instructions are categorized into scalar, vector, and logic instructions, which handle single data elements, multiple data elements, and control flow operations such as conditional judgments, respectively. They significantly impact processor resource usage and execution efficiency, encoded through semantic analysis.
Required Hardware Resources	The processor execution units occupied during instruction execution. Different instructions use different execution units, and some instructions can be executed on different units. Obtained through semantic analysis and hardware manuals, used to match hardware resources for optimal performance.

nature of the instruction dependencies, the adjacency matrix A takes an lower triangular form. This is due to the linear order of instruction execution, where subsequent instructions depend on preceding instructions, but preceding instructions do not depend on subsequent instructions. This results in the upper triangle of the matrix (including the diagonal) being all zeros, while the lower triangle indicates the dependencies.

4.4   Mechanism Design

In the reinforcement learning framework, the agent continuously interacts with the environment, which necessitates the design of a series of mechanisms to support these interactions. In this method, these mechanisms can be specifically categorized into the action generation function, state transition mechanism, and reward function.

Action Generation Functions. The action generation function is used to generate a set of feasible actions $\{a_1, a_2 \ldots a_k\}$ from the current state $s = G(X, A)$. Each action a represents the agent's choice of a set of assembly instructions to be scheduled in the current time cycle, while ensuring that the instruction selection conforms to data dependencies and hardware resource constraints. The implementation of the action generation function begins by generating a ready queue $ReadyQueue$, which contains all the assembly instructions that can be scheduled in the current time cycle and do not depend on any instructions that have not yet completed execution.

The adjacency matrix A describes the dependency relationships between instructions. For each column j in the matrix, when all the elements of the column are zero (for all i, $A_{ij} = 0$) it indicates that the in-degree of instruction j is zero. To handle the issue of instruction execution time, a $RemainingTime$ list is introduced to record the remaining waiting time of each instruction. An instruction is considered schedulable and added to $ReadyQueue$ only if its in-degree is zero and its corresponding value in $RemainingTime$ is zero. After generating $ReadyQueue$, the action generation function needs to select a group of instructions to form the action for the current time cycle. In the specific implementation, considering the hardware resource limitation (a maximum of C instructions can be scheduled per time cycle), k instructions are selected from $ReadyQueue$ in a combinatorial manner (where $k \leq C$ to form the action a, and finally, the set of all possible actions A is generated.

State Transfer Mechanism. The state transition mechanism defines how the agent transitions from the current state s to the next state s' by executing the action a. The core objective of state transition is to update the feature matrix X, the adjacency matrix A, and the $RemainingTime$ according to the action a selected by the agent, thereby accurately reflecting the state changes after instruction scheduling. The specific implementation involves three update operations:

Firstly, the feature matrix X is updated: after executing the action a, for each instruction i contained in the action, the corresponding row in the feature matrix is set to zero, indicating that the instruction has been scheduled. The updated feature matrix X' then serves as the state for the next cycle.

Next, update the adjacency matrix A: for each instruction i in action a, set its corresponding column in the adjacency matrix to zero (indicating that it is no longer dependent on any other instruction), and set its corresponding row to zero (indicating that no other instruction is dependent on it). The updated adjacency matrix A' then serves as the basis for the next state.

Finally, update the remaining time list $RemainingTime$: after executing the action, for each instruction i in the action (let its execution period be t_i), update the remaining time of all the successor instructions j that depend on i: if j already has a remaining time value, set the new value to the maximum of t_i and the existing value; if j has no remaining time value yet, set it directly to t_i. Meanwhile, adjust the positions of the values in the list to maintain correspondence with the feature matrix and the adjacency matrix, and ultimately generate a new remaining time list $RemainingTime'$ for the next state.

Reward Function. The reward function $R(s, a)$ is used to measure whether the action a chosen by the agent in the current state s is conducive to optimizing scheduling performance and to quantify this effect. The design of the reward function must consider multiple optimization objectives and plays a decisive role in whether the agent can learn an efficient scheduling strategy. The reward function is designed as follows:

$$
\begin{aligned}
R(s, a) = {}& \alpha \cdot R_{\text{num}}(s, a) + \beta \cdot R_{\text{path}}(s, a) + \lambda \cdot R_{\text{time}}(s, a) \\
& + \mu \cdot R_{\text{type}}(s, a) + \nu \cdot R_{\text{end}}(s, a)
\end{aligned}
\tag{3}
$$

The number of actions component $R_{\text{num}}(s, a)$ directly takes the number of instructions in an action k and is designed to incentivize the agent to make full use of hardware parallelism. This component quantifies the number of instructions scheduled in the current cycle by $R_{\text{num}}(s, a) = k$, which motivates the agent to maximize resource utilization.

The critical path component $R_{\text{path}}(s, a)$ counts the number of longest path instructions contained in the action, and is calculated as $\sum_{i \in a} X[i][j]$, where j corresponds to the position of "whether it is on the longest path" in the feature matrix. This component guides the agent to prioritize the scheduling of bottleneck instructions on the critical path, thus reducing the total execution time of the program.

The execution time component $R_{\text{time}}(s, a)$ calculates the total execution period of instructions in an action, denoted as $\sum_{i \in a} X[i][j]$ (j corresponds to the position of the execution time feature). This component motivates the agent to schedule long-cycle instructions as early as possible to optimize pipeline utilization.

The instruction type equalization component $R_{\text{type}}(s, a)$ computes the type distribution uniformity using the sum of the absolute values of the negative deviations $-\sum_{c \in C} |f_c(a) - |a|/|C||$, where C is the set of instruction types and $f_c(a)$ is the number of instructions of type c. This mechanism avoids overloading specific hardware resources and maintains scheduling flexibility.

The scheduling completion component $R_{\mathrm{end}}(s, a)$ gives a positive incentive when the action completes the scheduling of all instructions, and zero otherwise. This component ensures that the agent completes the scheduling task with minimal operations, improving overall efficiency.

The sub-rewards are weighted by the coefficients $\alpha, \beta, \lambda, \mu, \nu$ to jointly optimize the instruction scheduling efficiency.

4.5 GNN-Based State Compression

The graph neural network (GNN) learns the low-dimensional embedding representation of each node by aggregating the feature information of the nodes and the information of their neighboring nodes, and ultimately generates a global vector z. This process not only significantly reduces the dimensionality of the state space but also retains the key information in the graph structure. In this way, GNN provides a compact and information-rich state representation for the reinforcement learning agent, thereby enhancing learning efficiency and decision-making performance. The global vector z of the GNN can be expressed as:

$$z = \mathrm{GNN}(X, A) \tag{4}$$

The GNN framework is shown in the Fig. 5. The following section will provide a detailed description of the components depicted in the diagram, including the Encoder ϵ, the Readout Function R, the Discriminator D, and the specific design of the Self-Supervised Contrastive Learning framework.

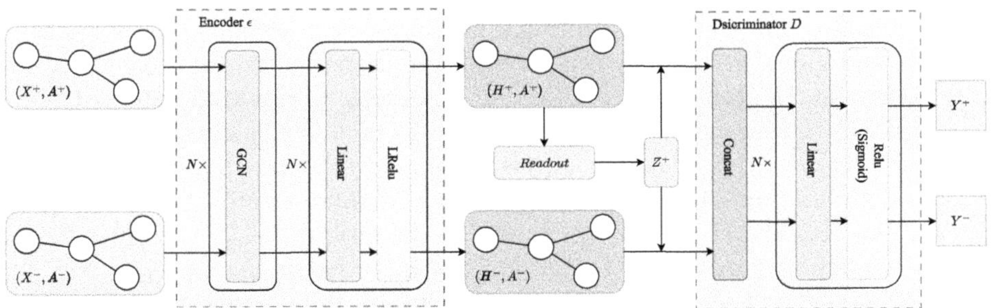

Fig. 5. GNN Framework.

Encoder ϵ. The role of the encoder ϵ is to take the input state space (X, A) and produce the transformed state space (H, A) after performing feature extraction and message passing. This process can be expressed as follows:

$$(H, A) = \epsilon(X, A) \tag{5}$$

The encoder ϵ comprises multiple graph convolutional layers, and the specific formulas for the network structure are as follows:

$$H^{l+1} = \text{contact}\left(\sigma\left(\hat{A}H^l\Theta^l\right), \sigma\left(\hat{A}_{\text{in}}H^l\Theta^l\right), \sigma\left(\hat{A}_{\text{out}}H^l\Theta^l\right)\right) \tag{4}$$

The formula 4 indicates how the data is specifically computed at different graph convolutional layers. H^l represents the result calculated at layer l, with $H^0 = X$. Θ^l is the updatable and shared weight matrix at layer l. σ denotes the activation function. The specific definitions of $\tilde{A}$, $\tilde{A}_{in}$ and $\tilde{A}_{out}$ are provided next.

$$\hat{A} = D^{-\frac{1}{2}}\tilde{A}D^{-\frac{1}{2}}$$

$$\hat{A}_{in} = D_{in}^{-\frac{1}{2}}\tilde{A}_{in}D_{in}^{-\frac{1}{2}} \tag{6}$$

$$\hat{A}_{out} = D_{out}^{-\frac{1}{2}}\tilde{A}_{out}D_{out}^{-\frac{1}{2}}$$

In the first line of Eq. 6, $\tilde{A}$ denotes the adjacency matrix with the addition of self-loops, which is computed as: $\tilde{A} = A + I$, where the matrix A is the original adjacency matrix of the graph. I is the identity matrix, which has the same dimensions as A, with diagonal elements of 1 and all other elements being 0. D denotes the degree matrix, which is a diagonal matrix whose diagonal elements $D(i, i)$ represent the degrees of the nodes. The specific formula is: $D = \text{diag}(\sum_j^n \tilde{A}(i, j))$. $D^{-\frac{1}{2}}$ is the inverse square root of the degree matrix, computed as $D^{-\frac{1}{2}} = diag(D_{ii}^{-\frac{1}{2}})$, representing the inverse square root of the diagonal elements of the degree matrix. The function of the formula in the first line is to normalize the adjacency matrix, which can mitigate the influence of high-degree nodes, allowing the graph convolutional neural network to better learn the local features and global structure of the nodes. The matrix obtained after the above transformation is often referred to as the symmetric normalized matrix.

Similarly, the second and third rows of the formula are calculated in the same manner. The definition of matrix $\tilde{A}_{in}$ is given below. Given an adjacency matrix $\tilde{A}$, $\tilde{A}_{in}(i, k) = 1$ and $\tilde{A}_{in}(k, i) = 1$ if and only if there exists at least one node j such that $\tilde{A}(j, i) = 1$ and $\tilde{A}(j, k) = 1$. The definition of $\tilde{A}_{out}$ is similar to $\tilde{A}_{in}$ but in the opposite direction. Given an adjacency matrix $\tilde{A}$, $\tilde{A}_{out}(i, k) = 1$ and $\tilde{A}_{out}(k, i) = 1$ if and only if there exists at least one node j such that $\tilde{A}(i, j) = 1$ and $\tilde{A}(k, j) = 1$.

The Fig. 6 illustrates the single-layer structure of graph convolution, focusing on the node at the center of the graph, which is highlighted in yellow. If we only follow the direction of the directed graph, the information on the yellow node cannot be propagated to its predecessor nodes. However, by constructing a second-order in-degree and out-degree matrix, the information on the yellow node can be efficiently propagated to the surrounding nodes, and the same applies to other nodes. Therefore, in addition to the basic graph convolutional layer, the computation of the second-order in-degree and out-degree matrix is incorporated, taking into account the directionality of the edges in the graph structure. This allows the model to extract more and deeper information from the graph structure. After passing through the multilayer graph convolution structure, the matrix is ultimately mapped to a hidden representation matrix H of the same dimension as the original adjacency matrix through a multilayer fully connected layer.

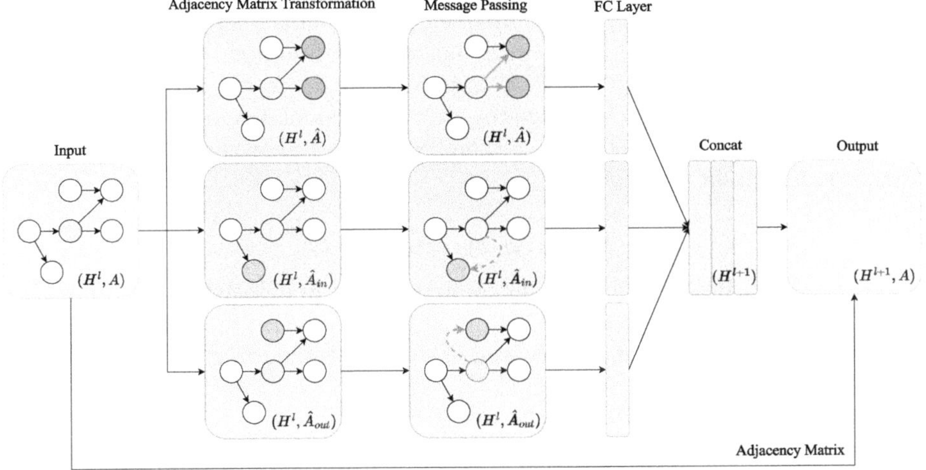

Fig. 6. Single Graph Convolutional Layer.

Readout Function R. To extract the global graph-level embedding vector z from the hidden representation H encoded by the multilayer graph neural network, a readout function R is introduced. This function serves to perform a weighted aggregation of the hidden representations H of all nodes. The formula is given below:

$$z = R(H) = \sum_{i=1}^{N} \left(\frac{\exp(\sqrt{\sum_{d=1}^{D} H_{id}^2})}{\sum_{j=1}^{N} \exp(\sqrt{\sum_{d=1}^{D} H_{jd}^2})} \right) H_i \tag{7}$$

where N denotes the total number of nodes in the graph, D denotes the dimension of each node's feature, H_{id} denotes the feature value of the i node in the d dimension, H_i denotes the hidden representation vector of the i node, α_i denotes the normalized attention weight of the i node, and z denotes the global graph-level embedding vector. This formulation first calculates the L_2 norm of the hidden representation of each node, then obtains the attention weight α_i for each node through an exponential function and normalization process. Finally, it multiplies each node's hidden representation H_i by its corresponding attention weight α_i and sums them to obtain the global graph-level embedding vector z.

Self-supervised Comparative Learning. Self-supervised contrastive learning is introduced to optimize the learning effect of graph neural networks. The basic idea is to bring similar samples closer together and push dissimilar samples farther apart, thereby enhancing the model's ability to understand the data distribution. In the training process, both positive and negative samples are randomly selected from the dataset, ensuring that they are distinct. The positive sample (X^+, A^+) is passed through the encoder ϵ for feature extraction, yielding the hidden representation $(H^+, A^+) = \epsilon(X^+, A^+)$. Similarly, the negative sample (X^-, A^-) is processed by the encoder ϵ to obtain the hidden representation $(H^-, A^-) = \epsilon(X^-, A^-)$. Subsequently, the hidden representation of the positive sample (H^+, A^+) is fed into the readout function R, resulting in the global embedding vector $z^+ = R(H^+, A^+)$. To measure the similarity between samples, a discriminator D is introduced, which calculates the degree of match between a given sample (H, A) and the global embedding vector z. The specific calculation formula is as follows:

$$D((H, A), z) = \sigma(H^T W z) \tag{8}$$

where, W is the learnable weight matrix and σ denotes the activation function. The formula computes the weighted inner product between the feature representation of the input sample and the global embedding vector to measure their similarity.

During the training process, the match for the positive sample is calculated as $Y^+ = D((H^+, A^+), z^+) = \sigma(H^{+T} W z^+)$, and the match for the negative sample is calculated as $Y^- = D((H^-, A^-), z^+) = \sigma(H^{-T} W z^+)$. To optimize the encoder ϵ and the discriminator D, the loss function is defined as follows:

$$\mathcal{L} = -\frac{1}{N} \left[0.5 \sum \log(Y_i^+) + 0.5 \sum \log(1 - Y_i^-) \right] \tag{9}$$

The design of this structure and function effectively prevents the encoder from falling into the problem of representation collapse, avoiding the convergence of the embedding representations of all samples to similar vectors, which would degrade the discriminative power.

4.6 Optimization of Instruction Scheduling Policies Driven by RL

DQN Model Structure. The structure of the DQN model is depicted in the fig.7. DQN overcomes the limitations of traditional Q-learning in large-scale problems by approximating the Q-function using a Deep Neural Network (DNN). The update formula for DQN is as follows:

$$Q(s, a; \theta) \leftarrow Q(s, a; \theta) + \alpha \left[R(s, a) + \gamma \max_{a'} Q(s', a'; \theta) - Q(s, a; \theta) \right] \tag{10}$$

where, θ denotes the network parameters, α is the learning rate, γ is the discount factor, $R(s, a)$ is the immediate reward, s and a are the current state and action, and s' and a' are the next state and action.

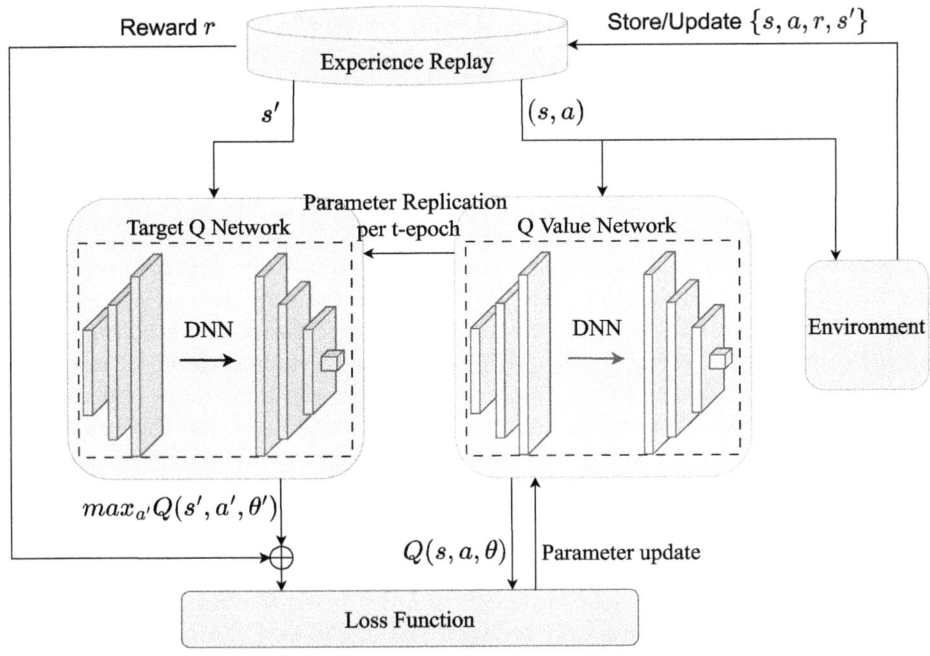

Fig. 7. DQN Framework.

The $Q(s, a; \theta)$ in Eq. 10 estimates the expected payoff for taking various actions in a given state. However, during model training, if only a single Q-value network is used for updating, the target value will change frequently, leading to oscillations in the learning process. To address this problem, a Target network is introduced, whose parameters are periodically copied from the Q network to provide stable target values. The structure of the two networks is identical, but their parameters differ during the same period of training. The Q-value network is responsible for estimating the expected payoff for taking various actions in a given state. The specific design of the networks is as follows:

$$Q(s, a; \theta) = \sigma\left(W_L h_{L-1} + b_L\right) \tag{11}$$

where, the hidden state h_i of each layer is calculated using the following recursive formula:

$$h_i = \sigma_i\left(W_i h_{i-1} + b_i\right), \quad i = 1, 2, \ldots, L - 1 \tag{12}$$

In Eqs. 11 and 12, s is the input state vector, a is the input action vector, W_i and b_i are the weight matrix and bias vector of the i layer, σ_i is the nonlinear activation function of the i layer, L is the total number of layers in the network, and $h_0 = s \cdot W_0 \cdot a + b_0$.

The parameters of the Target network θ_{target} are periodically copied from the parameters of the Q-value network θ. A soft update is used to gradually update the parameters of the target network by means of a smooth transition: $\theta_{target} \leftarrow \tau\theta + (1-\tau)\theta$. Here, τ is a small positive number used to control the rate of updating.

The loss function takes the form of a mean square error, defined by the formula:

$$\mathcal{L}(\theta) = \mathbb{E}_{(s,a,r,s')}\left[(Q(s,a;\theta) - y(s,a))^2\right] \tag{13}$$

where, $Q(s,a;\theta)$ is the Q-value predicted by the Q-value network, representing the expected cumulative reward for taking action a in state s; $y(s,a)$ is the target Q-value, which is used to provide the learning target and is computed as: $y(s,a) = r + \gamma \max_{a'} Q(s',a';\theta_{target})$. The target Q-value consists of two components: the immediate reward r, which represents the immediate benefit from the current action; and the discounted maximum Q-value for the next state $\gamma \max_{a'} Q(s',a';\theta_{target})$, which represents the maximum cumulative reward that may be obtained in the future.

Experience Replay Buffers and Sampling Strategies. The experience replay buffer is an important component in DQN, used to store the experiences generated during the interaction between the agent and the environment. Its introduction helps to break the correlation between data and improve the efficiency of data utilization. The quadruple $\{sars'\}$ is the basic unit in the experience replay buffer, and each quadruple contains the following information: s indicates the environment state that the agent is in before executing the action; a indicates the action that the agent chooses under the current state; r indicates the reward obtained after executing the action a; s' indicates the new state to which the agent transitions after executing the action. Together, this information describes a complete interaction process of the agent in the environment.

In the pre-population phase, the agent first collects a sufficient number of quadruples $\{sars'\}$ through its interaction with the environment and stores them in the experience replay buffer. During the training process of the model, the data update mechanism of the experience replay buffer is crucial to the performance of the algorithm. As the agent continues to interact with the environment, new experiences $\{sars'\}$ are continuously generated and stored in the experience replay buffer. When the experience replay buffer has not yet reached its preset capacity limit, new experiences are added directly to the buffer. When the buffer reaches its capacity limit, a First-In-First-Out (FIFO) strategy is employed to replace old experiences with new ones, thereby maintaining the dynamic nature of the buffer. During training, a random sampling strategy is used to uniformly select quadruples from the experience replay buffer without considering the weights or other features of each quadruple.

Exploration and Experience Utilization Strategy. During training, the agent selects actions and interacts with the environment using an $\epsilon - greedy$ strategy, which is a mechanism for balancing exploration and exploitation.

The parameter ϵ controls the probability of random exploration. When the agent interacts with the environment, if the randomly generated probability is less than ϵ, the agent will randomly select an action. This behavior helps the agent discover unknown aspects of the environment and avoid getting stuck in local optima. Conversely, if the randomly generated probability is greater than or equal to ϵ, the agent will choose the currently known optimal action, i.e., the action with the highest expected return. Additionally, ϵ should gradually decrease as training progresses. A linear decay approach is used to dynamically adjust ϵ, causing it to decay from a higher initial value to a lower final value. The calculation formula is as follows:

$$\epsilon_t = \max(\epsilon_{min}, \epsilon_{max} - k \cdot t) \tag{14}$$

where, ϵ_t denotes the exploration probability at training step t, ϵ_{max} is the initial exploration probability, ϵ_{min} is the minimum exploration probability, k is the decay coefficient, and t is the current training step. At the beginning of training, since the experience replay buffer has already been pre-populated, ϵ_{max} can be set to a lower value.

5 Experiment

5.1 Experiment Setup

Dataset. This experiment is developed and validated using a series of developed operators in the FT-Matrix accelerator, which extensively cover a wide range of typical computational tasks, including computation-intensive operations, data transfer operations, activation functions, normalization operations, and so on. These operators are not only functionally diverse but also contain complex control flow structures, with approximately 60% of the operators containing multiple layers of nested loops and conditional branches. Additionally, the operators in the dataset cover over 90% of the FT-Matrix instruction set, which fully reflects the diversity of the instruction set and the wide range of application scenarios. These features make the dataset an ideal testbed for validating and comparing the effectiveness of different optimization methods. Some statistical information of the dataset is shown in Table 2.

Table 2. Statistical Classification of Operators

Type	Num	Proportion	Operator Name
Computation-intensive	132	67.0%	gemm, sgemm, gelu, softmax, layernorm, rmsnorm, linear, etc.
Data Transfer and Auxiliary	45	22.8%	copy, dma_core, dma_transfer,etc.
Control and Special Functions	20	10.2%	div, dropout, embedding, set, transpose, mpid_barrier, etc.

Based on this dataset, a dataset for GNN training $\mathcal{G}$ and an experience replay pool for DQN training $\mathcal{R}$ are constructed in one step.

Network Performance Evaluation Metrics. In terms of model algorithm performance, the effectiveness of graph neural networks and reinforcement learning models needs to be considered. The evaluation metrics for graph neural networks include the following:

Positive and Negative Sample Discriminator Difference (ΔD): This metric measures the performance of the discriminator D in the graph neural network and reflects the model's ability to distinguish between feature representations of positive and negative samples. The formula is as follows:

$$\Delta D = \frac{1}{N} \sum_{i=1}^{N} \left[D((H_i^+, A_i^+), z_i^+) - D((H_i^-, A_i^-), z_i^+) \right] \tag{15}$$

where, H_i^+ denotes the embedding representation of the i positive sample, and H_i^- denotes the embedding representation of the i negative sample. $D((H^+, A^+), z^+))$ represents the score for the positive sample, while $D((H^-, A^-), z^+))$ represents the score for the negative sample. N is the number of sample pairs in the current batch.

ROC-AUC (Receiver Operating Characteristic - Area Under Curve) measures the model's ability to distinguish between positive and negative samples across different thresholds, reflecting the classification performance of the model's feature representation. The formula for calculating ROC-AUC is as follows:

$$ROC - AUC = \int_0^1 TPR(FPR) \, dFPR \tag{16}$$

The True Positive Rate (TPR) and False Positive Rate (FPR) in the formula are defined as follows:

$$TPR = \frac{TP}{TP + FN}, \quad FPR = \frac{FP}{FP + TN} \tag{17}$$

Here, TP denotes the number of true positives, FN denotes the number of false negatives, FP denotes the number of false positives, and TN denotes the number of true negatives.

Embedding Space Separation ($\Delta Dist$) measures the extent to which the distance between positive and negative samples in the embedding space is effectively separated, reflecting the model's ability to distinguish between positive and negative samples in the feature space. The formula for calculating $\Delta Dist$ is as follows:

$$\Delta Dist = \frac{1}{N} \sum_{i=1}^{N} \left(||H_i^- - Z_i^+|| - ||H_i^+ - Z_i^+|| \right) \tag{18}$$

where, $||H^+ - Z^+||$ denotes the distance between the matrix representation H^+ of the positive sample X^+ after passing through the encoder ϵ and the matrix representation Z^+ obtained by expanding the global vector of the positive samples. $||H^- - Z^+||$ denotes the distance between the matrix representation H^- of the negative sample X^- after passing through the encoder ϵ and Z^+. N represents the number of sample pairs in the current batch. This design incorporates the global vector z to evaluate the model.

Cosine Similarity Difference (Δcos) measures the difference in similarity between positive and negative samples, which similarly reflects the model's ability to distinguish between feature representations. The formula for calculating Δcos is as follows:

$$\Delta \cos = \frac{1}{N} \sum_{i=1}^{N} \left(\frac{H_i^+ \cdot H_i^-}{||H_i^+|| \cdot ||H_i^-||} \right) \tag{19}$$

where, $\frac{H_i^+ \cdot H_i^-}{||H_i^+|| \cdot ||H_i^-||}$ denotes the cosine similarity between the matrix representations of the positive and negative samples, and N is the number of sample pairs in the current batch. This design differs from the previous formula in that it directly compares the similarity difference between positive and negative samples without incorporating the global vector z.

t-SNE (t-Distributed Stochastic Neighbor Embedding) is a nonlinear dimensionality reduction method for visualizing high-dimensional data. Its core idea is to minimize the difference in probability distributions between the data in the high-dimensional space and the data in the low-dimensional space, thereby preserving as much of the local structure of the high-dimensional data as possible in the low-dimensional space. t-SNE constructs two probability distributions: the high-dimensional data distribution P_{ij} and the low-dimensional data distribution Q_{ij}. It uses KL divergence to measure the difference between these two distributions, with the final optimization goal being to minimize this divergence:

$$KL = \sum_{i} \sum_{j} P_{ij} \log \frac{P_{ij}}{Q_{ij}} \tag{20}$$

The high-dimensional data distribution P_{ij} and the low-dimensional data distribution Q_{ij} in the formula are defined as follows:

$$P_{ij} = \frac{\exp\left(-\frac{|H_i - H_j|^2}{2\sigma_i^2}\right)}{\sum_{k \neq i} \exp\left(-\frac{|H_i - H_k|^2}{2\sigma_i^2}\right)}, Q_{ij} = \frac{\left(1 + |Y_i - Y_j|^2\right)^{-1}}{\sum_{k \neq i} \left(1 + |Y_i - Y_k|^2\right)^{-1}} \tag{21}$$

where, H_i and H_j represent high-dimensional data points, Y_i and Y_j represent low-dimensional data points, and σ_i is a scale parameter that controls the local distribution of high-dimensional data points (determined by setting the perplexity).

Regarding the reinforcement learning model, the objective is to learn an optimal instruction scheduling strategy, which can be evaluated based on Q-value stability and decision optimality.

Q-value Convergence (σ_Q): This metric assesses the stability of Q-values, specifically whether the fluctuations in Q-values converge as training progresses. The formula for calculating σ_Q is as follows:

$$\sigma_Q = \sqrt{\frac{1}{N} \sum_{i=1}^{N} (Q_i - \bar{Q})^2} \tag{22}$$

where, Q_i denotes the Q-value at the i training iteration, $\bar{Q}$ represents the mean Q-value, and N is the number of sampling steps.

The Optimal Action Rate (O_t) measures the extent to which the model increasingly selects the optimal action during the reinforcement learning training process, reflecting whether the agent's action choices at different time steps are stable and consistent with the optimal strategy. The formula for calculating O_t is as follows:

$$O_t = \frac{1}{N} \sum_{i=1}^{N} e^{-\beta \cdot (Q^* - Q_i)} \tag{23}$$

Here, Q^* represents the theoretically optimal Q-value in the current state, Q_i is the Q-value of the action selected by the agent during the i training iteration, β is a tuning parameter, and N is the number of training rounds.

Optimization of Performance Evaluation Metrics. The Instruction Execution Speedup Ratio (*Speedup*) measures the reduction in the time period required for code execution after optimization compared to before optimization. This metric reflects the effectiveness of the optimization algorithm in enhancing program execution speed. The formula for calculating *Speedup* is as follows:

$$Speedup = \frac{T_{baseline}}{T_{optimized}} \tag{24}$$

Here, $T_{baseline}$ denotes the time period required for code execution before optimization, while $T_{optimized}$ denotes the time period required for code execution after optimization.

Pipeline Utilization measures the reduction in the idle rate of the pipeline following optimization and assesses the extent to which hardware resources are fully utilized to enhance instruction-level parallelism. The formula for calculating Pipeline Utilization is as follows:

$$PipelineUtilization = \frac{\sum_{i=1}^{N} IPC_i}{Max\_IPC \times TotalExecutionTimeCycles} \tag{25}$$

Here, IPC_i denotes the number of instructions actually executed in the i cycle, while Max_IPC represents the theoretical maximum number of instructions that the pipeline can execute per cycle.

5.2   Experiment Result

Performance Evaluation of GNN. Figure 8 illustrates the impact of the number of graph convolution layers on four distinct evaluation metrics. In sequence, the subplots depict ΔD, $ROC-AUC$, $\Delta Dist$ and Δcos across varying numbers of graph convolution layers. During the experiments, all hyperparameters remained constant except for the number of graph convolution layers, which was incremented from 1 to 4. Ten-fold cross-validation was employed to calculate the mean value of each metric. In the figure, the x-axis represents the number of graph convolution layers, while the y-axis represents the specific values of the evaluation metrics. The data across all subplots indicate that the model achieves optimal performance on all metrics when the number of graph convolution layers is set to 3, with the model's discriminative accuracy reaching approximately 86.8%. Additionally, each subplot includes shaded areas representing the standard deviation, which indicates the uncertainty or range of variation in the metric values.

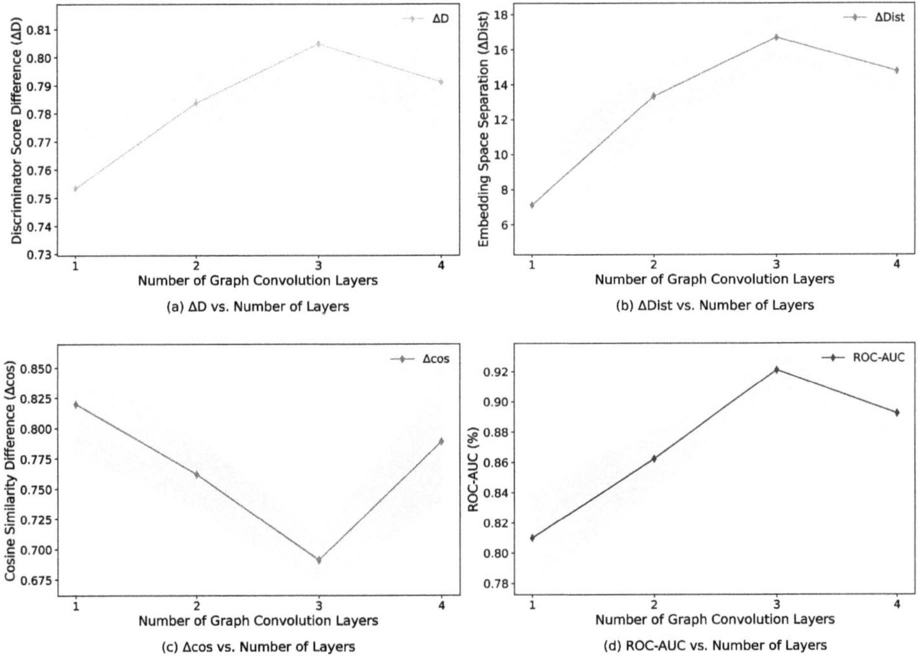

Fig. 8. GNN Performances.

Figure 9 illustrates the distribution of some data points during the training process, following t-SNE dimensionality reduction. Through t-SNE visualization, one can more intuitively observe how the encoder ϵ in the graph neural network learns the distribution of data within the dataset. As training progresses, the distribution of data points after t-SNE dimensionality reduction gradually changes, becoming more pronounced over time. This indicates that the model is progressively learning the feature distribution of the data, with the encoded feature representations forming a clearer and more distinct structure.

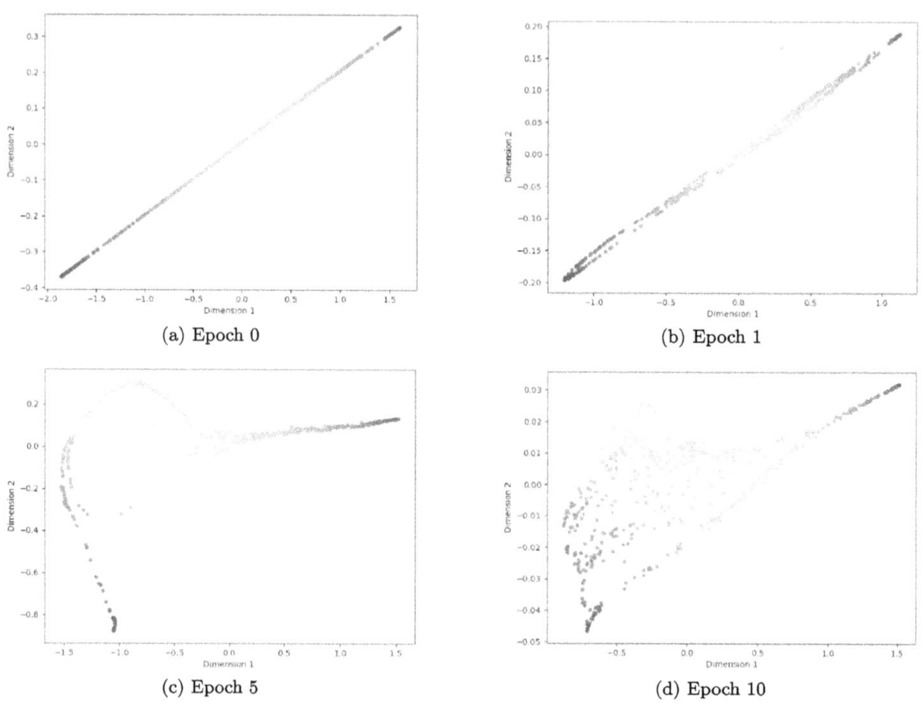

(a) Epoch 0   (b) Epoch 1

(c) Epoch 5   (d) Epoch 10

Fig. 9. t-SNE Visualization of Partial Data.

Performance Evaluation of DQN. Figure 10 displays σ_Q and O_t during model training. The training process consists of 500 steps. $\bar{Q}$ is calculated as the average Q-value per training round, Q^* is set to the maximum Q-value per round, and β is fixed at 0.1 to emphasize overall model stability. The right-hand figure illustrates how the Q-value distribution evolves over training. Initial refers to the first 50 rounds, Middle to rounds 100–200, and Final to rounds 300–400. The left figure shows that σ_Q decreases as training progresses, indicating more consistent Q-value estimation and stable decision-making. Meanwhile, O_t increases, demonstrating the model's ability to learn from the experience replay pool, select optimal actions, and refine its strategy. The right figure's Q-value distribution changes reveal the model's dynamic learning process. Initially, the Q-value distribution is dispersed, reflecting high environmental uncertainty. As training continues, the distribution stabilizes, showing enhanced environmental modeling and more accurate action value assessment.

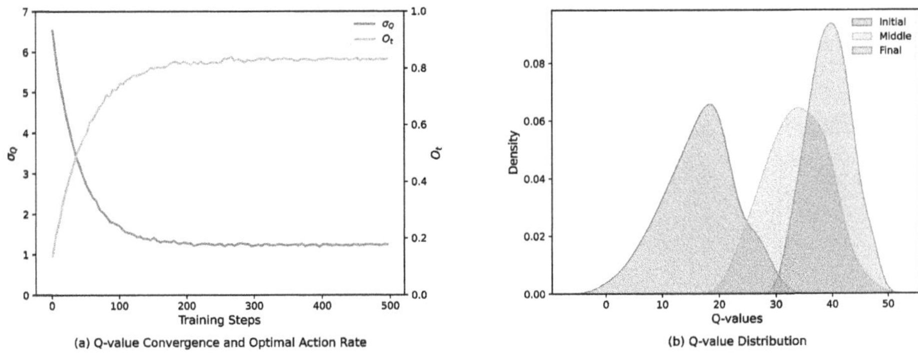

Fig. 10. DQN Performances.

Comparative Experimental Results. Figure 11 compares the execution time speedup ratio and pipeline utilization of different optimization methods across three operator classes: Computation Intensive (CI), Data Transfer & Auxiliary (DT&AF), and Control Flow & Special Functions (CF&SF). In the left figure, the execution time speedup ratio is evaluated, using manually optimized operators as the baseline. Various optimization methods were tested, including heuristic algorithms, reinforcement learning, manual rules, manual rules combined with heuristic algorithms, and manual rules combined with reinforcement learning. The results show that the combination of manual rules and reinforcement learning achieved the best performance improvements across all operator classes, with execution time speedup ratios of 1.096 for CI operators, 1.089 for DT&AF operators, and 1.072 for CF&SF operators. The right figure evaluates pipeline utilization, where the same combination of manual rules and reinforcement learning demonstrated the best performance in all operator classes, particularly increasing pipeline utilization from 36.2% to 41.8% in CF&SF operators. Heuristic algorithms, which locally adjust instruction order (e.g., by advancing schedulable instructions), can effectively reduce pipeline idle cycles and improve parallelism. However, their core limitation is the inability to break free from the initial instruction order for global optimization. This can lead to early scheduling imbalances that become fixed, and delays in critical instructions that are not prioritized can cause idle cycles and dependency blockages on the critical path, severely limiting subsequent optimization opportunities and overall performance.

In contrast, reinforcement learning-driven optimization offers significant advantages. It transcends the limitations of local rules and dynamically adjusts instruction scheduling from a global perspective. Reinforcement learning prioritizes instructions without dependencies and evaluates the potential impact of each instruction on overall execution time in real time. It actively identifies and prioritizes critical path instructions, even if it means adjusting other instructions or temporarily sacrificing local optimality. It also balances data dependencies, resource utilization, and execution priorities comprehensively. This approach allows reinforcement learning to fully utilize idle resources, accelerate the

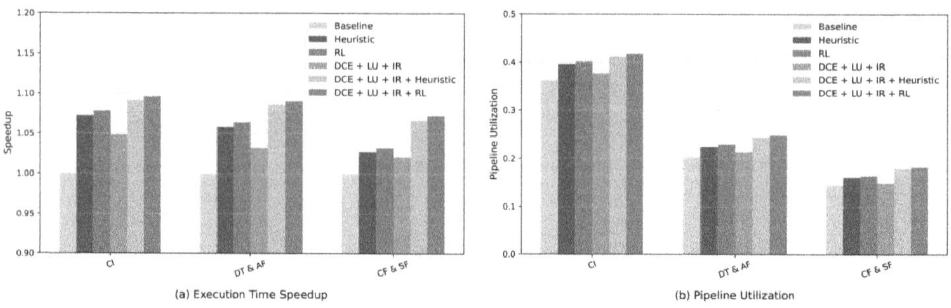

Fig. 11. Comparison of the Combination of Different Optimization Methods.

execution of critical instructions, and break free from the constraints of the initial instruction order. As a result, reinforcement learning provides more powerful optimization capabilities and achieves better outcomes in reducing overall execution time and improving pipeline utilization and throughput compared to heuristic algorithms.

6 Conclusion

In this paper, we present an automated solution based on GNN and RL for the problem of assembly instruction scheduling optimization within the FT-Matrix base block of a high-performance heterogeneous accelerator. The method models instructions and their dependencies as DAGs and leverages the powerful graph structure information extraction and state compression capabilities of GNNs to generate low-dimensional global embedding representations that contain rich dependency and instruction feature information. Building on this foundation, we construct a DQN-based reinforcement learning framework that drives the agent to learn optimal scheduling policies through a well-designed action generation mechanism, state transition mechanism, and multi-objective fusion reward function (covering instruction parallelism, critical path, execution cycle, type balance, and task completion). This approach effectively overcomes the inherent limitations of traditional heuristic algorithms, such as the tendency to fall into local optima, high computational overhead, and parameter sensitivity, achieving end-to-end, adaptive, and efficient instruction scheduling.

Experimental results demonstrate that the proposed method achieves notable performance improvements on the operator set of the FT-Matrix accelerator, which includes computationally intensive operations, data transfer and auxiliary operations, and control flow and special functions. Compared to the manually optimized baseline, the method shows superior performance in reducing instruction execution cycle time (up to approximately 9.6% improvement) and increasing pipeline utilization (particularly from 36.2% to 41.8% in control flow operators), thereby validating its effectiveness in maximizing instruction-level parallelism and hardware resource utilization. This study offers a solution for addressing the complex and dynamic instruction scheduling problem in AI-accelerated hardware, aiming to provide a valuable reference for research and practice in related fields.

References

1. Chen, S., et al.: Ft-matrix: A coordination-aware architecture for signal processing. IEEE Micro **34**(6), 64–73 (2014)
2. Chen, T., et al.: Tvm: an automated end-to-end optimizing compiler for deep learning. In: Proceedings of the 13th USENIX Conference on Operating Systems Design and Implementation, pp. 579–594. OSDI'18, USENIX Association, USA (2018)
3. Cummins, C., et al.: CompilerGym: Robust. CGO, Performant Compiler Optimization Environments for AI Research. In (2022)
4. Gou, J., Yu, B., Maybank, S.J., Tao, D.: Knowledge distillation: a survey. Int. J. Comput. Vision **129**(6) (2021)
5. Guo, D., et al.: Deepseek-coder: when the large language model meets programming – the rise of code intelligence (2024). https://arxiv.org/abs/2401.14196
6. Han, K., et al .: A survey on vision transformer. IEEE Trans. Pattern Anal. Mach. Intell. **45**(1), 87–110 (2023)
7. Islam, S., Elmekki, H., Pedrycz, R.W.: A comprehensive survey on applications of transformers for deep learning tasks. Expert Syst. Appl. **241**(May), 122666.1–122666.48 (2024)
8. Kool, W., Van Hoof, H., Welling, M.: Attention, learn to solve routing problems! arXiv preprint arXiv:1803.08475 (2018)
9. Lee, H., Cho, S., Jang, Y., Lee, J., Woo, H.: A global dag task scheduler using deep reinforcement learning and graph convolution network. IEEE Access **9**, 158548–158561 (2021)
10. Li, M., Liu, Y., Liu, X., Sun, Q., Qian, D.: The deep learning compiler: a comprehensive survey. IEEE Trans. Parallel Distrib. Syst. **32**(3), 708–727 (2021)
11. Nagel, M., Fournarakis, M., Amjad, R.A., Bondarenko, Y., Baalen, M.V., Blankevoort, T.: A white paper on neural network quantization (2021)
12. Song, W., Chen, X., Li, Q., Cao, Z.: Flexible job-shop scheduling via graph neural network and deep reinforcement learning. IEEE Trans. Ind. Inf. **19**(2), 1600–1610 (2023)
13. Stuckmann, F., Payá-Vayá, G.: A graph neural network approach to improve list scheduling heuristics under register-pressure. In: 2024 13th International Conference on Modern Circuits and Systems Technologies (MOCAST), pp. 01–06. IEEE (2024)
14. Talib, M.A., Majzoub, S., Nasir, Q., Jamal, D.: A systematic literature review on hardware implementation of artificial intelligence algorithms. J. Supercomput. **77**(12) (2020)
15. Trofin, M., Qian, Y., Brevdo, E., Lin, Z., Choromanski, K., Li, D.: Mlgo: a machine learning guided compiler optimizations framework (2021). https://arxiv.org/abs/2101.04808
16. Wang, Z., O'Boyle, M.: Machine learning in compiler optimization. Proc. IEEE **106**(11), 1879–1901 (2018)
17. Zoph, B., Le, Q.V.: Neural architecture search with reinforcement learning. arXiv preprint arXiv:1611.01578 (2016)

Tiling-Aware Vectorization Framework for Perfect Loop Nests in MLIR

Chenchen Hong[1,2], Zhong Liu[1,2(✉)], Hongli Zhong[1,2], and Hongbin Zhang[3]

[1] College of Computer Science and Technology, National University of Defense Technology, Changsha, Hunan, China
[2] Key Laboratory of Advanced Microprocessor Chips and Systems, National University of Defense Technology, Changsha, Hunan, China
`zhongliu@nudt.edu.cn`
[3] Institute of Software, Chinese Academy of Sciences, Beijing, China

Abstract. Automatic vectorization in traditional compilers is often unreliable, as the required high-level program structure is lost in low-level intermediate representations (IRs). This forces compilers to rely on fragile analyses to rediscover opportunities for SIMD execution, leading to unpredictable performance. This paper presents a novel, robust vectorization framework built upon the Multi-Level Intermediate Representation (MLIR) that avoids this pitfall by leveraging high-level semantics preserved in the linalg dialect. This paper introduces a **tiling-aware optimization** strategy, embodied in a three-phase, model-driven methodology. This strategy is guided by an **analytical cost model** that, by leveraging hardware-specific performance parameters, determines the lowest-cost vectorization plan that synergistically combines data locality from tiling with robust, masked vectorization to ensure high performance. Experimental evaluations demonstrate the framework's effectiveness. On compute-intensive kernels like **GEMM**, our method achieves speedups of **over 18x** for data sizes that fit within the L3 cache, significantly outperforming industry compilers like GCC and Clang. Notably, on large-scale **Conv2D**, the synergistic combination of tiling and vectorization delivers a speedup of over **30x**, validating our tiling-aware vectorization method.

Keywords: Automatic Vectorization · MLIR · Compilers · SIMD · High-Performance Computing

1 Introduction

The architectural evolution of modern processors has shifted from increasing clock speeds to integrating more on-chip parallel units. **Single Instruction, Multiple Data (SIMD)** [1,2] execution is a key mechanism for accelerating computation on these parallel units. For instance, Intel's Advanced Vector Extensions (AVX) [3] support 256-bit vectors to process eight single-precision floating-point numbers in one cycle, while ARM's NEON [4] handles 128-bit

H. Liu et al. (Eds.): ICA3PP 2025, LNCS 16381, pp. 568–584, 2026.
https://doi.org/10.1007/978-981-95-8399-7_30

vectors for mobile applications. However, manually leveraging SIMD is difficult and error-prone, making automatic vectorization essential. Traditional compilers, operating on low-level Intermediate Representations (IRs) such as **LLVM IR** [5], face a fundamental challenge: these IRs lack the high-level structural information of the original program. As a result, vectorization relies on brittle and complex analyses to recover this information, often failing unpredictably due to minor, functionally irrelevant changes in the source code.

Achieving consistent high performance through automatic vectorization thus remains a significant challenge. Mainstream compilers like GCC and Clang, while capable of handling simple loops, often exhibit unpredictable and sharply degraded performance in the face of minor or complex code changes. This performance "lottery" has led to a reliance on highly-tuned, hand-written kernel libraries (e.g., Intel's oneMKL [6]). While these libraries offer excellent performance, they cover a limited set of operations, and any new manual optimization demands substantial engineering effort, deep expertise, and long development cycles.

One promising direction to mitigate these challenges, a paradigm shift in compiler technology is underway, moving toward an infrastructure that retains and leverages high-level program semantics rather than rediscovering them through fragile analyses late in the optimization process. The Multi-Level Intermediate Representation (MLIR) [7] epitomizes this philosophy, enabling reliable transformations through its progressive lowering design. Within the MLIR ecosystem, the **linalg** dialect [8] provides a structured computational model for tensor operations. Specifically, the **linalg.generic** operation represents perfect loop nests declaratively with explicit indexing maps and iterator types, fundamentally eliminating the need for the complex and brittle dependency analyses that plague traditional approaches.

Although MLIR and the linalg dialect provide a solid foundation, sophisticated and automated vectorization strategies are still nascent within its compilation flows, leaving significant room for improvement. To bridge this gap, we propose a novel vectorization framework that operates directly on the high-level semantic model of linalg.generic. By exploiting its explicit structural information, our approach systematically transforms computations into efficient vectorized forms using the **affine** and **vector** dialects, paving the way for more reliable and high-performance automatic vectorization.

The main contributions of this work are as follows.

- We propose a **tiling-aware vectorization strategy** guided by an analytical cost model. This model analyzes linalg.generic semantics and memory access patterns to automate the selection of a high-performance transformation plan.
- We develop a framework that implements this strategy, automating the entire process from analysis to code generation. It systematically analyzes an operation and, based on the cost model's decision, generates efficient and legal vectorized code using the affine and vector dialects.
- We evaluate our framework on a diverse set of linear algebra kernels. The results show that our fully automated approach generates robust code achiev-

ing performance that is competitive with, and in key cases superior to, highly-tuned industry compilers like GCC and Clang.

The remainder of this paper is organized as follows. Section 2 introduces the necessary background on the MLIR framework, and discusses related work in compiler vectorization. Section 3 details the design and implementation of our proposed method. Section 4 presents our experimental setup and evaluates the performance of our approach. Finally, Sect. 5 concludes the paper and discusses directions for future work.

2   Background and Motivation

2.1   Perfect Loop Nests and the MLIR Infrastructure

To understand the core challenges in automatic vectorization, it is essential to first examine the loop structures that compilers confront. The effectiveness of optimizations hinges on a loop's internal characteristics, such as its memory access patterns and data dependencies. Among the various structures, one is particularly foundational for achieving high performance.

A particularly crucial structure for optimization is the **perfect loop nest** [9]. A loop nest is defined as "perfect" when all of its computational statements are located exclusively in the innermost loop. This structure, common in domains like dense linear algebra, provides a regular, predictable pattern of computation and memory access, making it an ideal candidate for advanced transformations like vectorization. A classic example is matrix multiplication:

```
for (int i = 0; i < M; i++)
  for (int j = 0; j < N; j++)
    for (int k = 0; k < K; k++)
      C[i][j] += A[i][k] * B[k][j];
```

However, manually programming for these architecture-specific instruction sets via intrinsics is complex, non-portable, and error-prone. Consequently, **automatic vectorization** has become crucial for bridging this gap. The success of this automation hinges on two critical tasks: legality and profitability analysis.

In stark contrast, many loops contain dependencies that fundamentally inhibit parallelism. The most challenging of these are **loop-carried dependencies**, where one iteration of a loop computationally depends on the result of a previous iteration. Consider the following simple loop:

```
for (int i = 1; i < n; i++) {
  a[i] = a[i-1] + b[i];
}
```

This loop cannot be naively vectorized because the calculation of a[i] requires the value of a[i−1], which was computed in the prior iteration. This creates a sequential chain that SIMD execution cannot break. For a traditional compiler operating on low-level IR, distinguishing the vectorizable structure of the preceding matrix multiplication example from the non-vectorizable loop with its dependency requires complex and often fragile data dependency analysis.

This is precisely the challenge that the Multi-Level Intermediate Representation (MLIR) framework is engineered to solve. It overcomes the representational limitations of traditional, single-level IR systems by introducing two core principles: an ecosystem of extensible **dialects** and a process of **progressive lowering**. Dialects allow the compiler to represent code at multiple levels of abstraction, while progressive lowering systematically transforms the representation from a higher level to a lower one. This design ensures that high-level semantic information is preserved throughout the optimization pipeline, rather than being discarded early.

Within this framework, the linalg dialect provides a first-class, declarative model for perfect loop nests, separating the specification of a computation ("what") from its concrete implementation as a loop nest ("how"). The linalg.generic operation defines a computation as a single, atomic unit through a declarative contract. Instead of implicit control flow, it uses two powerful attributes to make the loop's properties explicit:

- **indexing_maps:** A set of affine expressions that precisely describe the data access pattern for each operand.
- **iterator_types:** A list of semantic keywords (e.g., "parallel", "reduction") that define the nature of each loop dimension.

By construction, this model makes dependency information explicit and guarantees that the operation is a perfect loop nest, making subsequent transformations like tiling significantly simpler and more robust. This high-level linalg model is then lowered using supporting dialects: the affine dialect, with its mathematical rigidity, acts as a crucial intermediate step for materializing structured loops, and the vector dialect provides a target-independent abstraction for the final SIMD computations (e.g., vector.load, vector.addf, vector.store). This principled, model-driven approach forms the bedrock of our vectorization strategy.

2.2 Prior Works on Automatic Vectorization

A review of automatic vectorization techniques, particularly as they apply to the perfect loop nests that are foundational to scientific computing and machine learning, reveals several distinct paradigms. The core of our discussion centers on the dominant analysis-driven approach, contrasting it with model-driven paradigms that have proven successful in more specialized contexts. Critically, we find that each of these established methods has significant limitations in providing robust, general-purpose vectorization for perfect loop nests, and these shortcomings collectively form the motivation for our work.

The Analysis-Driven Paradigm. The most widespread approach to vectorization is exemplified by mainstream compilers such as GCC and LLVM/Clang. These systems typically employ a two-pronged strategy consisting of Loop Vectorization and **Superword-Level Parallelism (SLP)** vectorization [10]. The fundamental challenge for this paradigm stems from its reliance on a low-level

IR, such as GCC's GIMPLE or LLVM IR. At this level of abstraction, the high-level structural simplicity of a perfect loop nest is lost, forcing the compiler into a difficult process of reverse-engineering. This process, involving fragile legality and profitability analyses, often fails unpredictably due to minor code changes. Consequently, even for the structurally regular case of a perfect loop nest, this paradigm fails to generate consistently high-performance code.

The Model-Driven Paradigm in DSLs. A compelling alternative is the model-driven approach pioneered by Domain-Specific Languages (DSLs) like Halide [11] and TVM [12]. Halide introduced the core philosophy of decoupling the **algorithm** from the **schedule**, while TVM uses a **learning-based cost model** to find optimal schedules. The success of these DSLs demonstrates the power of preserving high-level semantics. However, their primary limitation is their domain-specific nature. To gain their benefits, a programmer must rewrite their algorithms, including any perfect loop nests, in a new, specialized language, making this approach unsuitable as a general-purpose solution for existing codebases.

The Polyhedral Transformation Framework [13]. Another powerful paradigm for loop optimization is the polyhedral model, a mathematically rigorous framework for transforming **Static Control Parts (SCoPs)** [14]. Implementations like **Pluto** [15] and LLVM's **Polly** [16] can effectively transform many perfect loop nests, which often fit the SCoP criteria. However, this model has two key limitations. First, its applicability is restricted to the domain of SCoPs, leaving many loops unhandled. Second, the complex code it generates can sometimes hinder, rather than help, the backend compiler's own auto-vectorizer, leading to unpredictable final performance.

In summary, each existing paradigm falls short. Traditional compilers are general-purpose but fragile and unreliable for vectorizing perfect loop nests. DSLs are powerful but domain-specific and not generally applicable. The polyhedral model is rigorous but limited in scope and can have unpredictable interactions with vectorization. This absence of a general-purpose yet semantically robust solution creates a clear opportunity for a new approach. We build our framework upon MLIR because, while it represents the state-of-the-art in compiler infrastructure, it currently lacks dedicated automatic vectorization strategies for perfect loop nests.

3   Framework Design

This section provides a detailed exposition of our novel tiling-aware optimization methodology, which is structured as a cohesive, three-phase framework.

Our tiling-aware optimization framework operates within the MLIR infrastructure, leveraging its progressive lowering capabilities to transform high-level linalg.generic operations into efficient machine code. As illustrated in Fig. 1, the process begins with linalg dialect representations of computations originating from various sources, including frontends like PyTorch [17] and TensorFlow [18] / XLA [19], as well as manually coded linalg operations. Our framework then analyzes and transforms these linalg operations, primarily utilizing the affine dialect for structured loop transformations and the vector dialect for generating efficient SIMD instructions with masking. Finally, the lowered code in the affine and vector dialects is further translated into LLVM IR, which serves as a common backend for generating machine code targeting diverse architectures such as x86, RISC-V, and ARM.

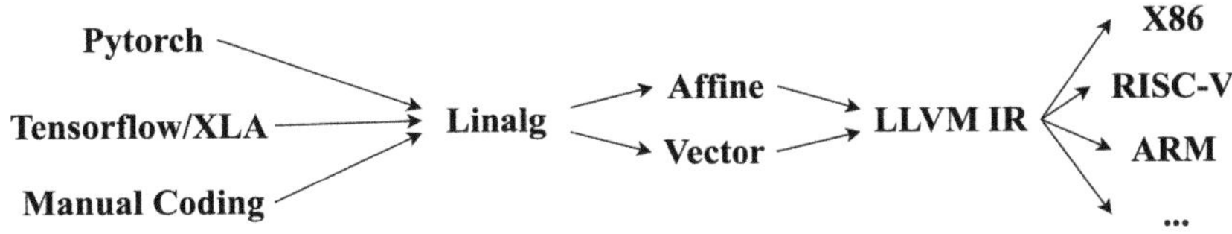

Fig. 1. Progressive lowering flow in our framework.

The first two phases, Comprehensive Analysis and Cost-Model-Based Decision, are tightly integrated to form a model-driven decision engine. The primary objective of this engine is to systematically deconstruct a high-level linalg.generic operation and derive a detailed **OptimizationBlueprint**, a plan that inherently captures a tiling-aware vectorization strategy. This blueprint serves as an unambiguous, prescriptive guide for the final Transformation and Code Generation phase. This integrated framework is distinguished by its reliance on the explicit semantics of the linalg dialect, allowing it to bypass the fragile analyses inherent to traditional compilers.

To make the process concrete, we will use the General Matrix Multiplication (GEMM) kernel, $C(m, n) + = A(m, k) \cdot B(k, n)$, as a running example to illustrate how the framework navigates complex optimization trade-offs.

```
#map_gemm = {
  indexing_maps = [
    affine_map<(m, n, k) -> (m, k)>, // Operand A
    affine_map<(m, n, k) -> (k, n)>, // Operand B
    affine_map<(m, n, k) -> (m, n)>  // Operand C (out)
  ],
  iterator_types = ["parallel", "parallel", "reduction"]
  // Loop order: m, n, k
}
linalg.generic #map_gemm
  ins(%A: tensor<128x256xf32>, %B: tensor<256x512xf32>)
  outs(%C: tensor<128x512xf32>) {
^bb0(%in1: f32, %in2: f32, %out: f32):
  %mul = arith.mulf %in1, %in2 : f32
  %add = arith.addf %out, %mul : f32
  linalg.yield %add : f32
}
```

Listing 1.1. MLIR representation of the GEMM kernel

3.1   Comprehensive Analysis

The first phase of the engine is dedicated to information gathering. It systematically inspects the linalg.generic operation and its hardware context to collect a set of objective facts, making no premature optimization decisions. This process culminates in a comprehensive **AnalysisResult** data structure that forms the immutable foundation for all subsequent cost model evaluations. The analysis is performed in three sequential steps.

First, a **Feasibility Analysis** acts as a prerequisite check. It verifies that the operation contains at least one loop dimension (iterator_types is not empty) and that all computations within its body (e.g., arith.mulf, arith.addf) are primitives that can be legally mapped to vector equivalents. For our GEMM example, this check passes, as it contains three loop iterators and its arithmetic operations are fully vectorizable.

Second, a **Hardware Capability Analysis** determines the theoretical performance ceiling. It calculates the maximum vectorization factor by dividing the target's vector register width by the bit-width of the operation's element type. For our running GEMM example, the maxVectorFactor is computed as $256/32 = 8$. This calculation is based on a target machine with 256-bit AVX2 registers operating on 32-bit floating-point numbers (f32).

Third, and most critically, a **Memory Pattern Analysis** builds a precise model of data access. This analysis is paramount as memory bandwidth is often the primary performance bottleneck.

Contiguity Analysis: By inspecting the indexing_maps, the engine determines which dimension corresponds to contiguous memory access for each operand. In the GEMM example, operand $A(m, k)$ is contiguous along the k dimension, while $B(k, n)$ and $C(m, n)$ are contiguous along the n dimension. This information is vital for identifying the most efficient vector load and store opportunities.

Dependency Analysis: The engine checks, for each parallel dimension, which input operands are invariant. For GEMM's parallel dimension n, the access to operand A, (m, k), does not depend on n. This identifies A as a candidate for a highly efficient **broadcast** operation if the n dimension is chosen for vectorization.

Upon completion, this phase yields an AnalysisResult containing all these facts: isVectorizable = true, maxVectorFactor = 8, and a detailed memory access model.

3.2   Cost-Model-Based Decision Making

With the objective facts from the AnalysisResult in hand, the decision phase begins. It moves beyond abstract heuristics to an analytical cost model, aiming to quantitatively estimate the performance of each potential vectorization strategy. This approach is grounded in the philosophy that the optimal choice is the one with the lowest estimated execution cost, a value derived from concrete hardware characteristics.

The Hardware Cost Parameter Library. The core of our model is a configurable library of hardware cost parameters. This library abstracts the performance of a target CPU by assigning an estimated cost, typically in clock cycles, to fundamental vector operations. This makes the framework extensible and portable: to target a new architecture, one primarily needs to provide a new cost library. An example library for a hypothetical x86 CPU with AVX2 might include:

Cost_VecLoad_Contiguous: 1 cycle. The cost for loading a vector from contiguous memory, which is highly efficient.

Cost_VecLoad_Stride: 8 cycles. The significantly higher cost for a strided load, which may require multiple micro-operations or gather instructions.

Cost_Broadcast: 3 cycles. The cost of broadcasting a single scalar value into a full vector register.

Cost_HorizontalReduce: 15 cycles. The substantial overhead for summing elements within a vector register, which involves costly shuffle and permute operations.

Candidate Evaluation. The process starts by generating a set of viable candidate strategies (e.g., {Strategy(VectorizeDim=m), Strategy(VectorizeDim=n), ...}). Then, for each candidate, the engine calculates a total estimated cost by summing the costs of the memory and computational operations required.

For our GEMM example, the cost estimation for the core operations within a single iteration of the vectorized loop proceeds as follows. We analyze the access pattern for each operand under each strategy:

$$\begin{aligned}
&\textbf{Cost(m):} \quad \text{Vectorizing along the } m \text{ dimension} \\
&\qquad\qquad\quad \text{A: Strided Load} + \text{B: Broadcast} \\
&\qquad\qquad\quad \approx 8 + 3 = \mathbf{11} \text{ cycles} \\
&\textbf{Cost(n):} \quad \text{Vectorizing along the } n \text{ dimension} \\
&\qquad\qquad\quad \text{A: Broadcast} + \text{B: Contiguous Load} \\
&\qquad\qquad\quad \approx 3 + 1 = \mathbf{4} \text{ cycles} \\
&\textbf{Cost(k):} \quad \text{Vectorizing along the } k \text{ dimension (a reduction)} \\
&\qquad\qquad\quad \text{A: Contiguous Load} + \text{B: Strided Load} + \text{Overhead} \\
&\qquad\qquad\quad \approx 1 + 8 + 15 \text{ (for Horizontal Reduce)} = \mathbf{24} \text{ cycles}
\end{aligned}$$

The analysis clearly shows that vectorizing along the n dimension yields the lowest estimated cost (4 cycles), making it the unequivocal choice.

On the Nature of the Cost Model. It is important to clarify the nature of these cost parameters and the model itself. The framework is designed to ingest a **Hardware Cost Parameter Library** where values are derived from concrete performance analysis, such as referencing CPU architecture manuals, using static analysis tools like llvm-mca, or conducting targeted micro-benchmarks. For clarity in this paper's example, the specific costs used are *illustrative values* based on typical x86 architectures, chosen to demonstrate the decision-making logic of the model.

Furthermore, we acknowledge that this cost model represents a **first-order approximation** of performance. It intentionally simplifies total cost as a linear sum of individual operation costs, focusing on the dominant effect of memory access patterns. This design choice prioritizes model generality and tractability over capturing more complex, second-order micro-architectural effects like instruction-level parallelism or latency hiding. Integrating more sophisticated performance models, such as those based on instruction throughput or dependency graph analysis [20], remains a promising direction for future work.

The engine finalizes this phase by constructing the OptimizationBlueprint. This blueprint is a comprehensive record that captures the winning decision: it prescribes vectorizing the n dimension with a factor of 8, specifies a **Broadcast** strategy for operand A and **VectorLoad/VectorStore** for B and C respectively, and includes a multi-dimensional tiling plan to optimize data reuse in caches.

Algorithm 1. Cost-Model-Based Vectorization Decision Engine

Input: Op (a linalg.generic operation), **HardwareCostParams** (cost library)
Output: OptimizationBlueprint or Failure
  1: // — *Phase 1: Comprehensive Analysis* —
  2: Create AnalysisResult struct
  3: AnalysisResult.isVectorizable ← IsVectorizationFeasible(Op);
  4: **if** not AnalysisResult.isVectorizable **then**
  5:     return Failure
  6: **end if**
  7: AnalysisResult.maxVectorFactor ← HardwareCostParams.vectorBits / GetElementBits(Op);
  8: AnalysisResult.memoryAccessInfo ← AnalyzeMemoryAccessPatterns(Op);
  9: // — *Phase 2: Cost-Model-Based Decision Making* —
 10: Candidates ← GenerateCandidateStrategies(AnalysisResult);
 11: **for** each candidate in Candidates **do**
 12:     candidate.cost ← CalculateStrategyCost(candidate, AnalysisResult.memoryAccessInfo, HardwareCostParams);
 13: **end for**
 14: BestStrategy ← SelectLowestCostCandidate(Candidates);
 15: Blueprint ← BuildOptimizationBlueprint(BestStrategy, AnalysisResult);
 16: return Blueprint;

3.3   Transformation and Code Generation

The final phase of our framework consumes the OptimizationBlueprint generated by the cost model and executes the prescribed transformations, lowering the high-level linalg.generic operation into an efficient, hardware-aware implementation. This process, outlined in Algorithm 2, integrates loop restructuring for data locality with kernel vectorization into a single, cohesive pass.

Loop Structuring via Tiling. Guided by the Blueprint.TilingPlan, this stage first restructures the computation. Using the affine dialect, whose mathematical rigidity is ideal for reasoning about structured loops, it generates a new loop nest that iterates over tiles of the original data. This serves the dual purpose of enhancing data locality by operating on cache-friendly data chunks, and preparing a perfectly structured inner kernel for vectorization.

Kernel Vectorization. Within the tiled loop structure, the kernel is vectorized according to the blueprint. The loop along the chosen vectorization dimension is transformed to step by the vector factor. The per-operand strategies defined in the blueprint guide the generation of memory operations in the vector dialect, such as vector.splat for broadcasted operands and vector.transfer_read for contiguous loads.

Unified Tail Handling with Masking. A key feature of our approach is the robust and efficient handling of loop tails. Instead of generating a separate, scalar epilogue loop – which increases code size and introduces complex control flow – we employ a unified, mask-based strategy. For any iteration that is not a multiple of the vector width, a vector.create_mask operation computes a bitmask indicating which SIMD lanes are active. This mask is then used by the vector.transfer_read and vector.transfer_write operations to ensure only valid memory elements are accessed. This approach maintains high performance for the entire loop nest and results in cleaner, more maintainable code.

Algorithm 2. Tiling and Vectorization Code Generation

Input: Op (original linalg.generic op), **Blueprint** (the optimization plan)
Output: Transformed, vectorized MLIR code
1: // *1. Apply Tiling for Locality*
2: TiledOp ← TILE(Op, Blueprint.TilingPlan);
3: // *2. Vectorize the Innermost Kernel*
4: InnerLoop ← getInnermostLoop(TiledOp);
5: VectorLoop ← setLoopStep(InnerLoop, Blueprint.vectorFactor);
6: // *3. Map Operands and Operations to Vector Dialect*
7: **for** each operand in Op **do**
8:     strategy ← Blueprint.OperandStrategies[operand];
9:     GENERATE_VECTOR_TRANSFER(operand, strategy); // *Emits masked reads/writes*
10: **end for**
11: REPLACE_SCALAR_OPS(VectorLoop.body); // *Maps arith ops to vector ops*
12: REPLACE(Op, TiledOp);

In summary, this section has detailed a complete, three-phase, model-driven vectorization framework. It begins with a comprehensive analysis of the high-level semantics of linalg operations, proceeds to make an optimal decision via an analytical cost model, and finally, in the code generation phase, materializes the abstract optimization blueprint into a concrete implementation that integrates tiling for data locality with efficient masked vectorization. The core advantage of this design is its ability to perform reliable and predictable optimizations based on explicit semantics, fundamentally avoiding the fragility of the analysis-recovery techniques upon which traditional compilers depend.

The ultimate measure of a robustly designed framework, however, is its empirical performance. Accordingly, the next section presents a comprehensive experimental evaluation. We will assess the performance and robustness of our generated code against established, state-of-the-art compilers across a variety of computational kernels to validate the practical efficacy of our approach.

4 Experimental Evaluation

To assess the effectiveness and robustness of our proposed vectorization framework, we conducted a series of experiments designed to evaluate its performance against both an unoptimized baseline and state-of-the-art traditional compilers. This section details our experimental setup, presents the performance results from two distinct sets of experiments, and provides an in-depth analysis of the findings, linking them back to the design principles of our method.

4.1 Experimental Setup

Hardware and Software Environment. All experiments were conducted on a server equipped with an Intel Xeon Platinum 8369B processor, capable of reaching speeds up to 3.500 GHz. The operating system was Ubuntu 22.04.5 LTS. Crucially, this CPU supports the **AVX2** instruction set, and all compilations were specifically targeted to utilize its 256-bit vector capabilities. For all performance measurements, program execution was pinned to a single core to ensure a fair and focused comparison of single-threaded code generation quality.

Methodology and Baselines. We evaluated a suite of common linear algebra kernels [21], including AXPY, GEMM, Dot-Product, Outer-Product, Batch-Mat-Vec, Batch-GEMM, Conv2D, and Softmax. For our scalability analysis, we focused on the compute-intensive GEMM and Conv2D kernels. We compare the performance of four different versions for each kernel:

- **baseline:** The MLIR implementation of each kernel as a linalg.generic operation, without our optimization pass applied. Its performance is normalized to 1x and serves as the reference for calculating speedup.
- **g++:** Equivalent C++ implementations compiled using GCC (g++) with its highest standard optimization level (-O3).

- **clang:** Equivalent C++ implementations compiled using Clang with aggressive optimizations enabled (-O3), including targeting the native CPU architecture (-march=native) and allowing fast-math computations (-ffast-math).
- **ours:** The MLIR implementation compiled with our proposed vectorization pass enabled.

We deliberately chose the -O3 optimization level as it represents the highest standard setting that guarantees strict adherence to language standards and numerical precision. While more aggressive, non-standard options like -Ofast exist, our choice ensures a fair and rigorous comparison against stable, industry-standard optimizations that do not alter the program's floating-point semantics. This provides a robust baseline to evaluate our structured, semantics-driven approach.

Tiling Strategy. To isolate the analysis on our vectorization decision framework, we employed a fixed, heuristic-based tile size for all experiments. Specifically, the dimension chosen for vectorization was tiled by the vector width (8 in our case), while other parallel dimensions were tiled by a factor of 16 to improve data reuse in the L1 cache. We acknowledge that this fixed tiling strategy, while simplifying the analysis of our vectorization decision engine, means the observed speedups reflect the combined effect of both tiling and vectorization. A precise decoupling of these two effects would require further experimentation, which we leave for future work. Determining optimal, kernel-specific tile sizes is a complex research problem in its own right; thus, a more sophisticated auto-tiling mechanism is another clear avenue for future work.

It is also worth noting that our proposed framework is implemented as a single, lightweight pass within the MLIR infrastructure. While a formal analysis of compilation overhead was not the primary focus of this study, we observed that the additional time introduced by our pass was negligible in the context of the entire compilation pipeline, which includes multiple stages of MLIR lowering and backend code generation by LLVM. This ensures that the significant performance gains are achieved without imposing a substantial penalty on the compilation time.

The primary metric for evaluation is **speedup**, calculated as the execution time of the baseline version divided by the execution time of the optimized version. Each benchmark was run multiple times, and the average execution time was used to ensure stable and reliable results.

4.2   Performance on Various Operator Kernels

The first experiment evaluates the breadth and effectiveness of our method across the diverse set of operator kernels. The results are presented in Fig. 2.

As the figure clearly shows, our method ('ours') achieves significant and robust speedups across all tested kernels. For compute-intensive operations such as GEMM, Batch-GEMM, and Conv2D, our approach not only vastly outperforms the baseline but is also significantly faster than both g++ and clang.

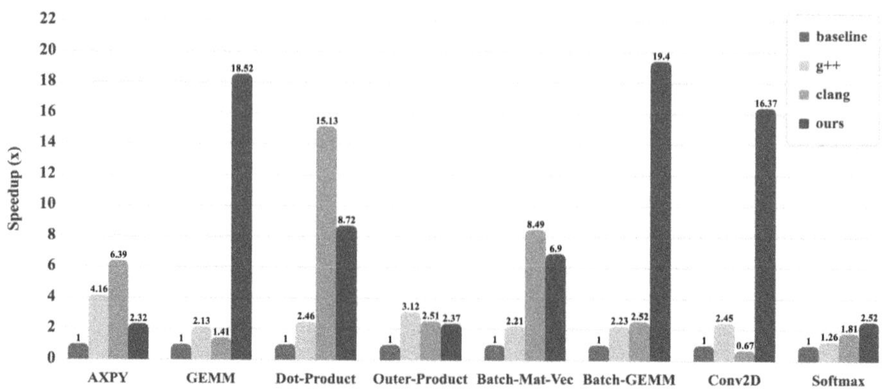

Fig. 2. Performance speedup for various operator kernels.

This success can be directly attributed to the design of our analysis and decision engine from Sect. 3. For these complex operations, the engine's model-aware analysis of memory access patterns allows it to identify the truly optimal vectorization dimension – the one offering contiguous memory access – which our analytical cost model correctly identifies as the lowest-cost strategy. This guarantees a transformation strategy that maximizes memory bandwidth utilization.

For kernels with different architectural properties, the results are also insightful. For AXPY, a classic memory-bound operation, all three optimized versions ('ours', g++, clang) show similar, significant speedups. This is expected, as performance is ultimately limited by memory bandwidth rather than computation, and all competent compilers converge on a similar, optimal instruction sequence. For Dot-Product, a reduction operation, the performance of our method is strong but comparable to clang. This reflects our engine's correct modeling of reduction overhead; while vectorization is beneficial, the final horizontal sum across the vector lanes incurs a cost that prevents the kind of extreme speedups seen in purely parallel kernels.

4.3   Scalability and Theoretical Performance Analysis

The second experiment investigates the performance scalability of our method on the GEMM and Conv2D kernels as the input data size increases. The data scale on the x-axis is calculated as the product of all data dimensions (e.g., $M \times N \times K$ for GEMM). To ground our analysis, we first establish the theoretical peak speedup for this platform. Given the target instruction set is AVX2 (256-bit) and the data type is 32-bit float, the theoretical peak speedup from vectorization alone is $256/32 = \mathbf{8x}$.

The performance trends are shown in Fig. 3. For GEMM, our method achieves a peak speedup of nearly 20x at small-to-medium data scales. However, as the data size becomes very large, the speedup gracefully degrades and stabilizes around the theoretical peak of 8x. This behavior clearly illustrates the interplay

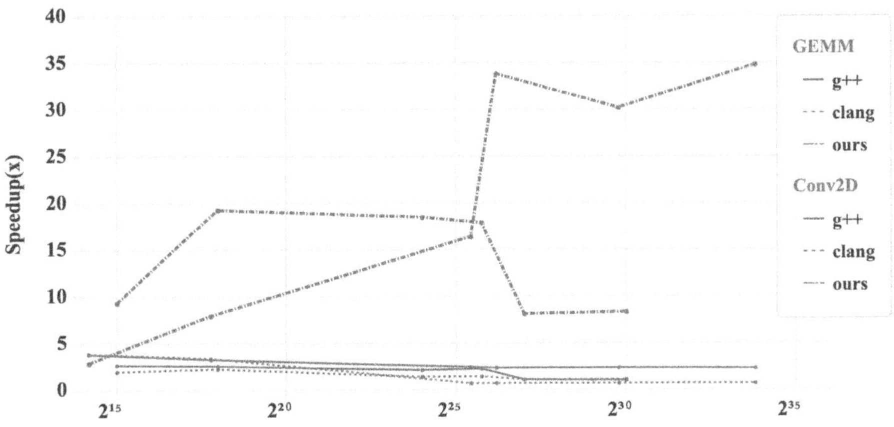

Fig. 3. Speedup trend as data scale increases for GEMM and Conv2D.

between tiling and vectorization. At smaller scales, the working set fits within the CPU cache, and the performance benefits from both the improved data locality of our tiling strategy and the computational throughput of vectorization. As the data size exceeds the cache capacity, the operation becomes memory-bound, effectively nullifying the benefit of tiling. The remaining 8x speedup is the pure, direct benefit of the AVX2 vector instructions, demonstrating that our framework extracts the maximum possible performance under memory-bound conditions.

Conversely, the performance of Conv2D demonstrates the powerful impact of our tiling-aware optimization. Its speedup continues to climb with the data scale, reaching over 30x – far exceeding the theoretical vectorization peak. This result clearly shows how a tiling-aware strategy creates a synergistic effect: by effectively exploiting the high data reuse in Conv2D, our framework's tiling strategy ensures the operation remains compute-bound even at large scales by keeping frequently accessed data in the cache. This drastically reduces memory latency, allowing the 8x computational speedup from vectorization to be fully realized and amplified, resulting in exceptional overall performance. This result highlights our framework's ability to adapt its strategy to the intrinsic properties of the algorithm.

5 Conclusion

In this paper, we presented a novel framework for the high-performance vectorization of perfect loop nests, driven by a tiling-aware optimization strategy within the MLIR compiler infrastructure. Our approach fundamentally differs from traditional compilers by operating on the high-level, semantic-rich representation of the 'linalg' dialect, thereby bypassing the fragile and complex analyses required to reverse-engineer program structure from low-level IR. We designed a cohesive, three-phase methodology that automates the entire optimization process: from a comprehensive analysis of memory patterns and hardware capabilities, through a principled decision engine guided by an analytical cost model, to a unified code generation pass that integrates tiling and robust masked vectorization.

Our experimental results validate the effectiveness of this model-driven approach. Across a diverse set of linear algebra kernels, our framework consistently generates high-performance code, outperforming industry-standard compilers like g++ and clang in compute-intensive scenarios such as GEMM and Conv2D. The scalability analysis further revealed our method's ability to effectively leverage cache locality through tiling in synergy with vectorization, achieving speedups that significantly exceed the theoretical peak of vectorization alone.

While our current implementation demonstrates strong performance on a key CPU architecture, the principles of our framework are designed for portability. Future work will focus on extending this methodology to more diverse hardware targets. One clear avenue is to enhance the generality of our analytical cost model, which currently relies on a hardware cost parameter library requiring manual calibration for new architectures. To address this limitation, we plan to develop an automated mechanism for constructing this library, potentially using micro-benchmarks or static analysis tools to profile a target's instruction latencies and throughputs. This would make the framework more readily adaptable to new hardware platforms. For instance, extending the framework to GPU architectures [22] presents a more significant challenge. Beyond interfacing with a new code generation backend (e.g., to produce the **gpu** dialect), our cost model would require a fundamental redesign. It would need to account for the unique memory hierarchy of GPUs (shared vs. global memory) and their **Single Instruction, Multiple Threads (SIMT)**execution model, which differs substantially from the SIMD model of CPUs in terms of parallelism granularity and memory access patterns. Ultimately, this work demonstrates that by preserving and leveraging high-level semantics, compilers can achieve more robust, predictable, and powerful automatic optimizations.

References

1. Franchetti, F., Kral, S., Lorenz, J., Ueberhuber, C.: Efficient utilization of SIMD extensions. Proc. IEEE **93**(2), 409–425 (2005). https://doi.org/10.1109/JPROC.2004.840491
2. Slotnick, D.L., Borck, W.C., McReynolds, R.C.: The Solomon computer. In: Proceedings of the 4–6 December 1962, Fall Joint Computer Conference, AFIPS 1962, pp. 97–107 (Fall). Association for Computing Machinery, New York, NY, USA (1962). https://doi.org/10.1145/1461518.1461528
3. Lomont, C.: Introduction to intel advanced vector extensions. Intel White Paper **23**(23), 1–21 (2011)
4. Jang, M., Kim, K., Kim, K.: The performance analysis of arm neon technology for mobile platforms. In: Proceedings of the 2011 ACM Symposium on Research in Applied Computation, pp. 104–106 (2011)

5. Lattner, C., Adve, V.: LLVM: a compilation framework for lifelong program analysis & transformation. In: International Symposium on Code Generation and Optimization, CGO 2004, pp. 75–86. IEEE (2004)
6. Krainiuk, M., Goli, M., Pascuzzi, V.R.: oneAPI open-source math library interface. In: 2021 International Workshop on Performance, Portability and Productivity in HPC (P3HPC), pp. 22–32. IEEE (2021)
7. Lattner, C., et al.: MLIR: scaling compiler infrastructure for domain specific computation. In: 2021 IEEE/ACM International Symposium on Code Generation and Optimization (CGO), pp. 2–14. IEEE (2021)
8. MLIR Community: Linalg dialect - MLIR documentation (2023). https://mlir.llvm.org/docs/Dialects/Linalg/. Accessed 13 July 2025
9. Li, H., Yan, P., Liu, H., Han, L., Nie, K., Chen, M.: Research on uncovering parallelism in perfect loop nests for GCC compiler. J. Phys. Conf. Ser. **2906**, 012013 (2024)
10. Chen, Y., Mendis, C., Amarasinghe, S.: All you need is superword-level parallelism: systematic control-flow vectorization with SLP. In: Proceedings of the 43rd ACM SIGPLAN International Conference on Programming Language Design and Implementation, pp. 301–315 (2022)
11. Ragan-Kelley, J., Barnes, C., Adams, A., Paris, S., Durand, F., Amarasinghe, S.: Halide: a language and compiler for optimizing parallelism, locality, and recomputation in image processing pipelines. SIGPLAN Not. **48**(6), 519–530 (2013). https://doi.org/10.1145/2499370.2462176
12. Chen, T., et al.: TVM: an automated end-to-end optimizing compiler for deep learning. In: Proceedings of the 13th USENIX Conference on Operating Systems Design and Implementation, OSDI 2018, pp. 579–594. USENIX Association, USA (2018)
13. Zhao, R., Cheng, J.: Phism: Polyhedral high-level synthesis in MLIR. arXiv preprint arXiv:2103.15103 (2021)
14. Bondhugula, U., Hartono, A., Ramanujam, J., Sadayappan, P.: A practical automatic polyhedral parallelizer and locality optimizer. SIGPLAN Not. **43**(6), 101–113 (2008). 22.1375595. https://doi.org/10.1145/1379022.1375595
15. Ferreira, J.D., et al.: pLUTo: enabling massively parallel computation in dram via lookup tables. In: 2022 55th IEEE/ACM International Symposium on Microarchitecture (MICRO), pp. 900–919. IEEE (2022)
16. Grosser, T., Groesslinger, A., Lengauer, C.: Polly – performing polyhedral optimizations on a low-level intermediate representation. Parallel Process. Lett. **22**(04), 1250010 (2012). https://doi.org/10.1142/S0129626412500107
17. Imambi, S., Prakash, K.B., Kanagachidambaresan, G.: Pytorch. In: Programming with TensorFlow: Solution for Edge Computing Applications, pp. 87–104. Springer (2021)
18. Developers, TensorFlow: Tensorflow. Zenodo (2022)
19. Bradbury, J., et al.: JAX: Autograd and XLA. Astrophysics Source Code Library, ascl–2111 (2021)

20. Abel, A., Reineke, J.: uiCA: accurate throughput prediction of basic blocks on recent intel microarchitectures. In: Proceedings of the 36th ACM International Conference on Supercomputing, pp. 1–14 (2022)
21. Psarras, C., Barthels, H., Bientinesi, P.: The linear algebra mapping problem. Current state of linear algebra languages and libraries. ACM Trans. Math. Softw. (TOMS) **48**(3), 1–30 (2022)
22. Kim, J.S., McCaskey, A., Heim, B., Modani, M., Stanwyck, S., Costa, T.: CUDA quantum: the platform for integrated quantum-classical computing. In: 2023 60th ACM/IEEE Design Automation Conference (DAC), pp. 1–4. IEEE (2023)

MLog: Achieving Low-Latency, Scalable Shared Log Writes via RDMA Multicast Protocol

Yanan Tao[1], Miao Cai[2(✉)], and Baoliu Ye[3]

[1] Hohai University, Nanjing 211100, China
[2] Nanjing University of Aeronautics and Astronautics, Nanjing 211106, China
`miaocai@nuaa.edu.cn`
[3] Nanjing University, Nanjing 210023, China

Abstract. The share log offers an efficient means to design reliable, coherent, and scalable cloud and distributed systems. However, current shared-log-based systems suffer from poor write performance, severely degrading overall throughput and impacting scalability. Specifically, traditional replication protocols, such as chain replication and primary-backup replication, lead to high network management overheads and excessive network round trips. Besides, existing shared log approaches use a centralized sequencer to realize the total ordering. Unfortunately, concurrent client requests inevitably render the centralized sequencer to be a scalability bottleneck.

We present MLog, a shared log system that leverages the RDMA multicast protocol to achieve low-latency, scalable write paths. We propose a broadcast-based replication mechanism, which achieves parallel, fault-tolerant data replication upon scalable RDMA multicast. Next, we design a decentralized ordering protocol. Leveraging the metadata embedded in the replication packet, every replica node performs localized, total ordering without expensive global coordination. We also propose a lightweight endpoint-driven recovery technique to overcome the reliability issue in the RDMA multicast mechanism. Compared to the state-of-the-art shared log system (i.e., LazyLog), MLog reduces the write latency by 19% and improves the throughput by 3.4× for various real-world workloads.

Keywords: Shared Log · Distributed System · RDMA Multicast

1 Introduction

A shared log is an abstraction that provides a totally ordered sequence of records, allowing multiple clients to append and retrieve records concurrently [1–4]. While conceptually simple, this abstraction is highly expressive and can serve as a fundamental building block for constructing distributed systems. In recent years, shared logs have become a key infrastructure component in cloud computing and data center systems. Architectures based on shared logs have emerged as the dominant paradigm for building large-scale, high-throughput messaging systems. This trend is particularly prevalent in commercial cloud services, with almost all major cloud platforms offering shared log-based messaging services, including Amazon Kinesis [5], Google Pub/Sub [6], and Microsoft Azure Event Hubs [7].

While the shared log abstraction offers a simple yet powerful model, its performance in practice is fundamentally shaped by the efficiency of the write path. This path governs both the per-operation latency and the system's throughput under concurrent workloads. As such, optimizing the write path is central to building high-performance shared log systems. However, a closer exploration of existing designs reveals that this critical path is often hindered by two core bottlenecks: the overhead of replication and the cost of establishing a total order.

First, shared log systems that employ traditional replication protocols such as chain replication and primary-backup replication face performance limitations in terms of latency and scalability. Chain replication, as used in CORFU [1], requires a record to traverse all replicas in sequence before being committed. This propagation causes write latency to increase with the number of replicas, making it difficult to scale without incurring significant performance penalties. In contrast, the primary-backup model, adopted by systems like Scalog [3], addresses the latency issue by allowing the primary to replicate to backups in parallel. However, this comes at the cost of scalability. In this model, all client writes must go through a single primary. More critically, the primary must manage and maintain a large number of connections, which becomes a substantial burden under high concurrency. This per-connection overhead effectively turns the primary into the scalability bottleneck of the shard. Thus, traditional replication methods inherently suffer from excessive network round trips and high connection management overheads, ultimately limiting both latency and scalability.

Second, most systems enforce total ordering through a centralized sequencing layer, which limits scalability and degrades performance under high write concurrency. In CORFU [1] and LazyLog [4], each write must obtain a global log position from a centralized sequencer. However, there are limits to the capacity of a centralized ordering layer to process records. As the number of concurrent clients increases, this centralized component becomes a throughput bottleneck. In summary, the high overhead of traditional replication results from serial data copies or network management, while centralized ordering services limit scalability. These problems reflect insufficient parallelism and the lack of a one-to-many delivery mechanism in the write and ordering paths.

To address both bottlenecks, we explore redesigning the write and ordering paths around RDMA multicast. This one-to-many transport sends records to all replicas in parallel, eliminating the sequential delay in chain replication. It also uses direct client-to-replica communication, avoiding network connection overhead of primary-backup replication. The inherent parallelism of multicast aligns well with our goal of decentralizing the ordering process, providing a natural foundation for rethinking shared log design. These features form the basis of a more scalable and efficient shared log design.

We propose MLog, a shared log system that leverages RDMA multicast to enable a low-latency and scalable write path. In MLog, a client performs each write by multicasting the record along with ordering metadata to all replicas in a single round of unreliable multicast. This design replaces excessive network round trips in chain replication with parallel, one-to-many delivery, thereby eliminating cumulative latency. At the same time, it avoids the scalability bottlenecks of primary-backup schemes by removing the overheads to maintain a large number of point-to-point connections. Although RDMA multicast uses unreliable transport, MLog ensures fault tolerance through lightweight endpoint logic: replicas detect dropped messages by tracking sequence number gaps and request retransmissions when needed, maintaining high performance even under packet loss.

Crucially, decentralized ordering over unreliable multicast is challenging. Packets may be lost or arrive out of order, and replicas cannot coordinate with each other to agree on a consistent order. To solve this, each client adds ordering metadata to every record it sends. This metadata includes the client ID, a sequence number, and a timestamp. Upon receiving a record, each replica applies a simple local rule to determine its position. This rule typically performs a lexicographical comparison of the metadata. As a result, all replicas in the shard reach the same local order without any coordination or a leader. A lightweight global layer then merges these local logs to form a consistent total order. This design eliminates the need for a centralized sequencer and allows MLog to scale under heavy concurrency.

We implement the client-side RDMA multicast transmission of MLog as a C-based dynamic library, while the data and storage layers are implemented in Go. We use CGO to enable interoperability between the Go and C code. For communication between the data and ordering layers, we adopt gRPC and Google Protocol Buffers [8] for message efficient serialization.

We have conducted an extensive evaluation on CloudLab. The results demonstrate that MLog significantly outperforms existing systems. When operating at similar throughput levels, MLog reduces append latency by up to 5x compared to CORFU and by 19% compared to the state-of-the-art LazyLog (Erwin). More importantly, MLog showcases superior scalability where prior systems falter. Under high multi-client concurrency, MLog's throughput scales nearly linearly, achieving over 3.4x the throughput of LazyLog, while maintaining consistently low, microsecond-level latency. In contrast, LazyLog's throughput stagnates, and its latency rises sharply. We further demonstrate that MLog's throughput continues to scale with the number of shards, a critical property for large-scale deployments.

In summary, this paper makes the following contributions:

We identify and analyze two performance bottlenecks in traditional shared log replication architectures: the linear latency scaling of chain replication and the connection-handling scalability limitations of primary-backup models, in addition to the undoubted bottleneck of centralized sequencers.

We present MLog, a shared log system that achieves low-latency, scalable replication via a single-round, unreliable RDMA multicast broadcast, made fault tolerant through endpoint-driven recovery. By embedding ordering metadata in each record, MLog enables replicas to independently perform consistent local ordering, which is finalized into a global total order by a lightweight global ordering layer, thus eliminating the need for a centralized sequencer.

We conduct a comparative evaluation of MLog against representative shared log systems on CloudLab. The results show that MLog significantly outperforms state-of-the-art systems, reducing write latency by 19% and improving throughput by 3.4x, confirming the effectiveness of our design in achieving both low latency and high scalability.

2   Background and Motivation

2.1   Shared Log Design

A shared log exposes a simple API that enables clients to concurrently append and read records. The client can use the append interface to add an entry to the log and return its location. The shared log also provides a total order of records, all of which can each be bound to a log position. The read interface accepts a position in the log and returns an entry for that position.

As a simple yet powerful abstraction, shared log has garnered significant attention and application in both academia and industry. Shared logs have been used by QuickSilver [9] and Camelot [10] for failure atomicity and node recovery since the late 1980s, and subsequent studies have focused on their technical optimization. In industry, this model is foundational to numerous cloud services. Notably, even systems that relax the total ordering guarantee to per-shard ordering have demonstrated excellent performance [11,12], showcasing the versatility of the shared log concept.

2.2   Performance Bottlenecks in Existing Shared Logs

While the shared log abstraction provides an elegant model for distributed applications, its practical implementations often struggle to deliver both low latency and high scalability simultaneously. A close examination of existing systems reveals that this challenge stems from two primary sources of overhead on the critical write path: the replication protocol used to distribute records, and the sequencing mechanism used for total ordering.

Bottlenecks in Replication Approaches. To ensure data durability, shared log systems usually adopt one of two common replication methods, each with its own trade-offs between latency and scalability. The first, chain replication, used in systems like CORFU [1] (Fig. 1a), is conceptually simple but incurs high write latency due to excessive network round trips. In this model, a record must be forwarded from one replica to the next along the chain, and the write is only acknowledged after it reaches the tail. As a result, the latency grows linearly with the number of replicas, making it unsuitable for scenarios requiring low latency. The second method, primary-backup replication, as used in Scalog [3]

(Fig. 1b), reduces latency by allowing the primary to replicate to all backups in parallel. However, it introduces a serious scalability issue: all client writes must go through a single primary. The primary becomes a central point for handling all client requests and must maintain connections with every client. As the number of clients grows, this connection overhead puts heavy pressure on the primary, eventually becoming a performance bottleneck for the whole shard. In short, current systems must choose between high latency (with chain replication) and poor scalability (with the primary-backup model).

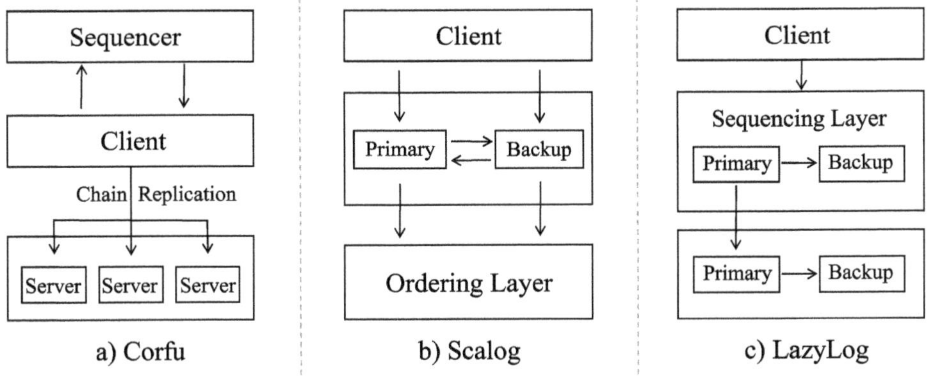

Fig. 1. Write Paths of Representative Shared Log Systems.

Scalability Limitations of Centralized Ordering. The second major bottleneck arises from the reliance on a centralized entity to establish a total order, which fundamentally limits system scalability. Systems ranging from CORFU [1], with its classic central sequencer, to more advanced designs like LazyLog [4], which employs a stateful Sequencing Layer, all share this characteristic (Fig. 1a, 1c). This centralized design creates a single point of serialization for all append requests from all clients. As client concurrency increases, this central coordinator inevitably becomes saturated, creating a throughput bottleneck that caps the entire system's performance, regardless of the resources allocated to the distributed data replicas. This serialization point is the root cause of poor scalability in high-concurrency workloads, preventing these systems from fully leveraging the power of modern multi-core servers and high-speed networks.

2.3   Opportunity: RDMA Multicast Mechanisms

The performance bottlenecks discussed above suggest that existing designs may have reached their limits. To break through these limits, we revisit the design of shared log write paths. In particular, we consider whether changing the communication model itself could offer a simpler and faster solution. Instead of relying on multi-round, point-to-point replication and centralized ordering, we wonder a single communication primitive handle both replication and ordering more efficiently.

In our search for such a primitive, RDMA multicast emerges as a strong candidate. It allows one sender to transmit data to many receivers at once, eliminating the sequential delay caused by chain replication. It also avoids the connection overhead of primary-backup schemes, as it does not require a per-client state. These properties make it attractive for building a faster and more scalable shared log. However, RDMA multicast operates on an unreliable transport [13]. Packets may be dropped, arrive out of order, or be duplicated. This makes it difficult to use directly in systems that require reliable and consistent behavior.

To make use of RDMA multicast in this setting, we must rethink where reliability and ordering are enforced. Rather than depending on the network to provide strong guarantees, we explore moving these responsibilities to the endpoints. This idea forms the basis of MLog. By redesigning the replication and ordering logic at the replica side, MLog aims to achieve both low-latency writes and scalable ordering over unreliable multicast. The next section describes how this is done in detail.

3   Design

3.1   Overview

MLog builds on a simple but powerful design principle. It replicates records in a single round using RDMA multicast. It also handles ordering and recovery at the endpoints, instead of relying on the network or a centralized service.

To realize this principle, MLog makes two key design decisions. First, it adopts scalable multicast to send each record to all replicas at once. This avoids extra network hops and reduces client-side connection overhead. Second, it embeds ordering metadata into each multicast packet. This allows replicas to independently determine the order of records without coordination or leader election. Together, these two decisions turn an unreliable transport into a low-latency and scalable write path. Reliability and total ordering are ensured by lightweight end-point logic, not by the network or a centralized sequencer.

We now describe how MLog puts this design into practice. As illustrated in Fig. 2, the MLog system architecture consists of three primary components: a client library, a distributed storage layer, and a global ordering layer.

Client Library. The client library is responsible for preprocessing records for the consolidated write path. Before transmission, it embeds each record with a crucial metadata tuple: (clientID, sequenceNumber, timestamp). This pre-assigned ordering tuple is multicast along with the record, enabling deterministic, coordination-free ordering at the storage replicas. This design is essential for imposing order on the inherently unordered and lossy RDMA Unreliable Datagram (UD) transport.

Storage Layer. The storage layer is partitioned into multiple logical shards for scalability. Each shard comprises $2f + 1$ replicas, allowing it to tolerate up to

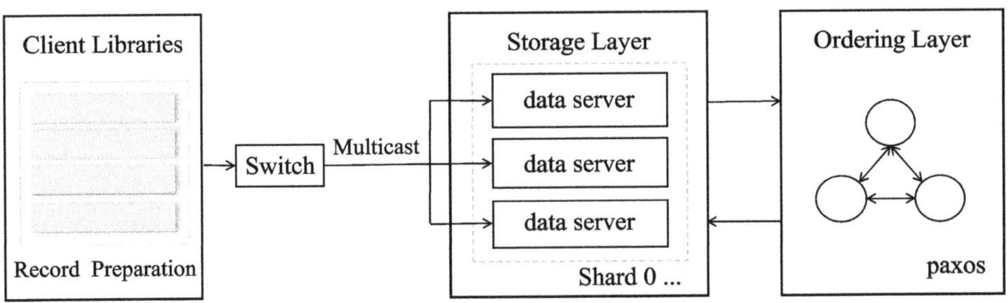

Fig. 2. MLog Architecture.

f node failures. Upon receiving a record via RDMA multicast, a storage server immediately sends an acknowledgement (ACK) back to the client. This receive-and-acknowledge approach allows the client to confirm a write quickly, while the more complex task of ordering is deferred to a background process, significantly minimizing latency on the critical write path. To ensure correctness over unreliable transport, all replicas within a shard apply an identical deterministic ordering algorithm to the received records. This is complemented by a lightweight loss detection and recovery protocol, ensuring that all replicas converge to an identical state.

Table 1. The MLog API. This provides a summary of the core functions available to clients.

API	Description
`Append(r, ack)`	Appends record r and returns an acknowledgment once it is durably stored
`Check()`	Returns the highest committed global sequence number
`Subscribe(l)`	Subscribes to a stream of records starting from global sequence number l
`Read(g, s)`	Reads the record with global sequence number g from shard s
`Trim(l)`	Deletes all records with global sequence numbers less than l

Global Ordering Layer. The global ordering layer aggregates the locally established orders from each shard. It then employs a consensus protocol, such as Paxos, to establish a globally consistent sequence for all records. This definitive global order is subsequently disseminated back to the individual shards to facilitate reads and further processing.

API. MLog provides a simple yet powerful API, as shown in Table 1. Append confirms persistence with an acknowledgment, Check reports commit progress, Subscribe and Read support continuous and direct record access, and Trim enables garbage collection. These interfaces offer a minimal yet sufficient abstraction for building high-performance shared log systems.

3.2 The Write Mechanism

MLog uses a broadcast-based write path to solve a coherence problems in shared log systems: high replication overhead. MLog replicates records in a single round using RDMA multicast. This removes the sequential delay of chain replication. It also avoids the connection overhead of primary-backup models. As a result, MLog achieves both low latency and high scalability. However, multicast over RDMA is unreliable and unordered. To make this primitive usable for strongly consistent logging, MLog pushes ordering and recovery logic to the endpoints.

To implement this mechanism, each client embeds an ordering tuple in every multicast packet. The tuple contains a timestamp, a sequence number, and a client ID. Replicas use this metadata to assign a logical position to the record on arrival. They do so without coordination or leader election. MLog uses RDMA's Unreliable Datagram (UD) mode, which supports one-to-many communication. Compared with the Reliable Connection (RC) mode, it requires far fewer client connection states. This allows MLog to scale well under high concurrency.

Unlike traditional designs, the client does not wait for acknowledgments or perform retransmissions. Instead, it sends each record once and each replica handles reliability independently. Upon receiving a record, a replica stores it locally and sends an acknowledgment immediately. Further steps, such as detecting loss and applying ordering, are handled asynchronously at the replica. This design keeps the client-side write path extremely lightweight and enables MLog to maintain high throughput even under packet loss. Sections 3.3 and 3.4 describe these mechanisms in detail.

3.3 Intra-shard Ordering

MLog adopts a decentralized ordering mechanism within each shard to support high-throughput writes. The goal is simple: let every replica decide the order of records on its own, yet still reach the same result as all others. This avoids dependence on a central coordinator or primary node, reducing delays and coordination overhead. The key enabler is the ordering tuple that each client attaches to a record before sending it. This tuple contains all the information needed for ordering, making the process fully deterministic. As long as replicas receive the same set of records and their metadata, they will produce an identical log sequence, even without any inter-replica communication.

To establish a stable and unambiguous total order within the shard, MLog employs a multi-field, lexicographical comparison logic. Figure 3 illustrates this process, in which multiple replicas independently order the same set of records into an identical sequence using client-assigned metadata. Records are first sorted by timestamp in ascending order. If two timestamps are equal, sequenceNumber breaks the tie to keep each client's records in original order. If both are still equal, clientID acts as the final tie-breaker. This ensures a consistent total order across all replicas.

The correctness of MLog's ordering comes from a simple principle: identical inputs produce identical outputs. The client attaches the ordering tuple to each

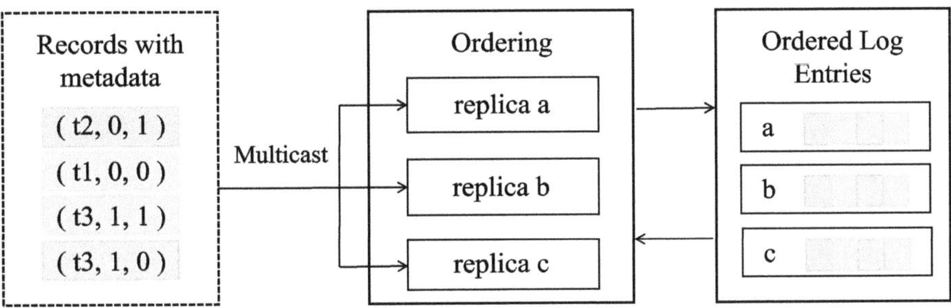

Fig. 3. Intra-Shard Ordering.

record before sending it, and this tuple never changes in transit. As a result, all replicas receive the same set of records with the same metadata. Each server then applies the same well-defined ordering rules. The process is deterministic, so network delays or out-of-order delivery do not affect the final result. Every replica ends up with the same ordered log segment. This removes the need for cross-server communication for ordering, such as metadata exchange or consensus, and greatly reduces system complexity and overhead.

3.4 Reliable Logging over Unreliable Multicast

Reliable logging is essential for a shared log system, even when built on top of an unreliable multicast transport. This section explains how MLog ensures data integrity and correctness despite packet loss, reordering, and node failures.

MLog uses RDMA's Unreliable Datagram (UD) mode for multicast communication. This transport does not offer a guarantee of reliability or in-order delivery. Packet loss and reordering can occur.

While such unreliability raises concerns about dependability, prior studies have shown that UD is highly robust in practice. For instance, FaSST [14] reported no packet losses over 50 PB of UD traffic, thanks to link-layer flow control, retransmissions, and forward error correction, which together ensure high reliability at the physical and link layers. This evidence suggests that building reliable protocols over UD is not only feasible but also well supported in practice. Building on this empirical foundation, MLog enhances robustness through endpoint-based detection and recovery mechanisms, ensuring correctness even under rare packet loss or node failure.

MLog integrates loss detection directly into its ordering process. This adds almost no extra overhead. Each client prepends every record with an ordering tuple: (timestamp, sequenceNumber, clientID). The sequenceNumber comes from a local counter that increases monotonically. As a storage server processes these tuples for intra-shard ordering, it also checks for missing sequence numbers.

The mechanism is simple: for a stream of records from a given clientID, a server detects packet loss by identifying a gap greater than one between the sequenceNumbers of consecutive records. For example, if a server receives a

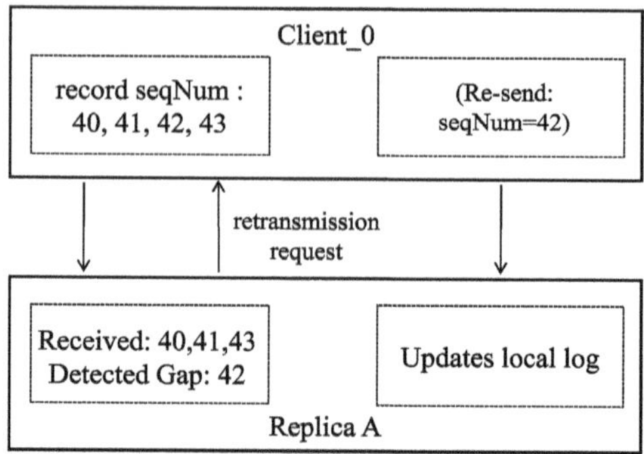

Fig. 4. Per-Client Sequence Tracking and Loss Recovery in MLog.

record with sequenceNumber = 43 immediately after one with 41, it infers that the record with sequenceNumber = 42 was lost. Upon detection, the server directly sends a retransmission request to client_0 for the missing record. This entire process of loss detection and recovery initiation is performed independently by each replica, improving both the robustness and responsiveness of failure handling. Figure 4 illustrates this sequence tracking and recovery process at the replica level.

Each MLog shard uses a $2f + 1$ replica architecture. This design allows the system to tolerate temporary packet loss and up to f node failures. It works through several coordinated mechanisms. First, a retransmission mechanism lets clients resend missing records when requested. These retransmitted records are merged into each server's log using deterministic ordering. Second, majority-based recovery ensures that even if a server permanently loses a record, it can be reconstructed during reads. This works as long as at least $f + 1$ replicas still hold the record. Finally, an acknowledgment and duplication control protocol ensures that servers acknowledge every received record. This allows clients to avoid redundant retransmissions after a timeout. Together, these mechanisms let MLog handle the unreliability of RDMA multicast while keeping high concurrency.

A final key part of MLog's reliability is repairing replica state after a crash. When a replica restarts, its local log may be behind other replicas. To rejoin the shard safely, the recovering node runs a state sync process. First, it compares its last stable log position with the current state of the shard. For example, it may ask a peer for its latest sequence numbers. This step helps it find the exact set of missing records. Next, it fetches these missing records from a healthy peer. It applies them to its local log using the same deterministic ordering rules. Once done, its state is up to date. This peer-to-peer recovery is fully decentralized. It restores the shard's full fault tolerance level without needing a central coordinator.

3.5 Read Operations

MLog's read path is designed to return only safe and consistent data, even under high concurrency and failures. The goal is to ensure that every record observed by a client has been globally ordered and durably persisted, avoiding any intermediate or inconsistent states.

The client can use Subscribe(l) or Read(g, s) to retrieve specific records. MLog enforces a strict policy: any record returned to a client must be part of the global cut, meaning it has been both globally ordered and durably persisted. Before reading, the client checks the status of the requested records with the storage layer. Only records that meet both conditions are returned. This approach ensures that clients never observe intermediate or inconsistent states, thereby preserving the determinism of read results.

To implement this, MLog adopts the global ordering mechanism of Scalog. It relies on periodically reported local cuts to track the progress of each shard. Each local cut reports the number of locally ordered and persisted records in its shard. MLog reuses the same global aggregation and consensus framework but adapts the content of the local cut to fit its single-round multicast replication. The global ordering layer periodically collects these local cuts and runs a consensus protocol (e.g., Paxos) to agree on them, thereby producing the global cut. The global cut marks the minimum safe progress across all shards and is disseminated to storage servers, allowing each server to identify which records are durably replicated and therefore safe to serve to readers.

Fault tolerance and high availability for read operations are built into this design. If a primary storage server fails, the client can retry the request. It can fetch the globally ordered records from any other available replica in the shard. This ensures that clients never observe records out of order or in uncommitted states, even under highly concurrent write workloads.

4 Evaluation

4.1 Experimental Setup

We evaluate the performance of MLog in six aspects: (1) comparison with the append latency of Corfu and Erwin, which are representative existing systems (Sect. 4.2); (2) MLog's ability to handle high concurrency with multiple clients (Sect. 4.3); (3) verifying that MLog has good horizontal scalability (Sect. 4.3); (4) analyzing the MLog's adaptability (Sect. 4.4); (5) evaluating the impact of the number of replicas on MLog's write performance (Sect. 4.5); and (6) read latency (Sect. 4.6). In our experiments, we compare MLog against two representative systems: Corfu and Erwin, the open-sourced implementation of LazyLog [4]. For brevity, we refer to Erwin simply as LazyLog throughout the rest of this paper.

To evaluate the performance of MLog, we conducted experimental evaluations on 16 c6525-25g servers on cloudlab Utah. Each server is configured with a 16-core AMD EPYC 7302P processor (3.00 GHz) and 128 GB of DDR4 ECC error-correcting memory (8 × 16 GB, 3200 MT/s). The storage system consists of two 480 GB SATA SSDs. To meet the high-bandwidth network requirements of the experiments, the server is equipped with two Mellanox ConnectX-5 dual-port 25GbE high-speed NICs (PCIe 4.0 interface), and two of the ports are designated for experimental communication. The server is running Ubuntu 22.04 LTS with Linux kernel version 5.15.0.

4.2 Append Latency Comparison

We first measure MLog's append latency, reporting both the mean and the 99th-percentile (P99) latency, in comparison with existing systems. We ran a single client to test append latency at 50k appends/s. We deployed a shard with three replicas for each system, and a single client continuously appends writes at 50k appends/s. Figure 5 shows the append latency for different systems.

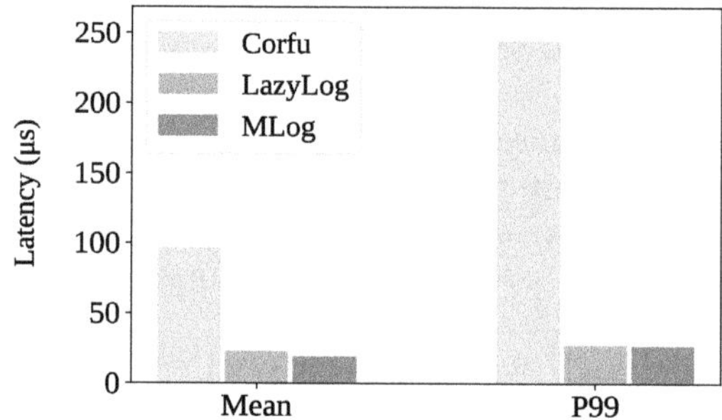

Fig. 5. Append Latency: Corfu vs LazyLog vs MLog

MLog achieves a nearly 5x reduction in latency compared to Corfu, mainly due to the fact that it abandons the traditional chained replication and centralized sequencer mechanism, and adopts lightweight multicast and local negotiation between replicas to quickly confirm the persistent state of the record, which effectively reduces the network and synchronization overhead in the write path.

MLog's append latency is also reduced by nearly 20% compared to the state-of-art implementation of Lazylog-Erwin, which indicates that MLog inherits LazyLog's idea of delayed global ordering while further utilizing RDMA's multicast optimization to further compress the write latency.

4.3 Multi-client Scalability

How does MLog perform against multiple clients? We measured the variation of latency and throughput under different client loads.

We used different numbers of client processes for append operations, and the latency results are shown in Fig. 6(a). The append latency of MLog remains stable. In contrast, the latency of LazyLog increases with the number of client processes, reaching 76 µs with 6 clients. This suggests that LazyLog has a bottleneck when handling highly concurrent writes, mainly because its records must pass through a centralized sequencing layer.

Figure 6(b) shows the corresponding throughput. MLog's throughput scales nearly linearly with the number of client processes, reaching 262K ops/s with 6 clients. In contrast, LazyLog's throughput stagnates at approximately 77k ops/s, demonstrating that MLog's decentralized, leaderless design effectively eliminates the single-point-of-contention bottleneck present in centralized sequencing architectures.

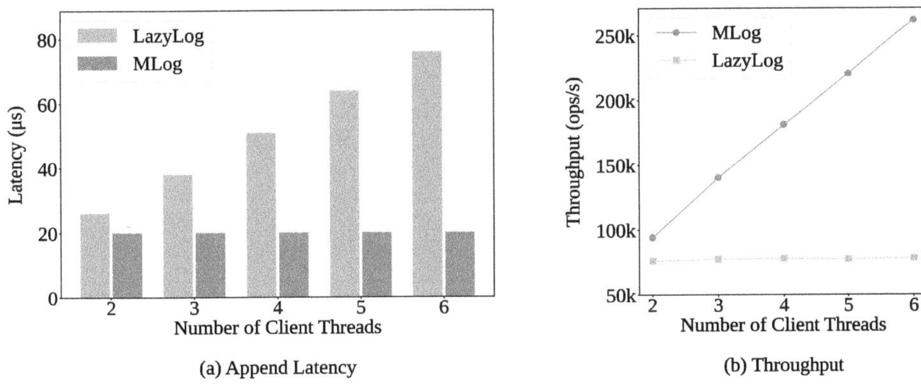

(a) Append Latency   (b) Throughput

Fig. 6. Performance and Scalability Comparison of MLog and LazyLog-Erwin

4.4 Sharding Scalability

Whether a system has good horizontal scalability is a key indicator of its deployability and long-term performance. We analyze the scalability of MLog from two dimensions:

First, we measure the append latency under different number of shards, and the results are shown in Fig. 7. MLog maintains a low and stable latency, while Corfu's latency fluctuates slightly as the number of shards grows. This indicates that MLog's write path can effectively scale horizontally to multiple shards without introducing additional synchronization overhead.

Second, we fixed the record size to 4 KB and measured the overall throughput and latency variation of the system for different number of shards (Fig. 8).

Figure 8(a) shows that MLog throughput scales approximately linearly with the number of shards, up to 250k ops/s, while LazyLog grows to about 77k and then stops improving;

Figure 8(b) shows the change of latency when the throughput increases, and MLog's additional latency is still stable even under high loads, while LazyLog's latency rises rapidly with the load, and the latency exceeds 100 µs when throughput reaches about 78k.

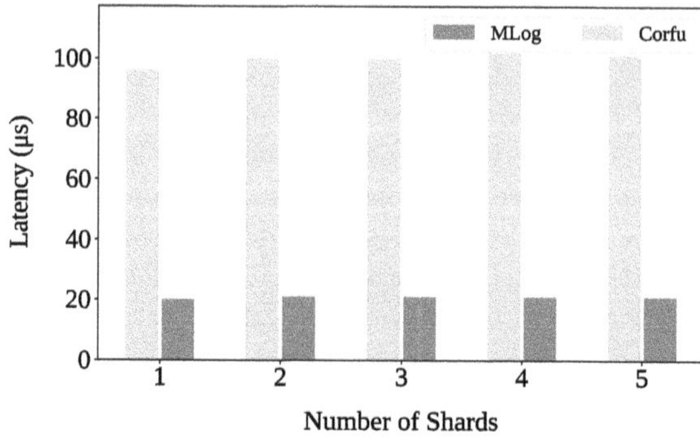

Fig. 7. Latency vs. Shard Count

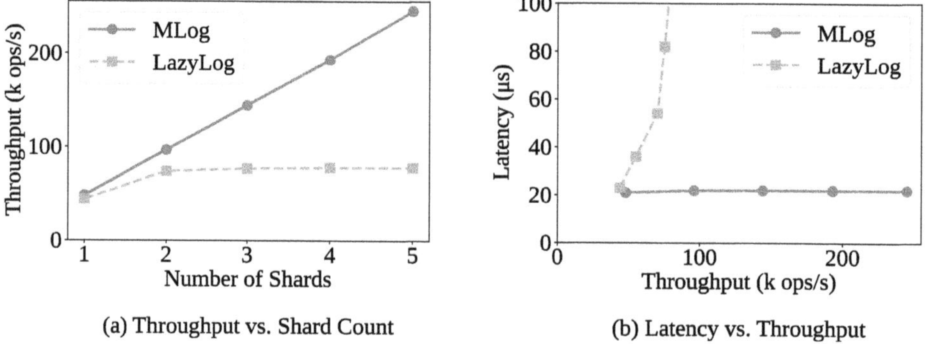

(a) Throughput vs. Shard Count

(b) Latency vs. Throughput

Fig. 8. Scalable Throughput with MLog

4.5   Record Size Impact

How does MLog's append throughput change with different record sizes? In real-world applications, log record sizes are often variable, so the system needs to be able to handle multiple record sizes.

Figure 9 shows the append throughput of MLog compared to LazyLog with record sizes ranging from 100 B to 8 KB. The experimental results show that MLog consistently outperforms LazyLog regardless of record size, with MLog throughput reaching 1.2M ops/s in a small record size scenario (e.g., 100 B), compared to LazyLog's 462k. The performance difference is primarily attributed to MLog's RDMA multicast-based write path. This design effectively avoids redundant memory copies and kernel-stack switching overhead, which become more pronounced with larger record sizes.

4.6   Replica Count Sensitivity

Distributed log systems often require replication of each record to multiple replica nodes in order to achieve high availability. However, traditional

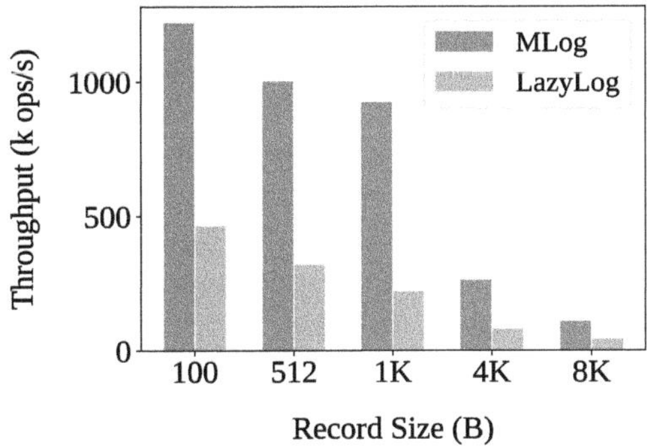

Fig. 9. Throughput vs. Record Size

replication mechanisms (e.g., chained replication) often bring significant performance degradation when the number of replicas increases. To this end, we measure the change in write latency of MLog and Corfu as the number of replicas increases from 2 to 6, as shown in Fig. 10.

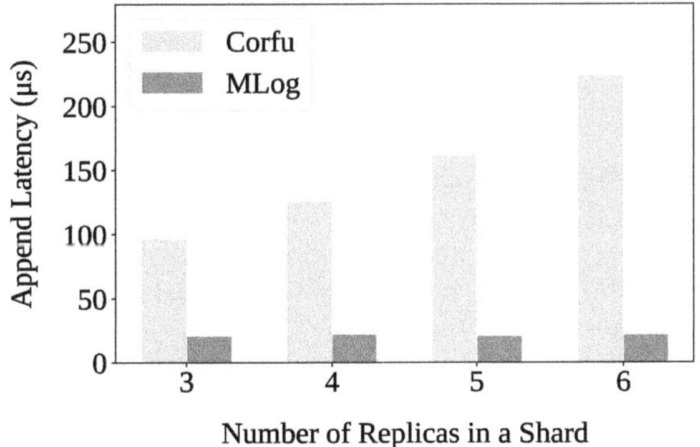

Fig. 10. Append Latency vs. Replica Number

The results show that MLog's append latency remains within 24 μs and hardly varies with the number of replicas, while Corfu's latency increases linearly with the number of replicas. This difference is due to the fundamental difference in architectural design: CORFU adopts serial chain replication, where each additional replica introduces an extra network hop and acknowledgment delay. In contrast, MLog leverages hardware-level RDMA multicast to perform data dissemination and concurrent acknowledgments in a single round, significantly shortening the replication path and reducing CPU overhead. MLog's hardware layer RDMA multicast capability enables a single-round communication pattern that combines a broadcast send with concurrent acknowledgements, significantly reducing the replication path length and CPU burden, and verifying MLog's stable low latency performance under high replica fault tolerance.

4.7   Read Performance

How is the read performance of MLog? Under the background load of 50k appends/s, we measure the read latency of each system with a delay of 4 ms after the appending is completed, and the results are shown in Fig. 11.

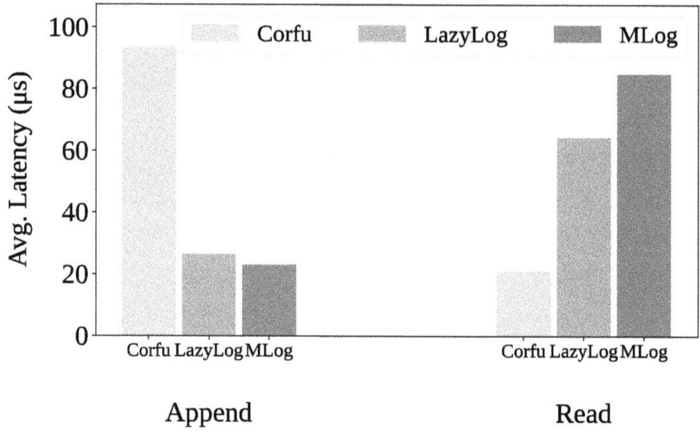

Fig. 11. Reads Lagging Behind Appends

MLog still shows the latency advantage of the write end, and its read latency is slightly higher than that of LazyLog's, which is mainly due to two points: (1) MLog uses the gRPC communication protocol, which has a certain overhead compared to the high-performance eRPC [15] protocol used by Erwin; (2) MLog enforces a strict visibility policy: records are exposed to clients only after they have been included in a finalized global cut. This prevents clients from reading data in intermediate states, such as records that have been replicated but are not yet globally ordered.

5   Related Work

5.1   Shared Log Designs

The design of high-performance shared logs is a central topic in distributed systems. Early and influential systems like CORFU [1] established a common architecture. It uses a centralized sequencer for total ordering, followed by a client-driven chain replication protocol for durability. This design, however, suffers from both of the key performance bottlenecks: the centralized sequencer inherently limits throughput scalability, while the $O(f)$ sequential network round trips in its replication path lead to high write latency that scales poorly with the number of replicas.

To address the scalability limitations of a central sequencer, Scalog [3] introduced a "persist-first, order-later" architecture. It first persists the records and establishes global order through a later aggregation step. Scalog's intra-shard replication still relies on a multi-round, primary-backup model, resulting in high append latencies that make it unsuitable for ultra-low-latency applications. Conversely, LazyLog [4] focuses on minimizing write-path latency with its "lazy ordering" abstraction. It defers global ordering until records are read, allowing for a faster initial acknowledgement. However, LazyLog's write path still relies on a stateful, centralized Sequencing Layer. This layer becomes the new scalability bottleneck for concurrent writes, fundamentally limiting the system's overall throughput.

In summary, existing state-of-the-art systems present a difficult choice in the design space. Systems like Scalog [3] optimize for ordering scalability at the cost of introducing a replication bottleneck (the primary node). Conversely, systems like LazyLog [4] optimize for low replication latency but re-introduce a centralized sequencing bottleneck. MLog introduces a novel architecture that simultaneously addresses the dual challenges of replication overhead and centralized ordering bottlenecks. By leveraging RDMA multicast, it replaces traditional point-to-point replication with a single parallel broadcast, enhanced by endpoint-side reliability mechanisms and embedded ordering metadata. This method forms the basis of a decentralized ordering framework, allowing MLog to combine the low latency of streamlined replication with the high scalability of a fully distributed design.

5.2 RDMA-Based Distributed Systems

RDMA has become a key enabler for high-performance distributed systems. It is widely used in key-value stores (e.g., FaRM [16], Pilaf [17], HERD [18], Orion [19]) and file systems (e.g., Octopus [20]) to achieve ultra-low latency and high throughput. Most of these systems leverage RDMA's Reliable Connection (RC) mode to provide point-to-point, reliable communication, which is a natural fit for client-server or primary-backup interactions. However, managing a large number of RC connections incurs significant memory and CPU overhead, posing a scalability challenge.

Some works have sought to apply RDMA to accelerate shared logs, for instance, by replacing the TCP transport in systems like Kafka with RDMA's point-to-point capabilities [21]. While improving throughput, these approaches act primarily as a transport-layer optimization. They accelerate the individual messages of the existing, decomposed replication and ordering protocols but do not alter the fundamental multi-round communication pattern.

MLog distinguishes itself from prior works by using RDMA multicast as more than just a transport replacement. It treats multicast as a foundational primitive for a new, lightweight write mechanism. Rather than speeding up the individual point-to-point messages of a multi-round protocol, MLog integrates the protocol design with the multicast mechanism itself. To enable decentralized ordering, it embeds ordering metadata in every multicast packet. This allows replicas to establish a consistent local order on their own, without a central coordinator. The protocol logic and hardware primitives work in close concert. As a result, replication and ordering finish in a single round of communication. This is the core innovation of MLog, which brings performance advantages.

6   Conclusion

This paper addresses two primary performance bottlenecks in modern shared log systems: the high write latency caused by traditional replication protocols and the scalability limitations of centralized sequencing. These challenges arise from excessive communication steps, connection management overhead, and the inherent serialization bottleneck of a centralized sequencer under high concurrency. We presented MLog, a shared log system that leverages the RDMA multicast protocol to construct a low-latency, scalable write path. To overcome the inefficiencies of conventional replication, MLog introduces a single-round, broadcast-based replication mechanism. Reliability is ensured through a lightweight, endpoint-driven loss recovery protocol. To eliminate the sequencing bottleneck, MLog embeds ordering metadata into the replication packets, allowing each replica to independently establish a consistent local order. These local orders are later finalized by a lightweight, decentralized global ordering layer. Our evaluation on CloudLab shows the effectiveness of MLog in delivering both low latency and high scalability for modern shared log applications.

Acknowledgments. We thank the reviewers for their helpful feedback. This paper is supported by the Natural Science Foundation of Jiangsu Province (Grant No. BK20220973), the Fundamental Research Funds for the Central Universities (Grant No. NS2024057), and 2024 Yangtze River Delta Science and Technology Innovation Community Joint Research Project (Grant No. 2024CSJZN00400)

References

1. Balakrishnan, M., Malkhi, D., Prabhakaran, V., Wobbler, T., Wei, M., Davis, J.D.: CORFU: a shared log design for flash clusters. In: 9th USENIX Symposium on Networked Systems Design and Implementation (NSDI 12), pp. 1–14. USENIX Association, San Jose, CA, April 2012
2. Lockerman, J., et al.: The FuzzyLog: a partially ordered shared log. In: 13th USENIX Symposium on Operating Systems Design and Implementation (OSDI 18), pp. 357–372. USENIX Association, Carlsbad, CA, October 2018
3. Ding, C., Chu, D., Zhao, E., Li, X., Alvisi, L., Van Renesse, R.: Scalog: seamless reconfiguration and total order in a scalable shared log. In: 17th USENIX Symposium on Networked Systems Design and Implementation (NSDI 20), pp. 325–338. USENIX Association, Santa Clara, CA, February 2020

4. Luo, X., Bhat, S.G., Hu, J., Alagappan, R., Ganesan, A.: LazyLog: a new shared log abstraction for low-latency applications. In: Proceedings of the ACM SIGOPS 30th Symposium on Operating Systems Principles, SOSP 2024, pp. 296–312. Association for Computing Machinery, New York, NY, USA (2024)

5. Amazon. Kinesis. aws.amazon.com/kinesis

6. Google. Pub/sub. cloud.google.com/pubsub

7. Microsoft. Event hubs. azure.microsoft.com/en-us/services/event-hubs

8. Google. Protocol buffers. developers.google.com/protocol-buffers

9. Haskin, R., Malachi, Y., Chan, G.: Recovery management in quicksilver. ACM Trans. Comput. Syst. **6**(1), 82–108 (1988)

10. Spector, A.Z., Pausch, R.F., Bruell, G.: Camelot: a flexible, distributed transaction processing system. In: Digest of Papers. COMPCON Spring 88 Thirty-Third IEEE Computer Society International Conference, pp. 432–437 (1988)

11. The Apache Software Foundation. Zookeeper programmer's guide. zookeeper.apache.org/doc/current/zookeeperProgrammers.html

12. Lamport, L.: Paxos made simple. ACM SIGACT News (Distrib. Comput. Column) **32**, 4 (Whole Number 121, December 2001), 51–58 (2001)

13. Kalia, A., Kaminsky, M., Andersen, D.G.: Design guidelines for high performance RDMA systems. In: Proceedings of the 2016 USENIX Conference on Usenix Annual Technical Conference, USENIX ATC 2016, pp. 437–450. USENIX Association, USA (2016)

14. Kalia, A., Kaminsky, M., Andersen, D.G.: FaSST: fast, scalable and simple distributed transactions with two-sided (RDMA) datagram RPCs. In: Proceedings of the 12th USENIX Conference on Operating Systems Design and Implementation, OSDI 2016, pp. 185–201. USENIX Association, USA (2016)

15. Kalia, A., Kaminsky, M., Andersen, D.G.: Datacenter RPCs can be general and fast. In: Proceedings of the 16th USENIX Conference on Networked Systems Design and Implementation, NSDI 2019, pp. 1–16. USENIX Association, USA (2019)

16. Dragojević, A., Narayanan, D., Hodson, O., Castro, M.: FaRM: fast remote memory. In: Proceedings of the 11th USENIX Conference on Networked Systems Design and Implementation, NSDI 2014, pp. 401–414. USENIX Association, USA (2014)

17. Mitchell, C., Geng, Y., Li, J.: Using one-sided RDMA reads to build a fast, CPU-efficient key-value store. In: Proceedings of the 2013 USENIX Conference on Annual Technical Conference, USENIX ATC 2013, pp. 103–114. USENIX Association, USA (2013)

18. Kalia, A., Kaminsky, M., Andersen, D.G.: Using RDMA efficiently for key-value services. SIGCOMM Comput. Commun. Rev. **44**(4), 295–306 (2014)

19. Yang, J., Izraelevitz, J., Swanson, S.: Orion: a distributed file system for non-volatile main memories and RDMA-capable networks. In: Proceedings of the 17th USENIX Conference on File and Storage Technologies, FAST 2019, pp. 221–234. USENIX Association, USA (2019)

20. Zhu, B., Chen, Y., Wang, Q., Lu, Y., Shu, J.: Octopus+: an RDMA-enabled distributed persistent memory file system. ACM Trans. Storage **17**(3) (2021)

21. Taranov, K., Byan, S., Marathe, V., Hoefler, T.: KafkaDirect: zero-copy data access for Apache Kafka over RDMA networks. In: Proceedings of the 2022 International Conference on Management of Data, SIGMOD 2022, pp. 2191–2204. Association for Computing Machinery, New York, NY, USA (2022)

Sumeru: An Efficient Hybrid-Granularity Cache Management Scheme for CXL-SSDs

Junjie Li, Xuchao Xie, Xinyu Xu, Qiulin Wu, Qi Xingyun,
and Zhenlong Song[✉]

College of Computer Science and Technology, National University of Defense
Technology, Changsha, China
{ljj707,xiexuchao,xuxinyu,qiulin_wu,qi_xingyun}@nudt.edu.cn,
songzhl@sina.com

Abstract. The Compute Express Link (CXL) protocol introduces memory semantics into solid-state drives (SSDs), enabling the construction of a unified memory-storage architecture. However, CXL's byte-addressable access pattern is inherently mismatched with the page- or block-level granularity of conventional SSD cache mechanisms. Specifically, the large cache granularity of existing policies causes significant I/O amplification and inefficient cache space utilization when handling fine-grained requests, ultimately diminishing CXL-SSDs' performance advantages. In contrast, simply reducing the cache granularity compromises spatial locality and increases write-back overhead.

To address this dilemma, we propose Sumeru, a hybrid-granularity cache management strategy in CXL-SSDs. Sumeru designs a multi-granularity hybrid policy to handle I/O requests of different sizes, proposes a selective eviction mechanism to capture data locality within a single cache unit, and introduces a differential probation method to reduce write-back overhead. Experimental results on five real-world workloads show that Sumeru improves the cache hit rate by an average of 9.4%, reduces access latency by 76.4%, and decreases write-back frequency by 46.4% compared to existing caching strategies. These gains fully unlock the performance potential of CXL-SSDs.

Keywords: CXL-SSD · Cache Management · Granularity

1 Introduction

For decades, the evolution of computer architecture has sought to break down the performance barrier between main memory and storage devices to achieve unified memory-style access [1–9]. The emergence of the Compute Express Link (CXL) protocol [10] has made this dream a reality. By bridging the block semantics of PCIe storage to memory-compatible byte semantics [11], CXL-SSDs allow upper-level applications to directly access storage at byte granularity via Load/Store instructions, presenting themselves to the operating system as a memory region. This paradigm shift brings unprecedented flexibility and performance potential

H. Liu et al. (Eds.): ICA3PP 2025, LNCS 16381, pp. 604–623, 2026.
https://doi.org/10.1007/978-981-95-8399-7_32

for data-intensive applications such as in-memory databases, large-scale graph analysis, and real-time data processing.

However, while the host-side access pattern has shifted to the byte level, the cache management units inside CXL-SSD devices still predominantly use the traditional page-level granularity designed for block I/O. This mismatch between external access and internal management causes severe read/write amplification and significantly limits the performance potential of CXL-SSDs. Existing DRAM caches in SSDs, whether their management policies are page-based or block-based, operate at a fixed granularity of 4KB or larger. When a byte-level request for only 1KB of data arrives, these traditional schemes are forced to read and cache a full 4KB page from NAND flash. This not only generates three times the required I/O traffic but also consumes valuable DRAM cache space with a large amount of invalid data, severely impeding the advantages of CXL [12,13].

A straightforward solution is to uniformly reduce the management granularity, for example to 256B. But this approach introduces two problems. First, for large requests like sequential scans, splitting a full page into multiple independent, fine-grained cache entries destroys the spatial locality between data. This prevents the cache policy from recognizing their association, leading to the premature eviction of data that still has potential value. Second, a finer granularity causes dirty data to be scattered across more cache entries. Each eviction might then require writing back multiple non-contiguous and fine-grained items, significantly increasing write amplification.

The real challenge lies not in choosing between coarse- and fine-grained caching, but in effectively combining them. To this end, we propose an adaptive CXL-SSD cache management scheme, Sumeru, which captures the combinatorial relationships of data within a cache unit. Our contributions can be summarized as follows:

- We design a multi-granularity hybrid policy that divides request granularities into three levels: page (4KB), L1 subpage (1KB), and L2 subpage (256B). The cache is dynamically partitioned into a page region and a subpage region to handle I/O requests of different sizes.
- We propose a selective eviction mechanism. For each page unit in the page region, we count the access frequency of combinations of L1 subpages within the page unit, rather than the frequency of individual L1 subpages. This better captures the locality between L1 subpages, thereby improving the cache hit rate. The design is also applied to L1 subpage units within the subpage region.
- We introduce a differential probation method. It classifies data into six categories based on hotness and dirty status, assigning different probation chances to each category. This effectively reduces write costs and improves system performance.

Sumeru is evaluated by a simulator. The simulation results show that, compared to three existing cache strategies under five real-world workloads, Sumeru improves the average hit rate by 9.4%, reduces average latency by 76.4%, and decreases the average number of write-backs by 46.4%.

The remainder of this paper is organized as follows. Section 2 introduces the background and motivation. Section 3 details the core design of Sumeru, including the multi-granularity hybrid, selective eviction, and differential probation mechanisms. Section 4 evaluates the performance of Sumeru. Section 5 summarizes related work, and Sect. 6 concludes the paper.

2   Background and Motivation

2.1   CXL-SSD

With the explosive growth of global data, the performance gap between memory and storage in traditional computer architectures has increasingly become a bottleneck. Although NVMe SSDs [14] offer superior performance, their nature of block devices limits how they interact with the CPU. The CPU cannot directly access the SSD with byte granularity using Load/Store instructions [15]. Instead, all data transfers must be mediated by main memory, which makes the process cumbersome and inefficient. Furthermore, the lack of cache coherence mechanisms between CPU caches and PCIe-connected devices leads to additional synchronization overhead.

To address these issues, the industry has proposed interconnect protocols such as OpenCAPI [16], GenZ [17], CCIX [18], NVLINK [19], and CXL. Among them, CXL, with its openness and technical advantages, has become the most promising next-generation interconnect standard.

CXL is a dynamic multi-protocol technology that unifies communication between CPUs, accelerators and memory devices. CXL not only provides a low-latency, high-bandwidth path for accelerators to access the system and for the system to access memory connected to CXL devices [20], but it also brings new possibilities to PCIe storage. Storage devices can be accessed via a byte-addressable interface, thereby enhancing data access flexibility.

CXL-SSD is a new type of storage device based on the CXL protocol. From the operating system's perspective, a CXL-SSD is not limited to traditional block-device access. Instead, it can appear as a directly addressable memory extension, accessible via memory semantic. This paradigm shift greatly simplifies the software stack, allowing the CPU to perform Load/Store operations directly on the CXL-SSD. This eliminates most of the overhead in the traditional I/O path, bringing unprecedented performance potential for data-intensive applications like in-memory databases and large-scale graph computing.

Figure 1 shows the typical architecture of a CXL-SSD [21]. When the host system issues a memory access request, the CXL interface receives it. The Indexing module checks if the requested data exists in the DRAM cache. If it's a cache hit, the data are returned directly. If it's a miss, the MSHR (Miss Status Holding Register) manages the request, tracking outstanding requests to avoid redundant processing. The request is then sent to the FTL (Flash Translation Layer), which performs address translation, transaction scheduling, and other functions. Finally, the corresponding data from the flash memory is transferred to the host.

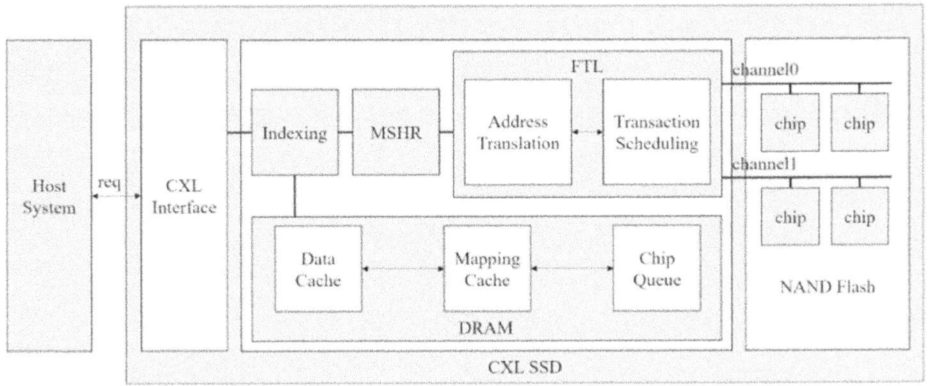

Fig. 1. Architecture of a CXL-SSD.

2.2 SSD Cache Management

Inside an SSD, the DRAM cache plays a crucial role. It acts as a high-speed buffer between the host and flash memory, and its management strategy directly affects the overall performance and lifespan of the device. An efficient cache scheme mainly aims to achieve two goals: first, to reduce the response time of the SSD, and second, to reduce the traffic written to flash memory to increase its lifespan [22].

SSDs use flash pages as the unit for reading and writing, and flash blocks as the unit for erasing. Research on SSD cache management can be divided into two categories based on granularity: page-based policies and block-based policies. Figure 2 illustrates these two types of policies using the most typical LRU strategy. In a page-based LRU policy, each node in the list stores a single page, so only one page is inserted or evicted at a time. In a block-based LRU policy, each node stores a block, and all pages within that block can be inserted or evicted at once.

Page-based policies, such as CFLRU [23], LRU-WSR [24], and CCF-LRU [25], manage individual pages as independent cache entries. They primarily optimize eviction decisions by considering the dirty status of pages to reduce expensive write operations.

In contrast, block-based policies, including FAB [26] and BPLRU [27], manage the pages of an entire data block as a collective unit. These policies are designed to better leverage spatial locality in sequential workloads and reduce garbage collection overhead.

Page-based cache management methods can directly identify the location of pages, improving the hit rate. However, they cannot leverage the spatial locality of pages; pages from a large request are scattered across multiple physical blocks, increasing the overhead of garbage collection. Block-based cache management methods utilize the spatial locality of multiple pages within the same block, resulting in lower garbage collection overhead. However, pages within the

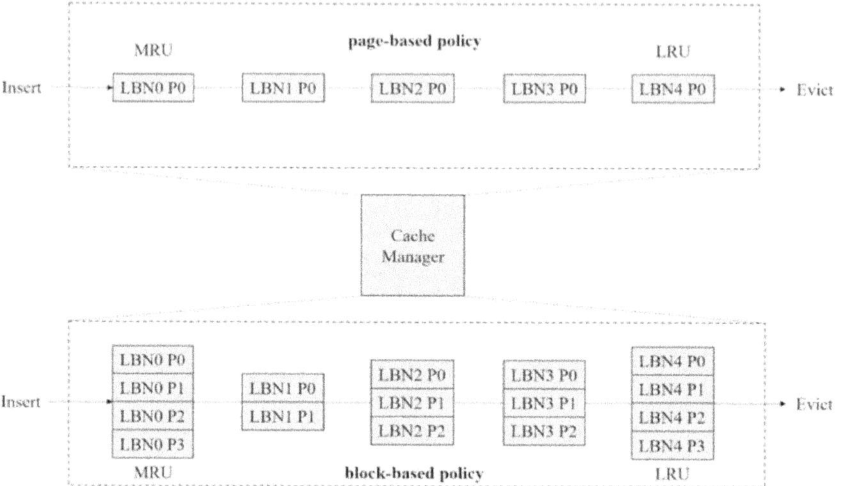

Fig. 2. Page-level vs. block-level policies.

same logical block may have varying hotness levels (hot blocks may contain cold pages), leading to reduced cache space utilization.

2.3   Motivation

CXL-SSDs have byte-addressability and can be accessed flexibly with memory semantics. However, this flexibility introduces a mismatch between access granularity and cache granularity inside existing SSDs. As shown in Fig. 3, traditional NVMe SSDs are designed to accommodate block I/O, whereas CXL-SSDs need to efficiently handle access requests with byte-level precision. This mismatch prevents current cache schemes from realizing the full potential of CXL.

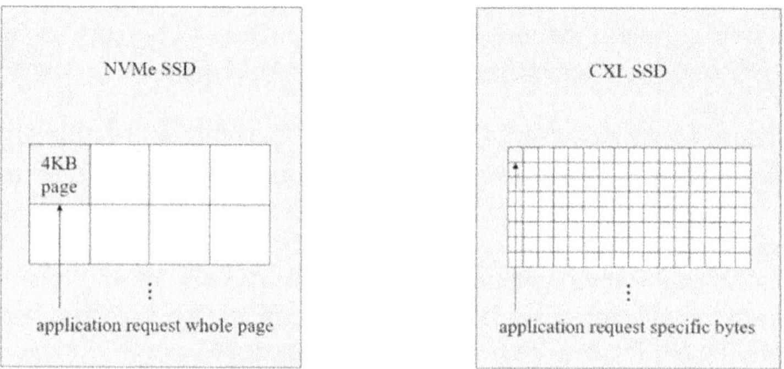

Fig. 3. Data access comparison between NVMe SSD and CXL-SSD.

Existing SSD cache policies, whether page-based or block-based, operate at a fixed granularity of 4KB or larger. In scenarios with dense small requests (such as real-time data analysis, index lookups), traditional page-level caching causes severe resource waste. For example, when an application only needs to access 1KB of data within a page, the entire 4KB page must be loaded into the cache. This not only wastes 3KB of precious cache space but also generates three times the actual required backend I/O traffic.

To solve the above problems, a seemingly obvious solution is to directly reduce the cache management granularity. However, this solution brings two new problems:

First, there is a loss of spatial locality. A large sequential request would be broken down into multiple independent small cache entries. This causes the cache manager to lose its holistic view of data access, failing to recognize the relationships between them. It might evict parts of a large request before it is even completed, severely reducing the cache hit rate.

Second, there is an increase in write amplification and management overhead. A smaller cache granularity means an exponential increase in metadata. More seriously, when evicting dirty data, what was once a single 4KB write-back operation could become 16 separate 256B write-back operations. This not only dramatically increases write amplification but also significantly raises the management burden on the FTL.

In summary, CXL-SSD cache management faces a critical trade-off: sticking with coarse-grained policies fails to fully utilize byte-addressability of CXL, leading to low resource utilization; directly adopting fine-grained policies sacrifices spatial locality and increases management overhead, harming overall performance. Therefore, the core challenge of this research is to design a cache management scheme that can dynamically adapt to different access patterns and combine the advantages of both coarse and fine granularities.

3   Design

We propose Sumeru, a cache management scheme specifically designed for the DRAM cache integrated in CXL-SSDs. The core idea of Sumeru is to intelligently match different I/O patterns through dynamic, multi-level mechanisms. Its overall design consists of three core mechanisms working in concert: multi-granularity hybrid, selective eviction, and differential probation. The multi-granularity hybrid adapts to requests of different sizes, selective eviction precisely retains hot data within a page, and differential probation reduces write-back costs.

As illustrated in Fig. 4, when a request enters the cache, it is routed to either the page region or the subpage region based on its size, with each region corresponding to a different management granularity. Each cache item tracks the access frequency of its internal subpage combinations. Based on these combination frequencies, subpages can be selectively evicted. Concurrently, each item is granted a certain number of probation chances, and it will only be evicted when this count becomes zero.

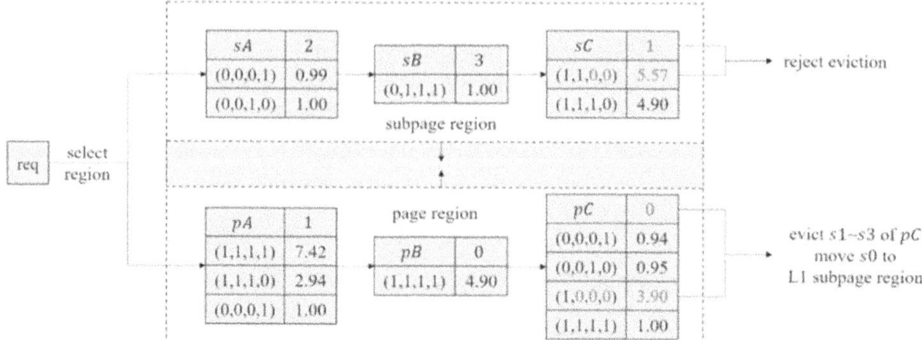

Fig. 4. Overall design of Sumeru.

3.1 Multi-Granularity Hybrid

Different applications, and even the same application at different stages, exhibit vastly different memory access localities. To efficiently handle the diverse workloads brought by CXL-SSDs, a single-granularity cache policy is clearly inadequate. Therefore, the primary design of Sumeru is a multi-granularity hybrid mechanism.

To avoid the resource rigidity of static partitioning, Sumeru treats the entire cache space as a unified resource pool. On top of this, we establish two logical partitions: a page region and a subpage region. The page region uses a top-down stack-based management [28] and primarily serves coarse-grained requests larger than an L1 subpage. Its design goal is to maximize the capture of spatial locality in large requests and sequential accesses. The subpage region uses a bottom-up heap-based management and is dedicated to handling fine-grained requests no larger than an L1 subpage. By managing smaller cache units, it focuses on capturing temporal locality in random small requests. These two regions are not rigidly separated but instead compete for resources through a global eviction policy. This allows their partition sizes to adapt dynamically to the workload's access characteristics.

We choose three different granularity levels to address different locality patterns:

- Page (4KB): Exists only in the page region. This is the standard page size widely used in existing systems, compatible with current page-level management policies, and suitable for large sequential requests, primarily capturing spatial locality.
- L1 Subpage (1KB): Exists in both the page region and the subpage region. For some medium-sized requests, using a 1KB granularity can reduce the amount of data transferred. It can capture both spatial and temporal locality.
- L2 Subpage (256B): Exists only in the subpage region. The 256B granularity can more efficiently handle scenarios with high-frequency small read/write requests, primarily capturing temporal locality.

This design allows the cache policy to be flexibly adjusted according to the request size, maximizing the use of locality and avoiding unnecessary waste of cache resources. Figure 5 illustrates the specific structure of the multi-granularity hybrid policy.

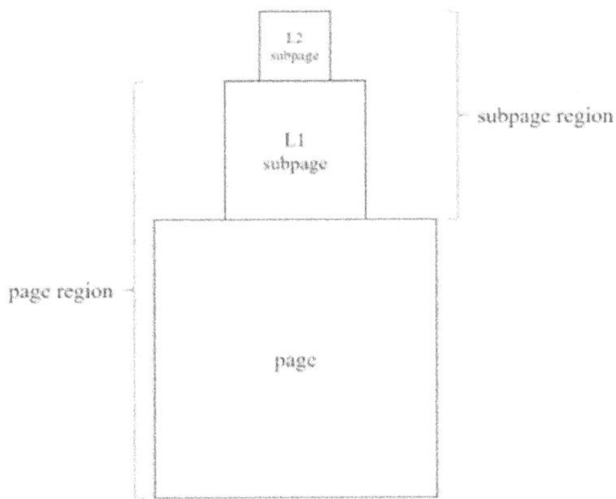

Fig. 5. Multi-granularity hybrid structure.

When a new I/O request requires data to be cached, Sumeru will make a decision based on its request size. Requests larger than 1KB are considered large requests with potential spatial locality, and the data will be cached in the page region; conversely, small requests will enter the subpage region.

To allow the ratio of the two regions to dynamically adapt to workload changes, Sumeru adopts a global perspective during eviction. When cache space is insufficient, it will simultaneously compare the data item at the LRU end of the page region and the data item at the LRU end of the subpage region, selecting the colder of the two as the victim for eviction. This design allows the subpage region to naturally encroach on the page region's space when small requests are dense, and the page region to reclaim space when large requests become more frequent, thus achieving dynamic adaptation of partition sizes.

3.2   Selective Eviction

Traditional cache policies use a coarse-grained eviction method, which cannot adapt to the access pattern where data within a page has non-uniform hotness. Under CXL's byte-level access, only some subpages within a 4KB page may be frequently accessed hot data. If the entire page is still treated as an indivisible unit for eviction, it will cause the accidental loss of this hot data, thereby reducing cache efficiency. Sumeru's selective eviction mechanism is designed to solve this problem. Its core idea is to gain insight into locality based on the combined access frequency of subpages within a page, thereby achieving precise partial eviction.

Unlike traditional methods that only track the access frequency of individual subpages, Sumeru tracks the access combinations of all L1 subpages within a page. For a page containing four subpages s0, s1, s2, s3, we are not only interested in whether s0 was accessed, but also whether s0, s1 were accessed together, or whether s1, s2, s3 were accessed as a group. This method can more accurately capture the spatial correlation between subpages. To balance efficiency and precision, we only record recently accessed combinations with non-zero frequency and introduce a time decay factor γ, giving more weight to recent accesses than historical ones.

Let all combinations of L1 subpages be S_p. Let x be one of the L1 subpage combinations in page unit p, denoted as $x \in \bigcup_{p \in P} S_p$. Then

$$Freq_x = \sum_{\substack{t=0 \\ x \in \bigcup_{p \in P} S_p}}^{n-1} \gamma^t \cdot I(x_t = x) \tag{1}$$

Here, n is the number of accesses tracked, set to 100. γ is the discount factor, giving higher weight to recent accesses, set to 0.95. $I(x_t = x)$ is an indicator function, which is 1 if x_t equals x, and 0 otherwise.

Algorithm 1: Selective Eviction of Victim Page

Input: victim page
Output: evicted L1 subpage(s)
maxFrequency = 0;
for *each non-zero frequency of L1 subpage combinations* **do**
  | maxFrequency = max(maxFrequency, frequency);
end
for *each non-zero frequency of L1 subpage combinations* **do**
  if *frequency == maxFrequency* **then**
    if *the combination contains all the L1 subpages in the page* **then**
      | return all the subpages;
    else
      if *the combination only contains one L1 subpage* **then**
        | move the L1 subpage to the subpage region;
      end
      return L1 subpage(s) not in the combination;
    end
  end
end

When a page unit p is selected as a victim, Sumeru executes the selective eviction logic shown in Algorithm 1. First, it finds the subpage combination with the highest access frequency within that page. If the most frequent combination includes all subpages in the page, it proves that the page has extremely strong spatial locality, and Sumeru will choose to evict the entire page. If the most frequent combination only includes some of the subpages, Sumeru will selectively evict the cold subpages that are not in this combination and retain the hot ones. When a page is left with only one subpage after multiple selective evictions, we migrate it to the subpage region for more fine-grained management to improve space utilization.

This mechanism is critical to Sumeru's high hit rate. By lowering the eviction granularity from page to subpage, Sumeru maximizes the retention of valid hot data in the cache. Units in the subpage region use a similar mechanism to manage their internal L2 subpages.

3.3   Differential Probation

Given that the cost and latency of write operations in NAND flash are much higher than read operations, reducing the write-back frequency must be a core optimization goal for SSD cache management. To this end, Sumeru introduces a differential probation mechanism. The core idea of this mechanism is to provide different levels of eviction immunity to a cache item based on its value.

We evaluate the value of a cache item from two dimensions: hotness and dirty status. A hot dirty page is undoubtedly the most valuable [15], as it is likely to be accessed again soon, and the cost of evicting it is the highest; a cold clean page has the lowest value.

Based on this, we set a probation chance counter for each cache item. When a cache item is selected as a victim, if its probation chance is greater than 0, it will skip the current eviction, and the counter is decremented by 1; if it is 0, it is evicted. The initial value of the probation chance is set according to the following empirical formula, which aims to quantify the value of the cache item:

$$Count_{probation} = C_{temp} + D_{dirty} \qquad (2)$$

Here, C_{temp} represents the hotness level. D_{dirty} represents the dirty state, taking the value 1 for a dirty page and 0 for a clean page. The hotness level is determined by a dual-threshold system based on access frequency:

$$C_{temp} = \begin{cases} 2, & \text{if } Freq > 2 \\ 1, & \text{if } 1 < Freq \leq 2 \\ 0, & \text{if } Freq \leq 1 \end{cases}$$

These thresholds and weights are empirically set based on a preliminary analysis of various typical workloads. As shown in Fig. 6, a newly generated hot dirty page will receive the maximum of 3 probation chances, while a cold clean page will have no probation chances and will be evicted immediately. Through this mechanism, Sumeru greatly reduces the probability of high-value data being mistakenly evicted, thus effectively reducing the number of write-backs to NAND, improving system performance, and extending SSD lifespan.

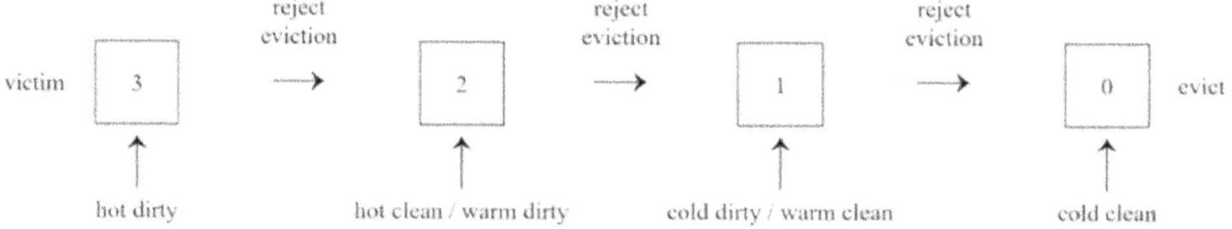

Fig. 6. Differential probation.

4   Experimental Evaluation

This chapter aims to comprehensively evaluate the performance of Sumeru through a series of experiments. We will compare Sumeru with several baseline policies under different workloads to verify its effectiveness.

4.1   Experimental Setup

To evaluate Sumeru's performance, we implemented an event-driven simulator based on Python. The simulation environment is configured with a 16MB cache. The specific parameters are shown in Table 1.

Table 1. Experimental Parameters

Parameter	Value
data cache capacity	16 MB
page capacity	4 KB
subpage capacity	1 KB
cache hit latency	100 ns
read latency	5 us
program latency	100 us
erase latency	1 ms

We selected five publicly available dataset traces with different characteristics for the experiments, each containing 5,000,000 requests, to ensure a comprehensive evaluation. These traces include:

- BERT [29]: An I/O trace from running the BERT model, representing the access patterns of deep learning applications.
- PageRank [30]: A trace of the PageRank algorithm, featuring typical memory access characteristics of graph computing.
- Radiosity [31]: The Radiosity application from the SPLASH-2 benchmark suite, simulating scientific computing scenarios.

- XZ [32,33]: A trace from the XZ data compression tool, exhibiting a mixed pattern of streaming and repetitive access.
- YCSB [34]: Workload F from the Yahoo! Cloud Serving Benchmark, dominated by read-modify-write operations, simulating database-intensive applications.

In addition to the classic LRU algorithm, we also selected two other policies, CFLRU and FAB, as the control group for this experiment.

- LRU: The most classic cache replacement policy, which replaces the cache page that has not been accessed for the longest time.
- CFLRU: Builds on the LRU policy by adding a check for whether a page in the cache is clean and prioritizes replacing clean pages to reduce write costs. CFLRU divides the LRU list into two regions: a working region and a clean-first region. The working region contains recently used pages, while the clean-first region contains candidate pages for replacement. It first tries to select a clean page from the clean-first region for replacement. If there are none, it selects the dirty page at the end of the working region.
- FAB: A policy for flash cache buffer management in PMP systems. This policy selects a victim block based on the proportion of valid pages within the block, prioritizing the eviction of the block with the highest "completeness" in the buffer. If multiple blocks have the same completeness, the LRU policy is used. During eviction, all dirty pages of the selected block are written to flash as a whole, which significantly reduces the overhead of copying valid pages during garbage collection.

4.2 Performance Results

Cache Hit Rate. The cache hit rate is a core metric for measuring cache efficiency. A high hit rate means that most memory accesses can be satisfied by the cache, thus avoiding access to the SSD's flash memory.

As can be seen from Fig. 7, the Sumeru algorithm achieved the highest hit rate under all workloads. Through adaptive multi-granularity management and precise selective eviction, Sumeru can maximize the utilization efficiency of the cache space. Traditional cache policies, whether the page-based LRU and CFLRU or the block-based FAB, use a fixed management granularity. When faced with the byte-level fine-grained access brought by CXL-SSDs, these traditional methods are still forced to read and cache entire pages. Sumeru, however, can perform adaptive caching based on the request size, thus precisely matching the workload's needs and more accurately identifying and retaining hot data. This allows Sumeru to achieve hit rates of 98.62% and 99.1% under workloads with high locality features like BERT and YCSB F, respectively. In contrast, other algorithms are limited in performance because their coarse-grained eviction policies cannot make such fine-grained judgments.

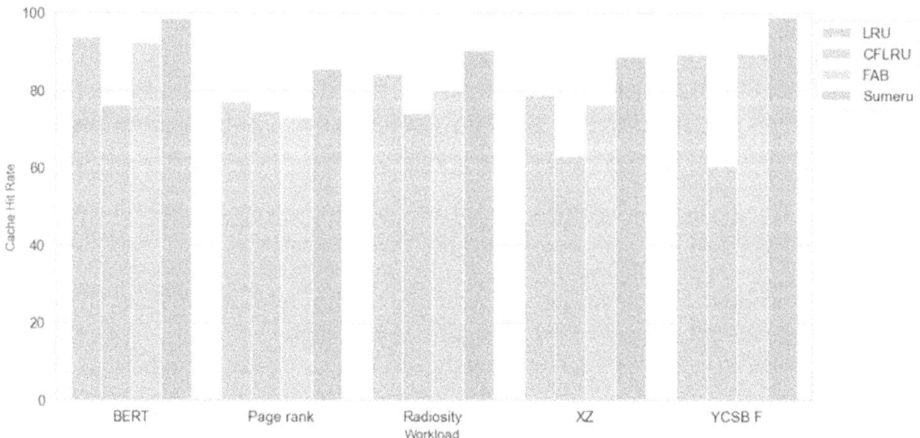

Fig. 7. Cache hit rate.

Latency. The average access latency is directly related to the execution efficiency of an application. This metric combines the latency of both cache hits and misses.

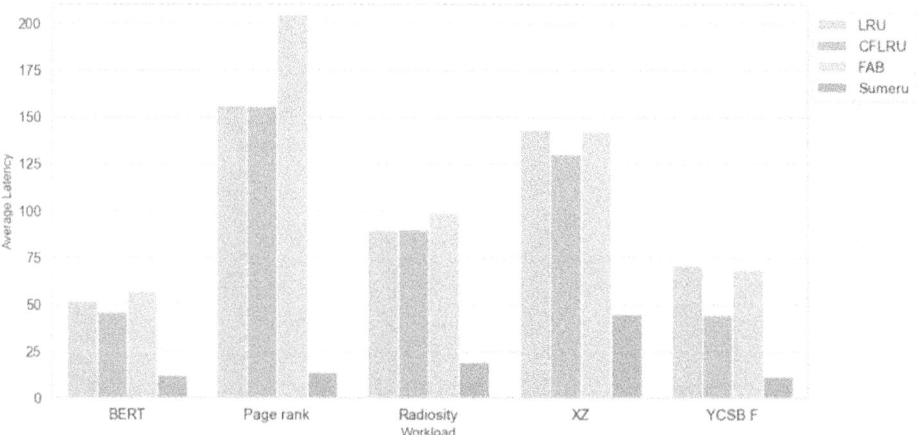

Fig. 8. Average latency.

As shown in Fig. 8, the Sumeru algorithm achieved the lowest average access latency in all cases. This advantage is most pronounced under the PageRank workload, where Sumeru reduces average latency to 13.98μs—significantly lower than that of other algorithms. Behind this performance leap is Sumeru's efficient collaborative mechanism. The multi-granularity hybrid policy avoids extra I/O caused by granularity mismatch at the source, while the selective eviction mechanism ensures that the data in the cache has the highest reuse value.

Together, they minimize the probability of a cache miss. Every successful cache hit avoids a time-consuming flash access, ultimately translating into significant performance gains and improved response speed at the application level. In contrast, LRU, CFLRU, and FAB, due to their relatively high miss rates, cause the system to access the underlying NAND flash more frequently, thus pushing up the overall average access latency.

Flush Count. The flush count represents the number of times dirty pages are evicted from the cache to persistent storage. This metric reflects the algorithm's degree of optimization for write operations.

As shown in Fig. 9, Sumeru significantly reduced the number of flushes under the BERT, XZ, and YCSB F workloads. This is thanks to the differential probation mechanism, which tends to retain dirty pages because the cost of evicting them is higher. To this end, Sumeru sets up a value assessment system for cache items based on their hotness and dirty status, giving the most valuable hot dirty pages up to 3 probation chances, allowing them to likely skip eviction and remain in the cache. This strong protection for high-value dirty data effectively reduces unnecessary write-back operations. In addition, the selective eviction policy also contributes to this, as it avoids having to write back an entire dirty page just to evict a small amount of cold data within it. In comparison, although CFLRU has a design that prioritizes evicting clean pages, its simple partitioning strategy is not adaptive enough for complex read-modify-write workloads like YCSB F, resulting in a relatively high number of flushes. Sumeru's strategy is clearly more intelligent and adaptive, thus achieving a significant reduction in write-back costs.

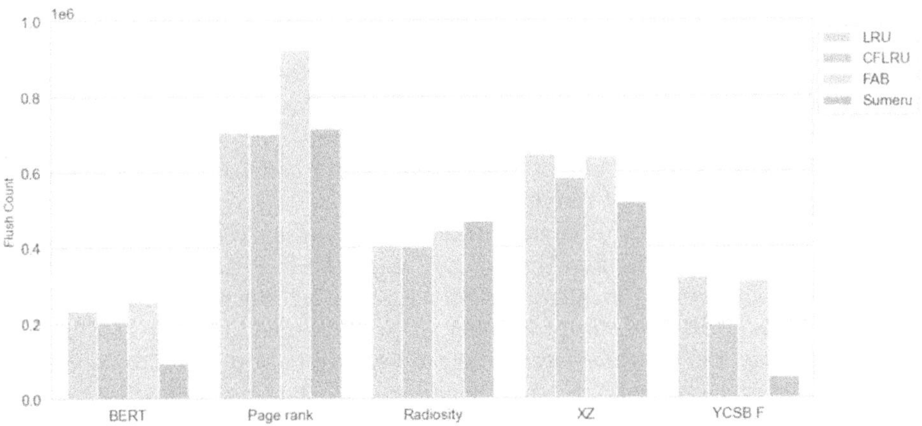

Fig. 9. Flush count.

4.3   Sensitivity Analysis

To gain a deeper understanding of the performance characteristics of the Sumeru scheme, we conducted a sensitivity analysis on the cache size and the hotness threshold in the differential probation mechanism.

Cache Size We varied the cache size from 2MB to 128MB under the five workloads (BERT, PageRank, Radiosity, XZ, and YCSB F) and observed Sumeru's performance in terms of hit rate, average access latency, and flush count.

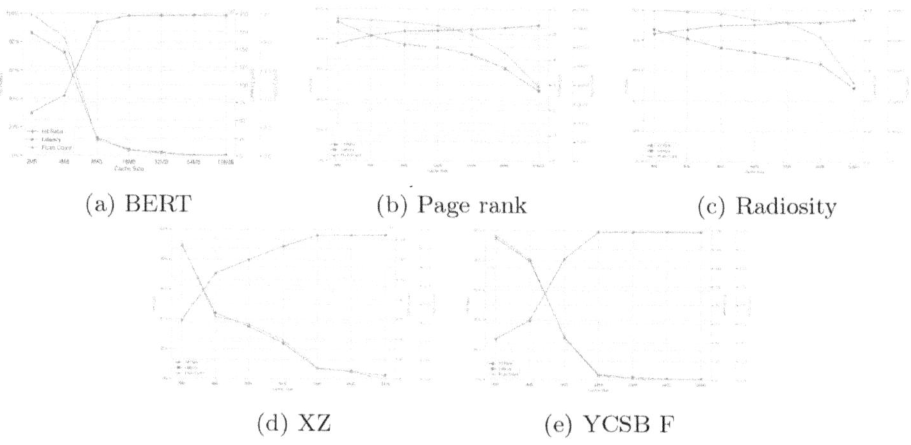

(a) BERT   (b) Page rank   (c) Radiosity

(d) XZ   (e) YCSB F

Fig. 10. Impact of cache size on Sumeru.

As the cache capacity increases, the performance of Sumeru improves significantly across all workloads (see Fig. 10). Specifically, the cache hit rate increases with capacity, while the average access latency and flush count decrease accordingly. This is expected, as a larger cache space means more data can be accommodated, thereby reducing evictions due to capacity constraints and giving hot data a higher probability of retention.

The performance under different workloads also reveals their respective access characteristics. Workloads like BERT, XZ, and YCSB F have relatively small working sets with strong locality. When the cache size increases from 2MB to 16MB, there is a dramatic improvement in performance. The hit rate quickly climbs to nearly 100%, while latency and flush count drop precipitously. Once the cache size exceeds 16MB, the performance curve flattens, indicating that the cache is now large enough to hold the core working set of these applications, and further increases in capacity yield diminishing marginal returns. This proves that Sumeru can efficiently utilize cache space; once the resources meet the working set demand, it can achieve extremely high performance. The PageRank and Radiosity workloads have larger working sets or poorer locality. Therefore, their performance shows a more gradual and continuous improvement as the cache capacity increases. Even with a 128MB cache, the hit rate is still slowly increasing, while latency and flush count continue to decrease. This indicates that for applications with large working sets, Sumeru also has good scalability and can steadily improve performance as resources increase.

The cache size sensitivity analysis shows that Sumeru's performance responds positively to cache resources. For applications with smaller working sets, a moderately sized cache is sufficient to unleash most of its potential; for applications with larger working sets, Sumeru can also continuously benefit from an increased cache size. This result justifies using a 16MB cache for our comparative experiments in Sect. 4.1 and demonstrates the effectiveness and robustness of Sumeru under different hardware resource configurations.

Hotness Threshold. The differential probation mechanism is key to Sumeru's reduction of write-back overhead, and its performance is closely related to the hotness level threshold settings used to judge data hotness. To evaluate the impact of this parameter, we conducted a sensitivity analysis by varying the threshold for determining hotness levels while keeping other conditions fixed. As shown in Fig. 11, we tested three different threshold configurations, as well as the case where the probation mechanism was completely disabled, and observed the average performance of each metric across the five workloads.

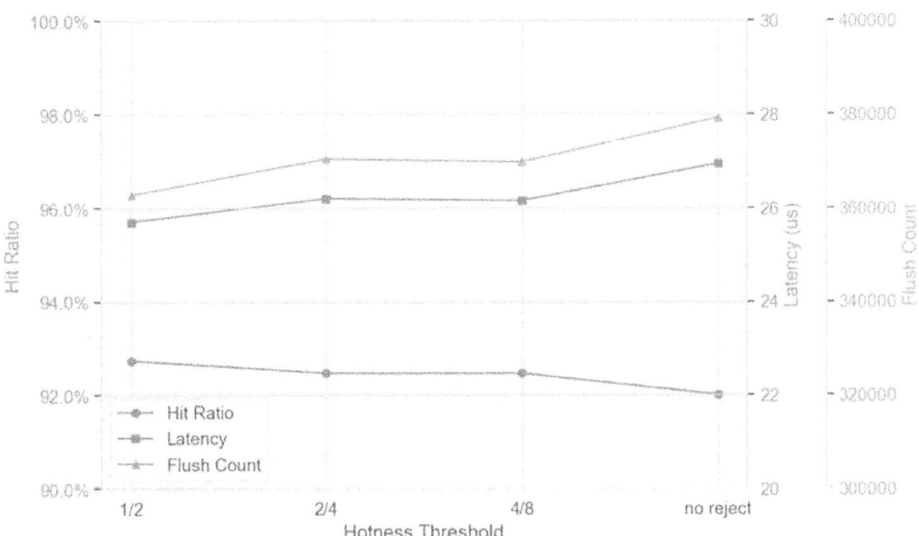

Fig. 11. Impact of hotness threshold on Sumeru.

The experimental results show that Sumeru's performance exhibits a certain sensitivity to the hotness threshold. As the threshold increases, the condition for a cache item to be judged as hot or warm becomes stricter. This leads to a decrease in the number of cache items that receive probation chances, especially dirty data, which in turn causes the flush count to rise. At the same time, because some data that should have been retained is evicted prematurely, the cache hit rate also shows a slight decline, ultimately leading to an increase in average access latency.

When the differential probation mechanism was completely disabled ("no probation"), the flush count and average latency reached their highest points. This demonstrates the critical role of this mechanism in protecting high-value dirty data and reducing expensive write-back operations.

Overall, the empirically set thresholds we chose in Sect. 3 achieved the best combined performance across the three metrics of hit rate, latency, and flush count in our tests. This analysis validates the reasonableness of our chosen parameters and highlights the contribution of the differential probation mechanism to Sumeru's overall performance.

5   Related Work

Cache policies within SSDs have undergone considerable development over the past decade. Their evolution has always revolved around the traditional block I/O model, forming two main paradigms: page-based policies and block-based policies. Page-based policies manage data at the granularity of a single flash page. To adapt to the flash medium, researchers have proposed optimization algorithms such as CFLRU and CCF-LRU, which consider the dirty status of pages and prioritize evicting clean pages to reduce the overhead associated with write operations. These types of policies can capture temporal locality well but perform poorly when handling large sequential reads and writes because they cannot perceive the spatial correlation between pages.

In contrast, block-based policies manage data at the granularity of an entire flash block. They can effectively utilize the spatial locality in sequential accesses and help reduce the cost of garbage collection. However, when the access pattern becomes random reads and writes, this strategy leads to a decrease in cache space utilization, as cold data within a block is forced to occupy precious cache space simply because it is adjacent to hot data. In our experiments, we chose LRU, CFLRU, and FAB as baselines, as they are classic representatives of general-purpose page-level, flash-aware page-level, and block-level cache designs, respectively. By comparing against these policies, we can clearly demonstrate that Sumeru's adaptive granularity design has a fundamental advantage over traditional fixed-granularity policies when dealing with the byte-level access introduced by CXL.

In recent years, the academic community has proposed more refined optimization schemes for traditional NVMe SSDs. For example, ECR [35] considers the state of the I/O waiting queue in its eviction decisions; Co-Active [36] introduces a collaborative write-back mechanism based on chip load; GCaR [37] considers the penalty factor of garbage collection; and VBBMS [16] partitions different cache regions for random and sequential requests.

Although these advanced policies show strong performance in traditional block I/O environments, their foundational models are predicated on fixed page or block granularities. Consequently, their core logic does not account for the fine-grained, byte-level access patterns native to CXL-SSDs. Adapting these complex algorithms to a byte-addressable, multi-granularity paradigm would require substantial architectural redesign, placing such a comparison beyond the scope of this work.

6 Conclusion

The byte-addressable nature of CXL-SSDs presents enormous opportunities for storage systems, but it also poses severe challenges to traditional cache management. This paper provides an in-depth analysis of the limitations of existing schemes in handling byte-granularity I/O and proposes a high-performance cache management scheme specifically designed for CXL-SSDs, named Sumeru. Sumeru's core contribution lies in its innovative selective eviction mechanism, which, by gaining insight into the combined access patterns of sub-blocks within a page, achieves precise retention of hot data and selective eviction of cold data. Combined with an adaptive multi-granularity hybrid architecture and an SSD-optimized differential probation policy, Sumeru achieves comprehensive performance leadership in hit rate, access latency, and write-back cost compared to various traditional and advanced cache policies across five real-world workloads. This paper validates that a fine-grained, multi-granularity cache management scheme can effectively overcome the core contradiction between CXL's external byte access and the SSD's internal page-level management. The design and evaluation results of Sumeru show that in the CXL era, cache decisions that delve inside the page, based on subpages combination access patterns, are essential to fully realizing its performance potential. This work provides a valuable reference and an efficient solution for the design of future CXL storage devices.

References

1. Zhang, J., Kwon, M., Gouk, D., et al.: Revamping storage class memory with hardware automated memory-over-storage solution. In: 2021 ACM/IEEE 48th Annual International Symposium on Computer Architecture (ISCA), pp. 762–775. IEEE (2021)
2. Lee, C., Shin, W., Kim, D.J., et al.: NVDIMM-C: a byte-addressable non-volatile memory module for compatibility with standard DDR memory interfaces. In: 2020 IEEE International Symposium on High Performance Computer Architecture (HPCA). pp. 502–514. IEEE (2020)
3. Liao, X., Lu, Y., Xu, E., et al.: Write dependency disentanglement with HORAE. In: 14th USENIX Symposium on Operating Systems Design and Implementation (OSDI 20), pp. 549–565 (2020)
4. Ruan, Z., Schwarzkopf, M., Aguilera, M.K., et al.: AIFM: high-performance, application-integrated far memory. In: 14th USENIX Symposium on Operating Systems Design and Implementation (OSDI 20), pp. 315–332 (2020)

5. Papagiannis, A., Xanthakis, G., Saloustros, G., et al.: Optimizing memory-mapped I/O for fast storage devices. In: 2020 USENIX Annual Technical Conference (USENIX ATC 20), pp. 813–827 (2020)
6. Badam, A., Pai, V.S.: SSDAlloc: hybrid SSD/RAM memory management made easy. In: 8th USENIX Symposium on Networked Systems Design and Implementation (NSDI 11) (2011)
7. Wu, K., Guo, Z., Hu, G., et al.: The storage hierarchy is not a hierarchy: optimizing caching on modern storage devices with orthus. In: 19th USENIX Conference on File and Storage Technologies (FAST 21), pp. 307–323 (2021)
8. Kazama, S., Gokita, S., Kuwamura, S., et al.: Memory expansion technology using software-controlled SSD. Proc. Flash Memory Summit 1–10 (20170
9. Tsai, S.Y., Shan, Y., Zhang, Y.: Disaggregating persistent memory and controlling them remotely: An exploration of passive disaggregated Key-Value stores. In: 2020 USENIX Annual Technical Conference (USENIX ATC 20), pp. 33–48 (2020)
10. CXL consortium. https://www.computeexpresslink.org/
11. Jung, M.: Hello bytes, bye blocks: Pcie storage meets compute express link for memory expansion (cxl-ssd). In: Proceedings of the 14th ACM Workshop on Hot Topics in Storage and File Systems, pp. 45–51 (2022)
12. Abulila, A., Mailthody, V.S., Qureshi, Z., et al.: Flatflash: Exploiting the byte-accessibility of ssds within a unified memory-storage hierarchy. In: Proceedings of the Twenty-Fourth International Conference on Architectural Support for Programming Languages and Operating Systems, pp. 971–985 (2019)
13. Mohan, J., Kadekodi, R., Chidambaram, V.: Analyzing IO amplification in Linux file systems. arXiv preprint arXiv:1707.08514 (2017)
14. NVM Express Base Specification (2024). https://nvmexpress.org/specifications/
15. Zhang D.: GenZ, CXL, NVLINK, OpenCAPI, CCIX Battle!. https://blog.csdn.net/TV8MbnO2Y2RfU/article/details/89308357, Accessed 18 Aug 2024
16. OpenCAPI Consortium. https://github.com/opencapi/
17. Gen-Z consortium. https://genzconsortium.org/specifications/
18. CCIX consortium. https://www.ccixconsortium.com/
19. NVLink consortium. https://www.nvidia.com/en-us/data-center/nvlink/
20. CXL Consortium. CXL Specification. https://computeexpresslink.org/cxl-specification/
21. Yang, S.P., Kim, M., Nam, S., et al.: Overcoming the memory wall with CXL-EnabledSSDs. In: 2023 USENIX Annual Technical Conference (USENIX ATC 23), pp. 601–617 (2023)
22. Tripathy, S., Satpathy, M.: SSD internal cache management policies: a survey. J. Syst. Architect. **122**, 102334 (2022)
23. Park, S., Jung, D., Kang, J., et al.: CFLRU: a replacement algorithm for flash memory. In: Proceedings of the 2006 International Conference on Compilers, Architecture and Synthesis for Embedded Systems, pp. 234–241 (2006)
24. Jung, H., Shim, H., Park, S., et al.: LRU-WSR: integration of LRU and writes sequence reordering for flash memory. IEEE Trans. Consum. Electron. **54**(3), 1215–1223 (2008)
25. Li, Z., Jin, P., Su, X., et al.: CCF-LRU: A new buffer replacement algorithm for flash memory. IEEE Trans. Consum. Electron. **55**(3), 1351–1359 (2009)
26. Jo, H., Kang, J.U., Park, S.Y., et al.: FAB: Flash-aware buffer management policy for portable media players. IEEE Trans. Consum. Electr. **52**(2), 485–493 (2006)
27. Kim H, Ahn S. BPLRU: A Buffer Management Scheme for Improving Random Writes in Flash Storage[C]//FAST. 2008, 8: 1-14

28. Zhang, X., Lu, T., Chang, Y., et al.: Morpheus: an adaptive DRAM Cache with Online Granularity Adjustment for Disaggregated Memory. In: 2023 IEEE 41st International Conference on Computer Design (ICCD), pp. 134–141. IEEE (2023)
29. Turc, I., Chang, M.W., Lee, K., et al.: Well-read students learn better: On the importance of pre-training compact models, arXiv preprint arXiv:1908.08962 (2019)
30. Beamer, S., Asanović, K., Patterson, D.: The GAP benchmark suite. arXiv preprint arXiv:1508.03619 (2015)
31. Sakalis, C., Leonardsson, C., Kaxiras, S., et al.: Splash-3: a properly synchronized benchmark suite for contemporary research. In: 2016 IEEE International Symposium on Performance Analysis of Systems and Software (ISPASS), pp. 101–111. IEEE (2016)
32. SPEC CPU (2017). https://www.spec.org/cpu2017/
33. SPEC Redistributable Sources Archive. https://www.spec.org/sources/
34. Cooper, B.F., Silberstein, A., Tam, E., et al.: Benchmarking cloud serving systems with YCSB. In: Proceedings of the 1st ACM symposium on Cloud computing, pp. 143–154 (2010)
35. Chen, H., Pan, Y., Li, C., et al.: ECR: Eviction-cost-aware cache management policy for page-level flash-based SSDs. Concurrency Comput. Practi. Experience **33**(15), e5395 (2021)
36. Sun, H., Dai, S., Huang, J., et al.: Co-active: a workload-aware collaborative cache management scheme for NVMe SSDs. IEEE Trans. Parallel Distrib. Syst. **32**(6), 1437–1451 (2021)
37. Wu, S., Lin, Y., Mao, B., et al.: GCaR: Garbage collection aware cache management with improved performance for flash-based SSDs. In: Proceedings of the. International Conference Supercomputing 2016, pp. 1–12 (2016)
38. Du, C., Yao, Y., Zhou, J., et al.: VBBMS: A novel buffer management strategy for NAND flash storage devices. IEEE Trans. Consum. Electron. **65**(2), 134–141 (2019)

Author Index

GPSR Compliance
The European Union's (EU) General Product Safety Regulation (GPSR) is a set
of rules that requires consumer products to be safe and our obligations to
ensure this.

If you have any concerns about our products, you can contact us on

ProductSafety@springernature.com

In case Publisher is established outside the EU, the EU authorized
representative is:

Springer Nature Customer Service Center GmbH
Europaplatz 3
69115 Heidelberg, Germany